Electromagnetic Waves and Transmission Lines

BASED ON JNTU SYLLABUS 2014

Electromagnetic Waves and Transmission Lines

Y Mallikarjuna Reddy

Principal
Vasireddy Venkatadri Institute of Technology, Guntur, A.P.

UNIVERSITIES PRESS (INDIA) PRIVATE LIMITED

Registered Office
3-6-747/1/A & 3-6-754/1, Himayatnagar
Hyderabad 500 029 (Telangana), India
E-mail: info@universitiespress.com; www.universitiespress.com

Distributed by
Orient Blackswan Private Limited

Registered Office
3-6-752 Himayatnagar, Hyderabad 500 029, Telangana, India

Other Offices
Bengaluru / Bhopal / Chennai / Ernakulam / Guwahati / Hyderabad /
Jaipur / Kolkata / Lucknow / Mumbai / New Delhi / Noida / Patna

First published 2015

ISBN 978 81 7371 948 6

Typeset in Times New Roman 10.6pt by
Techastra Solutions Pvt. Ltd., Hyderabad 500 029

Printed in India by
Graphica Printers, Hyderabad 500 013

Published by
Universities Press (India) Private Limited
3-6-747/1/A & 3-6-754/1, Himayatnagar
Hyderabad 500 029, Telangana, India

Contents

3. Electric Fields in Conductors and Dielectrics

4. Magnetostatics

5. Magnetic Force and Inductance

8. Electromagnetic Wave Characteristics

9. Transmission Lines - I

Preface

This book is prepared for the use of second year undergraduate engineering students specialising in electronics and communication engineering. Students taking the AMIE and IETE courses as well as those preparing for GATE, UPSC and other similar competitive examinations will also find the book very useful. Even though the book is designed according to the latest JNTU syllabus, it will meet the requirement of most technical universities in India.

At low frequencies, the analysis of voltages and currents is used for efficient signal transmission through electrical circuits. As frequency increases, the characteristics of physical parameters of the medium also play an important role. Here the voltages and currents exist in the form of electromagnetic waves. It is therefore important to study electromagnetic waves under two parts, namely static electromagnetics and dynamic electromagnetics. Static fields are best analysed with the help of Coloumb's law, Gauss' law, Ampere's law and Faraday's law. Dynamic or time-varying fields and waves can be comprehensively studied using Maxwell's equations. Transmission lines play a significant role in communications as carriers of electromagnetic waves through different media. Therefore, it is necessary for students pursuing courses in the discipline of electronics to learn the fundamentals of electromagnetic theory.

The main objective of this book is to impart a thorough understanding of electromagnetic waves and transmission lines. The material covered here is divided into four main parts. The first part, comprising Chapter 1, focuses on the basic mathematical analysis of vector fields required for the study of electromagnetic fields. The second part, comprising Chapters 2 to 6, introduces the concepts of static electric fields, static magnetic fields and time-varying fields. The third part, Chapters 7 and 8, discusses electromagnetic waves and their characteristics. The fourth part comprising Chapters 9 and 10 provides a detailed analysis of transmission lines.

The book includes many solved and unsolved problems. Emphasis is given to solving problems that appeared in JNTU examinations, as the text has been prepared according to the JNTU syllabus. Objective-type questions and the answers for unsolved problems are provided at the end of each chapter.

My lecture notes on the subject proved to be very useful in preparing the manuscript. The suggestions offered by students during the course of my teaching the subject over many semesters helped in bringing about better clarity to the structure and presentation of contents. More illustrations and solutions to questions from the latest JNTU papers were added on with the help of my colleagues and students. Great care has been taken to eliminate errors and inconsistencies in the book, and I am indebted to the faculty members who helped in this task with their critical reviews.

I sincerely thank Sri V Vidya Sagar, Chairman, Sri Badari Prasad, Secretary, Sri M Srikrishna, Joint Secretary, Sqn Ldr S R K Paramahamsa, Joint Secretary, and the management and staff of Vasireddy Venkatadri Institute of Technology, Nambur, Guntur, for their encouragement and support of this book. Especially, I wish to especially thank my friends and colleagues, Dr I A Pasha, Dr P Nageswara Rao, Dr E. Srinivasa Reddy, Dr Mohd. Moazzam, Dr K Giribabu, P Ammi Reddy, M R N Tagore, Dr M V Raghuram, N Sivarama Krishna and K Suresh Babu for their invaluable suggestions and reviews of the manuscript. My sincere thanks are also due to the publishers for their interest in the book. Finally, I thank my family members Srujana Devi, Sravan Kumar Reddy and Srinivasula Reddy for their patience and affection.

It is my sincere hope that readers' expectation of a textbook catering to their academic requirements is well met with this endeavour of mine. I welcome suggestions for improving the book in its next edition.

Outline of the book

Chapter 1 reviews mathematical tools like vector algebra and vector calculus, which are needed to express vector quantities in different coordinate systems. It also discusses del operators and theorems related to EM fields.

Chapter 2 discusses the basics of electrostatic fields—Coulomb's law, force, electric field intensity, charge distribution, Guass' law, application of Guass' law, work done, electric potential, electric dipole and energy density.

Chapter 3 explains conducting and dielectric materials, Ohm's law, conductivity, types of currents, current density, polarization, Laplace's and Poisson's equations for solving simple problems on static fields, boundary conditions, capacitance and energy stored in static electric fields.

Chapters 4 introduces the basics of static magnetic fields, Biot–Savart's law, magnetic field intensity, Amperes circuital law and its applications, Maxwells equations, field due to solenoids and toroids.

Chapters 5 details scalar and vector magnetic potentials, force in magnetic fields, Lorentz force equation, Ampere's force law, magnetic dipole and dipole moment, boundary conditions, self and mutual inductances, magnetic energy and magnetic circuits.

Chapters 6 elaborates the concepts of time-varying fields, Faraday's laws of electromagnetic induction, induced EMFs, Maxwell's equations for time-varying fields, displacement current and boundary conditions.

Chapters 7 and 8 explain electromagnetic waves, wave equations, uniform plain waves, propagation characteristics, wave propagation in different media, polarization, reflection and refraction, Poynting theorem, Poynting vector and average power.

Chapters 9 and 10 cover transmission lines, their types, line equations, line parameters, line characteristics, loading, impedances, OC and SC lines, reflections, impedance matching, UHF lines, stub matching, Smith charts and applications.

An appendix is provided at the end, which lists the symbols of quantities and material properties.

Y Mallikarjuna Reddy

Introduction to Vector Fields 1

Learning Objectives

The main objective of this chapter is to understand thoroughly the basic mathematical analysis of vector fields related to electromagnetic fields. Vector analysis in three dimensional space is a preamble for learning these fields. By reading this chapter, students will be able to relate to the following topics.

- Scalar and vector fields
- Addition, subtraction and multiplications of fields and their properties
- Orthogonal coordinate systems in three-dimensional space
- Transformation of vectors between Cartesian, cylindrical and spherical coordinate systems
- Differentiation and integration of vector fields in multi-dimensions
- Gradient, divergence and curl operators on vector fields and their properties
- Some useful del operators and theorems

1.1 Introduction

Analysis of vector fields require mathematical tools like vector algebra and vector calculus which are needed to express quantities in different coordinate systems with magnitude and direction. The study of electromagnetic fields (EM fields) involve electric and magnetic field vectors. Most of the applications of electromagnetic fields require the analysis, physical interpretation and synthesis of vector quantities like electric field intensity, magnetic field intensity and flux density. Vector analysis simplifies the formulation of various laws of EM fields. This chapter explains the basic concepts of vector algebra and vector calculus in different orthogonal coordinate systems like the rectangular coordinate system, the cylindrical coordinate system and the spherical coordinate system. Vector algebra deals with two quantities—vectors and scalars. It also explains basic vector operations such as addition, subtraction and multiplication. To make the analysis of EM fields simple, different types of orthogonal coordinate systems are required. For example, fields propagating through coaxial cables or circular wave guides can be easily analysed using cylindrical coordinate systems, whereas it is a laborious task to express them in rectangular coordinates. We will also explain how to transform vectors from one coordinate system to another.

Vector calculus is used to express fields with differential and integral operators. This chapter also discusses del operators such as gradient, divergence and curl. Finally, we explore some important theorems like divergence theorem, Stokes' theorem, Green's theorem and Helmholtz's theorem, which are essential for EM fields.

Therefore, we will study the basic concepts of vector analysis before we enter into a discussion on EM fields.

1.2 Basic Definitions

Scalar A scalar is a quantity which has only magnitude and is denoted by a single quantity with appropriate units. Time, mass, potential, distance etc., are scalar quantities. A scalar is represented by normal letters like A, B, C etc. For example, a mass of 10 kg is a scalar quantity.

Vector A vector is a quantity that has both magnitude and direction. It is denoted by its magnitude with appropriate units and its direction of application. Velocity, force, current, electric field intensity, magnetic field intensity etc., are vector quantities. A vector can be represented in many ways, the common ones being either a letter with a superscribed arrow such as $\vec{A}$, $\vec{B}$ etc., or with boldface type such as **A, B** etc. In this book, we have used the former convention.

For example, a velocity of 10 m/s in the North–East direction is a vector quantity.

Field A field is a function that describes a quantity at all points within a specified region. A quantity may have either a scalar or a vector field. In general, a field varies with both position and time. For example, charge distribution in a volume, sound intensity in a class room, temperature distribution in a furnace etc., are scalar fields. Wind velocity, current flow, electric forces, magnetic forces etc., are vector fields.

Unit vector In a three-dimensional space, a vector point can be located by the intersection of three surfaces. Generally, the well-known Cartesian or rectangular coordinate system is used to represent a three-dimensional space where the surfaces $x-y$, $y-z$ and $x-z$ are mutually perpendicular to each other.

The unit vectors along the axes in the Cartesian coordinate system can be represented as $\vec{a}_x, \vec{a}_y$ and $\vec{a}_z$ as shown in Fig. 1.1.

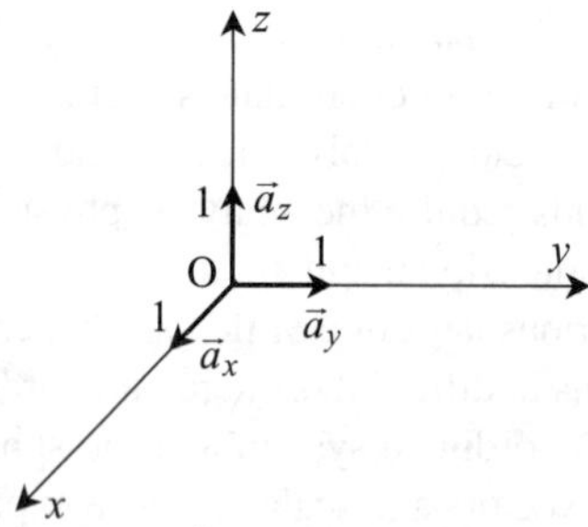

Fig. 1.1 *Unit vectors in the Cartesian coordinate system*

A vector can be represented as

$$\vec{A} = A_x\vec{a}_x + A_y\vec{a}_y + A_z\vec{a}_z, \tag{1.1}$$

where A_x, A_y and A_z are the scalar components of the vector $\vec{A}$ along the x, y and z axes respectively.

The magnitude of $\vec{A}$ is $|\vec{A}| = A = \sqrt{A_x^2 + A_y^2 + A_z^2}$.

For a given vector $\vec{A}$, a unit vector $\vec{a}_A$ is defined as a vector whose magnitude is unity and whose direction is along $\vec{A}$. It is given by

$$\vec{a}_A = \frac{\vec{A}}{|\vec{A}|} = \frac{\vec{A}}{A} \text{ and } |\vec{a}_A| = 1. \tag{1.2}$$

We can also write

$$\vec{A} = A\vec{a}_A. \tag{1.3}$$

Position vector A vector drawn from an arbitrary origin to the point in space whose coordinates are specified is called a position vector. In other words, an alternate way to denote the location of a point in space is by a position vector. If the point P in Cartesian coordinates is at (x, y, z), then the position vector is

$$\vec{A} = x\vec{a}_x + y\vec{a}_y + z\vec{a}_z \tag{1.4}$$

It also represents the distance vector from the origin to the point P as shown in Fig. 1.2(a).

Thus $\vec{A} = \overrightarrow{OP} = x\vec{a}_x + y\vec{a}_y + z\vec{a}_z$.

Its unit vector is

$$\vec{a}_{OP} = \frac{\vec{A}}{A} = \frac{x\vec{a}_x + y\vec{a}_y + z\vec{a}_z}{\sqrt{x^2 + y^2 + z^2}}. \tag{1.5}$$

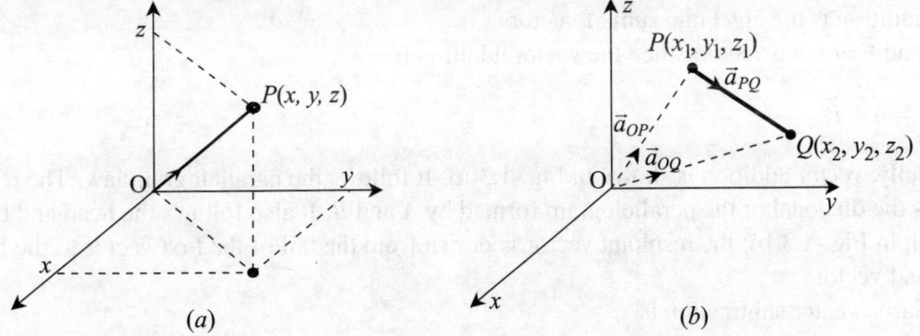

Fig. 1.2 *(a) Position vector; (b) distance vector and unit vectors*

Note: A point P is a scalar quantity which represents a position in space, whereas a position vector is either a variable depending on space variables x, y and z or a constant vector at a given position. For example, a position vector $\vec{A} = 2x\vec{a}_x + xy\vec{a}_y + y^2z\vec{a}_z$ is a variable depending on the space variables x, y and z. However, at a given point say (2, 1, 3), the vector becomes a constant vector, i.e., $\vec{A} = 4\vec{a}_x + 2\vec{a}_y + 3\vec{a}_z$.

Distance Vector It is a vector which represents displacement from one point to another point. If two points P and Q are position vectors at (x_1, y_1, z_1) and (x_2, y_2, z_2) as shown in Fig. 1.2(b), the distance vector from point P to point Q can be written as

$$\overrightarrow{PQ} = \overrightarrow{OQ} - \overrightarrow{OP} = (x_2 - x_1)\vec{a}_x + (y_2 - y_1)\vec{a}_y + (z_2 - z_1)\vec{a}_z.$$

From point Q to point P it is

$$\overrightarrow{QP} = \overrightarrow{OP} - \overrightarrow{OQ} = (x_1 - x_2)\vec{a}_x + (y_1 - y_2)\vec{a}_y + (z_1 - z_2)\vec{a}_z.$$

The distance between P and Q is

$$|\overrightarrow{PQ}| = |\overrightarrow{QP}| = \sqrt{(x_1 - x_2)^2 + (y_1 - y_2)^2 + (z_1 - z_2)^2}.$$

The unit vectors are

$$\vec{a}_{OP} = \frac{x_1\vec{a}_x + y_1\vec{a}_y + z_1\vec{a}_z}{\sqrt{x_1^2 + y_1^2 + z_1^2}}$$

$$\vec{a}_{OQ} = \frac{x_2\vec{a}_x + y_2\vec{a}_y + z_2\vec{a}_z}{\sqrt{x_2^2 + y_2^2 + z_2^2}}$$

and $$\vec{a}_{PQ} = \frac{(x_2 - x_1)\vec{a}_x + (y_2 - y_1)\vec{a}_y + (z_2 - z_1)\vec{a}_z}{\sqrt{(x_2 - x_1)^2 + (y_2 - y_1)^2 + (z_2 - z_1)^2}}. \quad (1.6)$$

1.3 Vector Addition and Subtraction

Vector addition is the algebraic sum of vectors.

If $\vec{A}$ and $\vec{B}$ are two vectors, then the vector addition is

$$\vec{C} = \vec{A} + \vec{B} \quad (1.7)$$

Graphically, vector addition is shown in Fig. 1.3(a). It follows the parallelogram law. The resultant vector is the diagonal of the parallelogram formed by $\vec{A}$ and $\vec{B}$. It also follows the head and tail rule as shown in Fig. 1.3(b); the resultant vector is drawn from the tail of the first vector to the head of the second vector.

Similarly, vector subtraction is

$$\vec{D} = \vec{A} - \vec{B} = \vec{A} + (-\vec{B}) \quad (1.8)$$

As shown in Fig. 1.3 (c), the resultant vector is the diagonal of the parallelogram formed by $\vec{A}$ and $-\vec{B}$.

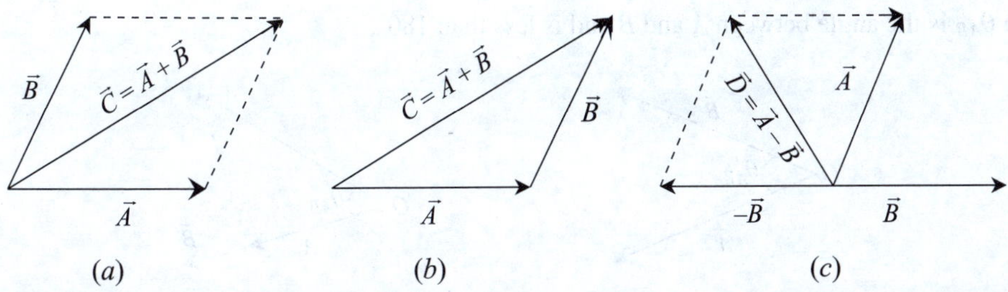

Fig. 1.3 *(a) Vector addition: parallelogram law; (b) vector addition: head and tail rule; (c) vector subtraction: parallelogram law*

Vector addition obeys commutative, associative and distributive laws.

Commutative law: $\vec{A}+\vec{B}=\vec{B}+\vec{A}$, $\alpha\vec{A}=\vec{A}\alpha$

Associative law: $\vec{A}+(\vec{B}+\vec{C})=(\vec{A}+\vec{B})+\vec{C}$, $\alpha(\beta\vec{A})=(\alpha\beta)\vec{A}$

Distributive law: $\alpha(\vec{A}+\vec{B})=\alpha\vec{A}+\alpha\vec{B}$, where α and β are scalar quantities.

Example 1.1 If two points are located in space at $A(3,\ -4,\ 5)$ and $B(-2,\ 0,\ 1)$, find (a) their position vectors, (b) the distance vector $\overrightarrow{AB}$, (c) a unit vector along $\overrightarrow{AB}$.

Solution Given points $A(3,\ -4,\ 5)$ and $B(-2,\ 0,\ 1)$.

(a) The position vectors are $\vec{A}=3\vec{a}_x-4\vec{a}_y+5\vec{a}_z$ and $\vec{B}=-2\vec{a}_x+\vec{a}_z$.

(b) Distance vector $\overrightarrow{AB}$ is

$$\overrightarrow{AB}=\vec{B}-\vec{A}=(-2\vec{a}_x+\vec{a}_z)-(3\vec{a}_x-4\vec{a}_y+5\vec{a}_z)=-5\vec{a}_x+4\vec{a}_y-4\vec{a}_z.$$

(c) The distance between $\vec{A}$ and $\vec{B}$ is $\sqrt{5^2+4^2+4^2}=7.55$.

(d) A unit vector along $\overrightarrow{AB}$, $\vec{a}_{AB}=\dfrac{\overrightarrow{AB}}{|AB|}=\dfrac{-5\vec{a}_x+4\vec{a}_y-4\vec{a}_z}{7.55}=0.66\vec{a}_x+0.53\vec{a}_y-0.53\vec{a}_z$.

1.4 Vector Multiplication

There are two types of multiplications

1. Scalar multiplication (dot product)
2. Vector multiplication (cross product)

1.4.1 Dot Product

The dot product of two vectors $\vec{A}$ and $\vec{B}$ is defined as the product of the magnitudes of $\vec{A}$, $\vec{B}$ and the cosine of the angle between them as shown in Fig. 1.4(a). The dot product is a scalar quantity.

Thus

$$\vec{A}\cdot\vec{B}=|\vec{A}||\vec{B}|\cos\theta_{AB}, \tag{1.9}$$

where θ_{AB} is the angle between A and B and is less than $180°$.

Fig. 1.4 *(a) Dot product of $\vec{A}$ and $\vec{B}$; (b) projection of $\vec{A}$ along $\vec{B}$*

If $\vec{A}$ has components (A_x, A_y, A_z) and $\vec{B}$ has components (B_x, B_y, B_z), then the dot product is

$$\vec{A}\cdot\vec{B} = (A_x\vec{a}_x + A_y\vec{a}_y + A_z\vec{a}_z)\cdot(B_x\vec{a}_x + B_y\vec{a}_y + B_z\vec{a}_z).$$

$$\vec{A}\cdot\vec{B} = A_xB_x + A_yB_y + A_zB_z. \tag{1.10}$$

Note 1: If two vectors are said to be orthogonal to each other, i.e., $\theta_{AB} = 90°$,

$$\text{then } \vec{A}\cdot\vec{B} = AB\cos 90° = 0. \tag{1.11}$$

If two vectors are parallel to each other, i.e., $\theta_{AB} = 0°$,

$$\text{then } \vec{A}\cdot\vec{B} = AB\cos 0° = AB, \text{ also } \vec{A}\cdot\vec{A} = A^2. \tag{1.12}$$

Note 2: For unit vectors,

$$\vec{a}_x\cdot\vec{a}_y = \vec{a}_y\cdot\vec{a}_z = \vec{a}_z\cdot\vec{a}_x = 0 \text{ (since the axes are orthogonal to each other).}$$

$$\vec{a}_x\cdot\vec{a}_x = \vec{a}_y\cdot\vec{a}_y = \vec{a}_z\cdot\vec{a}_z = 1 \text{ (since the angle is zero).} \tag{1.13}$$

Note 3: The angle between two vectors is $\theta_{AB} = \cos^{-1}\left(\dfrac{\vec{A}\cdot\vec{B}}{AB}\right)$ (1.14)

Note 4: The vector components along the axis are $A_x = \vec{A}\cdot\vec{a}_x,\ A_y = \vec{A}\cdot\vec{a}_y$ and $A_z = \vec{A}\cdot\vec{a}_z$ (1.15)

Note 5: *Components of a vector along another vector*
Dot product is useful in determining the projection (or component) of a vector in a given direction. Consider two vectors $\vec{A}$ and $\vec{B}$ with angle θ between them.

As shown in Fig.1.4(b), a scalar component A_B of $\vec{A}$ along vector $\vec{B}$ or projection of $\vec{A}$ along $\vec{B}$ is given by

$$A_B = A\cos\theta = |\vec{A}||\vec{a}_B|\cos\theta = \vec{A}\cdot\vec{a}_B \tag{1.16}$$

A vector component $\vec{A}_B$ of $\vec{A}$ along vector $\vec{B}$ is given by

$$\vec{A}_B = (A_B)\vec{a}_B = (A\cdot\vec{a}_B)\,\vec{a}_B \tag{1.17}$$

Note 6: The dot product obeys commutative and distributive laws.

Commutative law: $\vec{A} \cdot \vec{B} = \vec{B} \cdot \vec{A}$

Distributive law: $\vec{A} \cdot (\vec{B} + \vec{C}) = \vec{A} \cdot \vec{B} + \vec{A} \cdot \vec{C}$ (1.18)

1.4.2 Cross Product

The cross product of two vectors $\vec{A}$ and $\vec{B}$ is defined as the product of the magnitudes of $\vec{A}$ and $\vec{B}$ and sine of the angle between them. It is a vector quality whose direction is normal to the plane containing the vectors $\vec{A}$ and $\vec{B}$.

$$\therefore \quad \vec{A} \times \vec{B} = |\vec{A}||\vec{B}| \sin \theta_{AB} \vec{a}_n, \tag{1.19}$$

where $\vec{a}_n$ is a unit vector perpendicular to the plane of $\vec{A}$ and $\vec{B}$.

The direction of the unit vector follows the right-handed screw law. That is, if the screw is turned from $\vec{A}$ towards $\vec{B}$, the direction of the screw moment gives the direction of the cross product. Fig. 1.5(a) shows $\vec{A} \times \vec{B}$ with direction downwards and Fig. 1.5(b) shows $\vec{B} \times \vec{A}$ with direction upwards.

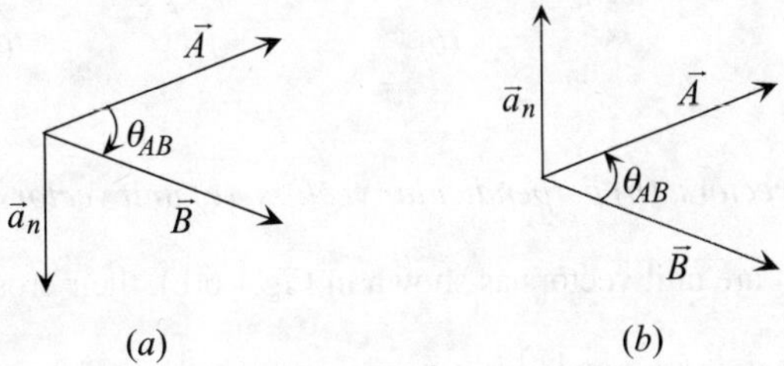

Fig. 1.5 *Cross product of $\vec{A}$ and $\vec{B}$: (a) $\vec{A} \times \vec{B}$ direction downwards; (b) $\vec{B} \times \vec{A}$ direction upwards*

If two vectors in Cartesian coordinates are given by

$$\vec{A} = A_x\vec{a}_x + A_y\vec{a}_y + A_z\vec{a}_z \quad \text{and} \quad \vec{B} = B_x\vec{a}_x + B_y\vec{a}_y + B_z\vec{a}_z,$$

then the cross product is

$$\vec{A} \times \vec{B} = (A_x\vec{a}_x + A_y\vec{a}_y + A_z\vec{a}_z) \times (B_x\vec{a}_x + B_y\vec{a}_y + B_z\vec{a}_z).$$

In matrix form,

$$\vec{A} \times \vec{B} = \begin{vmatrix} \vec{a}_x & \vec{a}_y & \vec{a}_z \\ A_x & A_y & A_z \\ B_x & B_y & B_z \end{vmatrix} \tag{1.20}$$

Note 1: The angle between the two vectors is

$$\sin \theta_{AB} = \frac{|\vec{A} \times \vec{B}|}{|\vec{A}||\vec{B}|} \tag{1.21}$$

and the unit vector is $\vec{a}_n = \dfrac{\vec{A}\times\vec{B}}{|\vec{A}||\vec{B}|\sin\theta_{AB}}$ (1.22)

The magnitude of the cross product $|\vec{A}\times\vec{B}|$ is equal to the area of the parallelogram having A and B as its sides.

Note 2: If two vectors are parallel, that is $\theta_{AB} = 0$ as shown in Fig. 1.6(a), the cross product is zero i.e., $\vec{A}\times\vec{B} = 0$, also $\vec{A}\times\vec{A} = 0$. (1.23)

If two vectors are perpendicular to each other as shown in Fig. 1.6(b), the cross product is given by $\vec{C} = \vec{A}\times\vec{B} = AB\vec{a}_n$. (1.24)

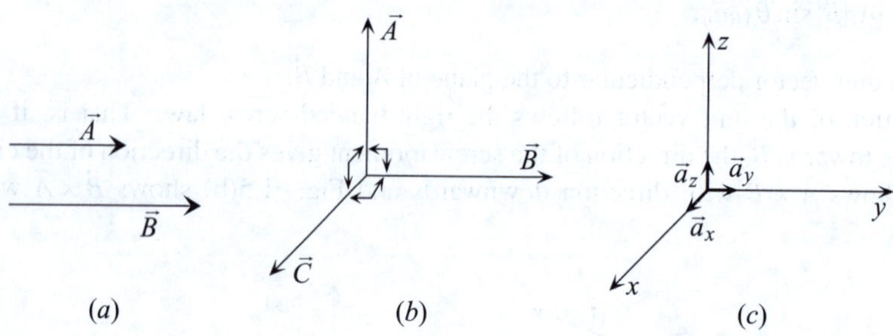

Fig. 1.6 *(a) Parallel vectors, (b) perpendicular vectors, (c) unit vectors*

Note 3: If $\vec{a}_x, \vec{a}_y$ and $\vec{a}_z$ are unit vectors as shown in Fig.1.6(c), their cross products are

$$\vec{a}_x\times\vec{a}_y = \vec{a}_z, \quad \vec{a}_y\times\vec{a}_z = \vec{a}_x \quad \text{and} \quad \vec{a}_z\times\vec{a}_x = \vec{a}_y. \tag{1.25}$$

Cross product follows cyclic permutation i.e., $\vec{a}_x \to \vec{a}_y \to \vec{a}_z \to \vec{a}_x \to \vec{a}_y$.

In the reverse order,

$$\vec{a}_y\times\vec{a}_x = -\vec{a}_z, \quad \vec{a}_x\times\vec{a}_z = -\vec{a}_y \quad \text{and} \quad \vec{a}_z\times\vec{a}_y = -\vec{a}_x$$

also $\quad \vec{a}_x\times\vec{a}_x = \vec{a}_y\times\vec{a}_y = \vec{a}_z\times\vec{a}_z = 0.$ (1.26)

Note 4: Commutative and associative laws are not valid for cross product since $\vec{A}\times\vec{B} = -(\vec{B}\times\vec{A})$ and

$$(\vec{A}\times\vec{B})\times\vec{C} \neq \vec{A}\times(\vec{B}\times\vec{C}).$$

However, it follows distributive law, that is

$$\vec{A}\times(\vec{B}+\vec{C}) = \vec{A}\times\vec{B}+\vec{A}\times\vec{C}. \tag{1.27}$$

Example 1.2 Given $\vec{A} = 5\vec{a}_x + 4\vec{a}_y - 3\vec{a}_z$ and $\vec{B} = 3\vec{a}_x - 2\vec{a}_y - 4\vec{a}_z$. Find unit vectors of $\vec{A}$, and $\vec{B}$, $\vec{A}\cdot\vec{B}$ and $\vec{A}\times\vec{B}$.

Solution Given $\vec{A} = 5\vec{a}_x + 4\vec{a}_y - 3\vec{a}_z$ and $\vec{B} = 3\vec{a}_x - 2\vec{a}_y - 4\vec{a}_z$.

$$\text{Unit vector of } \vec{A}, \quad \vec{a}_A = \frac{\vec{A}}{A} = \frac{5\vec{a}_x + 4\vec{a}_y - 3\vec{a}_z}{\sqrt{5^2+4^2+3^2}} = 0.707\vec{a}_x + 0.5656\vec{a}_y - 0.4242\vec{a}_z.$$

$$\text{Unit vector of } \vec{B}, \quad \vec{a}_B = \frac{\vec{B}}{B} = \frac{3\vec{a}_x - 2\vec{a}_y - 4\vec{a}_z}{\sqrt{3^2+2^2+4^2}} = 0.557\vec{a}_x + 0.3713\vec{a}_y - 0.743\vec{a}_z.$$

$$\text{Dot product,} \qquad \vec{A}\cdot\vec{B} = 5\times 3 + 4\times(-2) + (-3)\times(-4) = 19.$$

$$\text{Cross product,} \qquad \vec{A}\times\vec{B} = \begin{vmatrix} \vec{a}_x & \vec{a}_y & \vec{a}_z \\ 5 & 4 & -3 \\ 3 & -2 & -4 \end{vmatrix} = -22\vec{a}_x - 11\vec{a}_y - 22\vec{a}_z.$$

1.5 Triple Products

There are two types of triple products commonly used.

1. Scalar triple product
2. Vector triple product

1.5.1 Scalar Triple Product

If three vectors $\vec{A}, \vec{B}$ and $\vec{C}$ are given then, their scalar triple product is defined as

$$\vec{A}\cdot(\vec{B}\times\vec{C}) = \vec{B}\cdot(\vec{C}\times\vec{A}) = \vec{C}\cdot(\vec{A}\times\vec{B}). \tag{1.28}$$

It follows cyclic permutation, i.e., $\vec{A}\to\vec{B}\to\vec{C}\to\vec{A}\to\vec{B}$.

If $\quad \vec{A} = A_x\vec{a}_x + A_y\vec{a}_y + A_z\vec{a}_z$

$\vec{B} = B_x\vec{a}_x + B_y\vec{a}_y + B_z\vec{a}_z$

and $\vec{C} = C_x\vec{a}_x + C_y\vec{a}_y + C_z\vec{a}_z$,

then the scalar triple product in matrix form is

$$\vec{A}\cdot(\vec{B}\times\vec{C}) = \begin{vmatrix} A_x & A_y & A_z \\ B_x & B_y & B_z \\ C_x & C_y & C_z \end{vmatrix}. \tag{1.29}$$

Scalar triple product is a scalar quantity. It gives the volume of a parallelopiped structure having A, B and C as its sides.

Fig. 1.7 shows that the scalar triple product is equal to the volume of a parallelopiped with vectors $\vec{A}$, $\vec{B}$ and $\vec{C}$. That is

$$\vec{A}\cdot(\vec{B}\times\vec{C}) = |ABC\sin\theta_1\cos\theta_2|. \tag{1.30}$$

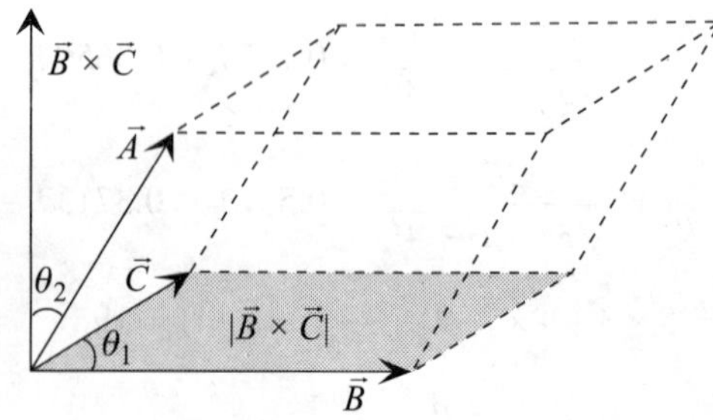

Fig. 1.7 *Scalar triple product*

1.5.2 Vector Triple Product

If three vectors $\vec{A}, \vec{B}$ and $\vec{C}$ are given, then the vector triple product is

$$\vec{A} \times (\vec{B} \times \vec{C}) = \vec{B}(\vec{A} \cdot \vec{C}) - \vec{C}(\vec{A} \cdot \vec{B})$$

(Bac−cab rule for easy identification)

Also $\vec{B} \times (\vec{C} \times \vec{A}) = \vec{C}(\vec{B} \cdot \vec{A}) - \vec{A}(\vec{B} \cdot \vec{C})$,

$$\vec{C} \times (\vec{A} \times \vec{B}) = \vec{A}(\vec{C} \cdot \vec{B}) - \vec{B}(\vec{C} \cdot \vec{A}) \tag{1.31}$$

and $(\vec{A} \cdot \vec{B})\vec{C} = \vec{C}(\vec{A} \cdot \vec{B})$. (1.32)

Vector triple product is a vector quantity.

Note: Vector division is not defined. Thus, $\alpha/\vec{A}$ or $\vec{A}/\vec{B}$ does not exist.

Example 1.3

Given three vectors $\vec{A} = 2\vec{a}_x - \vec{a}_z$ and $\vec{B} = \vec{a}_y + 2\vec{a}_z$ and $\vec{C} = \vec{a}_x - 2\vec{a}_y$. Find scalar and vector triple products.

Solution Given three vectors $\vec{A} = 2\vec{a}_x - \vec{a}_z$ and $\vec{B} = \vec{a}_y + 2\vec{a}_z$ and $\vec{C} = \vec{a}_x - 2\vec{a}_y$.

The scalar triple product in matrix form is

$$\vec{A} \cdot (\vec{B} \times \vec{C}) = \begin{vmatrix} A_x & A_y & A_z \\ B_x & B_y & B_z \\ C_x & C_y & C_z \end{vmatrix} = \begin{vmatrix} 2 & 0 & -1 \\ 0 & 1 & 2 \\ 1 & -2 & 0 \end{vmatrix} = 7.$$

The vector triple product is

$$\vec{A} \times (\vec{B} \times \vec{C}) = \vec{B}(\vec{A} \cdot \vec{C}) - \vec{C}(\vec{A} \cdot \vec{B}) = (2\vec{a}_x - \vec{a}_z)(2) - (\vec{a}_x - 2\vec{a}_y)(-2) = \vec{a}_x - 2\vec{a}_y + 4\vec{a}_z.$$

1.6 Orthogonal Coordinate System

So far, we represented vectors in Cartesian coordinate system. However, in general, a vector can be represented in any coordinate system. For easy analysis, we mostly use three-dimensional orthogonal coordinate systems. Here the three surfaces are mutually perpendicular to each other. The most commonly used coordinate systems for EM fields are

1. Cartesian or rectangular coordinate system
2. Circular cylindrical coordinate system and
3. Spherical coordinate system

It should be noted that the laws governing EM fields are invariant with the coordinate system. That is, we can transform the EM laws from one coordinate system to another for convenient representation.

1.6.1 Cartesian or Rectangular Coordinate System

In the Cartesian coordinate system, three coordinates x, y, and z axes are mutually at right angles to each other. For convenience, these axes are chosen as shown in Fig. 1.8.

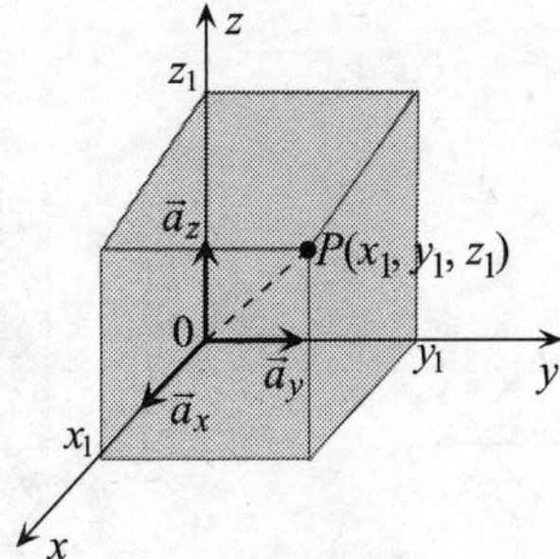

Fig. 1.8 *Cartesian coordinate system*

The axes of rotation in the anti-clockwise direction is taken as positive—this is called a right-handed system. The point O is taken as the origin. The three surfaces are

$y-z$ plane represented by $x=0$,
$x-z$ plane represented by $y=0$ and
$x-y$ plane represented by $z=0$.

The range of variables are $-\infty \leq x \leq \infty, -\infty \leq y \leq \infty$ and $-\infty \leq z \leq \infty$.

A point P on the rectangular coordinate system is represented as $P(x_1,\ y_1,\ z_1)$,

where x_1 = distance of point P along the x-axis from the origin,
y_1 = distance of point P along the y-axis from the origin,
z_1 = distance of point P along the z-axis from the origin.

The unit vectors along the axes x, y and z are $\vec{a}_x$, $\vec{a}_y$ and $\vec{a}_z$ respectively.

The vector $\vec{A}$ for point P can be represented as

$$\vec{A} - \overrightarrow{OP} - x_1\vec{a}_x + y_1\vec{a}_y + z_1\vec{a}_z.$$

The magnitude of vector $\vec{A}$ is

$$A = \sqrt{x_1^2 + y_1^2 + z_1^2}$$

and the unit vector of $\vec{A}$ is $\vec{a}_A = \dfrac{x_1\vec{a}_x + y_1\vec{a}_y + z_1\vec{a}_z}{\sqrt{x_1^2 + y_1^2 + z_1^2}}$. (1.33)

1.6.2 Circular Cylindrical Coordinate System

The circular cylindrical coordinate system uses polar coordinates on any one plane of the rectangular coordinates.

In general, the $x-y$ plane is taken as the polar coordinates $(\rho,\ \phi)$ and the z-axis is taken as it is.

A cylinder is formed with a circle of radius ρ_1, centre at the origin on the $x-y$ plane and the height z along the z-axis as shown in Fig. 1.9(a). The z-axis is the axis of the cylinder.

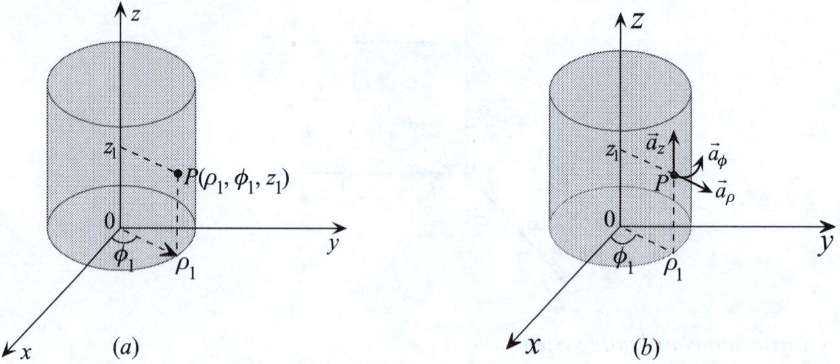

Fig. 1.9 *(a) Cylindrical coordinate system; (b) unit vectors in the cylindrical coordinate system*

A point $P(\rho_1,\ \phi_1,\ z_1)$ on the cylinder is the intersection of the radius ρ_1, the height z_1 and the angle ϕ_1, where ρ_1 is the radius of the point P from origin, z_1 is the height along the z-axis and ϕ_1 is the angle from the x-axis in the anti-clockwise direction.

The range of the variables are $0 < \rho \leq \infty, 0 \leq \phi \leq 2\pi$ and $-\infty \leq z \leq \infty$.

The unit vectors are $\vec{a}_\rho$ along the radial direction, $\vec{a}_\phi$ along the tangent to the surface of the cylinder and $\vec{a}_z$ along the z-axis upwards as shown in Fig. 1.9(b). These vectors are perpendicular to each other.

The vector $\vec{A}$ in cylindrical coordinates can be represented as

$$\vec{A} = A_\rho\vec{a}_\rho + A_\phi\vec{a}_\phi + A_z\vec{a}_z, \tag{1.34}$$

where A_ρ is the radial component of $\vec{A}$ along $\vec{a}_\rho$, A_ϕ is the phase component of $\vec{A}$ along $\vec{a}_\phi$ and A_z is the height component of $\vec{A}$ along $\vec{a}_z$.

The magnitude of vector $\vec{A}$ is

$$A = \sqrt{A_\rho^2 + A_\phi^2 + A_z^2}.$$

The relationships between Cartesian and Cylindrical coordinates are given by

$$\rho = \sqrt{x^2+y^2}, \ \phi = \tan^{-1}(y/x), \quad z = z, \tag{1.35}$$

and $\quad x = \rho\cos\phi, \ y = \rho\sin\phi, \quad z = z.$

Note 1: For unit vectors, the dot products are

$$\vec{a}_\rho \cdot \vec{a}_\phi = \vec{a}_\phi \cdot \vec{a}_z = \vec{a}_z \cdot \vec{a}_\rho = 0 \quad \text{(since the axes are orthogonal to each other).}$$

$$\vec{a}_\rho \cdot \vec{a}_\rho = \vec{a}_\phi \cdot \vec{a}_\phi = \vec{a}_z \cdot \vec{a}_z = 1 \text{ (since the angle is zero).} \tag{1.36}$$

Note 2: The cross products are

$$\vec{a}_\rho \times \vec{a}_\phi = \vec{a}_z, \quad \vec{a}_\phi \times \vec{a}_z = \vec{a}_\rho \quad \text{and} \quad \vec{a}_z \times \vec{a}_\rho = \vec{a}_\phi. \tag{1.37}$$

It follows cyclic permutation, i.e., $\vec{a}_\rho \to \vec{a}_\phi \to \vec{a}_z \to \vec{a}_\rho \to \vec{a}_\phi$.

Note 3: If two vectors are given by $\vec{A} = A_\rho\vec{a}_\rho + A_\phi\vec{a}_\phi + A_z\vec{a}_z$ and $\vec{B} = B_\rho\vec{a}_\rho + B_\phi\vec{a}_\phi + B_z\vec{a}_z$, then the dot product is

$$\begin{aligned} \vec{A}\cdot\vec{B} &= (A_\rho\vec{a}_\rho + A_\phi\vec{a}_\phi + A_z\vec{a}_z)\cdot(B_\rho\vec{a}_\rho + B_\phi\vec{a}_\phi + B_z\vec{a}_z) \\ &= A_\rho B_\rho + A_\phi B_\phi + A_z B_z \end{aligned} \tag{1.38}$$

and the cross product is

$$\vec{A}\times\vec{B} = (A_\rho\vec{a}_\rho + A_\phi\vec{a}_\phi + A_z\vec{a}_z)\times(B_\rho\vec{a}_\rho + B_\phi\vec{a}_\phi + B_z\vec{a}_z).$$

In matrix form

$$\vec{A}\times\vec{B} = \begin{vmatrix} \vec{a}_\rho & \vec{a}_\phi & \vec{a}_z \\ A_\rho & A_\phi & A_z \\ B_\rho & B_\phi & B_z \end{vmatrix}. \tag{1.39}$$

1.6.3 Spherical Coordinate System

The spherical coordinate system represents a sphere of radius r from the origin. The coordinates are $(r, \ \theta, \ \phi)$, where r is the radius of the sphere, ϕ is the horizontal angle (azimuth angle) measured from the x-axis and θ is the vertical angle (elevation) measured from the z-axis as shown in Fig. 1.10.

A point $P(r_1, \ \theta_1, \ \phi_1)$ on the sphere is the intersection of the radius r_1, horizontal angle ϕ_1 and the vertical angle θ_1, where r_1 is the radius of the point P from origin, ϕ_1 is the horizontal angle (azimuth angle) from the x-axis in the anti-clockwise direction and θ_1 is the vertical angle (elevation) from the z-axis in the clock-wise direction.

The range of the variables are $0 \le r \le \infty, \ \ 0 \le \theta \le \pi$ and $0 \le \phi \le 2\pi$.

The unit vectors $\vec{a}_r$, $\vec{a}_\theta$ and $\vec{a}_\phi$ are mutually perpendicular to each other.

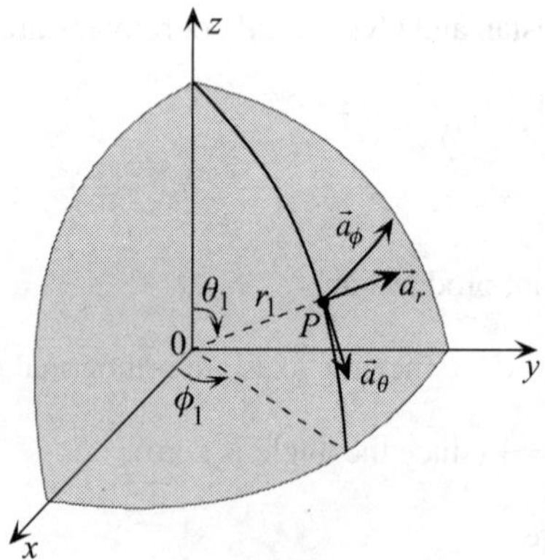

Fig. 1.10 *Unit vectors in the spherical coordinate system*

The vector A for point P is given by

$$\vec{A} = A_r\vec{a}_r + A_\theta\vec{a}_\theta + A_\phi\vec{a}_\phi, \tag{1.40}$$

where A_r is the radial component of A along $\vec{a}_r$, A_θ is the tangential component of A along $\vec{a}_\theta$ and A_ϕ is the tangential component of A along $\vec{a}_\phi$.

The magnitude of vector $\vec{A}$ is

$$A = \sqrt{= A_r^2 + A_\theta^2 + A_\phi^2}.$$

The relationships between spherical coordinates and Cartesian coordinates are given by

$$x = r\sin\theta\cos\phi, \quad y = r\sin\theta\sin\phi \quad \text{and} \quad z = r\cos\theta$$

$$\text{and} \quad r = \sqrt{x^2+y^2+z^2}, \quad \phi = \tan^{-1}(y/x) \quad \text{and} \quad \theta = \cos^{-1}(z/r). \tag{1.41}$$

Note 1: For unit vectors, the dot products are

$\vec{a}_r \cdot \vec{a}_\theta = \vec{a}_\theta \cdot \vec{a}_\phi = \vec{a}_\phi \cdot \vec{a}_r = 0$ (since the axes are orthogonal to each other).

$\vec{a}_r \cdot \vec{a}_r = \vec{a}_\theta \cdot \vec{a}_\theta = \vec{a}_\phi \cdot \vec{a}_\phi = 1$ (since the angle is zero). (1.42)

Note 2: The cross products are

$$\vec{a}_r \times \vec{a}_\theta = \vec{a}_\phi, \quad \vec{a}_\theta \times \vec{a}_\phi = \vec{a}_r \quad \text{and} \quad \vec{a}_\phi \times \vec{a}_r = \vec{a}_\theta. \tag{1.43}$$

It follows cyclic permutation, i.e., $\vec{a}_r \to \vec{a}_\theta \to \vec{a}_\phi \to \vec{a}_r \to \vec{a}_\theta$.

Note 3: If two vectors are given by $\vec{A} = A_r\vec{a}_r + A_\theta\vec{a}_\theta + A_\phi\vec{a}_\phi$ and $\vec{B} = B_r\vec{a}_r + B_\theta\vec{a}_\theta + B_\phi\vec{a}_\phi$, then the dot product is

$$\begin{aligned}\vec{A}\cdot\vec{B} &= (A_r\vec{a}_r + A_\theta\vec{a}_\theta + A_\phi\vec{a}_\phi)\cdot(B_r\vec{a}_r + B_\theta\vec{a}_\theta + B_\phi\vec{a}_\phi) \\ &= A_rB_r + A_\theta B_\theta + A_\phi B_\phi,\end{aligned} \tag{1.44}$$

and the cross product is

$$\vec{A}\times\vec{B}=(A_r\vec{a}_r+A_\theta\vec{a}_\theta+A_\phi\vec{a}_\phi)\times(B_r\vec{a}_r+B_\theta\vec{a}_\theta+B_\phi\vec{a}_\phi).$$

In matrix form

$$\vec{A}\times\vec{B}=\begin{vmatrix}\vec{a}_r & \vec{a}_\theta & \vec{a}_\phi\\ A_r & A_\theta & A_\phi\\ B_r & B_\theta & B_\phi\end{vmatrix}. \qquad (1.45)$$

Example 1.4 Express the point $P(-2,\ 6,\ 3)$ in cylindrical and spherical coordinate systems.

Solution Given a point $P(-2,\ 6,\ 3)$ in rectangular coordinates.

In cylindrical coordinates,

$$\rho=\sqrt{x^2+y^2}=\sqrt{2^2+6^2}=6.324.$$

$$\phi=\tan^{-1}(y/x)=\tan^{-1}(-6/2)=108.43^\circ.$$

$$z=z=3.$$

The point is $P(6.324,\ 108.43^\circ,\ 3)$.

In spherical coordinates,

$$r=\sqrt{x^2+y^2+z^2}=\sqrt{2^2+6^2+3^2}=7.$$

$$\theta=\cos^{-1}(z/r)=\cos^{-1}(3/7)=64.62^\circ.$$

$$\phi=\tan^{-1}(y/x)=\tan^{-1}(-6/2)=108.43^\circ.$$

The point is $P(6.324,\ 64.62^\circ,\ 108.43^\circ)$.

1.7 Transformation of Vectors

The transformation of a vector from one coordinate system to another coordinate system can be achieved by performing dot product of the vector with the unit vectors of the other coordinate system. Here we will discuss only the transformations of Cartesian coordinate system, cylindrical coordinate system and spherical coordinate system, since these are very useful in the applications of EM fields.

Transformation of vectors from Cartesian to cylindrical and vice versa

Let a vector in the Cartesian coordinate system be

$$\vec{A}=A_x\vec{a}_x+A_y\vec{a}_y+A_z\vec{a}_z$$

and in the cylindrical coordinate system be

$$\vec{A}=A_\rho\vec{a}_\rho+A_\phi\vec{a}_\phi+A_z\vec{a}_z.$$

The vector component along the radial direction is

$$A_\rho=A\cdot\vec{a}_\rho=(A_x\vec{a}_x+A_y\vec{a}_y+A_z\vec{a}_z)\cdot\vec{a}_\rho.$$

$$A_\rho = A_x\vec{a}_x\cdot\vec{a}_\rho + A_y\vec{a}_y\cdot\vec{a}_\rho + A_z\vec{a}_z\cdot\vec{a}_\rho.$$

Similarly, other vector components are

$$A_\phi = A\cdot\vec{a}_\phi = A_x\vec{a}_x\cdot\vec{a}_\phi + A_y\vec{a}_y\cdot\vec{a}_\phi + A_z\vec{a}_z\cdot\vec{a}_\phi$$

and $A_z = A_z.$ (1.46)

The relationships between unit vectors can be obtained from Fig. 1.11. When we transform the unit vectors from the Cartesian coordinate system to the cylindrical coordinate system and vice versa, it is observed that the unit vector $\vec{a}_z$ remains unchanged, i.e., $\vec{a}_z = \vec{a}_z$.

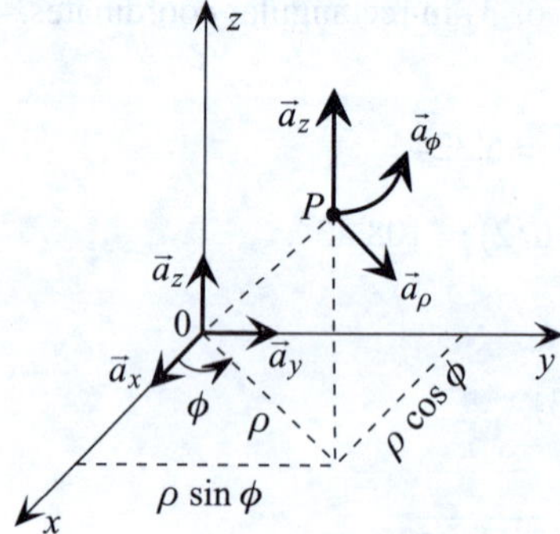

Fig. 1.11 *Directions of unit vectors*

To calculate the unit vectors $\vec{a}_\rho$ and $\vec{a}_\phi$, resolve the unit vectors $\vec{a}_x$ and $\vec{a}_y$ onto the $\rho-\phi$ plane by calculating the projections in the ρ and ϕ directions as shown in Fig. 1.12(a).

$$\vec{a}_x = \cos\phi\vec{a}_\rho - \sin\phi\vec{a}_\phi \text{ and } \vec{a}_y = \sin\phi\vec{a}_\rho + \cos\phi\vec{a}_\phi.$$

Similarly, to calculate the unit vectors $\vec{a}_x$ and $\vec{a}_y$, resolve the unit vectors $\vec{a}_\rho$ and $\vec{a}_\phi$ on the $x-y$ plane by calculating the projections in the x, y directions as shown in Fig. 1.12(b).

$$\vec{a}_\rho = \cos\phi\vec{a}_x + \sin\phi\vec{a}_y \text{ and } \vec{a}_\phi = -\sin\phi\vec{a}_x + \cos\phi\vec{a}_y.$$

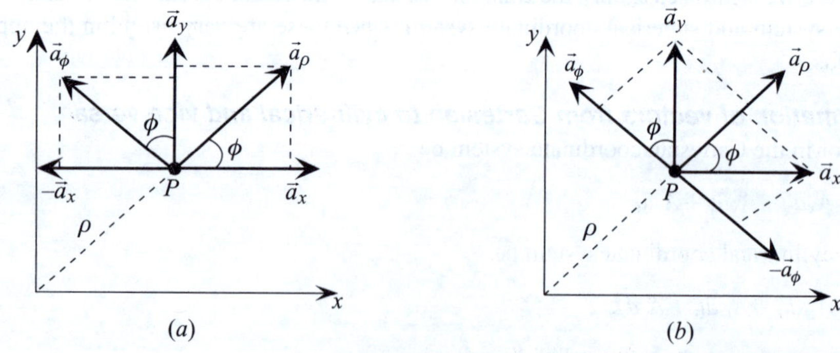

Fig. 1.12 *(a) Projections of unit vectors on the $\rho-\phi$ plane, (b) projections of unit vectors on the $x-y$ plane.*

Thus, the dot products of the unit vectors are given in Table 1.1.

Table 1.1 *Dot products of Cartesian and cylindrical unit vectors*

$\cdot$	$\vec{a}_\rho$	$\vec{a}_\phi$	$\vec{a}_z$
$\vec{a}_x$	$\cos\phi$	$-\sin\phi$	0
$\vec{a}_y$	$\sin\phi$	$\cos\phi$	0
$\vec{a}_z$	0	0	1

Now the vector components are

$$A_\rho = A_x \cos\phi + A_y \sin\phi,$$

$$A_\phi = -A_x \sin\phi + A_y \cos\phi,$$

$$A_x = A_\rho \cos\phi - A_\phi \sin\phi,$$

$$A_y = A_\rho \sin\phi + A_\phi \cos\phi,$$

and $A_z = A_z$.

The transformation of vectors from Cartesian to cylindrical coordinates in matrix form is

$$\begin{bmatrix} A_\rho \\ A_\phi \\ A_z \end{bmatrix} = \begin{bmatrix} \cos\phi & \sin\phi & 0 \\ -\sin\phi & \cos\phi & 0 \\ 0 & 0 & 1 \end{bmatrix} \begin{bmatrix} A_x \\ A_y \\ A_z \end{bmatrix}. \tag{1.47}$$

The inverse matrix gives the transformation of vectors from cylindrical to Cartesian coordinates.

$$\begin{bmatrix} A_x \\ A_y \\ A_z \end{bmatrix} = \begin{bmatrix} \cos\phi & \sin\phi & 0 \\ -\sin\phi & \cos\phi & 0 \\ 0 & 0 & 1 \end{bmatrix}^{-1} \begin{bmatrix} A_\rho \\ A_\phi \\ A_z \end{bmatrix}.$$

Since the matrix is symmetric

$$\begin{bmatrix} \cos\phi & \sin\phi & 0 \\ -\sin\phi & \cos\phi & 0 \\ 0 & 0 & 1 \end{bmatrix}^{-1} = \begin{bmatrix} \cos\phi & \sin\phi & 0 \\ -\sin\phi & \cos\phi & 0 \\ 0 & 0 & 1 \end{bmatrix}^{T} = \begin{bmatrix} \cos\phi & -\sin\phi & 0 \\ \sin\phi & \cos\phi & 0 \\ 0 & 0 & 1 \end{bmatrix}$$

$$\begin{bmatrix} A_x \\ A_y \\ A_z \end{bmatrix} = \begin{bmatrix} \cos\phi & -\sin\phi & 0 \\ \sin\phi & \cos\phi & 0 \\ 0 & 0 & 1 \end{bmatrix} \begin{bmatrix} A_\rho \\ A_\phi \\ A_z \end{bmatrix}. \tag{1.48}$$

Example 1.5 Express the vector $\vec{A} = x^2yz\vec{a}_x + xy^2z\vec{a}_y + xyz^2\vec{a}_z$ at the point $P(6.324,\ 108.43^\circ,\ 3)$ in the cylindrical coordinate system.

Solution Given the vector $\vec{A} = x^2yz\vec{a}_x + xy^2z\vec{a}_y + xyz^2\vec{a}_z$.

Transformation of the vector from Cartesian to cylindrical coordinates in matrix form is

$$\begin{bmatrix} A_\rho \\ A_\phi \\ A_z \end{bmatrix} = \begin{bmatrix} \cos\phi & \sin\phi & 0 \\ -\sin\phi & \cos\phi & 0 \\ 0 & 0 & 1 \end{bmatrix} \begin{bmatrix} A_x \\ A_y \\ A_z \end{bmatrix}$$

$$= \begin{bmatrix} \cos\phi & \sin\phi & 0 \\ -\sin\phi & \cos\phi & 0 \\ 0 & 0 & 1 \end{bmatrix} \begin{bmatrix} x^2yz \\ xy^2z \\ xyz^2 \end{bmatrix}$$

At the point $P(6.324,\ 108.43°,\ 3)$

$$\cos\phi = -0.316,\ \ \sin\phi = 0.9487.$$

$$x = \rho\cos\phi = 6.324\cos(108.43°) = -2.$$

$$y = \rho\sin\phi = 6.324\sin(108.43°) = 6.$$

$$z = z = 3.$$

$$\begin{bmatrix} A_\rho \\ A_\phi \\ A_z \end{bmatrix} = \begin{bmatrix} -0.316 & 0.9487 & 0 \\ -0.9487 & -0.316 & 0 \\ 0 & 0 & 1 \end{bmatrix} \begin{bmatrix} x^2yz \\ xy^2z \\ xyz^2 \end{bmatrix}$$

$$= \begin{bmatrix} -0.316 & 0.9487 & 0 \\ -0.9487 & -0.316 & 0 \\ 0 & 0 & 1 \end{bmatrix} \begin{bmatrix} 72 \\ -216 \\ -108 \end{bmatrix}.$$

$$A_\rho = -227.65,\ A_\phi = -0.0504,\ A_z = -108.$$

$$\vec{A} = -227.65\vec{a}_\rho - 0.0504\vec{a}_\phi - 108\vec{a}_z.$$

Transformation of vectors from Cartesian to spherical and vice versa

Let a vector $\vec{A}$ in Cartesian coordinate system be

$$\vec{A} = A_x\vec{a}_x + A_y\vec{a}_y + A_z\vec{a}_z$$

and in spherical coordinate system

$$\vec{A} = A_r\vec{a}_r + A_\theta\vec{a}_\theta + A_\phi\vec{a}_\phi.$$

The vector component along the radial direction is

$$A_r = A\cdot\vec{a}_r$$

$$= (A_x\vec{a}_x + A_y\vec{a}_y + A_z\vec{a}_z)\cdot\vec{a}_r.$$

$$\therefore\quad A_r = A_x\vec{a}_x\cdot\vec{a}_r + A_y\vec{a}_y\cdot\vec{a}_r + A_z\vec{a}_z\cdot\vec{a}_r.$$

The other components are

$$A_\theta = A_x\vec{a}_x \cdot \vec{a}_\theta + A_y\vec{a}_y \cdot \vec{a}_\theta + A_z\vec{a}_z \cdot \vec{a}_\theta.$$

$$A_\phi = A_x\vec{a}_x \cdot \vec{a}_\phi + A_y\vec{a}_y \cdot \vec{a}_\phi + A_z\vec{a}_z \cdot \vec{a}_\phi. \tag{1.49}$$

The relationships between unit vectors are obtained by resolving the components along the directions of the unit vectors as shown Fig. 1.13.

$$\vec{a}_x = \sin\theta\cos\phi\vec{a}_r + \cos\theta\cos\phi\vec{a}_\theta - \sin\phi\vec{a}_\phi,$$

$$\vec{a}_y = \sin\theta\sin\phi\vec{a}_r + \cos\theta\sin\phi\vec{a}_\theta + \cos\phi\vec{a}_\phi,$$

$$\vec{a}_z = \cos\theta\vec{a}_r - \sin\theta\vec{a}_\theta$$

and $\vec{a}_r = \sin\theta\cos\phi\vec{a}_x + \sin\theta\sin\phi\vec{a}_y + \cos\theta\vec{a}_z,$

$$\vec{a}_\theta = \cos\theta\cos\phi\vec{a}_x + \cos\theta\sin\phi\vec{a}_y - \sin\theta\vec{a}_z,$$

$$\vec{a}_\phi = -\sin\phi\vec{a}_x + \cos\phi\vec{a}_y.$$

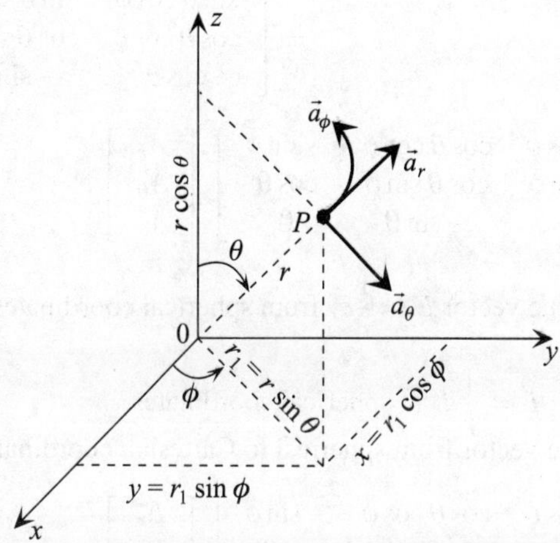

Fig. 1.13 *Directions of unit vectors*

The dot product of the unit vectors is given in Table 1.2.

Table 1.2 *Dot products of Cartesian and spherical unit vectors*

$\cdot$	$\vec{a}_r$	$\vec{a}_\theta$	$\vec{a}_\phi$
$\vec{a}_x$	$\sin\theta\cos\phi$	$\cos\theta\cos\phi$	$-\sin\phi$
$\vec{a}_y$	$\sin\theta\sin\phi$	$\cos\theta\sin\phi$	$\cos\phi$
$\vec{a}_z$	$\cos\theta$	$-\sin\theta$	0

From Table. 1.2, $A_r = A_x\sin\theta\cos\phi + A_y\sin\theta\sin\phi + A_z\cos\theta.$

Similarly, $A_\theta = A_x \cos\theta \cos\phi + A_y \cos\theta \sin\phi - A_z \sin\theta$ and $A_\phi = -A_x \sin\phi + A_y \cos\phi$.

The transformation in matrix form is

$$\begin{bmatrix} A_r \\ A_\theta \\ A_\phi \end{bmatrix} = \begin{bmatrix} \sin\theta\cos\phi & \sin\theta\sin\phi & \cos\theta \\ \cos\theta\cos\phi & \cos\theta\sin\phi & -\sin\theta \\ -\sin\phi & \cos\phi & 0 \end{bmatrix} \begin{bmatrix} A_x \\ A_y \\ A_z \end{bmatrix}. \tag{1.50}$$

The inverse matrix gives the transformation of vectors from spherical to Cartesian coordinates.

$$\begin{bmatrix} A_x \\ A_y \\ A_z \end{bmatrix}^{1} = \begin{bmatrix} \sin\theta\cos\phi & \sin\theta\sin\phi & \cos\theta \\ \cos\theta\cos\phi & \cos\theta\sin\phi & -\sin\theta \\ -\sin\phi & \cos\phi & 0 \end{bmatrix}^{-1} \begin{bmatrix} A_r \\ A_\theta \\ A_\phi \end{bmatrix}.$$

Since the matrix is symmetric,

$$\begin{bmatrix} \sin\theta\cos\phi & \sin\theta\sin\phi & \cos\theta \\ \cos\theta\cos\phi & \cos\theta\sin\phi & -\sin\theta \\ -\sin\phi & \cos\phi & 0 \end{bmatrix}^{-1} = \begin{bmatrix} \sin\theta\cos\phi & \sin\theta\sin\phi & \cos\theta \\ \cos\theta\cos\phi & \cos\theta\sin\phi & -\sin\theta \\ -\sin\phi & \cos\phi & 0 \end{bmatrix}^{T}$$

$$= \begin{bmatrix} \sin\theta\cos\phi & \sin\theta\sin\phi & -\sin\phi \\ \cos\theta\cos\phi & \cos\theta\sin\phi & \cos\phi \\ \cos\theta & -\sin\theta & 0 \end{bmatrix}.$$

$$\begin{bmatrix} A_x \\ A_y \\ A_z \end{bmatrix} = \begin{bmatrix} \sin\theta\cos\phi & \cos\theta\cos\phi & -\sin\phi \\ \sin\theta\sin\phi & \cos\theta\sin\phi & \cos\theta \\ \cos\theta & -\sin\theta & 0 \end{bmatrix} \begin{bmatrix} A_r \\ A_\theta \\ A_\phi \end{bmatrix}. \tag{1.51}$$

Example 1.6 Transform the vector $\vec{E} = \frac{1}{r}\vec{a}_r$ from spherical coordinates to Cartesian coordinates.

Solution Given the vector $\vec{E} = \frac{1}{r}\vec{a}_r$ in spherical coordinates.

The transformation of the vector from spherical to Cartesian coordinates is

$$\begin{bmatrix} E_x \\ E_y \\ E_z \end{bmatrix} = \begin{bmatrix} \sin\theta\cos\phi & \cos\theta\cos\phi & -\sin\phi \\ \sin\theta\sin\phi & \cos\theta\sin\phi & \cos\theta \\ \cos\theta & -\sin\theta & 0 \end{bmatrix} \begin{bmatrix} E_r \\ E_\theta \\ E_\phi \end{bmatrix}.$$

$$= \begin{bmatrix} \sin\theta\cos\phi & \cos\theta\cos\phi & -\sin\phi \\ \sin\theta\sin\phi & \cos\theta\sin\phi & \cos\theta \\ \cos\theta & -\sin\theta & 0 \end{bmatrix} \begin{bmatrix} 1/r \\ 0 \\ 0 \end{bmatrix}.$$

$$E_x = \frac{1}{r}(\sin\theta\cos\phi), \quad E_y = \frac{1}{r}(\sin\theta\sin\phi) \text{ and } E_z = \frac{1}{r}(\cos\theta)$$

From the relationships between coordinates, we know that

$$r = \sqrt{x^2 + y^2 + z^2}, \quad \cos\theta = \frac{z}{r} \text{ and } \tan\phi = \frac{y}{x}.$$

$$\therefore \quad \sin\theta = \frac{\sqrt{x^2+y^2}}{r}, \quad \cos\phi = \frac{x}{\sqrt{x^2+y^2}} \quad \text{and} \quad \sin\phi = \frac{y}{\sqrt{x^2+y^2}}.$$

$$\therefore \quad E_x = \frac{1}{r}\left(\frac{\sqrt{x^2+y^2}}{r}\right)\left(\frac{x}{\sqrt{x^2+y^2}}\right) = \frac{x}{r^2} = \frac{x}{x^2+y^2+z^2}.$$

$$\therefore \quad E_y = \frac{1}{r}\left(\frac{\sqrt{x^2+y^2}}{r}\right)\left(\frac{y}{\sqrt{x^2+y^2}}\right) = \frac{y}{r^2} = \frac{y}{x^2+y^2+z^2}.$$

$$E_z = \frac{1}{r}\left(\frac{z}{r}\right) = \frac{z}{x^2+y^2+z^2}.$$

$$\vec{E} = E_x\vec{a}_x + E_y\vec{a}_y + E_z\vec{a}_z$$

$$= \frac{1}{x^2+y^2+z^2}(x\vec{a}_x + y\vec{a}_y + z\vec{a}_z).$$

1.8 Differential Elements

Differential lengths, surfaces and volumes in the orthogonal coordinate system are useful in vector calculus to analyse EM fields.

Differential elements in Cartesian coordinates

Let dx, dy and dz be differential increments of point $P(x, y, z)$ as shown in Fig. 1.14(a).

The differential length vector is

$$d\vec{l} = dx\vec{a}_x + dy\vec{a}_y + dz\vec{a}_z. \tag{1.52}$$

Magnitude of the differential length vector is

$$dl = \sqrt{(dx)^2 + (dy)^2 + (dz)^2}. \tag{1.53}$$

The differential surface vectors are

$$d\vec{S}_x = dydz\vec{a}_x, \quad d\vec{S}_y = dxdz\vec{a}_y \quad \text{and} \quad d\vec{S}_z = dxdy\vec{a}_z. \tag{1.54}$$

The differential volume is a scalar quantity and is given by $dv = dxdydz$. (1.55)

Differential elements in cylindrical coordinates

If the differential increments in cylindrical coordinates are $d\rho$ in the radial direction, $\rho d\phi$ tangential to the surface and dz in the z direction from point $P(\rho, \phi, z)$ as shown in Fig. 1.14(b).

The differential length vector is

$$d\vec{l} = d\rho\vec{a}_\rho + \rho d\phi\vec{a}_\phi + dz\vec{a}_z. \tag{1.56}$$

Magnitude of the differential length vector is

$$dl = \sqrt{(d\rho)^2 + (\rho d\phi)^2 + (dz)^2}. \tag{1.57}$$

The differential surface vectors are

$$d\vec{S}_\rho = \rho d\phi dz \vec{a}_\rho,\ d\vec{S}_\phi = d\rho dz \vec{a}_\phi \text{ and } d\vec{S}_z = \rho d\phi d\rho \vec{a}_z. \tag{1.58}$$

The differential volume is

$$dv = \rho d\rho d\phi dz \tag{1.59}$$

Differential elements in spherical coordinates

Consider a point $P(r,\ \theta,\ \phi)$ in the spherical coordinate system as shown in Fig. 1.14(c). Let dr, $d\theta$ and $d\phi$ be the differential increments of point P along their directions.

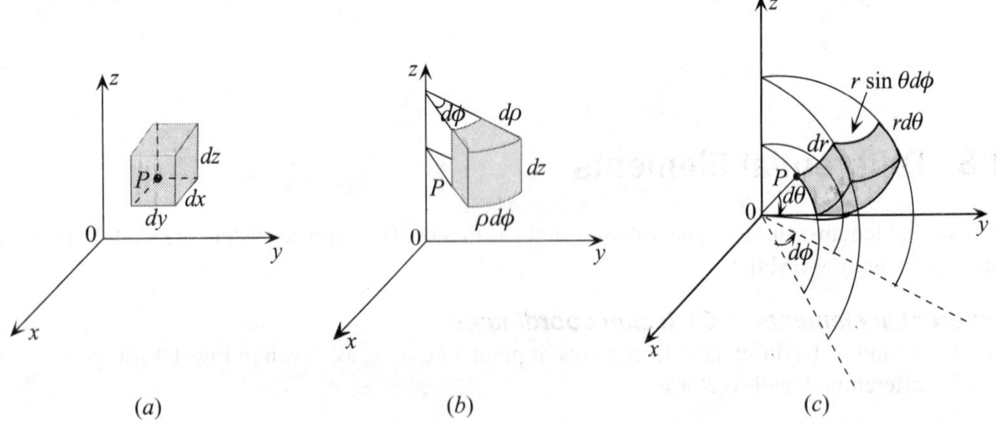

Fig. 1.14 *Differential elements in (a) Cartesian coordinates; (b) cylindrical coordinates and (c) spherical coordinates*

The differential length, in the radial direction is dr, in the θ direction is $rd\theta$ and in the ϕ direction is $r\sin\theta d\phi$.

The differential length vector is

$$d\vec{l} = dr\vec{a}_r + rd\theta\vec{a}_\theta + r\sin\theta d\phi\vec{a}_\phi. \tag{1.60}$$

Magnitude of the differential length vector is

$$dl = \sqrt{(dr)^2 + (rd\theta)^2 + (r\sin\theta d\phi)^2}. \tag{1.61}$$

The differential surface vectors are

$$d\vec{S}_r = r^2\sin\theta d\theta d\phi\vec{a}_r,\quad d\vec{S}_\theta = r\sin\theta dr d\phi\vec{a}_\theta \quad \text{and} \quad d\vec{S}_\phi = r dr d\theta\vec{a}_\phi. \tag{1.62}$$

The differential volume is

$$dv = r^2\sin\theta dr d\theta d\phi. \tag{1.63}$$

Note: The differential lengths and differential surfaces are vector quantities whereas the differential volume is a scalar quantity. The surface area of the differential surface vector is always normal to its direction.

1.9 Line, Surface and Volume Integrals

The integration of vector fields is often needed to find the desired quantities from the given fields. Line, surface and volume integrals are useful in vector analysis. These integrals are used in many applications to obtain electric and magnetic forces on charges, charge and current densities in materials etc.

Line integrals

Consider a vector field $\vec{A}$ and an arbitrary path C in the field as shown Fig. 1.15(a).

The line integral of the vector $\vec{A}$ over the path C is given by

$$B = \int_C \vec{A} \cdot d\vec{l} \text{ or } B = \int_l \vec{A} \cdot d\vec{l}, \tag{1.64}$$

where $d\vec{l}$ is the differential length vector, l in integration indicates line integral and B is a scalar quantity.

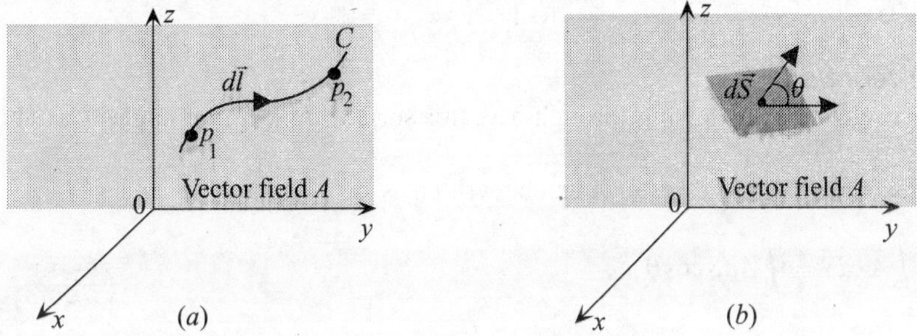

Fig. 1.15 *(a) Line integral, (b) surface integral*

If the integration is between two points P_1 and P_2, then the line integral is

$$B = \int_{p_1}^{p_2} \vec{A} \cdot d\vec{l}. \tag{1.65}$$

If the integration is around a closed path, the line integral becomes

$$B = \oint_l \vec{A} \cdot d\vec{l}. \tag{1.66}$$

This integration is called closed contour integral.

Note: If a closed contour integral of a vector is zero, then the vector field is called conservative field otherwise the field is non-conservative. For example, consider a force field—the line integral is work done. If the field is conservative, the total work done on any closed path in the field is zero. That is, for a conservative field,

$$B = \oint_l \vec{A} \cdot d\vec{l} = 0. \tag{1.67}$$

Example 1.7 Calculate the line integral from point (1, 1, 0) to point (3, 3, 0) for the force field $\vec{F} = xy\vec{a}_x + 4xy\vec{a}_y$.

Solution Given the force field $\vec{F} = xy\vec{a}_x + 4xy\vec{a}_y$ in Cartesian coordinates.

The line integral is $B = \int_l \vec{F} \cdot d\vec{l}$.

$$d\vec{l} = dx\vec{a}_x + dy\vec{a}_y + dz\vec{a}_z.$$

$$B = \int_l (xy\vec{a}_x + 4xy\vec{a}_y) \cdot (dx\vec{a}_x + dy\vec{a}_y + dz\vec{a}_z)$$

$$= \int_l (xydx + 4xydy) = \int_1^3 (xydx)\Big|_{y=1} + \int_1^3 (4xydy)\Big|_{x=3} = \int_1^3 xdx + \int_1^3 12ydy = 52.$$

Surface integrals

Consider a vector field $\vec{A}$ flowing through a vector surface $d\vec{S}$ with an angle θ as shown in Fig. 1.15(b).

The surface integral of the vector $\vec{A}$ can be written as

$$B = \int_s \vec{A} \cdot d\vec{S} = \int_s AdS\cos\theta. \tag{1.70}$$

Since $d\vec{S}$ is the product of two variables, the surface integral is a double integral.

If the surface area is a closed surface, the integral is called a closed surface integral and is denoted as

$$B = \oint_s \vec{A} \cdot d\vec{S}. \tag{1.71}$$

For example, if $\vec{A}$ is a charge density, the surface integral gives the net charge through the surface area.

Note: The line integral or surface integral of a vector results in a scalar quantity.

Example 1.8 Find the surface integral for the vector field $\vec{A} = \frac{5}{\rho}\cos\phi\vec{a}_\phi$ bounded by the surface $-\pi/4 \le \phi \le \pi/4,\ 0 \le z \le 5$.

Solution Given the vector $\vec{A} = \frac{5}{\rho}\cos\phi\vec{a}_\rho$ in cylindrical coordinates.

We know that the surface integral is

$$B = \int_s \vec{A} \cdot d\vec{S}.$$

$$d\vec{S} = \rho d\phi dz \vec{a}_\rho.$$

Given the surface is $-\pi/4 \le \phi \le \pi/4,\ 0 \le z \le 5$.

$$B = \int_s \vec{A} \cdot d\vec{S} = \int_0^5 \int_{-\pi/4}^{\pi/4} \left(\frac{5\cos\phi}{\rho} \rho d\phi \right) dz = 5 \times 5 \int_{-\pi/4}^{\pi/4} \cos\phi d\phi = 25\sqrt{2} = 35.35.$$

Volume integrals

There are two types of volume integrals: (i) scalar volume integral and (ii) vector volume integral.

Scalar volume integral

Consider a scalar field A within a volume dv. The volume integral of the scalar field A is given by

$$B = \int_v A \cdot dv. \tag{1.72}$$

Vector volume integral

Consider a vector field $\vec{A}$ within a volume dv. The volume integral of the vector field $\vec{A}$ is given by

$$\vec{B} = \int_v \vec{A} \cdot dv. \tag{1.73}$$

The vector volume integral in a given coordinate system should be evaluated over each component separately.

That is, in the Cartesian coordinate system, if $\vec{A} = A_x \vec{a}_x + A_y \vec{a}_y + A_z \vec{a}_z$, than

$$\vec{B} = \vec{a}_x \int_v A_x dv + \vec{a}_y \int_v A_y dv + \vec{a}_z \int_v A_z dv. \tag{1.74}$$

Since dv is the product of three variables, the volume integral is a triple integral.

1.10 Gradient, Divergence and Curl of a Vector

Differentiation of scalar and vector quantities are also needed in vector analysis. The useful types of operators are gradient, divergence and curl. These operators can be expressed in different types of orthogonal coordinate systems.

Cartesian coordinates

Gradient: The gradient of a scalar field V is a vector function which gives both magnitude and direction of the maximum rate of change of the scalar quantity V. Its magnitude and direction depends on the position in space where the gradient is to be evaluated.

It is denoted as

$$\text{grad } V = \nabla V = \left.\frac{dV}{dl}\right|_{\max}$$

or

$$\nabla V = \frac{\partial V}{\partial x}\vec{a}_x + \frac{\partial V}{\partial y}\vec{a}_y + \frac{\partial V}{\partial z}\vec{a}_z, \tag{1.75}$$

where del '∇' is a differential vector operator or gradient operator defined as

$$\nabla = \frac{\partial}{\partial x}\vec{a}_x + \frac{\partial}{\partial y}\vec{a}_y + \frac{\partial}{\partial z}\vec{a}_z. \tag{1.76}$$

Properties of the gradient of a scalar field *V*

1. The gradient of a scalar field is a vector field.
2. The gradient is perpendicular to a constant value surface.
3. The magnitude of the gradient is equal to the maximum rate of change in the scalar field per unit distance.
4. The direction of the gradient is in the direction of the maximum rate of change in scalar field. That is, the gradient gives the maximum slope of the scalar field at a given point.
5. The projection of the gradient of V in the direction of $d\vec{l} = dl\vec{a}_l$ is called the directional derivative of V along the direction $\vec{a}_l$.
6. The differential vector dV can be obtained as

$$dV = \nabla V \cdot d\vec{l} = \frac{\partial V}{\partial x}dx + \frac{\partial V}{\partial y}dy + \frac{\partial V}{\partial z}dz. \tag{1.77}$$

7. The gradient of the product of two scalar fields U and V is given by

$$\nabla(UV) = U\nabla V + V\nabla U. \tag{1.78}$$

Divergence: The divergence of a vector quantity $\vec{A}$ at a given point is the net outward flux from a small volume, through the closed surface enclosing the volume as the volume shrinks to zero about the point.

Mathematically, it can be expressed as

$$\text{div } \vec{A} = \nabla \cdot \vec{A} = \lim_{\Delta v \to 0} \frac{\oint_S \vec{A} \cdot d\vec{S}}{\Delta v}.$$

If $\vec{A} = A_x\vec{a}_x + A_y\vec{a}_y + A_z\vec{a}_z$,

$$\nabla \cdot \vec{A} = \text{div}\vec{A} = \frac{\partial A_x}{\partial x} + \frac{\partial A_y}{\partial y} + \frac{\partial A_z}{\partial z}, \tag{1.79}$$

where div $\vec{A}$ is a scalar quantity.

Divergence of a field indicates that there is a net flow out of the volume through the closed surface or the flow diverges from the point. Divergence represents the distribution or spread of a given vector quantity from a point. The divergence operator is useful in determining whether there is a source or a sink at locations in space where a vector field exists. In electromagnetic fields, the distribution of positive electric charges (sources) and negative electric charges (sinks) in space can be identified by divergence.

Properties of the divergence of a vector field

1. The divergence of a vector field gives a scalar quantity.
2. The divergence of a scalar field does not exist.
3. The divergence follows distributive law, i.e.,

$$\nabla \cdot (\vec{A}+\vec{B}) = \nabla \cdot \vec{A} + \nabla \cdot \vec{B}. \tag{1.80}$$

4. If V is a scalar and $\vec{A}$ is a vector, then

$$\nabla \cdot (V\vec{A}) = V(\nabla \cdot \vec{A}) + \vec{A} \cdot (\nabla V). \tag{1.81}$$

5. The divergence of a gradient V is

$$\nabla \cdot \nabla V = \nabla^2 V, \tag{1.82}$$

where $\nabla^2 = \dfrac{\partial^2}{\partial x^2}\vec{a}_x + \dfrac{\partial^2}{\partial y^2}\vec{a}_y + \dfrac{\partial^2}{\partial z^2}\vec{a}_z.$

∇^2 is called a Laplacian operator.

$$\nabla^2 V = \frac{\partial^2 V}{\partial x^2} + \frac{\partial^2 V}{\partial y^2} + \frac{\partial^2 V}{\partial z^2}. \tag{1.83}$$

6. If a vector field $\vec{A}$ is divergence free or $\nabla \cdot \vec{A} = 0$, the field is called a solenoidal field. And the field is source free. Otherwise, the field is non-solenoidal.

Curl: The curl of a vector quantity $\vec{A}$ is the net circulation of the vector $\vec{A}$ per unit area as the area tends to zero and its direction is normal to the area when the area is oriented such that the circulation is maximum.

Mathematically, it can be expressed as

$$\text{curl}\,\vec{A} = \nabla \times \vec{A} = \lim_{\Delta S \to 0} \left. \frac{\vec{a}_n \oint_l \vec{A} \cdot d\vec{l}}{\Delta S} \right|_{\max}.$$

If $\vec{A} = A_x\vec{a}_x + A_y\vec{a}_y + A_z\vec{a}_z$,

$$\nabla \times \vec{A} = \text{curl}\,\vec{A} = \begin{vmatrix} \vec{a}_x & \vec{a}_y & \vec{a}_z \\ \dfrac{\partial}{\partial x} & \dfrac{\partial}{\partial y} & \dfrac{\partial}{\partial z} \\ A_x & A_y & A_z \end{vmatrix}. \tag{1.84}$$

Curl of a vector field represents the rate of rotation or twist of a vector quantity at any point in space. It provides the maximum value of the circulation of the field per unit area and also indicates the direction of the maximum value.

Properties of the curl of a vector field V

1. The curl of a vector field is another vector field.
2. The curl of a scalar field does not exist.
3. The curl follows distributive law, i.e.,

$$\nabla \times (\vec{A}+\vec{B}) = \nabla \times \vec{A} + \nabla \times \vec{B}. \tag{1.85}$$

4. If V is a scalar and $\vec{A}$ is a vector, then

$$\nabla \times (V\vec{A}) = V(\nabla \times \vec{A}) + (\nabla V) \times \vec{A}. \tag{1.86}$$

5. The curl of a gradient of a scalar field is always zero, i.e., $\nabla \times (\nabla V) = 0$. (1.87)
6. The divergence of a curl of a vector field is also zero, i.e., $\nabla \cdot (\nabla \times A) = 0$. (1.88)
7. If a vector field $\vec{A}$ is curl free, that is, there no circulation, $\nabla \times \vec{A} = 0$, then the field is called irrotational or conservative.

Cylindrical coordinates

In the cylindrical coordinate system, the operator quantities are

Gradient: $$\nabla V = \frac{\partial V}{\partial \rho}\vec{a}_\rho + \frac{1}{\rho}\frac{\partial V}{\partial \phi}\vec{a}_\phi + \frac{\partial V}{\partial z}\vec{a}_z, \tag{1.89}$$

Divergence: $$\nabla \cdot \vec{A} = \frac{1}{\rho}\frac{\partial(\rho A_\rho)}{\partial \rho} + \frac{1}{\rho}\frac{\partial A_\phi}{\rho\partial \phi} + \frac{\partial A_z}{\partial z}, \tag{1.90}$$

Curl: $$\nabla \times A = \frac{1}{\rho}\begin{vmatrix} \vec{a}_\rho & \rho\vec{a}_\phi & \vec{a}_z \\ \dfrac{\partial}{\partial \rho} & \dfrac{\partial}{\partial \phi} & \dfrac{\partial}{\partial z} \\ A_\rho & \rho A_\phi & A_z \end{vmatrix}, \tag{1.91}$$

Laplacian operator: $$\nabla^2 V = \frac{1}{\rho}\frac{\partial}{\partial \rho}\left(\rho\frac{\partial V}{\partial \rho}\right) + \frac{1}{\rho^2}\frac{\partial^2 V}{\partial \phi^2} + \frac{\partial^2 V}{\partial z^2}. \tag{1.92}$$

Spherical coordinates

Similarly, in the spherical coordinate system, the operator quantities are

Gradient: $$\nabla V = \frac{\partial V}{\partial r}\vec{a}_r + \frac{1}{r}\frac{\partial V}{\partial \theta}\vec{a}_\theta + \frac{1}{r\sin\theta}\frac{\partial V}{\partial \phi}\vec{a}_\phi, \tag{1.93}$$

Divergence: $$\nabla \cdot \vec{A} = \frac{1}{r^2}\frac{\partial}{\partial r}(r^2 A_r) + \frac{1}{r\sin\theta}\frac{\partial}{\partial \theta}(\sin\theta A_\theta) + \frac{1}{r\sin\theta}\frac{\partial A_\phi}{\partial \phi}, \tag{1.94}$$

Curl: $$\nabla \times \vec{A} = \frac{1}{r^2\sin\theta}\begin{vmatrix} \vec{a}_r & r\vec{a}_\theta & r\sin\theta\vec{a}_\phi \\ \dfrac{\partial}{\partial r} & \dfrac{\partial}{\partial \theta} & \dfrac{\partial}{\partial \phi} \\ A_r & rA_\theta & r\sin\theta A_\phi \end{vmatrix} \tag{1.95}$$

Laplacian operator: $$\nabla^2 V = \frac{1}{r^2}\frac{\partial}{\partial r}\left(\frac{r^2\partial V}{\partial r}\right) + \frac{1}{r^2\sin\theta}\frac{\partial}{\partial\theta}\left(\sin\theta\frac{\partial V}{\partial\theta}\right) + \frac{1}{r^2\sin^2\theta}\frac{\partial^2 V}{\partial\phi^2}. \quad (1.96)$$

Example 1.9 Find the divergence of the vector $\vec{A} = \rho\sin\phi\vec{a}_\rho + 2\rho\cos\phi\vec{a}_\phi + 2z^2\vec{a}_z$ at the point $P(1,\ \phi,\ 5)$.

Solution Given the vector $\vec{A} = \rho\sin\phi\vec{a}_\rho + 2\rho\cos\phi\vec{a}_\phi + 2z^2\vec{a}_z$.

We know that the divergence in cylindrical coordinates is

$$\nabla\cdot\vec{A} = \frac{1}{\rho}\frac{\partial(\rho A_\rho)}{\partial\rho} + \frac{1}{\rho}\frac{\partial A_\phi}{\partial\phi} + \frac{\partial A_z}{\partial z}.$$

$$\nabla\cdot\vec{A} = \frac{1}{\rho}\frac{\partial}{\partial\rho}(\rho^2\sin\phi) + \frac{1}{\rho}\frac{\partial}{\partial\phi}(2\rho\cos\phi) + \frac{\partial}{\partial z}(2z^2)\vec{a}_z$$

$$= 2\sin\phi - 2\sin\phi + 4z = 4z.$$

At the point $P(1,\ \phi,\ 5)$, $\nabla\cdot\vec{A} = 20$.

Example 1.10 Find the $\nabla\times\vec{A}$ for the vector $\vec{A} = 5\sin\theta\vec{a}_\theta$ at the point $P(1,\ \pi/4,\ \pi)$.

Solution Given the vector $\vec{A} = 5\sin\theta\vec{a}_\theta$ in spherical coordinates.

We know that the curl is

$$\nabla\times\vec{A} = \frac{1}{r^2\sin\theta}\begin{vmatrix} \vec{a}_r & r\vec{a}_\theta & r\sin\theta\vec{a}_\phi \\ \frac{\partial}{\partial r} & \frac{\partial}{\partial\theta} & \frac{\partial}{\partial\phi} \\ A_r & rA_\theta & r\sin\theta A_\phi \end{vmatrix} = \frac{1}{r^2\sin\theta}\begin{vmatrix} \vec{a}_r & r\vec{a}_\theta & r\sin\theta\vec{a}_\phi \\ \frac{\partial}{\partial r} & \frac{\partial}{\partial\theta} & \frac{\partial}{\partial\phi} \\ 0 & 5r\sin\theta & 0 \end{vmatrix}$$

$$= \frac{1}{r^2\sin\theta}\frac{-\partial}{\partial\phi}(5r\sin\theta)\vec{a}_r + \frac{1}{r}\frac{\partial}{\partial r}(5r\sin\theta)\vec{a}_\phi = \frac{5\sin\theta}{r}\vec{a}_\phi.$$

At the point $P(1,\ \pi/4,\ \pi), \nabla\times\vec{A} = 5\sin(\pi/4) = 3.575\vec{a}_\phi$.

1.11 Del Operators

Consider two vector quantities $\vec{A}$ and $\vec{B}$ and two scalar quantities U and V. Some useful del operators are listed below.

$$\nabla(UV) = U(\nabla V) + V(\nabla U).$$

$$\nabla\cdot(V\vec{A}) = V(\nabla\cdot\vec{A}) + \vec{A}\cdot(\nabla V).$$

$$\nabla\cdot(\vec{A}+\vec{B}) = \nabla\cdot\vec{A} + \nabla\cdot\vec{B}.$$

$$\nabla\times(\vec{A}+\vec{B}) = \nabla\times\vec{A} + \nabla\times\vec{B}.$$

$$\nabla \times (\nabla V) = 0.$$

$$\nabla \cdot (\nabla \times \vec{A}) = 0.$$

$$\nabla \cdot (\nabla V) = \nabla^2 V.$$

$$\nabla \times (V\vec{A}) = V(\nabla \times \vec{A}) + (\nabla V) \times \vec{A}.$$

$$\nabla \cdot (\vec{A} \times \vec{B}) = \vec{B} \cdot (\nabla \times \vec{A}) - \vec{A} \cdot (\nabla \times \vec{B}).$$

$$\nabla \times (\vec{A} \times \vec{B}) = \vec{A}(\nabla \cdot \vec{B}) - \vec{B}(\nabla \cdot \vec{A}) + (\vec{B} \cdot \nabla)\vec{A} - (\vec{A} \cdot \nabla)\vec{B}.$$

$$\nabla \times \nabla \times \vec{A} = \nabla(\nabla \cdot \vec{A}) - \nabla^2 \vec{A}. \tag{1.97}$$

Vector laws and del operators are very useful in solving EM wave field equations.

1.12 Useful Theorems

1. **Divergence Theorem**: For a given vector field $\vec{A}$, the divergence theorem states that the volume integral of the divergence of a vector field $\vec{A}$ is equal to the total outward flux of the vector field $\vec{A}$ through the closed surface S that bounds the volume.

   It can be expressed as

$$\int_v \nabla \cdot \vec{A} dv = \oint_s \vec{A} \cdot d\vec{S}, \tag{1.98}$$

   where $d\vec{S} = dS\vec{a}_n$ is a differential area directed outward normal to S.

   It converts a volume integral of the divergence of a vector to a closed surface integral of the vector and vice versa.

   The divergence theorem is valid if the vector field $\vec{A}$ as well as its first derivatives exist. It applies to any volume that is bounded by a closed surface. The direction of the surface is always away from the volume and normal to the surface area. The divergence theorem is not valid for open surfaces like discs or planes.

2. **Stokes' Theorem**: For a given vector field $\vec{A}$, Stokes' theorem states that the surface integral of the curl of a vector field $\vec{A}$ over an open surface S is equal to the circulation (line integral) of a vector field $\vec{A}$ around a closed path l bounded by the surface S.

   It can be expressed as

$$\int_S (\nabla \times \vec{A}) \cdot d\vec{S} = \oint_l \vec{A} \cdot d\vec{l}. \tag{1.99}$$

   The equation converts a surface integral of the curl of a vector to a closed line integral of the vector and vice versa.

   Stokes' theorem is valid if the vector field $\vec{A}$ as well as its first derivatives exist and are continuous both on surface and the path. It applies to any surface that is bounded by a closed path. The surface should be an open surface like a two-dimensional disc or plane.

3. **Green's Theorem**: Green's theorem is applied to two-dimensional space. It states that if a vector field $\vec{A} = A_x\vec{a}_x + A_y\vec{a}_y$ is specified in a region S and enclosed by a contour C in the $x-y$ plane, then

$$\oint_C \vec{A} \cdot d\vec{l} = \int_S \left(\frac{\partial A_x}{\partial x} - \frac{\partial A_y}{\partial y} \right) \cdot dxdy, \tag{1.100}$$

where $d\vec{l}$ is the differential length vector,

$$d\vec{l} = dx\vec{a}_x + dy\vec{a}_y.$$

The equation converts a surface integral of a vector to a closed line integral of the vector and vice versa.

It is observed that Green's theorem is equal to Stoke's theorem in two-dimensional space.

4. **Helmholtz's Theorem**: Helmholtz' theorem states that a vector field $\vec{A}$ is uniquely defined within an additive constant if its divergence and curl are specified everywhere in the space.

That is, any vector field which is continuous and vanishes at infinity can be decomposed into two fields, one is the gradient of a scalar field and the other is the curl of a vector field.

If a scalar field V and a vector field $\vec{B}$ exist and are continuous within a bounded region, Helmholtz's theorem can be written as

$$\vec{A} = -\nabla V + \nabla \times \vec{B}. \tag{1.101}$$

According to Helmholtz's theorem, vector fields can be classified into four groups in terms of their divergence and curl.

1. Solenoidal and irrotational fields: These have no scalar and vector sources, that is, their divergence and curl vanishes, $\nabla \cdot \vec{A} = 0,\ \nabla \times \vec{A} = 0$.
2. Solenoidal and rotational fields: These have only a vector source and no scalar source, that is, their divergence vanishes, $\nabla \cdot \vec{A} = 0,\ \nabla \times \vec{A} \neq 0$.
3. Non-solenoidal and irrotational fields: These have only a scalar source and no vector source, that is, their curl vanishes. $\nabla \cdot \vec{A} \neq 0,\ \nabla \times \vec{A} = 0$.
4. Non-solenoidal and rotational fields: These have both a scalar and a vector source, that is, their divergence and curl exist, $\nabla \cdot \vec{A} \neq 0,\ \nabla \times \vec{A} \neq 0$.

   For example, in EM fields, a static electric field in a charge-free region is a solenoidal and irrotational field, whereas in a charged region, it becomes non-solenoidal and irrotational. A steady magnetic field in a material is a solenoidal and rotational field whereas a time-varying EM field in a charged region is a non-solenoidal and rotational field.

Example 1.11 Given two vectors $\vec{A} = 2\vec{a}_x - 5\vec{a}_y - 4\vec{a}_z$ and $\vec{B} = 3\vec{a}_x + 5\vec{a}_y + 2\vec{a}_z$, find the projection of $\vec{A}$ along $\vec{B}$, the angle between $\vec{A}$ and $\vec{B}$.

Solution Given $\vec{A} = 2\vec{a}_x - 5\vec{a}_y - 4\vec{a}_z$ and $\vec{B} = 3\vec{a}_x + 5\vec{a}_y + 2\vec{a}_z$.

Unit vector of $\vec{B}$, $\vec{a}_B = \dfrac{\vec{B}}{B} = \dfrac{3\vec{a}_x + 5\vec{a}_y + 2\vec{a}_z}{\sqrt{3^2 + 5^2 + 2^2}} = 0.487\vec{a}_x + 0.811\vec{a}_y - 0.325\vec{a}_z$.

Projection of $\vec{A}$ along $\vec{B} = \vec{A} \cdot \vec{a}_B = 0.487 \times 2 + 0.811 \times (-5) + 0.325 \times (-4) = -0.43.$

Angle between $\vec{A}$ and $\vec{B}$, $\cos\theta = \dfrac{\vec{A} \cdot \vec{B}}{|A||B|} = \dfrac{3 \times 2 + 5 \times (-5) + 2 \times (-4)}{\sqrt{45} \cdot \sqrt{38}} = -0.673.$

$$\theta = 130.76^\circ.$$

Example 1.12 If two vectors are given by $\vec{A} = 7\vec{a}_r + 5\vec{a}_\theta + 3\vec{a}_\phi$ and $\vec{B} = 2\vec{a}_r + 3\vec{a}_\theta + 4\vec{a}_\phi$, find $\vec{A} \cdot \vec{B}$ and $\vec{A} \times \vec{B}$.

Solution Given two vectors $\vec{A} = 7\vec{a}_r + 5\vec{a}_\theta + 3\vec{a}_\phi$ and $\vec{B} = 2\vec{a}_r + 3\vec{a}_\theta + 4\vec{a}_\phi$.

$$\vec{A} \cdot \vec{B} = 7 \times 2 + 5 \times (3) + (-3) \times (-4) = 19.$$

$$\vec{A} \times \vec{B} = \begin{vmatrix} \vec{a}_r & \vec{a}_\theta & \vec{a}_\phi \\ 7 & 5 & 3 \\ 2 & 3 & 4 \end{vmatrix} = 11\vec{a}_r - 22\vec{a}_\theta + 11\vec{a}_\phi.$$

Example 1.13 Express the points $P(2,\ 5\pi/6,\ 3)$ and $Q(2,\ 5\pi/6,\ \pi/2)$ in the rectangular coordinate system.

Solution Given a point $P(2,\ 5\pi/6,\ 3)$ in cylindrical coordinates.
In rectangular coordinates,

$$x = \rho\cos\phi = 2\cos(5\pi/6) = -1.732.$$

$$y = \rho\sin\phi = 2\sin(5\pi/6) = 1.$$

$$z = z = 3.$$

The point is $P(-1.732,\ 1,\ 3)$.

Given a point $Q(2,\ 5\pi/6,\ \pi/3)$ in spherical coordinates.
In rectangular coordinates,

$$x = r\sin\theta\cos\phi = 2\sin(5\pi/6)\cos(\pi/3) = 0.5.$$

$$y = r\sin\theta\sin\phi = 2\sin(5\pi/6)\sin(\pi/3) = 0.866.$$

$$z = r\cos\theta = 2\cos(5\pi/6) = -1.732.$$

The point is $Q(0.5,\ 0.866,\ -1.732)$.

Example 1.14 If two points are located in space at $P(2,\ \pi/6,\ 3\pi/8)$ and $Q(3,\ \pi/4,\ 2\pi/3)$, find (a) position vectors and (b) distance vector PQ.

Solution Given points $P(2,\ \pi/6,\ 3\pi/8)$ and $Q(3,\ \pi/4,\ 2\pi/3)$ in spherical coordinates.

(a) Conversion of the spherical coordinates into rectangular coordinates
We know that $x = r\sin\theta\cos\phi$.

$$y = r\sin\theta\sin\phi.$$

$$z = r\cos\theta.$$

$$\therefore \quad x_1 = 2\sin(\pi/6)\cos(3\pi/8) = 0.3826.$$

$$y_1 = 2\sin(\pi/6)\sin(3\pi/8) = 0.9238.$$

$$z_1 = 2\cos(\pi/6) = 1.732.$$

$$x_2 = 3\sin(\pi/4)\cos(2\pi/3) = -1.06.$$

$$y_2 = 3\sin(\pi/4)\sin(2\pi/3) = 1.837.$$

$$z_2 = 3\cos(\pi/4) = 2.12.$$

The position vectors are $\vec{P} = 0.3826\vec{a}_x + 0.9238\vec{a}_y + 1.732\vec{a}_z$ and $\vec{Q} = -0.16\vec{a}_x + 1.837\vec{a}_y + 2.12\vec{a}_z$.

(b) Distance vector $\overrightarrow{PQ}$ is

$$\overrightarrow{PQ} = \vec{Q} - \vec{P} = (-1.06\vec{a}_x + 1.837\vec{a}_y + 2.12\vec{a}_z) - (0.3826\vec{a}_x + 0.9238\vec{a}_y + 1.732\vec{a}_z)$$

$$= -1.4426\vec{a}_x + 0.9132\vec{a}_y + 0.388\vec{a}_z.$$

(d) A unit vector along $\overrightarrow{PQ}$ is $\vec{a}_{AB} = \dfrac{\overrightarrow{AB}}{|AB|} = \dfrac{-5\vec{a}_x + 4\vec{a}_y - 4\vec{a}_z}{\sqrt{5^2+4^2+4^2}} = \dfrac{-5\vec{a}_x + 4\vec{a}_y - 4\vec{a}_z}{7.55}$

$$= 0.66\vec{a}_x + 0.53\vec{a}_y - 0.53\vec{a}_z.$$

Example 1.15 Express the vector $\vec{A} = y\vec{a}_x + (x+z)\vec{a}_y$ at the point $P(-2,\ 6,\ 3)$ in spherical coordinates system.

Solution Given the vector $\vec{A} = y\vec{a}_x + (x+z)\vec{a}_y$.

Transformation of the vector from Cartesian to spherical coordinates in matrix form:

$$\begin{bmatrix} A_r \\ A_\theta \\ A_\phi \end{bmatrix} = \begin{bmatrix} \sin\theta\cos\phi & \sin\theta\sin\phi & \cos\theta \\ \cos\theta\cos\phi & \cos\theta\sin\phi & -\sin\theta \\ -\sin\theta & \cos\theta & 0 \end{bmatrix} \begin{bmatrix} A_x \\ A_y \\ A_z \end{bmatrix}.$$

At the point $P(-2,\ 6,\ 3)$ in spherical coordinates,

$$r = \sqrt{x^2+y^2+z^2} = \sqrt{2^2+6^2+3^2} = 7.$$

$$\theta = \cos^{-1}(z/r) = \cos^{-1}(3/7) = 64.62^\circ.$$

$$\phi = \tan^{-1}(y/x) = \tan^{-1}(-6/2) = 108.43^\circ.$$

$$\cos\phi = -0.316,\ \sin\phi = 0.9487,\ \cos\theta = 0.4286,\ \sin\theta = 0.903.$$

$$\begin{bmatrix} A_r \\ A_\theta \\ A_\phi \end{bmatrix} = \begin{bmatrix} -0.285 & 0.856 & 0.4286 \\ -0.1354 & 0.406 & -0.903 \\ -0.903 & -0.316 & 0 \end{bmatrix} \begin{bmatrix} y \\ x+z \\ 0 \end{bmatrix}$$

$$= \begin{bmatrix} -0.285 & 0.856 & 0.4286 \\ -0.1354 & 0.406 & -0.903 \\ -0.903 & -0.316 & 0 \end{bmatrix} \begin{bmatrix} 6 \\ 1 \\ 0 \end{bmatrix}.$$

$$A_r = -0.857, \ A_\theta = -0.41, \ A_\phi = -6.$$

$$\therefore \quad \vec{A} = -0.857\vec{a}_r - 0.41\vec{a}_\theta - 6\vec{a}_\phi.$$

Example 1.16 Transform the vector $\vec{E} = \sec\phi \vec{a}_\phi$ from cylindrical coordinates to Cartesian coordinates.

Solution Given the vector $\vec{E} = \sec\phi \vec{a}_\phi$ in cylindrical coordinates. The transformation of the vector from cylindrical to Cartesian coordinates is given below.

$$\begin{bmatrix} E_x \\ E_y \\ E_z \end{bmatrix} = \begin{bmatrix} \cos\phi & -\sin\phi & 0 \\ \sin\phi & \cos\phi & 0 \\ 0 & 0 & 1 \end{bmatrix} \begin{bmatrix} E_\rho \\ E_\phi \\ E_z \end{bmatrix}$$

$$= \begin{bmatrix} \cos\phi & -\sin\phi & 0 \\ \sin\phi & \cos\phi & 0 \\ 0 & 0 & 1 \end{bmatrix} \begin{bmatrix} 0 \\ \sec\phi \\ 0 \end{bmatrix}.$$

$$E_x = \sec\phi \cdot (-\sin\phi) = -\tan\phi.$$

$$E_y = \sec\phi \cdot (\cos\phi) = 1.$$

$$E_z = \sec\phi \cdot (0) = 0.$$

From the relationships, we know that,

$$\tan\phi = \frac{y}{x}.$$

$$E_x = -y/x, \ E_y = 1, \ E_z = 0.$$

$\therefore$ The vector in Cartesian coordinates is

$$\vec{E} = \frac{-y}{x}\vec{a}_x + \vec{a}_y.$$

Example 1.17 Transform the vector $\vec{B} = \frac{10}{r}\vec{a}_r + r\cos\theta\vec{a}_\theta + \vec{a}_\phi$ from spherical coordinates to Cartesian coordinates and find the vector at $(-3, 4, 0)$.

Solution Given the vector $\vec{B} = \frac{10}{r}\vec{a}_r + r\cos\theta\vec{a}_\theta + \vec{a}_\phi$ in spherical coordinates.

Transformation of the vector from spherical to Cartesian coordinates:

$$\begin{bmatrix} B_x \\ B_y \\ B_z \end{bmatrix} = \begin{bmatrix} \sin\theta\cos\phi & \cos\theta\cos\phi & -\sin\phi \\ \sin\theta\sin\phi & \cos\theta\sin\phi & \cos\phi \\ \cos\theta & -\sin\theta & 0 \end{bmatrix} \begin{bmatrix} B_r \\ B_\theta \\ B_\phi \end{bmatrix}.$$

$$\begin{bmatrix} B_x \\ B_y \\ B_z \end{bmatrix} = \begin{bmatrix} \sin\theta\cos\phi & \cos\theta\cos\phi & -\sin\phi \\ \sin\theta\sin\phi & \cos\theta\sin\phi & \cos\phi \\ \cos\theta & -\sin\theta & 0 \end{bmatrix} \begin{bmatrix} 10/r \\ r\cos\theta \\ 0 \end{bmatrix}.$$

$$B_x = \frac{10}{r}(\sin\theta\cos\phi) + r\cos^2\theta\cos\phi.$$

$$B_y = \frac{10}{r}(\sin\theta\sin\phi) + r\cos^2\theta\sin\phi.$$

$$B_z = \frac{10}{r}(\cos\theta) - r\cos\theta\sin\theta.$$

From the relationships, at point $(-3, 4, 0)$, we know that

$$r = \sqrt{x^2+y^2+z^2} = \sqrt{3^2+4^2} = 5, \quad \cos\theta = \frac{z}{r} = 0, \quad \sin\theta = 1,$$

and $\tan\phi = \dfrac{y}{x} = \dfrac{-4}{3} = -1.33, \ \cos\phi = 0.6, \sin\phi = -0.8.$

$$B_x = \frac{10}{5}(0.6) = 1.2.$$

$$B_y = \frac{10}{r}(\sin\theta\sin\phi) + r\cos^2\theta\sin\phi = \frac{10}{5}(-0.8) + 0 = -1.6.$$

$$B_z = \frac{10}{r}(\cos\theta) - r\cos\theta\sin\theta = 0.$$

$$\therefore \quad \vec{B} = 1.2\vec{a}_x - 1.6\vec{a}_y.$$

Example 1.18 Find the divergence of the vector $\vec{A} = 10\sin\theta\vec{a}_\theta + 10\cos\phi\vec{a}_\phi$ at the point $P(1, \pi/4, \pi/6)$.

Solution Given vector

$\vec{A} = 10\sin\theta\vec{a}_\theta + 10\cos\phi\vec{a}_\phi$ in spherical coordinates.

$A_r = 0$, $A_\theta = 10\sin\theta$, and $A_\phi = 10\cos\phi$

We know that the divergence is

$$\nabla\cdot\vec{A} = \frac{1}{r^2}\frac{\partial}{\partial r}(r^2A_r) + \frac{1}{r\sin\theta}\frac{\partial}{\partial\theta}(\sin\theta A_\theta) + \frac{1}{r\sin\theta}\frac{\partial A_\phi}{\partial\phi}$$

$$= \frac{1}{r\sin\theta}\frac{\partial}{\partial\theta}(10\sin^2\theta) + \frac{1}{r\sin\theta}\frac{\partial}{\partial\phi}(10\cos\phi)$$

$$= \frac{20\cos\theta}{r} - \frac{10\sin\phi}{r\sin\theta}.$$

At the point $P(1,\ \pi/4,\ \pi/6)$, $\nabla\cdot\vec{A} = \frac{20\cos(\pi/4)}{1} - \frac{10\sin(\pi/6)}{\sin(\pi/4)} = 7.071.$

Summary

1. Vector algebra and vector calculus provide basic tools to analyse electromagnetic fields in material space
2. A scalar is a quantity represented by a real number. It has only magnitude, whereas a vector has both magnitude and direction.
3. A field is a function that describes a quantity at all points within a specified region. It may be a scalar or vector.
4. A unit vector along a vector $\vec{A}$ is defined as a vector whose magnitude is unity and its direction is along A.
5. Multiplying two or more vectors results in either a scalar or a vector.
6. A dot product gives a scalar quantity whereas a cross product gives a vector quantity.
7. Dot product is used to find the component or projection of a vector in a given direction.
8. The magnitude of the cross product of two vectors is equal to the area of the parallelogram generated by the vectors.
9. Scalar triple product of three vectors, A, B, C gives the volume of a parallelopiped structure having A, B and C as its sides.
10. The most commonly used orthogonal coordinate systems for EM fields are
    1. Cartesian or rectangular coordinate system
    2. Circular cylindrical coordinate system and
    3. Spherical coordinate system
11. Transformation of a vector from one coordinate system to another coordinate system can be obtained by dot product of the vector with the unit vectors of the other coordinate system.
12. Transformation of vectors helps us to analyse problems in a convenient way. A hard problem in one coordinate system may become simple in another system by applying symmetrical conditions.
13. The fundamental laws on vectors do not vary with vector transformation.
14. For a given vector field $\vec{A}$, the divergence theorem can be expressed as

    $$\int_v \nabla\cdot\vec{A}\,dv = \oint_s \vec{A}\cdot d\vec{S}.$$

    It converts a volume integral of the divergence of a vector to a closed surface integral of the vector and vice versa.

15. For a given vector field $\vec{A}$, Stokes' theorem can be expressed as

$$\int_s (\nabla \times \vec{A}) \cdot d\vec{S} = \oint_l \vec{A} \cdot d\vec{l}.$$

It converts a surface integral of the curl of a vector to a closed line integral of the vector and vice versa.

16. Green's theorem: Green's theorem for a vector field $\vec{A} = A_x \vec{a}_x + A_y \vec{a}_y$ can be expressed as

$$\oint_C \vec{A} \cdot d\vec{l} = \int_s \left(\frac{\partial A_x}{\partial x} - \frac{\partial A_y}{\partial y} \right) \cdot dxdy.$$

It converts a surface integral of a vector to a closed line integral of the vector and vice versa.

17. Helmholtz's theorem for a vector field $\vec{A}$ can be written as

$$\vec{A} = -\nabla V + \nabla \times \vec{B}.$$

18. A summary of the unit vectors, the differential lengths, the differential surfaces, and the differential volumes for the three coordinate systems is given in Table 1.3. A summary of the transformations of the variables between coordinate systems is given in Table 1.4.
19. Three differential operations of fields, the ***gradient*** of scalar field, the ***divergence*** of vector field, and the ***curl*** of the vector field are discussed in this chapter. Cartesian, cylindrical and spherical coordinates, their expressions together with the Laplacian operation are summarised in Table 1.5.

Table 1.3 *Three orthogonal coordinate systems*

Coordinate system	**Cartesian**	**Cylindrical**	**Spherical**
Coordinate variables	(x, y, z)	(ρ, ϕ, z)	(r, θ, ϕ)
Unit vector properties	$\vec{a}_x \cdot \vec{a}_y = \vec{a}_y \cdot \vec{a}_z = \vec{a}_z \cdot \vec{a}_x = 0$ $\vec{a}_x \cdot \vec{a}_x = \vec{a}_y \cdot \vec{a}_y = \vec{a}_z \cdot \vec{a}_z = 1$ $\vec{a}_x \times \vec{a}_y = \vec{a}_z, \vec{a}_y \times \vec{a}_z = \vec{a}_x$ $\vec{a}_z \times \vec{a}_x = \vec{a}_y$	$\vec{a}_\rho \cdot \vec{a}_\phi = \vec{a}_\phi \cdot \vec{a}_z = \vec{a}_z \cdot \vec{a}_\rho = 0$ $\vec{a}_\rho \cdot \vec{a}_\rho = \vec{a}_\phi \cdot \vec{a}_\phi = \vec{a}_z \cdot \vec{a}_z = 1$ $\vec{a}_\rho \times \vec{a}_\phi = \vec{a}_z, \vec{a}_\phi \times \vec{a}_z = \vec{a}_\rho,$ $\vec{a}_z \times \vec{a}_\rho = \vec{a}_\phi$	$\vec{a}_r \cdot \vec{a}_\theta = \vec{a}_\theta \cdot \vec{a}_\phi = \vec{a}_\phi \cdot \vec{a}_r = 0$ $\vec{a}_r \cdot \vec{a}_r = \vec{a}_\theta \cdot \vec{a}_\theta = \vec{a}_\phi \cdot \vec{a}_\phi = 1$ $\vec{a}_r \times \vec{a}_\theta = \vec{a}_\phi, \vec{a}_\theta \times \vec{a}_\phi = \vec{a}_r,$ $\vec{a}_\phi \times \vec{a}_r = \vec{a}_\theta$
Vector representation	$A_x\vec{a}_x + A_y\vec{a}_y + A_z\vec{a}_z$	$A_\rho\vec{a}_\rho + A_\phi\vec{a}_\phi + A_z\vec{a}_z$	$A_r\vec{a}_r + A_\theta\vec{a}_\theta + A_\phi\vec{a}_\phi$
Dot product $\vec{A} \cdot \vec{B}$	$A_xB_x + A_yB_y + A_zB_z$	$A_\rho B_\rho + A_\phi B_\phi + A_z B_z$	$A_rB_r + A_\theta B_\theta + A_\phi B_\phi$
Cross product $\vec{A} \times \vec{B}$	$\begin{vmatrix} \vec{a}_x & \vec{a}_y & \vec{a}_z \\ A_x & A_y & A_z \\ B_x & B_y & B_z \end{vmatrix}$	$\begin{vmatrix} \vec{a}_\rho & \vec{a}_\phi & \vec{a}_z \\ A_\rho & A_\phi & A_z \\ B_\rho & B_\phi & B_z \end{vmatrix}$	$\begin{vmatrix} \vec{a}_r & \vec{a}_\theta & \vec{a}_\phi \\ A_r & A_\theta & A_\phi \\ B_x & B_\theta & B_\phi \end{vmatrix}$
Differential length $d\vec{l}$	$dx\vec{a}_x + dy\vec{a}_y + dz\vec{a}_z$	$d\rho\vec{a}_\rho + \rho d\phi\vec{a}_\rho + dz\vec{a}_z$	$dr\vec{a}_r + rd\theta\vec{a}_\theta + r\sin\theta d\phi\vec{a}_\phi$
Differential surface areas	$d\vec{S}_x = dydz\vec{a}_x$ $d\vec{S}_y = dxdz\vec{a}_y$ $d\vec{S}_z = dxdy\vec{a}_z$	$d\vec{S}_\rho = \rho d\phi dz\vec{a}_\rho$ $d\vec{S}_\phi = d\rho dz\vec{a}_\phi$ $d\vec{S}_z = \rho d\phi d\rho\vec{a}_z$	$d\vec{S}_r = r^2\sin\theta d\theta d\phi\vec{a}_r$ $d\vec{S}_\theta = r\sin\theta drd\phi\vec{a}_\theta$ $d\vec{S}_\phi = rdrd\theta\vec{a}_\phi$
Differential volume dv	$dx\,dy\,dz$	$\rho d\rho d\theta dz$	$r^2\sin\theta dr\,d\theta\,d\phi$

Table 1.4 *Summary of transformation between coordinate systems*

Transformation	Coordinate variables	Vector components
Cylindrical to Cartesian	$x=\rho\cos\phi$ $y=\rho\sin\phi$ $z=z$	$\begin{bmatrix}A_x\\A_y\\A_z\end{bmatrix}=\begin{bmatrix}\cos\phi & -\sin\phi & 0\\ \sin\phi & \cos\phi & 0\\ 0 & 0 & 1\end{bmatrix}\begin{bmatrix}A_\rho\\A_\phi\\A_z\end{bmatrix}$
Cartesian to cylindrical	$\rho=\sqrt{x^2+y^2}$ $\phi=\tan^{-1}(y/x)$ $z=z$	$\begin{bmatrix}A_\rho\\A_\phi\\A_z\end{bmatrix}=\begin{bmatrix}\cos\phi & \sin\phi & 0\\ -\sin\phi & \cos\phi & 0\\ 0 & 0 & 1\end{bmatrix}\begin{bmatrix}A_x\\A_y\\A_z\end{bmatrix}$
Spherical to cartesian	$x=r\sin\theta\cos\phi$ $y=r\sin\theta\sin\phi$ $z=r\cos\theta$	$\begin{bmatrix}A_x\\A_y\\A_z\end{bmatrix}=\begin{bmatrix}\sin\theta\cos\phi & \cos\theta\cos\phi & -\sin\phi\\ \sin\theta\sin\phi & \cos\theta\sin\phi & \cos\phi\\ \cos\theta & -\sin\theta & 0\end{bmatrix}\begin{bmatrix}A_r\\A_\theta\\A_\phi\end{bmatrix}$
Cartesian to spherical	$r=\sqrt{x^2+y^2+z^2}$ $\phi=\tan^{-1}(y/x)$ $\theta=\cos^{-1}(z/r)$	$\begin{bmatrix}A_r\\A_\theta\\A_\phi\end{bmatrix}=\begin{bmatrix}\sin\theta\cos\phi & \sin\theta\sin\phi & \cos\theta\\ \cos\theta\cos\phi & \cos\theta\sin\phi & -\sin\theta\\ -\sin\phi & \cos\phi & 0\end{bmatrix}\begin{bmatrix}A_x\\A_y\\A_z\end{bmatrix}$

Table 1.5 *Differential operations in three orthogonal coordinate systems*

Coordinate system	Cartesian	Cylindrical	Spherical
Gradient of scalar field ∇V	$\frac{\partial V}{\partial x}\vec{a}_x+\frac{\partial V}{\partial y}\vec{a}_y+\frac{\partial V}{\partial z}\vec{a}_z$	$\frac{\partial V}{\partial \rho}\vec{a}_\rho+\frac{1}{\rho}\frac{\partial V}{\partial \phi}\vec{a}_\phi+\frac{\partial V}{\partial z}\vec{a}_z$	$\frac{\partial V}{\partial r}\vec{a}_r+\frac{1}{r}\frac{\partial V}{\partial \theta}\vec{a}_\theta+\frac{1}{r\sin\theta}\frac{\partial V}{\partial \phi}\vec{a}_\phi$
Divergence of vector field $\nabla\cdot\vec{A}$	$\frac{\partial A_x}{\partial x}+\frac{\partial A_y}{\partial y}+\frac{\partial A_z}{\partial z}$	$\frac{1}{\rho}\frac{\partial(\rho A_\rho)}{\partial \rho}+\frac{1}{\rho}\frac{\partial A_\phi}{\partial \phi}+\frac{\partial A_z}{\partial z}$	$\frac{1}{r^2}\frac{\partial}{\partial r}(r^2A_r)+\frac{1}{r\sin\theta}\frac{\partial A_\phi}{\partial \phi}+\frac{1}{r\sin\theta}\frac{\partial}{\partial \theta}(\sin\theta A_\theta)$
Curl of vector field $\nabla\times\vec{A}$	$\begin{vmatrix}\vec{a}_x & \vec{a}_y & \vec{a}_z\\ \frac{\partial}{\partial x} & \frac{\partial}{\partial y} & \frac{\partial}{\partial z}\\ A_x & A_y & A_z\end{vmatrix}$	$\frac{1}{\rho}\begin{vmatrix}\vec{a}_\rho & \rho\vec{a}_\phi & \vec{a}_z\\ \frac{\partial}{\partial \rho} & \frac{\partial}{\partial \phi} & \frac{\partial}{\partial z}\\ A_\rho & \rho A_\phi & A_z\end{vmatrix}$	$\frac{1}{r^2\sin\theta}\begin{vmatrix}\vec{a}_r & r\vec{a}_\theta & r\sin\theta\vec{a}_\phi\\ \frac{\partial}{\partial r} & \frac{\partial}{\partial \theta} & \frac{\partial}{\partial \phi}\\ A_r & rA_\theta & r\sin\theta A_\phi\end{vmatrix}$
Laplacian of scalar field $\nabla^2 V$	$\frac{\partial^2 V}{\partial x^2}+\frac{\partial^2 V}{\partial y^2}+\frac{\partial^2 V}{\partial z^2}$	$\frac{1}{\rho}\frac{\partial}{\partial \rho}\left(\rho\frac{\partial V}{\partial \rho}\right)+\frac{1}{\rho^2}\frac{\partial^2 V}{\partial \phi^2}+\frac{\partial^2 V}{\partial z^2}$	$\frac{1}{r^2}\frac{\partial}{\partial r}\left(\frac{r^2\partial V}{\partial r}\right)+\frac{1}{r^2\sin^2\theta}\frac{\partial^2 V}{\partial \phi^2}+\frac{1}{r^2\sin\theta}\frac{\partial}{\partial \theta}\left(\sin\theta\frac{\partial V}{\partial \theta}\right)$

Review Questions

1. Define scalar and vector fields with examples.
2. Explain scalar and vector products.
3. Explain vector addition and describe its properties.
4. Discuss dot product in cylindrical coordinates.
5. Discuss dot product in rectangular coordinates.
6. Define and explain orthogonal coordinate systems.

7. Explain the procedure to transform rectangular coordinates to cylindrical coordinates.
8. Explain the procedure to transform rectangular coordinates to spherical coordinates.
9. State and explain divergence theorem.
10. State and explain Stokes' theorem.
11. State and explain Helmholtz's theorem.
12. Classify the vector field according to Helmholtz's theorem.
13. Explain the properties of the curl of a vector field.
14. List various del operators.
15. Define line, surface and volume integrals used in vector analysis.

Problems

1. Given three vectors $\vec{A} = \vec{a}_x + 2\vec{a}_y + 3\vec{a}_z$, $\vec{B} = 3\vec{a}_x + 4\vec{a}_y + 5\vec{a}_z$ and $\vec{C} = 2\vec{a}_x - 2\vec{a}_y + 7\vec{a}_z$. Compute
   (a) the scalar product $\vec{A} \cdot \vec{B}$
   (b) the angle between $\vec{A}$ and $\vec{B}$
   (c) the scalar projection of $\vec{A}$ on $\vec{B}$
   (d) the vector product $\vec{A} \times \vec{B}$
   (e) the area of the parallelogram whose sides are specified by A and B, i.e., $|\vec{A} \times \vec{B}|$
   (f) the volume of a parallelopiped defined by vectors A, B and C, i.e., $|\vec{A} \cdot \vec{B} \times \vec{C})|$
2. The Cartesian coordinates for a point P is (1, 2, 3). Find the cylindrical coordinates.
3. A point is located at $P(-2, 4, -1)$. Express it in cylindrical coordinates system and spherical coordinate system.
4. Express the vector field $\vec{A} = 5\sin\varphi\vec{a}_\rho + 7\rho\vec{a}_\varphi - 3\cos\varphi\vec{a}_z$ in Cartesian coordinates at $(-3, 4, 5)$.
5. Calculate the line integral along the circular path from point (4, 0, 0) to point (0, 4, 0) for the force field $\vec{F} = 3xy\vec{a}_x + 4x\vec{a}_y$.
6. Find the total charge within a volume defined by $1 \le r \le 5, \pi/6 \le \theta \le \pi/2, \pi/4 \le \theta \le 3\pi/4$ if the charge density is given as $\rho_v = r\cos\varphi$.
7. If $\vec{A} = 2xy\vec{a}_x + (x^2 - z^2)\vec{a}_y - 3xz^2\vec{a}_z$, evaluate the line integral along the straight path from the origin to a point (2, 1, 3).
8. Given two points $P(2, 1, 3)$ and $Q(4, -50°, 2)$, find the distance between P and Q.
9. A surface is modelled with the equation $z = e^{-(2x^2+y^2)}$. Calculate the gradient of z along z-axis.
10. Find the divergence of the vector $\vec{A} = 4x\vec{a}_x + xy^2\vec{a}_y - 2xye^{-z}\vec{a}_z$ at the point $P(1, -2, 0)$.
11. Find the divergence of the vector $\vec{A} = \rho e^{-(\rho/5)^2}\vec{a}_\rho$ at the point $P(1, -2, 5)$.
12. Evaluate both sides of the divergence theorem for the vector field $\vec{A} = 5xy\vec{a}_x + 2xy^2\vec{a}_y + 5z\vec{a}_z$ defined in the region $1 \le x \le 2, -1 \le y \le 3, -2 \le z \le 2$.
13. Verify the divergence theorem for the vector field $\vec{A} = 2\rho z\vec{a}_\rho + 3z\sin\varphi\vec{a}_\varphi - 4\rho\cos\varphi\vec{a}z$ defined in the region $1 \le \rho \le 2,\ 0 \le \varphi \le 45^0,\ 0 \le z \le 5$.
14. Show that $\nabla \times \vec{A} = 0$ if $\vec{A} = \frac{1}{\rho}\vec{a}_\rho$ in cylindrical coordinates.

15. Show that $\nabla \times \vec{A} = 0$ if $\vec{A} = r^2\vec{a}_r$ in spherical coordinates.
16. Verify Stokes' theorem for the vector field
$\vec{A} = (2x - y)\vec{a}_x + 2yz^2\vec{a}_y - 3zy^2\vec{a}_z$
on the lower half-surface of the sphere defined in the region $0 \leq \theta \leq \pi/2,\ 0 \leq \varphi \leq 2\pi$ with radius $r = 2$. The contour C on the $x - y$ plane is the rim of the half-sphere.

Answers

1. 20, 40.89°, 2.8284, $-2\vec{a}_x + 14\vec{a}_y - 10\vec{a}_z$, 17.32, 102
2. $P(2.23,\ 63.43°,\ 3)$
3. $P(4.49,\ -63.4°,\ -1), P(4.18,\ -102.6°,\ -63.4°)$
4. $\vec{A} = 25.6\vec{a}_x + 24.2\vec{a}_y - 1.8\vec{a}_z$
5. $-64 + 16\pi$.
6. 95.53
7. −39.5
8. 6.78
9. 0
10. 4.
11. 1.31.

Multiple-choice Questions

1. The vector product obeys the following law.
(a) commutative (b) associative (c) distributive (d) parallelogram
2. Which product is governed by the right-hand screw rule?
(a) vector (b) scalar (c) simple mathematical (d) none of these
3. If two vectors are orthogonal, then their scalar product is
(a) 1 (b) zero (c) infinite (d) 100
4. Multiplication of two vectors is
(a) vector (b) scalar (c) either vector or scalar (d) cannot say
5. ——— of two vectors uses the law of parallelogram.
(a) multiplication (b) division (c) addition (d) all of these
6. A unit vector has the ——— direction as that of the main vector.
(a) same (b) opposite (c) normal upwards (d) normal downwards.
7. A scalar quantity has
(a) magnitude (b) direction (c) both (a) and (b) (d) none of these
8. Which of the following is/are scalar quantity(s)?
(a) distance (b) density (c) temperature (d) all of these
9. Which of the following is not a scalar quantity?
(a) entropy (b) displacement (c) volume (d) mass.
10. A vector quantity has
(a) magnitude (b) direction (c) both (a) and (b) (d) none of these
11. Which of the following is/are vector quantity(s)?
(a) force (b) electric field intensity (c) acceleration (d) all of these

12. ——— product of two unit vectors is the third unit vector.
(a) vector (b) scalar (c) both (a) and (b) (d) none of these

13. A point in Cartesian coordinates is (3, 4, 5). In cylindrical coordinates, the radius is
(a) 12 (b) 7 (c) 10 (d) 5

14. Given $\vec{A} = 5\vec{a}_x + 2\vec{a}_y - 6\vec{a}_z$, the projection of $\vec{A}$ along $\vec{a}_z$ is
(a) 7 (b) -6 (c) 2 (d) 5

15. Given $\vec{A} = \alpha\vec{a}_x + \vec{a}_y + \vec{a}_z$ and $\vec{B}^{=\vec{a}_x} + \alpha\vec{a}_y + \vec{a}_z$, if $\vec{A}$ and $\vec{B}$ are normal to each other, then α is
(a) 1/4 (b) $-1/2$ (c) 2 (d) 1

16. The volume of a parallelopiped with vectors $\vec{A}$, $\vec{B}$ and $\vec{C}$ is given by
(a) vector triple product (b) gradient triple product
(c) scalar triple product (d) curl triple product

17. If a closed contour integral of a vector field is zero, then the vector field is called
(a) conservative (b) rotational (c) irrotational (d) non-conservative field

18. The maximum slope of the scalar field at a given point is called
(a) gradient (b) divergence (c) curl (d) del

19. If there are no sources and no sinks in a vector field, then the divergence of the field is always
(a) 1 (b) a scalar (c) a vector (d) zero

20. The curl of a gradient of a scalar field is always
(a) zero (b) infinite (c) vector (d) 1

21. For a vector field $\vec{A}$, Stoke's theorem can be written as
(a) $\oint_S (\nabla \times \vec{A}) \cdot d\vec{S} = \int_l \vec{A} \cdot d\vec{l}$ (b) $\int_S (\nabla \cdot \vec{A}) \times d\vec{S} = \oint_l \vec{A} \cdot d\vec{l}$
(c) $\int_S (\nabla \times \vec{A}) \cdot d\vec{S} = \oint_l \vec{A} \times d\vec{l}$ (d) $\int_S (\nabla \times \vec{A}) \cdot d\vec{S} = \oint_l \vec{A} \cdot d\vec{l}$

22. For a vector field $\vec{A}$, the divergence theorem can be written as
(a) $\int_v \nabla\vec{A} dv = \int_S \vec{A} \cdot d\vec{S}$ (b) $\int_v \nabla \cdot \vec{A} dv = \oint_S \vec{A} \cdot d\vec{S}$
(c) $\int_v \nabla \times \vec{A} dv = \oint_S \vec{A} \cdot d\vec{S}$ (d) $\int_v \nabla \cdot \vec{A} dv = \int_S \vec{A} \times d\vec{S}$

23. For a vector field $\vec{A}$, the condition for a solenoidal and irrotational field is
(a) $\nabla . \vec{A} \neq 0,\ \nabla \times \vec{A} = 0$ (b) $\nabla . \vec{A} = 0,\ \nabla \times \vec{A} = 0$
(c) $\nabla . \vec{A} = 0,\ \nabla \times \vec{A} \neq 0$ (d) $\nabla . \vec{A} \neq 0,\ \nabla \times \vec{A} \neq 0$

24. For a vector field $\vec{A}$, the condition for a solenoidal and rotational field is

    (a) $\nabla.\vec{A} \neq 0,\ \nabla \times \vec{A} = 0$ (b) $\nabla.\vec{A} = 0,\ \nabla \times \vec{A} = 0$

    (c) $\nabla.\vec{A} = 0,\ \nabla \times \vec{A} \neq 0$ (d) $\nabla.\vec{A} \neq 0,\ \nabla \times \vec{A} \neq 0$

25. For a scalar field V and a vector field $\vec{B}$, Helmholtz's theorem can be written as

    (a) $-\nabla \cdot V + \nabla \times \vec{B}$ (b) $-\nabla V + \nabla \times \vec{B}$ (c) $-\nabla \cdot V - \nabla \times \vec{B}$ (d) $\nabla V + \nabla \vec{B}$

Answers

1. (c)	2. (a)	3. (b)	4. (c)	5. (c)	6. (a)	7. (a)	8. (d)	9. (b)	10. (c)
11. (d)	12. (a)	13. (d)	14. (b)	15. (b)	16. (c)	17. (a)	18. (a)	19. (d)	20. (a)
21. (d)	22. (b)	23. (b)	24. (c)	25. (b)					

Electrostatics | 2

Learning Objectives

The main objectives of this chapter are to understand thoroughly:

- Coulomb's law and its applications, electric field intensity
- Fields due to different charge distributions.
- Electric flux density, Gauss' law and its applications
- Divergence of flux density, divergence theorem
- Work done, electric potential, equipotential surface
- Potential due to different charge distributions
- Electric dipole, field due to dipole, torque on an electric dipole
- Energy density, energy stored in electrostatic fields

2.1 Introduction to Electrostatic Fields

This chapter introduces the basic concepts of electrostatic fields produced by static electric charges. Electrostatic fields exist due to the force of attraction or repulsion among the spatial distribution of electric charges (cloud of electrons) at rest in a medium. Since a static charge does not change with time, the study mainly refers to the magnitude and position of charges in a given medium. Electrostatic field is conservative in nature, i.e., the charge can neither be created nor destroyed. This principle is a law of nature and must be satisfied always. Thus the total work done in moving a charge around any closed path within the field is always zero. Static charge distribution is a very common natural phenomenon. Some examples are cloud formation, lightning, and corona.

Electrostatic fields play an important role in many industrial devices like ink jet printers, X-ray machines, medical scanners, microphones, spray painting machines, etc. Electronic solid state devices such as FETs, MOSFETs, LCDs and touchpads too work based on the principles of electrostatic fields. In order to understand the working of these devices, it is important to know the fundamental theory of electrostatic fields. This chapter explains the laws governing the behaviour of static electric fields. Coulomb's law, Gauss' law, Maxwell's first law, electric potential, potential gradient, work done in electrostatic field, etc., are some of the topics that will be covered here. In the next chapter, we will study Laplace's and Poisson's equations for solving boundary value problems to obtain fields in the surrounding space.

Electromagnetic field theory was developed from the basic principles of electrostatics and magnetic fields. It is the study of the inter relationships between time-varying electric charges at rest and in motion. It deals directly with electric (E) and magnetic (H) field vectors. The field vectors in three dimensions depend on time and distance, and their analysis is more complex than electrical signals. A knowledge of vector analysis is a prerequisite for learning electromagnetic field theory. Since the field is distributive in nature, Ohm's law and Kirchhoff's laws are not valid in the analysis of electromagnetic field vectors. The concepts of magnetic fields and electromagnetic fields will be discussed in the following chapters.

It was the experiments performed on static charges by French army engineer, Colonel Charles A de Coulomb, that laid the foundation for formulating the principles of electrostatics. The fundamental Coulomb's law was formulated in 1785.

2.2 Coulomb's law

Coulomb's law states that *the force between two point charges, Q_1 and Q_2, separated by a distance R_{12}, is proportional to the product of the charges and inversely proportional to the square of the distance between them. The direction of the force is along the line joining the charges.*

Coulomb also postulated that like charges repel and unlike charges attract each other. Mathematically, Coulomb's law in vector form can be expressed as

$$\vec{F}_{12} = K\frac{Q_1Q_2}{R_{12}^2}\vec{a}_{R_{12}} \text{ newtons (N)}, \tag{2.1}$$

where Q_1 and Q_2 are positive or negative quantities of charge in coulombs (C) and R_{12} is the magnitude of the radial distance between the charges in metres (m), $\vec{a}_{R_{12}}$ is the unit vector in the radial direction from Q_1 to Q_2 and K is the proportionality constant.

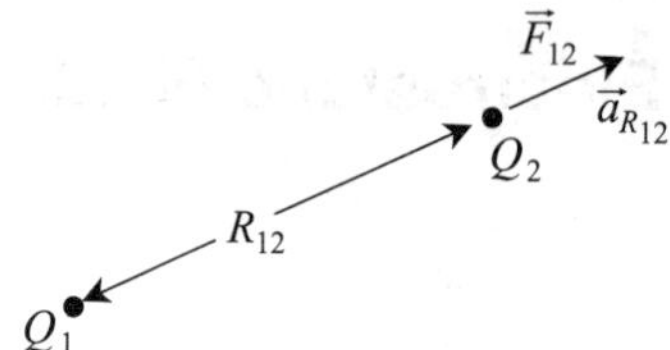

Fig. 2.1 *Coulomb's law*

Coulomb's law is illustrated in Fig. 2.1. The point charges located at a point in the medium are very small in size, negligible compared to the distance between them.

Charge of an electron e is equal to 1.6×10^{-19} coulombs.

The proportionality constant K can be determined from practical experiments. It depends on the dielectric constant (permittivity) of the medium.

For convenience, it can be expressed as

$$K = \frac{1}{4\pi\varepsilon}, \tag{2.2}$$

where $\varepsilon = \varepsilon_0\,\varepsilon_r$ is the dielectric constant or permittivity of the medium,

ε_r = relative permittivity of the medium (dimensionless quantity) and

ε_0 = permittivity of vacuum (absolute permittivity) measured in farads/metre (F/m).

The permittivity of vacuum is given by

$$\varepsilon_0 = \frac{1}{36\pi} \times 10^{-9} = 8.854 \times 10^{-12}\ \text{F/m}. \tag{2.3}$$

The relative permittivity is for vacuum, $\varepsilon_r = 1$, for air or free space, $\varepsilon_r \approx 1$, and for dielectric materials, $\varepsilon_r > 1$.

The proportionality constant for free space or vacuum is

$$K = \frac{1}{4\pi\varepsilon_0} = \frac{1}{4\pi \times \dfrac{1}{36\pi} \times 10^{-9}} = 9 \times 10^9\ \text{m/F}. \tag{2.4}$$

Therefore, the force between two charges Q_1 and Q_2 in free space is

$$\vec{F}_{12} = \frac{9 \times 10^9 Q_1 Q_2}{R_{12}^2} \vec{a}_{R_{12}} \tag{2.5}$$

and the magnitude of force is $F = \dfrac{9 \times 19^9 Q_1 Q_2}{R_{12}^2}$ newtons.

The unit vector $\vec{a}_{R_{12}}$ can be determined by considering two point charges, Q_1 and Q_2, located at two points at distance R_1 and R_2 respectively from the origin.

Let $\vec{a}_{R_1}$ and $\vec{a}_{R_2}$ be unit vectors as shown in Fig. 2.2.

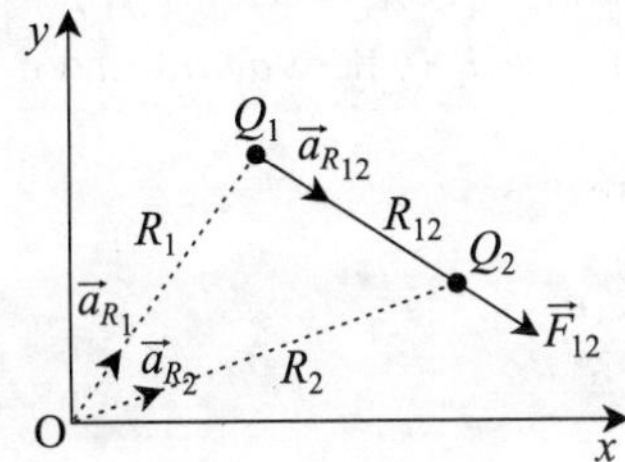

Fig. 2.2 *Force between two charges*

The distance vector is $\vec{R}_{12} = \vec{R}_2 - \vec{R}_1$.

The unit vector along the distance vector $\vec{R}_{12}$ is given by

$$\vec{a}_{R_{12}} = \frac{\vec{R}_{12}}{|R_{12}|} = \frac{\text{Distance vector}}{\text{Magnitude of distance vector}}. \tag{2.6}$$

The force on Q_2 due to Q_1 is

$$\vec{F}_{12} = \frac{Q_1 Q_2}{4\pi\varepsilon\, |R_{12}|^2} \vec{a}_{R_{12}} = \frac{Q_1 Q_2}{4\pi\varepsilon\, |R_{12}|^2} \times \frac{\vec{R}_{12}}{|R_{12}|} = \frac{Q_1 Q_2}{4\pi\varepsilon\, |R_{12}|^3} \vec{R}_{12}. \tag{2.7}$$

The magnitude of the force is

$$F_{12} = |\vec{F}_{12}| = \frac{Q_1 Q_2}{4\pi\varepsilon\, R_{12}^2}.$$

Since $\vec{a}_{R_{12}} = -\vec{a}_{R_{21}}$, i.e, $\vec{R}_{12} = -\vec{R}_{21}$, the force on Q_1 due to Q_2 is

$$\therefore \quad \vec{F}_{21} = -\vec{F}_{12}. \tag{2.8}$$

2.3 Force in Terms of Rectangular Coordinates

Consider two point charges, Q_1 and Q_2, located at positions $A(x_1, y_1, z_1)$ and $B(x_2, y_2, z_2)$ respectively, as shown in Fig. 2.3.

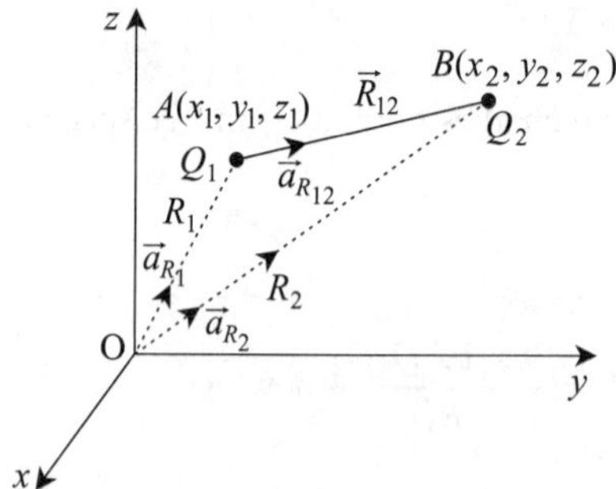

Fig. 2.3 *Force between two point charges*

The distance vector of A from the origin is $\vec{R}_1 = x_1\vec{a}_x + y_1\vec{a}_y + z_1\vec{a}_z$ and the distance vector of B from the origin is $\vec{R}_2 = x_2\vec{a}_x + y_2\vec{a}_y + z_2\vec{a}_z$ (where $\vec{a}_x, \vec{a}_y$, and $\vec{a}_z$ are unit vectors along x, y and z-axes respectively).

The unit vectors of R_1 and R_2 are

$$\vec{a}_{R_1} = \frac{\vec{R}_1}{|R_1|} = \frac{x_1\vec{a}_x + y_1\vec{a}_y + z_1\vec{a}_z}{\sqrt{x_1^2 + y_1^2 + z_1^2}},$$

$$\vec{a}_{R_2} = \frac{\vec{R}_2}{|R_2|} = \frac{x_2\vec{a}_x + y_2\vec{a}_y + z_2\vec{a}_z}{\sqrt{x_2^2 + y_2^2 + z_2^2}}. \tag{2.9}$$

The distance vector from A to B is $\vec{R}_{12} = \vec{R}_2 - \vec{R}_1$.

$$\vec{R}_{12} = (x_2 - x_1)\vec{a}_x + (y_2 - y_1)\vec{a}_y + (z_2 - z_1)\vec{a}_z. \tag{2.10}$$

The unit vector along $\vec{R}_{12}$ is

$$\vec{a}_{R_{12}} = \frac{\vec{R}_{12}}{|R_{12}|} = \frac{(x_2 - x_1)\vec{a}_x + (y_2 - y_1)\vec{a}_y + (z_2 - z_1)\vec{a}_z}{\sqrt{(x_2 - x_1)^2 + (y_2 - y_1)^2 + (z_2 - z_1)^2}}. \tag{2.11}$$

From Eq. (2.7), the force on Q_2 due to Q_1 is

$$\vec{F}_{12} = \frac{Q_1 Q_2}{4\pi\varepsilon} \frac{(x_2 - x_1)\vec{a}_x + (y_2 - y_1)\vec{a}_y + (z_2 - z_1)\vec{a}_z}{((x_2 - x_1)^2 + (y_2 - y_1)^2 + (z_2 - z_1)^2)^{3/2}}. \tag{2.12}$$

Note 1: If one charge is at the origin and the other at (x, y, z), then the force between them is

$$\vec{F}_{12} = \frac{Q_1 Q_2}{4\pi\varepsilon} \frac{x\vec{a}_x + y\vec{a}_y + z\vec{a}_z}{(x^2 + y^2 + z^2)^{3/2}}.$$

Note 2: For like charges (having same signs), the force is repulsive. It is indicated with a positive sign. For unlike charges (having opposite signs), the force is attractive. It is indicated with a negative sign.

Example 2.1 Two point charges of 10 μC each and of the same sign are situated 1.5 cm apart. Calculate the magnitude of the repulsive force?

Solution Given $Q_1 = Q_2 = 10\ \mu\text{C} = 10 \times 10^{-6}$ C and $R = 1.5$ cm $= 0.015$ m. Assume that the charges are in free space ($\varepsilon = \varepsilon_0$).

We know that $$F = \frac{Q_1 Q_2}{4\pi\varepsilon_0 R^2} = \frac{10 \times 10 \times 10^{-12}}{4\pi \times 8.854 \times 10^{-12} \times (0.015)^2} = 4 \times 10^3 \text{ N}.$$

Since charges are of the same sign, the force is repulsive.

Example 2.2 Find the force on charge $Q_1 = 450\ \mu$C located at (5, 0, 0) due to charge $Q_2 = 250\ \mu$C at $(0, 4, 5)$ in free space.

Solution Given that

$$Q_1 = 450 \times 10^{-6} \text{ at } (5,\ 0,\ 0), \quad \vec{R}_1 = 5\vec{a}_x \text{ and}$$

$$Q_2 = 250 \times 10^{-6} \text{ C at } (0,\ 4,\ 5), \quad \vec{R}_2 = 4\vec{a}_y + 5\vec{a}_z.$$

The distance vector $\vec{R}_{21} = \vec{R}_1 - \vec{R}_2$.

$$\vec{R}_{21} = (5-0)\vec{a}_x + (0-4)\vec{a}_y + (0-5)\vec{a}_z = 5\vec{a}_x - 4\vec{a}_y - 5\vec{a}_z.$$

Magnitude $|R_{21}| = \sqrt{(5)^2 + (-4)^2 + (-5)^2} = 8.12$.

The unit vector is $$\vec{a}_{21} = \frac{\vec{R}_{21}}{R_{21}} = \frac{5\vec{a}_x - 4\vec{a}_y - 5\vec{a}_z}{8.12} = 0.615\vec{a}_x - 0.5\vec{a}_y - 0.615\vec{a}_z.$$

Therefore, the force on Q_1 due to Q_2 is

$$\vec{F} = \frac{Q_1 Q_2}{4\pi\varepsilon_0 R_{21}^2}\vec{a}_{R_{21}} = \frac{450 \times 10^{-6} \times 250 \times 10^{-6}}{4\pi \times 8.854 \times 10^{-12} \times 8.12^2}(0.615\vec{a}_x - 0.5\vec{a}_y - 0.615\vec{a}_z)$$

$$= 15.3(0.615\vec{a}_x - 0.5\vec{a}_y - 0.615\vec{a}_z) = 9.4\vec{a}_x - 7.65\vec{a}_y - 9.4\vec{a}_z \text{ N}.$$

2.4 Force due to *N* Number of Charges

Consider N number of charges, $Q_1, Q_2, \ldots, Q_N$ as shown in Fig. 2.4, with associated vectors $\vec{r}_1, \vec{r}_2, \ldots, \vec{r}_N$. Let another point charge Q be at position $\vec{r}$.

The total force exerted on a point charge Q due to N number of charges satisfies the principle of superposition, i.e., it is equal to the vector sum of all the forces exerted on that charge due to each of the other point charges.

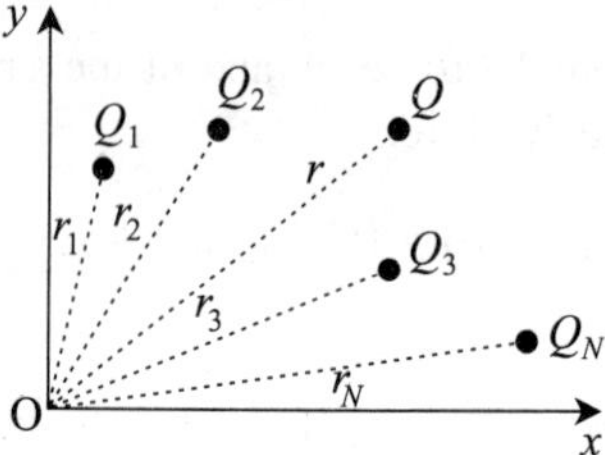

Fig. 2.4 *Point charges at different positions*

Let the vectors $\vec{R}_1, \vec{R}_2, \ldots, \vec{R}_N$ be the distance vectors from Q to $Q_1, Q_2, \ldots, Q_N$ respectively.

Then the forces $\vec{F}_1, \vec{F}_2, \ldots., \vec{F}_N$ exerted on Q due to $Q_1, Q_2, \ldots, Q_N$ respectively are

$$\vec{F}_1 = \frac{QQ_1}{4\pi\varepsilon R_1^2}\vec{a}_{R_1},$$

$$\vec{F}_2 = \frac{QQ_2}{4\pi\varepsilon R_2^2}\vec{a}_{R_2},$$

$$\vdots$$

and $\vec{F}_N = \dfrac{QQ_N}{4\pi\varepsilon\, R_N^2}\vec{a}_{R_N}.$

The respective unit vectors are

$$\vec{a}_{R_1} = \frac{\vec{R}_1}{|R_1|} = \frac{\vec{r}-\vec{r}_1}{|\vec{r}-\vec{r}_1|},$$

$$\vec{a}_{R_2} = \frac{\vec{R}_2}{|R_2|} = \frac{\vec{r}-\vec{r}_2}{|\vec{r}-\vec{r}_2|},$$

$$\vdots$$

and $\vec{a}_{R_N} = \dfrac{\vec{R}_N}{|R_N|} = \dfrac{\vec{r}-\vec{r}_N}{|\vec{r}-\vec{r}_N|}.$

Hence the total force on Q, according to the principle of superposition, is

$$\vec{F} = \vec{F}_1 + \vec{F}_2 + \vec{F}_3 + \ldots. + \vec{F}_N$$

$$= \frac{QQ_1}{4\pi\varepsilon\, R_1^2}\vec{a}_{R_1} + \frac{QQ_2}{4\pi\varepsilon\, R_2^2}\vec{a}_{R_2} + \ldots. + \frac{QQ_N}{4\pi\varepsilon\, R_N^2}\vec{a}_{R_N}$$

$$= \frac{Q}{4\pi\varepsilon}\sum_{n=1}^{N}\frac{Q_n\vec{a}_{R_n}}{R_n^2}. \qquad (2.13)$$

Example 2.3 Four concentrated charges, $Q_1 = 0.3\ \mu$C, $Q_2 = 0.2\ \mu$C, $Q_3 = 0.3\ \mu$C and $Q_4 = 0.2\ \mu$C are located at the vertices of a rectangle. The length of the rectangle is 5 cm and the breadth of the rectangle is 2 cm. Find the magnitude and direction of the resultant force on Q_1.

Solution Let the rectangle with length 0.05 m and breadth 0.02 m be placed on the $x-y$ plane.
The four charges are located at the vertices of the rectangle in free space as shown in Fig. 2.5.

Given $Q_1 = 0.3\ \mu$C at (0.05, 0.02, 0), $Q_2 = 0.2\ \mu$C at (0, 0.02, 0),
$Q_3 = -0.3\ \mu$C at (0, 0, 0) and $Q_4 = 0.2\ \mu$C at (0.05, 0, 0).

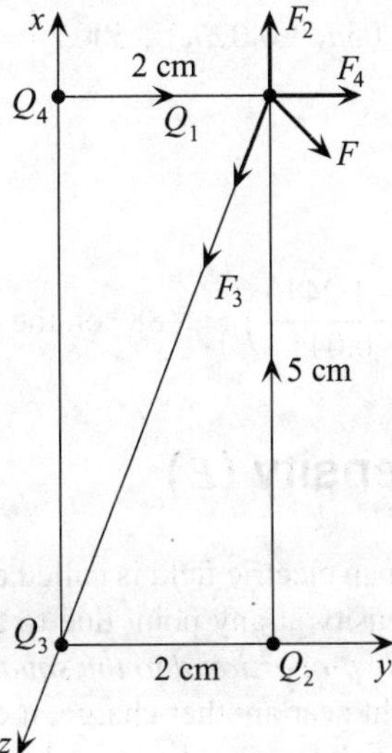

Fig. 2.5 *Four charges located at the vertices of a rectangle*

Using Coulomb's law and the superposition principle, the force on Q_1 is

$$\vec{F} = \frac{1}{4\pi\varepsilon_0}\left[\frac{Q_1Q_2}{R_{21}^2}\vec{a}_{R_{21}} + \frac{Q_1Q_3}{R_{31}^2}\vec{a}_{R_{31}} + \frac{Q_1Q_4}{R_{41}^2}\vec{a}_{R_{41}}\right].$$

Now the distance vectors

$\vec{R}_{21} = (0.05,\ 0.02,\ 0) - (0,\ 0.02,\ 0) = 0.05\vec{a}_x$ and $R_{21} = 0.05.$

$$\vec{a}_{R_{21}} = \frac{0.05\vec{a}_x}{0.05} = \vec{a}_x.$$

$\vec{R}_{31} = (0.05,\ 0.02,\ 0) - (0,\ 0,\ 0) = 0.05\vec{a}_x + 0.02\vec{a}_y$ and

$R_{31} = \sqrt{0.05^2 + 0.02^2} = \sqrt{0.29} = 0.0538.$

$$\vec{a}_{R_{31}} = \frac{1}{\sqrt{0.29}}(0.05\vec{a}_x + 0.02\vec{a}_y).$$

$$\vec{R}_{41} = (0.05,\ 0.02) - (0.05,\ 0,\ 0) = 0.02\ \vec{a}_y \text{ and } R_{41} = 0.02.$$

$$\vec{a}_{R_{41}} = \frac{0.02}{0.002}\vec{a}_y.$$

Substituting these values in the equation for $\vec{F}$,

$$\vec{F} = \frac{Q_1}{4\pi\varepsilon_0}\left[\frac{Q_2}{R_{21}^2}\vec{a}_{R_{21}} + \frac{Q_3}{R_{31}^2}\vec{a}_{R_{31}} + \frac{Q_4}{R_{41}^2}\vec{a}_{R_{41}}\right]$$

$$= 0.3\times10^{-6}\times9\times10^{+9}\left[\frac{0.2\times10^{-6}}{0.05^2}\vec{a}_x + \frac{-0.3\times10^{-6}}{0.0538^2}\left(\frac{0.05\vec{a}_x+0.02\vec{a}_y}{0.0538}\right) + \frac{0.2\times10^{-6}}{0.02^2}\vec{a}_y\right]$$

$$= 2.7\times10^{-3}\left[80\vec{a}_x - 1926.5(0.05\vec{a}_x+0.02\vec{a}_y)+500\vec{a}_y\right] = 2.7\times10^{-3}\left[-16.325\vec{a}_x+461.47\vec{a}_y\right]$$

$$= -0.044\vec{a}_x + 1.246\vec{a}_y.$$

$$|\vec{F}| = 1.2467 \text{ N}.$$

The direction of the force is $\tan^{-1}\left(\frac{-1.2446}{0.044}\right) = -88^\circ$ on the $x-y$ plane.

2.5 Electric Field Intensity ($\vec{E}$)

The force exerted on a unit charge in an electric field is called electric field intensity ($\vec{E}$) or electric field strength. The electric field intensity at any point due to a point charge is proportional to the *magnitude of the charge and inversely proportional to the square of the distance from the charge to that point.* If a point charge is brought near another charge, it experiences a force. The field where a charge exerts a force on any another charge is called the electric field of that charge.

Consider a single point charge Q_2 located at any point near another fixed point charge Q_1, as shown in Fig. 2.6. The electric field strength due to the point charge Q_1 at the point charge Q_2 is defined as the force per unit charge at Q_2. That is,

$$\vec{E} = \frac{\vec{F}}{Q_2} = \frac{Q_1}{4\pi\varepsilon\, R_{12}^2}\vec{a}_{R_{12}} \text{ newtons/coulomb (N/C) or volts/metre(V/m)}, \tag{2.14}$$

where $\vec{R}_{12} = \vec{R}_2 - \vec{R}_1$ is the radial distance between Q_1 and Q_2.

Here, the fixed charge Q_1 becomes the source of the force acting on Q_2. The vector field $\vec{E}$ gives the force per unit charge anywhere in space.

In general, the electric field intensity is defined as

$$\vec{E} = \lim_{Q\to0}\frac{\vec{F}}{Q} \text{ N/C}.$$

The vector field, electric field intensity is proportional to the force and the direction is along the force.

It can be seen that the electric field intensity is symmetrical in the radial direction and varies inversely with square of the distance.

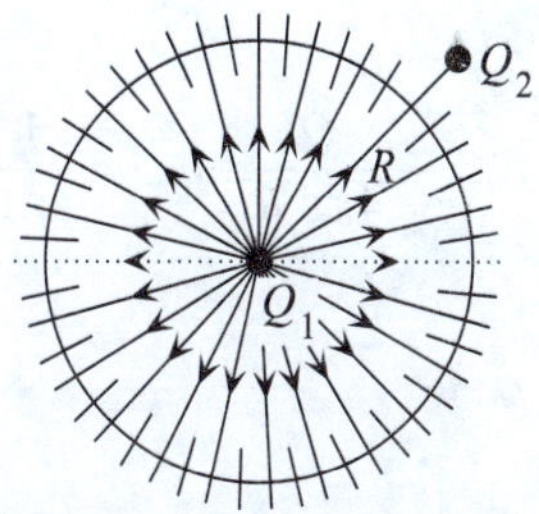

Fig. 2.6 *Field of point charge Q_1*

If a positive charge Q is located at the centre of an imaginary sphere of radius R with the radial unit vector $\vec{a}_R$ directed radially outwards from the centre, the electric field intensity $\vec{E}$ at any point on the spherical surface is given by

$$\vec{E} = \frac{Q}{4\pi\varepsilon R^2}\vec{a}_R \text{ V/m}. \tag{2.15}$$

The direction of $\vec{E}$ is along the radial distance from the positive charge Q_1, i.e., the electric field has the same intensity at all points on the spherical surface.

The magnitude of the electric field strength is

$$E = \frac{Q}{4\pi\varepsilon R^2}.$$

In Cartesian coordinates, from Eq. (2.12), the electric field intensity at a point (x_2, y_2, z_2) due to a charge Q at (x_1, y_1, z_1) is

$$\vec{E} = \frac{Q}{4\pi\varepsilon}\frac{(x_2-x_1)\vec{a}_x + (y_2-y_1)\vec{a}_y + (z_2-z_1)\vec{a}_z}{((x_2-x_1)^2 + (y_2-y_1)^2 + (z_2-z_1)^2)^{3/2}}. \tag{2.16}$$

If the charge Q is at the origin, the electric field intensity at a point (x, y, z) is

$$\vec{E} = \frac{Q}{4\pi\varepsilon}\frac{x\vec{a}_x + y\vec{a}_y + z\vec{a}_z}{(x^2+y^2+z^2)^{3/2}}.$$

Note: The force on a stationary charge Q in an electric field $\vec{E}$ is given by

$$\vec{F} = Q\vec{E} \text{ newtons}. \tag{2.17}$$

Electric field intensity due to N number of charges

Consider N charges, $Q_1, Q_2, Q_3, \ldots, Q_N$, located at different points defined by position vectors $\vec{r}_1, \vec{r}_2, \vec{r}_3, \ldots, \vec{r}_N$ with unit vectors $\vec{a}_{R_1}, \vec{a}_{R_2}, \vec{a}_{R_3}, \ldots, \vec{a}_{R_N}$. Since the field is linear, the principle of superposition holds good. The total electric field intensity $\vec{E}$ at any point P at distance r from the origin is the vector sum of the individual electric field intensities caused by all the individual charges at point P. It is expressed as

$$\vec{E} = \vec{E}_1 + \vec{E}_2 + \vec{E}_3 + \ldots\ldots + \vec{E}_N. \tag{2.18}$$

From Eq. 2.15,

$$\vec{E} = \frac{Q_1}{4\pi\varepsilon\, R_1^2}\vec{a}_{R_1} + \frac{Q_2}{4\pi\varepsilon\, R_2^2}\vec{a}_{R_2} + \ldots + \frac{Q_N}{4\pi\varepsilon\, R_N^2}\vec{a}_{R_N} = \frac{1}{4\pi\varepsilon}\sum_{n=1}^{N}\frac{Q_n}{R_n^2}\vec{a}_{R_n},$$

where, $\vec{R}_n = \vec{r} - \vec{r}_n$ and $\vec{a}_{R_n} = \dfrac{\vec{r} - \vec{r}_n}{|r - r_n|}$.

$$\therefore \quad \vec{E} = \frac{1}{4\pi\varepsilon}\sum_{n=1}^{N}\frac{Q_n}{|r - r_n|^2}\vec{a}_{R_n}. \tag{2.19}$$

Example 2.4 Determine the electric field $\vec{E}$ at the origin due to a point charge of 54.9 nC located at $(-4,\ 5,\ 3)$ in air.

Solution Given $Q = 54.9$ nC at point $(-4,\ 5,\ 3)$, $\vec{E} = \dfrac{Q}{4\pi\varepsilon\, R^2}\vec{a}_R$.

In air, $\varepsilon = \varepsilon_0$.

$$\vec{E} = \frac{54.9 \times 10^{-9}}{4\pi\varepsilon_0\, R^2}\vec{a}_R.$$

Now the distance vector from the point charge to the origin is

$$\vec{R} = (0,0,0) - (-4,5,3) = 4\vec{a}_x - 5\vec{a}_y - 3\vec{a}_z.$$

$$R = \sqrt{4^2 + 5^2 + 3^2} = \sqrt{50}.$$

$$\vec{a}_R = \frac{4\vec{a}_x - 5\vec{a}_y - 3\vec{a}_z}{\sqrt{50}}.$$

$$\therefore \quad \vec{E} = \frac{54.9 \times 10^{-9}}{4\pi\varepsilon_0\,(\sqrt{50})^2}\left(\frac{4\vec{a}_x - 5\vec{a}_y - 3\vec{a}_z}{\sqrt{50}}\right) = \frac{54.9 \times 10^{-9} \times 9 \times 10^9}{50}\left(\frac{4\vec{a}_x - 5\vec{a}_y - 3\vec{a}_z}{\sqrt{50}}\right)$$

$$= 5.6\vec{a}_x - 6.77\vec{a}_y - 4.19\vec{a}_z \text{ V/m}.$$

2.6 Charge Distributions

The electric charge in a medium can be of any of the following forms.

1. Point charge
2. Line charge
3. Surface charge
4. Volume charge

The total charge caused by a charge distribution can be obtained by integrating the charge element over the charge distribution.

Point charge A small amount of charge Q occupying an infinitesimal volume is called a point charge. Its unit is coulomb (C). Its value can be either positive $(+Q)$ or negative $(-Q)$.

Line charge A charge uniformly distributed along a line is called line charge. Line charge density is the charge per unit length. It is expressed as

$$\rho_l = \frac{\text{Total line charge}}{\text{Line length}} = \frac{Q}{l} \text{ C/m.} \tag{2.20}$$

If the line charge density is a function of length, then the differential charge dQ on a differential length dl is given by

$$dQ = \rho_l dl. \tag{2.21}$$

The total charge $Q = \int dQ = \int_l \rho_l dl$ coulombs.

Integration along a line is called line integration. The line may be of infinite length.

If the line is in the form of a closed path, then

$$Q = \oint \rho_l dl \text{ coulombs.} \tag{2.22}$$

Surface charge A charge uniformly distributed over a surface or a sheet is called surface charge or a sheet of charge. The surface charge density is defined as the charge per unit surface area. It is expressed as

$$\rho_s = \frac{\text{Total surface charge}}{\text{Surface area}} = \frac{Q}{S} \text{ C/m}^2. \tag{2.23}$$

If the differential charge on a differential surface area dS is given by

$$dQ = \rho_s dS,$$

then the total charge is

$$Q = \int_s dQ = \int_s \rho_s dS \text{ coulombs.} \tag{2.24}$$

Integration along a surface is called surface integration. The surface may be of infinite area.

For a closed surface, the charge is

$$Q = \oint_s \rho_s dS \text{ coulombs.}$$

Volume charge A volume charge is a charge uniformly distributed throughout a volume. The volume charge density, ρ_v, is defined as the charge per unit volume. It is expressed as

$$\rho_v = \frac{\text{Total volume charge}}{\text{Volume}} = \frac{Q}{v} \text{ C/m}^3. \tag{2.25}$$

If the differential charge on a differential volume dv is given by

$$dQ = \rho_v dv,$$

the total charge is $$Q = \int_v dQ = \int_v \rho_v \, dv \text{ coulombs.} \tag{2.26}$$

Integration over a volume is called volume integration.

Example 2.5 Find the charge in the volume defined by $0 \leq x \leq 2$ m, $0 \leq y \leq 2$ m and $0 \leq z \leq 2$ m if $\rho_v = 30y^2x \ \mu C/m^3$.

Solution Given $\rho_v = 30y^2x \ \mu C/m^3$ and the volume is $0 \leq x \leq 2,\ 0 \leq y \leq 2,\ 0 \leq z \leq 2$. We know that the charge in the volume is

$$Q = \int \rho_v dv = \int_0^2 \int_0^2 \int_0^2 30y^2 x dx \, dy \, dz \ \mu C$$

$$= 30 \left[\frac{x^2}{2}\right]_0^2 \left[\frac{y^3}{3}\right]_0^2 [z]_0^2 = 30 \times \frac{2^2}{2} \times \frac{2^3}{3} \times 2 = 320 \ \mu C.$$

2.7 Electric Field Intensity due to Charge Distributions

Electric field intensity due to a line charge Consider a line charge of charge density ρ_l C/m along a line length l as shown in Fig. 2.7.

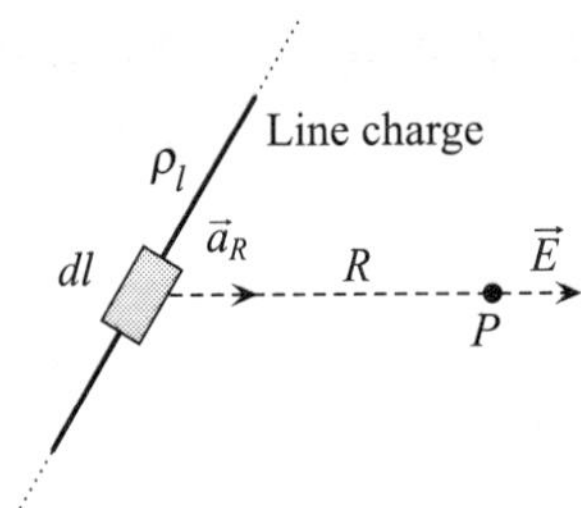

Fig. 2.7 *Line charge*

The differential charge dQ on a differential length dl is

$$dQ = \rho_l dl.$$

The differential electric field at a point P having unit vector $\vec{a}_R$ from dl is, using Eq. (2.15),

$$d\vec{E} = \frac{dQ}{4\pi\varepsilon R^2}\vec{a}_R = \frac{\rho_l dl}{4\pi\varepsilon R^2}\vec{a}_R.$$

Therefore, the electric field intensity at point P due to the line charge is

$$\vec{E} = \int_l \frac{\rho_l dl}{4\pi\varepsilon R^2}\vec{a}_R = \frac{1}{4\pi\varepsilon}\int_l \frac{\rho_l dl}{R^2}\vec{a}_R \ \text{V/m}. \tag{2.27}$$

Electric field intensity due to a surface charge Consider a surface charge of charge density ρ_s C/m^2 over the surface area S as shown in Fig. 2.8

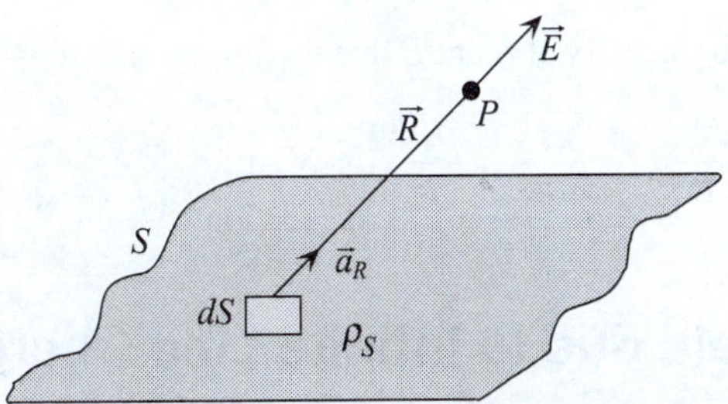

Fig. 2.8 *Surface charge*

The differential charge dQ on the differential surface dS is

$$dQ = \rho_s dS.$$

The differential electric field intensity at point P having unit vector $\vec{a}_R$ from dS is

$$d\vec{E} = \frac{dQ}{4\pi\varepsilon\, R^2}\vec{a}_R = \frac{\rho_s dS}{4\pi\varepsilon\, R^2}\vec{a}_R.$$

Therefore, the electric field intensity at point P due to surface charge is

$$\vec{E} = \int_s d\vec{E} = \int_s \frac{\rho_s dS}{4\pi\varepsilon\, R^2}\vec{a}_R = \frac{1}{4\pi\varepsilon}\int_s \frac{\rho_s dS}{R^2}\vec{a}_R \ \text{V/m}. \tag{2.28}$$

Electrical field intensity due to a volume charge Consider a volume charge of charge density ρ_v C/m^3 in a volume v as shown in Fig. 2.9.

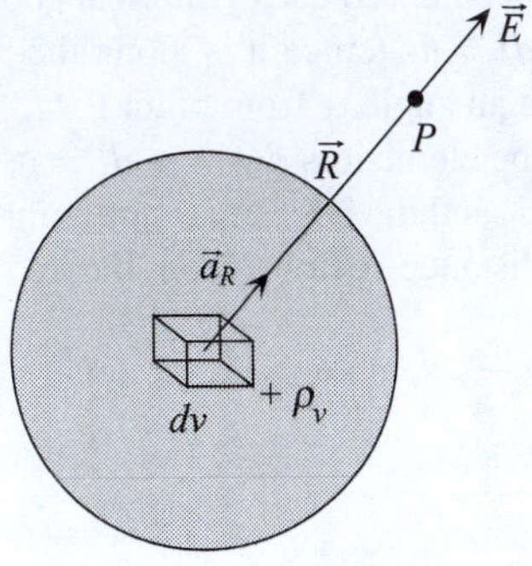

Fig. 2.9 *Volume charge*

The differential charge dQ in the differential volume dv is

$$dQ = \rho_v dv.$$

The differential electric field intensity at point P at a distance R is

$$d\vec{E} = \frac{dQ}{4\pi\varepsilon R^2}\vec{a}_R = \frac{\rho_v dv}{4\pi\varepsilon R^2}\vec{a}_R.$$

Therefore, the electric field intensity at point P due to volume charge is

$$\vec{E} = \int_v d\vec{E} = \int_v \frac{\rho_v dv}{4\pi\varepsilon R^2}\vec{a}_R = \frac{1}{4\pi\varepsilon}\int_v \frac{\rho_v dv}{R^2}\vec{a}_R \text{ V/m.} \tag{2.29}$$

2.8 Electric Field due to Infinite Line Charge

Consider an infinitely long straight line carrying charge with a uniform charge density ρ_l C/m along the z-axis.

Let P be a point on the y-axis at a distance 'ρ' from the line as shown in Fig. 2.10.

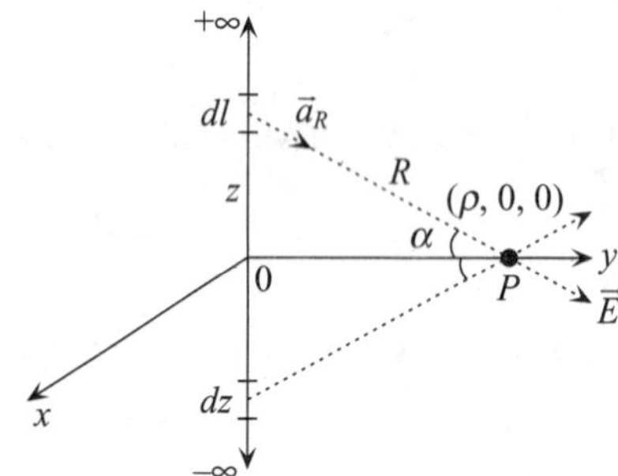

Fig. 2.10 *Field due to a line charge of infinite length*

Since the problem has cylindrical symmetry, that is, the electric field depends only on distance ρ from the line, it would be most convenient to use cylindrical coordinates.

Let a differential line element $dl = dz$ (since it is along the z-axis) be at an arbitrary distance z from the origin. Also, let dl makes an angle α from point P.

The differential charge in the line element is $dQ = \rho_l\, dl = \rho_l\, dz$.

The position vector for dQ is $z\vec{a}_z$ and the coordinates of point P is $(\rho,\ 0,\ 0)$.

In cylindrical coordinates, the distance vector R from the line element to the point P is

$$\vec{R} = (\rho\vec{a}_\rho + 0\vec{a}_\phi + 0\vec{a}_z) - z\vec{a}_z = \rho\vec{a}_\rho - z\vec{a}_z$$

and the unit vector $\vec{a}_R = \dfrac{\rho\vec{a}_\rho - z\vec{a}_z}{\sqrt{\rho^2 + z^2}}$.

The differential electric field intensity due to a differential line charge at point $+z$ is

$$d\vec{E}_+ = \frac{\rho_l dl}{4\pi\varepsilon R^2}\vec{a}_R = \frac{\rho_l dz}{4\pi\varepsilon\,(\rho^2+z^2)}\frac{(\rho\vec{a}_\rho - z\vec{a}_z)}{\sqrt{\rho^2+z^2}} = \frac{\rho_l(\rho\vec{a}_\rho - z\vec{a}_z)dz}{4\pi\varepsilon\,(\rho^2+z^2)^{3/2}}$$

$$d\vec{E}_+ = \frac{\rho_l\rho\vec{a}_\rho dz}{4\pi\varepsilon\,(\rho^2+z^2)^{3/2}} - \frac{\rho_l z\vec{a}_z dz}{4\pi\varepsilon\,(\rho^2+z^2)^{3/2}}. \tag{2.30}$$

Similarly, elemental charge $\rho_l dz$ in the line at a point $(0,\ 0,\ -z)$ is a mirror-image of the charge $\rho_l dz$ at $(0,\ 0,\ z)$ and would produce a differential field intensity

$$d\vec{E}_- = \frac{\rho_l\rho\vec{a}_\rho dz}{4\pi\varepsilon\,(\rho^2+z^2)^{3/2}} + \frac{\rho_l z\vec{a}_z dz}{4\pi\varepsilon\,(\rho^2+z^2)^{3/2}}.$$

As every point $(0,\ 0,\ z)$ has a mirror-image charge point $(0,\ 0,\ -z)$, in the process of integration, the $\vec{a}_z$ component will get cancelled.

$$\therefore\quad E = \int_l d\vec{E} = \int_{-\infty}^{\infty} \frac{\rho_l\rho\vec{a}_\rho dz}{4\pi\varepsilon\,(\rho^2+z^2)^{3/2}} = \frac{\rho_l\rho\vec{a}_\rho}{4\pi\varepsilon}\int_{-\infty}^{\infty}\frac{dz}{(\rho^2+z^2)^{3/2}}. \tag{2.31}$$

To evaluate the integral, from Fig. 2.10, let $z = \rho\tan\alpha,\ dz = \rho\sec^2\alpha d\alpha$.

The limits for α are $-\pi/2$ to $\pi/2$.

$$\therefore\quad \vec{E} = \frac{\rho_l\rho\vec{a}_\rho}{4\pi\varepsilon}\int_{-\pi/2}^{\pi/2}\frac{\rho\sec^2\alpha d\alpha}{(\rho^2+\rho^2\tan^2\alpha)^{3/2}}$$

$$\vec{E} = \frac{\rho_l\rho\vec{a}_\rho}{4\pi\varepsilon}\int_{-\pi/2}^{\pi/2}\frac{\rho\sec^2\alpha d\alpha}{\rho^3\sec^3\alpha} = \frac{\rho_l\vec{a}_\rho}{4\pi\varepsilon\rho}\int_{-\pi/2}^{\pi/2}\cos\alpha d\alpha$$

$$\vec{E} = \frac{\rho_l\vec{a}_\rho}{4\pi\varepsilon\rho}[1-(-1)]$$

$$\vec{E} = \frac{\rho_l\vec{a}_\rho}{2\pi\varepsilon\rho}\quad \text{F/m}. \tag{2.32}$$

Since no practical line is infinitely long, the above expression gives the approximate value of $\vec{E}$. It is observed that the field depends only on the radial component ρ, which is the perpendicular distance between the line charge and the point P.

Example 2.6 A uniform line charge, infinite in extent, with $\rho_l = 25$ nC/m, is lying on the line $x = -3$ m and $z = 4$ m in free space. Find the electric field intensity at point (2, 5, 3).

Solution Given $\rho_l = 25\ \text{nC/m} = 25 \times 10^{-9}\ \text{C/m}$.
The location of the line charge is along $x = -3,\ z = 4$ that is at $(-3, y, 4)$. We have to find the electric field intensity at the point $P(2, 5, 3)$.

The distance vector between point (2, 5, 3) and $(-3, y, 4)$ is

$$R = 5\vec{a}_x - \vec{a}_z.$$

(Since the line charge is along the y-axis, there is no $\vec{a}_y$ component.)

$$R = \sqrt{5^2+1} = \sqrt{26};\ \vec{a}_R = \frac{5\vec{a}_x - \vec{a}_z}{\sqrt{26}}.$$

We know that the electric field at a point P due to line charge in free space is

$$\vec{E} = \frac{\rho_l}{2\pi\varepsilon_0 R}\vec{a}_R = \frac{25\times 10^{-9}(5\vec{a}_x - \vec{a}_z)}{2\pi\varepsilon_0 \times \sqrt{26}\sqrt{26}}$$

$$= \frac{25\times 10^{-9}\times 8\times 10^{9}}{26}(5\vec{a}_x - \vec{a}_z) = 17.3(5\vec{a}_x - \vec{a}_z) = 86.5\vec{a}_x - 17.3\vec{a}_z.$$

2.9 Electric Field due to A Finite Line Charge

Consider a line of finite length l with uniform charge density ρ_l placed along the z-axis. Let the end points of the line be at distances a and $-b$ from the origin. Consider a differential charge dQ of length dl at a distance z from the origin, as shown in Fig. 2.11.

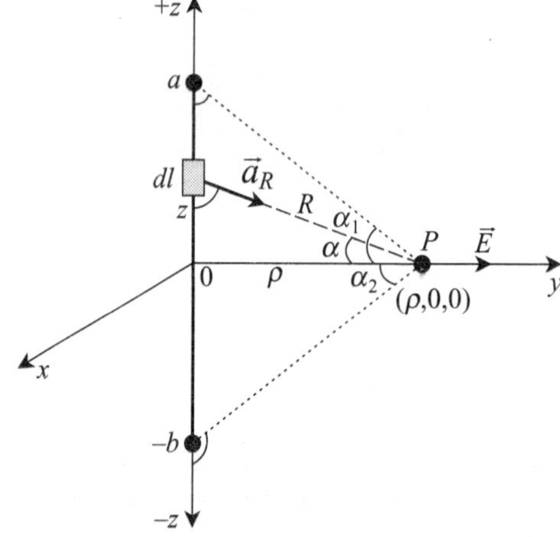

Fig. 2.11 *Field due to line charge of finite length*

Then, $dQ = \rho_l dl = \rho_l dz$.

Let point P be on the y-axis, which is at a distance ρ from the line charge.

Consider cylindrical coordinates. The position of dQ is given by coordinates $(0, 0, z)$ and that of point P by $(\rho, 0, 0)$.

The distance vector between the points $(\rho,\ 0,\ 0)$ and $(0,\ 0,\ z)$ is

$$\vec{R}=(\rho-0)\vec{a}_\rho+0\vec{a}_\phi+(0-z)\vec{a}_z=\rho\vec{a}_\rho-z\vec{a}_z. \tag{2.33}$$

The unit vector is $\vec{a}_R=\dfrac{\vec{R}}{|R|}=\dfrac{\rho\vec{a}_\rho-z\vec{a}_z}{\sqrt{\rho^2+z^2}}.$

We know that the differential field strength is

$$d\vec{E}=\frac{\rho_l dl}{4\pi\varepsilon R^2}\vec{a}_R=\frac{\rho_l dz}{4\pi\varepsilon\,(\rho^2+z^2)^{3/2}}(\rho\vec{a}_\rho-z\vec{a}_z)$$

$$=\frac{\rho_l}{4\pi\varepsilon}\left[\frac{\rho dz\vec{a}_\rho}{(\rho^2+z^2)^{3/2}}-\frac{zdz\vec{a}_z}{(\rho^2+z^2)^{3/2}}\right]. \tag{2.34}$$

The total electric field intensity is

$$\vec{E}=\int_l d\vec{E}=\frac{\rho_l}{4\pi\varepsilon}\int_{-b}^{a}\frac{\rho\vec{a}_\rho dz}{(\rho^2+z^2)^{3/2}}-\frac{\rho_l}{4\pi\varepsilon}\int_{-b}^{a}\frac{z\vec{a}_z dz}{(\rho^2+z^2)^{3/2}}. \tag{2.35}$$

To evaluate the integral, let α be the angle of $\vec{a}_R$ at point P. Also, let α_1 and α_2 be the angles of the end points a and $-b$ of the line charge makes with point P respectively.

From Fig. 2.11, $z=\rho\tan\alpha$, $a=\rho\tan\alpha_1$, and $-b=\rho\tan\alpha_2$,

$$\alpha_1=\tan^{-1}\left(\frac{a}{\rho}\right),\quad \alpha_2=\tan^{-1}\left(\frac{-b}{\rho}\right)$$

$$dz=\rho\ \sec^2\alpha d\alpha.$$

Substituting the values in Eq. (2.35),

$$\vec{E}=\frac{\rho_l}{4\pi\varepsilon}\int_{\alpha_2}^{\alpha_1}\frac{\rho^2\sec^2\alpha d\alpha\vec{a}_\rho}{(\rho^2+\rho^2\tan^2\alpha)^{3/2}}-\frac{\rho_l}{4\pi\varepsilon}\int_{\alpha_2}^{\alpha_1}\frac{\rho^2\tan\alpha\ \sec^2\alpha d\alpha\vec{a}_z}{(\rho^2+\rho^2\tan^2\alpha)^{3/2}}$$

$$=\frac{\rho_l}{4\pi\rho\varepsilon}\int_{\alpha_2}^{\alpha_1}\frac{\rho^2\sec^2\alpha d\alpha\vec{a}_\rho}{\rho^3\sec^3\alpha}-\frac{\rho_l}{4\pi\varepsilon}\int_{\alpha_2}^{\alpha_1}\frac{\rho^2\tan\alpha.\ \sec^2\alpha d\alpha\vec{a}_z}{\rho^3\sec^3\alpha}$$

$$=\frac{\rho_l}{4\pi\rho\varepsilon}\int_{\alpha_2}^{\alpha_1}\cos\alpha\, d\alpha\,\vec{a}_\rho-\frac{\rho_l}{4\pi\rho\varepsilon}\int_{\alpha_2}^{\alpha_1}\sin\alpha d\alpha\vec{a}_z.$$

$$\therefore\quad \vec{E}=\frac{\rho_l}{4\pi\rho\varepsilon}\left[(\sin\alpha_1-\sin\alpha_2)\vec{a}_\rho+(\cos\alpha_1-\cos\alpha_2)\vec{a}_z\right]. \tag{2.36}$$

Note 1: For an infinite line, $\alpha_1=\dfrac{\pi}{2}$, $\alpha_2=\dfrac{-\pi}{2}$.

$$\therefore \quad \vec{E} = \frac{\rho_l}{4\pi\varepsilon\rho}\left[(\sin\frac{\pi}{2} + \sin\frac{\pi}{2})\vec{a}_\rho + (\cos\frac{\pi}{2}) - \cos\frac{\pi}{2})\vec{a}_z\right].$$

$$\vec{E} = \frac{\rho_l}{4\pi\varepsilon\rho}2\vec{a}_\rho = \frac{\rho_l\vec{a}_\rho}{2\pi\varepsilon\rho},$$

which is the same as Eq. (2.32).

Note 2: If the point P is on the perpendicular bisector of the line charge, i.e., $a = b$, then $\alpha_2 = -\alpha_1$.

$$\vec{E} = \frac{\rho_l}{4\pi\varepsilon\rho}\left[(\sin\alpha_1 - \sin(-\alpha_1))\vec{a}_\rho + \cos\alpha_1 - \cos(-\alpha_1)\vec{a}_z\right]$$

$$\vec{E} = \frac{\rho_l}{4\pi\varepsilon\rho}\left[(\sin\alpha_1 + \sin\alpha_1)\vec{a}_\rho + (\cos\alpha_1 - \cos\alpha_1)\vec{a}_z\right]$$

$$\vec{E} = \frac{\rho_l \sin\alpha_1}{2\pi\varepsilon\rho}\vec{a}_\rho. \tag{2.37}$$

Note 3: For semi-infinity line, that is the line extends from 0 to infinity, and if the point P is along the y-axis, then

$$\alpha_1 = \frac{\pi}{2}, \; \alpha_2 = 0,$$

$$\vec{E} = \frac{\rho_l}{4\pi\varepsilon\rho}\left[(\sin\frac{\pi}{2} - \sin 0)\vec{a}_\rho - (\cos\frac{\pi}{2} - \cos 0)\vec{a}_z\right].$$

$$\vec{E} = \frac{\rho_l}{4\pi\varepsilon\rho}(\vec{a}_\rho + \vec{a}_z).$$

2.10 Electric Field Strength due to a Circular Ring of Charge

Consider a circular ring of radius ρ placed in the $x-y$ plane with its centre at the origin, carrying charge Q. Let the charge density be ρ_l C/m. Since we are dealing with cylindrical symmetry, the cylindrical coordinate system is chosen.

Let a point $P(0, \phi, z)$ be on the z-axis, perpendicular to the ring plane as shown in Fig. 2.12.

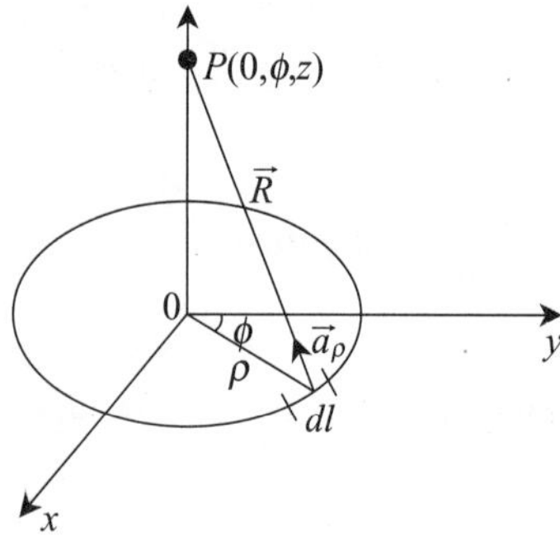

Fig. 2.12 Circular path charge

Let dQ be the differential charge on a differential length dl on the circular path. $dQ = \rho_l dl$.

The position vector of dQ has coordinates $(\rho, \phi, 0)$. From Fig. 2.12,

$$dl = \rho d\phi.$$

The distance vector is $\vec{R} = (0-\rho)\vec{a}_\rho + (z-0)\vec{a}_z = -\rho\vec{a}_\rho + z\vec{a}_z$.

$$|R| = \sqrt{\rho^2 + z^2}.$$

The unit vector is $\quad \vec{a}_R = \dfrac{-\rho\vec{a}_\rho + z\vec{a}_z}{\sqrt{\rho^2+z^2}}.$

The differential field strength is

$$d\vec{E} = \frac{\rho_l dl}{4\pi\varepsilon R^2}\vec{a}_R$$

$$= \frac{\rho_l \rho d\phi}{4\pi\varepsilon(\rho^2+z^2)} \frac{(-\rho\vec{a}_\rho + z\vec{a}_z)}{\sqrt{\rho^2+z^2}}.$$

$$d\vec{E} = \frac{-\rho_l \rho^2 d\phi \vec{a}_\rho}{4\pi\varepsilon(\rho^2+z^2)^{3/2}} + \frac{\rho_l \rho z d\phi}{4\pi\varepsilon(\rho^2+z^2)^{3/2}}\vec{a}_z.$$

The radial components of $\vec{E}$ at point P are symmetrical about the $x-y$ plane and are cancelled out. Thus, the radial component, $\vec{a}_\rho$, is zero.

$$\therefore \quad d\vec{E} = \frac{\rho_l \rho z \vec{a}_z}{4\pi\varepsilon(\rho^2+z^2)^{3/2}} d\phi.$$

The total field strength is therefore

$$\vec{E} = \frac{\rho_l \rho z \vec{a}_z}{4\pi\varepsilon(\rho^2+z^2)^{3/2}} \int_0^{2\pi} d\phi = \frac{\rho_l \rho z \vec{a}_z}{4\pi\varepsilon(\rho^2+z^2)^{3/2}} \times 2\pi$$

$$= \frac{\rho_l \rho z}{2\varepsilon(\rho^2+z^2)^{3/2}}\vec{a}_z \text{ V/m.} \tag{2.38}$$

Note: If $z = 0$, the electric field strength $\vec{E} = 0$. That is, if a point charge is kept at the centre of the ring on the same plane, the force on the point charge is zero.

2.11 Electric Field Strength due to an Infinite Sheet of Charge

Consider an infinite sheet of charge placed in the *x-y* plane with uniform charge density ρ_s C/m^2. Let the differential charge on a differential surface area dS on the *x-y* plane be dQ, and let point P be a point at a distance z on the z-axis, as shown in Fig. 2.13.

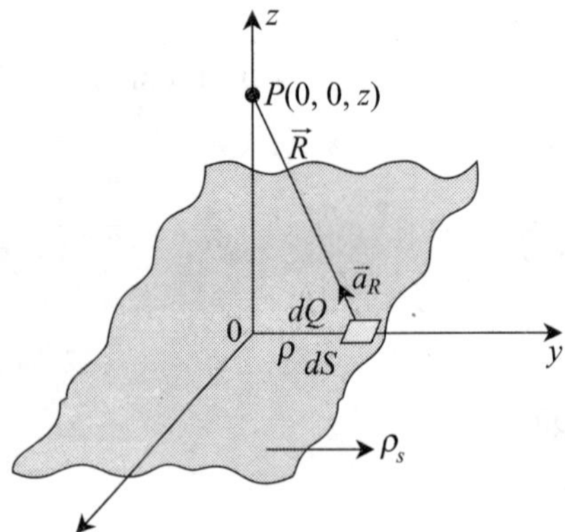

Fig. 2.13 *Infinite sheet of charge*

Using cylindrical coordinates, the position vector of dQ has coordinates $(\rho,\ 0,0)$ and that of P is $(0,0,z)$.

The differential charge is $dQ = \rho_s dS = \rho_s\ \rho d\rho\ d\phi$.

The distance vector is

$$\vec{R} = -\rho\vec{a}_\rho + z\vec{a}_z;\ |R| = \sqrt{\rho^2+z^2}$$

and the unit vector is $\vec{a}_R = \dfrac{R}{|R|} = \dfrac{-\rho\vec{a}_\rho + z\vec{a}_z}{\sqrt{\rho^2+z^2}}$.

We know that the differential field strength is

$$d\vec{E} = \frac{\rho_s dS}{4\pi\varepsilon\, R^2}\vec{a}_R = \frac{\rho_s \rho d\rho\ d\phi}{4\pi\varepsilon\ (\rho^2+z^2)^{3/2}}(-\rho\vec{a}_\rho + z\vec{a}_z)$$

$$= \frac{\rho_s \rho d\rho\ d\phi}{4\pi\varepsilon\ (\rho^2+z^2)^{3/2}}(-\rho\vec{a}_\rho) + \frac{\rho_s \rho d\rho\ d\phi}{4\pi\varepsilon\ (\rho^2+z^2)^{3/2}}z\vec{a}_z.$$

Since the sheet is symmetrical about the radial distance, the $\vec{a}_\rho$ components will be cancelled.

$$\therefore\quad d\vec{E} = \frac{\rho_s \rho d\rho\ d\phi}{4\pi\varepsilon\ (\rho^2+z^2)^{3/2}}(z\vec{a}_z).$$

The total field strength is

$$\vec{E} = \int_s d\vec{E} = \int_0^{2\pi}\int_0^{\infty} \frac{\rho_s \rho d\rho\ d\phi}{4\pi\varepsilon\ (\rho^2+z^2)^{3/2}}z\vec{a}_z. \tag{2.39}$$

(For an infinite sheet charge, the limits for ρ are from 0 to ∞ and for ϕ, they are from 0 to 2π.)

$$\therefore\quad \vec{E} = \frac{\rho_s z\vec{a}_z}{4\pi\varepsilon}\left[\phi\right]_0^{2\pi}\int_0^{\infty}\frac{\rho d\rho}{(\rho^2+z^2)^{3/2}} = \frac{\rho_s z\vec{a}_z}{2\varepsilon}\int_0^{\infty}\frac{\rho d\rho}{(\rho^2+z^2)^{3/2}}.$$

To evaluate the integral, let $\rho^2 + z^2 = t^2$. Then $2\rho\, d\rho = 2t dt$. The limits are from z to ∞.

$$\vec{E} = \frac{\rho_s z \vec{a}_z}{2\varepsilon} \int_z^\infty \frac{t dt}{t^3}$$

$$= \frac{\rho_s z \vec{a}_z}{2\varepsilon} \int_z^\infty \frac{1}{t^2} dt = \frac{\rho_s z \vec{a}_z}{2\varepsilon} \left(\frac{1}{z}\right)$$

$$\vec{E} = \frac{\rho_s}{2\varepsilon} \vec{a}_z \ \text{V/m}. \tag{2.40}$$

Note 1: The field strength has component normal to the plane of sheet charge. In general, if $\vec{a}_n$ is the normal component and if the sheet charge is in any plane, then the electric field strength is $\vec{E} = \frac{\rho_s}{2\varepsilon}\vec{a}_n$ V/m.

Note 2: If the point P is considered on the negative axis, then the field strength becomes

$$\vec{E} = \frac{-\rho_s}{2\varepsilon} \vec{a}_z \ \text{V/m}.$$

It is observed that the electric field strength at any point due to an infinite sheet charge is always normal to the plane of the sheet and is independent of the position (distance) of the point.

2.12 Electric Field Strength due to a Sheet of Circular Disc

Consider a sheet of circular disc radius a laying in the $x-y$ plane. Let the disc be carrying a charge Q with surface charge density ρ_s C/m^2. Also, let a differential charge dQ be on a differential surface area dS at a distance ρ on the y-axis, and let point P be at a distance z on the z-axis as shown in Fig. 2.14.

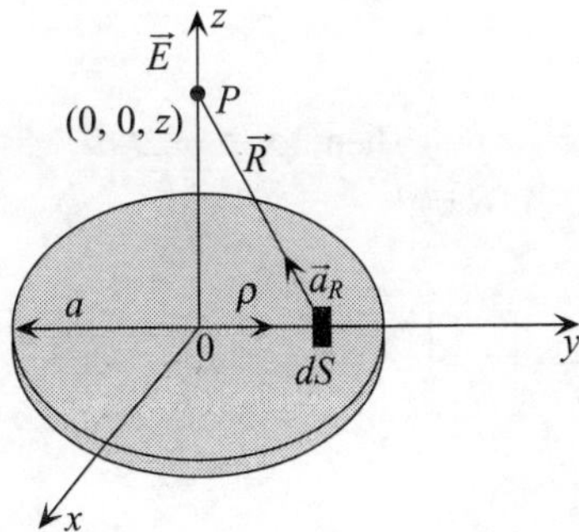

Fig. 2.14 *Field due to sheet of circular disc*

Due to the symmetry, we will use cylindrical coordinates. The position vector at dQ is $\rho\vec{a}_\rho$ and the coordinates of P is $(0, 0, z)$.

The differential charge is

$$dQ = \rho_s dS = \rho_s \rho d\rho\, d\phi.$$

The distance vector is $\vec{R} = (0-\rho)\vec{a}_\rho + 0\vec{a}_\phi + (z-0)\vec{a}_z = -\rho\vec{a}_\rho + z\vec{a}_z$.

$$|R| = \sqrt{R^2 + z^2}$$

and the unit vector $\vec{a}_R = \dfrac{|R|}{\vec{R}} = \dfrac{-\rho\vec{a}_\rho + z\vec{a}_z}{\sqrt{\rho^2+z^2}}$.

The differential field strength is

$$d\vec{E} = \frac{\rho_s dS}{4\pi\varepsilon R^2}\vec{a}_R = \frac{\rho_s \rho d\rho\, d\phi}{4\pi\varepsilon\,(\rho^2+z^2)^{3/2}}(-\rho\vec{a}_\rho + z\vec{a}_z)$$

$$= \frac{\rho_s \rho d\rho\, d\phi}{4\pi\varepsilon\,(\rho^2+z^2)^{3/2}}(-\rho\vec{a}_\rho) + \frac{\rho_s \rho d\rho\, d\phi}{4\pi\varepsilon\,(\rho^2+z^2)^{3/2}} z\vec{a}_z.$$

Since the sheet is symmetrical about the radial distance, the $\vec{a}_\rho$ components will be cancelled.

$$\therefore \quad d\vec{E} = \frac{\rho_s \rho d\rho\, d\phi}{4\pi\varepsilon\,(\rho^2+z^2)^{3/2}} z\vec{a}_z. \tag{2.41}$$

The limits of the circular disc for ρ is 0 to a, and for θ it is 0 to 2π.

The total field strength is

$$\vec{E} = \int_s d\vec{E} = \int_0^{2\pi}\int_0^{a} \frac{\rho_s \rho d\rho\, d\phi}{4\pi\varepsilon\,(\rho^2+z^2)^{3/2}} z\vec{a}_z = \frac{\rho_s z\vec{a}_z}{4\pi\varepsilon}[\phi]_0^{2\pi}\int_0^a \frac{\rho d\rho}{(\rho^2+z^2)^{3/2}}$$

$$= \frac{\rho_s z\vec{a}_z}{2\varepsilon}\int_0^a \frac{\rho d\rho}{(\rho^2+z^2)^{3/2}}.$$

To evaluate the integral, let $\rho^2 + z^2 = t^2$. Then $2\rho d\rho = 2t\, dt$.
The limits are: at $\rho = 0$, $t = z$; at $\rho = a$, $t = \sqrt{a^2+z^2}$.

$$\therefore \quad \vec{E} = \frac{\rho_s z\vec{a}_z}{2\varepsilon}\int_z^{\sqrt{a^2+z^2}} \frac{tdt}{t^3} = \frac{\rho_s z\vec{a}_z}{2\varepsilon}\left[\frac{-1}{t}\right]_a^{\sqrt{a^2+z^2}}$$

$$= \frac{\rho_s z\vec{a}_z}{2\varepsilon}\left[\frac{-1}{\sqrt{a^2+z^2}} + \frac{1}{z}\right]$$

$$= \frac{\rho_s}{2\varepsilon}\left[1 - \frac{z}{\sqrt{a^2+z^2}}\right]\vec{a}_z \tag{2.42}$$

Note 1: If the point P is on the centre of the disc, i.e., $z = 0$, the electric field strength is

$$\vec{E} = \frac{\rho_s \vec{a}_z}{2\varepsilon}.$$

Note 2: If the circular disc is an infinite sheet charge i.e., $a = \infty$, the electric field strength is

$$\vec{E} = \frac{\rho_s \vec{a}_z}{2\varepsilon},$$

which is same as Eq. (2.40).

Example 2.7 A circular disc of 10 cm radius is charged uniformly with a total charge of 100 μC. Find E at a point 20 cm on its axis.

Solution Given radius of a circular disc $a = 10$ cm $= 0.1$m; $Q = 100\ \mu\text{C} = 10^{-4}$ C.

Let the disc be placed in the $x-y$ plane in space. We have to calculate $\vec{E}$ on its axis at a point where $z = 20$ cm $= 0.2$ m.

We know that

$$\vec{E} = \frac{\rho_s}{2\varepsilon_0}\left[1 - \frac{z}{\sqrt{a+z^2}}\right]\vec{a}_z.$$

Now, $\rho_s = Q/A = \dfrac{10^{-4}}{\pi a^2} = 3.184 \times 10^{-3}\text{C/m}^2$

$$\vec{E} = \frac{3.1.84 \times 10^{-3}}{2 \times 8854 \times 10^{-12}}\left[1 - \frac{0.2}{\sqrt{0.1^2 + 0.2^2}}\right]$$

$$= 18.97 \times 10^6\ \text{V/m}.$$

Example 2.8 A line charge located parallel to the y-axis at $x = 2$ and $z = 2$ carries a charge density $\rho_l = 10\pi$ nC/m, and two sheet charges at $x = 2$ and at $y = -4$ carry charge densities of 20 nC/m^2 and 20 nC/m^2 respectively. Calculate the electric field intensity $\vec{E}$ at (1, 1, 1) due to the three charge distributions.

Solution Given: $\rho_l = 10\pi$ nC/m at $x = 2$, $z = 2$,

$$\rho_{s_1} = 20\ \text{nC/m}^2 \text{ at } x = 2,$$
$$\rho_{s_2} = 15\ \text{nC/m}^2 \text{ at } y = -4$$

The field at (1, 1, 1) due to the line charge is

$$\vec{E}_1 = \frac{\rho_l}{2\pi\varepsilon_0 R}\vec{a}_R.$$

The distance vector between the coordinates (1, 1, 1) and (2, 1, 2).

$$\vec{R} = (1-2)\vec{a}_x + (1-1)\vec{a}_y + (1-2)\vec{a}_z$$
$$= -\vec{a}_x - \vec{a}_z = \sqrt{1+1} = \sqrt{2}.$$

$$\vec{E}_1 = \frac{10\pi \times 10^{-9}}{2\pi \times 8.854 \times 10^{-12} \times \sqrt{2}} \times \frac{-\vec{a}_x - \vec{a}_z}{\sqrt{2}}$$

$$= -90\pi(\vec{a}_x + \vec{a}_z) = -282.7(\vec{a}_x + \vec{a}_z).$$

The field at point $x_1 = 1$ due to sheet charge at $x = 2$ is

$$\vec{E}_2 = \frac{\rho_{s_1}}{2\varepsilon_0}(-\vec{a}_x).$$

Since $(x > x_1)$, the field direction is along the $-x$-axis.

$$\therefore \quad \vec{E}_2 = \frac{-20 \times 10^{-9}}{2 \times 8.854 \times 10^{-12}}\vec{a}_x = -1130.9\vec{a}_x.$$

The field at point $y_1 = 1$ due to sheet charge at $y = -3$ is

$$\vec{E}_3 = \frac{\rho_{s3}}{2\varepsilon_0}\vec{a}_y = \frac{15 \times 10^{-9}}{2 \times 8.854 \times 10^{-12}} = 848.23\vec{a}_y.$$

Since $(y < y_1)$, the field direction is along the positive y-axis. Therefore, the total electric field intensity is

$$\vec{E} = \vec{E}_1 + \vec{E}_2 + \vec{E}_3$$

$$= -282.7(\vec{a}_x + \vec{a}_z) - 1130.9\vec{a}_x + 848.23\vec{a}_y.$$

$$\vec{E} = -1413.6\vec{a}_x + 848.23\vec{a}_y - 282.7\vec{a}_z \text{ V/m}.$$

2.13 Electric Flux and Flux Density

Michael Faraday in 1837, while conducting experiments on static electric fields, used a pair of concentric conducting spheres as shown in Fig. 2.15. A positive charge, $+Q$, was placed on the inner sphere. Insulating material or air was filled in between the spheres. There is no net charge on the surface of the outer sphere. The outer sphere was earthed momentarily through a closed switch. The inner sphere was carefully removed and the charge on the outer sphere was then measured. Faraday found that:

1 On the outer sphere, a negative charge $(-Q)$ equal in magnitude to the positive charge was observed.

2 The charge on the outer surface was independent of the size of the spheres and the dielectric medium between them.

3 As the charge $(+Q)$ on the inner surface increases, the charge $(-Q)$ on the outer surface also increases with equal magnitude.

Faraday postulated that there was an electric displacement from the positive charge to the negative charge through the medium. This displacement is called electric flux. It is the total number of lines of force in the electric field and is represented by the symbol ' ψ '. The lines of force are called flux lines or stream lines.The amount of dispacement of flux is equal in magnitude to the positive charge.

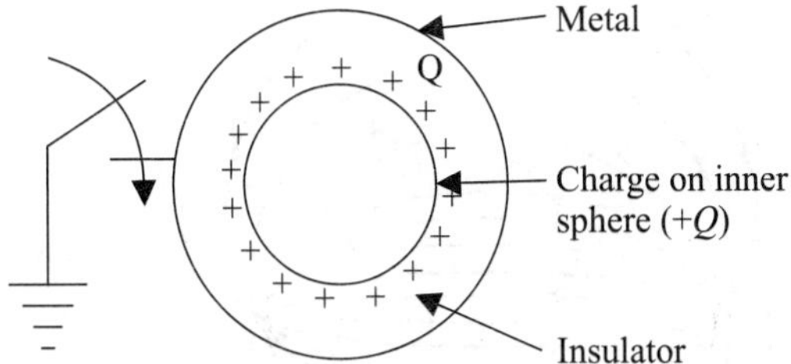

Fig. 2.15 *A schematic diagram of Faraday's experiment with concentric spheres*

Electric flux (ψ) An electric charge produces electric flux and as the charge increases, the electric flux also increases. Therefore, electric flux is numerically equal to the electric charge producing it. If an electric charge produces Q coulombs, then the electric flux, associated with it is given by

$$\psi = Q \text{ coulombs.}$$

For an isolated positive charge Q, the electric flux originates radially from the positive charge and terminates at infinity (see Fig. 2.16(a)).

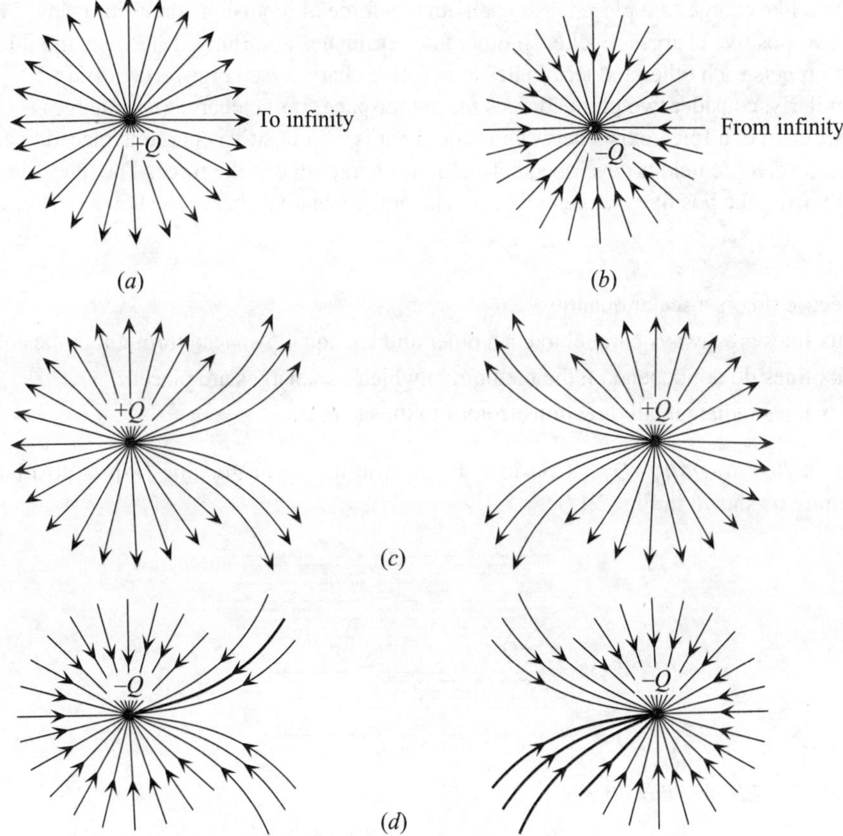

Fig. 2.16 *Flux lines of (a) an isolated positive charge and (b) an isolated negative charge. Flux lines between (c) two positive charges, (d) two negative charges.*

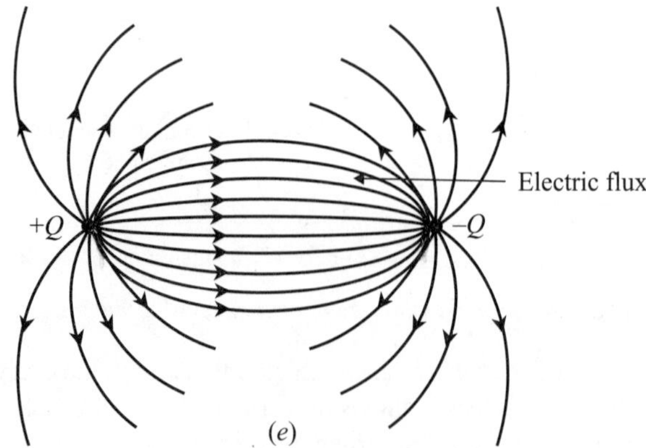

Fig. 2.16 contd *Flux lines of (e) one positive and one negative charge.*

Similarly, for an isolated negative charge, the electric flux from infinity radially terminates at the negative charge as shown in Fig. 2.16(b).

If two like charges are placed near each other, a force of repulsion comes into play. The flux lines from two positive charges repel each other and terminates at infinity. Likewise, the flux lines from infinity repels each other and terminates at negative charges (see Fig. 2.16(c) and (d)).

Similarly, consider two point charges placed in space. If one charge (test charge) is brought near another charge, a force exists between them. That is, if a positive charge is placed near a negative charge, an electric field is developed between the charges due to the force. The lines force originates radially from the positive charge and terminate at the negative charge as shown in Fig. 2.16(e).

Note:

1. Electric flux is a scalar quantity.
2. Flux lines are always parallel to each other and are equally spaced throughout the solid angle.
3. Flux lines do not depend on the medium in which the charges are placed.
4. Flux lines radiate in all directions from/into the charges.

Electric flux density $(\vec{D})$ Consider the electric flux ψ originating from a positive charge, Q coulombs, as shown in Fig. 2.17.

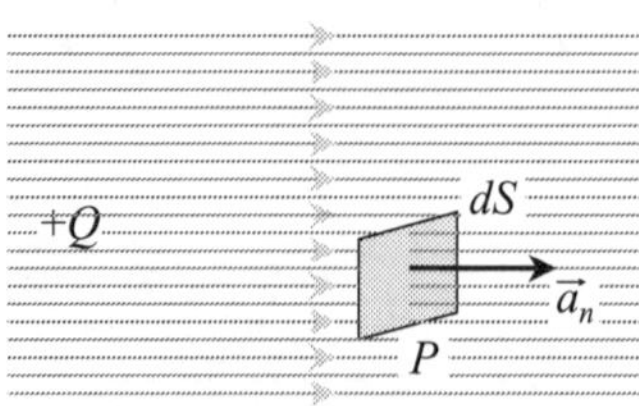

Fig. 2.17 *Electric flux lines*

Let dS be a differential surface area at a point P perpendicular to the lines of flux. The electric flux crossing through this surface area is $d\psi$.

The net electric flux passing through the unit surface area is called electric flux density $\vec{D}$. The direction flux density is normal to the surface area. It is expressed as

$$\vec{D} = \frac{d\psi}{dS}\vec{a}_n \text{ C/m}^2 \tag{2.43}$$

The flux density $\vec{D}$ at any point depends on the direction of the differential surface area. If the flux density makes an angle θ with the normal at dS, then

$$d\psi = DdS\cos\theta = \vec{D}\cdot d\vec{S}.$$

That is, for an electric flux density $\vec{D}$, the flux passing through the surface $d\vec{S}$ is the dot product of $\vec{D}$ and $d\vec{S}$.

If S is the total area, the total flux is $\psi = \vec{D}\cdot\vec{S}$.

2.14 Electric Flux Density due to Point Charge *Q*

Consider an imaginary sphere of radius r and a point charge $+Q$ placed at its centre as shown in Fig. 2.18. The flux lines originating from the point charge distribute radially over the surface of the sphere.

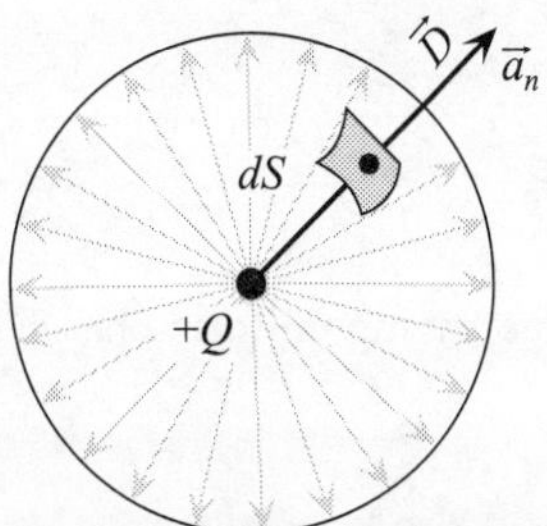

Fig. 2.18 *Flux from point charge*

The flux density at a differential surface dS is

$$\vec{D} = \frac{d\psi}{dS}\vec{a}_n.$$

If the total surface area through which the flux ψ passes is considered, due to symmetry in the radial direction, $\vec{D}$ is always normal to the surface area.

Also, $\vec{D} = \dfrac{\text{Total flux}}{\text{Surface area of the sphere}} = \dfrac{\psi}{4\pi r^2}\vec{a}_n.$

Since $\psi = Q$, $\vec{D} = \dfrac{Q}{4\pi r^2}\vec{a}_n$. C/m^2

Since $\vec{a}_n = \vec{a}_r$, the radial direction normal to the surface area,

$$\vec{D} = \frac{Q}{4\pi r^2}\vec{a}_r \tag{2.44}$$

The electric flux density is also called displacement flux density. It can be seen from the above that electric flux density depends on the charge and the radial distance, and is independent of the characteristics of the medium.

Relationship between $\vec{E}$ and $\vec{D}$ From Eq. (2.15), we know that the electric field strength $\vec{E}$ at a distance r from a positive point charge Q is given by

$$\vec{E} = \frac{Q}{4\pi\varepsilon r^2}\vec{a}_r = \frac{1}{\varepsilon}\cdot\vec{D}.$$

$$\therefore \quad \vec{D} = \varepsilon\vec{E}. \tag{2.45}$$

Therefore, $\vec{D}$ and $\vec{E}$ are related through the permittivity of the medium. E is a function of the permittivity of the medium, while the flux density is not.

For a medium where permittivity is constant, the direction of displacement flux density is along the direction of E.

2.15 Electric Flux Density due to Charge Distributions

Case 1 For an infinite line of charge density ρ_l, we know from Eq. (2.32) that

$$\vec{E} = \frac{\rho_l}{2\pi\varepsilon\rho}\vec{a}_\rho.$$

$$\text{Then } \vec{D} = \varepsilon\vec{E} = \varepsilon\frac{\rho_l}{2\pi\varepsilon\rho}\vec{a}_\rho = \frac{\rho_l}{2\pi\rho}\vec{a}_\rho. \tag{2.46}$$

Case 2 For an infinite sheet charge having uniform charge density ρ_s C/m^2, we know from Eq. (2.40) that

$$\vec{E} = \frac{\rho_s}{2\varepsilon}\vec{a}_n.$$

$$\text{Then } \vec{D} = \varepsilon\vec{E} = \frac{\varepsilon\,\rho_s}{2\varepsilon}\vec{a}_n = \frac{\rho_s}{2}\vec{a}_n, \tag{2.47}$$

where $\vec{a}_n$ is a unit vector normal to the plane of the sheet.

Case 3 For a volume charge enclosed by a sphere of radius r, having a uniform charge density ρ_v C/m^3,

$$\vec{E} = \frac{Q}{4\pi\varepsilon r^2}\vec{a}_r.$$

The flux density is $\quad \vec{D} = \varepsilon\,\vec{E} = \dfrac{Q}{4\pi r^2}\vec{a}_r$, where $\quad Q = \displaystyle\int_v \rho_v dv.$

$$\text{Then, } \quad D = \int_v \frac{\rho_v dv}{4\pi r^2}\vec{a}_r. \tag{2.48}$$

2.16 Gauss' Law (Integral Form)

Gauss' law states that *the net electric flux passing through any closed surface is equal to the total charge enclosed by that surface.*

Mathematically, it can be expressed as

$$\psi = Q_{\text{encl}} = \oint_S \vec{D} \cdot d\vec{S}, \tag{2.49}$$

where Q_{encl} = charge enclosed by the surface, $\vec{D}$ = electric flux density and $d\vec{S}$ = differential surface area.

Proof Consider an arbitrary surface enclosing a charge Q. Let dS be a differential surface area through which the flux passes, as shown in Fig. 2.19. The differential surface area vector is always chosen normal to the surface area.

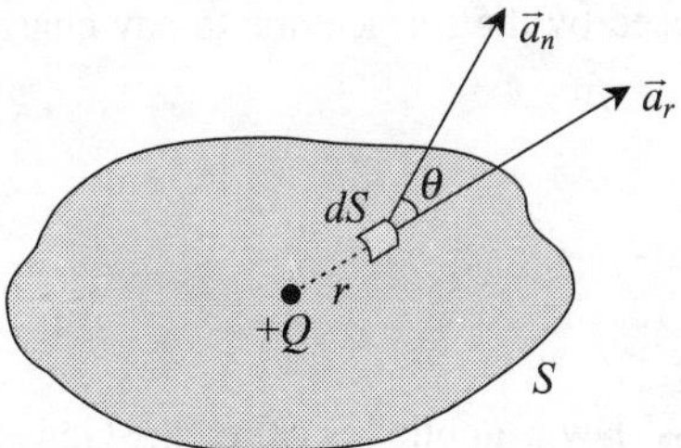

Fig. 2.19 *Charge enclosed by the surface S*

$$d\vec{S} = dS\vec{a}_n.$$

We know that the electric flux density at a distance r from a point charge is

$$\vec{D} = \frac{Q}{4\pi r^2}\vec{a}_r.$$

Taking the dot product with surface area $d\vec{S}$ on both sides,

$$\vec{D} \cdot d\vec{S} = \frac{Q\vec{a}_r}{4\pi r^2} \cdot dS\vec{a}_n.$$

Let θ be the angle between the direction of $\vec{a}_r$ and the normal to the surface, $\vec{a}_n$.

$$\vec{D} \cdot d\vec{S} = \frac{Q}{4\pi r^2} dS\vec{a}_r . \vec{a}_n = \frac{Q}{4\pi r^2} dS\cos\theta.$$

The net flux passing through the closed surface is obtained by taking closed surface integration on both sides.

$$\oint_S \vec{D} \cdot d\vec{S} = \oint_S \frac{Q}{4\pi r^2} dS\cos\theta = \frac{Q}{4\pi} \oint_S \frac{dS\cos\theta}{r^2}.$$

The quantity $dS\cos\theta$ is the component of dS in the radial direction, and $\frac{dS\cos\theta}{r^2}$ is the projection of dS on a spherical surface of unit radius. This represents the solid angle of the surface area dS at the point charge.

Therefore, the solid angle of the sphere is $\oint_s \frac{dS\cos\theta}{r^2} = 4\pi$.

So $\oint_s \vec{D}\cdot d\vec{S} = \frac{Q}{4\pi} \times 4\pi = Q$.

Therefore, the total displacement through the closed surface is

$$\oint_s \vec{D}\cdot d\vec{S} = Q = \psi.$$

If Q_{encl} is the total charge enclosed by the surface due to any charge distribution, then Gauss' law states that

$$\oint_s \vec{D}\cdot d\vec{S} = Q_{\text{encl}}$$ (Hence proved.)

Note:

1. The surface over which Gauss' law is applied is called the Gaussian surface.
2. The Gaussian surface is always a closed surface. It may be of any shape, but must have a constant electric field.
3. The surface is chosen such that at each point on the Gaussian surface, the flux density $\vec{D}$ is either normal or tangential to the surface.
4. When D is normal to the Gaussian surface, $\vec{D}\cdot d\vec{S} = DdS$ is a constant. If $\vec{D}$ is tangential to the surface, $\vec{D}\cdot d\vec{S} = 0$.
5. If there are more than one charge or charge distribution enclosed by the Gaussian surface, the net charge is the algebraic sum of all the charges.

For a volume charge with charge density ρ_v C/m^3, the total flux is

$$\psi = \int_v \rho_v dv = \oint_s \vec{D}_s \cdot d\vec{S}.$$

For a line charge with charge density $\rho_l \mathrm{C/m}$,

$$\psi = \int_l \rho_l dl = \oint_s \vec{D}_s \cdot d\vec{S}.$$

For a surface charge with charge density ρ_s C/m^2,

$$\psi = \int_s \rho_s dS = \oint_s \vec{D}_s \cdot d\vec{S}.$$

Gauss' law can be applied to the above described charge distributions when symmetry conditions exist. It is easy for symmetrical distributions because the problem of integration always disappears when a proper closed surface is chosen. Gauss' law is not useful if there is no suitable Gaussian surface.

Example 2.9 A circular disc of radius 20 m with a charge density $\rho_s = 100\sin\phi\ \text{nC/m}^2$ is enclosed by surface S. What net flux crosses the surface?

Solution Given: $\rho_s = 100\sin\phi\ \text{nC/m}^2$; radius $a = 20$ m.

Flux crossing the surface is $\psi = Q = \int \rho_s dS$

$$= \int_{\phi=0}^{2\pi}\int_{r=0}^{20} (100\sin\phi \times 10^{-9})\rho d\rho\ d\phi$$

$$= 100\times 10^{-9}\int_0^{20}\rho\ d\rho\int_0^{2\pi}\sin\phi d\phi$$

$$= 100\times 10^{-9}\frac{10^2}{2}\times 0 = 0.$$

Example 2.10 A point charge, $Q = 60$ nC, is located at the origin of a Cartesian coordinate system. Find the electric flux density D at (4, 7, −8).

Solution Given: $Q = 60$ nC at origin and a point at (4, 7, −8).
The distance vector is

$$\vec{R} = 4\vec{a}_x + 7a_y - 8\vec{a}_z.$$

$$R = \sqrt{4^2+7^2+8^2} = \sqrt{129}.$$

$$\vec{a}_R = \frac{4\vec{a}_x + 7\vec{a}_y - 8\vec{a}_z}{\sqrt{4^2+7^2+8^2}} = \frac{4\vec{a}_x + 7\vec{a}_y - 8\vec{a}_z}{\sqrt{129}}.$$

We know that the flux density is

$$\vec{D} = \frac{Q}{4\pi R^2}\vec{a}_R = \frac{60\times 10^{-9}}{4\pi\times 129}\cdot\frac{4\vec{a}_x + 7\vec{a}_y - 8\vec{a}_z}{\sqrt{129}}$$

$$= 3.25\times 10^{-12}(4\vec{a}_x + 7\vec{a}_y - \vec{a}_z) = (13\vec{a}_x + 22.8\vec{a}_y - 26\vec{a}_z)\ \text{pC/m}^2.$$

2.17 Gauss' Law in Point Form (Maxwell's First Law)

Gauss' law in point form states that *the divergence of the electric flux density* ($\vec{D}$) *is equal to the volume charge density* ρ_v *at any point in a given medium.*

It can be expressed as

$$\nabla \cdot \vec{D} = \rho_v, \text{ or div } \vec{D} = \rho_v. \tag{2.50}$$

Proof Let us consider a differential Gaussian surface in the form of a rectangular box with dimensions Δx, Δy and Δz as shown in Fig. 2.20.

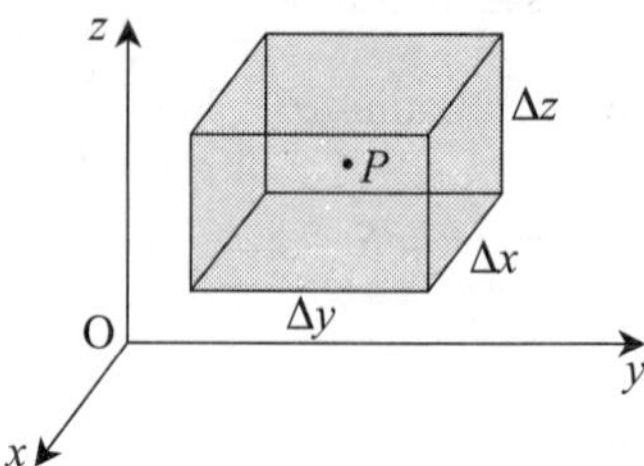

Fig. 2.20 *Differential rectangular Gaussian surface*

Let P be a point at the centre of the box and let charge Q be enclosed by the surface.

Let the flux density at the centre of the box be

$$\vec{D} = D_{x0}\vec{a}_x + D_{y0}\vec{a}_y + D_{z0}\vec{a}_z.$$

The integral form of Gauss' law is

$$Q = \oint_s \vec{D} \cdot d\vec{S}.$$

The integration over the closed surface of the box is

$$Q = \int_{\text{front}} \vec{D}_1 \cdot d\vec{S}_1 + \int_{\text{back}} \vec{D}_2 \cdot d\vec{S}_2 + \int_{\text{left}} \vec{D}_3 \cdot d\vec{S}_3 + \int_{\text{right}} \vec{D}_4 \cdot d\vec{S}_4 + \int_{\text{top}} \vec{D}_5 \cdot d\vec{S}_5 + \int_{\text{bottom}} \vec{D}_6 \cdot d\vec{S}_6, \tag{2.51}$$

where D_1 to D_6 are the flux densities on the box surfaces with differential surface areas dS_1 to dS_6 respectively.

Since point P is at the centre and the box surfaces are very small, the flux density at the front surface is approximately equal to (using Taylor series expansion for D_x and neglecting higher order terms)

$$D_1 \approx D_{x0} + \frac{\partial D_x}{\partial x}\frac{\Delta x}{2}$$

So, $$\int_{\text{front}} \vec{D}_1 \cdot d\vec{S}_1 \approx \left(D_{x0} + \frac{\partial D_x}{\partial x}\frac{\Delta x}{2}\right)\Delta y \Delta z,$$

and for the back surface,

$$\int_{\text{back}} \vec{D}_2 \cdot d\vec{S}_2 \approx \left(-D_{x0} + \frac{\partial D_x}{\partial x}\frac{\Delta x}{2}\right)\Delta y\, \Delta z.$$

Then, $\int\limits_{\text{front}} \vec{D}_1 \cdot d\vec{S}_1 + \int\limits_{\text{back}} \vec{D}_2 \cdot d\vec{S}_2 \approx \frac{\partial D_x}{\partial x}\Delta x\,\Delta\, y\Delta z \approx \frac{\partial D_x}{\partial x}\Delta v.$

Similarly, for the left and the right surfaces,

$$\int\limits_{\text{left}} \vec{D}_3 \cdot d\vec{S}_3 + \int\limits_{\text{right}} \vec{D}_4 \cdot d\vec{S}_4 \approx \frac{\partial D_y}{\partial y}\Delta x\,\Delta y\,\Delta z \approx \frac{\partial D_y}{\partial y}\Delta v,$$

and for the top and bottom surfaces,

$$\int\limits_{\text{top}} \vec{D}_5 \cdot d\vec{S}_5 + \int\limits_{\text{bottom}} \vec{D}_6 \cdot d\vec{S}_6 \approx \frac{\partial D_z}{\partial z}\Delta x\,\Delta y\,\Delta z \approx \frac{\partial D_z}{\partial z}\Delta v,$$

where $\Delta x\,\Delta y\,\Delta z = \Delta v$, is the differential volume.

From Eq. (2.51), $Q \approx \frac{\partial D_x}{\partial x}\Delta v + \frac{\partial D_y}{\partial y}\Delta v + \frac{\partial D_z}{\partial z}\Delta v$ (2.52)

$$\approx \left(\frac{\partial D_x}{\partial x} + \frac{\partial D_y}{\partial y} + \frac{\partial D_z}{\partial z}\right)\Delta v.$$

The charge enclosed in a volume is given by

$$\frac{Q}{\Delta v} \approx \frac{\partial D_x}{\partial x} + \frac{\partial D_y}{\partial y} + \frac{\partial D_z}{\partial z}.$$

Substituting the value of Q,

$$\frac{\oint_s \vec{D} \cdot d\vec{S}}{\Delta v} \approx \frac{\partial D_x}{\partial x} + \frac{\partial D_y}{\partial y} + \frac{\partial D_z}{\partial z}.$$

For better approximation, let the value Δv tend to zero by taking limits,

$$\lim_{\Delta v \to 0} \oint\limits_s \frac{\vec{D} \cdot d\vec{S}}{\Delta v} = \lim_{\Delta v \to 0} \frac{Q}{\Delta v} = \frac{\partial D_x}{\partial x} + \frac{\partial D_y}{\partial y} + \frac{\partial D_z}{\partial z}. \tag{2.53}$$

We know that the volume charge density is defined as

$$\rho_v = \lim_{\Delta v \to 0} \frac{Q}{\Delta v}.$$

Also, the divergence of $\vec{D}$ is

$$\rho_v = \frac{\partial D_x}{\partial x} + \frac{\partial D_y}{\partial y} + \frac{\partial D_z}{\partial z} = \nabla \cdot \vec{D}$$

or $\quad \rho_v = \nabla \cdot \vec{D}$ (Hence proved.) (2.54)

Note: The point form of Gauss' law is also called Maxwell's first law. This relationship is useful to determine the volume charge density when a flux density is given irrespective of the medium properties.

2.18 Divergence of Electric Flux Density

From Eq. (2.54), the divergence of electric flux density can be expressed as

$$\nabla\cdot\vec{D} = \frac{\partial D_x}{\partial x} + \frac{\partial D_y}{\partial y} + \frac{\partial D_z}{\partial z} = \lim_{\Delta v\to 0} \oint \frac{\vec{D}\cdot d\vec{S}}{\Delta v} = \rho_v \tag{2.55}$$

i.e., the divergence of flux density is the outflow of electric flux from a closed surface per unit volume as the volume shrinks to zero. This quantity is equal to the volume charge density. In physical terms, the divergence of the flux density is the extent to which the flux density behaves like a source or a sink at a given point. If the divergence is non zero at some point, there must be a source or a sink at that point.

The divergence of electric flux density in various coordinte systems is as given below.

Cartiesian coordinates

$$\nabla\cdot\vec{D} = \frac{\partial D_x}{\partial x} + \frac{\partial D_y}{\partial y} + \frac{\partial D_z}{\partial z} \tag{2.56}$$

Cylindrical coordinates

$$\nabla\cdot\vec{D} = \frac{1}{\rho}\frac{\partial}{\partial \rho}(\rho D_\rho) + \frac{1}{\rho}\frac{\partial D_\phi}{\partial \phi} + \frac{\partial D_z}{\partial z} \tag{2.57}$$

Spherical coordinates

$$\nabla\cdot\vec{D} = \frac{1}{r^2}\frac{\partial}{\partial r}(r^2 D_r) + \frac{1}{r\sin\theta}\frac{\partial}{\partial\theta}(\sin\theta D_\theta) + \frac{1}{r\sin\theta}\frac{\partial D_\phi}{\partial\phi} \tag{2.58}$$

Example 2.11 Find the total charge in a volume defined by six planes for which $1 \le x \le 2,\ 2 \le y \le 3,\ 3 \le z \le 4$, $\vec{D} = 4x\vec{a}_x + 3y^2\vec{a}_y + 2z^3\vec{a}_z$ C/m^2

Solution Given $\vec{D} = 4x\vec{a}_x + 3y^2\vec{a}_y + 2z^3\vec{a}_z$ C/m^2.
The volume is enclosed by : $1 \le x \le 2,\ 2 \le y \le 3,\ 3 \le z \le 4$.

We know from Maxwell's first equation that the volume charge density is

$$\begin{aligned}\rho_v &= \nabla\cdot\vec{D}\\ &= \frac{\partial D_x}{\partial x} + \frac{\partial D_y}{\partial y} + \frac{\partial D_z}{\partial z} = \frac{\partial(4x)}{\partial x} + \frac{\partial(3y^2)}{\partial y} + \frac{\partial}{\partial z}(2z^3)\\ &= 4 + 6y + 6z^2 \text{ C/m}^3.\end{aligned}$$

The total charge $Q = \int_v \rho_v dv$

$$= \int_{1}^{2}\int_{2}^{3}\int_{3}^{4}(4+6y+6z^2)dx\,dy\,dz = [x]_1^2\int_{2}^{3}\int_{3}^{4}(4+6y+6z^2)dy\,dz$$

$$= \int_{3}^{4}\left[4y+\frac{6y^2}{2}+6z^2y\right]_2^3 dz = \int_{3}^{4}\left[12+\frac{54}{2}+18z^2-8-12-12z^2\right] dz$$

$$= \int_{3}^{4}(19+6z^2)\,dz = \left[19z+\frac{6z^3}{3}\right]_3^4 = 19+2\,(4^3-3^3) = 93 \text{ C.}$$

2.19 Divergence Theorem of Electric Flux Density

The divergence theorem states that *the integral of the normal component of electric flux density over a closed surface is equal to the integral of the divergence of this flux density throughout the volume enclosed by that closed surface.*

Mathematically,

$$\oint_s \vec{D}\cdot d\vec{S} = \int(\nabla\cdot\vec{D})dv. \qquad (2.59)$$

Proof Consider a volume of charge Q with volume charge density ρ_v C/m^3 enclosed by a Gaussian surface S. Let dS be a differential surface area on the surface through which the flux density D is radially outwards in the $\vec{a}_r$ direction as shown in Fig. 2.21.

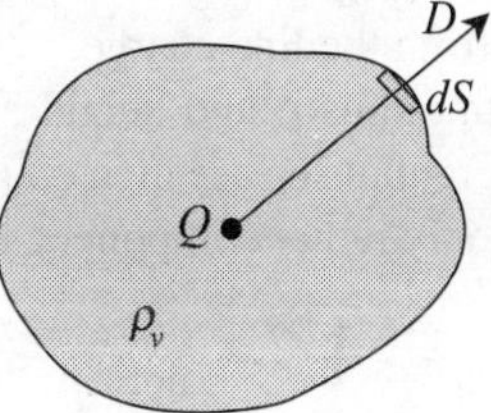

Fig. 2.21 *Volume charge distribution*

From Gauss' law, we know that

$$Q = \oint_s \vec{D}\cdot d\vec{S}.$$

Since the charge is distributed uniformly within the volume with charge density ρ_v C/m^3,

$$Q = \int_v \rho_v dv.$$

$$\therefore \quad \oint_s \vec{D}\cdot d\vec{S} - \int_v \rho_v dv.$$

But according to the point form of Gauss' law

$$\text{div } \vec{D} = \nabla \cdot \vec{D} = \rho_v$$

$$\therefore \quad \oint_s \vec{D} \cdot d\vec{S} = \int_v (\nabla \cdot \vec{D}) dv. \qquad \text{Hence proved.}$$

Note: The divergence of the flux density is the net flux crossing the surface per unit volume. The divergence theorem relates a volume integration to a surface integration when the volume of charge is enclosed by the closed surface.

Applications of Gauss' law Gauss' law and divergence theorem were formulated by Karl Friedrich Gauss, a German mathematician of the 18th century. Gauss' law is similar to Coulomb's law. It can be applied to many practical problems where the electric field of charge distributions has symmetrical geometry, such as sphere, cylinder or plane. We can easily obtain the net flux density, electric field and the charge enclosed by the surfaces for such symmetrical charge distributions. Gauss' law allows us to find electric fields without needing to integrate.

Coulomb's law is more powerful and enables us to solve problems for which Gauss' law cannot be applied. However, while solving problems, it requires complex integations.

2.20 Electric Flux Density due to an Infinite Line Charge Using Gauss' Law

The electric field of an infinitely long line charge at any point P can be solved by using Gauss' law.

Consider an infinite line charge of density ρ_l C/m lying along the z-axis extending from $-\infty$ to $+\infty$. Let a point P be at a distance ρ from the line charge.

Consider a cylindrical surface with radius ρ and height z, as shown in Fig. 2.22. Let it be the Gaussian surface, and let dS be a differential surface area on it. Since the line charge is symmetrical about the z-axis, the flux density has only radial component, a_ρ, normal to the surface.

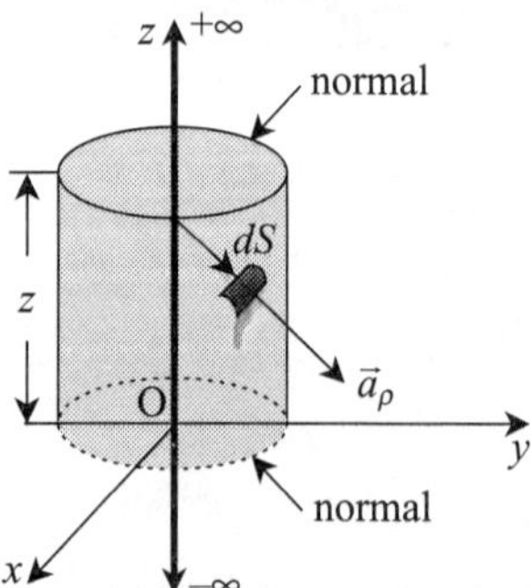

Fig. 2.22 *Infinite line charge*

From Gauss' law, the net charge enclosed by the cylindrical surface is

$$Q = \oint_s \vec{D} \cdot d\vec{S}.$$

The integration is evaluated over the closed surface of the cylinder.

$$Q = \int_{\text{side}} \vec{D} \cdot d\vec{S} + \int_{\text{top}} \vec{D} \cdot d\vec{S} + \int_{\text{bottom}} \vec{D} \cdot d\vec{S}. \tag{2.60}$$

Since $\vec{D}$ has only radial component, there is no component along the z-axis.

Thus, $\int_{\text{top}} \vec{D} \cdot d\vec{S} = \int_{\text{bottom}} \vec{D} \cdot d\vec{S} = 0$

$$\therefore \quad Q = \int_{\text{side}} \vec{D} \cdot d\vec{S}$$

$$d\vec{S} = \rho \; d\phi \; dz \, \vec{a}_\rho \;\text{ and }\; \vec{D} = D\vec{a}_\rho.$$

$$Q = \int_0^z \int_0^{2\pi} D\vec{a}_\rho \cdot \rho \; d\phi \; dz \, \vec{a}_\rho = \int_0^z \int_0^{2\pi} D \, \rho \; d\phi \; dz$$

$$= D(2\pi\rho z) \quad \text{or,} \quad D = \frac{Q}{2\pi\rho z}.$$

Since $Q = \rho_l z$,

$$D = \frac{\rho_l}{2\pi\rho}.$$

The electric flux density

$$\vec{D} = \frac{\rho_l}{2\pi\rho}\vec{a}_p \;\text{C/m}^2, \tag{2.61}$$

and the electric field strength

$$\vec{E} = \frac{\vec{D}}{\varepsilon} = \frac{\rho_l}{2\pi\varepsilon\rho}\vec{a}_p, \tag{2.62}$$

which is the same as Eq. (2.32).

Example 2.12 Find the flux density at $(2, -2, 1)$ due to a point charge 10 mC at $(1, -4, 0)$ and a line charge 0.3 mC/m along the y-axis.

Solution Given: $Q = 10 \times 10^{-3}$ C at $(1, -4, 0)$ and $\rho_l = 30 \times 10^{-3}$ C/m.

To calculate the flux density D at $(2, -2, 1)$ due to the point charge, we know that

$$\vec{D}_1 = \varepsilon\vec{E} = \frac{Q}{4\pi R^2}\vec{a}_R.$$

The distance vector is $\vec{R} = (2, -2, 1) - (1, -4, 0)$.

$$\vec{R} = (2-1)\vec{a}_x + (-2+4)\vec{a}_y + (1-0)\vec{a}_z = \vec{a}_x + 2\vec{a}_y + \vec{a}_z.$$

$$R = \sqrt{1+2^2+1^2} = \sqrt{6}.$$

$$\therefore \quad \vec{D}_1 = \frac{10 \times 10^{-3}}{4\pi 6} \times \frac{\vec{a}_x + 2\vec{a}_y + \vec{a}_z}{\sqrt{6}}$$

$$= 5.4(\vec{a}_x + 2\vec{a}_y + \vec{a}_z) \times 10^{-6}\ \mathrm{C/m^2}.$$

Flux density D at $(2,\ -2,1)$ due to the line charge along the y-axis is

$$\vec{D}_2 = \varepsilon\vec{E} = \frac{\rho_l}{2\pi R}\vec{a}_R.$$

The point on the line charge along the y-axis is $(0,\ -2,0)$.

Distance vector $\vec{R} = (2,\ -2,1) - (0,\ -2,0) = 2\vec{a}_x + \vec{a}_z$.

$$R = \sqrt{2^2+1} = \sqrt{5}.$$

$$\therefore \quad \vec{D}_2 = \frac{0.3 \times 10^{-3}}{2\pi \times \sqrt{5}} \times \frac{2\vec{a}_x + \vec{a}_z}{\sqrt{5}} = 9.5(2\vec{a}_x + \vec{a}_z) \times 10^{-6}\ \mathrm{C/m^2}$$

Total flux density $\vec{D} = \vec{D}_1 + \vec{D}_2 = (24.4\vec{a}_x + 10.8\vec{a}_y + 14.9\vec{a}_z)\ \mu\mathrm{C/m^2}$.

2.21 Flux Density due to an Infinite Sheet Charge Using Gauss' Law

Consider an infinite sheet of charge with charge density ρ_s $\mathrm{C/m^2}$ lying along the $x-y$ plane at $z=0$ as shown in Fig. 2.23.

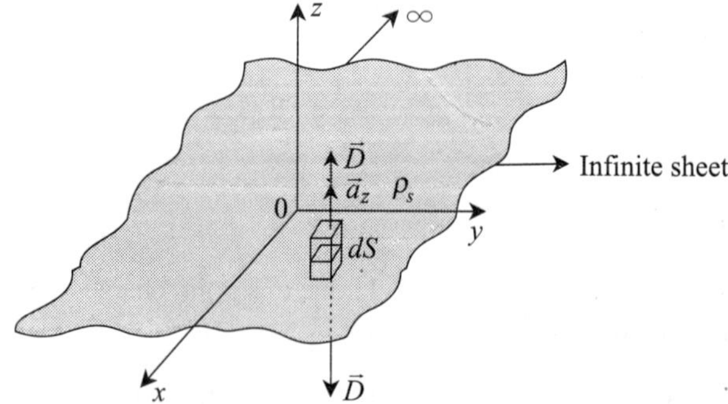

Fig. 2.23 *Field due to an infinite sheet of charge*

Choose a rectangular box as a Gaussian surface parallel to the z-axis which cuts the sheet of charge with differential surface area dS at an arbitrary point.

The charge enclosed by the surface is

$$Q = \oint_s \vec{D}\cdot d\vec{S} = \int_{\text{sides}} \vec{D}\cdot d\vec{S} + \int_{\text{top}} \vec{D}\cdot d\vec{S} + \int_{\text{bottom}} \vec{D}\cdot d\vec{S}. \tag{2.63}$$

Since $\vec{D}$ has no component in the x- and y- directions,

$$\int_{\text{sides}} \vec{D}\cdot d\vec{S} = 0 \text{ and } d\vec{S} = dx\,dy\,\vec{a}_z.$$

$$\therefore \quad \int_{\text{top}} \vec{D}\cdot d\vec{S} = \int D\vec{a}_z\cdot dx\,dy\,\vec{a}_z = DS;$$

$$\int_{\text{bottom}} \vec{D}\cdot d\vec{S} = \int D(-\vec{a}_z)\,dx\,dy(-\vec{a}_z) = DS.$$

$$\therefore \quad Q = DS + DS = 2DS.$$

Since $\rho_s = \dfrac{Q}{S}$, $D = \dfrac{Q}{2S} = \dfrac{\rho_s}{2}$.

Therefore, $\vec{D} = \dfrac{\rho_s}{2}\,\vec{a}_n$

and the field strength $$\vec{E} = \frac{\vec{D}}{\varepsilon} = \frac{\rho_s}{2\varepsilon}\,\vec{a}_n, \tag{2.64}$$

which is the same as Eq. (2.43).

2.22 Flux Density for a Coaxial Cable

Consider a coaxial cable with outer conductor radius b and inner conductor radius a as shown in Fig. 2.24. Let L be the length of the cable, ρ_s be the charge density on the inner conductor and $-\rho_s$ be the charge density on the outer conductor. Consider a Gaussian surface of radius ρ inside the coaxial cable.

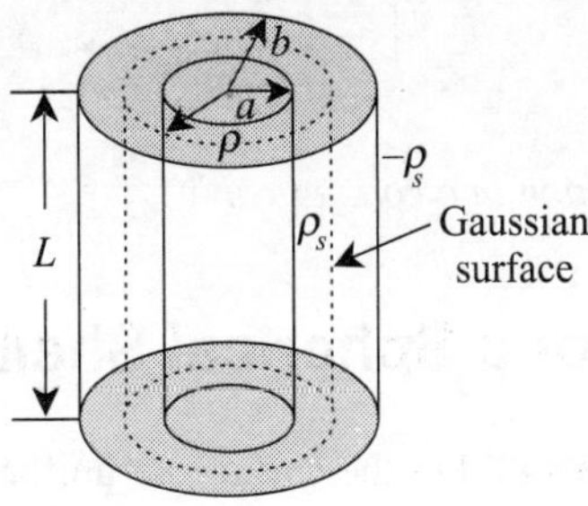

Fig. 2.24 *Flux density for a coaxial cable*

If $\rho < a$, the charge enclosed by the Gaussian surface is zero. That is, $D = 0$ for $\rho < a$.

If $a < \rho < b$, the charge enclosed by the Gaussian surface is the sheet charge with inner conductor of charge density ρ_s C/m^2.

$$\vec{Q} = \oint_s \vec{D} \cdot d\vec{S} = D\, 2\pi\, \rho L.$$

$$\therefore \quad \vec{D} = \frac{Q}{2\pi\rho L}\, \vec{a}_\rho.$$

But the charge density $\rho_s = \dfrac{Q}{S}$ or, $\rho_S = \dfrac{Q}{2\pi a L}$.

$$\therefore \quad \vec{D} = \frac{a\rho_s}{\rho}\, \vec{a}_\rho \quad a < \rho < b.$$

At $\rho = a$, $\vec{D} = \rho_s \vec{a}_\rho$ and at $\rho = b$, $\vec{D} = \dfrac{a\rho_s \vec{a}_\rho}{b}$. (2.65)

For $\rho > b$, both sheet charges of inner and outer conductors are enclosed by the Gaussian surface. Since equal and opposite charges exist on the conductors, the net charge enclosed by the surface is zero.

Therefore, $\vec{D} = 0$ for $\rho > b$.

Hence, the flux density distribution is shown in Fig. 2.25 and is given by

$$\vec{D} = \begin{cases} 0 & \text{for } \rho < a \text{ and } \rho > b \\ \dfrac{a\rho_s}{\rho} & \text{for } a < \rho < b \\ \rho_s & \text{for } \rho = a \\ \dfrac{a\rho_s}{b} & \text{for } \rho = b \end{cases} \qquad (2.66)$$

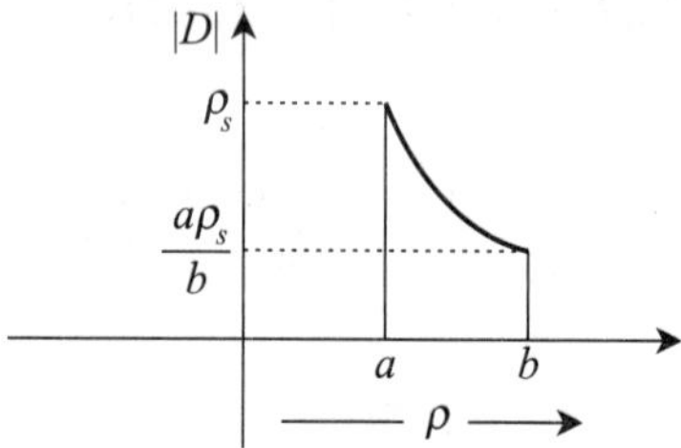

Fig. 2.25 *Flux density distribution in a coaxial cable*

2.23 Flux Density for a Spherical Shell of Charge

Consider a spherical shell of radius a. Let the surface be uniformly charged with charge density ρ_s C/m^2. Also consider an imaginary spherical Gaussian surface of radius r from the centre of the sphere as shown in Fig. 2.26.

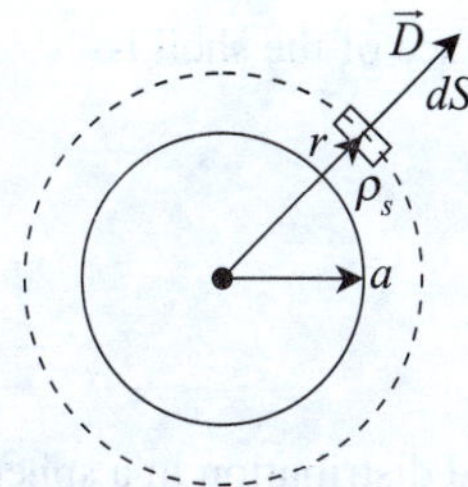

Fig. 2.26 *Flux density for a spherical surface*

(1) If $r < a$, there is no charge enclosed by the surface and $\vec{D} = 0$.

Note: The charge $Q, \vec{D}$ and $\vec{E}$ at any point inside a spherical shell is zero.

(2) If $r > a$, the flux density is directed radially outwards and normal to the surface.

Let dS be a differential surface through which the flux is passing.
In spherical coordinates,

$$d\vec{S} = r^2 \sin\theta \, d\theta \, d\phi \, \vec{a}_r \quad \text{and} \quad \vec{D} = D \cdot \vec{a}_r.$$

$$\therefore \quad \text{The flux } \psi = \int_s \vec{D} \cdot d\vec{S} = \int_{\phi=0}^{2\pi} \int_0^{\pi} Dr^2 \sin\theta \, d\theta \, d\phi \, \vec{a}_r \cdot \vec{a}_r.$$

$$Q = \psi = \int_0^{\pi} Dr^2 \sin\theta \, d\theta \, (2\pi)$$

$$= 2\pi Dr^2 \left[-\cos\theta \right]_0^{\pi} = 4\pi r^2 D.$$

$$D = \frac{Q}{4\pi r^2}.$$

$$\therefore \quad \vec{D} = \frac{Q}{4\pi r^2} \vec{a}_r \ \text{C/m}^2. \tag{2.67}$$

We know that the charge density $\rho_s = \dfrac{Q}{S}$.

$$\rho_s = \frac{Q}{4\pi a^2}.$$

$$\therefore \quad \vec{D} = \frac{\rho_s a^2}{r^2} \vec{a}_r. \tag{2.68}$$

The field strength $\quad \vec{E} = \vec{D}/\varepsilon$,

$$\vec{E} = \frac{Q}{4\pi\varepsilon r^2} \vec{a}_r = \frac{\rho_s a^2}{\varepsilon r^2} \vec{a}_r. \tag{2.69}$$

(3) If $r = a$, the flux density on the surface of the shell is

$$\vec{D} = \frac{Q}{4\pi a^2}\vec{a}_r = \rho_s \tag{2.70}$$

$$\text{and} \quad \vec{E} = \frac{Q}{4\pi\varepsilon a^2}\vec{a}_r \ \text{V/m}. \tag{2.71}$$

Fig. 2.27 shows the plot of electric field distribution in a spherical shell.

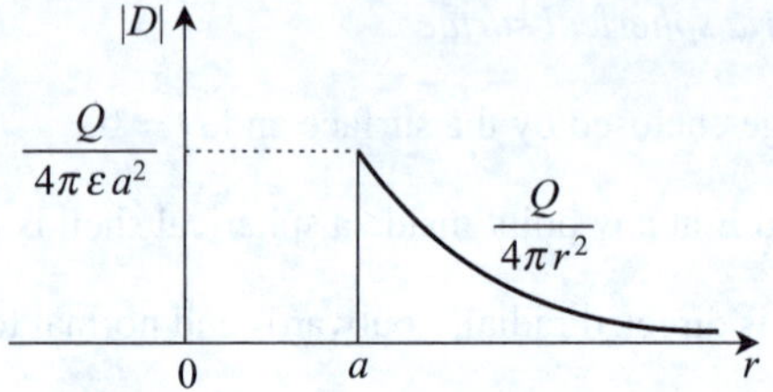

Fig. 2.27 *Electric field distribution in a spherical shell*

2.24 Flux Density for a Uniformly Charged Sphere

Consider a sphere of radius a as shown in Fig. 2.28. Let charge Q be uniformly distributed in the sphere with volume charge density ρ_v C/m^3. Also consider an imaginary Gaussian surface of radius r from the centre of the sphere. Let the flux density $\vec{D}$ be radially outward along $\vec{a}_r$ through a differential surface area dS.

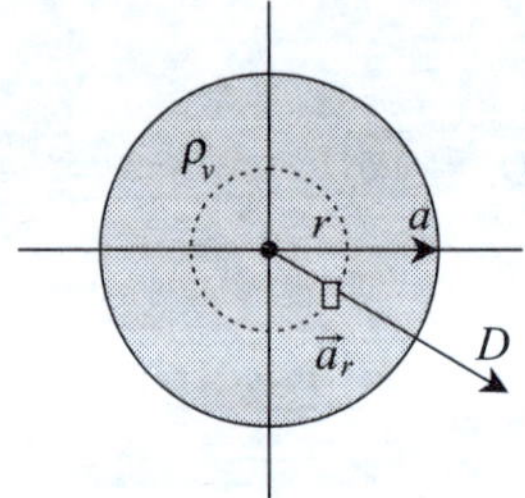

Fig. 2.28 *Sphere of volume charge*

In spherical coordinates,

$$d\vec{S} = r^2 \sin\theta \, d\theta \, d\phi \, \vec{a}_r \quad \text{and} \quad \vec{D} = D \cdot \vec{a}_r. \tag{2.72}$$

(1) For $r < a$, the charge enclosed by the surface is

$$Q = \oint_s \vec{D} \cdot d\vec{S} = \int_{\phi=0}^{2\pi} \int_{\theta=0}^{\pi} D\, r^2 \sin\theta \, d\theta \, d\phi \, \vec{a}_r \cdot \vec{a}_r$$

$$= Dr^2 2\pi \left[-\cos\theta\right]_0^{\pi} = 4\pi r^2 D.$$

The flux density $\vec{D} = \dfrac{Q}{4\pi r^2}\,\vec{a}_r$ C/ m^2.

For volume charge density ρ_v, the charge within the sphere of radius r is

$$Q = \int_v \rho_v dv = \frac{4\pi r^3}{3} \cdot \rho_v.$$

$$\vec{D} = \frac{Q}{4\pi r^2}\,\vec{a}_r.$$

The flux density $\vec{D} = \dfrac{4\pi r^3 \rho_v}{12\pi r^2}\vec{a}_r = \dfrac{r}{3}\rho_v\vec{a}_r.$ (2.73)

(2) For $r > a$, the charge enclosed by the sphere of radius a is

$$Q = \frac{4}{3}\pi a^3\,\rho_v.$$

$$\vec{D} = \frac{Q}{4\pi r^2} = \frac{4\pi a^3 \rho_v}{12\pi r^2} = \frac{a^3\rho_v}{3r^2}\vec{a}_r. \tag{2.74}$$

(3) At $r = a$, the flux density is

$$\vec{D} = \frac{a^3\rho_v}{3a^2}\vec{a}_r = \frac{a\rho_v}{3}\vec{a}_r. \tag{2.75}$$

Thus the flux density at all regions is:

$$\vec{D} = \begin{cases} \dfrac{r}{3}\rho_v\,\vec{a}_r & r < a \\ \dfrac{a\rho_v}{3}\,\vec{a}_r & r = a \\ \dfrac{a^3\rho_v}{3r^2}\,\vec{a}_r & r > a \end{cases}. \tag{2.76}$$

The electric field intensity is $\vec{E} = \vec{D}/\varepsilon$.

$$\vec{E} = \begin{cases} \dfrac{r}{3\varepsilon}\rho_v\,\vec{a}_r & r < a \\ \dfrac{a\rho_v}{3\varepsilon}\,\vec{a}_r & r = a \\ \dfrac{a^3\rho_v}{3\varepsilon r^2}\,\vec{a}_r & r > a \end{cases}. \tag{2.77}$$

Fig. 2.29 shows the variation of electric flux density in a charged sphere.

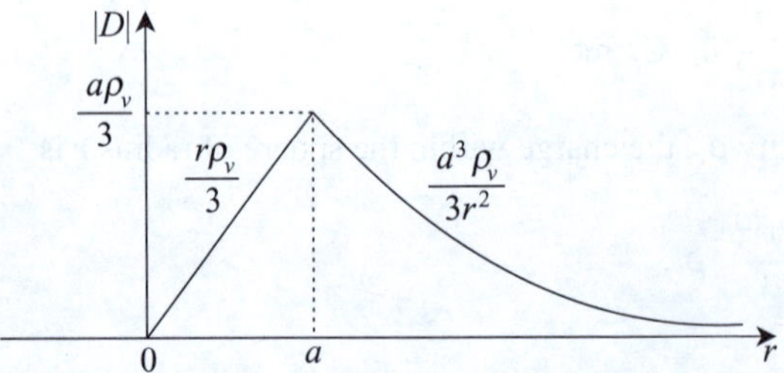

Fig. 2.29 *Field distribution in a charged sphere*

Example 2.13 A volume enclosed between cylindrical coordinates $\rho = 2$ m and $\rho = 4$ m has a uniform charge density of ρ_v C/m^3. Use Gauss' law to find D in all the regions.

Solution Given the volume enclosed by cylindrical coordinates $\rho = 2$ m and $\rho = 4$ m, and charge density ρ_v C/m^3. Assume that the length of the cylinder is L metres. The surface area of the cylinder is $2\pi\rho L$ m^2.

For $0 < \rho < 2$,

$$\int \vec{D} \cdot d\vec{S} = Q.$$

$D \cdot S = D(2\pi\rho L) = 0$ since no charge is enclosed.

$$D = 0.$$

For $2 \leq \rho \leq 4$,

$$\oint D \cdot dS = Q = \int \rho_v dv.$$

$$D \cdot S = \int_0^L \int_{\phi=0}^{2\pi} \int_{\rho=2}^{\rho} \rho_v \rho \, d\rho \, d\phi \, dz.$$

$$D \, 2\pi\rho L = 2\pi \, \rho_v L \left(\frac{\rho^2}{2}\right)_2^{\rho}.$$

$$D = \frac{\rho_v}{2\rho}(\rho^2 - 4) \ \text{C/m}^2$$

$$\vec{D} = \frac{\rho_v}{2\rho}(\rho^2 - 4) \ \text{C/m}^2$$

For $\rho > 4$,

$$D(2\pi\rho L) = \int_0^L \int_0^{2\pi} \int_2^4 \rho_v \rho \, d\rho \, d\phi \, dz = 2\pi\rho_v \frac{(4^2 - 2^2)}{2}$$

$$D = \frac{6\rho_v}{\rho} \quad \text{and} \quad \vec{D} = \frac{6\rho_v}{\rho}\vec{a}_\rho \ \text{C/m}^2.$$

2.25 Work Done in Moving a Point Charge in an Electrostatic Field

Consider an electric field $\vec{E}$ due to a positive charge Q in free space. If a test charge Q_t is placed at any point in this field, it experiences a force and tends to move along the direction of the force. Let Q_t be a positive charge. Due to the repulsive force between Q and Q_t, the charge Q_t will move away from Q as shown in Fig. 2.30.

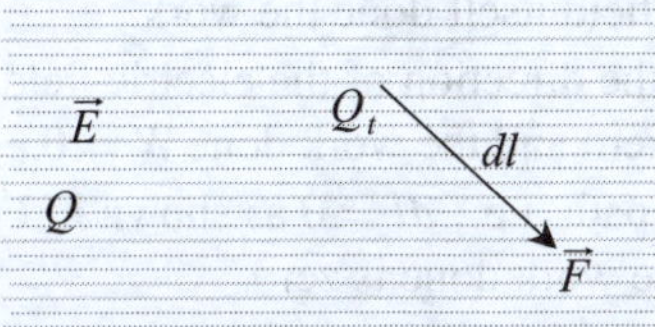

Fig. 2.30 *Repulsive force on Q_t due to Q*

Let dl be the distance moved by Q_t along the direction $\vec{a}_l$. The force exerted by the electric field is

$$\vec{F} = Q_t\vec{E}.$$

The component of $\vec{F}$ in the direction $\vec{a}_l$ is

$$\vec{F}_l = (Q_t\vec{E} \cdot \vec{a}_l)\vec{a}_l$$

To keep the test charge in equilibrium, an equal and opposite force is to be applied on Q_t.

The required applied force is

$$\vec{F}_{\text{app}} = -\vec{F}_l = -Q_t(\vec{E} \cdot \vec{a}_l)\vec{a}_l$$

We know that the differential work done by an external (applied) source in moving the test charge Q_t through a differential distance dl opposite to the electric field $\vec{E}$ is

Work done = Force × distance

Since $\vec{a}_l$ is the unit vector along the distance dl,

$$dW = \vec{F}_{\text{app}} \cdot d\vec{l} = -Q_t\vec{E} \cdot d\vec{l} \text{ joules (J),} \tag{2.78}$$

where $\vec{E} \cdot d\vec{l}$ is the dot product of electric vector $\vec{E}$ with distance vector $d\vec{l}$.

In general, the total work done by an applied force in moving a unit positive charge from point A to point B in the electric field $\vec{E}$ is given by

$$W = \int_A^B dW = \int_A^B -Q_t\vec{E} \cdot d\vec{l} = -Q_t \int_A^B \vec{E} \cdot d\vec{l} \text{ joules.} \tag{2.79}$$

Note:

1. The integral expression in Eq. (2.79) is called a line integral.
2. Since the field is conservative, the work done in moving a charge from one point A to another point B in the electric field is independent of the path chosen. Also the line integral depends only on end points of the path, as shown in Fig. 2.31 (a)
3. When the work done, W is positive, the charge is moving against the direction of the electric field $\vec{E}$. This indicates that the applied force has done the work.
   When the work done, W is negative, the charge is moving along the direction of the electric field $\vec{E}$. This indicates that the field itself done the work.
4. If the path is perpendicular to the direction of the electric field, i.e., the angle between $\vec{E}$ and $d\vec{l}$ is 90°, then the dot product is zero and the work done $W = 0$.
5. If the path selected is a closed path, i.e., $dl = 0$ as shown in Fig. 2.31 (b), the net movement of the charge is zero and hence the work done is zero.

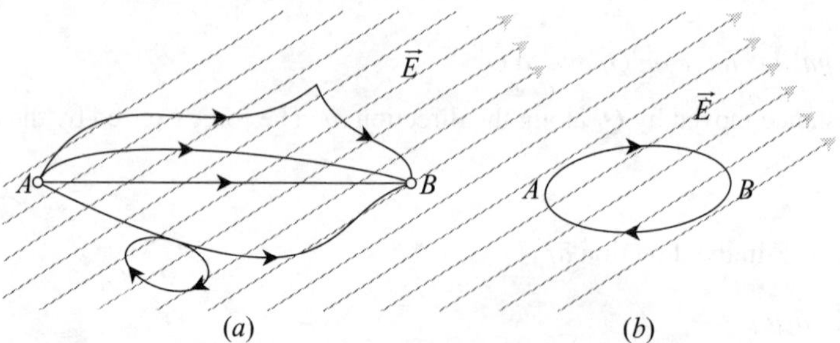

***Fig. 2.31** (a) Paths between points A and B in field $\vec{E}$, (b) circular path*

6. If the charge is moving in different fields $\vec{E}_1, \vec{E}_2, \ldots, \vec{E}_N$, and covering distances $d\vec{l}_1$, $d\vec{l}_2$, $d\vec{l}_3, \ldots,$ $d\vec{l}_N$, respectively, then the total work done is

$$W = -Q\sum_{i=1}^{N} \vec{E}_i \cdot d\vec{l}_i. \tag{2.80}$$

7. The differential length vector in Cartesian coordinate system is

$$d\vec{l} = dx\,\vec{a}_x + dy\,\vec{a}_y + dz\,\vec{a}_z. \tag{2.81}$$

In cylindrical coordinate system, it is

$$d\vec{l} = d\rho\,\vec{a}_\rho + \rho d\phi\,\vec{a}_\phi + dz\,\vec{a}_z. \tag{2.82}$$

In spherical coordinate system, it is

$$d\vec{l} = dr\,\vec{a}_r + rd\theta\,\vec{a}_\theta + r\,\sin\theta\,d\phi\vec{a}_\phi. \tag{2.83}$$

2.26 Electric Potential

Electric potential at a point P due to a fixed charge Q is defined as *the work done in moving one coulomb of charge from infinity (∞) to the point P in an electric field against the force created by* *the fixed charge*. The work must be done against the field. Electric potential is also defined as the work done per unit charge.

Mathematically, the potential is

$$V_p = \frac{W}{Q} = -\int_{\infty}^{P} \vec{E}\cdot d\vec{l} \quad \text{J/C or volts (V).} \tag{2.84}$$

The potential is a scalar quantity produced at any point due to the charge Q. It is also called the scalar potential.

Potential difference Potential difference is defined as the *work done in moving a unit charge from a point A to another point B against field $\vec{E}$. It is denoted by V_{BA}.*

$$V_{BA} = \frac{W}{Q} = -\int_{A}^{B} \vec{E}\cdot d\vec{l}. \tag{2.85}$$

Here, A is the initial point and B, the final point. The negative sign indicates loss in potential energy in moving Q from A to B, i.e., the work is being done by the field.

Potential difference is also defined as the difference between potentials at given points.

$$V_{BA} = V_B - V_A,$$

where $$V_A = -\int_{\infty}^{A} \vec{E}\cdot d\vec{l} \tag{2.86}$$

and $$V_B = -\int_{\infty}^{B} \vec{E}\cdot d\vec{l} \quad \text{volts.} \tag{2.87}$$

V_A and V_B are the absolute potentials at A and B, respectively.

The work done is

$$W = QV_{BA} = Q(V_B - V_A).$$

Note:

1. The potential difference depends only on the terminal points. It is independent of the path chosen.
2. The potential at infinity is zero.
3. The potential around a closed path is equal to zero, that is, $V = \oint \vec{E}\cdot d\vec{l} = 0$.
4. One volt potential at a point is equal to one joule of work done in moving a unit charge from infinity to that point in the electric field $\vec{E}$.

Example 2.14 Calculate the work done in moving a 40 μC charge from a point A (3, 45°, 135°) to B(4, 120°, 75°) in a field having a potential given by $V = 25r\cos^2\theta\sin\phi$ volts.

Solution Given: $V = 25r\cos^2\theta\sin\phi$ volts, $Q = 40$ μC and the points A (3, 45°, 135°) and B(4, 120°, 75°) in spherical coordinates.

Work done $W = QV_{BA} = Q(V_B - V_A)$.

But the potential at point B is

$$V_B = 25(4)\cos^2(120)\sin(75) = 24.14 \text{ volts}$$

and that at point A is $V_A = 25 \times 3\cos^2(45)\sin(135)$

$$= 75 \times 0.5 \times 0.207 = 26.516 \text{ volts}.$$

Work done $W = Q(V_B - V_A) = 40 \times 10^{-6}(24.14 - 26.516) = -95\ \mu\text{J}$.

The minus sign indicates that the work is being done by the field.

2.27 Potential due to a Point Charge

Consider a point charge Q located at the origin of spherical coordinates. The field at a distance r is given by

$$\vec{E} = \frac{Q}{4\pi\varepsilon r^2}\vec{a}_r. \tag{2.88}$$

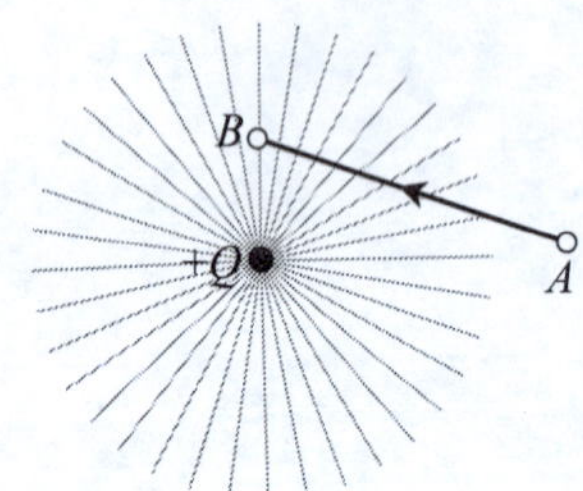

Fig. 2.32 *Potential difference*

Let a unit charge be placed at point A with radius r_A from the origin. It is moved to point B at radius r_B against the direction of the field.

The potential difference V_{BA} is

$$V_{BA} = -\int_A^B \vec{E} \cdot d\vec{l}.$$

The differential length in spherical coordinates is

$$d\vec{l} = dr\vec{a}_r + rd\theta\vec{a}_\theta + r\sin\theta d\phi\vec{a}_\phi.$$

$$\therefore \quad V_{BA} = -\int_A^B \frac{Q}{4\pi\varepsilon r^2}\vec{a}_r \cdot (dr\,\vec{a}_r + rd\theta\,\vec{a}_\theta + r\sin\theta\,d\phi\,\vec{a}_\phi)$$

$$= -\frac{Q}{4\pi\varepsilon}\int_{r_A}^{r_B}\frac{[dr(\vec{a}_r\cdot\vec{a}_r)+rd\theta(\vec{a}_r\cdot\vec{a}_\theta)+r\sin\theta\, d\phi(\vec{a}_r\cdot\vec{a}_\phi)]}{r^2} = -\frac{Q}{4\pi\varepsilon}\int_{r_A}^{r_B}\frac{1}{r^2}\cdot dr.$$

$$V_{BA} = \frac{-Q}{4\pi\varepsilon}\left[\frac{-1}{r}\right]_{r_A}^{r_B}$$

$$= \frac{Q}{4\pi\varepsilon}\left[\frac{1}{r_B}-\frac{1}{r_A}\right] = V_B - V_A. \tag{2.89}$$

Note 1: If the charge is moved from infinity to the point B, i.e., $\frac{1}{r_A} = 0$, then the potential at point B is

$$V_B = \frac{Q}{4\pi\varepsilon r_B}. \tag{2.90}$$

Potential at infinity is zero.

In general, the potential at radius r from the point charge at the origin in the electric field is called absolute potential, i.e.,

$$V = \frac{Q}{4\pi\varepsilon r} \text{ volts.} \tag{2.91}$$

In practical situations, the absolute potentials are measured with respect to the zero potential at the reference point (infinity point). This reference point is considered as ground point, i.e., the potential at a ground point is zero.

Note 2: If any other point is choosen as the reference point, the potential at point B is

$$V_B = \frac{Q}{4\pi\varepsilon r} + C \tag{2.92}$$

where C is a constant that can be determined from the potential at the reference point.

Note 3: If the point charge Q is not located at the origin but say at the position radius $\vec{r}_1$, then the potential at point B is

$$V_B = \frac{Q}{4\pi\varepsilon\,|\vec{r}-\vec{r}_1|} + C.$$

$C = 0$ if reference point is at infinity.

Example 2.15 Find the potential at $(-2, 4, 3)$ due to point charge $(0, -5, 1)$ in the electric field. Assume that potential at infinity is zero.

Solution Given $Q = 2.5$ nC $= 2.5\times10^{-9}$ C at $(0, -5, 1)$; the point where potential is to be found is $(-2, 4, 3)$.

We know that the potential at a point in the field is

$$V = \frac{Q}{4\pi\varepsilon_o R} + C.$$

If potential at infinity is zero, $C = 0$.

$$\vec{R} = \vec{r} - \vec{r}_1 = (-2,\ 4,\ 3) - (0,\ -5,\ 1)$$

$$\vec{R} = (-2+0)\vec{a}_x + (4+5)\vec{a}_y + (3-1)\vec{a}_z$$

$$\vec{R} = -2\vec{a}_x + 9\vec{a}_y + 2\vec{a}_z$$

$$R = \sqrt{2^2 + 9^2 + 2^2} = \sqrt{89}$$

$$V = \frac{Q}{4\pi\varepsilon_0\sqrt{89}} = \frac{2.5 \times 10^{-9}}{4\pi\varepsilon_0\sqrt{89}} = 2.38 \text{ volts.}$$

2.28 Potential due to Point Charges

Consider N point charges located at distances $r_1,\ r_2 \ldots r_N$ from the origin as shown in Fig. 2.33. Consider a unit charge at point A, a distance r from the origin. The potential due to all the N number of point charges at point A is the algebraic sum of the individual potentials acting at that point. Assume that the reference point is at the infinity has zero potential.

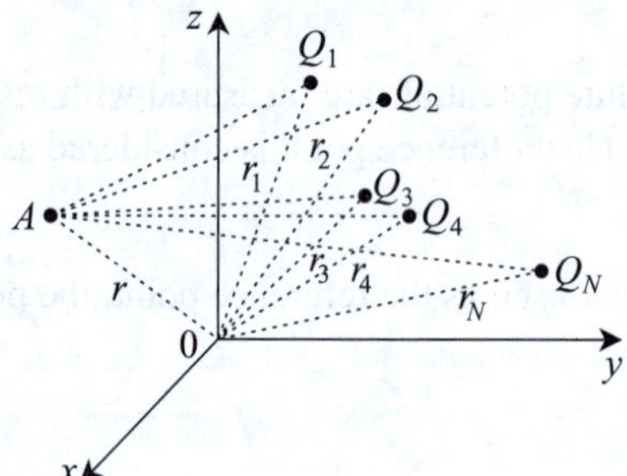

Fig. 2.33 *Potential due to N point charges*

The potential due to Q_1 at point A is

$$V_{A1} = \frac{Q_1}{4\pi\varepsilon\, R_{A1}} = \frac{Q_1}{4\pi\varepsilon\,|\vec{r} - \vec{r}_1|} \text{ volts,} \tag{2.93}$$

where $R_{A1} = |r - r_1|$ = distance between point A and Q_1.

The potential due to Q_2 at point A is

$$V_{A2} = \frac{Q_2}{4\pi\varepsilon\, R_{A2}} = \frac{Q_2}{4\pi\varepsilon|\vec{r} - \vec{r}_2|} \text{ volts.}$$

Similarly, the potential due to Q_N at point A is

$$V_{AN} = \frac{Q_N}{4\pi\varepsilon\, R_{AN}} = \frac{Q_N}{4\pi\varepsilon|\vec{r} - \vec{r}_N|} \text{ volts.}$$

The total potential at A is

$$V_A = V_{A1} + V_{A2} + V_{A3} + \ldots\ldots + V_{AN}$$

$$V_A = \frac{1}{4\pi\varepsilon}\left[\frac{Q_1}{|\vec{r}-\vec{r}_1|} + \frac{Q_2}{|\vec{r}-\vec{r}_2|} + \frac{Q_3}{|\vec{r}-r_3|} + \ldots + \frac{Q_N}{|\vec{r}-\vec{r}_N|}\right]$$

$$V_A = \frac{1}{4\pi\varepsilon}\sum_{i=1}^{N}\frac{Q_i}{|\vec{r}-\vec{r}_i|} \text{ volts.} \tag{2.94}$$

Hence the potential due to N number of point charges at a point A is the sum of the individual potentials at that point.

Note : Constant C is added if reference point (zero potential) is not at infinity.

Example 2.16 A point charge of 10 nC is located at $(-1, 2, 0)$, while a line charge at $y = 1$, $z = 2$ carries a uniform line charge of charge density $\rho_l = 4$ nC/m. Find the potential at point $A(1, 0, 2)$. Assume the potential at origin is zero.

Solution Given point charge $Q = 10$ nC at point $(-1, 2, 0)$.
Line charge $\rho_l = 4$ nC/m at $y = 1$, $z = 2$, i.e., at $(x, 1, 2)$.
Let potential at point $A(1,0,2)$ be V.
Assume that the potential at the origin (reference point) is zero, $V(0) = 0$.
Also, let potential V_1 be due to the point charge and potential V_2 be due to the line charge.

$\therefore$ The total potential is $V = V_1 + V_2$.

(1) The potential at A due to point charge is

$$V_1 = \frac{Q}{4\pi\varepsilon R_1} - \frac{Q}{4\pi\varepsilon R_0}$$

$$V_1 = \frac{10\times 10^{-9}\times 36\pi}{4\pi\times 10^{-9}}\left[\frac{1}{R_1} - \frac{1}{R_0}\right]$$

where the distance vector from the point charge to the point A is

$$\vec{R}_1 = (1, 0, 2) - (-1, 2, 0)$$

$$\vec{R}_1 = (1+1)\vec{a}_x + (0-2)\vec{a}_y + (2-0)\vec{a}_z$$

$$\vec{R}_1 = 2\vec{a}_x - 2\vec{a}_y + 2\vec{a}_z$$

$$R_1 = \sqrt{2^2+2^2+2^2} = \sqrt{12},$$

and the distance vector from the point charge to the origin is

$$\vec{R}_0 = (0, 0, 0) - (-1, 2, 0)$$

$$\vec{R}_0 = (0+1)\vec{a}_x + (0-2)\vec{a}_y + (0-0)\vec{a}_z$$

$$\vec{R}_0 = \vec{a}_x - 2\vec{a}_y$$

$$R_0 = \sqrt{1+4} = \sqrt{5}.$$

$$\therefore \quad V_1 = 90\left[\frac{1}{\sqrt{12}} - \frac{1}{\sqrt{5}}\right] = 90[0.288 - 0.447]$$

$$V_1 = -14.31 \text{ volts.}$$

(2) Potential due to a line charge is

$$V_2 = \frac{\rho_l}{2\pi\varepsilon} \ln\left(\frac{\rho_0}{\rho_2}\right)$$

where the distance vector from the line charge to the point A is

$$\vec{\rho}_1 = (x, 0, 2) - (x,\ 1,\ 2)$$

$$\vec{\rho}_1 = (0-1)\vec{a}_y + (2-2)\vec{a}_z$$

$$\vec{\rho}_1 = -\vec{a}_y$$

$$\rho_1 = 1$$

and the distance vector from the line charge to the origin is

$$\vec{\rho}_0 = (x,\ 0,\ 0) - (x,\ 1,\ 2)$$

$$\vec{\rho}_0 = (0-1)\vec{a}_y + (0-2)\vec{a}_z$$

$$\vec{\rho}_0 = -\vec{a}_y - 2\vec{a}_z$$

$$\rho_0 = \sqrt{5}.$$

$$\therefore \quad V_2 = \frac{4\times 1^{-9}\times 36\pi}{2\pi\times 10^{-9}} \ln\left(\frac{\sqrt{5}}{1}\right) = 72 \ln(\sqrt{5}) = 72\times 0.804$$

$$V_2 = 58 \text{ volts.}$$

$\therefore$ The potential is $V = V_1 + V_2$

$$V = 58 - 14.31 = 43.69 \text{ volts.}$$

2.29 Potential due to Charge Distribution

Line charge Consider a line charge of charge density ρ_l C/m. Let differential charge dQ exist on a differential length dl at a distance $\vec{r}_1$ from the origin.

The potential on a unit charge located at point A at distance $\vec{r}$ from the origin

$$dV_A = \frac{dQ}{4\pi\varepsilon|\vec{r}-\vec{r}_1|} = \frac{\rho_1 dl}{4\pi\varepsilon|\vec{r}-\vec{r}_1|} = \frac{\rho_l dl}{4\pi\varepsilon R},$$

where $R = |\vec{r}-\vec{r}_1|$ is the distance between the differential length element and point A.

The net potential is

$$V_A = \int_l \frac{\rho_l dl}{4\pi\varepsilon R} \text{ volts.} \tag{2.95}$$

Sheet of charge Consider a sheet of charge density ρ_s C/m^2 distributed uniformly over the surface S. Let the differential charge dQ exist on the differential area dS at distance $\vec{r}_1$ from the origin.

The potential on the unit charge located at point A at distance $\vec{r}$ from the origin is

$$dV_A = \frac{dQ}{4\pi\varepsilon|\vec{r}-\vec{r}_1|} = \frac{\rho_s dS}{4\pi\varepsilon|\vec{r}-\vec{r}_1|} = \frac{\rho_s dS}{4\pi\varepsilon R}. \tag{2.96}$$

The net potential is

$$V_A = \int_s \frac{\rho_s dS}{4\pi\varepsilon R} \text{ volts.} \tag{2.97}$$

Volume charge Consider a volume of charge density ρ_v V/m^3 distributed uniformly. Let the differential charge dQ exist on the differential volume dv at a distance $\vec{r}_1$ from the origin.

The potential on the unit charge located at point A at distance $\vec{r}$ from the origin is

$$dV_A = \frac{dQ}{4\pi\ \varepsilon|\vec{r}-\vec{r}_1|} = \frac{\rho_v dv}{4\pi\ \varepsilon|\vec{r}-\vec{r}_1|} = \frac{\rho_v dv}{4\pi\varepsilon R}. \tag{2.98}$$

The net potential is

$$V_A = \int_v \frac{\rho_v dv}{4\pi\varepsilon\ R} \text{ volts.} \tag{2.99}$$

Note: Constant C is added to the Eq. (2.95), Eq. (2.97) and Eq. (2.99) if the reference point (zero potential) is not at infinity.

2.30 Potential Difference due to an Infinite Line Charge

Consider an infinite line charge of charge density ρ_l lying along the z-axis.

Let two points A and B be located at distances ρ_A and ρ_B radially from the line charge as shown in Fig. 2.34. Also, let us use the cylindrical coordinate system.

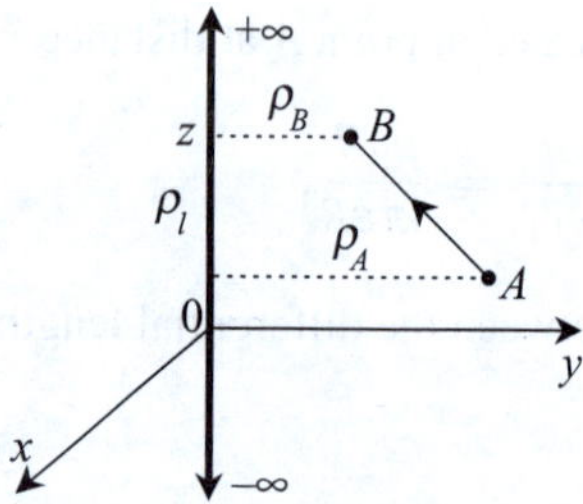

Fig. 2.34 *Infinite line charge*

We know that the field intensity $\vec{E}$ due to an infinite line charge along the z-axis at a distance ρ is

$$\vec{E} = \frac{\rho_l}{2\pi\,\varepsilon\,\rho}\vec{a}_\rho,$$

and the differential length along the radial direction is

$$d\vec{l} = d\rho\,\vec{a}_\rho.$$

We also know that the potential difference is

$$V_{BA} = -\int_A^B \vec{E}\cdot d\vec{l} = -\int_{\rho_A}^{\rho_B} \frac{\rho_l}{2\pi\varepsilon\,\rho}\vec{a}_\rho \cdot d\rho\,\vec{a}_\rho$$

$$= \frac{-\rho_l}{2\pi\varepsilon\,\rho}\int_{\rho_A}^{\rho_B} \frac{1}{\rho}d\rho = \frac{\rho_l}{2\pi\varepsilon}\left[-\ln(\rho)\right]_{\rho_A}^{\rho_B}$$

$$= \frac{\rho_l}{2\pi\varepsilon}\left[\ln(\rho_A) - (\rho_B)\right] = \frac{\rho_l}{2\pi\varepsilon}\ln\left(\frac{\rho_A}{\rho_B}\right) \text{ volts.} \tag{2.100}$$

Note 1: If point A is on the line charge, then $r_A = 0$.

$$\therefore\quad V_B = \frac{\rho_l}{2\pi\varepsilon}\ln\left(\frac{1}{\rho_B}\right) = \frac{-\rho_l}{2\pi\varepsilon}\ln(\rho_B) \text{ volts.} \tag{2.101}$$

Note 2: If there is a constant potential (V_A) at point A, then

$$V_B = \frac{-\rho_l}{2\pi\varepsilon}\ln(\rho_B) + V_A \text{ volts.} \tag{2.102}$$

2.31 Potential due to a Line Charge of Finite Length

Consider a uniform line charge of charge density ρ_l C/m^2 of finite length $2L$ along the x-axis, centred at the origin as shown in Fig. 2.35. Let a differential length dl be considered at a distance x from the origin on the line charge. Let a unit charge be located at a point A on the y-axis at a distance a from the origin i.e., the point A is on the perpendicular bisector of the line charge.

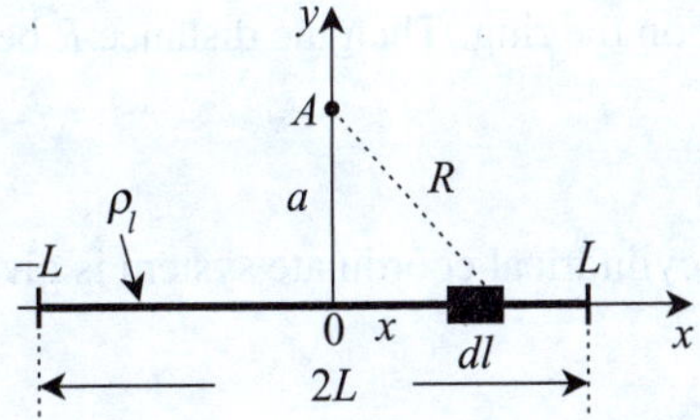

Fig. 2.35 *Finite line charge*

$$R = \sqrt{x^2 + a^2};\ dl = dx.$$

We know that for line charge distribution, the potential at point A is

$$V_A = \int \frac{\rho_l dl}{4\pi\varepsilon R} = \int_{-L}^{L} \frac{\rho_l dx}{4\pi\varepsilon\sqrt{(a^2 + x^2)}}$$

or $$V_A = \frac{2\rho_l}{4\pi\varepsilon}\int_{0}^{L} \frac{dx}{\sqrt{a^2 + x^2}}$$

We know that $\displaystyle\int \frac{dx}{\sqrt{a^2 + x^2}} = \ln[x + \sqrt{x^2 + a^2}].$

$$\therefore \quad V_A = \frac{\rho_l}{2\pi\varepsilon}\ln\left[x + \sqrt{x^2 + a^2}\right]\Big|_0^L = \frac{\rho_l}{2\pi\varepsilon}\left[\ln(L + \sqrt{L^2 + a^2}) - \ln a\right]$$

$$= \frac{\rho_l}{2\pi\varepsilon}\ln\left[\frac{L + \sqrt{L^2 + a^2}}{a}\right] \text{ volts.} \qquad (2.103)$$

2.32 Potential due to a Circular Ring

Consider a uniform line charge with charge density ρ_l C/m in the form of a circular ring radius of a located in the $x - y$ plane, centred at the origin as shown in Fig. 2.36. Let a unit charge be located at point A on the z-axis at a distance z from the origin.

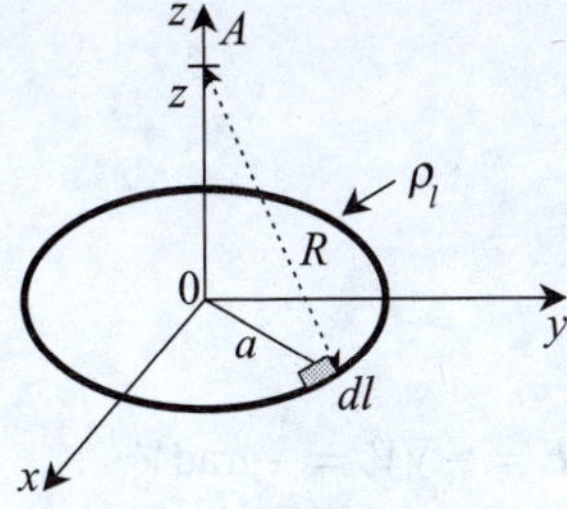

Fig. 2.36 *Circular ring charge*

Consider a differential length dl on the ring. Then the distance R between dl and the point A is

$$R = \sqrt{a^2 + z^2}.$$

The differential length dl in the cylindrical coordinate system is given by

$$dl = a d\phi.$$

We know that for a line charge distribution, the potential at point A is

$$V_A = \int_l \frac{\rho_l dl}{4\pi\varepsilon R} = \int_0^{2\pi} \frac{\rho_l a d\phi}{4\pi\varepsilon\sqrt{a^2+z^2}}$$

$$= \frac{\rho_l a}{4\pi\varepsilon\sqrt{a^2+z^2}}[\phi]_0^{2\pi} = \frac{\rho_l a 2\pi}{4\pi\varepsilon\sqrt{a^2+z^2}}$$

$$= \frac{\rho_l a}{2\varepsilon\sqrt{a^2+z^2}} \text{ volts.} \tag{2.104}$$

Note: If the point A is at the centre of the circular ring on the same plane as the ring charge, i.e., $z = 0$, the potential is

$$V_A = \frac{\rho_l}{2\varepsilon} \text{ volts.}$$

2.33 Potential Gradient

In a given electric field, the maximum value of rate of change of potential with respect to the distance is called potential gradient. It is a differential change of potential.

Mathematically, it can be expressed as

$$\nabla V = \left.\frac{dV}{dl}\right|_{\max} = \text{grad } V = \text{Gradient of potential.} \tag{2.105}$$

We know that $V = -\int \vec{E} \cdot d\vec{l}$. Differential voltage is $dV = -\vec{E} \cdot d\vec{l}$. If θ is the angle between $\vec{E}$ and $d\vec{l}$,

$$dV = -Edl \cos\theta.$$

The maximum value is at $\cos\theta = 1$.

$$\left.\frac{dV}{dl}\right|_{\max} = -E,$$

or $$E = -\left.\frac{dV}{dl}\right|_{\max} = -\nabla V. \text{ Thus, } E = -\nabla V = -\text{grad} V.$$

The electric field intensity is equal to the negative of the potential gradient at a given point. The field is directed towards the maximum gradient.

Del operator (∇) The ∇ symbol in the equation, $E = -\nabla V$, is called the del operator. The del operator in Cartesian coordinates is given as

$$\nabla = \frac{\partial}{\partial x}\vec{a}_x + \frac{\partial}{\partial y}\vec{a}_y + \frac{\partial}{\partial z}\vec{a}_z. \tag{2.106}$$

Therefore, the gradient of potential ∇V in Cartesian coordinates is

$$\nabla V = \frac{\partial V}{\partial x}\vec{a}_x + \frac{\partial V}{\partial y}\vec{a}_y + \frac{\partial V}{\partial z}\vec{a}_z. \tag{2.107}$$

In the cylindrical coordinate system,

$$\nabla V = \frac{\partial V}{\partial \rho}\vec{a}_\rho + \frac{1}{\rho}\frac{\partial V}{\partial \phi}\vec{a}_\phi + \frac{\partial V}{\partial z}\vec{a}_z, \tag{2.108}$$

and in the spherical coordinate system,

$$\nabla V = \frac{\partial V}{\partial r}\vec{a}_r + \frac{1}{r}\frac{\partial V}{\partial \theta}\vec{a}_\theta + \frac{1}{r\sin\theta}\frac{\partial V}{\partial \phi}\vec{a}_\phi. \tag{2.109}$$

2.34 Relationship between ($\vec{E}$) and V: Maxwell's Second Equation

We know that in a given electric field $\vec{E}$, the potential difference between points A and B is independent of the path chosen, i.e.,

$$V_{AB} = -\int_B^A \vec{E}.d\vec{l} \text{ and } V_{BA} = -\int_A^B \vec{E}.d\vec{l} = -V_{AB}.$$

$$\therefore \quad V_{AB} + V_{BA} = \oint \vec{E}.d\vec{l} = 0. \tag{2.110}$$

Thus the potential difference around a closed path must be zero.

Applying Stokes' theorem,

$$\oint \vec{E}.d\vec{l} = \int (\nabla \times \vec{E}).d\vec{S} = 0$$

or $\quad \nabla \times \vec{E} = 0, \qquad (2.111)$

that is, the curl of the electric field must be zero. Any electric field that satisfies the above equation is called a conservative field or an irrotational one.

Now, the potential at any point in the field is

$$V = \int \vec{E}.d\vec{l}.$$

In Cartesian coordinates, $dV = -\vec{E} \cdot d\vec{l} = -E_x dx - E_y dy - E_z dz.$

Also, $dV = \frac{\partial V}{\partial x}dx + \frac{\partial V}{\partial y}dy + \frac{\partial V}{\partial z}dz.$

Therefore, $E_x = -\frac{\partial V}{\partial x}$; $E_y = -\frac{\partial V}{\partial y}$; and $E_z = -\frac{\partial V}{\partial z}$.

We know that $\vec{E} = E_x\vec{a}_x + E_y\vec{a}_y + E_z\vec{a}_z$

$$= -\frac{\partial V}{\partial x}\vec{a}_x - \frac{\partial V}{\partial y}\vec{a}_y - \frac{\partial V}{\partial z}\vec{a}_z \tag{2.112}$$

or, $\vec{E} = -\nabla V.$

Thus the electric field intensity is equal to the gradient of the potential. The negative sign indicates that the field direction is opposite to the direction of the potential increment.

2.35 Potential Function (V)

An electric potential at any point in the field which depends on the coordinates (or position) is called potential function.

The electrostatic field E can be described completely by means of a potential function $V(x, y, z)$. The potential gradient of V is given by

$$E = -\nabla V,$$

where V is a scalar potential function. This equation gives a simple method to find the electric field intensity from the potential.

It should be noted that if two points in the electric field have non-zero potential difference, electric lines of force should be present. These lines of force travel from a point of higher potential to a point of lower potential.

Properties of potential function

Potential function has the following properties.

1. It has a single value at any point in the electrostatic field.
2. It is a continuous function.
3. The difference in potential between any two points does not depend on the path of integration.
4. An electrostatic field is conservative in nature, since $\nabla \times (\nabla V)$, which is the work done in moving a point charge around any closed path in the field, is zero.

2.36 Equipotential Surfaces

An equipotential surface is defined as *a surface in an electric field on which all points on the surface are at the same potential. In other words, the potential difference between any two points on an equipotential surface is always zero.* Hence the work done in moving a charge from one point to another point on an equipotential surface is always zero.

The gradient of the potential and the equipotential surface are always orthogonal to each other.

Many equipotential surfaces may exist in a given electric field distribution. For example, consider a point charge located at the centre of a sphere. The potential at a distance r from the point charge depends only on the radial distance r, not on θ and ϕ. Therefore, the surface of a sphere with a radius r forms an equipotential surface, and concentric spherical surfaces at $r = r_1, r_2, r_3, \ldots, r_n$ are all equipotential surfaces.

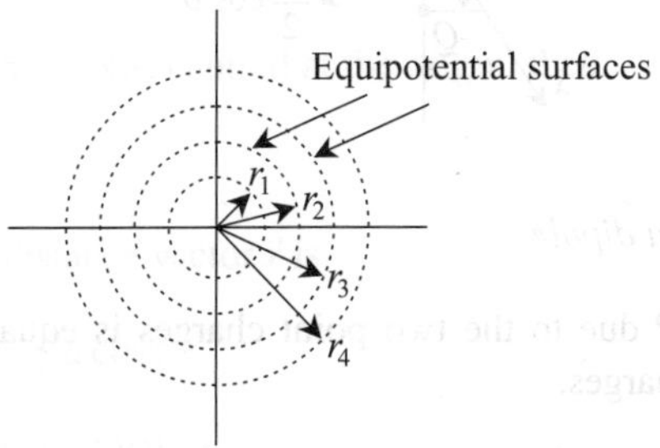

Fig. 2.37 *Equipotential surfaces*

The potential $V = \dfrac{Q}{4\pi\varepsilon r}$ is constant on an equipotential surface as shown in Fig. 2.37. The equipotential surfaces are normal to the direction of $\vec{E}$. For a uniform electric field, equipotential exist at the same increments of distance. For a non-uniform field, equipotentials are not at the same increments of distance but are still normal to $\vec{E}$.

2.37 Electric Dipole

An electric dipole is defined as *a pair of point charges with equal magnitude and opposite polarity separated by a small distance in a medium* as shown in Fig. 2.38.

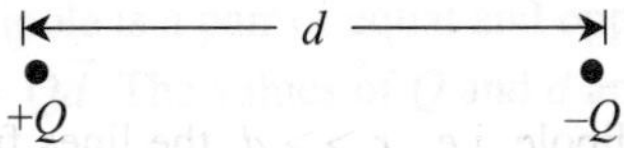

Fig. 2.38 *Electric dipole*

Materials are made up of atoms. Dipoles are common in atoms and molecule. They exist whenever electrons are unevenly distributed around the nuclei of atoms, or whenever they are shared unevenly among atoms in the covalent bond of a molecule. Antennas radiate because of oscillating electric dipoles.

The potential due to dipole moment is

$$V = \frac{\vec{p}\cdot\vec{a}_r}{4\pi\varepsilon_0 r^2}.$$

The distance vector $\vec{r} = (5,0,0) - (3,0,0) = 2\vec{a}_x$.

$$\vec{a}_r = \frac{\vec{r}}{r} = \frac{2\vec{a}_x}{2} = \vec{a}_x.$$

$$\therefore \quad V = \frac{10\times10^{-9}\vec{a}_x\cdot\vec{a}_x}{4\pi\varepsilon_0 2^2} = \frac{10\times10^{-9}\times9\times10^9}{4} = 22.5 \text{ V}.$$

Example 2.18 Two point charges of 1 μC and -1μC are located at (0, 0, 1) and $(0,0,-1)$ respectively, in free space.

(i) Find the potential at (0, 3, 4).

(ii) Recalculate the same potential treating the charges as a pure dipole.

Solution Given, a point charge 1 μC located at $(0,0,1)$, another point charge -1 μC located at $(0, 0, -1)$. We have to calculate the potential at the point $(0,3,4)$.

(i) The potential due to the two point charges is

$$V = \frac{Q_1}{4\pi\varepsilon_0 R_1} + \frac{Q_2}{4\pi\varepsilon_0 R_2}.$$

The distance vector $\vec{R}_1 = (0,3,4) - (0,0,1) = 3\vec{a}_y + 3\vec{a}_z$.

$$R_1 = \sqrt{3^2+3^2} = \sqrt{18}$$

and the distance vector $\vec{R}_2 = (0,3,4) - (0,0,\ -1) = 3\vec{a}_y + 5\vec{a}_z$.

$$R_2 = \sqrt{3^2+5^2} = \sqrt{34}.$$

$$\therefore \quad V = \frac{1\times10^{-6}}{4\pi\varepsilon_0\sqrt{18}} + \frac{-1\times10^{-6}}{4\pi\varepsilon_0\sqrt{34}}$$

$$= 9\times10^9\times10^{-6}\left[\frac{1}{\sqrt{18}} - \frac{1}{\sqrt{34}}\right] = 9000\,[0.235 - 0.171] = 576 \text{ volts}.$$

(ii) If the charges constitute a pure dipole, then

$$V = \frac{\vec{p}\cdot\vec{a}_R}{4\pi\varepsilon\, R^2}.$$

The dipole moment $\vec{p} = Q\cdot\vec{d}$.

$$\vec{d} = (0,0,1) - (0,0,-1) = 2\vec{a}_z.$$

$$\therefore \quad \vec{p} = 2\times10^{-6}\vec{a}_z.$$

The distance vector $\vec{R} = (0,3,4) = 3\vec{a}_y + 4\vec{a}_z$.

$$R = \sqrt{3^2+4^2} = 5.$$

$\therefore$ The potential due to the dipole is

$$V = \frac{2\times 10^{-6}\vec{a}_z\cdot(3\vec{a}_y+4\vec{a}_z)}{4\pi\varepsilon_0(5^2)(5)}$$

$$= \frac{2\times 10^{-6}\times 9\times 10^9\times 4}{125} = 576 \text{ volts.}$$

Note: Both methods yield the same result.

2.40 Electric Field due to a Dipole

We know that the potential due to a dipole is

$$V = \frac{Qd\cos\theta}{4\pi\varepsilon r^2} \text{ volts.}$$

Since the field due to the dipole is symmetrical along the radial direction, it is convenient to use spherical coordinates. In spherical coordinates, the electric field intensity in terms of potential is given by

$$\vec{E} = -\nabla V$$

$$= -\left[\frac{\partial V}{\partial r}\vec{a}_r + \frac{\partial V}{r\partial\theta}\vec{a}_\theta + \frac{\partial V}{r\sin\theta\partial\phi}\vec{a}_\phi\right]. \tag{2.121}$$

Since for the dipole equipotential surfaces exist at different values of ϕ, the potential is not a function of ϕ.

$$\therefore \quad E_\phi = 0$$

Then, $$\vec{E} = -\left[\frac{\partial}{\partial r}\left(\frac{Qd\cos\theta}{4\pi\varepsilon r^2}\right)\vec{a}_r + \frac{1}{r}\frac{\partial}{\partial\theta}\left(\frac{Qd\cos\theta}{4\pi\varepsilon r^2}\right)\vec{a}_\theta\right]$$

$$= \frac{-Qd\cos\theta\vec{a}_r}{4\pi\varepsilon}\frac{\partial}{\partial r}\left(\frac{1}{r^2}\right) - \frac{Qd\vec{a}_\theta}{4\pi\varepsilon r^3}\frac{\partial}{\partial\theta}(\cos\theta)$$

$$= \frac{Qd}{4\pi\varepsilon r^3}[2\cos\theta\vec{a}_r + \sin\theta\vec{a}_\theta] \text{ V/m.} \tag{2.122}$$

We know that the magnitude of dipole movement is $|p| = Qd$.

$$\therefore \quad \vec{E} = \frac{|p|}{4\pi\varepsilon r^3}[2\cos\theta\vec{a}_r + \sin\theta\vec{a}_\theta] \text{ V/m.} \tag{2.123}$$

Note: It is to be noted that for a given dipole, the potential is inversely proportional to the square of the distance, whereas the electric field intensity is inversely proportional to the cube of the distance.

Example 2.19 What is the electric field at $(0,0,5)$ due to a pure dipole $1\vec{a}_z$ nC-m at the origin?

Solution Given dipole $\vec{p} = 1\vec{a}_z$ μC-m located at origin.
We know that the electric field due to a dipole is

$$\vec{E} = \frac{|p|}{4\pi\varepsilon\, r^3}(2\cos\theta\vec{a}_r + \sin\theta\vec{a}_\theta)\ \text{V/m in spherical coordinates.}$$

The point $(0,0,5)$ in spherical coordinates:

$$r = \sqrt{0^2+0^2+5^2} = 5\ \text{m}$$

$$\theta = \cos^{-1}\left(\frac{z}{r}\right) = \cos^{-1}\left(\frac{5}{5}\right) = 0$$

Then, $$\vec{E} = \frac{10^{-6}}{4\pi\varepsilon r^3}(2\cos(0)\vec{a}_r + \sin(0)\vec{a}_\theta)$$

$$= \frac{10^{-6}\times 9\times 10^9}{125}2\vec{a}_r = 144\vec{a}_r\ \text{V/m.}$$

2.41 Torque on an Electric Dipole in an Electric Field

The torque on a dipole is equal to the cross product of the dipole moment $\vec{p}$ and the electric field intensity $\vec{E}$. It can be expressed as

$$\vec{T} = \vec{p}\times\vec{E}\ \text{N-m.} \tag{2.124}$$

Proof Consider an electric dipole in a uniform electric field $\vec{E}$ as shown in Fig. 2.40. Let the dipole axis make an angle θ with the field. d is the distance between the charges.

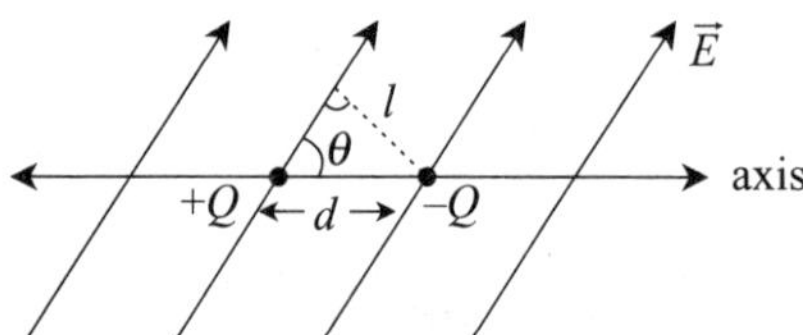

Fig. 2.40 *Torque on a dipole*

The force on the dipole is $\vec{F} = Q\vec{E}$.

We know that the torque is $T = \text{Force}\times\text{Distance} = \vec{F}\,l = Q\vec{E}\,l,$ (2.125)

where l is the perpendicular distance between E and Q.

From Fig. 2.40, $\sin\theta = \dfrac{l}{d}$ or $l = d\sin\theta$.

$$\therefore \quad \vec{T} = Q\vec{E}d\sin\theta = Qd\vec{E}\sin\theta.$$

We know that the dipole moment, $\vec{p} = Q.\vec{d}$.

$$\vec{T} = \vec{p}\,\vec{E}\sin\theta = \vec{p}\times\vec{E}. \qquad \text{(Proved).}$$

Thus the torque on the dipole is equal to the cross product of the dipole moment and the electric field intensity. The direction of the torque is perpendicular to the dipole plane.

Example 2.20 Compute the torque for a dipole comprising 1μC charges in an electric field $E = 10^3(z\vec{a}_x - \vec{a}_y - a_z)$ separated by 1 mm and located on the z-axis at the origin.

Solution Given $Q = 1\ \mu\text{C} = 1\times10^{-6}$C; $d = 1$ mm; $\vec{d} = 10^{-3}\vec{a}_z$; $\vec{E} = 10^3(z\vec{a}_x - \vec{a}_y - \vec{a}_z)$.
The torque on the dipole is given by

$$\vec{T} = \vec{p}\times\vec{E}$$

where $\vec{p} = Q\vec{d} = Qd\vec{a}_z = 10^{-6}\times10^{-3}\vec{a}_z = 10^{-9}\vec{a}_z$

$$\therefore \quad \vec{T} = 10^{-9}\vec{a}_z\times10^3(z\vec{a}_x - \vec{a}_y - \vec{a}_z) = 10^{-6}\begin{vmatrix} \vec{a}_x & \vec{a}_y & \vec{a}_z \\ 0 & 0 & 1 \\ z & -1 & -1 \end{vmatrix}.$$

$$= 10^{-6}[\vec{a}_x + z\vec{a}_y] = (\vec{a}_x + z\vec{a}_y)\ \mu\text{N-m}.$$

Example 2.21 Find the potential due to an electric dipole consisting of $+Q$ at $(a/2,0,0)$ and $-Q$ at $(-a/2,0,0)$ at a distance point (r,θ,ϕ) in spherical coordinates.

Solution Given: The dipole $+Q$ at $(a/2,0,0)$ and $-Q$ at (a/2, 0,0).
The distance between the charges in the dipole is

$$\vec{d} = \frac{a\vec{a}_x}{2} - \left(\frac{-a\vec{a}_x}{2}\right) = a\vec{a}_x.$$

The dipole moment is $\vec{p} = Q\vec{d} = Q\,a\,\vec{a}_x$.
The field point is given at (r,θ,ϕ) in spherical coordinates. We know that the potential at point of distance r is

$$V = \frac{\vec{p}\cdot\vec{a}_r}{4\pi\varepsilon r^2} = \frac{Qa\vec{a}_x\cdot\vec{a}_r}{4\pi\varepsilon r^2}$$

The dot product $\vec{a}_x\cdot\vec{a}_r = \cos\phi\sin\theta$ (Refer to Chapter 1)

$$\therefore \quad V = \frac{Qa\cos\phi\sin\theta}{4\pi\varepsilon r^2}.$$

2.42 Energy Stored in an Electrostatic Field

The energy stored in an electric field due to N point charges is expressed as

$$W_E = \frac{1}{2}\sum_{i=1}^{N} V_i Q_i \text{ joules} \tag{2.126}$$

Proof Consider a free space where there is no electric field. Let a point charge Q_1 be moved from infinity to a point 1 in the free space as shown in Fig. 2.41.

Since there is no field, the work done is zero, i.e., $W_1 = 0$. Now due to Q_1 in the free space, an electric field is established.

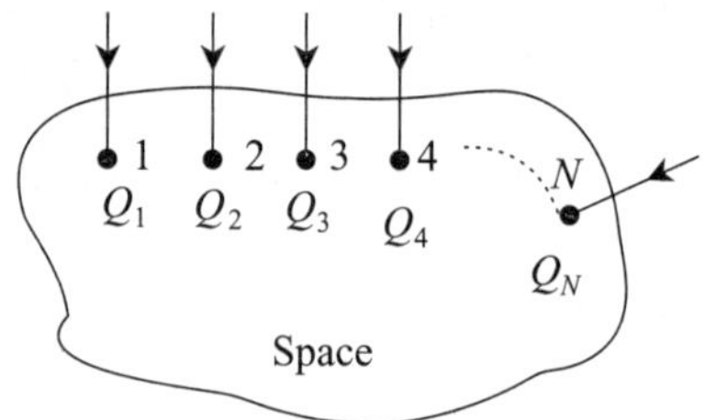

Fig. 2.41 *Point changes in space*

If a point charge Q_2 is moved from infinity to a point 2 in this field, then the work done in keeping Q_2 is

$$W_2 = V_{21}Q_2, \tag{2.127}$$

where V_{21} is the potential at point 2 due to Q_1. Now the field increases due to Q_1 and Q_2.

If another point charge Q_3 is moved from infinity to a point 3 in this field, then the work done in keeping Q_3 is

$$W_3 = V_{31}Q_3 + V_{32}Q_3,$$

where V_{31} is the potential at point 3 due to Q_1, and V_{32} is the potential at point 3 due to Q_2.

Similarly, if the point charges Q_4, $Q_5, \ldots\ldots Q_{N-1}$, etc., are moved into this field, the field strength increases.

Finally, if the point charge Q_N is moved from infinity to a point N in this field, then the work done in keeping Q_N is

$$W_E = V_{N1}Q_N + V_{N2}Q_N + V_{N3}Q_N + \ldots. + V_{NN-1}Q_N.$$

The total work done in positioning all the point charges at their respective points is

$$\begin{aligned} W_E &= W_1 + W_2 + W_3 + \ldots.. + W_N \\ &= 0 + (V_{21}Q_2) + (V_{31}Q_3 + V_{32}Q_3) + (V_{41}Q_4 + V_{42}Q_4 + V_{43}Q_4) \\ &\quad + (V_{51}Q_5 + V_{52}Q_5 + V_{53}Q_5 + V_{54}Q_5) + \cdots\cdot \\ &\quad + (V_{N1}Q_N + V_{N2}Q_N + V_{N3}Q_N + \ldots\ldots + V_{NN-1}Q_N). \end{aligned} \tag{2.128}$$

The total work done in positioning all the point charges in a space is called the potential energy or the energy stored in an electrostatic field.

To simplify the equation, let us consider that all the point charges are placed in reverse order. That is,

$$Q_iV_{ij} = Q_jV_{ji}.$$

The energy stored in the field can be written as

$$\begin{aligned} W_E &= (V_{12}Q_1) + (V_{13}Q_1 + V_{23}Q_2) + (V_{14}Q_1 + V_{24}Q_2 + V_{34}Q_3) \\ &\quad + (V_{15}Q_1 + V_{25}Q_2 + V_{35}Q_3 + V_{45}Q_4) + \ldots\ldots. \\ &\quad + V_{1N}Q_1 + V_{2N}Q_2 + V_{3N}Q_3 + \ldots\ldots + V_{N-1N}Q_N \end{aligned} \tag{2.129}$$

Adding Eqs (2.128) and (2.129), we get

$$\begin{aligned} 2W_E &= Q(V_{12} + V_{13} + V_{14} + V_{15} + \ldots. + V_{1N}) + Q_2(V_{21} + V_{23} + V_{24} + \ldots. + V_{2N}) \\ &\quad + Q_3(V_{31} + V_{32} + V_{34} + \ldots.. + V_{3N}) \\ &\quad + Q_4(V_{41} + V_{42} + V_{43} + V_{44} + \ldots + V_{4N}) \cdots\cdots \\ &\quad + Q_N(V_{N1} + V_{N2} + N_{N3} + \ldots. + V_{NN}). \end{aligned}$$

Let the potential at Q_1 due to all the other charges be

$$V_1 = V_{12} + V_{13} + V_{14} + \ldots. + V_{1N},$$

and that at Q_2 be $V_2 = V_{21} + V_{23} + V_{24} + \ldots. + V_{2N}$,

.

.

.

and at Q_N $\quad V_N = V_{N1} + V_{N2} + V_{N3} + \ldots. + V_{NN}.$

$$\therefore \quad 2W_E = Q_1V_1 + Q_2V_2 + Q_3V_3 + \ldots. + Q_NV_N.$$

or $\quad W_E = \dfrac{1}{2}(Q_1V_1 + Q_2V_2 + Q_3V_3 + \ldots. + Q_NV_N).$

Thus the potential energy stored in the field is

$$W_E = \frac{1}{2}\sum_{i=1}^{N} Q_iV_i \text{ joules.} \qquad \text{(Proved)}$$

Similarly, for a line charge of charge density ρ_l,

$$W_E = \frac{1}{2}\int_l V\,\rho_l\,dl \text{ joules;} \tag{2.130}$$

for a surface charge of charge density ρ_s,

$$W_E = \frac{1}{2}\int_s V\,\rho_s\,dS \text{ joules;} \tag{2.131}$$

and for a volume charge of charge density ρ_v

$$W_E = \frac{1}{2}\int_v V\,\rho_v\,dv \text{ joules.} \tag{2.132}$$

2.43 Energy Stored in Terms of $\vec{E}$ and $\vec{H}$

Consider a volume charge density ρ_v existing in a volume with applied potential, V volts.

We know that the energy stored is

$$W_E = \frac{1}{2}\int_v V\,\rho_v\,dv \text{ joules.}$$

From Gauss' law, $\rho_v = \nabla\cdot\vec{D} = \nabla\cdot\varepsilon\vec{E}$.

$$\therefore \quad W_E = \frac{\varepsilon}{2}\int_v V(\nabla\cdot E)dv.$$

Using the vector identity

$$(\nabla\cdot\vec{E})V = \nabla\cdot V\vec{E} - \vec{E}\cdot\nabla V.$$

$$W_E = \frac{\varepsilon}{2}\left[\int_v(\nabla\cdot V\vec{E})dv - \int_v \vec{E}\cdot\nabla V dv\right].$$

From the divergence theorem, since no closed surface encloses $\vec{E}$,

$$\int_v(\nabla\cdot V\vec{E})dv = \oint_s V\vec{E}\cdot\vec{dS} = 0.$$

$$\therefore \quad W_E = \frac{-\varepsilon}{2}\int_v \vec{E}\cdot\nabla V dv.$$

We know that $\vec{E} = -\nabla V$.

$$\text{So} \quad W_E = \frac{\varepsilon}{2}\int_v \vec{E}\cdot\vec{E}\cdot dv = \frac{\varepsilon}{2}\int_v E^2 dv$$

$$\text{or} \quad W_E = \int_v \frac{1}{2}\varepsilon E^2 dv = \int_v \frac{1}{2}\vec{D}\cdot\vec{E}dv \text{ joules} \tag{2.133}$$

$$\text{or} \quad W_E = \int_v \frac{1}{Q\varepsilon} D^2 dv \text{ joules.} \tag{2.134}$$

Therefore the energy density in an electrostatic field is

$$w_E = \frac{W_E}{v} = \frac{1}{2}\varepsilon E^2 = \frac{1}{2}\vec{D}\cdot\vec{E} = \frac{1}{2\varepsilon}D^2 \text{ joules/m}^3 \text{ or J/m}^3 \tag{2.135}$$

Additional Solved Problems

Example 2.23 A point charge at $Q_1 = 600\ \mu$C located at $(2, -4, -3)$ experiences a force $\vec{F} = 4\vec{a}_x - 4\vec{a}_y + 2\vec{a}_z$ due to a point charge Q_2 at $(4, -6, -4)$. Determine the magnitude of Q_2.

Solution Given: $Q_1 = 600\ \mu$C at $(2, -4, -3)$; $\vec{F} = 4\vec{a}_x - 4\vec{a}_y + 2\vec{a}_z$; Q_2 at $(4, -6, -4)$.

We know that $\vec{F} = \dfrac{Q_1 Q_2}{4\pi\varepsilon_0 R_{21}^2}\vec{a}_{R_{21}}$.

The distance vector is

$$\vec{R}_{21} = (2,-4,-3) - (4,-6,-4)$$

$$= (2-4)\vec{a}_x + (-4+6)\vec{a}_y + (-3+4)\vec{a}_z = -2\vec{a}_x + 2\vec{a}_y + \vec{a}_z$$

$$|\vec{R}_{21}| = \sqrt{2^2+2^2+1} = 3.$$

Since the force is in the direction of the distance vector, magnitude of force is

$$|\vec{F}_{21}| = \frac{Q_1 Q_2}{4\pi\varepsilon_0 R_{21}^2}|\vec{a}_{R_{21}}|.$$

But $|\vec{F}_{21}| = \sqrt{4^2+4^2+2^2} = 6$.

$$\therefore \quad 6 = \frac{600\times 10^{-6}\times Q_2}{4\pi\times 8.854\times 10^{-12}\times 3^2}.$$

$$Q_2 = 10\ \mu\text{C}.$$

Example 2.24 Two point charges, $Q_1 = 35\ \mu$C and $Q_2 = 60\ \mu$C, are located at $(-4, 6, -8)$ and $(3, 5, 2)$ respectively. Find the force on Q_1.

Solution Given $Q_1 = 35\ \mu$C at $(-4, 6, 8)$ and $Q_2 = 60\ \mu$C at $(3, 5, 2)$.

$$\vec{R}_{21} = (-4-3)\vec{a}_x + (6-5)\vec{a}_y + (-8+2)\vec{a}_z = -7\vec{a}_x + \vec{a}_y - 6\vec{a}_z.$$

$$\vec{a}_{R_{21}} = \frac{\vec{R}_{21}}{|R_{21}|} = \frac{-7\vec{a}_x + \vec{a}_y - 6\vec{a}_z}{\sqrt{49+1+36}} = \frac{-7\vec{a}_x + \vec{a}_y - 6\vec{a}_z}{\sqrt{86}}.$$

We know that

$$\vec{F} = \frac{Q_1 Q_2}{4\pi\varepsilon_0 R_{21}^2}\vec{a}_{R_{21}}.$$

$$\therefore \quad \text{Force on } Q_1 \text{ is } \vec{F} = \frac{35\times10^{-6}\times60\times10^{-6}}{4\pi\times8.8854\times10^{-12}\times86}\left(\frac{-7\vec{a}_x+\vec{a}_y-6\vec{a}_z}{\sqrt{86}}\right)$$

$$= (0.0236)(-7\vec{a}_x+\vec{a}_y-6\vec{a}_z)$$

$$= -0.164\vec{a}_x+0.02\vec{a}_y-0.142\vec{a}_z \text{ N}.$$

Example 2.25 Four point charges are located in free space as shown in Fig. 2.42. Find the force experienced by the 1μC charge.

Solution Given $Q_1 = 1\ \mu$C at $(0,0,0)$; $Q_2 = -4\ \mu$C at $(0,0.1,0)$; $Q_3 = 4\ \mu$C at $(0,0,0.1)$; $Q_4 = -8\sqrt{2}\ \mu$C at $(0,0.1,0.1)$.

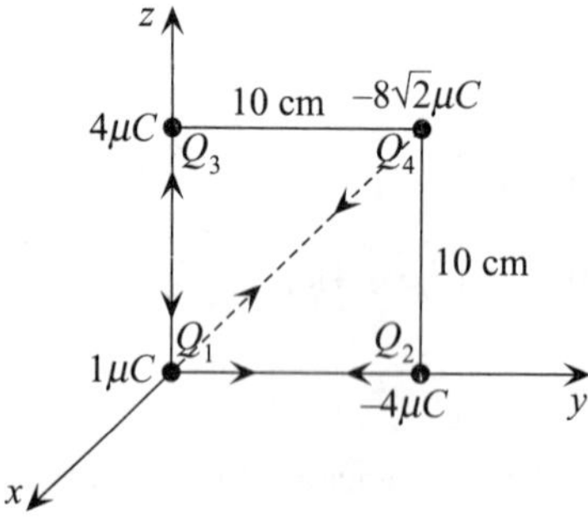

Fig. 2.42 *Four point charges on a square*

The force experienced by $Q_1 = 1\ \mu$C is obtained from the superposition principle as

$$\vec{F} = \vec{F}_1+\vec{F}_2+\vec{F}_3$$

$$= \frac{1}{4\pi\varepsilon_0}\left[\frac{Q_1Q_2}{R_{21}^2}\vec{a}_{R_{21}}+\frac{Q_1Q_3}{R_{31}^2}\vec{a}_{R_{31}}+\frac{Q_1Q_4}{R_{41}^2}\vec{a}_{R_{41}}\right].$$

The distance vectors are

$$\vec{R}_{21} = (0,0,0)-(0,0.1,0) = -0.1\vec{a}_y.$$

$$R_{21} = \sqrt{(-01)^2} = 0.1,\ \ \vec{a}_{21} = \frac{-0.1\vec{a}_y}{0.1} = -\vec{a}_y.$$

$$\vec{R}_{31} = (0,0,0)-(0,0.1,0) = -0.1\vec{a}_z.$$

$$R_{31} = 0.1,\ \ \vec{a}_{31} = -a_z \text{ and}$$

$$\vec{R}_{41} = (0,0,0)-(0,0.1,0.1) = -0.1\vec{a}_y-0.1\vec{a}_z.$$

$$\vec{R}_{41} = \sqrt{(-0.1)^2 + (-0.1)^2} = 0.1\sqrt{2}, \quad \vec{a}_{41} = \frac{-0.1\vec{a}_y - 0.1\vec{a}_z}{0.1\sqrt{2}} = \frac{-\vec{a}_y - \vec{a}_z}{\sqrt{2}}.$$

Also, $\dfrac{1}{4\pi\varepsilon_0} = 9 \times 10^9$.

$$\therefore \quad \vec{F} = 10^{-12} \times 9 \times 10^9 \left[\frac{-4}{0.1^2}(-\vec{a}_y) + \frac{4}{0.1^2}(-\vec{a}_z) - \frac{8\sqrt{2}}{(0.1\sqrt{2})^2}\left(\frac{-\vec{a}_y - \vec{a}_z}{\sqrt{2}}\right)\right]$$

$$= 9 \times 10^{-3}[400\vec{a}_y - 400\vec{a}_z + 400\vec{a}_y + 400\vec{a}_z]$$

$$\vec{F} = 9 \times 10^{-3} \times 800\vec{a}_y = 7.2\vec{a}_y \text{ N}.$$

Example 2.26 Four point charges of 500 μC each are placed at the comers of a $3\sqrt{2}$m side square. The square is located in the $z = 0$ plane between $x = \pm\dfrac{3}{\sqrt{2}}$ and $y = \pm\dfrac{3}{\sqrt{2}}$m in free space. Find the force on a point charge 30μC located at $(0, 0, 4)$.

Solution: Given four point charges:

$$Q_1 = 500\ \mu\text{C at } \left(\frac{-3}{\sqrt{2}}, \frac{-3}{\sqrt{2}}\right); \ Q_2 = 500\ \mu\text{C at } \left(\frac{3}{\sqrt{2}}, \frac{-3}{\sqrt{2}}\right);$$

$$Q_3 = 500\ \mu\text{C at } \left(\frac{3}{\sqrt{2}}, \frac{3}{\sqrt{2}}\right); \text{ and } Q_4 = 500\ \mu\text{C at } \left(\frac{-3}{\sqrt{2}}, \frac{3}{\sqrt{2}}\right).$$

The force experienced by a point charge $Q = 30\ \mu$C located at $(0, 0, 4)$ is shown in Fig. 2.43.

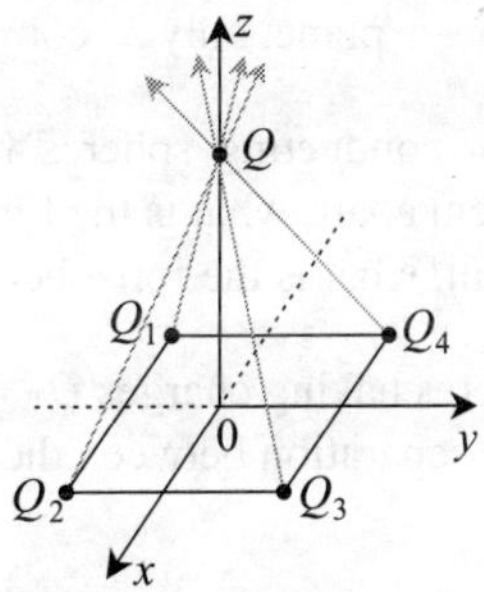

Fig. 2.43 *Force on Q due to four point charges located on a square*

Using superposition principle, the force is

$$\vec{F} = \vec{F}_1 + \vec{F}_2 + \vec{F}_3 + \vec{F}_4$$

$$= \frac{1}{4\pi\varepsilon_0}\left[\frac{QQ_1}{R_1^2}\vec{a}_{R_1} + \frac{QQ_2}{R_2^2}\vec{a}_{R_2} + \frac{QQ_3}{R_3^2}\vec{a}_{R_3} + \frac{QQ_4}{R_4^2}\vec{a}_{R_4}\right].$$

The distance vectors are

$$\vec{R}_1 = (0,0,4) - \left(\frac{-3}{\sqrt{2}}, \frac{-3}{\sqrt{2}}, 0\right) = \frac{3}{\sqrt{2}}\vec{a}_x + \frac{3}{\sqrt{2}}\vec{a}_y + 4\vec{a}_z.$$

$$R_1 = \sqrt{\left(\frac{3}{\sqrt{2}}\right)^2 + \left(\frac{3}{\sqrt{2}}\right)^2 + 4^2} = 5; \ \vec{a}_{R_1} = \frac{1}{5}\left(\frac{3}{\sqrt{2}}\vec{a}_x + \frac{3}{\sqrt{2}}\vec{a}_y + 4\vec{a}_z\right).$$

Similarly, $R_2 = R_3 = R_4 = 5$ and

$$\vec{a}_{R_2} = \frac{1}{5}\left(\frac{-3}{\sqrt{2}}\vec{a}_x + \frac{3}{\sqrt{2}}\vec{a}_y + 4\vec{a}_z\right)$$

$$\vec{a}_{R_3} = \frac{1}{5}\left(\frac{-3}{\sqrt{2}}\vec{a}_x - \frac{3}{\sqrt{2}}\vec{a}_y + 4\vec{a}_z\right)$$

$$\vec{a}_{R_4} = \frac{1}{5}\left(\frac{3}{\sqrt{2}}\vec{a}_x - \frac{3}{\sqrt{2}}\vec{a}_y + 4\vec{a}_z\right).$$

$$\therefore \quad \vec{F} = \frac{500\times10^{-6}\times30\times10^{-6}\times9\times10^{9}}{5^2\times5}\left(\frac{3}{\sqrt{2}}\vec{a}_x + \frac{3}{\sqrt{2}}\vec{a}_y + 4\vec{a}_z\right.$$

$$\left. -\frac{3}{\sqrt{2}}\vec{a}_x + \frac{3}{\sqrt{2}}\vec{a}_y + 4\vec{a}_z - \frac{3}{\sqrt{2}}\vec{a}_x - \frac{3}{2}\vec{a}_y + 4\vec{a}_z + \frac{3}{\sqrt{2}}\vec{a}_x - \frac{3}{\sqrt{2}}\vec{a}_y + 4\vec{a}_z\right)$$

$$\vec{F} = \frac{135}{125}\times4\vec{a}_z = 4.32\vec{a}_z.$$

Note: Due to the symmetry on the $x-y$ plane, only $\vec{a}_z$ component exists.

Example 2.27 Two small identical conducting spheres have charges of 2 nC and −0.5 nC respectively. When they are placed 4 cm apart, what is the force between them? If they are brought into contact and then separated by 4 cm, what is the force between them?

Solution Given: Two identical spheres having charges $Q_1 = 2$ nC $= 2\times10^{-9}$ C and $Q_2 = -0.5$ nC $= -0.5\times10^{-9}$ C. The separation between them is $R = 4$ cm $= 0.04$ m.

According to Coulomb's law,

$$F_1 = \frac{Q_1Q_2}{4\pi\varepsilon_0R^2} = \frac{2\times10^{-9}(-0.5\times10^{-9})\times9\times10^{9}}{0.04^2} = -5.62\ \mu\text{N}.$$

The negative sign indicates that the force is attractive. If the two charges are brought into contact, the charge on both spheres becomes their average value.

$$\therefore \quad Q_1 = Q_2 = \frac{2+(-0.5)}{2}\times10^{-9} = 0.75\times10^{-9}.$$

The force between the charges after they are separated by 4 cm is

$$F_2 = \frac{Q_1 Q_2}{4\pi\varepsilon_0 R^2} = \frac{0.75 \times 10^{-9} \times 0.75 \times 10^{-9} \times 9 \times 10^9}{0.04^2} = 3.16\ \mu\text{N}.$$

The positive sign indicates repulsive force.

Example 2.28 Concentrated charges of 0.25 C are located at the vertices of an equilateral triangle of 10 m side. Find the magnitude and direction of the force on one charge due to the other two charges.

Solution Given that charges 0.25 μC are located at the vertices of an equilateral triangle of 10 m side.

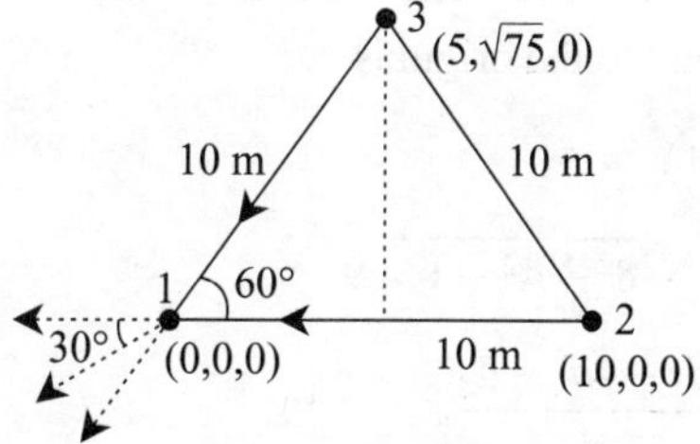

Fig. 2.44 *Point charges on a triangle*

Let the charges be located at $(0,\ 0,\ 0)$, $(10,\ 0,\ 0)$, $(5,\ \sqrt{75},\ 0)$.

The force on one charge due to the other two charges is obtained by using the superposition principle. Therefore the total charge at the origin due to other charges is

$$\vec{F} = \vec{F}_{21} + \vec{F}_{31},$$

where $\vec{F}_{21} = \dfrac{Q^2}{4\pi\varepsilon_0 R_{21}^2}\vec{a}_{21}$ and $\vec{F}_{31} = \dfrac{Q^2}{4\pi\varepsilon_0 R_{31}^2}\vec{a}_{R_{31}}$.

From Fig. 2.44, the distance vectors are

$$\vec{R}_{21} = (0,\ 0,\ 0) - (10,\ 0,\ 0) = -10\vec{a}_x;\ R_{21} = 10;\ \vec{a}_{R_{21}} = \frac{-10\vec{a}_x}{10} = -\vec{a}_x$$

and $\vec{R}_{31} = (0,\ 0,\ 0) - (5,\ \sqrt{75},\ 0) = -5\vec{a}_x - \sqrt{75}\vec{a}_y;\ R_{31} = \sqrt{5^2 + 75} = 10.$

$$\vec{a}_{R_{31}} = \frac{-5\vec{a}_x - \sqrt{75}\vec{a}_y}{10} = -0.5\vec{a}_x - 0.866\vec{a}_y.$$

$$\therefore\quad \vec{F} = \frac{Q^2}{4\pi\varepsilon_0}\left[\frac{\vec{a}_{R_{21}}}{R_{21}^2} + \frac{\vec{a}_{R_{31}}}{R_{31}^2}\right]$$

$$= 0.25^2 \times 10^{-12} \times 9 \times 10^9 \left[\frac{-\vec{a}_x}{100} + \frac{-0.5\vec{a}_x - 0.866\vec{a}_y}{100}\right]$$

$$= \frac{0.56 \times 10^{-3}}{100}\left[-a_x - 0.5\vec{a}_x - 0.866\vec{a}_y\right]$$

$$= 5.6 \times 10^{-6}[-1.5\vec{a}_x - 0.866\vec{a}_y] \text{ N}$$

$$= (-8.4\vec{a}_x - 4.85\vec{a}_y)\ \mu\text{N}.$$

The magnitude is $|F| = \sqrt{84^2 + 485^2} = 9.7\ \mu$N and direction is $\theta = \tan^{-1}\left(\frac{4.85}{8.4}\right) = 30^\circ$ negative of the x-axis.

Example 2.29 A point charge 75 nC is located at $(-8,\ 4,\ 1)$ in the Cartesian coordinate system. Find $\vec{E}$ at the origin.

Solution Given $Q = 75$ nC $= 75 \times 10^{-9}$ C located at $(-8, 4, 1)$.
The distance vector from the point to the origin is

$$R = (0,\ 0,\ 0) - (-8,\ 4,\ 1)$$

$$\vec{R} = 8\vec{a}_x - 4\vec{a}_y - \vec{a}_z;\ \ R = \sqrt{8^2 + 4^2 + 1} = 9.$$

$$\vec{a}_R = \frac{8\vec{a}_x - 4\vec{a}_y - \vec{a}_z}{\sqrt{8^2 + 4^2 + 1^2}} = \frac{8\vec{a}_x - 4\vec{a}_y - \vec{a}_z}{9}.$$

We know that $\vec{E} = \dfrac{Q}{4\pi\varepsilon_0 R^2}\vec{a}_R$

$$= \frac{75 \times 10^{-9}}{4\pi \times 8.854 \times 10^{-12}}\left(\frac{8\vec{a}_x - 4\vec{a}_y - \vec{a}_z}{9^3}\right) = 0.925(8\vec{a}_x - 4\vec{a}_y - \vec{a}_z)$$

$$= 7.4\vec{a}_x - 3.7\vec{a}_y - 0.925\vec{a}_z.$$

Example 2.30 Find the force on a point charge of 70 μC at $(0,\ 0,\ 8)$ m due to a charge of 700 π μC uniformly distributed over the circular disc of radius $r \leq 8$ m, located at $z = 0$ m as shown in Fig. 2.45.

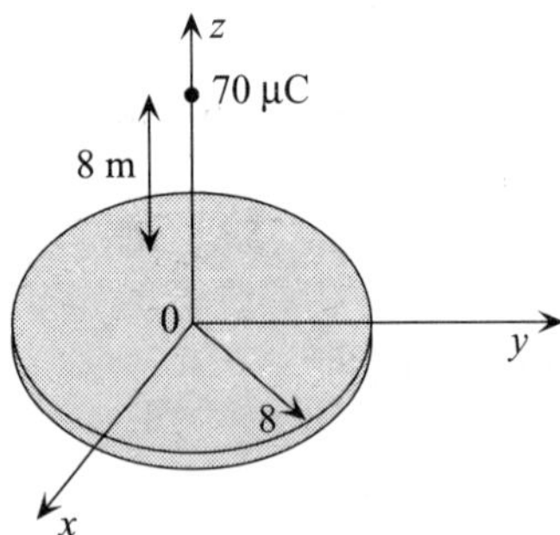

Fig. 2.45 *Force on point charge due to disc charge*

Solution: Given disc of radius $a = 8$ m located at $z = 0$; $Q = 700\pi\ \mu$C.

$$\therefore \quad \text{Charge density } \rho_s = \frac{Q}{A} = \frac{700\pi \times 10^{-6}}{\pi(8^2)} = 10.9\ \mu\text{C/m}^2$$

The point charge is located on the z-axis at $z = 8$ m. We know that the electric field intensity at a point on the z-axis due to a circular disc in the $x-y$ plane is

$$\vec{E} = \frac{\rho_s}{2\varepsilon_0}\left[1 - \frac{z}{\sqrt{a^2+z^2}}\right]\vec{a}_z$$

$$= \frac{10.9\times10^{-6}}{2\times 8854\times10^{-12}}\left[1 - \frac{8}{\sqrt{8^2+8^2}}\right]\vec{a}_z = 0.18\times10^6\vec{a}_z \text{ V/m}.$$

Force on point charge of 70 μC is $\vec{F} = \vec{E}\cdot Q = 0.18\times10^6\times70\times10^{-6} = 12.626\vec{a}_z$ N.

Example 2.31 Find E at (0, 0, 8) due to $Q_1 = 3.5$ μC (0, 7, 0) and $Q_2 = -5.5$ μ C at (6, 0, 0).

Solution Given two point charges as shown in Fig. 2.46.

$Q_1 = 3.5\times10^{-6}$ C at (0, 7, 0) and $Q_2 = -5.5\times10^{-6}$ C at (6, 0, 0).

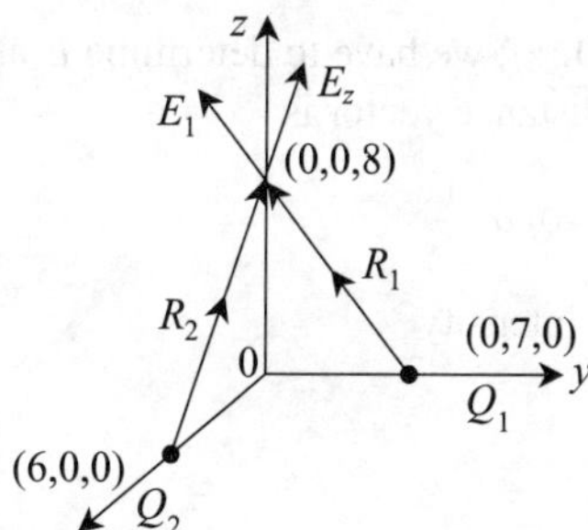

Fig. 2.46 *Field due to point charges*

The distance vectors are

$$\vec{R}_1 = (0,0,8)-(0,7,0); \quad \vec{R}_1 = -7\vec{a}_y + 8\vec{a}_z$$

and $\vec{R}_2 = (0,0,8)-(6,0,0); \quad \vec{R}_2 = -6\vec{a}_x + 8\vec{a}_z.$

Field intensity due to Q_1 is $\vec{E}_1 = \dfrac{Q_1}{4\pi\varepsilon_0 R_1^2}\vec{a}_{R_1} = \dfrac{3.5\times10^{-6}}{4\pi\times8.854\times10^{-12}}\left(\dfrac{-7\vec{a}_y+8\vec{a}_z}{(49+64)^{3/2}}\right)$

$$= -183.4\vec{a}_y + 209.6\vec{a}_z.$$

Field intensity due to Q_2 is $\vec{E}_2 = \dfrac{Q_2}{4\pi\varepsilon_o R_2^2}\vec{a}_{R_2} = \dfrac{-5.5\times10^{-6}}{4\pi\times8.854\times10^{-12}}\left(\dfrac{-6\vec{a}_x+8\vec{a}_z}{(36+64)^{3/2}}\right)$

$$= +296.74\vec{a}_x - 395.66\vec{a}_z.$$

Total electric field at $(0,0,8)$ is $\vec{E} = \vec{E}_1 + \vec{E}_2$

$$= -183.4\vec{a}_y + 209.6\vec{a}_z + 296.74\vec{a}_x - 395.66\vec{a}_z$$

$$= 296.74\vec{a}_x - 183.4\vec{a}_y - 186.06\vec{a}_z.$$

Example 2.32 As shown in Fig. 2.47, $\rho_l = 40$ nC/m. Determine $\vec{E}$ at point (8, 0, 0) in cylindrical coordinates.

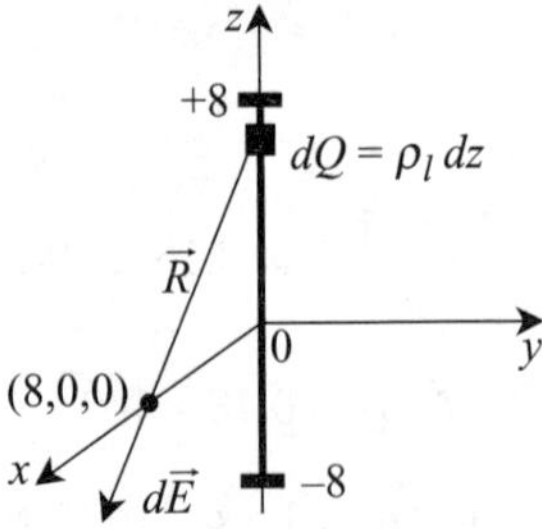

Fig. 2.47 *Field due to line charges*

Solution Given a line charge with charge density $\rho_l = 40 \times 10^{-9}$ C/m, lying on the z-axis between -8 m and 8 m.

Let a differential charge be at $(0,\ 0,\ z)$; we have to determine E at point $(8,\ 0,\ 0)$.

In cylindrical coordinates, the distance vector is

$$\vec{R} = 8\vec{a}_\rho - z\vec{a}_z; \quad \text{and} \quad dQ = \rho_l\, dz.$$

We know that the differential field intensity

$$d\vec{E} = \frac{dQ}{4\pi\varepsilon_o R^2}\vec{a}_R.$$

Then the field intensity $\vec{E} = \int\limits_{z=-8}^{8} \frac{\rho_l dz}{4\pi\varepsilon_o R^2}\vec{a}_R$

$$= \int\limits_{z=-8}^{8} \frac{40 \times 10^{-9} dz}{4\pi \times 8.854 \times 10^{-12}} \frac{8\vec{a}_\rho - z\vec{a}_z}{(8^2 + z^2)^{3/2}}.$$

Since the charge is distributed uniformly along the z-axis, the z components will be cancelled.

$$\therefore \quad \vec{E} = 360 \int\limits_{z=-8}^{8} \frac{8\vec{a}_\rho dz}{(8^2 + z^2)^{3/2}}$$

$$= 2880 \int\limits_{-8}^{8} \frac{dz}{(8^2 + z^2)^{3/2}}\vec{a}_\rho.$$

To evaluate the integral, let $z = 8\ \tan\theta$.

$$dz = 8\sec^2\theta d\theta.$$

The limits are at $z = \pm 8,\ \theta = \pm\pi/4$.

$$\vec{E} = 2880 \int_{-\pi/4}^{\pi/4} \frac{8\sec^2\theta\, d\theta}{(8^2 + 8^2\tan^2\theta)^{3/2}} \vec{a}_\rho$$

$$= 2880 \int_{-\pi/4}^{\pi/4} \frac{8\sec^2\theta\, d\theta}{(8^2)^{3/2}(\sec^2\theta)^{3/2}} = 45 \int_{-\pi/4}^{\pi/4} \cos\theta\, d\theta\, \vec{a}_\rho = 90 \int_{0}^{\pi/4} \cos\theta\, d\theta \vec{a}_\rho$$

$$= 90 \Big[\sin\theta\Big]_0^{\pi/4} \vec{a}_\rho = 63.64 \vec{a}_\rho \text{ V/m.}$$

Example 2.33 A circular ring of charge with radius 2 m lies in the $z = 0$ plane, with the centre at the origin. If it has uniform charge density $\rho_l = 10$ nC/m, find the point charge Q at the origin which would produce the same electric field E at (0, 0, 5).

Solution Given a circular ring of radius $\rho = 2$ m on the $z = 0$ plane as shown in Fig. 2.48. $\rho_l = 10$ nC/m. The electric field intensity due to the circular ring at (0, 0, 5) is calculated as follows.

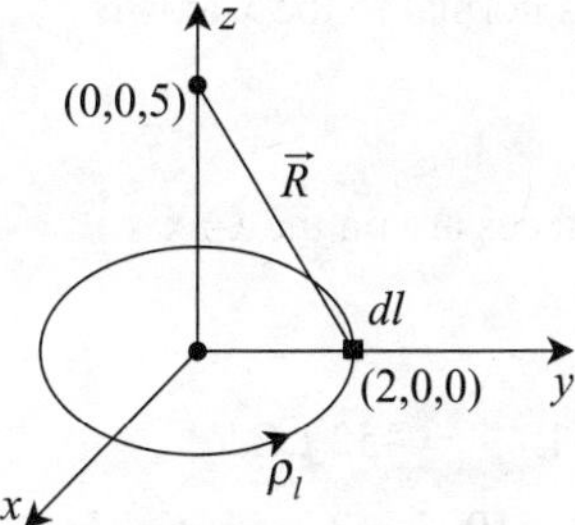

Fig. 2.48 *Field due to a ring*

We know that the differential field intensity $d\vec{E} = \dfrac{dQ}{4\pi\varepsilon_0 R^2}\vec{a}_R$.

Using cylindrical coordinates, distance vector $\vec{R} = (0, 0, 5) - (2, 0, 0)$.

$$\vec{R} = -2\vec{a}_\rho + 5\vec{a}_z,\ dl = 2d\phi;\ R = \sqrt{2^2 + 5^2} = \sqrt{29};\ \vec{a}_R = \frac{-2\vec{a}_\rho + 5\vec{a}_z}{\sqrt{2^2 + 5^2}}.$$

$$\therefore \quad \vec{E} = \int \frac{\rho_l dl}{4\pi\varepsilon_0 R^2}\vec{a}_R$$

$$= \int_{\phi=0}^{2\pi} \frac{10 \times 10^{-9}}{4\pi\varepsilon_0(4+25)} \left(\frac{-2\vec{a}_\rho + 5\vec{a}_z}{\sqrt{4+25}} 2d\phi \right).$$

The radial component $\vec{a}_\rho = 0$ due to symmetry about the z-axis.

$$\therefore \quad \vec{E} = \frac{10 \times 10^{-9}}{4\pi\varepsilon_o(29)} \left(\frac{10\vec{a}_z}{\sqrt{29}} \right)(2\pi) = 36.16\vec{a}_z \text{ V/m.}$$

Electric field intensity due to a point charge at the origin is

$$\vec{E} = \frac{Q}{4\pi\varepsilon_o R^2}\vec{a}_z.$$

If the electric field intensity is the same at $(0,\ 0,\ 5)$, then,

$$36.16\vec{a}_z = \frac{Q}{4\pi\varepsilon_o 8^2}\vec{a}_z \quad \text{or} \quad Q = 257 \text{ nC}.$$

Example 2.34 Given that $D = 500e^{-0.1x}\,\vec{a}_x\ \mu\text{C/m}^2$, find the electric flux crossing the surfaces of area 1 m^2 normal to the x-axis and located at $x = 1$ m, and $x = 5$ m.

Solution Given $\vec{D} = 500e^{-0.1x}\,\vec{a}_x \times 10^{-6}\ \text{C/m}^2$; surface area $S = 1\ \text{m}^2$.
We know that the flux crossing the surface is

$$\psi = \int_s \vec{D}\cdot d\vec{S},$$

where dS = differential surface area normal to the x-axis is

$$dS = dy\,dz\,\vec{a}_x.$$

Assuming that the centre of the surfaces are on the x-axis,

$$\psi = DS.$$

$$\text{At} \quad x = 1\text{ m}, \quad \psi = 500e^{-0.1} \times 10^{-6} = 452\ \mu\text{C}.$$

$$\text{At} \quad x = 5\text{ m}, \quad \psi = 500 \times e^{-0.5} \times 10^{-6} = 303.2\ \mu\text{C}.$$

Example 2.35 Charge in the form of a plane sheet with density $\rho_s = 200\ \mu\text{C/m}$ lies at $z = 0.5$m. A uniform line charge of $\rho_l = -400\ \mu\text{C/m}$ lies along the x-axis. What net flux crosses the surface of a cube of side 2 m, centred at the origin?

Solution Given $\rho_s = 200\ \mu\text{C/m}^2$ at $z = 0.5$ m and $\rho_l = -400\ \mu\text{C/m}$ along the x-axis.

Let the Gaussian surface be a cube of side 2 m, centred at the origin. The surface area $dS = 2^2 = 4\ \text{m}^2$; edge length = 2 m.

We know that the charge enclosed by the sheet charge is

$$\begin{aligned} Q_1 &= \rho_s dS \\ &= 200 \times 10^{-6} \times 4 = 800\ \mu\text{C}. \end{aligned}$$

The charge enclosed by the line is

$$Q_2 = \rho_l l = -400 \times 10^{-6} \times 2 = -800\ \mu\text{C}.$$

$\therefore$ The total charge enclosed by the cube is

$$Q_{\text{encl}} = Q_1 + Q_2 = \psi = (800 - 800)\ \mu\text{C/m}^2 = 0.$$

Example 2.36 Find the electric field at any point between two concentric spherical shells, the inner spherical shell having a charge Q_1 and the outer a charge Q_2.

Solution Given two concentric spherical shells as shown in Fig. 2.49.

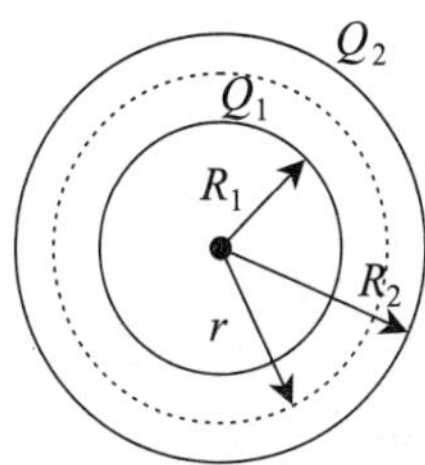

Fig. 2.49 *Two spherical shells*

Let the radius of the inner shell be R_1m and that of the outer shell be R_2m. The charge on the inner shell is Q_1 and the charge on the outer shell is Q_2. Consider a Gaussian surface of radius r inside the shell.

From Gauss' law, we know that,

$$\int_s D \cdot dS = Q_{\text{encl}}.$$

(1) If $r < R_1$, $\quad Q_{\text{encl}} = 0.$

$$\therefore \quad \vec{D} = 0 \text{ and } \vec{E} = 0.$$

(2) For $r = R_1$, $\quad Q_{\text{encl}} = Q_1.$

$$D 4\pi R_1^2 = Q_1.$$

$$\vec{D} = \frac{Q_1}{4\pi R_1^2}\vec{a}_r \quad \text{and} \quad \vec{E} = \frac{Q_1}{4\pi\varepsilon_0 R_1^2}\vec{a}_r.$$

(3) For $R_1 < r < R_2$, $\quad Q_{\text{encl}} = Q_1.$

$$\therefore \quad \vec{E} = \frac{Q_1}{4\pi\varepsilon_0 r^2}\,\vec{a}_r.$$

(4) For $r = R_2$, $\quad Q_{\text{encl}} = Q_1 + Q_2.$

$$\therefore \quad \vec{E} = \frac{Q_1 + Q_2}{4\pi\varepsilon_0 R_2^2}\vec{a}_r.$$

(5) For $r > R_2$, $\quad \vec{E} = \dfrac{Q_1 + Q_2}{4\pi\varepsilon_0 r^2}\vec{a}_r.$

Example 2.37 The flux density $\vec{D} = \frac{r}{3}\vec{a}_r$ nC/m^2 in free space. Find

(i) $\vec{E}$ at $r = 0.2$ m.

(ii) The total electric flux leaving a sphere of $r = 0.2$ m.
(iii) The total charge within a sphere of $r = 0.3$ m.

Solution Given the flux density $\vec{D} = \frac{r}{3}\vec{a}_r$ nC/m^2; for free space, $\varepsilon = \varepsilon_0$.

(i) E at $r = 0.2$ m:

We know that flux density $\vec{D} = \varepsilon_0 E$.

$$\vec{E} = \frac{\vec{D}}{\varepsilon_0} = \frac{r}{3\varepsilon_0} \times 10^{-9}\vec{a}_r$$

$$= \frac{0.2 \times 10^{-9}}{3 \times 8.854 \times 10^{-12}}\vec{a}_r = 7.53\vec{a}_r \text{ V/m.}$$

(ii) Given a sphere of radius $r = 0.2$ m, the flux $\psi = Q_{\text{encl}} = \oint_s \vec{D} \cdot d\vec{S}$.

$$Q_{\text{encl}} = \frac{r}{3}(4\pi r^2) = \frac{4\pi}{3}r^3 \text{ C.}$$

Therefore, the flux leaving the sphere is $\psi = \frac{4\pi}{3} \times 0.2^3 = 0.0335$ nC.

(iii) Total charge within a sphere of radius $r = 0.3$ m is

$$Q_{\text{encl}} = \frac{4\pi}{3} \times 0.3^3 = 0.113 \text{ nC.}$$

Example 2.38 Calculate the work done in moving a point charge of 10 μC from point $(4, 90°, 60°)$ to $(3, 30°, 120°)$ if $V = \frac{10}{r}\cos\theta\sin\phi$ volts.

Solution Given $V = \frac{10}{r}\cos\theta\sin\phi$; $Q = 10\ \mu$C; points $(4,\ 90°, 60°)$ and $(3,\ 30°, 120°)$ in spherical coordinates.

We know that work done is $W = Q(V_2 - V_1)$.

At point (4, 90°, 60°), $V_1 = \frac{10}{4}\cos 90\ \sin 60 = 0$ V

and at point (3, 30°, 120°), $V_2 = \frac{10}{3}\cos 30\ \sin 120 = 2.5$ V.

$\therefore$ Work done $= 10 \times 10^{-6}(2.5 - 0) = 25\ \mu$J.

Example 2.39 Six equal point charges, $Q = 100$ nC, are located at $x = 5, 6, 7, 8, 9$ and 10 metres. Find the potential at the origin.

Solution Given six point charges at $x = 5, 6, 7, 8, 9$ and 10 m, $Q = 100$ nC.

We know that the potential at any point due to several point charges is

$$V = \frac{1}{4\pi\varepsilon_o}\sum_{i=1}^{n}\frac{Q_n}{R_n}.$$

$$= \frac{100\times10^{-9}}{4\pi\times8.854\times10^{-12}}\left(\frac{1}{5}+\frac{1}{6}+\frac{1}{7}+\frac{1}{8}+\frac{1}{9}+\frac{1}{10}\right)$$

$$= 761 \text{ volts.}$$

Example 2.40 Three point charges $-6\ \mu$C, $2\ \mu$C and $4\ \mu$C, are located at $(-1,\ -2,\ -3)$, $(3,\ 2,\ 1)$ and $(1,\ 2,\ 3)$ respectively. Find the potential at $(1,\ 1,\ 1)$. Assume zero potential at infinity.

Solution Given $Q_1 = -6\ \mu C$ at $(-1,\ -2, -3)$; $Q_2 = 2\ \mu C$ at $(3, 2, 1)$; $Q_3 = 4\ \mu C$ at $(1, 2, 3)$
Assume $V(\infty) = 0$.
To find the potential at point $(1, 1, 1)$, compute the distance vectors:

$$\vec{R}_1 = (1+1)\vec{a}_x + (1+2)\vec{a}_y + (1+3)\vec{a}_z = 2\vec{a}_x + 3\vec{a}_y + 4\vec{a}_z;\ R_1 = \sqrt{2^2+3^2+4^2} = \sqrt{29}.$$

$$\vec{R}_2 = (1-3)\vec{a}_x + (1-2)\vec{a}_y + (1-1)\vec{a}_z = -2\vec{a}_x - \vec{a}_y;\ R_2 = \sqrt{4+1} = \sqrt{5}.$$

$$\vec{R}_3 = (1-1)\vec{a}_x + (1-2)\vec{a}_y + (1-3)\vec{a}_z = -\vec{a}_y - 2\vec{a}_z;\ R_3 = \sqrt{1+4} = \sqrt{5}.$$

The potential due to the three point charges is

$$V = V_1 + V_2 + V_3 + V(\infty)$$

$$= \frac{Q_1}{4\pi\varepsilon R_1} + \frac{Q_2}{4\pi\varepsilon R_2} + \frac{Q_3}{4\pi\varepsilon R_3} + V(\infty)$$

$$= \frac{36\pi}{4\pi\times10^{-9}}\left[\frac{-6}{\sqrt{29}} + \frac{2}{\sqrt{5}} + \frac{4}{\sqrt{5}}\right]\times10^{-6} + 0$$

$$V = 9[-1.114 + 0.89 + 1.789]\times10^3 = 14 \text{ kV}.$$

Example 2.41 The electric field in cylindrical coordinates is given by

$$E = \begin{cases} \dfrac{3\vec{a}_r}{r} & \text{V/m}\quad 0 \le r \le 2 \\ 3.5\vec{a}_r & \text{V/m}\quad r \ge 2 \end{cases}$$

Find the potential difference V_{AB} between points $A(1,\ 0,\ 0)$ and $B(4,\ 0,\ 0)$.

Solution Given the points A at $(1,\ 0,\ 0)$ and B at $(4,\ 0,\ 0)$, we have to find the potential difference.

$$V_{AB} = -\int_B^A \vec{E}\cdot d\vec{l}.$$

$$d\vec{l} = dr\,\vec{a}_r.$$

$$V_{AB} = -\int_4^1 \vec{E} \cdot d\vec{l}$$

$$= -\int_4^2 3.5\vec{a}_r \cdot dr\vec{a}_r - \int_2^1 \frac{3\vec{a}_r}{r} \cdot dr\vec{a}_r = -\int_4^2 3.5\, dr - \int_2^1 \frac{3}{r} dr$$

$$V_{AB} = -[3.5r]_4^2 - [3\ln r]_2^1 = -3.5(2-4) - 3\ln 1 + 3\ \ln 2 = 9.08 \text{ volts.}$$

Example 2.42 A 300 nC of charge is uniformly distributed around a circular disc of radius 4 m. Find the potential at a point on the axis 5 m from the plane of the ring.

Solution Given charge $Q = 300 \times 10^{-9}$ C uniformly distributed around a circular disc as shown in Fig. 2.50.

Circular disc area $S = \pi r^2 = \pi \times 4^2 = 16\pi$ m^2.

We know that surface charge density is

$$\rho_s = \frac{Q}{S} = \frac{300 \times 10^{-9}}{16\pi} = 5.97 \times 10^{-9} \text{ C/m}^2.$$

$$V = \int \frac{\rho_s d\vec{S}}{4\pi\varepsilon_o R}$$

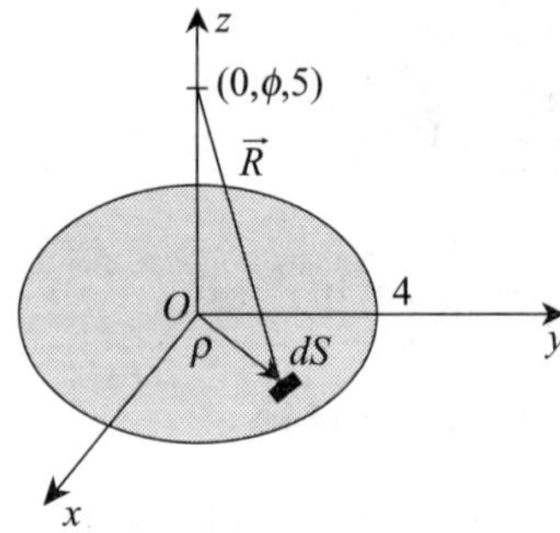

Fig. 2.50 *Field due to a circular disc*

Using cylindrical coordinates,

$$dS = \rho d\rho\ d\phi.$$

Let dS be at $(\rho,\ \phi,\ 0)$ on the disc.

Distance vector to a point 5 m away on the axis is $\vec{R} = (0,\ \phi,\ 5) - (\rho,\ \phi,\ 0)$.

$$\vec{R} = -\rho\vec{a}_\rho + 5\vec{a}_z;\ \ R = \sqrt{\rho^2 + 5^2};$$

$$\therefore \quad V = \int_s \frac{5.97 \times 10^{-9}}{4\pi\varepsilon_0} \left(\frac{\rho d\rho d\phi}{\sqrt{\rho^2 + 25}} \right)$$

$$V = 53.7 \int_{\phi=0}^{2\pi} \int_{r=0}^{4} \frac{\rho d\rho\, d\phi}{\sqrt{\rho^2+25}} = 53.7[\phi]_0^{2\pi} \int_0^4 \frac{\rho d\rho}{\sqrt{\rho^2+25}}$$

$$= 337.13 \int_0^4 \frac{\rho d\rho}{\sqrt{\rho^2+25}}$$

$$V = 337.13\left(\sqrt{\rho^2+25}\right)_0^4 = 337.13\left(\sqrt{4^2+25}-5\right) = 473 \text{ volts.}$$

Example 2.43 Find the electric flux density $\vec{D}$ at $(4, \pi/4, 0)$ if the potential is $V = \frac{15}{r^2}\sin\theta\cos\phi$.

Solution Given $V = \frac{15}{r^2}\sin\theta\cos\phi$, to find the flux density $\vec{D} = \varepsilon_0 E$ at point $(4, \pi/4, 0)$.

We know that $\vec{E} = -\nabla V$. In spherical coordinates,

$$\vec{E} = -\nabla V = -\left[\frac{\partial V}{\partial r}\vec{a}_r + \frac{1}{r}\frac{\partial V}{\partial \theta}\vec{a}_\theta + \frac{1}{r\sin\theta}\frac{\partial V}{\partial \phi}\vec{a}_\phi\right].$$

$$\frac{\partial V}{\partial r} = \frac{-2}{r^3} \times 15\sin\theta\cos\phi = \frac{-30}{r^3}\sin\theta\cos\phi.$$

$$\frac{\partial V}{\partial \theta} = \frac{15}{r^2}\cos\theta\cos\phi.$$

$$\frac{\partial V}{\partial \phi} = \frac{-15}{r^2}\sin\theta\sin\phi.$$

$$\therefore \quad \vec{E} = \frac{30}{r^3}\sin\theta\cos\phi\vec{a}_r - \frac{15}{r^3}\cos\theta\cos\phi\vec{a}_\theta + \frac{15}{r^3}\sin\phi\vec{a}_\phi.$$

Now at point $(4, \pi/4, 0)$, i.e., $r = 4$, $\theta = \pi/4$ and $\phi = 0$,

$$\vec{E} = \frac{30}{4^3}\sin(\pi/4)\cos(0)\vec{a}_r - \frac{15}{4^3}\cos(\pi/4)\cos(0)\vec{a}_\theta + \frac{15}{4^3}\sin(0)\vec{a}_\phi$$

$$= 0.33\vec{a}_r - 0.166\vec{a}_\theta \text{ V/m.}$$

The flux density is $\vec{D} = \varepsilon_0[0.33\vec{a}_r - 0.166\vec{a}_\theta]$

$$= (2.92\vec{a}_r - 1.47\vec{a}_\theta) \text{ pC/m}^2.$$

Example 2.44 The potential for a certain region is given by $V = \frac{300}{x}$ volts, where x is in metres. Find the electric field at point $P(x = 1 \text{ m})$.

Solution Given potential function $V = \frac{300}{x}$.

We know that the electric field intensity

$$\vec{E} = -\nabla V$$

$$= -\frac{\partial V}{\partial x}\vec{a}_x - \frac{\partial V}{\partial y}\vec{a}_y - \frac{\partial V}{\partial z}\vec{a}_z$$

$$= -\frac{\partial}{\partial x}\left(\frac{300}{x}\right)\vec{a}_x = \frac{300}{x^2}\vec{a}_x.$$

Now at $x = 1$ m, $\vec{E} = 300\vec{a}_x$ V/m.

Example 2.45 A dipole at the origin in free space has $\vec{P} = 95\pi\varepsilon_0\vec{a}_z$ C-m. Find (a) V at $P(x, y, z)$ and (b) E at $P(x, y, z)$ in Cartesian coordinates.

Solution Given

$$\vec{P} = 95\pi\varepsilon_0\vec{a}_z \text{ C-m.}$$

(a) We know that the potential at point $P(x, y, z)$ is

$$V = \frac{\vec{P}\cdot\vec{a}_r}{4\pi\varepsilon_0 r^2}.$$

The distance vector $r = x\vec{a}_x + y\vec{a}_y + z\vec{a}_z$

$$\vec{a}_r = \frac{x\vec{a}_x + y\vec{a}_y + z\vec{a}_z}{\sqrt{x^2+y^2+z^2}}$$

$$\therefore \quad V = \frac{95\pi\varepsilon_0\vec{a}_z\cdot(x\vec{a}_x + y\vec{a}_y + z\vec{a}_z)}{4\pi\varepsilon_0(x^2+y^2+z^2)\sqrt{x^2+y^2+z^2}}$$

$$V = \frac{23.75z}{(x^2+y^2+z^2)^{3/2}}$$

(b) The electric field intensity

$$\vec{E} = -\nabla V$$

$$\vec{E} = -\frac{\partial V}{\partial x}\vec{a}_x - \frac{\partial V}{\partial y}\vec{a}_y - \frac{\partial V}{\partial z}\vec{a}_z$$

$$\vec{E} = -23.75\left(\frac{\partial}{\partial x}\left(\frac{z\,\vec{a}_x}{(x^2+y^2+z^2)^{3/2}}\right) + \frac{\partial}{\partial y}\left(\frac{z\,\vec{a}_y}{(x^2+y^2+z^2)^{3/2}}\right) + \frac{\partial}{\partial z}\left(\frac{z\,\vec{a}_z}{(x^2+y^2+z^2)^{3/2}}\right)\right)$$

$$\vec{E} = -23.75\left[\frac{-z\left(\frac{3}{2}\right)(2x)}{(x^2+y^2+z^2)^{2.5}}\vec{a}_x + \frac{-z\left(\frac{3}{2}\right)(2y)}{(x^2+y^2+z^2)^{2.5}}\vec{a}_y + \frac{(x^2+y^2+2z^2)}{(x^2+y^2+z^2)^{2.5}}\vec{a}_z\right]$$

$$\therefore \quad \vec{E} = \frac{71.25xz\vec{a}_x + 71.25yz\vec{a}_y + 23.75(x^2+y^2+2z^2)\vec{a}_z}{(x^2+y^2+z^2)^{2.5}}.$$

Example 2.46 Calculate the potential due to a dipole of dipole moment 45×10^{-10} C/m at a distance 1 m from it (i) on its axis and, (ii) on its perpendicular bisector.

Solution Given dipole moment $P = 45 \times 10^{-10}$ C/m.

(i) Distance on its axis $r = 1$ m; $\theta = 0°$.

We know that the potential due to dipole is

$$V = \frac{P\cos\theta}{4\pi\varepsilon_0 r^2} = \frac{45 \times 10^{-10}}{4\pi\varepsilon_0} = 45 \times 10^{-10} \times 9 \times 10^9$$

$$V = 40.5 \text{ volts}$$

(ii) On its perpendicular bisector

$$\theta = 90°$$

$$V = 0 \text{ volts.}$$

Example 2.47 What is the electric field at (1, 2, 4) due to a pure dipole $10\vec{a}_z$ μCm at (1, −2, 3)?

Solution Given dipole $P = 10\vec{a}_z$ μC m located at (1, −2, 3) and a point at (1, 2, 4).

We know that the electric field due to dipole is

$$\vec{E} = \frac{|P|}{4\pi\varepsilon r^3}(2\cos\theta\vec{a}_r + \sin\theta\vec{a}_\theta) \text{ V/m in spherical coordinates.}$$

The distance vector is $r = (1, 2, 4) - (1, -2, 3)$.

Convert the point in spherical coordinates:

$$r = 4\vec{a}_y + \vec{a}_z$$

$$r = \sqrt{4^2+1} = \sqrt{17}$$

$$\theta = \cos^{-1}\left(\frac{z}{r}\right) = \cos^{-1}\left(\frac{1}{\sqrt{17}}\right) = 76°$$

$$\vec{E} = \frac{10 \times 10^{-6}}{4\pi\varepsilon r^3}(2\cos 76°\vec{a}_r + \sin(76°)\vec{a}_\theta)$$

$$\vec{E} = \frac{10^{-6} \times 9 \times 10^9}{(\sqrt{17})^3}(0.484\vec{a}_r + 0.97\vec{a}_\theta)$$

$$\therefore \quad \vec{E} = 321.46\vec{a}_r + 1245.5\vec{a}_\theta \text{ V/m.}$$

Example 2.48 Point charges of 30 μC and – 30 μC are located at (0, 0, 0.5) and (0, 0, −0.5) m respectively in free space. Find potential at (1, 8, 5), treating the charges as a pure dipole.

Solution Given point charges 30 μC and –30 μC. If the charges are pure dipole, then $Q = 30$ μC.

Separation is: $d = (0,\ 0,\ 0.5) - (0,\ 0,\ -0.5) = 1\vec{a}_z$

The dipole moment $\vec{P} = Q\vec{d} = 30\vec{a}_z$ μC-m.

$$\vec{P} = 30 \times 10^{-6}\vec{a}_z$$

$$\vec{R} = (1,\ 8,\ 5) - (0,\ 0,\ 0)$$

$$\vec{R} = \vec{a}_x + 3\vec{a}_y + 4\vec{a}_z$$

$$R = \sqrt{1+64+16} = 9$$

We know that potential due to dipole is

$$V = \frac{\vec{P}\cdot\vec{a}_R}{4\pi\varepsilon\cdot R^2}$$

$$V = \frac{(30\times10^{-6}\vec{a}_z)\times 9\times10^9(\vec{a}_x + 3\vec{a}_y + 4\vec{a}_z)}{9^3}$$

$$V = 370.37(\vec{a}_z\cdot 4\vec{a}_z)$$

$$V = 1481.8 \text{ volts.}$$

Example 2.49 An electric field $\vec{E} = 3\vec{a}_x + 4\vec{a}_y + 4\vec{a}_z - 12\vec{a}_z$ V/m exists in a dielectric material, which has relative permittivity 3. Obtain the energy density.

Solution Given $\vec{E} = 3\vec{a}_x + 4\vec{a}_y - 12\vec{a}_z$

$$\varepsilon_r = 3$$

$$|E| = \sqrt{3^2+4^2+12^2} = \sqrt{169}$$

We know that $w_e = \dfrac{1}{2}\varepsilon|E|^2$.

$$w_e = \frac{1}{2}\times 8.854\times10^{-12}\times 3\times 169$$

$$w_e = 2.244 \text{ nJ/m}^3$$

Example 2.50 Given the potential function $V = 3x^2 + 4y^2$. Find the energy stored in the volume described by $0 \le x \le 1$, $0 \le y \le 1$ and $0 \le z \le 1$ m.

Solution Given $V = 3x^2 + 4y^2$

$$\vec{E} = -\nabla V = -\left[\frac{\partial V}{\partial x}\vec{a}_x + \frac{\partial V}{\partial y}\vec{a}_y + \frac{\partial V}{\partial z}\vec{a}_z\right]$$

$$\vec{E} = -6x\vec{a}_x - 8y\vec{a}_y$$

Energy density $w_e = \frac{1}{2}\vec{D}\vec{E} = \frac{\varepsilon_0}{2}E^2 = \frac{\varepsilon_0}{2}[36x^2 + 64y^2]$ C/m^2

Energy stored $W_e = \int\limits_v w_e dv$

$$= \frac{\varepsilon_0}{2}\int_0^1\int_0^1\int_0^1 (36x^2 + 64y^2)dxdydz = \frac{\varepsilon_0}{2}\left(\frac{36}{3} + \frac{64}{3}\right)$$

$$W_e = \frac{50}{3}\varepsilon_0 \text{ joules} = 0.147 \text{ nJ}.$$

Summary

1. Coulomb's law in vector form can be expressed as

$$\vec{F} = \frac{Q_1Q_2}{4\pi\varepsilon_0 R_{21}^2}\vec{a}_{R_{21}} \text{ Newtons (N)}$$

where Q_1 and Q_2 are positive or negative quantities of charge in coulombs (C).

R_{12} = magnitude of the radial distance between the charges

$\vec{a}_{R_{12}}$ = unit vector in the radial direction from Q_1 to Q_2

ε = permittivity of the medium.

For N charges, $\vec{F} = \frac{Q}{4\pi\varepsilon}\sum\limits_{n=1}^{N}\frac{Q_n\vec{a}_{R_n}}{R_n^2}$

2. The force exerted on a unit charge in an electric field is called electric field intensity $\vec{E}$ or electric field strength.

$$\vec{E} = \frac{\vec{F}}{Q_2} = \frac{Q_1}{4\pi\varepsilon R_{12}^2}\vec{a}_{R_{12}} \text{ V/m.}$$

3. The total charge for different types of charge distributions is

$$Q = \int dQ = \int\limits_l \rho_l dl = \int\limits_S \rho_S ds = \int\limits_v \rho_v dv \text{ Coulombs.}$$

4. Electric field intensity at point P for different types of charge distributions is

$$\vec{E} = \frac{1}{4\pi\varepsilon}\int\limits_l \frac{\rho_l dl}{R^2}\vec{a}_R = \frac{1}{4\pi\varepsilon}\int\limits_S \frac{\rho_S ds}{R^2}\vec{a}_R = \frac{1}{4\pi\varepsilon}\int\limits_v \frac{\rho_v dv}{R^2}\vec{a}_R \text{ V/m.}$$

5. Electric field intensity at point P due to an infinite line charge is $\vec{E} = \frac{\rho_l\vec{a}_\rho}{2\pi\varepsilon\rho}$ F/m. The field depends only on the radial component ρ, which is the perpendicular distance between the line charge and point P.

6. Electric field intensity due to an infinite sheet of charge is $\vec{E} = \frac{\rho_s}{2\varepsilon}\vec{a}_z$ V/m.
7. The electric flux is numerically equal to the electric charge $\psi = Q$ Coulombs.
8. Gauss' law can be expressed as

$$\text{Flux } \psi = Q_{\text{encl}} = \oint_s \vec{D} \cdot d\vec{S},$$

where $\vec{D}$ = electric flux density.

9. $\vec{D}$ and $\vec{E}$ are related through the permittivity of the medium, $\vec{D} = \varepsilon\vec{E}$.
10. Gauss' law in point form or Maxwell's first equation can be expressed as

$$\nabla \cdot \vec{D} = \rho_v, \text{ or div } \vec{D} = \rho_v$$

11. The Gaussian surface is always a closed surface. The flux density $\vec{D}$ is either normal or tangential to the Gaussian surface.
12. The total work done by an applied force in moving a unit positive charge from point A to point B in electric field $\vec{E}$ is given by

$$W = -Q_t \int_A^B \vec{E} \cdot d\vec{l} \text{ joules.}$$

13. Electric potential at a point P due to a fixed charge Q can be expressed as

$$V_p = \frac{W}{Q} = -\int_\infty^P \vec{E} \cdot d\vec{l} \quad \text{J/C or volts.}$$

Electric potential at r due to a point charge Q at r_1 is

$$V = \frac{Q}{4\pi\varepsilon|\vec{r} - \vec{r_1}|} + C$$

where C is potential at reference point. $C = 0$ if potential at infinity is zero

14. Since the field is conservative, the potential around a closed path is equal to zero

that is $V = \oint \vec{E} \cdot d\vec{l} = 0$.

15. The potential at radius r from a point charge at the origin in the electric field is called absolute potential. Assume that the potential at infinity is zero.

The absolute potential is $V = \frac{Q}{4\pi\varepsilon r}$ volts.

The potential difference is defined as difference between potentials at given points A and B i.e,

$$V_{BA} = V_B - V_A$$

16. The potential due to different charge distributions

$$V = \int_{l} \frac{\rho_l dl}{4\pi\varepsilon R} = \int_{S} \frac{\rho_S dS}{4\pi\varepsilon R} = \int_{v} \frac{\rho_v dv}{4\pi\varepsilon R} \text{ volts.}$$

17. Given the potential, the electric field intensity is $E = -\nabla V = -\text{grad V}$.
18. Potential at a distance r due to dipole is

$$V = \frac{Qd\cos\theta}{4\pi\varepsilon r^2} \text{ volts.}$$

19. The potential at a distance $\vec{r}$ due to the dipole moment $\vec{P}$ is $V = \dfrac{\vec{P}\cdot\vec{r}}{4\pi\varepsilon r^3}$ volts.
20. The potential energy stored in the field due to different charge distributions is

$$W_E = \frac{1}{2}\sum_{i=1}^{N} Q_i V_i = \frac{1}{2}\int_{l} V\rho_l dl = \frac{1}{2}\int_{s} V\rho_s dS = \frac{1}{2}\int_{v} V\rho_v dv \text{ joules.}$$

21. The energy density in an electrostatic field is

$$\omega_E = \frac{W_E}{v} = \frac{1}{2}\varepsilon E^2 = \frac{1}{2}\vec{D}\vec{E} = \frac{1}{2\varepsilon}D^2 \text{ J/m}^3.$$

Review Questions

1. State and explain Coulomb's law of electrostatic field in vector form. *(April 2011, 2008, 2012)*
2. State Coulomb's law of force between any two point charges and state the units of quantities. *(May 2008, 2010, 2011, Aug 2014)*
3. Derive the concept of electric field intensity from Coulomb's law. *(Nov 2008, May 2012, Jan 2014)*
4. Obtain an expression for the total force experienced by a point charge due to an infinite number to point charges around it. *(May 2011)*
5. Obtain an expression for electrical field intensity at a point due to an infinite number of point charges. *(Nov 2010)*
6. Derive an expression for the electric field intensity due to an infinite length line charge along the z-axis at an arbitrary point $Q(x, y, z)$. *(Nov 2008, 2011, April 2013, Aug 2014)*
7. A line charge density is uniformly distributed over a length $2a$ with centre as origin, along the x-axis. Find E at a point P on the z-axis at a distance d. *(Nov 2010)*
8. Find the electric field at any point between the two concentric spherical shells, the inner spherical shell having a charge Q_1 and outer spherical shell a charge Q_2. *(May 2010)*
9. State and explain Gauss' law. *(Nov 2009, 2010, Aug 2014)*
10. Derive Maxwell's first equation as applied to the electrostatics using Gauss' law. *(May 2010, 2012)*
11. Derive the expression for D due to an infinite sheet of charge placed in the $z = 0$ plane using Gauss' law. *(May 2011, Jan 2014)*

12. Explain Gauss' law applied to the case of infinite line charge and derive the expression for D due to the infinite line charge. *(May 2009, 2011)*
13. It is required to hold four equal point charges, each in equilibrium at the corners of a square. Find the point charge which will do this if placed at the centroid of the square. *(May 2011)*
14. Prove that electric field intensity is equal to the negative gradient of the potential, i.e., $E = -\nabla V$, where V is the potential. *(May 2009, 2011)*
15. Define and explain the following terms.

    (a) electric flux density D (b) electric field intensity E

    (c) electrostatic energy (d) electric potential V *(May 2012)*
16. Show that the electric field intensity at any point inside a hollow charged spherical conductor is zero.
17. Deduce Gauss' law from Coulomb's law. *(Nov 2011, Nov 2013)*
18. Derive the expressions for potential and electric field intensity due to an electric dipole. *(May 2011, Nov 2013)*
19. Derive the expression for the r, θ and ϕ components of the electric field intensity at a point $P(r,\ \theta,\ \phi)$ due to an electric dipole. *(May 2011, May 2013)*
20. Show that the torque on a physical dipole $\vec{P}$ in a uniform electric field $\vec{E}$ is given by $\vec{P} \times \vec{E}$. Extend this result to a pure dipole. *(May 2012)*
21. Derive the expression for electric potential due to a dipole. *(May 2011, 2012)*
22. Show that the electric field intensity due to an electric dipole represents a conservative field (show that $\nabla \times \nabla V = 0$). *(May 2012)*
23. Derive the expression for energy stored in an electric field due to N point charges.
24. Show that the energy density in electrostatic field is

$$w_e = \frac{1}{2}\varepsilon E^2 = \frac{1}{2}\vec{D}\cdot\vec{E} = \frac{1}{2\varepsilon}D^2 \text{ J/m}^3.$$

(Nov 2010)

Problems

1. Point charges 1 mC and -2 mC are located at $(3,\ 2,\ -1)$ and $(-1,\ -1,\ 4)$, respectively. Calculate the electric force on a 10 nC charge located at $(0,\ 3,\ 1)$.
2. Point charges 5 nC and -2 nC are located at $(2,\ 0, 4)$ and $(-3,\ 0,\ 5)$, respectively. (a) Determine the force on a 1 nC point charge located at $(1,\ -3,\ 7)$. (b) Find the electric field E at $(1,\ -3,\ 7)$.
3. Planes $x = 2$ and $y = -3$, respectively, carry charges $10\ \text{nC/m}^2$ and $15\ \text{nC/m}^2$. If the line $x = 0$, $z = 2$ carries charge 10π nC/m, calculate E at $(1,\ 1, -1)$ due to the three charge distributions.
4. If $D = (2y^2 + z)\vec{a}_x + 4xy\vec{a}_y + x\vec{a}_z\ \text{C/m}^2$, find (a) the volume charge density at $(-1, 0, 3)$, (b) the flux through the cube defined by $0 \le x \le 1, 0 \le y \le 1, 0 \le z \le 1$, and (c) the total charge enclosed by the cube.
5. Two point charges, $-4\ \mu$C and $5\ \mu$C are located at $(2,\ -1,\ 3)$ and $(0,\ 4,\ -2)$, respectively. Find the potential at $(1,\ 0,\ 1)$, assuming zero potential at infinity.
6. A point charge of 5 nC is located at the origin. If $V = 2\ V$ at $(0,\ 6,\ -8)$, find (a) the potential at $A(-3,\ 2,\ 6)$, (b) the potential at $B(1,\ 5,\ 7)$ and (c) the potential difference V_{AB}.

7. Point charges of 50 nC each are located at $A(1,\ 0,\ 0)$, $B(-1,\ 0,\ 0)$, $C(0,\ -1,\ 0)$, and $D(0,\ 1,\ 0)$ in free space. Find the total force on the charge at A.
8. Let a point charge $Q_1 = 25$ nC be located at $P_1(4,\ -2,\ 7)$ and a charge $Q_2 = 60$ nC be at $P_2(-3,\ 4,\ -2)$. Find (a) $\vec{E}$ at $P_3(1,\ 2,\ 3)$. (b) At what point on the y-axis is $\vec{E}_x = 0$?
9. Given point charges of -1μC at $P_1(0,0,0.5)$ and $P_2(0,0,-0.5)$, and a charge of 2 μC at the origin, find $\vec{E}$ at $P(0,\ 2,\ 1)$ in spherical coordinates.
10. A spherical volume having a 2 μm radius contains a uniform volume charge density of 10^{15} C/m^3. What total charge is enclosed in the spherical volume?
11. Uniform line charges of 0.4 μC/m and -0.4μC/m are located in the $x = 0$ plane at $y = -0.6$ and $y = 0.6$ m respectively. Find E at $P(x,\ 0,\ z)$.
12. Given the surface charge density, $\rho_s = 2\ \mu$C/m^2 in the region $\rho < 0.2$ m, $z = 0$, and is zero elsewhere, find E at $P(\rho = 0,\ z = 0.5)$.
13. The cylindrical surfaces $\rho = 1$, 2, and 3 cm carry uniform surface charge densities of 20, -8, and 5 nC/m^2, respectively. How much electric flux passes through the closed surface $\rho = 5$ cm, $0 < z < 1$m?
14. Two point charges, 1 nC at $(0,\ 0,\ 0.1)$ and -1 nC at $(0,\ 0,\ -0.1)$ are in free space. Calculate the potential at $P(0.3,\ 0,\ 0.4)$.
15. Two concentrated charges, $Q_1 = 3\mu$C and $Q_2 = 3\mu$C are located 0.8 cm apart in free space. Find the electric field intensity at a point 0.6 m from both charges.
16. Four charges each of 5 μC are located in free space at the corners of a rectangle of 8m $\times$ 6m. Q_1 is located at $(0,\ 0,\ 0)$. The other charges are located as follows. Q_2 at $(8,\ 0,\ 0)$, Q_3 at $(8, 6, 0)$ and Q_4 at $(0,\ 6,\ 0)$. Determine the electric potential at (a) $(4, 3, 0)$ and (b) $(0,\ 3,\ 0)$.
17. A uniform line charge $\rho_l = 10$ nC/m lies in the $z = 0$ plane parallel to the x -axis at $y = 1$ m. Find the potential difference V_{AB} for the points $A(2,\ 0,\ 4)$ and $B(0,\ 0,\ 0)$.
18. Find the total charge in the specified volume of a sphere of radius 2 m and centred at the origin if $\rho_v = 5(1 + \cos\theta)\sin(\phi/2)$ C/m^3.
19. Three point charges each 5×10^{-9} C are located on the x-axis at $(-1,\ 0,\ 0)$, $(0,\ 0,\ 0)$ and $(1,\ 0,\ 0)$ in free space. Find (a) electric field intensity at $(5,\ 0,\ 0)$ and (b) determine the value and location of the equivalent single point charge which would produce the same field at very large distances.
20. Show that the electric potential at a distance d from the centre of the ring radius r and situated on the axis is $\dfrac{Q}{4\pi\varepsilon_0}(\sqrt{r^2 + d^2})$, where Q is the total charge on the ring.
21. A hemi-spherical surface is uniformly charged with a surface density ρ_s C/m^2. Show that the electric field intensity at the centre of the sphere is $\rho_s\vec{a}_y/4\varepsilon_0$.
22. Given the potential function $V = v_0(x+y)(x+z)$, find $\vec{E}$, ρ_v and $\vec{D}$ at $x = y = z = 1$.
23. A uniform charge density 2 C/m^2 exist in the volume 2 m $\leq x \leq$ 4 m. Using Gauss' law, find $\vec{D}$ in all the regions.
24. A dipole having a moment $\vec{P} = 3\vec{a}_x - 5\vec{a}_y + 10\vec{a}_z$ nC-m is located at $Q(1,\ 2,\ -4)$ in free space. Find V at $P(2,\ 3,\ 4)$.
25. A dipole having $Qd/(4\pi\varepsilon_0) = 100$ V/m^2 is located at the origin in free space and aligned so that its moment is in the $\vec{a}_z$ direction. Find V and $\vec{E}$.

26. Two point charges, 1 nC at (0, 0, 0.1) and −1 nC at (0, 0, −0.1), are in free space. (a) Calculate V at $P(0.3,\ 0,\ 0.4)$; (b) calculate $|E|$ at P. (c) Now treat the two charges as a dipole at the origin and find V at P.
27. A dipole, having a moment $\vec{P} = 2\vec{a}_z$ nC m, is located at the origin in free space. Give the magnitude of E and its direction in Cartesian components at $r = 100$ m, $\phi = 90°$, and θ as: (a) 0°; (b) 30°; (c) 90°.
28. A dipole with $\vec{P} = 0.1\vec{a}_z$ μCm is located at $A(1,\ 0,\ 0)$ in free space and the $x = 0$ plane is perfectly conducting. Find V at $P(2,\ 0,\ 1)$.
29. Find the energy stored in a system of three equal point charge ($Q = 20$ nC) arranged in a line with 0.5 m separation between them.
30. Given a potential function $V = 3x^2 + 4y^2$ volts, find the energy stored in the cube of side 1 m with one corner at the origin.
31. An electric field is given by $\vec{E} = \left(\frac{x}{2}\right)\vec{a}_x - 2x\vec{a}_y$ V/m. Find the work done in moving a point charge $Q = -20$ nC from the origin to (4, 0, 0).
32. If $E = 120a_\rho$ V/m, find the incremental amount of work done in moving a 50 μC charge a distance of 2 mm from a point $P(1,\ 2,\ 3)$ toward $Q(2,\ 1,\ 4)$.

Answers

1. $\vec{F} = -6.507\vec{a}_x + 3.817\vec{a}_y + 7.506\vec{a}_z$ mN
2. (a) $-1.004\vec{a}_x - 1.284\vec{a}_y + 1.4\vec{a}_z$ nN, , (b) $-1.004\vec{a}_x - 1.284\vec{a}_y + 1.4\vec{a}_z$ V/m.
3. $\vec{E} = -162\pi\vec{a}_x + 270\pi\vec{a}_y - 54\pi\vec{a}_z$ V/m   4. (a) −4 C/m^3; (b) 2 C; (c) 2 C
5. −5.872 kV   6. (a) 3.929 V (b) 2.696 V (c) −1.233 V.   7. $21.5\vec{a}_x$ μN
8. (a) $4.58\vec{a}_x - 0.15\vec{a}_y + 5.51\vec{a}_z$; (b) $y = -6.89,\ -2.21$ 9. $E_r = 160.9$, $E_\theta = -120.5$; $E_\phi = 0$
10. 3.35×10^{-2} C   11. $-625.8\vec{a}_x - 241.6\vec{a}_y$ V/m   12. 8. 1 kV/m   13. 1. 2 nC
14. 5. 78 V   15. 10000 V/m acts normal to the line joining the charges.
16. 36×10^3 V; 45.45×10^3 V   17. 273.52 V   18. 360 C
19. 5.85 V/m; 5×10^{-9} C
22. (a) $-2v_0(2\vec{a}_x + \vec{a}_y + \vec{a}_z)$; (b) $-2v_0\varepsilon_0$; (c) $-2v_0\varepsilon_0(2\vec{a}_x + \vec{a}_y + \vec{a}_z)$
23. $D = \begin{cases} -2\vec{a}_x & x < 2 \\ 2(x-3)\vec{a}_x & 2 \le x \le 4 \\ 2\vec{a}_x & x > 4 \end{cases}$ 24. 1.31 V
25. $V = |100\cos\theta|$ and $E = 100(1 + 3\cos^2\theta)^{1/2}$
26. (a) 5.78 V (b) $|E| = 25.2$ V/m, (c) V = 5.76 V
27. (a) 36.0 μV/m (b) 32.5 μV/m (c) 18.0 μV/m
28. 289.5 V   29. 18×10^{-6} J   30. 147 pJ   31. 80 μJ
32. 3.1 μJ

Multiple-choice Questions

1. If the force on a stationary charge Q is F then the electric field intensity E is
   (a) F/Q (b) FQ (c) F/Q^2 (d) P/Q.

2. Force on a stationary or moving electric charge Q in electric field E is given by
   (a) E/Q^2 (b) EQ (c) E/Q (d) E^2/Q

3. If $E = 167\vec{a}_z$ is given, the $\vec{E}$ in spherical coordinate system is
   (a) $\vec{E} = 167a_r$ (b) $\vec{E} = 167a_\phi$ (c) $\vec{E} = 167a_\theta$ (d) $\vec{E} = 167a_z$

4. Given $\vec{E} = 4\vec{a}_r + 6\vec{a}_\theta + 4\vec{a}_\phi$ at $(1,\ \pi/3,\ \pi/6)$, the component of $\vec{E}$ perpendicular to the surface $\theta = \pi/3$ is
   (a) $2\vec{a}_r$ (b) $6\vec{a}_\theta$ (c) $5\vec{a}_\phi$ (d) $2\ a_r + 5\vec{a}_\phi$

5. Electric field due to single point charge is given by
   (a) $\dfrac{Q_1}{4\pi\varepsilon r^2}$ (b) $\dfrac{Q_1}{4\pi\varepsilon r}$ (c) $\dfrac{Q_1}{4\pi r^2}$ (d) none of these

6. E at a distance of 0.2 meters from a charge of 2 nC in vacuum is
   (a) 25×10^2 kV/m (b) 4.5×10^2 kV/m (c) 2.5×10^2 V/m (d) none of these

7. If $\vec{E} = 3\vec{a}_x + 4\vec{a}_y + 7\vec{a}_z$, then the unit vector in the direction of $\vec{E}$ is
   (a) $\dfrac{3\vec{a}_x + 4\vec{a}_y + 7\vec{a}_z}{\sqrt{74}}$ (b) $\dfrac{3\vec{a}_x + 4\vec{a}_y + 7\vec{a}_z}{\sqrt{14}}$
   (c) $(3\vec{a}_x + 4\vec{a}_y + 7\vec{a}_z)$ (d) $\dfrac{(3\vec{a}_x + 4\vec{a}_y + 7\vec{a}_z)}{30}$

8. The component of the vector $\vec{E} = 6\vec{a}_x + 2\vec{a}_y - 3\vec{a}_z$ along $3\vec{a}_x - 4\vec{a}_y$ is
   (a) 2 (b) 3 (c) 5 (d) 7

9. $\rho_s dS/(4\pi\varepsilon r^2)$ is the expression for
   (a) electric field due to line charge (b) electric field due to surface charge
   (c) electric field due to volume charge (d) none of the above

10. If the plane $z = 15$m carries a charge 30 c/m^2, then the electric field intensity at the origin is
    (a) $\dfrac{15}{\varepsilon_0}\vec{a}_z$ (b) $\dfrac{-15}{\varepsilon_0}\vec{a}_z$ (c) $\dfrac{30}{\varepsilon_0}\vec{a}_z$ (d) $\dfrac{-30}{\varepsilon_0}\vec{a}_z$

11. If a conducting sphere of radius a carries a charge Q, then the electric field at a distance $r(r > a)$ from the centre of the sphere is
    (a) $\dfrac{Q}{4\pi\varepsilon_0 r^2}\vec{a}_r$ (b) $\dfrac{-Q}{4\pi\varepsilon_0 r^2}\vec{a}_r$ (c) $\dfrac{Q}{4\pi\varepsilon_0 r}\vec{a}_r$ (d) $\dfrac{-Q}{4\pi\varepsilon_0 r}\vec{a}_r$

12. A point charge Q is located at (0, 0, 0) and a line charge of ρ_l C/m is situated at (2, 0, 0) parallel to z-axis. The electric field intensity at (1, 0, 0) is
    (a) $\dfrac{\rho_l}{2\pi\varepsilon_0}\vec{a}_x$ (b) $\dfrac{Q+\rho_l}{4\pi\varepsilon_0}\vec{a}_x$ (c) $\dfrac{\vec{a}_z}{2\pi\varepsilon_0}\left(\dfrac{Q}{2}+\rho_l\right)$ (d) $\left[\dfrac{Q}{4\pi\varepsilon_0} - \dfrac{\rho_l}{2\pi\varepsilon_0}\right]\vec{a}_z$

13. Three point charges, $Q_1 = 30$ nC, $Q_2 = 150$ nC and $Q_3 = -70$ nC are enclosed by sur- face S. The net flux crossing the surface is

(a) -110 nC (b) 110 nC (c) 100 nC (d) none of these

14. The force between two point charges Q_1 and Q_2 which are separated by a distance d in air is F If the charges are placed in a dielectric with $\varepsilon_r = 4$ and other factors remain the same, then the force between the charges is

(a) F (b) $F/2$ (c) $2F$ (d) $F/4$

15. If the relative permittivity of a medium is 2, its dielectric constant is

(a) $1\varepsilon_0$ (b) $2\varepsilon_0$ (c) $4\varepsilon_0$ (d) $3\varepsilon_0$

16. A conducting sphere of charge 2 μC is placed inside a hollow sphere carrying a charge 3 μC. If the two bodies are connected by fine metal wire, then the charges on the outer sphere is

(a) 1μC (b) 2μC (c) 5μC (d) 3 μC

17. Electric flux originates at______ and terminate at_____

(a) $-$ve charge and $-$ve charge (b) $-$ve charge and $+$ve charge
(c) $+$ve charge and $-$ve charge (d) $+$ve charge and $+$ve charge

18. The expression, $\psi = \int \vec{D} \cdot d\vec{S} = Q$, is

(a) Coulomb's law (b) Gauss' law (c) Ohm's law (d) Faraday's law

19. Electric flux ψ is given by

(a) $\psi = D/A$ (b) $\psi = D\,A$ (c) $\psi = D^2\,A$ (d) $\psi = D\,A^2$

20. Voltage around a closed loop is given by

(a) $V = \int(E.dl)$ (b) $V = \int(E.dS)$ (c) $V = \int(E.dV)$ (d) $V = \int(QE.dl)$

21. The electric flux lines emit from a charged conducting body such that the angle between the surface of the conductor and the flux line is

(a) $\pi/2$ (b) $\pi/3$ (c) $\pi/6$ (d) $\pi/4$

22. In an electric field, the maximum value of rate change potential with respect to the distance is called

(a) Field gradient (b) Potential gradient
(c) Divergence potential (d) Divergence of electric field

23. Divergence theorem can be used to convert the

(a) surface into volume integral (b) line into surface integral
(c) line into volume integral (d) volume to surface integral

24. The potential difference of two concentric conducting spheres of radius a and b is given by

(a) $\dfrac{Q}{4\pi\varepsilon_0}\left[\dfrac{b-a}{ab}\right]$ (b) $\dfrac{Q}{4\pi\varepsilon_0}\left[\dfrac{a-b}{ab}\right]$ (c) $\dfrac{Q}{4\varepsilon_0}\left[\dfrac{b-a}{ab}\right]$ (d) $\dfrac{Q}{4\pi}\left[\dfrac{b-a}{ab}\right]$

25. The electric potential has

(a) only magnitude (b) only direction

(c) neither magnitude and direction (d) none of the above

26. If charge Q and electric flux density D are given, then the electric field intensity is

(a) $D = \varepsilon E$ (b) $E = Q/4\pi\varepsilon r$ (c) $E = -\nabla D$ (d) $E = \varepsilon D$

27. Charge is distributed along the z-axis between $z = \pm 5$ m with a uniform density $\rho_L = 20$ nC/m. To find E at (2, 0, 0) m, the integration limits for z are

(a) 0 to 5 (b) 5 to ∞ (c) -5 to $+5$ (d) $-\infty$ to -5

28. Charge is distributed along the z-axis from $z = +5$ m to ∞ and from $z = -5$ m to $-\infty$ with charge density p_L. To find E at a point, the integration limits for z is

(a) $z = 5$ to ∞ and $-\infty$ to -5 (b) $z = -5$ to $+5$ and ∞ to $-\infty$

(c) $z = -\infty$ to $+5$ and 0 to ∞ (d) none of these

29. The charge in the volume defined by $0 \le x \le 1$ m, $0 \le y \le 1$ m and $0 \le z \le 1$ m and charge density $\rho = 90x^2y$ μC is

(a) -5 μC (b) 15 μC (c) 10 μC (d) 11 μC

30. If $\vec{D} = 9\vec{a}_\rho - 8\vec{a}_\phi + 5\vec{a}_z$, at $(1, \pi/2, 0)$ the component of $\vec{D}$ parallel to the surface $\rho = 1$ is

(a) $4\vec{a}_\rho$ (b) $5\vec{a}_z$ (c) $-8\vec{a}_\phi$ (d) $-8\vec{a}_\phi - 5\vec{a}_z$

31. A line charge of 4 C/m is intercepted by a hollow sphere of diameter 3 m. If the line charge is passing through the centre of the sphere, then the amount of normal flux leaving the surface of the sphere is

(a) 4 C (b) 3 C (c) 12 C (d) 24 C

32. A conducting body carries a charge density of ρ_s C/m^2. The electric field intensity just outside the body is

(a) ρ_s (b) $\dfrac{\rho_s}{2\varepsilon_0}$ (c) $\dfrac{\rho_s}{\varepsilon_0}$ (d) $\dfrac{\rho_s}{\varepsilon_0}$

33. A hollow conductor arbitrary shape carries a charge with density ρ_s C/m^2. If a small section of a shell is removed from the hollow body, then the electric field intensity in the hole is

(a) $\rho_s\vec{a}_n$ (b) $\dfrac{\rho_s}{2\varepsilon_0}\vec{a}_n$ (c) $\dfrac{\rho_s}{\varepsilon_0}\vec{a}_n$ (d) $\dfrac{2\rho_s}{\varepsilon_0}\vec{a}_n$

34. The volume charge density associated with the field $\vec{D} = 2xy^2\vec{a}_x + yx^2\vec{a}_y + 2z\vec{a}_z$ at point (1, 1, 0) is

(a) 2 C/m^3 (b) 2 C/m^3 (c) 4 C/m^3 (d) 5 C/m^3

35. Electric flux lines cannot exist

(a) normal to equipotential surface (b) along the surface of the conductor (c) at singular points (d) both (b) and (c) are tme.

36. Electric flux lines and equipotential surfaces intersect at

(a) $\pi/4$ (b) $\pi/2$ (c) $\pi/6$ (d) $\pi/3$

37. The volume charge density in C/m^3 associated with the field $\vec{D} = \rho z^2 \sin^2\phi a_\rho + \rho z^2 \sin\phi\cos\phi a_\theta + \rho z^2 \sin^2\phi a_\phi$ C/m at (1, 1,1) is
(a) 1 (b) 3 (c) 2 (d) 5

38. If the electrostatic field is conservative, which of the following statements is not correct?
(a) $E = -\nabla V$ (b) $\oint \vec{E}\cdot\vec{dl} = 0$
(c) $\nabla\times\vec{E} = 0$ (d) Potential difference between two points is zero.

39. Two concentric cylinders of radius a and b carry charges of ρ_{l_1} C/m and ρ_{l_2} C/m respectively. Presuming $b > a$, the electric field intensity at a radius $r > b$ is
(a) $\frac{(\rho_{l_1}+\rho_{l_2})}{2\pi\varepsilon_0 l}$ (b) $\frac{\rho_{l_1}}{2\pi\varepsilon_0 l}$ (c) $\frac{\rho_{l_2}}{2\pi\varepsilon_0 l}$ (d) $\frac{(\rho_{l_1}-\rho_{l_2})}{2\pi\varepsilon_0 l}$

40. The condition that the electrostatic field is conservative is
(a) $\nabla\cdot\vec{D} = 0$ (b) $\nabla\times\vec{D} = 0$ (c) $\nabla\cdot\vec{F} = 0$ (d) $\nabla^2\vec{D} = 0$

41. The potential along the axis of the dipole is
(a) zero (b) $qd/4\varepsilon r^2$ (c) $qd/2\pi\varepsilon r^2$ (d) $2qd/\pi\varepsilon r^2$

42. For a given dipole, the potential at the point on the plane perpendicular to the middle of the dipole axis is
(a) maximum (b) zero (c) $qd/4\pi\varepsilon r^2$ (d) $qd/2\pi\varepsilon r^2$

43. If two charges with Q coulombs are separated by a distence d, the dipole moment is
(a) $Q\vec{d}$ (b) $\vec{d}/Q$ (c) Q/d (d) $Q^2\vec{d}$

44. The units of dipole moment is
(a) Coulomb-metre (b) Coulomb/m^2 (c) Coulomb/time (d) Farads/m^2

45. Electric field intensity at a point 'r' from the dipole varies as
(a) $1/r$ (b) $1/r^2$ (c) $1/r^3$ (d) none

46. For a given dipole, if the dipole moment is $\vec{P}$ and the electric field intensity is $\vec{E}$, then the torque on the dipole is
(a) $\vec{P}\cdot\vec{E}$ (b) $\vec{P}\vec{E}$ (c) $\vec{P}\times\vec{E}$ (d) $\vec{P}/\vec{E}$

47. The expression for energy density for two charges present in a system is
(a) $\frac{V_1Q_2+V_2Q_1}{2}$ (b) $\frac{(Q_1V_1+Q_2V_2)}{2}$ (c) $\frac{(Q_1V_1)}{2}$ (d) both (b) and (c)

48. The expression for energy density for electrostatic fields is
(a) $\int(D.V)/2$ (b) $\sum\frac{(D.E)}{2}$ (c) $\int\frac{(D^2.E)}{2}$ (d) $\int\frac{(D\times V)}{2}$

49. The expression for the energy density for electrostatic fields
(a) $\int\frac{(D\times E)}{2}$ (b) $\int\frac{(D^2E)}{2}$ (c) $\sum\frac{(\varepsilon E^2)}{2}$ (d) $\sum\frac{(D\times V)}{2}$

50. The energy stored in a volume with volume charge density ρ_v and applied potential V volts is

(a) $\frac{1}{2}\int_v V^2 \rho_v dv$ (b) $\frac{1}{2}\int_v V \rho_v dv$ (c) $\sum \frac{(\rho_v V^2)}{2}$ (d) $\int \frac{(\rho_v \times V)}{2}$

Answers

1. *a*	2. *c*	3. *b*	4. *b*	5. *a*	6. *b*	7. *a*	8. *a*	9. *b*	10. *b*
11. *a*	12. *d*	13. *b*	14. *d*	15. *b*	16. *c*	17. *a*	18. *b*	19. *b*	20. *a*
21. *a*	22. *b*	23. *a*	24. *a*	25. *a*	26. *a*	27. *c*	28. *a*	29. *b*	30. *c*
31. *c*	32. *b*	33. *b*	34. *d*	35. *d*	36. *b*	37. *c*	38. *d*	39. *a*	40. *b*
41. *b*	42. *b*	43. *a*	44. *a*	45. *c*	46. *c*	47. *b*	48. *b*	49. *c*	50. *b*

Electric Fields in Conductors and Dielectrics 3

Learning Objectives

The main objectives of this chapter are to understand thoroughly:

- Conductors, electric current, current density
- Behaviour of conductors in an electric field
- Conductivity, point form of Ohm's law
- Dielectric material, polarization, permittivity
- Continuity equation, relaxation time
- Resistance and power
- Boundary conditions between conductors and dielectrics
- Poisson's and Laplace's equations and its applications
- Capacitance—parallel plate, coaxial, spherical capacitors
- Energy stored in capacitors

3.1 Introduction

In the previous chapter, we saw how Coulomb's law and Gauss' law can be used to determine a static electric field when a known charge distribution is given. We also saw how the electric field can be calculated when the potential is given. In this chapter, we will discuss the characteristics of static electric fields in conductors and dielectric materials. We will also define the concepts of electric current in conductors and polarization in dielectric. We will explain Ohm's law, continuity equation, resistance, power and boundary conditions. Next, we will explore boundary value problems which can be solved using Poisson's or Laplace's equations. These equations are used to find the electric field throughout a region when only boundary values are known. Finally, the concepts of capacitance and energy will be discussed.

3.2 Conductors

Materials can be classified into two categories in terms of their electrical properties: current carrying materials, called conductors, and nonconducting materials, called insulators or dielectrics.

In a crystalline material, the electrons possess energy due to interaction forces between adjacent atoms. The energy levels possessed by electrons are grouped into energy bands at a given temperature. The electrons with low energy levels, called valence electrons, are located in the valence band. The high energy levels where the electrons move freely are called conduction band. For conductors, the valence band merges with the conduction band. There is no energy gap between them. When an external energy is applied, the electrons in the valance band enter the conduction band resulting in an electron flow. Materials such as silver, gold, aluminium, etc., have a very large number of free electrons at room temperature and are therefore good conductors of electricity.

If the forbidden gap, i.e., the gap between the valance band and the conduction band is large, the valance electrons cannot jump into the conduction band with the application of normal external energy. Materials with a large forbidden gap are called insulators. For example, wood, plastic, rubber, etc., do not posses free electrons at room temperature, and when an external energy is applied, electrons cannot move into the conduction band, resulting in zero current. But when a sufficiently large amount of external energy is applied, the valance electrons jump into the conduction band, resulting in the breakdown of the material. Thereafter, the material may not behave as an insulator anymore.

Materials with small energy gap between the valence band and the conduction band are called semiconductors. Only a small amount of external energy is sufficient to move the valence electrons into the conduction band. Semiconductors such as silicon and germanium have a relatively small number of free electrons.

The energy bands of materials are shown in Fig. 3.1. Students can refer to any standard textbook on material physics for additional information on this.

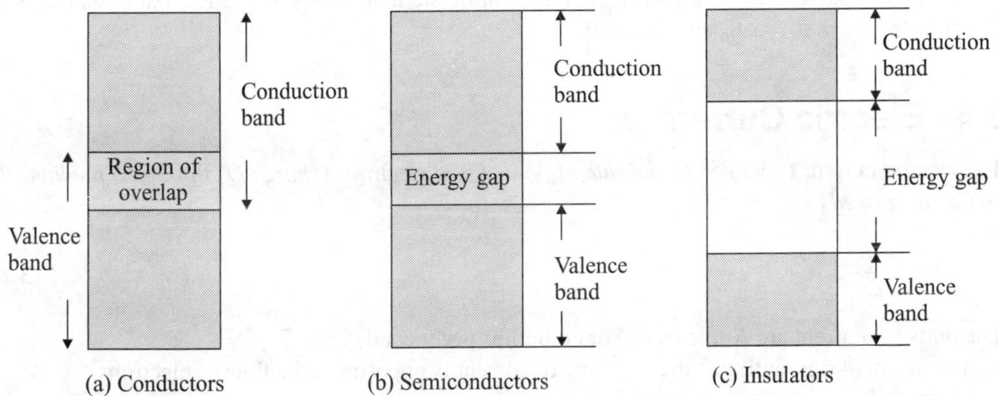

Fig. 3.1 *Energy bands of materials*

3.3 Behaviour of Conductors in an Electric Field

Conductors, such as metals, possess a large number of mobile charges. Under equilibrium condition, the free electrons move randomly within the conductor and constitute a static charge. The net flow of charge is zero. When an external electric field $\vec{E}$ is applied, the positive free charges move along the direction of the field while the negative free charges move in the opposite direction within the conductor. The free charges pile up on the exterior surfaces of the conductor. This charge is called induced surface charge, with surface charge density ρ_s. This induced charge between the surfaces develops an induced field $\vec{E}_i$, which cancels the applied field $\vec{E}$. Therefore, the net electric field inside the conductor is zero. The charges on the surface have equal potential, i.e., the conductor surfaces become equipotential surfaces ($\vec{E} = -\nabla V = 0$). According to the Gauss' law, since the electric field inside the conductor is zero, the charge density ρ_v is also zero under static conditions.

Therefore, under static conditions, for a perfect conductor,

$$\vec{E} = 0, \quad \rho_v = 0, \quad \text{and} \quad V = 0 \text{ inside the material.} \tag{3.1}$$

Properties of conductors under an electric field

1. Electronic charges move on to the surface of the conductor. This results in zero electric field inside the conductor.
2. The volume charge density ρ_v is zero within the conductor.
3. The charge exists only on the conductor surface and is called induced surface charge.
4. The conductor has an equipotential surface, i.e., the potential is the same everywhere on the surface.
5. The electric field intensity at any point on the surface of the conductor is normal to the surface.
6. The potential across the conductor is equal in magnitude to the external applied potential, but opposite in direction.
7. The flow of charge is proportional to the applied potential. The proportional constant is called resistance R.
8. The surface charge density ρ_s and the field on the conductor surface should satisfy the boundary conditions, namely only the normal component of the field exists and tangential component is zero, i.e., $D_n = \rho_s = \varepsilon E_n$ and $E_{\tan} = 0$.

3.4 Electric Current (*I*)

The electric current is defined as *the rate of flow of free electrons (charge Q) through a medium. It can be expressed as*

$$I = \frac{-dQ}{dt}. \tag{3.2}$$

The units for current are Amperes (A) or coulombs per second (C/s).

The negative sign indicates that the flow of current is opposite to the flow of electrons.

When a charge of one coulomb passes across a surface in one second, a current of one ampere flows through that surface.

There are three types of currents.

Conduction current or drift current: *The current flowing due to the flow of free electrons or the drifting of electrons in conductors under the influence of an applied voltage is called conduction current or drift current.*

The direction of conduction current is opposite to the direction of flow of free electrons.

Displacement current: *The current flowing due to the flow of charges in dielectrics under the influence of a time varying electric field is called displacement current.*

For example, the current flowing between the plates of a capacitor (through dielectric material or vacuum) when a time varying potential is applied constitutes displacement current.

Diffusion current: *The current flowing due to the movement of free electrons and free holes to make uniform charge density in semiconductor materials is called diffusion current.*

3.5 Current Density ($\vec{J}$)

The current density in a medium at a given point is defined as *the flow of current per unit area at that point*. The direction is normal to the surface area as shown in Fig. 3.2.

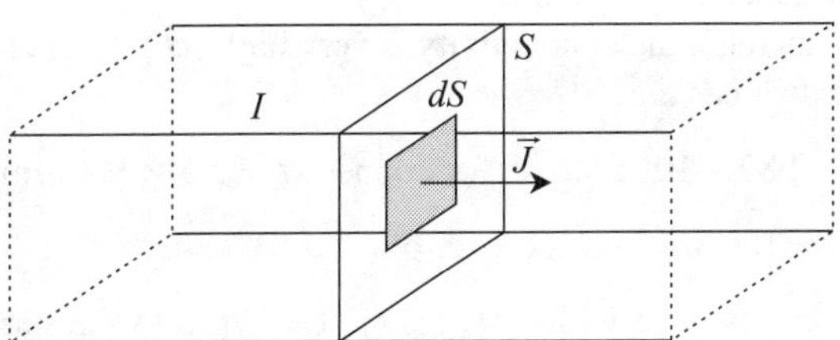

Fig. 3.2 *Current density in a medium*

Consider a differential current dI flowing through a differential surface area $d\vec{S}$.

The current density $\vec{J}$ can be expressed as

$$\vec{J} = \frac{dI}{dS} \text{ A/m}^2. \tag{3.3}$$

The differential current is

$$dI = \vec{J} \cdot d\vec{S}$$

The total current is $$I = \int_S \vec{J} \cdot d\vec{S} \text{ A}. \tag{3.4}$$

If $\vec{J}$ and $d\vec{S}$ are parallel (in the same direction) to each other, then

$$I = \int_S JdS \text{ A}. \tag{3.5}$$

For uniform surface area S, the current is $I = JS$ amperes.

3.6 Ohm's Law and Conductivity

Conductors are materials which allow the flow of electrons through it with very low resistance.

When a potential V is applied across a conductor from an external source, the free charges move and an electric field is created due to the force on the charge carriers. Here, $E \neq 0$ inside the conductor. The charge flow depends on the resistance of the conductor. According to Ohm's law, *flow of current is proportional to the applied potential and its direction is opposite to the direction of flow of electrons*. Mathematically, Ohm's law is stated as

$$V = RI \text{ volts.} \tag{3.6}$$

The proportionality constant R is called the *resistance* of the material, measured in ohms (Ω)

Let the conductor have a length l and uniform cross-sectional area S.

Then the resistance is $$R = \frac{V}{I} = \frac{El}{JS} = \frac{El}{\sigma ES} = \frac{l}{\sigma S} = \frac{\rho l}{S} \text{ ohms,} \tag{3.7}$$

where σ is the *conductivity* expressed in mhos/metre ($\mho/\text{m}$) or siemens/metre (S/m) and $\rho = \dfrac{1}{\sigma}$ is the resistivity of the material expressed in Ω-m. For example, a copper wire of 1.5 mm diameter and length 50 m, offers a resistance of 0.49 ohms.

For good conductors (metals), the conductivity is very high ($\sigma >> 1$) and is constant over weak electric fields. The typical values are:

for aluminium, $\sigma = 3.82 \times 10^7$ S/m, for copper, $\sigma = 5.8 \times 10^7$ S/m

for silver, $\sigma = 6.17 \times 10^7$ S/m and for lead, $\sigma = 5 \times 10^6$ S/m.

For insulators, the conductivity is very low ($\sigma << 1$). The typical values are:

for water $\sigma \approx 10^{-4}$ S/m and for paper $\sigma \approx 10^{-11}$ S/m.

For semiconductors, the conductivity is in between those of metals and insulators. The typical values are:

for silicon $\sigma \approx 10^{-3}$ S/m and for germanium $\sigma = 1.2$ S/m.

For perfect insulators, conductivity is zero. This requires that $J = 0$ and is independent of $\vec{E}$.

For perfect conductors, conductivity is infinite. This requires that $E = 0$ and is independent of J.

Conductivity a depends on temperature when it is very low. Generally, conductivity of metals increases as temperature decreases. The resistivity of some metals abruptly drops to zero and conductivity raises to infinity as temperature decreases to a few degrees kelvin. Such metals are called superconductors.

3.7 Relationship between $\vec{J}$ and ρ_v

Consider a volume charge distribution with charge density ρ_v C/m^3. Let dv be the differential volume across the surface dS at a point at distance dx, as shown in Fig. 3.3. The surface S is normal to the direction of current density $\vec{J}$.

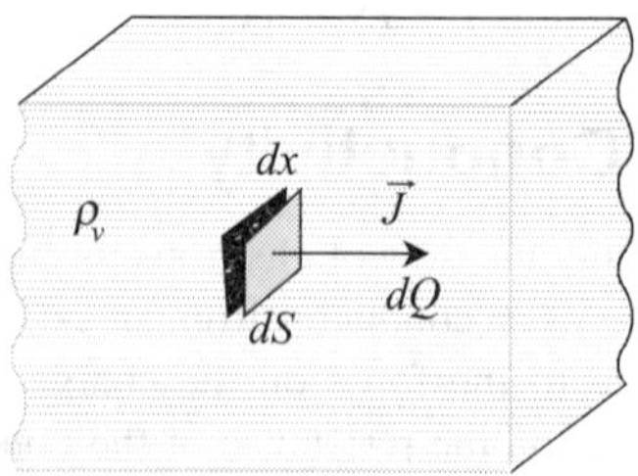

Fig. 3.3 *Volume charge distribution*

The differential charge within the differential volume is

$$dQ = \rho_v dv = \rho_v dS\, dx.$$

The differential current is

$$dI = \frac{dQ}{dt} = \rho_v\, dS \frac{dx}{dt} = \rho_v\, dS\, \vec{v}_d$$

$$\frac{dI}{dS} = \rho_v\, \vec{v}_d.$$

We know that the current density $\vec{J} = \dfrac{dI}{dS}$.

$$\therefore \quad \vec{J} = \rho_v\, \vec{v}_d. \tag{3.8}$$

The current density is linearly proportional to the drift velocity of electrons and the charge density.

3.8 Point Form of Ohm's Law (Relationship between $\vec{J}$ and $\vec{E}$)

Consider free electrons moving in a conductor under an electric field $\vec{E}$. We know that the force on the electrons having charge Q is

$$\vec{F} = Q\vec{E}. \tag{3.9}$$

Due to force $\vec{F}$, the electrons will move with an average velocity called drift velocity, v_d. The drift velocity is linearly proportional to the electric field intensity.

That is,

$$\vec{v}_d \;\alpha\; \vec{E} \quad \text{or} \quad \vec{v}_d = -\mu_e\, \vec{E}, \tag{3.10}$$

where μ_e is called the *mobility* of the electrons within the conductor. The negative sign indicates that the electron velocity is opposite to the direction of the electric field.

The unit of mobility is

$$\mu_e = \frac{-v_d}{E}\,\frac{\text{m/s}}{\text{V/ m}} \text{ or m}^2\text{ /V-s} \tag{3.11}$$

From Eq. (3.9), it is also known that when the current is flowing, the conduction current density J is related linearly to drift velocity as

$$\vec{J} = \rho_e \vec{v}_d \;\; \text{A/m}^2, \tag{3.12}$$

where $\rho_e = ne$ is the electron charge density $=$ no. of electrons per cubic metre $\times$ electric charge $= n \times 1.6 \times 10^{-19}$ C/m^3. It is also called volume charge density, ρ_v.

Substituting the value of v_d from Eq.(3.10) in Eq.(3.12),

$$\vec{J} = -\mu_e\, \rho_e\, \vec{E} = \rho_e(-\mu_e\, \vec{E}) \tag{3.13}$$

$$= -\rho_e \mu_e\, \vec{E} = -ne\mu_e\, \vec{E}. \tag{3.14}$$

The constant $\sigma = -\mu_e\, \rho_e$ is called the conductivity of the material.

$$\therefore \quad \vec{J} = \sigma\vec{E} = -\mu_e\, \rho_e\, \vec{E}\ \ \mathrm{A/m^2}. \tag{3.15}$$

This relationship is called the point form of Ohm's law. Thus conductivity σ is a material property which relates the applied electric field $\vec{E}$ to the electric current density $\vec{J}$ in the material due to free charge carriers.

Example 3.1 A sample of silver has $\sigma = 6.17 \times 10^7$ ℧/m and $\mu_e = 0.056$ m^2/V-s. Determine the electric field intensity $\vec{E}$ if (a) drift velocity $v_d = 1$ mm/s and (b) $J = 10^7$A/m.

Solution (a) Given $\sigma = 6.17 \times 10^7$ ℧/m; $\mu_e = 0.0056$ m^2/V-s; $v_d = 1 \times 10^{-3}$ m/s.

We know that the drift velocity $v_d = -\mu_e\, E$

$$\text{or the field intensity} \quad E = \frac{-\text{Drift velocity } (v_d)}{\text{Mobility } (\mu_e)} = \frac{-1 \times 10^{-3}}{0.0056}\ \mathrm{V/m}$$

$$= -0.1786\ \mathrm{V/m}.$$

(b) Given $J = 10^7$ A/m.
We know that $J = \sigma E$.

$$\therefore \quad E = \frac{J}{\sigma} = \frac{10^7}{6.17 \times 10^7} = 0.1621\ \mathrm{V/m}.$$

Example 3.2 A wire of diameter 1 mm, length 2 m and $\sigma = 5.8 \times 10^7$ S/ m has 10^{30} free electrons per cubic metre, when $E = 20$ mV/ m is applied. Find (a) current density, (b) drift velocity, (c) current in the wire and (d) resistance of the wire.

Solution Given $n = 10^{30}$ per m^3; $E = 20 \times 10^{-3}$ V/ m; $\sigma = 5.8 \times 10^7$ S/m.

We know that electron charge $e = -1.6 \times 10^{-19}$ C.

$\therefore$ Electron charge density $\rho_v = ne = -16 \times 10^{-19} \times 10^{30} = -1.6 \times 10^{11}\ \mathrm{C/m^3}$.

(a) Current density $J = \sigma E$

$$= 5.8 \times 10^7 \times 20 \times 10^{-3} = 116 \times 10^4\ \mathrm{A/m^2}.$$

(b) Also, we know that $J = \rho_v v_d$.

$$\therefore \quad \text{Drift velocity } v_d = \frac{J}{\rho_e} = \frac{116 \times 10^4}{1.6 \times 10^{11}} = 7.25\ \mu\mathrm{m/s}.$$

(c) Current in the wire $I = JS = 116 \times 10^4 \times \frac{\pi}{4}(10^{-3})^2 = 91 \times 10^{-2}\ \text{A} = 0.91\ \text{A}.$

(d) Resistance $R = \frac{l}{\sigma S} = \frac{2}{5.8 \times 10^7 \times \frac{\pi}{4} \times 10^{-6}} = 0.044\ \Omega.$

Voltage across the wire, $V = RI = 0.044 \times 0.91 = 0.04$ volts.

3.9 Dielectric Materials

A dielectric material is a non-conducting material (insulator), in which no free mobile charges exist. For example, plastic, glass, rubber, etc., are dielectric materials. The charges inside the dielectric are held by bonding forces and are not free to move, and they cannot contribute to conduction. In such materials, a large forbidden gap exists between the valence band and the conduction band. The conductivity of these materials is zero, and the resistance is infinite. Under the influence of electric fields, the bounded charges in dielectric materials shift their relative positions infinitesimally against the bonding forces. The positive and negative charges shift in opposite directions, resulting in pairs of positive and negative bounded charges that act like electric dipoles.

3.10 Polarization

Consider a dielectric material with no free electrons. If there is no electric field, the molecules/atoms within the material are in neutral positions since the electrons and photons are bounded. The atoms are said to be electrically neutral, with the positive and negative charges equal in magnitude. The negative charges form a cloud of electrons. No dipole exists in this situation and the atoms are in an unpolarized state.

When an electric field $\vec{E}$ is applied on the dielectric material, the atoms are polarized. The positive charges, under the influence of force $\vec{F} = Q\vec{E}$, shift in one direction, while the negative charges, under the force $\vec{F} = -Q\vec{E}$, shift in the opposite direction. The separation of the positive charge and the negative charge constitute an electric dipole. The dipole is oriented along the direction of the applied electric field. Such atoms are called polarized atoms and this process is called electric polarization.

With respect to polarization, dielectric materials are of two types: polar and non-polar. In polar dielectric materials, the molecules or atoms possess permanent dipoles which are randomly oriented as shown in Fig. 3.4 (a). Each pair of charges acts as a dipole with the same dipole moment. Since the dipoles are randomly oriented, the net dipole moment is zero. Under the influence of an electric field, the dipoles experiences a torque and gets oriented along the direction of the field, as shown in Fig. 3.4 (b). This process is called polarization of polar dielectric materials.

Water and sulphur dioxide are examples of polar dielectric materials.

In non-polar dielectric materials, the positive and negative charges are coincident and there is no separation between them as shown in Fig. 3.4 (c). The net dipole moment is hence zero. The atoms in non polar dielectric materials do not form dipoles until the application of an electric field. Under the influence of an electric field, the charges are separated and they form dipoles with some dipole moment. This process is called polarization of non-polar materials. Hydrogen and oxygen are examples of non-polar type of dielectric materials.

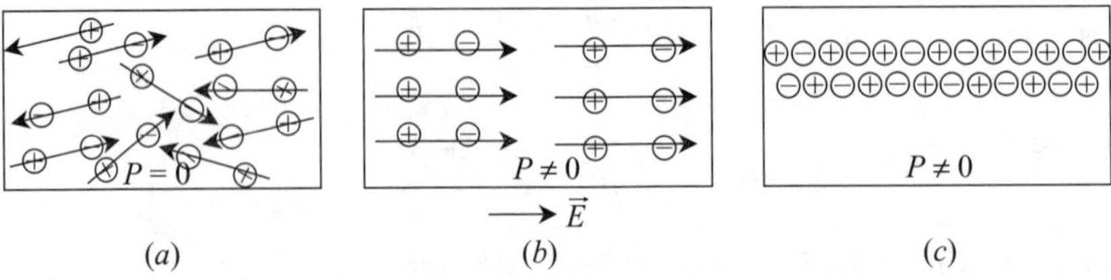

Fig. 3.4 *Polarization of dielectric material: (a) polar with no electric field, (b) polar with electric field and (c) non-polar with electric field*

Dipole moment For a polarized dielectric material, *if the distance between positive charge and negative charge is d, then the dipole moment is the product of charge Q and the distance d*

$$\text{i.e.,} \quad \vec{p} = Q\vec{d}. \tag{3.16}$$

If there are N dipoles in a polarized dielectric, the total dipole moment is

$$\vec{p} = \sum_{i=1}^{N} Q_i \vec{d}_i. \tag{3.17}$$

Polarization For a polarized dielectric material, polarization is defined as the *dipole moment per unit volume* (Δv) *as* Δv *tends to zero.*

$$\text{That is,} \quad \vec{P} = \lim_{\Delta v \to 0} \sum_{i=0}^{N} \frac{Q_i \vec{d}_i}{\Delta v} \ \text{C/m}^2. \tag{3.18}$$

Properties of dielectrics

Linear: A dielectric material is said to be linear if the flux density within the material, $\vec{D}$, varies linearly with $\vec{E}$, that is, the material properties like permittivity should be independent of the applied field. Otherwise, it is non-linear.

Homogeneous: A dielectric material is said to be homogenous if its material properties, permittivity ε and conductivity σ, does not vary from point to point in the material, i.e., ε and σ should be constant throughout the material. For an inhomogeneous or non-homogeneous material, its ε and σ depends on space coordinates.

Isotropic: A dielectric material is said to be isotropic if $\vec{D}$ and $\vec{E}$ are in the same direction; that is, the material properties are independent of direction, otherwise, the material is non-isotropic.

Dielectric constant For a linear and isotropic dielectric medium, polarization vector is proportional to the applied electric field $\vec{E}$,

$$\text{that is,} \quad \vec{P} = \chi_e \varepsilon_0 \vec{E}, \tag{3.19}$$

where χ_e is called the *electric susceptibility* of the dielectric material. It is a dimensionless quantity.

As mentioned before, when an electric field is applied, the dielectric material gets polarized. Due to polarization, the electric flux density increases. Therefore, the electric flux density in a dielectric medium can be expressed as

$$\vec{D} = \varepsilon_0 \vec{E} + \vec{P}. \tag{3.20}$$

Using Eq.(3.19),

$$\vec{D} = \varepsilon_0\,\vec{E} + \chi_e\,\varepsilon_0\,\vec{E} = (1+\chi_e)\varepsilon_0\,\vec{E}.$$

But, we know that $\vec{D} = \varepsilon_r\,\varepsilon_0\,\vec{E}.$ (3.21)

$$\therefore \quad \varepsilon_r = 1+\chi_e = \frac{\varepsilon}{\varepsilon_0}, \tag{3.22}$$

where ε_r is the relative permittivity or dielectric constant of the dielectric material.

Thus the dielectric constant, ε_r, is defined as the ratio of the permittivity of the dielectric material to that of free space.

Dielectric strength (K) When the applied electric field within a dielectric material is sufficiently large, the electrons with high acceleration will come out from the molecules and collide with other atoms. As a result an avalanche effect may occur and a very large number of free electrons will be generated. At this stage, the dielectric material starts conducting. This is known as dielectric breakdown. The maximum electric field that a dielectric material can tolerate without breakdown is called dielectric strength of the material.

If the applied field is more than the dielectric strength of a material, the material becomes conducting and short circuits between the sources. As a result, it destroys the system. For example, lightning in air ($K = 3\times 10^6$ V/m), oil breakdown in transformers ($K \approx 1.5\times 10^7$ V/m), etc.

Note: All dielectric materials have a dielectric strength more than that of air.

Example 3.3 A slab of dielectric material of relative dielectric constant 4 is placed in a uniform electric field of 240 V/m. If the dielectric material is lossless, find (a) electric flux density and (b) polarization.

Solution Given $E = 240\ \text{V/m},\ \varepsilon_r = 4.$

(a) We know that $D = \varepsilon E$

$$= \varepsilon_r\varepsilon_0\,E = 4\times 8.854\times 10^{-12}\times 240 = 8.5\times 10^{-9}\ \text{C/m}^2.$$

(b) Polarization $P = D - \varepsilon_0 E$

$$= \varepsilon_r\,\varepsilon_0\,E - \varepsilon_0 E = \varepsilon_0(\varepsilon_r - 1)E$$

$$= 8.854\times 10^{-12}\times 3\times 240 = 6.37\ \text{nC/m}^2.$$

Example 3.4 The polarization within a region having $\varepsilon_r = 2.7$ has the uniform value of $\vec{P} = -0.2\vec{a}_x + 0.7\vec{a}_y + 0.3\vec{a}_z\ \mu\text{C/m}^2$. Find (i) $\vec{E}$ and (ii) $\vec{D}$.

Solution Given $\varepsilon_r = 2.7$ and polarization $\vec{P} = -0.2\vec{a}_x + 0.7\vec{a}_y + 0.3\vec{a}_z\ \mu\text{C/m}^2.$

(i) We know that $\vec{P} = X_e\varepsilon_0\vec{E}$ but $\chi_e + 1 = \varepsilon_r = 2.7.$

$$\chi_e = 2.7 - 1 = 1.7$$

$$\vec{P} = 1.7\varepsilon_0\,\vec{E}$$

or $\quad \vec{E} = \dfrac{1.0^{-6}}{17\varepsilon_0}\left[-0.2\vec{a}_x + 0.7\vec{a}_y + 0.3\vec{a}_z\right] = -13.3\vec{a}_x + 46.5\vec{a}_y + 20\vec{a}_z \text{ kV/ m}^2.$

(ii) $\vec{D} = \varepsilon_0\varepsilon_r\vec{E} = 2.7\varepsilon_0(-13.3\vec{a}_x + 46.5\vec{a}_y + 20\vec{a}_z) \times 10^3$

$$= -0.318\vec{a}_x + 1.11\vec{a}_y + 0.48\vec{a}_z \ \mu\text{C/m}^2$$

3.11 Electric Displacement Vector in Dielectrics

Consider a non-polar material in which the dipole moment is zero. Let an incremental surface element ΔS be considered within the dielectric material.

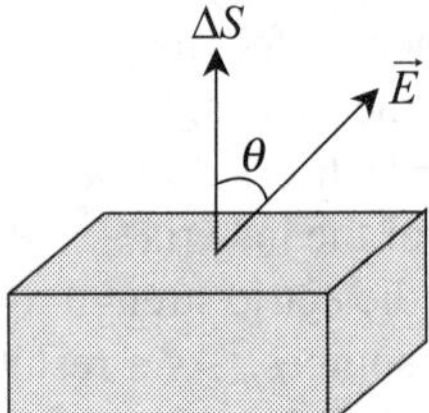

Fig. 3.5 *Field on dielectrics*

When an electric field $\vec{E}$ is applied at an angle θ with ΔS as shown in Fig. 3.5, the electric field produces a dipole moment $\vec{p} = q\vec{d}$ in each molecule and the direction of $\vec{p}$ is aligned with the electric field where d is the separation between charges and q is the electron change. If there are n molecules/m^3, the net bound charge crossing the surface ΔS is

$$\Delta Q_b = nqd \cdot \Delta S$$

or $$\Delta Q_b = \vec{P} \cdot \Delta\vec{S},$$

where $\vec{P} = n\vec{p}$, the polarization vector.

The total bound charge or polarization charge within the closed surface is

$$Q_b = -\oint_S \vec{P} \cdot d\vec{S}.$$

The negative sign indicates the charge is crossing the surface $\Delta\vec{S}$.

If free electrons of charge Q exists within the closed surface, the total charge now becomes $Q_t = Q_b + Q$. According to Gauss' law,

$$Q_{\text{encl}} = \oint_S \vec{D} \cdot d\vec{S} = Q_t$$

$$\therefore \quad Q_b + Q = \oint_S \vec{D} \cdot d\vec{S}$$

$$\text{or} \quad Q_b + Q = \oint_S \varepsilon_0 \vec{E} \cdot d\vec{S}. \tag{3.23}$$

The free charge

$$Q = \oint_S \varepsilon_0 \vec{E} \cdot d\vec{S} - Q_b = \oint_S \varepsilon_0 \vec{E} \cdot d\vec{S} + \oint_S \vec{P} \cdot d\vec{S}$$

$$= \oint_S (\vec{P} + \varepsilon_0 \vec{E}) \cdot d\vec{S}. \tag{3.24}$$

If $\vec{D}$ is the electric displacement vector in dielectrics, then

$$Q = \oint_S \vec{D} \cdot d\vec{S}$$

$$\therefore \quad \vec{D} = \varepsilon_0 \vec{E} + \vec{P}. \tag{3.25}$$

Thus, when an electric field is applied, the dielectric material gets polarized. Due to polarization, the electric flux density increases.

Note: Polarization in free space is $\vec{P} = 0$. So free space is a perfect dielectric.

Volume charge densities If ρ_b is the bounded charge volume density and ρ_v the free charge volume density, then

$$Q_b = \int_v \rho_b dv, \quad Q = \int_v \rho_v dv \quad \text{and} \quad Q_t = \int_v \rho_t \, dv, \tag{3.26}$$

where the total volume charge density is

$$\rho_t = \rho_b + \rho_v. \tag{3.27}$$

According to the divergence theorem,

$$\nabla \cdot \vec{D} = \rho_v$$

$$\nabla \cdot (\varepsilon_0 \vec{E} + \vec{P}) = \rho_v$$

$$\nabla \cdot \vec{P} + \nabla \cdot \varepsilon_0 \vec{E} = \rho_v$$

$$\nabla \cdot \varepsilon_0 \vec{E} = \rho_v - \nabla \cdot \vec{P}.$$

From Eq. (3.27) and since

$$\rho_t = \nabla . \varepsilon_0 \vec{E},$$

and the bounded charge density is

$$\rho_b = -\nabla \cdot \vec{P}. \tag{3.28}$$

In isotropic materials, the vectors $\vec{E}$ and $\vec{P}$ are always parallel. In non-isotropic materials, the dielectric constant depends on charge orientation. It is different for different directions of $\vec{P}$, and $\vec{E}$ may not be parallel. The components of $\vec{D}$ may be a function of respective components of $\vec{E}$.

Therefore, the displacement density along each direction is:

$$\vec{D}_x = \varepsilon_{xx}\,\vec{E}_x + \varepsilon_{xy}\,\vec{E}_y + \varepsilon_{xz}\,\vec{E}_z$$

$$\vec{D}_y = \varepsilon_{yx}\,\vec{E}_x + \varepsilon_{yy}\,\vec{E}_y + \varepsilon_{yz}\,\vec{E}_z$$

$$D_z = \varepsilon_{zx}\,\vec{E}_x + \varepsilon_{zy}\,\vec{E}_y + \varepsilon_{zz}\,\vec{E}_z$$

or
$$\begin{bmatrix} D_x \\ D_y \\ D_z \end{bmatrix} = \begin{bmatrix} \varepsilon_{xx} & \varepsilon_{xy} & \varepsilon_{xz} \\ \varepsilon_{yx} & \varepsilon_{yy} & \varepsilon_{yz} \\ \varepsilon_{zx} & \varepsilon_{zy} & \varepsilon_{zz} \end{bmatrix} \begin{bmatrix} \vec{E}_x \\ \vec{E}_y \\ \vec{E}_z \end{bmatrix}. \tag{3.29}$$

Note: For crystals, only the diagonal terms of the permittivity matrix exists.

Example 3.5 A condenser is composed of two parallel conducting plates separated by a sheet of insulation 2 mm thick and with relative permittivity 4. If external charges of 1 nC/m^2 and $-1\ \text{nC/m}^2$ are placed on the two conducting plates respectively, find the polarization.

Solution Given that $d = 2$ mm $= 2\times10^{-3}$ m, $\varepsilon_r = 4$, $\rho_s = 1\ \text{nC/m}^2$.

The displacement charge density at the boundary is equal to the surface charge density.

$$\therefore\quad D = \rho_s = 10^{-9}\ \text{C/m}^2.$$

We know that $D = \varepsilon_0 E + P$.

$$\therefore\quad \text{Polarization,}\quad P = D - \varepsilon_0 E$$

$$= D - \frac{\varepsilon_0 D}{\varepsilon_r \varepsilon_0} = \rho_s\left(1 - \frac{1}{\varepsilon_r}\right)$$

$$= 1\times10^{-9}\left(1 - \frac{1}{4}\right) = 0.75\ \text{nC/m}^2.$$

Example 3.6 A conducting sphere radius 20 cm is surrounded up to a radius of 1 m by a linear dielectric constant 2. An external charge of 1 nC is placed on the conductor. Find all the bound charge densities.

Solution Given a conducting sphere of radius $r_1 = 20$ cm $= 0.2$ m; radius dielectric material, $r_2 = 1$ m; $\varepsilon_r = 2$; external charge $Q = 1$ nC.

Surface charge density on the conductor is

$$\rho_s = \frac{Q}{4\pi r^2} = \frac{1\times10^{-9}}{4\pi(0.2)^2} = 1.989\times10^{-9}\ \text{C/m}^2.$$

On the surface of the conductor, $\vec{D} = \rho_s\,\vec{a}_r$

The electric field intensity applied on the dielectric material is

$$\vec{E} = \frac{\vec{D}}{\varepsilon} = \frac{\rho_s \vec{a}_r}{\varepsilon_0\,\varepsilon_r}$$

$$= \frac{1.989 \times 10^{-9}}{8.854 \times 10^{-12} \times 2}\vec{a}_r = 112.32\vec{a}_r.$$

The polarization $\vec{P} = \chi_e \varepsilon_0\,\vec{E} = (\varepsilon_r - 1)\varepsilon_0\,\vec{E}$

$$= \varepsilon_0\,\vec{E} = 9.947 \times 10^{-10}\,\vec{a}_r.$$

We know that the bound charge density

$$\rho_b = -\nabla \cdot \vec{P}$$

Since $\vec{P}$ has only a radial component, using the spherical coordinates system,

$$\rho_b = \frac{-1}{r^2 \sin\theta}\frac{\partial}{\partial r}(r^2 P \sin\theta)$$

$$= \frac{-1}{r^2 \sin\theta} 2r(\sin\theta)(9.947 \times 10^{-10})$$

$$= \frac{-2 \times 9.947}{r} \times 10^{-10}.$$

$\therefore$ Bound charge density $\rho_b = -\dfrac{1.98}{r}$ nC/ m^2.

Bound charge densities exist at different values of r.

Example 3.7

(a) Find the maximum charge that can be held on an isolated sphere 2 m in diameter, the sphere being in air with dielectric strength 40 kV/cm.

(b) What would be the maximum charge if this sphere is in oil of $\varepsilon_r = 3.5$ and dielectric strength of 75 kV/cm?

Solution Given diameter of the sphere $d = 2$ m and the field strengths

$$E_1 = 40 \text{ kV/ cm} = 40 \times 10^5 \text{ V/ m in air}$$

and $E_2 = 75 \text{ kV/ cm} = 75 \times 10^5 \text{ V/ m}$ in oil.

We know that for a sphere $E_1 = \dfrac{Q}{4\pi\varepsilon r^2}$.

(a) For air $\varepsilon = \varepsilon_0$.
The maximum charge is $Q = 4\pi\varepsilon_0\, E_1 r^2$

$$= 4\pi \times 8.854 \times 10^{-12} \times 40 \times 10^5 \times 2^2 = 1.78 \text{ mC}.$$

(b) For oil $\varepsilon = \varepsilon_0 \varepsilon_r = 3.5\varepsilon_0$.

$$Q = 4\pi \times 8.854 \times 10^{-12} \times 75 \times 10^5 \times 3.5 \times 2^2 = 11.68 \text{ mC}.$$

3.12 Continuity Equation

The continuity equation states that *the current density diverging from a small volume is equal to the time rate of decrease of charge per unit volume at any given point.*

Mathematically,

$$\nabla \cdot \vec{J} = \frac{-\partial \rho_v}{\partial t}. \tag{3.30}$$

This equation is called the point form of the continuity equation.

Proof Consider an arbitrary volume v bounded by a closed surface S and current density J is flowing through surface area S. Let a differential surface dS be normal to the direction of current density $\vec{J}$ as shown in Fig. 3.6.

The total current crossing the surface S is

$$I = \oint_S \vec{J} \cdot d\vec{S}.$$

The principle of conservation of energy says that *charge can neither be created nor be destroyed, that is, the outward rate of flow of positive charge is equal to the rate of decrease of charge inside the closed surface.*

Let Q be the charge within the closed surface. Then the rate of charge decreasing through the surface is $I = \dfrac{-dQ}{dt}$.

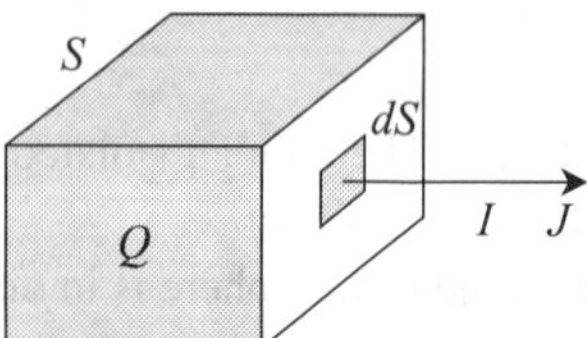

Fig. 3.6 *Current flow through a closed surface*

$$I = \oint_S \vec{J} \cdot d\vec{S} = \frac{-dQ}{dt}.$$

From the divergence theorem, we know that

$$\oint_S \vec{J} \cdot dS = \int_v \nabla \cdot \vec{J} \, dv$$

Also, $Q = \displaystyle\int_v \rho_v \cdot dv.$

$$\therefore \quad I = \int_v \nabla \cdot \vec{J} dv = \frac{-d}{dt} \int_v \rho_v \cdot dv$$

$$\text{or} \quad \int_v (\nabla \cdot \vec{J}) dv = \int_v \frac{-\partial \rho_v}{\partial t} \cdot dv.$$

Eliminating the volume integration on both sides,

$$\nabla \cdot \vec{J} = \frac{-\partial \rho_v}{\partial t}$$

$$\nabla \cdot \vec{J} = \dot{\rho}_v. \qquad \text{(Proved)}$$

Note 1: For steady current, the volume charge density is constant with respect to time.

$$\therefore \quad \nabla \cdot \vec{J} = 0 \tag{3.31}$$

Thus the study currents are solenoidal, that is, they are not diverged. The total charge entering a volume is equal to the charge leaving it.

Note 2: The integral form of the continuity equation states that *the current flow outward from a closed surface is equal to the rate of decrease of charge inside that surface.*

$$I = \oint_S \vec{J} \cdot d\vec{S} = \frac{-dQ}{dt}.$$

3.13 Relaxation Time (T_r)

The charge density through the closed surface within a conductor decreases exponentially with time. The relaxation time T_r is defined as *the time taken by the charge density to decay to 36.8% of its initial value at a given point.*

It is expressed as

$$T_r = \frac{\varepsilon}{\sigma} \text{ seconds,} \tag{3.32}$$

where ε = permittivity of the medium and σ = conductivity of the medium.

The relaxation time depends only on medium properties.

Proof Consider a linear and homogeneous medium with constants σ and ε through which current is flowing with current density $\vec{J}$.

We know that $\vec{J} = \sigma \vec{E} = \sigma \dfrac{\vec{D}}{\varepsilon}$.

From the continuity equation, we know that

$$\nabla \cdot \vec{J} = -\dot{\rho}_v.$$

$$\therefore \quad \nabla \cdot \left(\frac{\sigma}{\varepsilon}\right) \vec{D} = -\dot{\rho}_v$$

or $$\frac{\sigma}{\varepsilon}\left(\nabla \cdot \vec{D}\right) = -\dot{\rho}_v.$$

From Gauss' law, we know that

$$\nabla \cdot \vec{D} = \rho_v.$$

$$\therefore \quad \frac{\sigma}{\varepsilon}\rho_v = -\dot{\rho}_v = \frac{-\partial \rho_v}{\partial t}$$

or $$\frac{\partial \rho_v}{\partial t} + \frac{\sigma}{\varepsilon}\rho_v = 0.$$

This is a first order differential equation, whose solution is

$$\rho_v = Ae^{-\sigma t/\varepsilon}.$$

where A is a constant.

Let the initial value at $t = 0$ be $\rho_v = \rho_{v_0}$ so that

$$A = \rho_{v_0}.$$

$$\therefore \quad \rho_v = \rho_{v_0} e^{-\sigma t/\varepsilon}.$$

The time constant $\varepsilon/\sigma = T_r$ is the relaxation time in seconds.

$$\rho_v = \rho_{v_0} e^{-t/T_r} \tag{3.33}$$

At $$t = T_r,\ \rho_v = \frac{\rho_{v_0}}{e} = 0.368\ \rho_{v_0}. \tag{3.34}$$

Therefore, the relaxation time $T_r = \frac{\varepsilon}{\sigma}$, is the time taken for the charge density to decay to 36.8% or $(1/e)$ times its initial value. This is shown in Fig. 3.7.

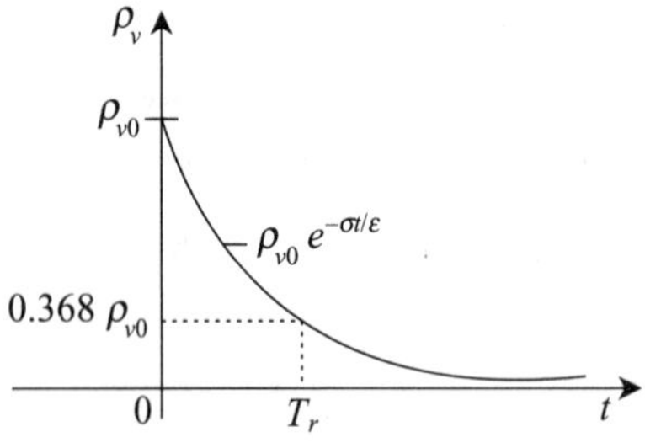

Fig. 3.7 *Charge density and relaxation time*

3.14 Resistance and Power

If a conductor has non-uniform cross-sectional area, Eq. (3.7) is not valid. However from basic definition, the resistance is given by

$$R = \frac{V}{I} = \frac{\int \vec{E} \cdot d\vec{l}}{\vec{J} \cdot d\vec{S}}. \tag{3.35}$$

Power is dissipated through the conductor due to resistance. This is expressed from Joule's law as

$$P = \int_v \vec{E} \cdot \vec{J} \cdot dv \quad \text{watts.} \tag{3.36}$$

and power density is $P = \vec{E} \cdot \vec{J} = \sigma |E|^2$ watts/m^3.

For a conductor having uniform cross-section,

$$P = \int_v \vec{E} \cdot \vec{J}.\ (dl\ dS) = \int_l \vec{E} \cdot d\vec{l} \int_s \vec{J} \cdot d\vec{S}$$

$$P = VI = I^2 R \quad \text{watts.} \tag{3.37}$$

Leakage resistance of a capacitor When the dielectric medium in the capacitor is lossy, a small amount of conduction current will flow between the plates. The expression for resistance and capacitance for a given lossy dielectric medium is

$$R = \frac{V}{I} \quad \text{and} \quad C = \frac{Q}{V}. \tag{3.38}$$

Then the product $RC = \frac{V}{I} \cdot \frac{Q}{V} = \frac{Q}{I}$. Let the medium be homogeneous and the field $\vec{E}$ established in it.

$$RC = \frac{\int \vec{D} \cdot d\vec{S}}{\int \vec{J} \cdot d\vec{S}} = \frac{\int \varepsilon \vec{E} \cdot d\vec{S}}{\int \sigma \vec{E} \cdot d\vec{S}} = \frac{\varepsilon (\int \vec{E} \cdot d\vec{S})}{\sigma (\int \vec{E} \cdot d\vec{S})} = \frac{\varepsilon}{\sigma}$$

$$\therefore \quad RC = \frac{\varepsilon}{\sigma} \quad \text{seconds,} \tag{3.39}$$

which represents the relaxation time of the medium.

The leakage resistance of capacitors can be obtained from the above equation. Consider a capacitor filled with dielectric medium having permittivity ε and conductivity σ. The resistance of the capacitor is the leakage resistance between the plates.

Note 1: The leakage resistance of the parallel plate capacitor is

$$R = \frac{\varepsilon}{C\sigma} = \frac{\varepsilon d}{\varepsilon S \sigma}$$

$$\therefore \quad R = \frac{d}{\sigma S} \quad \text{ohms,} \tag{3.40}$$

where d is the separation between the conductors and S is the surface area.

Note 2: For a single isolated sphere, the leakage resistance is $R = \dfrac{\varepsilon}{C\sigma} = \dfrac{\varepsilon}{4\pi\varepsilon a\sigma}$.

$$\therefore \quad R = \frac{1}{4\pi\sigma a} \text{ ohms,} \tag{3.41}$$

where a is the radius of the sphere.

Note 3: For a coaxial cable, the leakage resistance

$$R = \frac{\varepsilon}{C\sigma} = \frac{\varepsilon}{\dfrac{2\pi\varepsilon L}{\ln(b/a)}\sigma} = \frac{\ln(b/a)}{2\pi\sigma L} \text{ ohms,} \tag{3.42}$$

where b is the outer conductor radius, a is the inner conductor radius and L is the length of the cable.

3.15 Boundary Conditions

Consider an electric field $\vec{E}$ moving from one medium to another. The field conditions which exist at the boundary of the two media are called boundary conditions. The normal and tangential components of electric field intensity $(\vec{E})$ and electric flux density $\vec{D}$ are the important field conditions at the boundary (shown in Fig. 3.8).

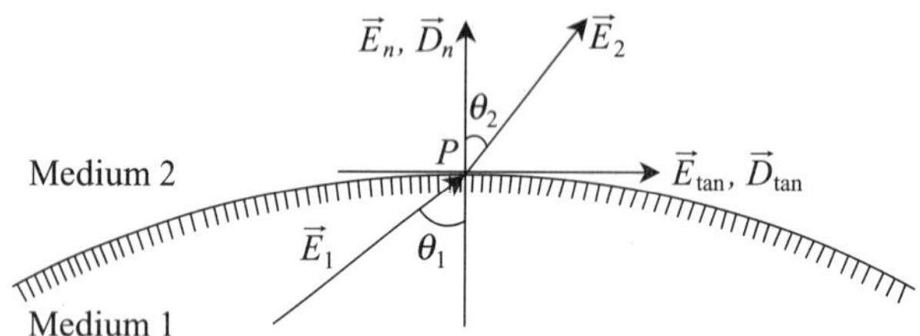

Fig. 3.8 *Boundary conditions for an electric field*

Consider that the field makes an angle θ_1 with the normal to the boundary at point P and is entering from medium 1 to medium 2. The fields normal to the boundary surface are called normal components ($\vec{E}_n$ and $\vec{D}_n$) and the fields tangential to the boundary surface are called tangential components ($\vec{E}_{\tan}$ and $D_{\tan}$) or simply $\vec{E}_t$ and $\vec{D}_t$.

In medium 1, the electric field is $\vec{E}_1 = \vec{E}_{n1} + \vec{E}_{\tan 1}$ and in medium 2, the electric field is $\vec{E}_2 = \vec{E}_{n2} + \vec{E}_{\tan 2}$.

Statement 1 *The tangential component of the electric field intensity $\vec{E}$ is continuous at the boundary. That is, the tangential components of $\vec{E}$ in both media are equal.*

$$\vec{E}_{\tan 1} = \vec{E}_{\tan 2}. \tag{3.43}$$

Statement 2 *If the surface charge density ρ_s exists on the boundary, then the normal components of the electric flux density $\vec{D}$ at the boundary between medium 1 and medium 2 are related as*

$$D_{n1} - D_{n2} = \rho_s. \tag{3.44}$$

For materials other than conductors $(\rho_s = 0)$, the normal components are continuous at the boundary, that is,

$$D_{n1} = D_{n2} \tag{3.45}$$

Proof Consider the boundary between two media where the field is crossing from medium 1 to medium 2.

Condition for electric field intensity $\vec{E}$ Let a small rectangular area of height Δh and width $\Delta\omega$ be considered on the boundary between the two media such that height $\Delta h/2$ lies in each media as shown in Fig. 3.9.

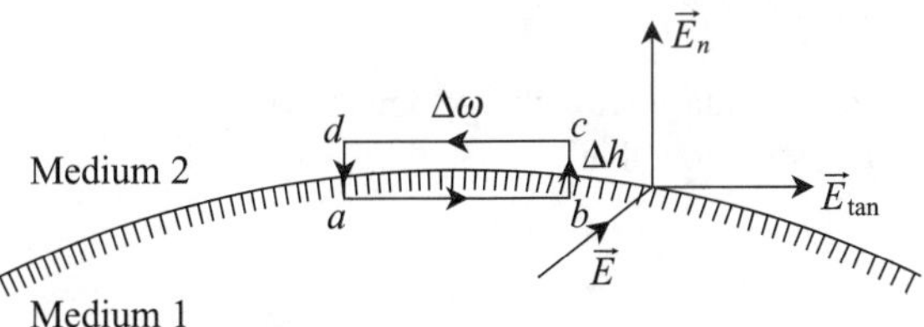

Fig. 3.9 *Boundary conditions of electric field* $\vec{E}$

Since the field is conservative, the line integral of $\vec{E}.d\vec{l}$ around a closed path is zero,

that is, $\oint \vec{E}.d\vec{l} = 0$.

From Fig. 3.9, the closed rectangular path in the counter-clockwise direction is

$$\int_a^b \vec{E}.d\vec{l} + \int_b^c \vec{E}.d\vec{l} + \int_c^d \vec{E}.d\vec{l} + \int_d^a \vec{E}.d\vec{l} = 0$$

$$= \vec{E}_{\tan 1}\Delta\omega + \vec{E}_{n1}\frac{\Delta h}{2} + \vec{E}_{n2}\frac{\Delta h}{2} - \vec{E}_{\tan 2}\,\Delta\omega - \vec{E}_{n2}\frac{\Delta h}{2} - \vec{E}_{n1}\frac{\Delta h}{2} = 0.$$

As $\Delta h \to 0$,

$$\oint \vec{E}.d\vec{l} = \vec{E}_{\tan 1}\Delta\omega - \vec{E}_{\tan 2}\Delta\omega = 0$$

or $\quad \vec{E}_{\tan 1} = \vec{E}_{\tan 2}$ or $\vec{E}_{t1} = \vec{E}_{t2}$. (Proved)

Let $\varepsilon_1 = \varepsilon_{r1}\varepsilon_0$ and $\varepsilon_2 = \varepsilon_{r2}\varepsilon_0$ be the permittivities of the media.

We know that $D_{\tan 1} = \varepsilon_1 E_{\tan 1}$ and $D_{\tan 2} = \varepsilon_2 E_{\tan 2}$.

$$\therefore \quad \frac{D_{\tan 1}}{D_{\tan 2}} = \frac{\varepsilon_1}{\varepsilon_2} = \frac{\varepsilon_{r1}}{\varepsilon_{r2}} \quad \text{or} \quad \frac{D_{t1}}{D_{t2}} = \frac{\varepsilon_{r1}}{\varepsilon_{r2}}.$$

The tangential components of flux density $\vec{D}$ are proportional to the relative permittivities of the two media.

Condition for electric flux density $\vec{D}$ Let a small cylinder of cross-sectional area ΔS with radius r and height Δh be placed across the boundary between the two media such that $\Delta h/2$ lies in each media as shown in Fig. 3.10.

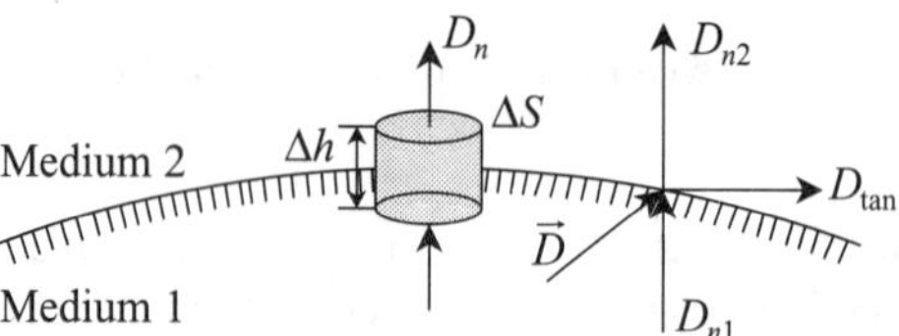

Fig. 3.10 *Boundary conditions for flux density $\vec{D}$*

Let the charge Q exist at the boundary with charge density ρ_s C/m^2.

According to Gauss' law, we know that

$$\oint \vec{D}\cdot d\vec{S} = Q.$$

From Fig. 3.10, the closed surface of the cylinder is

$$\int_{\text{bottom}} \vec{D}\cdot d\vec{S} + \int_{\text{top}} \vec{D}\cdot d\vec{S} + \int_{\text{lateral}} \vec{D}\cdot d\vec{S} = Q$$

$$D_{n1}\Delta S - D_{n2}\Delta S + \vec{D}(2\pi r)\Delta h = Q.$$

As $\Delta h \to 0$ and $Q = \rho_s \Delta S$, the above equation can be written as

$$(D_{n1} - D_{n2})\Delta S = \rho_s \Delta S$$

or $D_{n1} - D_{n2} = \rho_s$. (Proved)

Case 1 Boundary between conductor and dielectric materials

Since no charge exists within a conductor, the fields $\vec{E}, \vec{D}$ and ρ_v inside a conductor are zero. Only the surface charge density ρ_s C/m^2 appears on the surface of the conductor.

$$\therefore \quad E_{\tan 1} = E_{\tan 2} = 0. \tag{3.46}$$

That is, the tangential components of the electric field intensity $\vec{E}$ is zero at the boundary between the conductor and the dielectric.

Since $\vec{D} = \varepsilon \vec{E}$,

$$D_{\tan 1} = D_{\tan 2} = 0. \tag{3.47}$$

Since $\vec{D}$ is also zero inside a conductor, $D_{n2} = 0$.

We know that $D_{n1} - D_{n2} = \rho_s$

$$\therefore \quad D_{n1} = \rho_s. \tag{3.48}$$

The normal component of the electric flux density is equal to the surface charge density.

$\therefore \quad D_n = \rho_s.$

Since $D_n = \varepsilon E_n$,

$$E_n = \frac{\rho_s}{\varepsilon}. \tag{3.49}$$

Case 2 Boundary between two perfect dielectrics

For dielectric materials, fields exist inside the material and there is, no surface charge density.

Therefore, ρ_s on the surface at the boundary is zero, i.e., $\rho_s = 0$.

The boundary conditions are

$$E_{\tan 1} = E_{\tan 2}.$$

Since $D = \varepsilon\, E$, $\dfrac{D_{\tan 1}}{\varepsilon_0\, \varepsilon_{r1}} = \dfrac{D_{\tan 2}}{\varepsilon_0\, \varepsilon_{r2}}$

or $\varepsilon_{r2} D_{\tan 1} = \varepsilon_{r1} D_{\tan 2}$.

Also, $D_{n1} = D_{n2}$

$$\varepsilon_1 E_{n1} = \varepsilon_2 E_{n2}$$

or $\varepsilon_{r1} E_{n1} = \varepsilon_{r2} E_{n2}$

$$\frac{E_{n1}}{E_{n2}} = \frac{\varepsilon_{r2}}{\varepsilon_{r1}} \tag{3.50}$$

Thus the tangential components of $\vec{E}$ and the normal components of $\vec{D}$ are continuous at the boundary.

Note: If θ_1 and θ_2 are the angles that the field makes with the normal to the boundary in medium 1 and medium 2 respectively, and E_1, D_1 and E_2, D_2 are the fields in medium 1 and medium 2 respectively, at the boundary,

$$E_{\tan 1} = E_{\tan 2}, \text{ or } E_1 \sin\theta_1 = E_2 \sin\theta_2$$

and $D_{n1} = D_{n2}$ or $D_1 \cos\theta_1 = D_2 \cos\theta_2$.

$$\therefore \quad \frac{E_1 \sin\theta_1}{D_1 \cos\theta_1} = \frac{E_2 \sin\theta_2}{D_2 \cos\theta_2} \quad \text{or} \quad \frac{\tan\theta_1}{\varepsilon_1} = \frac{\tan\theta_2}{\varepsilon_2} \quad \text{or} \quad \frac{\tan\theta_1}{\tan\theta_2} = \frac{\varepsilon_{r_1}}{\varepsilon_{r_2}}.$$

Case 3 Boundary between conductor and free space

Consider a boundary between a conductor and the free space.

$\vec{E}, \vec{D}$ and ρ_v within the conductor are zero. The charge density ρ_s exists on the surface of the conductor.

$$E_{\tan 1} = E_{\tan 2} = 0.$$

Also, $D_{\tan 1} = \varepsilon_0 E_{\tan 1} = 0.$

Also, $D_{n2} = 0$ inside the conductor.

$$D_{n1} = \rho_s$$

or $$E_{n1} = \frac{\rho_s}{\varepsilon_0}.$$

Example 3.8 An electric field in a medium whose relative permittivity is 7 flows into another medium of relative permittivity 2. If $\vec{E}$ makes an angle of 60° with the boundary normal, what angle does the field makes with the normal in the second dielectric?

Solution Given two dielectric materials with $\varepsilon_{r_1} = 7$ and $\varepsilon_{r_2} = 2; \theta_1 = 60°$.

We know that at the boundary

$$\frac{\tan\theta_1}{\tan\theta_2} = \frac{\varepsilon_{r1}}{\varepsilon_{r2}}$$

$$\tan\theta_2 = \frac{\varepsilon_{r2}}{\varepsilon_{r1}} \times \tan\theta_1 = \frac{2}{7}\tan 60° = 0.4948.$$

$$\therefore \quad \theta_2 = 26.33°.$$

3.16 Poisson's and Laplace's Equations

Here we examine another useful approach for finding electric potentials and fields in free space or in a dielectric medium. The approach is to obtain solution to boundary value problems when the boundary conditions of the medium are given. The fields can be obtained by solving either Laplace's or Poisson's equation. The Poisson's equation states that *in a homogeneous medium, if there exists a volume charge density* ρ_v C/m^3 *at any point, then the potential at that point is given by*

$$\nabla^2 V = -\rho_v/\varepsilon, \tag{3.51}$$

where ∇ is the del operator and ε is the permittivity of the medium.

If the volume charge density is zero (charge-free region of space), and only either point or charge distribution exists in a medium, then

$$\nabla^2 V = 0. \tag{3.52}$$

This relationship is called Laplace's equation.

Proof Consider a homogeneous and linear medium where ε is constant throughout the region in which the field is defined. Let ρ_v be the volume charge density. Then, from the point form of Gauss' law, we know that

$$\nabla \cdot \vec{D} = \rho_v$$

Also, $\vec{D} = \varepsilon\vec{E}$.

Then, $\nabla \cdot \varepsilon \vec{E} = \rho_v$

$$\nabla \cdot \vec{E} = \rho_v/\varepsilon.$$

We also know that $\vec{E} = -\nabla V$.

$$\therefore \quad \nabla \cdot \nabla V = -\rho_v/\varepsilon$$

or $\nabla^2 V = -\rho_v/\varepsilon$.

If $\rho_v = 0$, then $\nabla^2 V = 0$ (Proved)

where ∇^2, called the del square operator, is given in different coordinate systems as:
Cartesian coordinate system:

$$\nabla^2 V = \frac{\partial^2 V}{\partial x^2} + \frac{\partial^2 V}{\partial y^2} + \frac{\partial^2 V}{\partial z^2}. \tag{3.53}$$

Cylindrical coordinate system:

$$\nabla^2 V = \frac{1}{\rho}\frac{\partial}{\partial \rho}\left(\rho \frac{\partial V}{\partial \rho}\right) + \frac{1}{\rho^2}\left(\frac{\partial^2 V}{\partial \phi^2}\right) + \frac{\partial^2 V}{\partial z^2}. \tag{3.54}$$

Spherical coordinate system:

$$\nabla^2 V = \frac{1}{r^2}\frac{\partial}{\partial r}\left(r^2 \frac{\partial V}{\partial r}\right) + \frac{1}{r^2 \sin\theta}\frac{\partial}{\partial \theta}\left(\sin\theta \frac{\partial V}{\partial \theta}\right) + \frac{1}{r^2 \sin^2\theta}\frac{\partial^2 V}{\partial \phi^2}. \tag{3.55}$$

Expressing the above equation in different coordinate systems to take advantage of the symmetry of a charge distribution helps in solving problems related to electric potential. Laplace's and Poisson's equations are useful for solving problems when the potential and charge are known at some boundary and it is required to find the electric field and potential throughout the volume.

Procedure for solving Poisson's or Laplace's equations

1. Solve the given Laplace's equation, $\nabla^2 V = 0$, or the Poisson's equation, $\nabla^2 V = \dfrac{\rho_V}{\varepsilon}$, to obtain the potential by using integration either for a single variable or for multiple variables in a given coordinate system. The solution has two unknown integration constants.
2. Apply boundary conditions to determine the integration constants such that the solution for potential becomes unique.
3. Find the potential by substituting the integration constants.
4. Find the electric field intensity using $\vec{E} = -\nabla V$ and the electric flux density using $\vec{D} = \varepsilon \vec{E}$.
5. The charge exists in a medium can be obtained from

$$Q = \int_s \vec{D} \cdot d\vec{S}$$

6. The capacitance between two conductors can be obtained from $C = Q/V$.

3.17 Uniqueness Theorem

The uniqueness theorem states that *Laplace's or Poisson's equation has only one solution, irrespective of the method adopted.* That is, if there are more than one solution, the solutions must be the same.

Proof Assume that Laplace's equation has two solutions, V_1 and V_2, at the boundary. Then Laplace equations are

$$\nabla^2 V_1 = 0 \text{ and } \nabla^2 V_2 = 0.$$

At the boundary, on the equipotential surface, the potentials at different points are equal.

$$\therefore \quad V_1 = V_2.$$

If a potential difference between V_1 and V_2 exists, then

$$V_d = V_1 - V_2.$$

Using Laplace's equation for potential difference V_d,

$$\nabla^2 V_d = \nabla^2 (V_1 - V_2) = 0.$$

On the boundary $V_d = 0$. Consider the vector identity

$$\nabla \cdot (\alpha A) = \alpha(\nabla A) + A \cdot \nabla \alpha \qquad \text{(Refer to Chapter 1)}$$

for any vector A and a constant α.

Let $\alpha = V_1 - V_2$ and $A = \nabla(V_1 - V_2)$.

Then, $\nabla \cdot (V_1 - V_2)\nabla(V_1 - V_2) = (V_1 - V_2)(\nabla \cdot \nabla(V_1 - V_2)) + \nabla(V_1 - V_2) \cdot \nabla(V_1 - V_2)$

$$\text{or} \quad \nabla \cdot (V_d \nabla V_d) = V_d \nabla^2 V_d + \nabla V_d \cdot \nabla V_d. \tag{3.23}$$

Taking volume integration on both sides,

$$\int_v \nabla \cdot (V_d \nabla V_d) dv = \int_v V_d \nabla^2 V_d dv + \int_v \nabla V_d \cdot \nabla V_d dv.$$

Applying divergence theorem, for any vector $\vec{B}$, we know that,

$$\int_v \nabla \cdot \vec{B} dv = \oint_s \vec{B}.\vec{dS}.$$

If $\vec{B} = V_d \nabla V_d$, then

$$\int_s V_d \nabla V_d \cdot \vec{dS} = \int_v V_d \nabla^2 V_d dv + \int_v \nabla V_d \cdot \nabla V_d dv.$$

At the boundary, we know $\nabla^2 V_d = 0$.

$$\therefore \quad \int_s V_d \nabla V_d \cdot \vec{ds} = \int_v \nabla V_d \cdot \nabla V_d dv.$$

Since V_d is also zero at the boundary,

$$\int_v \nabla V_d \cdot \nabla V_d dv = 0.$$

Also, we know that $\nabla V_d \cdot \nabla V_d = \left|\nabla V_d\right|^2$.

$$\therefore \quad \int_v \left|\nabla V_d\right|^2 dv = 0$$

$$\text{or} \quad \int_v \left|\nabla(V_1 - V_2)\right|^2 dv = 0. \tag{3.56}$$

This integral equation is possible if

1. the quantity in the integration is zero or
2. the quantity has equal and opposite regions.

Since the quantity is always real and non-negative,

$$\left|\nabla(V_1 - V_2)\right|^2 = 0 \quad \text{or} \quad \nabla(V_1 - V_2) = 0.$$

$\therefore \quad V_1 = V_2.$ (Hence proved.)

Applications of Laplace's and Poisson's equations

1. The surface charge density $\vec{D}$ and electric field intensity $\vec{E}$ at the given regions can be obtained directly without using Coulumb's or Gauss's law.
2. By applying boundary conditions on Laplace's and or Poisson's equations we can calculate the capacitance of different structures.

3.18 Electric Field between Two Concentric Conducting Spheres Using Laplace's Equation

Consider two concentric conducting spherical surfaces with radii a and b, $(b > a)$, as shown in Fig. 3.11. The space between the spheres is filled with a homogenous medium of dielectric constant ε.

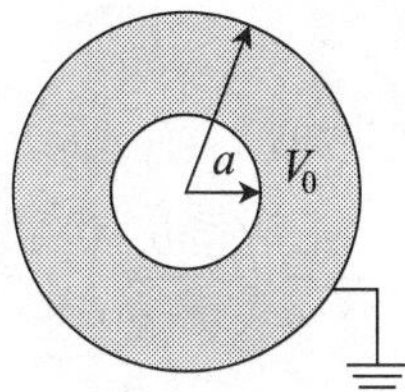

Fig. 3.11 *Spherical capacitor*

The problem involves solving second order partial differential equations for an electric scalar potential subject to known boundary conditions. Here we know that the potential on the inner sphere is, say, V_0. The outer sphere is grounded, i.e., it is at zero volts.

We know that the field intensity $\vec{E}$ for spherical distribution has only a radial component.

According to Laplace's equation for potential,

$$\nabla^2 V = 0.$$

Using spherical coordinates, since the potential distribution does not vary with respect to θ and ϕ,

$$\frac{1}{r^2}\frac{\partial}{\partial r}\left(r^2\frac{\partial V}{\partial r}\right) = 0. \tag{3.57}$$

Noting r^2 in the denominator, we can exclude $r = 0$ from the solution. Multiply both sides by r^2 and then integrate with respect to r. We get,

$$\int \frac{d}{dr}\left(r^2\frac{dV}{dr}\right) dr = 0$$

$$r^2\frac{dV}{dr} = A_1$$

or $$\frac{dV}{dr} = \frac{A_1}{r^2},$$

where A_1 is the constant of integration. (3.58)

Again integrating with respect to r, we get

$$V = \frac{-A_1}{r} + A_2,$$

where A_2 is another constant. The constants A_1 and A_2 can be found from the boundary conditions.

At $r = b$, $V = 0$ and at $r = a$, $V = V_0$,

$$V_0 = \frac{-A_1}{a} + A_2 \quad \text{and} \quad 0 = \frac{-A_1}{b} + A_2.$$

Solving the equations, we get

$$A_1 = \frac{-V_0}{\left(\frac{1}{a} - \frac{1}{b}\right)} \quad \text{and} \quad A_2 = \frac{A_1}{b}.$$

The potential between the two spheres is then

$$V = \frac{V_0}{r\left(\frac{1}{a} - \frac{1}{b}\right)} - \frac{V_0}{b\left(\frac{1}{a} - \frac{1}{b}\right)} = \frac{V_0}{b-a}\left(\frac{ab}{r-a}\right). \tag{3.59}$$

Now the electric field intensity is

$$\vec{E} = -\nabla V = \frac{-\partial V}{\partial r}\vec{a}_r = \frac{-\partial}{\partial r}\left[\frac{V_0}{r\left(\frac{1}{a}-\frac{1}{b}\right)} - \frac{V_0}{b\left(\frac{1}{a}-\frac{1}{b}\right)}\right]\vec{a}_r$$

$$= \frac{V_0}{r^2\left(\frac{1}{a}-\frac{1}{b}\right)}\vec{a}_r \text{ V/m.} \tag{3.60}$$

3.19 Electric Field due to a Coaxial Cable Using Laplace's Equation

Consider a long coaxial cable with two concentric conducting cylinders of radii a and b, $(b > a)$, as shown in Fig. 3.12. The space between the cylinders is filled with a homogeneous medium of dielectric constant ε. Let the potential on the inner conductor be V_0 and on the outer conductor be zero, i.e., it is grounded.

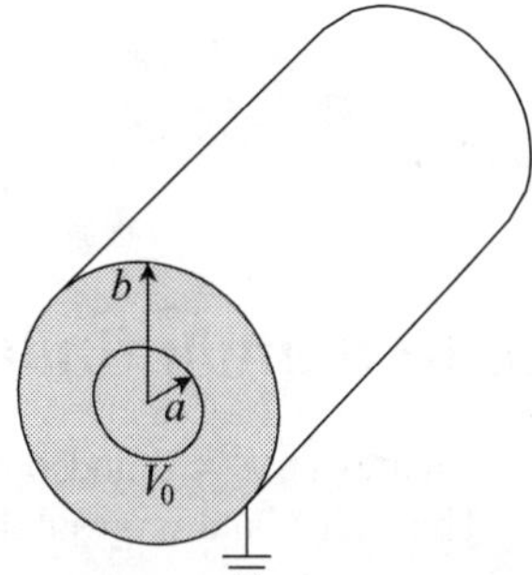

Fig. 3.12 *Potential due to a coaxial cable*

We know that the field intensity $\vec{E}$ of a spherical distribution has only the radial component.

Using Laplace's equation,

$$\nabla^2 V = 0.$$

Since the conductors are cylinders, using cylindrical coordinate system,

$$\frac{1}{\rho}\frac{\partial}{\partial \rho}\left(\rho\frac{\partial V}{\partial \rho}\right) = 0. \tag{3.61}$$

Noting ρ in the denominator, we can exclude $\rho = 0$ from the solution, multiply it by ρ and integrate to get

$$\rho\frac{\partial V}{\partial \rho} = A_1 \text{(a constant) or } \quad \frac{\partial V}{\partial \rho} = \frac{A_1}{\rho}.$$

Again integrating with respect to ρ,

$$V = A_1 \ln(\rho) + A_2,$$

where A_2 is a constant. Apply the following boundary conditions:

At $\rho = a,\ V = V_0$ and at $\rho = b,\ V = 0$.

Then $V_0 = A_1 \ln(a) + A_2$ and $0 = A_1 \ln(b) + A_2$.

Solving the equations,

$$A_1 = \frac{V_0}{\ln(a/b)} \quad \text{and} \quad A_2 = -A_1 \ln(b) = \frac{-V_0 \ln(b)}{\ln(a/b)}.$$

$$\text{Then, } V = \frac{1}{\ln(a/b)}(V_0 \ln(\rho) - V_0 \ln(b)) = V_0 \frac{\ln(\rho/b)}{\ln(a/b)}. \tag{3.62}$$

Now, the electric field intensity is

$$\vec{E} = -\nabla V = \frac{-\partial V}{\partial \rho}\vec{a}_\rho$$

$$= \frac{-V_0}{\ln(a/b)} \frac{\partial}{\partial \rho}(\ln(\rho) - \ln(b))\vec{a}_\rho$$

$$\vec{E} = \frac{-V_0}{\rho \ln(a/b)}\vec{a}_\rho = \frac{V_0}{\rho \ln(b/a)}\vec{a}_\rho \text{ V/m}. \tag{3.63}$$

3.20 Electric Field due to Semi Infinite Conducting Planes

Consider two semi infinite conducting planes at $\phi = 0$ and $\phi = \phi_0$, separated by an infinitesimal insulating gap along the $x - z$ plane. The first plane is grounded and the potential on the second plane is V_0, as shown in Fig. 3.13.

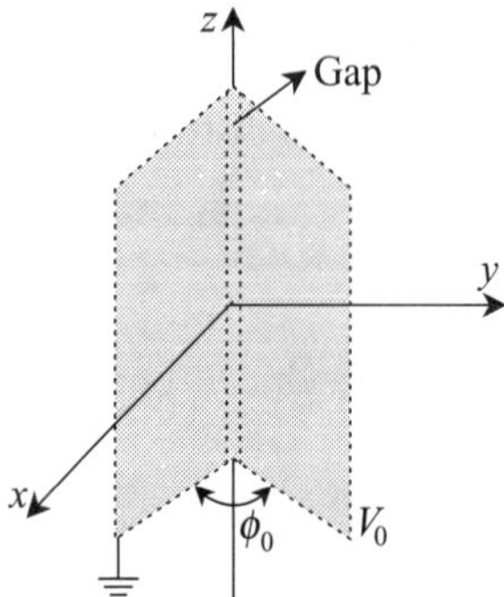

Fig. 3.13 *Potential due to semi infinite conducting planes*

In cylindrical coordinates, the Laplace's equation is given by

$$\frac{1}{\rho}\frac{\partial}{\partial \rho}\left(\rho \frac{\partial V}{\partial \rho}\right) + \frac{1}{\rho^2}\frac{\partial^2 V}{\partial \phi^2} + \frac{\partial^2 V}{\partial z^2} = 0.$$

Assume that the planes are equipotential surfaces.

Since V depends only on ϕ, Laplace's equation therefore becomes

$$\frac{1}{\rho^2}\frac{\partial^2 V}{\partial \phi^2} = 0 \quad \text{or} \quad \frac{d^2 V}{d\phi^2} = 0.$$

Integrating twice, we get

$$V = A\phi + B,$$

where A and B are constants. Applying the boundary conditions:

At $\phi = 0,\ V = 0;$ and at $\phi = \phi_0,\ V = V_0,$

we get, $0 = A(0) + B$ or $B = 0$

and $V_0 = A\phi_0$ or $A = \dfrac{V_0}{\phi_0}$.

Therefore the potential is $V = A\phi + B = \dfrac{V_0 \phi}{\phi_0}$. (3.64)

The electric field is $\vec{E} = -\nabla V = \dfrac{-dV}{\rho d\phi}\vec{a}_\phi$

$$= \frac{-1}{\rho}\frac{d}{d\phi}\left(\frac{V_0\phi}{\phi_0}\right)\vec{a}_\phi = \frac{-V_0}{\rho\phi_0}\vec{a}_\phi. \tag{3.65}$$

3.21 Electric Field due to Two Axial Conducting Cones

Consider two axial conducting cones of infinite extent with its axis along the z-axis. Let the cones be separated by an infinitesimal insulated gap at the origin as shown in Fig. 3.14.

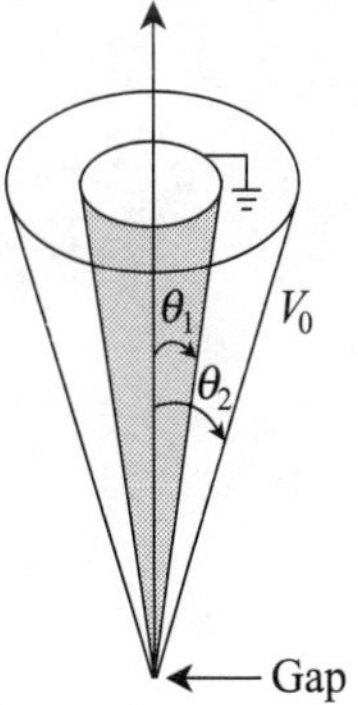

Fig. 3.14 *Potential due to axial conducting cones*

Consider spherical coordinates. Let the inner cone with angle $\theta = \theta_1$ be grounded and the outer cone with angle $\theta = \theta_2 (\theta_2 > \theta_1)$ has a potential V_0. Assume that the cones have equipotential surfaces. The potential depends only on θ.

Laplace's equation in spherical coordinates is

$$\nabla^2 V = \frac{1}{r^2 \sin\theta} \frac{\partial}{\partial\theta}\left[\sin\theta \frac{\partial V}{\partial\theta}\right] = 0$$

or $$\frac{d}{d\theta}\left[\sin\theta \frac{dV}{d\theta}\right] = 0.$$

Integrating with respect to θ,

$$\sin\theta \frac{dV}{d\theta} = A \quad (A \text{ is the constant of integration})$$

$$\frac{dV}{d\theta} = \frac{A}{\sin\theta}. \tag{3.66}$$

Integrating again, we get

$$V = \int \frac{A}{\sin\theta} d\theta = \frac{A}{2\sin(\theta/2)\cos(\theta/2)} d\theta$$

$$= A \int \frac{1}{\frac{2\sin(\theta/2)}{\cos(\theta/2)} \cos^2(\theta/2)} d\theta = A \int \frac{\frac{1}{2}\sec^2(\theta/2)}{\tan(\theta/2)} d\theta. \tag{3.67}$$

Let $\tan(\theta/2) = x$. Then $\frac{1}{2}\sec^2(\theta/2)\, d\theta = dx$.

$$\therefore \quad V = A \int \frac{dx}{x} = A \ln(x) + B$$

or $$V = A \ln(\tan(\theta/2)) + B, \tag{3.68}$$

where B is another constant.

Applying boundary conditions

(i) at $\theta = \theta_1$, $V = 0$

$$0 = A \ln(\tan(\theta_1/2)) + B$$

or $$B = -A \ln(\tan(\theta_1/2))$$

(ii) at $\theta = \theta_2$, $V = V_0$

Then $$V_0 = A \ln(\tan(\theta_2/2)) - A \ln(\tan(\theta_1/2))$$

$$V_0 = A \ln\left[\frac{\tan(\theta_2/2)}{\tan(\theta_1/2)}\right].$$

$$\therefore \quad A = \frac{V_0}{\ln\left[\dfrac{\tan(\theta_2/2)}{\tan(\theta_1/2)}\right]} \quad \text{and} \quad B = -\frac{V_0 \ln(\tan(\theta_1/2))}{\ln\left[\dfrac{\tan(\theta_2/2)}{\tan(\theta_1/2)}\right]}.$$

$$\text{Thus,} \quad V = \frac{V_0 \ln(\tan(\theta/2))}{\ln\left[\dfrac{\tan(\theta_2/2)}{\tan(\theta_1/2)}\right]} - \frac{V_0 \ln(\tan(\theta_1/2))}{\ln\left[\dfrac{\tan(\theta_2/2)}{\tan(\theta_1/2)}\right]}$$

$$= \frac{V_0 \ln\left[\dfrac{\tan(\theta/2)}{\tan(\theta_1/2)}\right]}{\ln\left[\dfrac{\tan(\theta_2/2)}{\tan(\theta_1/2)}\right]}. \tag{3.69}$$

The electric field is $\vec{E} = -\nabla V$

$$= -\frac{1}{r}\frac{dV}{d\theta}\vec{a}_\theta.$$

From Eq.(3.66), $\quad \vec{E} = \dfrac{-1}{r}\dfrac{A}{\sin\theta}\vec{a}_\theta$

$$\therefore \quad \vec{E} = \frac{-V_0}{r(\sin\theta)\ln\left[\dfrac{\tan(\theta_2/2)}{\tan(\theta_1/2)}\right]}\vec{a}_\theta. \tag{3.70}$$

Note: Potential between a cone and a plane

Consider that the outer cone of Fig. 3.14 becomes a plane, i. e., $\theta_2 = \pi/2$ with zero potential as shown in Fig. 3.15. Let the potential on the cone be V_0 and let θ_1 be the angle $(\theta_1 < \pi/2)$.

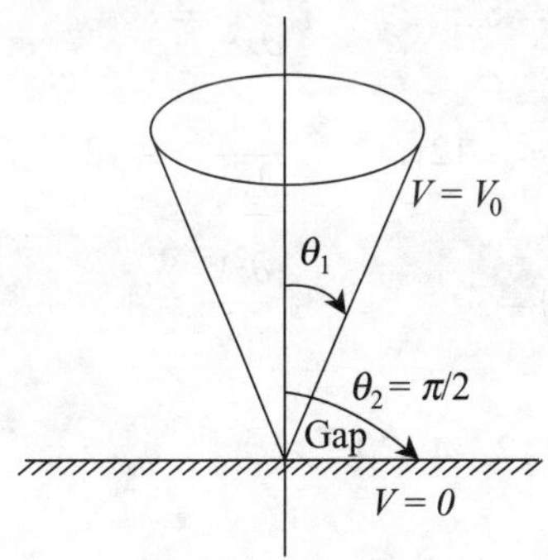

Fig. 3.15 *Potential between a cone and a plane*

The boundary conditions are:

at $\theta = \theta_1, \; V = V_0$; and at $\theta = \pi/2, \; V = 0$.

From Eq. (3.68), $\quad V = A\ln(\tan\theta/2) + B$.

Then $0 = A\ln(\tan\pi/4) + B$

or $B = 0$

and $V_0 = A\ln(\tan\theta_1/2)$.

$$\therefore \quad A = \frac{V_0}{\ln(\tan\theta_1/2)}.$$

The potential is $V = \dfrac{V_0\ln(\tan\theta/2)}{\ln(\tan\theta_1/2)}$. (3.71)

The electric field strength is

$$\vec{E} = -\nabla V = \frac{-1}{r}\frac{\partial V}{\partial\theta}\vec{a}_\theta = \frac{-A}{r\sin\theta}\vec{a}_\theta$$

$$= -\frac{V_0\vec{a}_\theta}{r(\sin\theta)\ln\left(\tan\dfrac{\theta_1}{2}\right)}. \tag{3.72}$$

Example 3.9 Verify whether the potential fields given below satisfy Laplace's equation.

(i) $V = 4x^2 - 6y^2 + 2z^2$; (ii) $V = \rho\cos\phi + 4z$

Solution (i) Given $V = 4x^2 - 6y^2 + 2z^2$. To satisfy Laplace's equation $\nabla^2 V = 0$, in Cartesian coordinates,

$$\nabla^2 V = \frac{\partial^2 V}{\partial x^2} + \frac{\partial^2 V}{\partial y^2} + \frac{\partial^2 V}{\partial z^2}$$

$$\frac{\partial V}{\partial x} = \frac{\partial}{\partial x}(4x^2 - 6y^2 + 2z^2) = 8x; \qquad \frac{\partial^2 V}{\partial x^2} = 8$$

$$\frac{\partial V}{\partial y} = \frac{\partial}{\partial y}(4x^2 - 6y^2 + 2z^2) = -12y; \qquad \frac{\partial^2 V}{\partial y^2} = -12$$

$$\frac{\partial V}{\partial z} = \frac{\partial}{\partial z}(4x^2 - 6y^2 6 + 2z^2) = 4z; \qquad \frac{\partial^2 V}{\partial z^2} = 4$$

Then $\dfrac{\partial^2 V}{\partial x^2} + \dfrac{\partial^2 V}{\partial y^2} + \dfrac{\partial^2 V}{\partial z^2} = 8 - 12 + 4 = 0$

i.e., $\nabla^2 V = 0$ (satisfies Laplace's equation).

(ii) $V = \rho\cos\phi + 4z$

Given the potential in cylindrical coordinates. To satisfy Laplace's equation, $\nabla^2 V = 0$,

$$\nabla^2 V = \frac{1}{\rho}\frac{\partial}{\partial\rho}\left(\rho\frac{\partial V}{\partial\rho}\right) + \frac{1}{\rho^2}\frac{\partial^2 V}{\partial\phi^2} + \frac{\partial^2 V}{\partial z^2}$$

Now, $\dfrac{\partial V}{\partial \rho} = \dfrac{\partial}{\partial \rho}(\rho \cos\phi + 4z) = \cos\phi,$

or $\dfrac{1}{\rho}\dfrac{\partial}{\partial \rho}\left(\rho \dfrac{\partial V}{\partial \rho}\right) = \dfrac{\cos\phi}{\rho}.$

And $\dfrac{\partial V}{\partial \phi} = \dfrac{\partial}{\partial \phi}(\rho \cos\phi + 4z) = -\rho \sin\phi,$

$$\frac{\partial^2 V}{\partial \phi^2} = \frac{\partial}{\partial \phi}(-\rho \sin\phi) = -\rho \cos\phi.$$

And $\dfrac{\partial V}{\partial z} = \dfrac{\partial}{\partial z}(\rho \cos\phi + 4z) = 4,$

$$\frac{\partial^2 V}{\partial z^2} = \frac{\partial}{\partial z}(4) = 0.$$

$$\therefore \quad \frac{1}{\rho}\frac{\partial}{\partial \rho}(\rho \frac{\partial V}{\partial \rho}) + \frac{1}{\rho^2}\frac{\partial^2 V}{\partial \phi^2} + \frac{\partial^2 V}{\partial z^2} = \frac{1}{\rho}(\cos\phi) + \frac{1}{\rho^2}(-\rho \cos\phi) + 0$$

$$= \frac{\cos\phi}{\rho} - \frac{\cos\phi}{\rho} = 0.$$

$\nabla^2 V = 0$ (satisfies Laplace's equation).

3.22 Capacitance

A capacitor consists of two conducting plates, A and B, separated by free space or a dielectric medium of thickness d, with permittivity ε. Let the conducting plates carry equal and opposite charge as shown in Fig. 3.16. Due to the force of attraction between the charges, an electric field normal to the conducting surface is established. The flux flows from the positive charge to the negative charge and a potential difference V is developed across the plates. The potential difference across the plates is proportional to the charge on the plates, i.e.,

$$Q \,\alpha\, V \quad \text{or} \quad Q = CV.$$

The proportional constant is called capacitance, and is defined as *the ratio of the charge Q on the conductor to the potential difference V between the conductors.*

It is denoted by $C = \dfrac{Q}{V}$ *farads or coulombs/volt.*

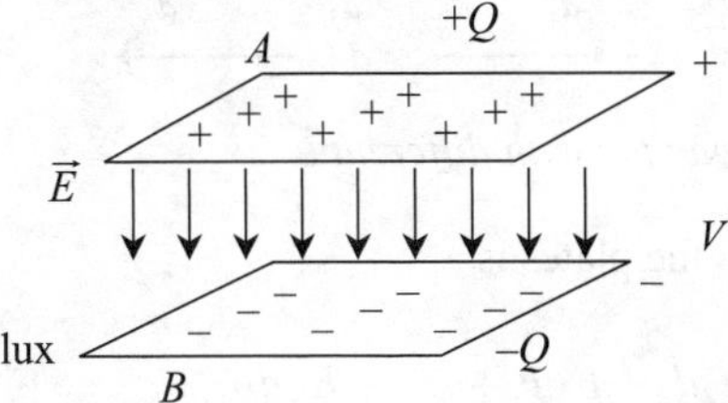

Fig. 3.16 *Parallel plate capacitor*

If the separation between the plates is d metres and the area of the conductor is S, then the capacitance of the parallel plate capacitor is

$$C = \frac{Q}{V} = \frac{\varepsilon S}{d} \text{ farads.} \tag{3.73}$$

Proof Let two parallel plates in the $x-y$ plane be separated by a distance d metres along the z-axis. Let ε be the permittivity of the dielectric medium between the plates and S the surface area of the conductor.

If the plates carry equal and opposite charges, an electric field develops between the plates, which can be obtained from the electric flux density as

$$\vec{D} = \rho_s \vec{a}_z = \frac{Q}{S}\vec{a}_z.$$

$$\vec{E} = \frac{\vec{D}}{\varepsilon} = \frac{Q}{\varepsilon S}\vec{a}_z.$$

The potential difference $V = \int\limits_l \vec{E} \cdot d\vec{l} = \frac{Q}{\varepsilon S}\vec{a}_z \cdot d \cdot \vec{a}_z = \frac{Qd}{\varepsilon S}\vec{a}_z \cdot \vec{a}_z = \frac{Qd}{\varepsilon S}$ volts.

Then the capacitance C is

$$C = \frac{Q}{V} = \frac{\varepsilon S}{d} \text{ farads.} \qquad \text{(Proved)}$$

Note: The capacitance depends only on the physical properties of the capacitor.

Capacitor with different dielectric materials Consider two parallel plate conductors separated by different dielectric materials as shown in Fig. 3.17. Let $\varepsilon_1, \varepsilon_2, \ldots, \varepsilon_N$ be the permittivities of the dielectric materials and $d_1, d_2, \ldots, d_N$ their thicknesses, all of the same area, S.

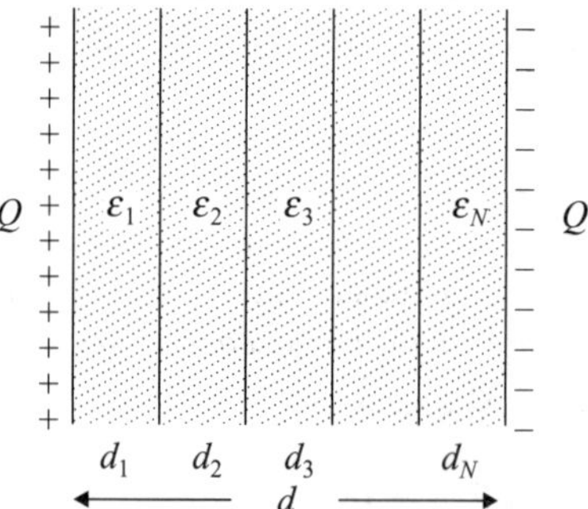

Fig. 3.17 *Parallel plate capacitor with different* ε

The potential difference across the plates is

$$V = \int\limits_l \vec{E} \cdot d\vec{l} = E_1 d_1 + E_2 d_2 + E_3 d_3 + \ldots. + E_N d_N.$$

Since $E = \dfrac{D}{\varepsilon} = \dfrac{Q}{S\varepsilon}$,

then, $V = \dfrac{D_1 d_1}{\varepsilon_1} + \dfrac{D_2 d_2}{\varepsilon_2} + \ldots.. + \dfrac{D_N d_N}{\varepsilon_N}$

$$V = \frac{Q}{S}\frac{d_1}{\varepsilon_1} + \frac{Qd_2}{S\,\varepsilon_2} + \ldots.. + \frac{Qd_N}{S\,\varepsilon_N}$$

$$V = \frac{Q}{S}\left(\frac{d_1}{\varepsilon_1} + \frac{d_2}{\varepsilon_2} + \frac{d_3}{\varepsilon_3} + \ldots.. + \frac{d_N}{\varepsilon_N}\right)$$

$\therefore$ The capacitance $C = \dfrac{Q}{V} = \dfrac{S}{\left(\dfrac{d_1}{\varepsilon_1} + \dfrac{d_2}{\varepsilon_2} + \dfrac{d_3}{\varepsilon_3} + \ldots.. + \dfrac{d_N}{\varepsilon_N}\right)}$ farads. (3.74)

Capacitors connected in series If N capacitors of different dimensions are connected in series as shown in Fig. 3.18, then

$$C_1 = \frac{\varepsilon_1 S_1}{d_1},\; C_2 = \frac{\varepsilon_2 S_2}{d_2}, \ldots.,\; C_N = \frac{\varepsilon_N S_N}{d_N}.$$

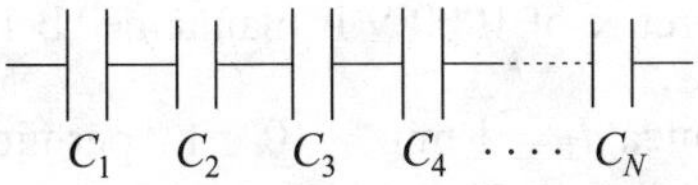

C_1 C_2 C_3 C_4 .... C_N

Fig. 3.18 *Capacitors connected in series*

The potential difference is

$$V = V_1 + V_2 + V_3 + \ldots\ldots + V_N$$

$$= \frac{Q}{C_1} + \frac{Q}{C_2} + \ldots.. + \frac{Q}{C_N}$$

$$= Q\left(\frac{1}{C_1} + \frac{1}{C_2} + \ldots.. + \frac{1}{C_N}\right).$$

$$C = \frac{Q}{V} = \frac{1}{\dfrac{1}{C_1} + \dfrac{1}{C_2} + \dfrac{1}{C_3} + \ldots.. + \dfrac{1}{C_N}}$$

or $\dfrac{1}{C} = \dfrac{1}{C_1} + \dfrac{1}{C_2} + \ldots.. + \dfrac{1}{C_N}$. (3.75)

Capacitors connected in parallel If N capacitors are connected in parallel as shown in Fig. 3.19, and the potential across the capacitors is V, then the the total charge on all the capacitors is

$$Q = Q_1 + Q_2 + Q_3 + \ldots.. + Q_N.$$

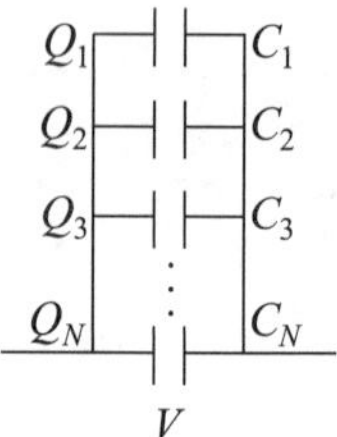

Fig. 3.19 *Capacitors connected in parallel*

Now, $Q = C_1V + C_2V + C_3V + \ldots. + C_NV.$

$$\frac{Q}{V} = C = C_1 + C_2 + \ldots.. + C_N.$$

$$\therefore \quad C = C_1 + C_2 + C_3 + \ldots.. + C_N. \tag{3.76}$$

Thus the total capacitance is equal to the sum of all the individual capacitances.

Example 3.10 A parallel plate capacitor has a separation of 2 mm between the plates. A sheet mica 1 mm thick, having a permittivity of 6, is placed between them. Find the charge density on each plate when a potential difference of 1000 V is maintained between them.

Solution Given thickness of mica $d_1 = 1$ mm $= 10^{-3}$ m; permittivity $\varepsilon_{r1} = 6$.
Assume that there is a free space of $d_2 = 1$ mm thickness between the mica and a plate.

Let E_1 and E_2 be the electric fields between the dielectrics.

Then, $V = E_1d_1 + E_2d_2.$

Let D be the charge density,

then, $$V = \frac{D}{\varepsilon_0\,\varepsilon_{r_1}}d_1 + \frac{D}{\varepsilon_0}d_2 = \frac{D}{\varepsilon_0}\left(\frac{d_1}{\varepsilon_{r_1}} + d_2\right)$$

$$1000 = \frac{D}{8.884 \times 10^{-12}}\left[1 + \frac{1}{6}\right] \times 10^{-3}$$

$$\therefore \quad D = 7.6 \times 10^{-6}\ \mathrm{C/m^2}.$$

3.23 Capacitance between Two Concentric Spheres

Consider two concentric spheres of conductors with radius a and b, where $b > a$, as shown in Fig. 3.20. The spheres are filled with dielectric material of permittivity ε. Let the inner conductor carry a positive charge Q and the outer conductor carry a negative charge $-Q$.

Consider a Gaussian spherical surface of radius r.

The field intensity in radial direction at radius r is

$$\vec{E} = \frac{Q}{4\pi\,\varepsilon r^2}\vec{a}_r \text{ V/m}.$$

The potential difference is

$$V = \int_a^b \vec{E}\cdot d\vec{l} = \int_a^b \frac{Q}{4\pi\,\varepsilon r^2}\vec{a}_r \cdot dr\,\vec{a}_r$$

$$= \frac{Q}{4\pi\,\varepsilon}\int_a^b \frac{1}{r^2}dr = \frac{Q}{4\pi\,\varepsilon}\left(\frac{1}{a} - \frac{1}{b}\right). \qquad (3.77)$$

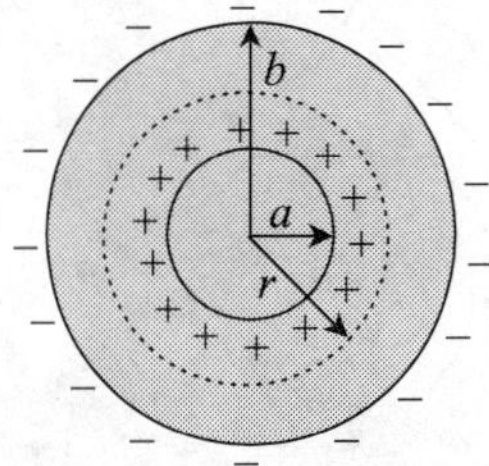

Fig. 3.20 *Capacitance between two concentric spheres*

The capacitance $C = \dfrac{Q}{V} = \dfrac{4\pi\,\varepsilon}{\left(\dfrac{1}{a} - \dfrac{1}{b}\right)}.$ (3.78)

Note: If there is only one sphere of radius a carrying positive charge Q and if the outer sphere is at infinity, i.e., $b = \infty$, then from Eq. (3.78),

$$C = \frac{4\pi\,\varepsilon}{\dfrac{1}{a} - \dfrac{1}{\infty}} = \frac{4\pi\varepsilon}{\dfrac{1}{a}} \qquad (3.79)$$

or $C = 4\pi\,\varepsilon a.$

Thus the capacitance of a single isolated sphere of radius a is $C = 4\pi\,\varepsilon\,a$.

Example 3.11 Obtain the capacitance of an isolated conducting sphere of radius 1 cm.

Solution Given $r = 1$ cm $= 0.01$ m

For an isolated sphere, $V = \dfrac{Q}{4\pi\varepsilon_0 r}$.

Capacitance $C = \dfrac{Q}{V} = 4\pi\varepsilon_0 r = \dfrac{0.01}{9\times 10^9} = 1.11$ pF.

3.24 Capacitance of a Coaxial Cable

Consider two coaxial conducting cylinders separated by a dielectric material of permittivity ε. The radius of the inner conductor is a, that of the outer conductor is b, and the height is L as shown in Fig. 3.21. Let ρ_s be the charge density on the inner conductor and $-\rho_s$, the charge density on the outer conductor. Let the Gaussian cylindrical surface be at radius ρ.

We know that the electric field intensity between the conductors is

$$\vec{E} = \frac{\rho_s}{\varepsilon}\vec{a}_\rho.$$

$$Q = \rho_s 2\pi\rho\, L.$$

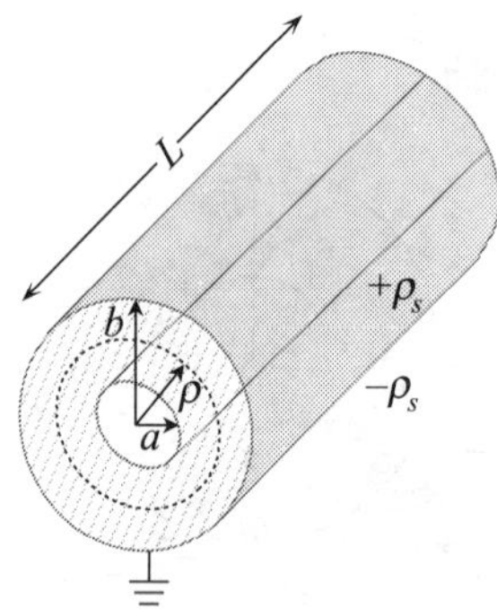

Fig. 3.21 Capacitance of a coaxial cable

The potential difference is $V = \int_a^b \vec{E}\cdot d\vec{l} = \int_a^b \frac{\rho_s}{\varepsilon}\vec{a}_\rho \cdot d\rho\,\vec{a}_\rho$

$$= \int_a^b \frac{Q}{2\pi\rho\varepsilon L}d\rho = \frac{Q}{2\pi\varepsilon L}\int_a^b \frac{1}{\rho}d\rho$$

$$V = \frac{Q}{2\pi\varepsilon L}\ln(b/a).$$

The capacitance is $C = \frac{Q}{V} = \frac{2\pi\varepsilon L}{\ln(b/a)}$. (3.80)

3.25 Capacitance of Two Parallel Wires (Single-phase Transmission Line)

Consider a single phase transmission line comprising two parallel conducting wires, each of radius a separated by a distance d from their axis as shown in Fig. 3.22. Assume that the conductors are infinitely long. ρ_l and $-\rho_l$ are the charge densities on the conducting wires A and B respectively. Let P be a point at a distance x from ρ_l along the x-axis.

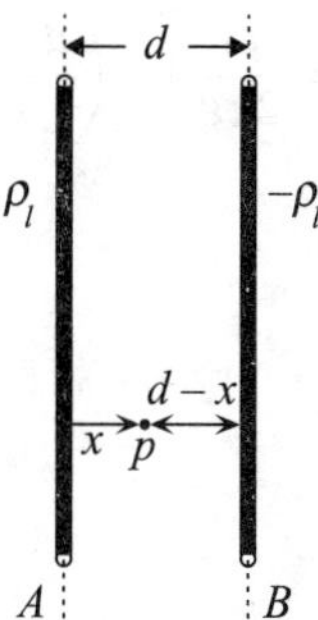

Fig. 3.22 *Two parallel wires*

The electric field intensity at P due to ρ_l is

$$\vec{E}_1 = \frac{\rho_l}{2\pi\,\varepsilon\,x}\vec{a}_x$$

and that due to $-\rho_l$ with a distance $(d-x)$ is

$$\vec{E}_2 = \frac{-\rho_l}{2\pi\varepsilon(d-x)}\vec{a}_x.$$

The total field at point P is

$$\vec{E} = \vec{E}_1 + \vec{E}_2 = \frac{\rho_l\vec{a}_x}{2\pi\varepsilon x} + \frac{-\rho_l\vec{a}_x}{2\pi\varepsilon(d-x)}$$

$$= \frac{\rho_l}{2\pi\varepsilon}\left[\frac{1}{x} - \frac{1}{d-x}\right]\vec{a}_x.$$

The potential difference between the two wires A and B is

$$V_{AB} = -\int_B^A \vec{E}\cdot d\vec{l} = -\int_B^A E\,dx$$

$$= -\int_B^A \frac{\rho_l}{2\pi\varepsilon}\left[\frac{1}{x} - \frac{1}{d-x}\right]dx$$

$$= -\int_B^A \frac{\rho_l}{2\pi\varepsilon x}dx + \int_B^A \frac{\rho_l}{2\pi\varepsilon(d-x)}dx.$$

Between the wires A and B, the variable x varies from $d-a$ to a against the field.

$$\therefore \quad V_{AB} = \frac{\rho_l}{2\pi\varepsilon}\left[\int_{d-a}^{a} \frac{1}{d-x}dx - \int_{d-a}^{a} \frac{1}{x}dx\right]$$

$$= \frac{\rho_l}{2\pi\varepsilon}\Big[\ln(d-x) - \ln(x)\Big]_{d-a}^{a} = \frac{\rho_l}{2\pi\varepsilon}\left[\ln\left(\frac{d-a}{a}\right) + \ln\left(\frac{d-a}{a}\right)\right]$$

$$V_{AB} = \frac{\rho_l}{\pi\varepsilon}\ln\left(\frac{d-a}{a}\right).$$

Since $d - a \approx d$ as a is very small,

$$V_{AB} = \frac{\rho_l}{\pi\varepsilon}\ln\left(\frac{d}{a}\right).$$

Capacitance per unit length $C = \dfrac{\rho_l}{V}$ F/m.

$$C = \frac{\pi\,\varepsilon}{\ln(d/a)} \text{ F/m.}$$

If L is the length of the wire, then the total capacitance is

$$C = \frac{\pi\,\varepsilon\,L}{\ln(d/a)} \text{ farads.} \tag{3.81}$$

Note: If C_n is the capacitance of a single line with neutral, then

$$C_n = C + C = 2C.$$

$$C_n = \frac{2\pi\varepsilon_0}{\ln\left(\dfrac{d}{a}\right)} \text{ F/m.}$$

3.26 Energy Stored in a Capacitor

When potential V volts is applied on a capacitor C, the energy stored in the electric field of a capacitor is given by

$$W_E = \frac{1}{2}CV^2 \text{ joules.} \tag{3.82}$$

Proof Consider a parallel plate capacitor as shown in Fig. 3.23. Let V be the potential applied. We know that the energy stored in the electrostatic field is

$$W_E = \frac{1}{2}\int \vec{D}\cdot\vec{E}\,dv.$$

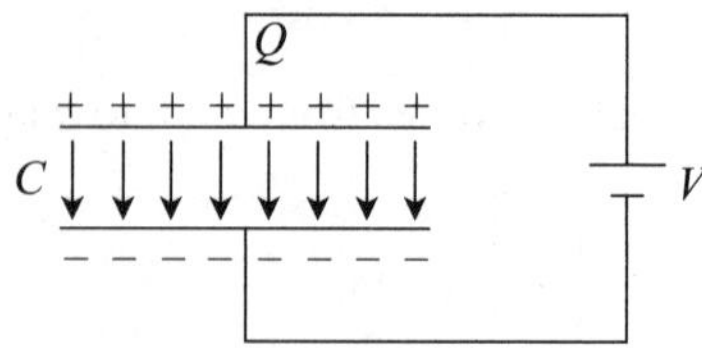

Fig. 3.23 *Energy stored in a capacitor*

Let S be the cross-sectional area, Q, the charge and d, the separation between plates.

Then the flux density $\vec{D} = \frac{Q}{S}\vec{a}_x$ and the electric field intensity $\vec{E} = \frac{V}{d}\vec{a}_x$.

$$W_E = \frac{1}{2}\int \frac{Q}{S}\vec{a}_x \cdot \frac{V}{d}\vec{a}_x dv = \frac{QV}{2Sd}\int_v dv$$

$$W_E = \frac{1}{2}QV\frac{v}{Sd}.$$

Since volume $v = Sd$,

$$W_E = \frac{1}{2}QV$$

or $W_E = \frac{1}{2}CV^2$ joules. (Proved)

Another method When potential V volts is applied on a capacitor C, the incremental energy stored in the electric field is the product of voltage applied and the incremental charge. Thus

$$dW_E = VdQ = \frac{Q}{C}dQ.$$

The total energy stored is

$$W_E = \frac{1}{C}\int_0^Q QdQ = \frac{Q^2}{2C}.$$

$\therefore$ $W_E = \frac{1}{2}CV^2$ joules.

Example 3.12 The construction of a paper capacitor is as follows. Aluminium foil of 100 cm^2 area is placed on both sides of paper thickness 0.03 mm. If the dielectric constant of paper is given as 3 and the dielectric breakdown strength is 200 kV/ cm, what is the rating of the capacitor?

Solution Given area: $S = 100$ cm$^2 = 100 \times 10^{-4} = 10^{-2}$ m^2; paper thickness $= d = 0.03$ mm $= 0.03 \times 10^{-3}$m; $\varepsilon_r = 3$; maximum breakdown strength $E_{\text{max}} = 200$ kV/ cm $= 200 \times 10^3 \times 10^2$ V/m $= 2 \times 10^7$ V/m.

We know that capacitance

$$C = \frac{\varepsilon S}{d} = \frac{8.854 \times 10^{-12} \times 3 \times 10^{-2}}{3 \times 10^{-5}} = 8.854 \text{ nF}.$$

Maximum operating voltage $V = E.d = 2 \times 10^7 \times 3 \times 10^{-5} = 600$ V.

Maximum energy stored $W_E = \frac{1}{2}CV^2 = \frac{1}{2} \times 8.854 \times 10^{-9}(600)^2 = 1.6$ mJ.

The rating of the capacitor is $V_{\text{max}} = 600$V and $W_{E\,\text{max}} = 1.6$ mJ.

3.27 Energy Stored in a Coaxial Cable

Consider a coaxial cable with two concentric conducting cylinders of radii a and b, $(b > a)$ and length L centred along the z-axis as shown in Fig. 3.24. The space between the cylinders is filled with material of dielectric constant ε.

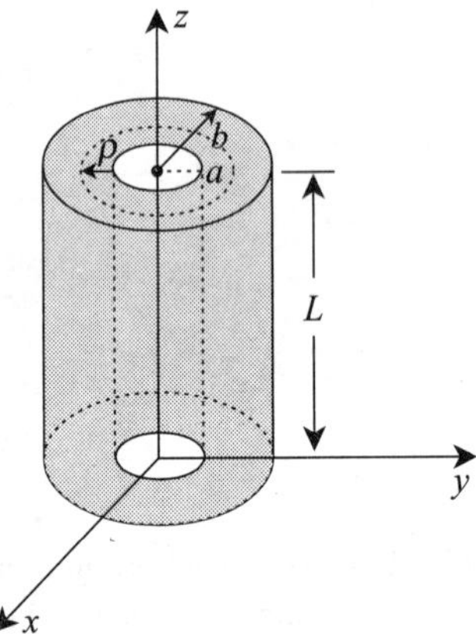

Fig. 3.24 *Energy in a coaxial cable*

Let Q be the charge on the inner conductor. Also, let the Gaussian surface be of radius ρ within the cable.

When $a \leq \rho \leq b$, since the Gaussian surface encloses the inner conductor, the total charge is

$$Q = \oint_s \vec{D} \cdot d\vec{S}$$

$$= \oint_s D\vec{a}_\rho \cdot \rho d\phi dz \vec{a}_\rho = \oint_s \rho D d\phi dz \vec{a}_\rho \cdot \vec{a}_\rho$$

$$= \int_0^L \int_\theta^{2\pi} \rho D \, d\phi dz = 2\pi\rho L D$$

or $$D = \frac{Q}{2\pi\rho L} \quad \text{and} \quad \vec{D} = \frac{Q}{2\pi\rho L}\vec{a}_\rho. \tag{3.83}$$

But we know that the charge density on the inner conductor is $\rho_s = \dfrac{Q}{2\pi a L}$ C/m^2.

$$\therefore \quad \vec{D} = \frac{a\rho_s}{\rho}\vec{a}_\rho. \tag{3.84}$$

The energy stored in the electric field between two conductors is

$$W_E = \frac{1}{2}\int_v \frac{|\vec{D}|^2}{\varepsilon} dv = \frac{1}{2\varepsilon}\int_v \frac{a^2\rho_s^2}{\rho^2} \rho d\rho \, d\phi dz$$

$$= \frac{a^2\rho_s^2}{2\varepsilon}\int_0^L\int_0^{2\pi}\int_a^b \frac{1}{\rho}d\rho\, d\phi\, dz = \frac{a^2\rho_s^2}{2\varepsilon}2\pi L[\ln\rho]_a^b.$$

$$W_E = \frac{\pi La^2\rho_s^2}{\varepsilon}\left(\frac{b}{a}\right) \text{ joules} \quad \text{for} \quad a \le \rho \le b \tag{3.85}$$

Since the charge enclosed in the regions $\rho < a$ and $\rho > b$ is zero, the energy stored is zero for $\rho < b$ and $\rho > b$.

Additional Solved Problems

Example 3.13 The current density is given by $\vec{J} = \frac{4}{r^2}\cos\theta\vec{a}_r + 20e^{-2r}\sin\theta\vec{a}_\theta - r\sin\theta\cos\phi\vec{a}_\rho$ A/m^2. Find current at $r = 3$, passing through the surface $0 \le \theta < 20,\ 0 < \phi < 2\pi$, the $\vec{a}_r$ direction.

Solution Given $\vec{J} = \frac{4}{r^2}\cos\theta\vec{a}_r + 20e^{-2r}\sin\theta\vec{a}_\theta - r\sin\theta\cos\phi\vec{a}_\phi.$

Now, $I = \oint_s J\cdot dS$ where $dS = r^2\sin\theta d\theta d\phi\ \vec{a}_r.$

$$I = \int_0^{20}\int_0^{2\pi}\left[\frac{4}{r^2}\cos\theta\vec{a}_r + 20e^{-2r}\sin\theta\vec{a}_\theta - r\sin\theta\cos\phi\vec{a}_\phi\right]r^2\sin\theta\, d\theta d\phi\vec{a}_r$$

$$= \int_0^{20}\int_{20}^{2\pi} 4\sin\theta\cos\theta d\theta d\phi = 4\pi\int_0^{2\pi}\sin 2\theta d\theta = \frac{4\pi}{2}(-\cos 2\theta)\Big|_0^{2\pi}$$

$= 1.47$ A.

Example 3.14 For a linear dielectric, show that $\vec{E}_{\text{total}} = \vec{E}_{\text{ext}}/\varepsilon_r$, where ε_r is the relative permittivity of the dielectric, $\vec{E}_{\text{total}}$ is the total electric field and $\vec{E}_{\text{ext}}$ is the electric field due to charges other than bound charges.

Solution Given $\vec{E}_{\text{total}}$ = total electric field, $\vec{E}_{\text{ext}}$ = electric field due to free charges.

Let $\vec{E}_{\text{in}}$ = electric field due to bound charges.

We know from the theory polarization of dielectric material,

$$\vec{D}_{\text{total}} = \varepsilon_0\vec{E}_{\text{total}} + \vec{P} \quad \text{and} \quad \vec{P} = \chi_e\varepsilon_0\vec{E}_{\text{total}}.$$

$$\therefore \quad \vec{D}_{\text{total}} = \varepsilon_0(1+\chi_e)\vec{E}_{\text{total}}$$

$$\vec{D}_{\text{total}} = \varepsilon\, E_{\text{total}}$$

Also, $\vec{D}_{\text{total}} = \varepsilon_0\vec{E}_{\text{ext}}$

$$\varepsilon\,\vec{E}_{\text{total}} = \varepsilon_0\vec{E}_{\text{ext}}$$

$$\varepsilon_0\, \varepsilon_r\, \vec{E}_{\text{total}} = \varepsilon_0\, \vec{E}_{\text{ext}}$$

$$\therefore\quad \vec{E}_{\text{total}} = \frac{\vec{E}_{\text{ext}}}{\varepsilon_r}. \qquad \text{(Proved)}$$

Example 3.15 A spherical dielectric shell has its centre at origin. Its internal and external radii are 49 cm and 50 cm respectively. The shell has a polarization of $3\vec{a}_r$ nC/m^2. Considering the bound surface density on the outer surface, find the potential of the dielectric shell if $\varepsilon_r = 4$.

Solution Given a spherical dielectric shell with centre at the origin.
Inner radius $a = 49$ cm $= 0.49$ m; Outer radius $b = 50$ cm $= 0.5$ m.
Polarization $\vec{P} = 3\vec{a}_r$ nC/m^2.

We know that $\vec{P} = \chi_e\, \varepsilon_0\, \vec{E}$

$$\varepsilon_r = 4$$

$$\varepsilon_r = (\chi_e + 1) = 4$$

$$\chi_e = 4 - 1 = 3$$

$$\vec{P} = 3\varepsilon_0\, \vec{E}$$

$$\vec{E} = \frac{10^{-9}\vec{a}_r}{\varepsilon_0}\ \text{F/m} = 113\vec{a}_r\ \text{F/m}.$$

Also, we know that the potential

$$V = -\int_a^b \vec{E}\cdot d\vec{l}$$

$$= -\int_{0.49}^{0.5} \frac{10^{-9}}{\varepsilon_0}\vec{a}_r \cdot dr\, \vec{a}_r = \frac{-1}{\varepsilon_0}\int_{0.49}^{0.5} dr = \frac{-10^{-9}}{\varepsilon_0}\big[r\big]_{0.49}^{0.5}$$

$$= \frac{-10^{-9}}{8.854\times 10^{-12}}[0.5 - 0.49] = -1.13\ \text{V}.$$

Example 3.16 Given that $15\vec{a}_x - 6\vec{a}_y + 20\vec{a}_z$ V/m at a charge free dielectric's interface as shown in Fig. 3.25, find D_2 and angles α_1 and α_2.

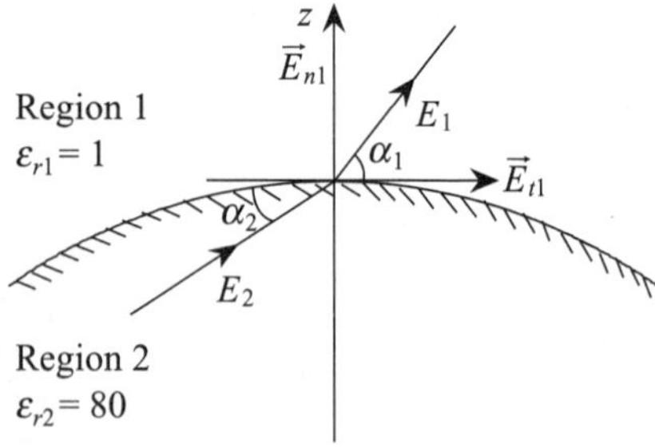

Fig. 3.25 *Boundary between dielectrics*

Solution Given $\vec{E} = 15\vec{a}_x - 6\vec{a}_y + 20\vec{a}_z$ V/m.

At the boundary,

$$E_{n1} = \vec{E}_1 \cdot \vec{a}_z = (15\vec{a}_x - 6\vec{a}_y + 20\vec{a}_z) \cdot \vec{a}_z = 20\vec{a}_z.$$

From the boundary conditions, $E_{t1} = E_{t2}$, since $\vec{E}_1 = \vec{E}_{t1} + \vec{E}_{n1}$

$$E_{t2} = E_{t1} = \vec{E}_1 - \vec{E}_n = 15\vec{a}_x - 6\vec{a}_y.$$

Also, $\vec{D}_{n2} = \vec{D}_{n1}$ or $\vec{E}_{n2} = \dfrac{\varepsilon_{r1}}{\varepsilon_{r2}}\vec{E}_{n1} = \dfrac{1}{80}(20\vec{a}_z) = 0.25\vec{a}_z.$

$$\therefore \quad \vec{E}_2 = \vec{E}_{n2} + \vec{D}_{n2} = 15\vec{a}_x - 6\vec{a}_y + 0.25\vec{a}_z.$$

$$\vec{D}_2 = \varepsilon\vec{E}_2 = 80\varepsilon_0(15\vec{a}_x - 6\vec{a}_y + 0.25\vec{a}_z)$$

$$= 1200\varepsilon_0\vec{a}_x - 480\varepsilon_0\vec{a}_y + 20\varepsilon_0\vec{a}_z.$$

To find the angles, from Fig. 3.25,

$$|\vec{E}_{n_1}| = |E_1|\sin\alpha_1$$

$$E_1 = \sqrt{15^2 + 6^2 + 20^2} = 25.71$$

$$|E_{n_1}| = 20.$$

$$\therefore \quad \sin\alpha_1 = \frac{20}{25.71} \quad \text{or} \quad \alpha_1 = 51°.$$

Also, $\sin\alpha_2 = \dfrac{|E_{n_2}|}{|E_2|} = \dfrac{0.25}{\sqrt{15^2 + 6^2 + 0.25^2}}$

$$\alpha_2 = 0.88°.$$

Example 3.17 Region 1 defined by $x < 0$ is free space, while region 2 defined by $x \geq 0$ is a dielectric material for which $\varepsilon_{r_2} = 4.4$, as shown in Fig. 3.26. Given $\vec{D}_1 = 4\vec{a}_x + 3\vec{a}_y - 6\vec{a}_z$ C/m^2, find E_2 and the angles θ_1 and θ_2.

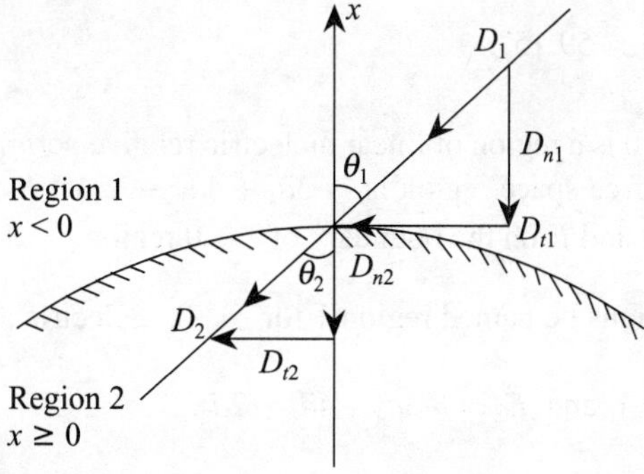

Fig. 3.26 *Boundary between two regions*

Solution Given $\vec{D}_1 = 4\vec{a}_x + 3\vec{a}_y - 6\vec{a}_z \text{ C/m}^2$; $\varepsilon_{r1} = 1$ and $\varepsilon_{r2} = 4.4$.
At the boundary, let $\vec{D}_{n1} = D_1 \cdot \vec{a}_x$

$$= (4\vec{a}_x + 3\vec{a}_y - 6\vec{a}_z) \cdot \vec{a}_x = 4\vec{a}_x.$$

Also, $\vec{D}_{n2} = \vec{D}_{n1} = 4\vec{a}_x.$

$$\vec{D}_{t1} = \vec{D}_1 - \vec{D}_{n1} = (4\vec{a}_x + 3\vec{a}_y - 6\vec{a}_x) - 4\vec{a}_x = 3\vec{a}_y - 6\vec{a}_z,$$

and $\vec{D}_{t2} = \dfrac{\varepsilon_{r_2}}{\varepsilon_{r_1}}\vec{D}_{t1} = \dfrac{4.4}{1}(3\vec{a}_y - 6\vec{a}_z) = 13.2\vec{a}_y - 26.4\vec{a}_z.$

Now, $\vec{D}_2 = \vec{D}_{t2} + \vec{D}_{n2}$

$$= 4\vec{a}_x + 13.2\vec{a}_y - 26.4\vec{a}_z.$$

$$\vec{E}_2 = \frac{\vec{D}_2}{\varepsilon_0 \varepsilon_{r_2}} = \frac{1}{4.4\varepsilon_0}(4\vec{a}_x + 13.2\vec{a}_y - 26.4\vec{a}_z)$$

$$= (10.27\vec{a}_x + 33.89\vec{a}_y - 67.76\vec{a}_z) \times 10^{10} \text{ V/m}.$$

To find the angles, from Fig. 3.26,

$$|D_{n2}| = |D_2| \cos\theta_2$$

$$4 = \left(\sqrt{(13.2)^2 + (26.4)^2 + (4)^2}\right)\cos\theta_2$$

$$\theta_2 = \cos^{-1}\left(\frac{4}{29.78}\right) = 82.27^\circ$$

and $|D_{n1}| = |D_1| \cos\theta_1.$

$$4 = \left(\sqrt{16 + 9 + 36}\right)\cos\theta_1$$

$$\theta_1 = \cos^{-1}\left(\frac{4}{7.8}\right) = 59.15^\circ.$$

Example 3.18 Given $z < 0$ is a region of linear dielectric relative permittivity 6.5 and $z > 0$ is free space. Electric field in the free space region is $(-3\vec{a}_x + 4\vec{a}_y - 2\vec{a}_z)$ V/m. Find (i) $\vec{D}$ for $z > 0$; (ii) tangential components of $\vec{D}$ and $\vec{E}$ on the boundary of $z < 0$ region.

Solution Let the two regions be named region 1 for $z < 0$, dielectric and region 2 for $z > 0$, free space.

Then, $\varepsilon_{r1} = 6.5$, $\varepsilon_{r2} = 1$ and $\vec{E}_2 = -3\vec{a}_x + 4\vec{a}_y - 2\vec{a}_z.$

(i) We know that for $z > 0$,

$$\vec{D}_2 = \varepsilon_2 \vec{E}_2 = 8.854 \times 10^{-12}(-3\vec{a}_x + 4\vec{a}_y - 2\vec{a}_z)$$

$$= (-26.56\vec{a}_x + 35.41\vec{a}_y - 17.7\vec{a}_z) \times 10^{-12} \text{ C/m}^2.$$

(ii) The tangent and normal components of $\vec{D}_1$, $\vec{E}_1, \vec{D}_2$ and $\vec{E}_2$ are shown in Fig. 3.27.

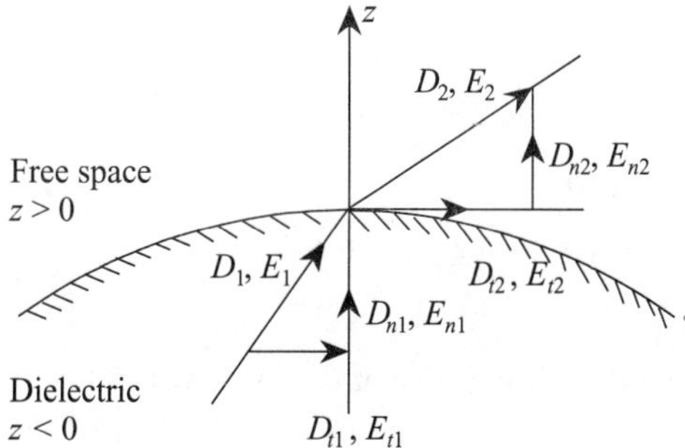

Fig. 3.27 *Boundary between dielectric and free space*

From Fig. 3.27, $\vec{E}_2 = \vec{E}_{t2} + \vec{E}_{n2}$.

$$\therefore \quad \vec{E}_{n2} = -2\vec{a}_z.$$

$$E_{t2} = \vec{E}_2 - \vec{E}_{n2} = -3\vec{a}_x + 4\vec{a}_y - 2\vec{a}_z + 2\vec{a}_z = -3\vec{a}_x + 4\vec{a}_y.$$

At the boundary, the tangential components are continuous.

$$\vec{E}_{t1} = \vec{E}_{t2} = -3\vec{a}_x + 4\vec{a}_y.$$

$$|E_{t1}| = \sqrt{3^2 + 4^2} = 5 \text{ V/m}.$$

The flux density is $D_{t1} = \varepsilon E_{t2} = \varepsilon(-3\vec{a}_x + 4\vec{a}_y)$

$$= 6.5 \times 8.854 \times 10^{-12}(-3\vec{a}_x + 4\vec{a}_y)$$

$$= (-17.2\vec{a}_x + 23.02\vec{a}_y) \times 10^{-11} \text{ C/m}^2.$$

$$|D_{t1}| = 0.287 \text{ nC/m}^2.$$

Example 3.19 Find V at $P(2,1,3)$ for the field of two axial conducting cones with $V = 50$ volts at $\theta = 30^\circ$ and $V = 20$ volts at $\theta = 50^\circ$.

Solution Consider two axial conducting cones. Given boundary conditions

$$\text{at } \theta = 30^\circ, \; V = 50 \text{ volts} \;\; ; \;\; \text{at } \theta = 50^\circ, \; V = 20 \text{ volts}.$$

Since V depends only on θ, in spherical coordinates, Laplace's equation is

$$\nabla^2 V = \frac{1}{r^2 \sin\theta} \frac{\partial}{\partial\theta}\left(\sin\theta \cdot \frac{\partial V}{\partial\theta}\right) = 0$$

or $\dfrac{d}{d\theta}\left(\sin\theta\cdot\dfrac{dV}{d\theta}\right)=0.$

Integrating, we get

$$\sin\theta\frac{dV}{d\theta}=A_1 \text{ (where } A_1 \text{ is a constant).}$$

$$\frac{dV}{d\theta}=\frac{A_1}{\sin\theta}.$$

Again integrating, $V=A_1\displaystyle\int\frac{1}{\sin\theta}d\theta$

$$=A_1\ln(\tan\theta/2)+A_2, \qquad \text{(From Eq. (3.68))}$$

where A_2 is another constant.

Now applying the boundary conditions that at $\theta=30^\circ$, $V=50$,

$$A_1\ln(\tan 15^\circ)+A_2=50,$$

$$-1.32A_1+A_2=50.$$

Also, at $\theta=50^\circ$, $V=20$ V.

$$20=A_1\ln(\tan 25^\circ)+A_2$$

$$-0.763A_1+A_2=20.$$

Solving for A_1 and A_2, we get

$$A_1=-53.86;\ A_2=-21.09.$$

$$\therefore \quad V=-53.86\ln(\tan\theta/2)-21.09 \text{ volts.}$$

Since in spherical coordinates, $\theta=\tan^{-1}\left(\dfrac{\sqrt{x^2+y^2}}{z}\right)$,

at point $P(2,1,3)$, $\theta=\tan^{-1}\left(\dfrac{\sqrt{5}}{3}\right)=36.7^\circ.$

$$\therefore \quad V=-53.86\ln\left(\tan\frac{36.7}{2}\right)-21.09=38.35 \text{ volts.}$$

Example 3.20 In cylindrical coordinates, $V=60$ V at $\rho=6$ mm and $V=0$ at $\rho=66$ mm. Find the voltage at $\rho=140$ mm if the potential depends only on ρ.

Solution Given that the potential depends only on ρ. So consider cylindrical coordinates. The boundary conditions are

$$\text{at} \quad \rho=6 \text{ mm}=6\times10^{-3} \text{ m}, \quad V=60 \text{ V};$$

$$\text{at} \quad \rho=66 \text{ mm}=66\times10^{-3} \text{ m}, \quad V=0 \text{ V}.$$

To find the potential, use Laplace's equation $\nabla^2 V = 0$.

In cylindrical coordinates, this is

$$\frac{1}{\rho}\frac{\partial}{\partial\rho}\left(\rho\frac{\partial V}{\partial\rho}\right) = 0 \quad \text{since } V \text{ depends only on } \rho.$$

Integrating, $\rho\dfrac{dV}{d\rho} = A_1$ (where A_1 is a constant).

$$\frac{dV}{d\rho} = \frac{A_1}{\rho}.$$

Again integrating,

$$V = A_1 \ln\rho + A_2 \quad \text{(where } A_2 \text{ is another constant)}.$$

Apply the boundary conditions that at $\rho = 66\times10^{-3}$ m, $V = 0$.

$$0 = A_1 \ln(66\times10^{-3}) + A_2$$

$$0 = -2.718A_1 + A_2, \quad \text{or} \quad A_2 = 2.718\,A_1$$

and at $\rho = 6\times10^{-3}$m, $V = 60$ V,

$$60 = A_1 \ln(6\times10^{-3}) + A_2 = -5.116\,A_1 + A_2.$$

On solving the equations, $A_1 = \dfrac{-60}{2.398} = -25,$

$$A_2 = 2.718\times(-25) = -67.95.$$

Substituting these values in the potential equation,

$$V = -25\ln\rho - 67.95.$$

Now at $\rho = 140$ mm $= 140\times10^{-3}$, the voltage is

$$V = -25\ln(140\times10^{-3}) - 67.95$$

$$= 49.15 - 67.95 = -18.8 \text{ volts}.$$

Example 3.21 In spherical coordinates, $V = 0$ at $r = 0.1$ m and $V = 100$ volts at $r = 2$ m. Assuming free space between concentric spherical shells, find $\vec{E}$ and $\vec{D}$.

Solution Given spherical coordinates and boundary conditions that at $r = 0.1$ m, $V = 0$, and at $r = 2$ m, $V = 100$ V.

From Laplace's equation, in spherical coordinates, we have

$$\nabla^2 V = \frac{1}{r^2}\frac{\partial}{\partial r}\left(r^2\frac{\partial V}{\partial r}\right) + \frac{1}{r^2\sin\theta}\frac{\partial}{\partial\theta}\left(\sin\theta\frac{\partial V}{\partial\theta}\right) + \frac{1}{r^2\sin^2\theta}\frac{\partial^2 V}{\partial\phi^2} = 0.$$

Since V depends only on r, $\frac{1}{r^2}\frac{\partial}{\partial r}\left(r^2\frac{\partial V}{\partial r}\right)=0.$

Integrating, $r^2\frac{dV}{dr}=A_1$ (where A_1 is a constant) or, $\frac{dV}{dr}=\frac{A_1}{r^2}$

Again integrating, $V=\frac{-A_1}{r}+A_2$ (A_2 is a constant).

Applying the boundary conditions that at $r=0.1\text{m},\ V=0$,

$$V=\frac{-A_1}{0.1}+A_2=0.$$

$$-10A_1+A_2=0 \text{ or } A_2=10\,A_1.$$

At $r=2\text{m},\ V=100.$

$$\frac{-A_1}{2}+A_2=100.$$

$$-A_1+2A_2=200.$$

Solving the equations for A_1 and A_2, we get

$$A_1=\frac{200}{19} \text{ and } A_2=\frac{2000}{19}.$$

$$\therefore \quad V=\frac{200}{19}\left(\frac{-1}{r}+10\right)=\frac{-10.52}{r}+105.2 \text{ volts.}$$

We know that the electric field intensity $\vec{E}=-\nabla V$.

$$\vec{E}=-\left[\frac{\partial V}{\partial r}\vec{a}_r+\frac{1}{r}\frac{\partial V}{\partial \theta}\vec{a}_\theta+\frac{\partial V}{\partial \phi}\vec{a}_\phi\right].$$

Since V is a function of r,

$$\vec{E}=-\frac{d}{dr}\left[\frac{200}{19}\left(\frac{-1}{r}+10\right)\right]\vec{a}_r=10.52\frac{d}{dr}\left(\frac{1}{r}-10\right)\vec{a}_r$$

$$=10.52\ln(r)\vec{a}_r \text{ V/m.}$$

The electric flux density

$$\vec{D}=\varepsilon_0\vec{E} \quad \text{(in free space)}$$

$$=10.52\varepsilon_0\ln(r)\vec{a}_r=93.2\ln(r)\vec{a}_r \text{ pC/m}^2.$$

Example 3.22 Show that the potential due to an electric dipole satisfies Laplace's equation.

Solution We know that the potential due to a dipole in spherical coordinates is

$$V=\frac{\vec{p}\cdot\vec{a}_r}{4\pi\varepsilon_0\,r^2}.$$

Converting into Cartesian coordinates, the distance vector

$$\vec{r} = x\vec{a}_x + y\vec{a}_y + z\vec{a}_z; \quad r = \sqrt{x^2+y^2+z^2}$$

and $\vec{a}_r = \dfrac{x\vec{a}_x + y\vec{a}_y + z\vec{a}_z}{(x^2+y^2+z^2)^{1/2}}.$

Assume $\vec{p} = p\,\vec{a}_z$ along the z-axis.

$$\therefore \quad V = \frac{p\vec{a}_z \cdot (x\,\vec{a}_x + y\,\vec{a}_y + z\,\vec{a}_z)}{4\pi\,\varepsilon_0 (x^2+y^2+z^2)^{1/2}} = \frac{pz}{4\pi\varepsilon_0 (x^2+y^2+z^2)^{3/2}}.$$

Let $\dfrac{p}{4\pi\varepsilon_0} = A$, a constant. Then $V = \dfrac{Az}{\left(x^2+y^2+z^2\right)^{3/2}}.$

We know that Laplace's equation in Cartesian coordinates is

$$\nabla^2 V = 0 \quad \text{or} \quad \nabla^2 V = \frac{\partial^2 V}{\partial x^2} + \frac{\partial^2 V}{\partial y^2} + \frac{\partial^2 V}{\partial z^2}.$$

Now,
$$\frac{\partial^2 V}{\partial x^2} = \frac{\partial}{\partial x}\left[\frac{-Az\left(\frac{3}{2}\right)2x}{(x^2+y^2+z^2)^{5/2}}\right]$$

$$= 3Az\left[\frac{1}{(x^2+y^2+z^2)^{5/2}} - \frac{x\left(\frac{5}{2}\right)(2x)}{(x^2+y^2+z^2)^{7/2}}\right]$$

$$= 3Az\left[\frac{x^2+y^2+z^2-5x^2}{(x^2+y^2+z^2)^{7/2}}\right]$$

or,
$$\frac{\partial^2 V}{\partial x^2} = A\left[\frac{-3z(z^2+y^2-4x^2)}{(x^2+y^2+z^2)^{7/2}}\right].$$

Similarly,
$$\frac{\partial^2 V}{\partial y^2} = A\left[\frac{-3z(x^2+z^2-4y^2)}{(x^2+y^2+z^2)^{7/2}}\right],$$

and,
$$\frac{\partial^2 V}{\partial z^2} = \frac{\partial}{\partial z}\left[\frac{A(x^2+y^2-2z^2)}{(x^2+y^2+z^2)^{5/2}}\right].$$

$$= \frac{-4Az}{(x^2+y^2+z^2)^{5/2}} + \frac{A(x^2+y^2+2z^2)\left(\frac{-5}{2}\right)(2z)}{(x^2+y^2+z^2)}$$

$$= \frac{-Az[4(x^2+y^2+z^2)+5(x^2+y^2-2z^2)]}{(x^2+y^2+z^2)^{7/2}}$$

or $\dfrac{\partial^2 V}{\partial z^2} = \dfrac{-A(9x^2z + 9y^2z - 6z^3)}{(x^2+y^2+z^2)^{7/2}}$

$$\therefore \quad \frac{\partial^2 V}{\partial x^2} + \frac{\partial^2 V}{\partial y^2} + \frac{\partial^2 V}{\partial z^2}$$

$$= \frac{-A}{(x^2+y^2+z^2)^{7/2}}[-3y^2z - 3z^3 + 12x^2z - 3x^2z - 3z^3 + 12y^2z - 9x^2z - 9y^2z + 6z^3]$$

$$= \frac{-A}{(x^2+y^2+z^2)^{7/2}}[9x^2z + 9y^2z - 6z^3 - 9x^2z - 9y^2z + 6z^3] = 0$$

$\therefore \quad \nabla^2 V = 0$, satisfying Laplace's equation.

Example 3.23 Find the capacitance of a parallel plate capacitor when $S = 1\ \text{m}^2$ and the distance between the plates is 1 mm. The voltage gradient is 10^5 V/m and the charge density on the plate is $2\ \mu\text{C/m}^2$.

Solution Given $S = 1\ \text{m}^2$; $d = 1\ \text{mm} = 1 \times 10^{-3}$ m; $E = 10^5$ V/m;

$$\rho_s = 2\ \mu\text{C/m}^2 = 2 \times 10^{-6}\ \text{C/m}^2$$

We know that the capacitance, $C = \dfrac{Q}{V}$.

The charge on the plates $Q = \rho_s S = 2 \times 10^{-6} \times 1 = 2\ \mu\text{C}$.

The potential $V = Ed = 10^5 \times 1 \times 10^{-3} = 100$ V.

$$\therefore \quad C = \frac{2 \times 10^{-6}}{100} = 0.02\ \mu\text{F}.$$

Example 3.24 Calculate the capacitance of a parallel plate capacitor with the following particulars: plate area $= 100\ \text{cm}^2$, $\varepsilon_{r_1} = 4$, $d_1 = 2$ mm, $\varepsilon_{r_2} = 3$, $d_2 = 3$ mm.

Solution Given $S = 10^{-2}$ m; $d_1 = 2 \times 10^{-3}$ m; $d_2 = 3 \times 10^{-3}$ m; $\varepsilon_{r_1} = 4$, $\varepsilon_{r_2} = 3$.
(a) Since the plates are connected in parallel,

$$\frac{1}{C} = \frac{1}{C_1} + \frac{1}{C_2} = \frac{d_1}{\varepsilon_1 S_1} + \frac{d_2}{\varepsilon_2 S_2}$$

$$\frac{1}{C} = \frac{1}{8.854 \times 10^{-12} \times 10^{-2}} \left(\frac{2 \times 10^{-3}}{4} + \frac{3 \times 10^{-3}}{3} \right)$$

$$C = \frac{8.8.54 \times 10^{-14}}{15 \times 10^{-3}} = 5.9 \times 10^{-11}\ \text{F} = 59\ \text{pF}.$$

Example 3.25 A parallel plate capacitor is considered with the following details: plate area $= 100\ \text{cm}^2$, $\varepsilon_{r_1} = 4$, $d_1 = 2$ mm, $\varepsilon_{r_2} = 3$ and $d_2 = 3$ mm. If 200 V is applied across the plates, what will be the voltage gradient across each dielectric?

Solution Given $V = 200$ V; $S = 100$ cm^2;

$$\varepsilon_{r_1} = 4,\ d_1 = 2\times 10^{-3}\text{ m};\ \varepsilon_{r_2} = 3,\ d_2 = 3\times 10^{-3}\text{ m}.$$

If V_1 and V_2 are the potential across the dielectrics, then the total potential is

$$V = V_1 + V_2 = 200\ V.$$

Also, if E_1 and E_2 are the potential gradients across the dielectrics, then

$$E_1 d_1 + E_2 d_2 = 200.$$

$$2\times 10^{-3} E_1 + 3\times 10^{-3} E_2 = 200.$$

$$2\,E_1 + 3E_2 = 200\times 10^3.$$

At the boundary between the dielectrics, we know that,

$$\varepsilon_1 E_1 = \varepsilon_2 E_2.$$

$$4\,E_1 = 3E_2 \quad \text{or} \quad E_1 = \frac{3}{4}E_2.$$

$$\therefore \quad \frac{6E_2}{4} + 3E_2 = 2\times 10^5.$$

$$4.5\,E_2 = 2\times 10^5.$$

$$E_2 = 44.44\times 10^3\ \text{V/m}$$

$$\text{and} \quad E_1 = \frac{3}{4}E_2 = 33.33\times 10^3\ \text{V/m}.$$

Therefore the voltage gradients are $E_1 = 33.33\times 10^3$ V/m and $E_2 = 44.44\times 10^3$ V/m.

Example 3.26 A parallel plate capacitor consists of two plates, each 30 mm ×30 mm, spaced 2 mm apart and 2 dielectrics, each 1 mm thick, having relative primitivities of 3 and 5 respectively. If the potential difference between the plates is 5000 V, calculate the voltage gradient in each dielectric.

Solution Consider a parallel plate capacitor as shown in Fig. 3.28.

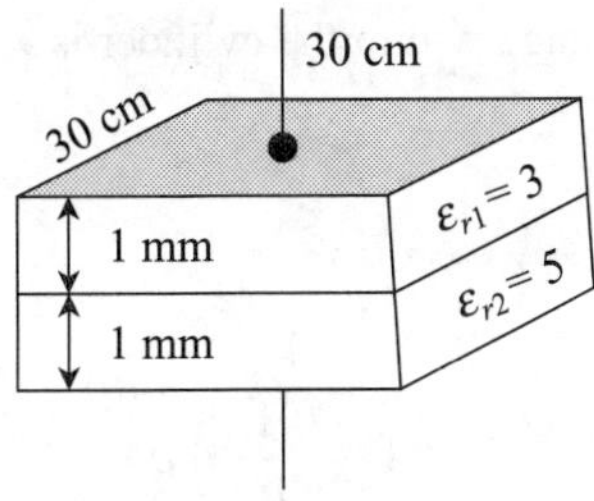

Fig. 3.28 *Parallel plate capacitor*

Given the area of a capacitor $S = 30 \times 30 = 900\ \text{cm}^2 = 9 \times 10^{-2}\ \text{m}^2$.

$$d_1 = d_2 = 1\ \text{mm} = 1 \times 10^{-3}\text{m};\ \varepsilon_{r_1} = 3;\ \varepsilon_{r_2} = 5;\ V = 5000\ \text{volts}.$$

Therefore, the capacitance of the dielectric are:

$$C_1 = \frac{\varepsilon_0 \varepsilon_1 S}{d} = \frac{3\varepsilon_0 \times 9 \times 10^{-2}}{10^{-3}} = 270\varepsilon_0$$

and $C_2 = \dfrac{\varepsilon_0 \varepsilon_2 S}{d} = \dfrac{5\varepsilon_0 \times 9 \times 10^{-2}}{10^{-3}} = 450\varepsilon_0.$

Total capacitance is $C = \dfrac{C_1 C_2}{C_1 + C_2} = \dfrac{270\varepsilon_0 \times 450\varepsilon_0}{270\varepsilon_0 + 450\varepsilon_0}$

$$= 168.75\ \varepsilon_0 = 1.5\ \text{nF}.$$

If V_1 and V_2 are the potentials across the dielectrics, then

$$V_1 = \frac{VC_2}{C_1 + C_2} = \frac{5000 \times 450}{720} = 3125\ \text{volts}.$$

$$V_2 = \frac{VC_1}{C_1 + C_2} = \frac{5000 \times 270}{720} = 1875\ \text{volts}.$$

The voltage gradient across the first dielectric is

$$E_1 = V_1/d = \frac{3125}{10^{-3}} = 3125\ \text{kV/m}.$$

The voltage gradient across the second dielectric is

$$E_2 = 1875\ \text{kV/m}.$$

Example 3.27 A cylindrical capacitor has radii $a = 2$ cm and $b = 3$ cm. The capacitor is filled with dielectric material with $\varepsilon_r = 1 + \dfrac{0.15}{\rho}$, where ρ is the radius. Find the capacitance if the length of the cylinder is 1.5 m.

Solution Given $a = 2\ \text{cm} = 2 \times 10^{-2}$ m; $b = 3\ \text{cm} = 3 \times 10^{-2}$ m; $L = 1.5$ m; $\varepsilon_r = 1 + \dfrac{0.15}{\rho}$.

We know that the differential voltage within the cylinder is

$$dV = \frac{-Qd\rho}{2\pi\varepsilon L\rho} = \frac{-Qd\rho}{2\pi\varepsilon_0\varepsilon_r L\rho}.$$

$$V = \frac{-Q}{2\pi L}\int_b^a \frac{1}{\varepsilon_r \varepsilon_0\ \rho} d\rho = \frac{-Q}{2\pi L\varepsilon_0}\int_b^a \frac{1}{\left(1 + \dfrac{0.15}{\rho}\right)\rho} d\rho$$

$$= \frac{-Q}{2\pi L\varepsilon_0} \int_b^a \frac{1}{(0.15+\rho)} d\rho = \frac{-Q}{2\pi\varepsilon_0 L} \ln(0.15+\rho)\Big|_b^a$$

$$V = \frac{Q}{2\pi\varepsilon_0 L} \ln\left(\frac{0.15+b}{0.15+a}\right).$$

Capacitance:

$$C = \frac{Q}{V} = 2\pi\varepsilon_0 L \ln\left(\frac{0.15+0.03}{0.15+0.02}\right) = 2\pi\varepsilon_0\, 1.5 \ln\left(\frac{0.18}{0.17}\right)$$

$$= \frac{1.5 \times 0.057 \times 10^{-9}}{18} = 4.77 \text{ pF}.$$

Example 3.28 Determine the capacitance of a capacitor consisting of two parallel metal plates of 30 cm ×30 cm surface area, separated by 5 mm gap in air. What is the total energy stored by the capacitor if the capacitor is charged with 500 V? What is the energy density?

Solution Given $S = 30 \text{ cm} \times 30 \text{ cm} = 900 \text{ cm}^2 = 9 \times 10^{-2} \text{ m}^2$;
$d = 5 \text{ mm} = 5 \times 10^{-3}\text{m}$; ε_r of air $= 1$; $V = 500$ V.

The capacitance $C = \dfrac{\varepsilon S}{d} = \dfrac{\varepsilon_0 \varepsilon_r S}{d} = \dfrac{\varepsilon_0 9 \times 10^{-2}}{5 \times 10^{-3}} = 0.16 \text{ nC}.$

The total energy stored by the capacitor is

$$W_E = \frac{1}{2}CV^2 = \frac{1}{2} \times 0.16 \times 10^{-9} \times 500^2 = 20\ \mu\text{J}.$$

The energy density is

$$w_E = \frac{W_E}{Sd} = \frac{20 \times 10^{-6}}{9 \times 10^{-2} \times 5 \times 10^{-3}} = 44 \text{ mJ/m}^3.$$

Example 3.29 If $C_1 = 100\ \mu$F, $C_2 = 50\ \mu$F, calculate the joint capacitance (i) for series connection, (ii) for parallel connection and (iii) the total energy stored with a steady applied potential difference of 1000 V in each case.

Solution Given $C_1 = 100\ \mu\text{F}; C_2 = 50\ \mu\text{F}$.
(i) If the capacitors are connected in series, the joint capacitance is

$$C_S = \frac{C_1 C_2}{C_1 + C_2} = 33.33\ \mu\text{F}.$$

(ii) If the capacitors are connected in parallel, the joint capacitance is

$$C_P = C_1 + C_2 = (100 + 50)\ \mu\text{F} = 150\ \mu\text{F}.$$

(iii) Given $V = 1000$ V.

We know that energy stored $W = \frac{1}{2}CV^2$

For series connection,

$$W_S = \frac{1}{2} \times 33.33 \times 10^{-6} \times 1000^2 = 16.67 \text{ J.}$$

For parallel connection,

$$W_P = \frac{1}{2} \times 150 \times 10^{-6} \times 1000^2 = 75 \text{ J.}$$

Example 3.30 A parallel plate capacitor with a large plate area is situated in air. When a potential difference of 100 V between the plates is applied, the stored energy is 44.21 μJ per unit area. Find the distance of separation between the plates.

Solution Given $V = 100$ V; stored energy $= 44.21$ μJ per unit area.
Let $S =$ area of the plates. Then $W_E = 44.21 \times 10^{-6} \times S$.

We know that

$$W_E = \frac{1}{2}CV^2.$$

$$\text{The capacitance} \quad C = \frac{2W_E}{V^2} = \frac{2 \times 44.21 \times 10^{-6} \times S}{100^2} = 88.42 \times 10^{10} \times S.$$

$$\text{or} \quad \frac{C}{S} = 88.42 \times 10^{-10}.$$

$$\text{We know that} \quad C = \frac{\varepsilon S}{d} \quad \text{or} \quad \frac{C}{S} = \frac{\varepsilon}{d} = 88.42 \times 10^{-10}.$$

$$\therefore \quad d = \frac{8.8.54 \times 10^{-12}}{8842 \times 10^{-10}} = 1 \text{ mm.}$$

Example 3.31 A parallel plate capacitor with air as dielectric has a plate area of 36 πcm^2 and a separation of 1 mm between the plates. It is charged to 100 volts by connecting it across a battery. If the battery is disconnected and plate separation is increased to 2 mm, calculate the change in (i) potential difference across the plates and (ii) energy stored.

Solution Given $S = 36\pi$ cm^2 $= 36\pi \times 10^{-4}$ m^2; $d_1 = 1$ mm $= 10^{-3}$; ε_r for air $= 1$;
$V_1 = 100$ V.

$$\text{The capacitance} \quad C_1 = \frac{\varepsilon S}{d_1} = \frac{\varepsilon_0 \varepsilon_r S}{d_1} = \frac{\varepsilon_0 \times 36\pi \times 10^{-4}}{10^{-3}} = 100 \text{ pF.}$$

The charge on the capacitor $Q = C_1V_1 = 100 \times 100 \times 10^{-12} = 10$ nC.

If the plate separation is increased to 2 mm, $d_2 = 2 \times 10^{-3}$ m.

The capacitance is changed to

$$C_2 = \frac{\varepsilon S}{d_2} = \frac{8.854 \times 10^{-12} \times 36\pi \times 10^{-4}}{2 \times 10^{-3}} = 50 \text{ pF}.$$

When the battery is disconnected, the charge remains the same.

(i) The potential difference across plates is

$$Q = V_2 C_2.$$

$$\therefore \quad V_2 = \frac{Q}{C_2} = \frac{10 \times 10^{-9}}{50 \times 10^{-12}} = 200 \text{ volts}.$$

So the potential difference is increased from 100 V to 200 V.

(ii) Energy stored:

$$\text{When } V_1 = 100\text{V}, \quad W_1 = \frac{1}{2} C_1 V_1^2 = \frac{1}{2} \times 100 \times 10^{-12} \times 100^2 = 0.5\ \mu\text{J},$$

$$\text{and when } V_2 = 200\text{V}, \quad W_2 = \frac{1}{2} C_2 V_2^2 = 0.5\ \mu\text{J}.$$

There is no change in energy.

Example 3.32 Two capacitors each of capacitance C_1 and C_2 are charged to Q and zero respectively. If the capacitors are connected in parallel, find the potential and charges on each capacitor, and the loss in energy stored.

Solution Given capacitors of capacitances C_1 and C_2. The charge on C_1 is Q and the charge on C_2 = zero.

If the capacitors are connected in parallel, the charge is distributed on both capacitors. Let Q_1 and Q_2 be the charges on each capacitor.

Then the total charge $Q = Q_1 + Q_2$.

The potential across the capacitors is

$$V = \frac{Q}{C_1 + C_2}.$$

$$\text{Then} \quad Q_1 = C_1 V = \frac{QC_1}{C_1 + C_2} \quad \text{and} \quad Q_2 = C_2 V = \frac{QC_2}{C_1 + C_2}.$$

The energy stored on capacitor C_1 before connection is

$$W_1 = \frac{Q^2}{2C_1}$$

and that on capacitor C_2 is zero.

The energy stored after parallel connection is

$$W_2 = \frac{1}{2} QV = \frac{1}{2} Q \frac{Q}{C_1 + C_2} = \frac{Q^2}{2(C_1 + C_2)}.$$

Loss in energy $= W_1 - W_2 = \dfrac{Q^2}{2C_1} - \dfrac{Q^2}{2(C_1+C_2)} = \dfrac{Q^2}{2}\left[\dfrac{1}{C_1} - \dfrac{1}{C_1+C_2}\right]$ joules.

Example 3.33 Given the potential function $V = 3x^2 + 4y^2$, find the energy stored in the volume described by $0 \le x \le 1,\ 0 \le y \le 1$ and $0 \le z \le 1$ m.

Solution Given $V = 3x^2 + 4y^2$.

$$\vec{E} = -\nabla V = -\left[\frac{\partial V}{\partial x}\vec{a}_x + \frac{\partial V}{\partial y}\vec{a}_y + \frac{\partial V}{\partial z}\vec{a}_z\right] = -6x\vec{a}_x - 8y\vec{a}_y.$$

Energy density $\quad w_E = \dfrac{1}{2}\vec{D}\cdot\vec{E} = \dfrac{\varepsilon_0}{2}|E|^2 = \dfrac{\varepsilon_0}{2}[36x^2 + 64y^2].$

Energy stored $\quad w_E = \displaystyle\int_v w_E dv = \frac{\varepsilon_0}{2}\int_0^1\int_0^1\int_0^1 (36x^2 + 64y^2)dxdydz = \frac{\varepsilon_0}{2}\left(\frac{36}{3} + \frac{64}{3}\right).$

$$W_E = \frac{50}{3}\varepsilon_0 \text{ joules} = 0.147 \text{ nJ}.$$

Example 3.34 A spherical capacitor has an inner conductor of radius a, carrying a charge Q at zero potential. If the outer conductor contracts from a radius b to c under intemal forces, find the work done by the electric field as a result of the contraction.

Solution Given the charge on the inner conductor of radius a is Q, consider an arbitrary Gaussian surface of radius r $(a \le r \le b)$.

Applying Gauss' law,

$$\int_S \vec{D}\cdot d\vec{S} = Q_{\text{encl}}.$$

$$D\cdot 4\pi r^2 = -Q$$

since at zero potential, charge is negative

$$D = \frac{-Q}{4\pi r^2} \quad \text{or,} \quad E = \frac{-Q}{4\pi\varepsilon_0 r^2}.$$

Work done on contraction of radius from b to c is

$$W = W_E = \int_v w_E dv.$$

$$W = \int_v \frac{\vec{D}\cdot\vec{E}}{2}dv = \int_b^c \frac{\varepsilon_0|E|^2}{2}4\pi r^2\, dr$$

$$= \int_b^c \frac{\varepsilon_0}{2}\left(\frac{-Q}{4\pi\varepsilon_0 r^2}\right)^2 4\pi r^2\, dr = \int_b^c \frac{\varepsilon_0}{2}\frac{Q^2}{16\pi^2\varepsilon_0^2 r^4} \times 4\pi r^2\, dr$$

$$= \int_b^c \frac{Q^2}{8\pi\varepsilon_0 r^2} dr = \frac{Q^2}{4\pi\varepsilon_0}\left[\frac{-1}{r}\right]_b^c = \frac{Q^2}{8\pi\varepsilon_0}\left[\frac{1}{c}-\frac{1}{b}\right]$$

$$= \frac{Q^2}{8\pi\varepsilon_0}\left[\frac{1}{c}-\frac{1}{b}\right].$$

Summary

1. Materials can be classified into conductors, semi conductors and insulators or dielectrics according to their conductivity.
2. Electric current can be expressed as

   $$I = \frac{-dQ}{dt} = \int_s \vec{J}\cdot d\vec{S} \text{ amps}$$

   where $\vec{J}$ is the current density.
3. Resistance is $R = \frac{V}{I} = \frac{l}{\sigma A} = \frac{\rho l}{A}$ ohms

   where σ is the conductivity, and $\rho = \frac{1}{\sigma}$ is the resistivity of the material.
4. Point form of Ohm's law is given by

   $$J = \sigma E = -\mu_e \rho_e \vec{E} \text{ A/m}^2.$$

5. Inside the perfect conductor $\vec{E} = 0$, $\rho_v = 0$, and $V = 0$, the conductor surfaces are equipotential surfaces. The surface charge density is $\rho_s = \vec{D} = \varepsilon_0\vec{E}$, normal to the conductor surface.
6. For a linear and isotropic dielectric medium, polarization is given by

   $\vec{P} = \chi_e \varepsilon_0 \vec{E}$, where χ_e is called electric susceptibility.

   and the electric flux density is $\vec{D} = \varepsilon_0\vec{E} + \vec{P} = (1+\chi_e)\varepsilon_0\vec{E} = \varepsilon_r\varepsilon_0\vec{E}$.
7. The continuity equation can be expressed as $\nabla\cdot\vec{J} = \frac{-\partial\rho_v}{\partial t}$.

   For steady current, $\nabla\cdot\vec{J} = 0$.
8. Relaxation time depends only on the medium properties of the dielectric material and it can be expressed as

   $T_r = RC = \frac{\varepsilon}{\sigma}$ s, where R is the leakage resistance and C is the capacitance
9. Boundary conditions:

   (1) The tangential component of the electric field intensity is continuous at the boundary.

   $\vec{E}_{\tan 1} = \vec{E}_{\tan 2}$ and $\frac{D_{\tan_1}}{D_{\tan_2}} = \frac{\varepsilon_1}{\varepsilon_2} = \frac{\varepsilon_{r_1}}{\varepsilon_{r_2}}$

   (2) If the surface charge density ρ_s exists on the boundary, then

   $D_{n1} - D_{n2} = \rho_s$.

Between two perfect dielectrics, $\rho_s = 0$ and $D_{n_1} = D_{n_2}$.

Between a conductor and a dielectric, $E_{\tan 1} = E_{\tan 2} = 0,\ D_{n1} = \rho_s$.

10. Poisson's equation can be expressed as $\nabla^2 V = -\rho_v/\varepsilon$. If $\nabla^2 V = 0$, it is called Laplace's equation.
11. By solving Laplace's equation or Poisson's equation with the given boundary conditions, we can find electric field intensity from $\vec{E} = -\nabla V$, electric flux density from $\vec{D} = \varepsilon\vec{E}$, the charge from $Q = \int_S D \cdot dS$ and the capacitance from $C = Q/V$.
12. The capacitance of a parallel plate capacitor can be expressed as

$$C = \frac{Q}{V} = \frac{\varepsilon S}{d} \text{ farads.}$$

13. The energy stored in the electric field of a capacitor is given by

$$W = \frac{1}{2}CV^2 \text{ joules}$$

and the energy density is $w_E = \frac{1}{2}\vec{D}\cdot\vec{E} = \frac{1}{2}\varepsilon(E)^2$ joules/m^3.

Review Questions

1 Describe dielectric materials in brief. *(May 2008)*

2 Describe electric displacement vector in dielectrics. *(May 2008, 2010)*

3 State and explain the continuity equation of current in integral form and point form. *(May 2010, Nov 2011, Aug 2014)*

4 Explain and derive the boundary conditions for a conductor and free space interface. *(May 2011, 2012)*

5 State and explain the continuity equation of current in integral form and point from. *(May 2011, Jan 2014)*

6 State and derive boundary conditions between two perfect dielectrics. *(Nov 2011, May 2012)*

7 Derive Poisson's and Laplace's equations. *(May 2010, 2011, Aug 2014)*

8 Obtain the solution of Laplace's equation in spherical coordinates. *(May 2010)*

9 Briefly explain the behaviour of conductors in an electric field. *(May 2010, 2011)*

10 Show that electric field can be determined from the scalar potential V through the relation $E = -\nabla V$. Hence arrive at the Poisson's equation. *(May 2009, 2011, Nov 2013)*

11 Explain how the Laplace's and the Poisson's equations are useful in solving electrostatic potential and field distribution problems. *(Nov 2010, Jan 2014)*

12 State and prove the conditions on the tangential and normal components of electric flux density and electric field intensity at the boundary between dielectrics. *(May 2011, 2012)*

13 Explain the difference between conductors, insulators and dielectrics using the band energy gap model. (*Nov 2009, 2011*)

14 A conducting body is in the electric field of static charges. Explain why the net electric field at any point inside the conducting body will be zero. (*May 2011, May 2013*)

15 A uniform charge density of ρ_v C/m^2 exists throughout the volume of a sphere of radius b metres. Using Poisson's equation, find the value of the electric field intensity and potential at any point inside the sphere for which $0 \leq r \leq b$. (*May 2010*)

16. Define capacitance. Derive the expression for capacitance of a spherical type capacitor. (*Aug 2014*)

Problems

1 Let $S = 120$ cm^2, $d = 4$ mm, and $\varepsilon_r = 12$ for a parallel-plate capacitor. (a) Calculate the capacitance. (b) After connecting a 40 V battery across the capacitor, calculate E, D, Q, and the total stored electrostatic energy.

2 A dielectric circular cylinder used between the plates of a capacitor has a thickness of 0.2 mm and a radius of 1.4 cm. The dielectric properties are $\varepsilon_r = 400$ and $\sigma = 10^{-5}$S/ m. Calculate C.

3 Two coaxial conducting cylinders of radius 2 cm and 4 cm have a length of 1 m. The region between the cylinders contains a layer of dielectric from $\rho = 2$ cm to $\rho = 3$ cm with $\varepsilon_r = 4$. Find the capacitance.

4 A copper sphere of radius 4 cm carries a uniformly-distributed total charge of 5 μC in free space. (a) Use Gauss' law to find D external to the sphere. (b) Calculate the total energy stored in the electrostatic field.

5 Given the current density $\vec{J} = -10^4(\sin(2x)e^{-2y}\vec{a}_x + \cos(2x)e^{-2y}\vec{a}_y)$ KAm^2, (a) find the total current crossing the plane $y = 1$ in the $\vec{a}_y$ direction in the region $0 < x < 1, 0 < z < 2$ and (b) evaluate $\nabla.\vec{J}$ at $P(\rho = 2.4,\ \phi = 0.08,\ z = 6.05)$.

6 Determine the capacitance of a conducting sphere radius 5 cm deeply immersed in sea water $(\varepsilon_r = 80)$.

7 A conducting sphere of radius 2 cm is surrounded by a concentric conducting sphere of radius 5 cm. If the space between the spheres is filled with sodium chloride $(\varepsilon_R = 5.9)$, calculate the capacitance of the system.

8 A spherical capacitor with $a = 1.5$ cm, $b = 4$ cm, has an inhomogeneous dielectric $\varepsilon = 10\varepsilon_0/r$. Calculate the capacitance.

9 Estimate the current density and electric field intensity when the drift-velocity of the electrons is 5.3×10^4 m/s in an aluminium conductor. Take conductivity as 3.82×10^7 S/m and mobility as 1.4×10^{-3}m^2 / V-s.

10 In a certain region $\vec{J} = 3r^2 \cos\theta\vec{a}_r - r^2 \sin\theta\vec{a}_\theta$ A/m^2, find the current crossing the surface defined by $\theta = 30°; 0 < \phi < 2\pi,\ 0 < r < 2$.

11 Find the current in the circular wire of radius 2 mm, if the current density is $\vec{J} = 15(1 - e^{-1000r})\vec{a}_z$ A/m^2.

12 A current sheet width 5 m lies in the $z = 0$ planc and contains a total charge of 15 A in a direction from the origin to point (1, 3, 0). Find an expression for linear charge density.

13 Find the polarization P in the dielectric with $\varepsilon_2 = 2.8$, given the electric flux density $D = 3.0 \times 10^{-7}\vec{a}_x$ C/m^2

14 Find the dielectric constant for a material in which the electric flux density is four times the polarization.

15 Given that $\vec{D}_1 = 2\vec{a}_x - 4\vec{a}_y + 1.5\vec{a}_z$ C/m^2 in the region $x > 0$, which is free space, find the polarization $\vec{P}_2$ in the region $x < 0$ with a dielectric constant $\varepsilon_2 = 5.0$.

16 In a dielectric material with $\varepsilon_2 = 3.6$, $\vec{D} = 2.5 \times 10^{-9}$ C/m^2, find the magnitude of E, P and χ_c of the material.

17 Region 1 defined by $x < 0$ is a free space, while region 2 defined by $x > 0$ is a dielectric material for which $\varepsilon_r = 24$. Given $D_1 = 3\vec{a}_x - 4a_y + 6\vec{a}_z$ C/m^2, determine $\vec{E}_2$.

18 The plane interface between two dielectrics is given by $3x + z = 5$ m on the side including the origin $\vec{D}_1 = (4.5\vec{a}_x + 3.2\vec{a}_y) \times 10^{-7}$ C/m^2 and $\varepsilon_{r_1} = 4.3$. On the other side, $\varepsilon_{r2} = 18$. Find E_2 and θ_2, the angle between the field and the boundary.

19 Calculate the emerging angle by which the vector E changes its direction as it passes from a medium with $\varepsilon_r = 10$ into air making an angle of 45° with the interface as it enters.

20 In a one-dimensional device, the charge density is given by $\rho_v = \rho_0 x/a$. If $\vec{E} = 0$ at $x = 0$ and $V = 0$ at $x =$ a, find V and E.

21 Two conducting cones ($\theta = \pi/10$ and $\theta = \pi/6$) of infinite extent are separated by an infinitesimal gap at $r = 0$. If $V(\theta = \pi/10) = 0$ and $V(\theta = \pi/6) = 50$ V, find V and $\vec{E}$ between the cones.

22 In free space, $V = 6xy^2z + 8$. At point $P(1, 2, -5)$, find $\vec{E}$ and ρ_v.

23 Show that the following potentials satisfy Laplace's equation.

$$(a)V = e^{-5x}\cos(13y)\sinh(12z);\ (b)\mathrm{V} = \frac{z\cos\theta}{p};\ (c)\mathrm{V} = \frac{30\cos\theta}{r^2}.$$

24 If $\rho_v = 100$ in spherical coordinates, (a) find $\vec{D}$ and $\vec{E}_r$ using Gauss' law. (b) Verify the result by any suitable method. (c) Find the potential V from $\vec{E}_r$ taking $V = 0$ at $r = 0$.

25 Find the potential function subject to the boundary conditions $V = 100$ V at $z = 0.01$ and $V = -33$ V at $z = 0.02$.

26 Find voltage at $\theta = 20°$ for the field between two conduting cones with $V = 0$ at $\theta = 30°$ and $V = 100$ volts at $\theta = 10°$.

27 The voltages on two concentric cylinders with radii $\rho_1 = 2$ cm and $\rho_2 = 6$ cm are 60 V and -20 V respectively. Calculate the potential and electric field at $\rho = 4$ cm.

28 Two conducting cones of infinite extent have potentials $V = 100$ V at $\theta = 120°$ and $V = 0$ at 30°. Find V and $\vec{E}$ at $P(3, 0, 4)$.

29 Find the potential V at the point $P(2, 3, 4)$ for the field of two coaxial conducting cylinders, given $V = 60$ V at $r = 3$ m and $V = 10$ V at $r = 5$ m.

Answers

1. (a) 319 pF (b)104 V/m, 1.06 μC/m^2, 12.7 nC, 255 nJ
2. 10.9 nF
3. 143 pF
4. 2.81 J
5. -1.23 mA, 2.0 A/m^3
6. 444 pF
7. 21.85 pF
8. 1.13 nF
9. 1.45×10^7A/m^2; 0.379 V/m
10. $= -6.283$ A
11. 0.133 A
12. $\dfrac{3(\vec{a}_x + 3\vec{a}_r)}{\sqrt{10}}$ A/m
13. 1.93×10^{-7} C/m^2
14. 1.33
15. $8\vec{a}_x - 16\vec{a}_y + 6\vec{a}_z$ C/m^2
16. 8.99 kV/m; 206×10^{-9} C/m^2; 2.6
17. $0.125\vec{a}_x - 4\vec{a}_y + 4\vec{a}_z$ V/m
18. 3.37×10^4 V/m, 83.0°
19. 5.7° from normal
20. $\dfrac{\rho_0}{6\varepsilon a}(a^3 - x^3)$, $\dfrac{\rho_0 x^2}{2\varepsilon a}\vec{a}_x$
21. $95.1\left[\dfrac{\tan\theta/2}{0.1584}\right]$ V; $\vec{E} = \dfrac{-95.1}{r\sin\theta}a_\theta$V/m
22. $\vec{E} = 120\vec{a}_x + 120\vec{a}_y - 12\vec{a}_z$V/m, 530.5 pC/m^3
23.
24. $\vec{D}_r = 50, \vec{E}_r = 50/\varepsilon_0$, $V = \dfrac{50r}{\varepsilon_0}$
25. $V = -1.33 \times 10^{-4}z + 2.33 \times 10^2$ volts
26. 37.4 V
27. 9.52 V, $18.205\vec{a}_\rho$ V/m
28. $V = 11.7$ V, $-1786\vec{a}_\theta$ V/m
29. $V = -97.8r + 167.53$V, 42 V

Multiple-choice Questions

1 The polarization $\vec{P}$ C/m^2 in the dielectric material with $\varepsilon_r = 2.8$ when $\vec{D} = 3 \times 10^{-7}\vec{a}_x$C/m^2 is
(a) $2 \times 10^{-7}\vec{a}_x$ (b) $1.5 \times 10^{-7}\vec{a}_x$ (c) $1.93 \times 10^{-7}\vec{a}_x$ (d) $1.07 \times 10^{-7}\vec{a}_x$

2 The boundary condition for a perfect dielectric is
(a) $D_{n1} = D_{t1}$ (b) $E_{t1} = E_{t2}$ (c) $E_{n1} = E_{n2}$ (d) $D_{n1} = D_{t2}$

3 The boundary condition for a perfect dielectric is
(a) $\dfrac{D_{n1}}{D_{n2}} = \dfrac{\varepsilon_{r2}}{\varepsilon_{r1}}$ (b) $\dfrac{E_{n1}}{E_{n2}} = \dfrac{\varepsilon_{r2}}{\varepsilon_{r1}}$ (c) $\dfrac{E_{n1}}{E_{n2}} = \dfrac{\varepsilon_{r1}}{\varepsilon_{r2}}$ (d) $\dfrac{D_{n1}}{D_{n2}} = \dfrac{\varepsilon_{r1}}{\varepsilon_{r2}}$

4 The boundary condition for a perfect dielectric is
(a) $\dfrac{D_{t1}}{D_{t2}} = \dfrac{\varepsilon_{r1}}{\varepsilon_{r2}}$ (b) $\dfrac{E_{t1}}{E_{t2}} = \dfrac{\varepsilon_{r2}}{\varepsilon_{r1}}$ (c) $\dfrac{E_{t1}}{E_{t2}} = \dfrac{\varepsilon_{r1}}{\varepsilon_{r2}}$ (d) $\dfrac{D_{t1}}{D_{t2}} = \dfrac{\varepsilon_{r2}}{\varepsilon_{r1}}$

5 If $\vec{D}_1 = 13\vec{a}_x - 14\vec{a}_y + 6\vec{a}_z$, the incident angle θ_1 at the boundary on the *x*-*y* plane is
(a) 32.6° (b) 22.6^6 (c) 102.6 (d) 17.45

6 If $\vec{D}_2 = 13\vec{a}_x - 9.2\vec{a}_y + 14.4\vec{a}_z$, the emerging angle θ_2 at the boundary on the *x*-*y* plane is
(a) 9.83° (b) 22.6° (c) 102.6 (d) 42.12

7 Boundary conditions for a perfect dielectric material is derived from

(a) Ohm's law (b) Gauss' law

(c) Faraday's law (d) Coulomb's law

8 The tangential component of the electric field is always

(a) parallel to the boundary (b) perpendicular to the boundary

(c) it makes angle with boundary (d) all of the above

9 An air-filled parallel plate capacitor carries a charge of Q per plate and is isolated. If a dielectric slab is introduced into the air space, then the potential across the plates

(a) remains unaltered (b) increases (c) decreases (d) becomes zero

10 An air-filled parallel plate capacitor is connected to a voltage source V. If a dielectric slab is introduced into the air space, then the charge on the plates

(a) remains unaltered (b) increase (c) reduces to zero (d) decreases

11 Relaxation time for a copper conductor ($\sigma = 5.8 \times 10^7$ S/m) is

(a) 3×10^{-10} seconds (b) 10^{-16} seconds

(c) 2×10^{-5} seconds (d) 1.5×10^{-19} seconds

12 The region 1 defined by $x < 0$ is a free space while the region 2 $(x > 0)$ is a dielectric with $\varepsilon_{r_2} = 2$. Given $\vec{D}_1 = 3\vec{a}_x - 4\vec{a}_y + 6\vec{a}_z$ C/m^2, the value of D_{2x} is

(a) $3\vec{a}_x$ (b) $1.5\vec{a}_x$ (c) $6\vec{a}_x$ (d) $8\vec{a}_x$

13 If E_{n1} and E_{n2} are the normal electric field intensities at the dielectric boundary, then the bound charge density at the dielectric interfaces is

(a) $(E_{n1} - E_{n2})\varepsilon_0$ (b) $\varepsilon_0 E_{n1}$ (c) $\varepsilon_0 E_2$ (d) $(E_{n1} + E_{n2})\varepsilon_0$

14 A void is formed in a parallel plate capacitor which is filled with a dielectric of $\varepsilon_r = 4$. If the electric stress in the dielectric is 20 kV/cm, then the electric stress in the void is

(a) 20 kV/cm (b) 80 kV/cm (c) 5 kV/cm (d) 200 kV/ cm

15 A point charge Q is placed at the boundary ofa dielectric interface which is formed by two dielectrics of dielectric constants k_1 and k_2. The electric flux lines from Q in the medium with $\varepsilon_r = k_1$ is

(a) $\dfrac{Q}{k_1 + k_2}$ (b) $\dfrac{Qk_2}{k_1 + k_2}$ (c) $\dfrac{Qk_1}{k_1 + k_2}$ (d) $\dfrac{Qk_1}{k_2}$

16 The capacitance between two conductors depends upon

(a) shape of the conductors

(b) separation between the conductors

(c) medium in which conductors are placed

(d) all the above

17 A spherical capacitor is formed with inner radius a and outer radius b. If the charge on the inner conductor is $-Q$ when grounded, the charge on the outer conductor will be

(a) $\dfrac{Qb}{a}$ (b) $\dfrac{Qa}{b}$ (c) Q (d) $\dfrac{Qb^2}{a}$

18 The pressure acting at the dielectric interface between the media having dielectric constants $(k_1 < k_2)$ will

(a) pull the dielectric k_2 into k_1

(b) pull the dielectric k_1 into k_2

(c) pull along the interface

(d) none of the above

19 A capacitor has a capacitance $C = 2\ \mu$F. If a voltage of 100 V is applied on it, the maximum energy stored will be

(a) 10 J (b) 10 mJ (c) 100 mJ (d) 1 mJ

20 The following expression represents conservation of charge.

(a) $\nabla \cdot \vec{J} = \dfrac{-d\rho}{dt}$ (b) $\nabla \cdot \vec{J} = \sigma$ (c) $\vec{J} = \sigma\vec{E}$ (d) $\vec{J} = p\vec{u}$

21 In a battery, electrostatic field exists

(a) outside the battery terminals

(b) inside the battery terminals

(c) normal to the battery terminals

(d) all the above

22 Which of the following is not a convection current?

(a) An electron beam in a television tube

(b) Electron in a vacuum tube

(c) Electric current flow in aluminum wire

(d) Electron beam in CRO

23 When there is a current flow in a conductor, the following is true.

(a) Electric field exists along the surface

(b) Electric field exist normal to the surface

(c) Magnetic field is along the conductor

(d) Magnetic field is perpendicular to the conductor

24 Power dissipated per unit volume is

(a) $\dfrac{\vec{E} \cdot \vec{J}}{2}$ (b) $\vec{E} \cdot \vec{J}$ (c) $\vec{D} \cdot \vec{J}$ (d) $\dfrac{\vec{D}}{J}$

25 The boundary conditions at the interface of two conducting bodies are

(a) $J_{n1} = J_{n2}$ (b) $E_{t1} = E_{t2}$ (c) $\dfrac{\tan\theta_1}{\tan\theta_2} = \dfrac{\sigma_1}{\sigma_2}$ (d) All of the above

26 When the current flows across a boundary from a medium of higher conductivity to a medium of lower conductivity, the flow lines

(a) bend towards the normal at the boundary

(b) bend away from the normal at the boundary

(c) have no refraction at the boundary

(d) none of the above

27 The current flow in metallic conductors is due to

(a) motion of electrons (b) motion of positive charges
(c) both (a) and (b) (d) plasma

28 The number of electrons required to constitute a current of 1 A is

(a) 10×10^{18} (b) 5×10^{18} (c) 6.25×10^{18} (d) 15×10^{18}

29 For copper, the mobility $\mu = 0.0032$ m^2 /V-s and conductivity is 5.8×10^7 S/m. The corresponding volume charge density is

(a) 10^{10} C/m^3 (b) 1.81×10^{10} C/m^3 (c) 2×10^{18} C/m^3 (d) 2×10^{19} C/m^3

30 Ohm's law in point form is

(a) $\vec{E} = \sigma \vec{J}$ (b) $\vec{J} = \dfrac{\vec{E}}{\sigma}$ (c) $\vec{J} = \sigma \vec{E}$ (d) $\vec{J} = p\vec{u}$

31 The Laplace's equation for the vector $\vec{E}$ is

(a) $\nabla \times \vec{E} = 0$ (b) $\int \vec{E} \cdot d\vec{l} = 0$
(c) $\nabla \cdot \vec{E} = 0$ (d) $\nabla^2 \vec{E} = 0$

32 According to band theory of solids there is no energy gap between conduction band and valency band for

(a) conductors (b) insulators (c) semiconductors (d) none of the above

33 Two conducting bodies, A and B, with potentials V_1 and V_2 are placed inside a hollow body. If the electric flux lines start from A and end on B, then the relative potentials of the bodies will be as follows.

(a) $V_1 < V_2$ (b) $V_1 = V_2$ (c) $V_1 << V_2$ (d) $V_1 > V_2$

34 The Poisson's equation for the potential V is

(a) $\int V \cdot dv = 0$ (b) $-\nabla V = E$ (c) $\nabla^2 V = 0$ (d) $\nabla^2 V = \dfrac{-\rho_v}{\varepsilon}$

35 If the energy band gap between the conduction band and valence band is small, the material is

(a) metal (b) insulator (c) semiconductor (d) dielectric

36 For perfect conductors, the charge density inside the surfaces is

(a) maximum (b) zero (c) one (d) minimum

37 The electric field inside a perfect conductor is

(a) one (b) maximum (c) zero (d) minimum

38 The surface charge density (ρ_s) on a perfect conductor can be expressed as

(a) $\vec{E}$ (b) $\varepsilon \vec{E}$ (c) $\varepsilon_0 \rho_v$ (d) $\vec{E}/\varepsilon_0$

39 The potential due to two semi-infinite conducting plates with one plate at $(\rho = 0, V = 0)$ and the other plate at $(\rho = \phi_0, V = V_0)$ is

(a) $V_0 \phi / \phi_0$ (b) $V_0 \phi$ (c) $V_0 \phi \phi_0$ (d) $\phi_0 \phi / V_0$

40 The relaxation time of a medium with conductivity $\sigma = 2.8 \times 10^{-2}$ S/m and $\varepsilon_r = 2.4$ is

(a) 7.8 ps (b) 8 ns (c) 7.6 ns (d) 0.76 ns

Anwers

1. *c* 2. *b* 3. *b* 4. *a* 5. *d* 6. *d* 7. *b* 8. *a* 9. *c* 10. *d*

11. *d* 12. *a* 13. *a* 14. *b* 15. *c* 16. *d* 17. *a* 18. *a* 19. *b* 20. *a*

21. *d* 22. *c* 23. *a* 24. *b* 25. *d* 26. *a* 27. *a* 28. *c* 29. *b* 30. *c*

31. *d* 32. *a* 33. *a* 34. *d* 35. *c* 36. *b* 37. *c* 38. *b* 39. *a* 40. *d*

Magnetostatics | 4

Learning Objectives

The main objectives of this chapter are to understand thoroughly:

- Static magnetic field, magnetic field due to current-carrying conductors
- Magnetic flux, magnetic flux density, magnetic field intensity
- Biot–Savart's law, Biot–Savart's law for distributed currents
- Magnetic field intensity due to different current loops
- Maxwell's second equation for magnetic fields
- Differences between static electric field and magnetic fields
- Ampere's circuital law and its applications
- Differential form of Ampere's circuital law
- Stokes' theorem for magnetic field
- Magnetic field intensity due to a solenoid and a toroid

4.1 Introduction

We have seen in the previous chapters that electric charges are the sources of electric fields and an electric field exerts force on electric charges. Analogous to this, one can think of a magnetic charge (magnetic monopole) as being the source of magnetic fields, except that isolated magnetic monopoles do not exist in nature. They always occur in pairs or dipoles, comprising a magnetic north pole and a magnetic south pole. A region around a magnet (or a distribution of magnetic dipole) experiences a magnetic field. It is also observed that a steady magnetic field exists around a conductor when a steady current (direct current) flows through it. This phenomenon was first noted in 1820 by Hans Christian Oersted, who observed that a magnetic needle gets deflected in the region around a current-carrying conductor. It may then be inferred that the current-carrying conductor is the source of the magnetic field.

In magnetostatics, the sources of magnetic fields are associated with either current-carrying conductors or magnetic materials (dipoles). The magnetic effects observed can thus be described in terms of either the distribution of dipoles or distribution of current. In nature, magnetic materials produce magnetic fields by virtue of their electronic orientation. A permanent magnet is one that exhibits a persistent magnetic field. An electromagnet, on the other hand, acts as a magnet only when an electric current passes through it.

In this chapter, we will study the characteristics of static magnetic fields associated with steady, non-time varying currents (magnetostatics). We will also study Biot–Savart's law (corresponding to Coulomb's empirical law in electrostatics) to find the magnetic field intensity due to various current distributions. Ampere's circuital law, which can be used to find the magnetic field intensity more easily than Biot–Savart's law, is discussed. Important applications of Ampere's circuital law like computation of fields due to circular, rectangular and triangular loops are also discussed. Finally, the concept of point form of Ampere's circuital law, which is also called Maxwell's third equation, is explained.

4.2 Magnetic Field due to a Current-Carrying Conductor

When a conductor carries a current, it produces a magnetic field around it along the conductor. The direction of the magnetic flux lines (magnetic field) is given by the right-hand thumb rule (see Fig. 4.1), which is stated as follows.

When the right-hand thumb is opened and the fingers are closed, if the direction of the current is parallel to the thumb, then the direction of the fingers indicate the direction of magnetic field in a closed loop path.

Alternatively, the right-hand screw rule (Fig. 4.2) stated below also gives the direction of the magnetic field:

When a right-handed screw is turned, if the screw moves in the direction of the current flow, then the direction of the rotation of the screw gives the direction of magnetic field.

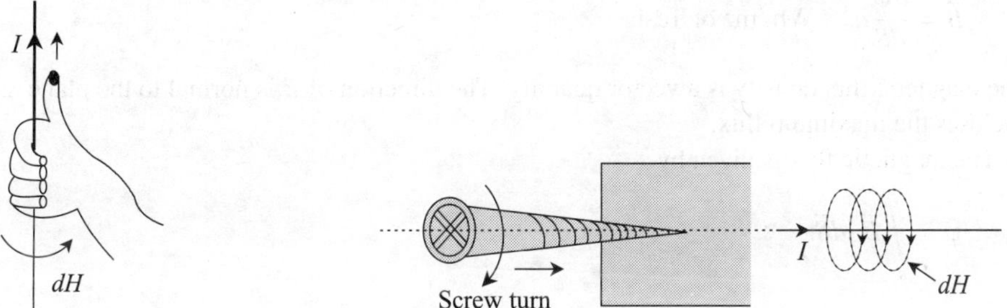

Fig. 4.1 *Right-hand thumb rule* **Fig. 4.2** *Right-hand screw rule*

In other words, we can say that if the current flows into the plane of paper (symbol ⊗), then the direction of the magnetic field is in the clockwise direction, and if the current flows outwards (symbol ⊙), the direction of the magnetic field is anti-clockwise (see Fig. 4.3). Both the right-hand screw rule and the right-hand thumb rule will give the same direction for magnetic field.

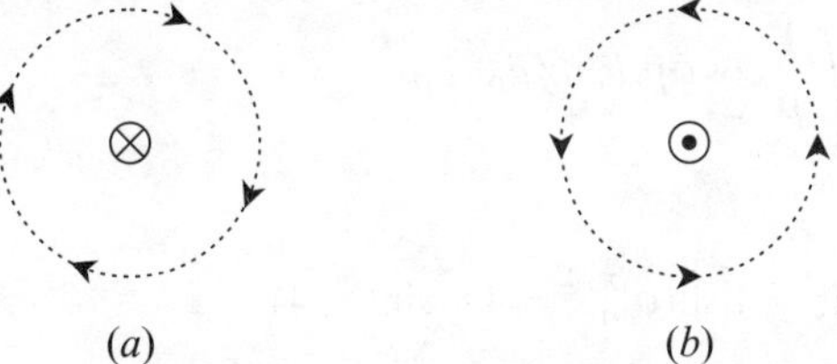

Fig. 4.3 (*a*) *Current is flowing inwards into the paper and the flux direction is clockwise.*
(*b*) *Current is flowing outwards from the paper and the flux direction is anti-clockwise.*

4.3 Magnetic Flux and Flux Density

Magnetic flux gives the distribution of magnetic field passing through a given surface. In other words, the total number of magnetic lines of force passing through a given surface is called magnetic flux. It is denoted by the symbol φ, and its unit is weber (Wb). In electrostatics, we saw how the electric field could be described on the basis of electric flux density D. In magnetostatics, the corresponding quantity is magnetic flux density, denoted by B, whose units are Wb/m^2, also known as Tesla (T).

Magnetic flux density is defined *as the magnetic flux crossing a unit area normal to the direction of the magnetic flux.*

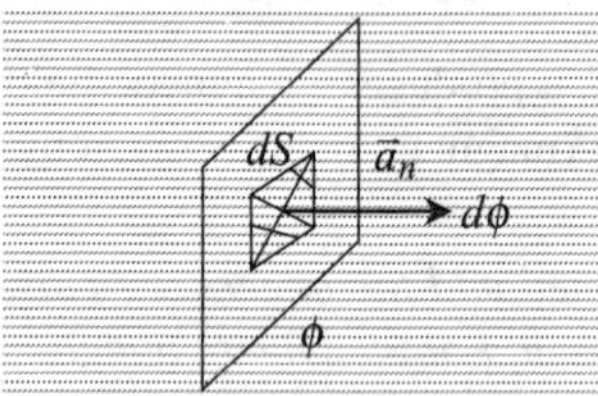

Fig. 4.4 *Magnetic flux density*

If the differential flux $d\varphi$ crosses the differential area dS normal to the plane of the loop as shown in Fig. 4.4, then the magnetic flux density is given by

$$\vec{B} = \frac{d\varphi}{dS}\vec{a}_n \quad \text{Wb/m}^2 \text{ or Tesla.} \tag{4.1}$$

The magnetic flux density is a vector quantity. The direction of B is normal to the plane area which encloses the maximum flux.

The magnetic flux is given by

$$\varphi = \int_s \vec{B}\cdot d\vec{S}. \tag{4.2}$$

Example 4.1 If the magnetic flux density in a medium is given by $\vec{B} = \frac{1}{\rho}\cos\phi\vec{a}_\rho$, find the flux crossing the surface defined by $0 \le z \le 4$ m and $0 < \phi < \pi/4$.

Solution Given $\vec{B} = \frac{1}{\rho}\cos\phi\vec{a}_\rho$. In cylindrical coordinates, $dS = \rho d\phi dz\,\vec{a}_\rho$. The flux is

$$\varphi = \int_s B\cdot dS = \int_0^{\pi/4}\int_0^4 \frac{1}{\rho}\cos\phi\rho d\phi dz\,\vec{a}_\rho\cdot\vec{a}_\rho$$

$$= \int_0^{\pi/4}\int_0^4 \cos\phi d\phi dz = 4[\sin\phi]_0^{\pi/4} = 4\times\sin(\pi/4)$$

$$= 4\times\frac{1}{\sqrt{2}} = 2.828 \text{ Wb.}$$

Magnetic field intensity or magnetic field strength

The magnetic field intensity at any point in a magnetic field is defined as *the magnetic force experienced by a unit pole at that point. It is also called magnetic field strength.*

It is a vector quantity, analogous to electric field intensity and is denoted by $\vec{H}$. Its units are newtons/weber (N/Wb) or ampere/metre (A/m).

When a current is flowing through a conducting wire, a magnetic field is produced around the wire. The field can be represented by concentric circles drawn around the wire with proper direction (flux line) as shown in Fig. 4.5. According the right-hand thumb rule, I indicates the current flowing out of the paper; the flux lines are in the anti-clockwise direction. The field strength decreases as the distance from the wire increases.

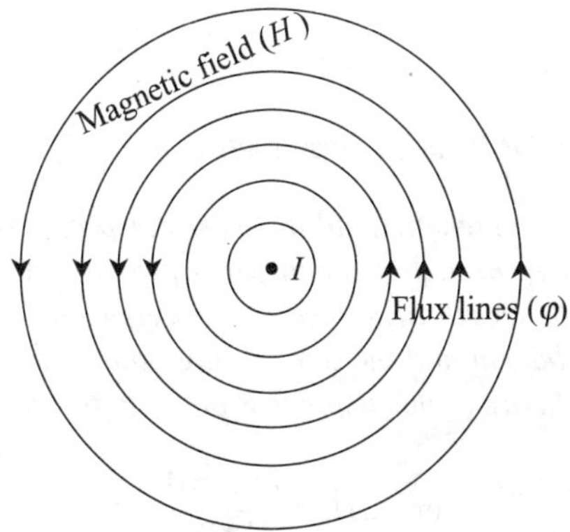

Fig. 4.5 *Flux lines around a current-carrying wire*

The magnetic flux density $\vec{B}$ is related to magnetic field intensity $(\vec{H})$ through the material property called permeability (μ), i.e.,

$$\vec{B} = \mu\vec{H} \tag{4.3}$$

or $\vec{B} = \mu_0\mu_r\vec{H}$

where μ_r = relative permeability and μ_0 = absolute permeability.

For free space, $\mu_r = 1$ and $\mu = \mu_0 = 4\pi \times 10^{-7}$ H/m.

For magnetic materials, μ_r is greater than 1 and for non-magnetic materials, μ_r is equal to 1.

The magnetic field strength is independent of the medium's permeability.

Example 4.2 Given $\vec{H} = 4\vec{a}_x + 2\vec{a}_y - 3\vec{a}_z$ A/m at a point in free space, what is the flux density?

Solution Given $\vec{H} = 4\vec{a}_x + 2\vec{a}_y - 3\vec{a}_z$.

We know that flux density $\vec{B} = \mu_0\vec{H} = 4\pi \times 10^{-7}(4\vec{a}_x + 2\vec{a}_y - 3\vec{a}_z)$

$$= 5.026\vec{a}_x + 2.513\vec{a}_y - 3.767\vec{a}_z\ \mu\text{Wb/m}^2.$$

4.4 Biot–Savart's Law

Consider a current-carrying conductor producing a steady magnetic field around it. Let dl be a differential length on the conductor with current element Idl. Let us consider a point P at a distance R from the current element, as shown in Fig. 4.6.

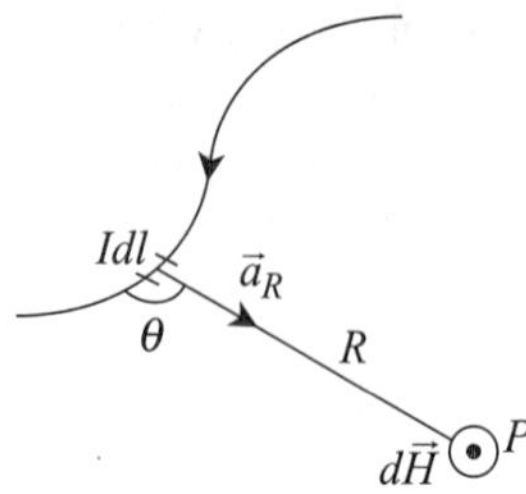

Fig. 4.6 *Magnetic field at point P due to current element*

Biot–Savart's law states that *the magnetic field intensity $d\vec{H}$ produced at a point P due to the differential current element flowing through a conductor is proportional to the product of the current I and the differential conductor length dl, called current element (Idl); and the sine of the angle between the direction of the element and the line joining point P to the current element. It is also inversely proportional to the square of the distance R between the current element and the point P.*

Mathematically, $d\vec{H} \propto \dfrac{Id\vec{l}\sin\theta}{R^2}$, or $d\vec{H} = \dfrac{Id\vec{l}\sin\theta}{4\pi R^2}$.

From the definition of the cross product, this can be written as

$$d\vec{H} = \frac{Id\vec{l} \times \vec{a}_R}{4\pi R^2} \quad \text{A/m} \tag{4.4}$$

where $1/4\pi$ is the proportionality constant and $\vec{a}_R = \dfrac{\vec{R}}{|R|}$ is the unit vector along the direction of $\vec{R}$ (line joining the current element and point P).

Then, $$d\vec{H} = \frac{Id\vec{l} \times \vec{R}}{4\pi R^3}. \tag{4.5}$$

∴ The total magnetic field intensity $\vec{H}$ is

$$\vec{H} = \oint \frac{Id\vec{l} \times \vec{a}_R}{4\pi R^2} \tag{4.6}$$

and the magnetic flux density $\vec{B}$ is

$$\vec{B} = \frac{\mu}{4\pi}\oint \frac{Id\vec{l} \times \vec{a}_R}{R^2}.$$

The direction of the magnetic field can be determined by the right-hand thumb rule or the right-hand screw rule. Biot–Savart's law is also called Ampere's law for a current element.

Example 4.3 A steady current element $10^{-3}\,\vec{a}_z$ A-m is located at the origin in free space. (i) What is the magnetic field $\vec{B}$ due to this element at the point $(1, 0, 0)$ in rectangular coordinates? (ii) What is the magnetic field at the point $(0,0,1)$?

Solution (i) Given current element, $Id\vec{l} = 10^{-3}\,\vec{a}_z$.

The magnetic field $d\vec{B}$ at the point (1, 0, 0) due to $Id\vec{l}$ located at origin (0, 0, 0) is,

$$d\vec{B} = \frac{\mu_0 Id\vec{l} \times \vec{R}}{4\pi\,|R|^3}.$$

The distance vector is $\vec{R} = (1-0)\vec{a}_x + (0-0)\vec{a}_y + (0-0)\vec{a}_z = \vec{a}_x$; $R = 1$.

$$\therefore \quad d\vec{B} = \frac{\mu_0}{4\pi}\frac{10^{-3}(\vec{a}_z \times \vec{a}_x)}{(1)^3}$$

$$= \frac{\mu_0}{4\pi}(10^{-3}\vec{a}_z \times \vec{a}_x) = \frac{4\pi}{4\pi} \times 10^{-7} \times 10^{-3}(\vec{a}_y)$$

$$= 10^{-10}\vec{a}_y \text{ Tesla}$$

(ii) The magnetic field at the point (0, 0, 1) due to current element Idl at (0, 0, 0) is

$$d\vec{B} = \frac{\mu_0}{4\pi}\frac{Id\vec{l} \times \vec{R}}{|R|^3}$$

Now, $\vec{R} = (0-0)\vec{a}_x + (0-0)\vec{a}_y + (1-0)\vec{a}_z = \vec{a}_z$; $R = 1$.

$$d\vec{B} = \frac{\mu_0}{4\pi}\frac{10^{-3}\vec{a}_z \times \vec{a}_z}{(1)^3} = 0.$$

Note: It is observed that the magnetic field is always perpendicular to the current-carrying element.

4.5 Biot–Savart's Law for Distributed Currents

Sheet Current Consider a sheet of conductor lying on the $x-y$ plane, carrying current I uniformly distributed over the surface as shown in Fig. 4.7. Let the surface current density be $\vec{K}$ A/m. If we consider a differential surface area dS, the current element with differential length dl having the current density is

$$Id\vec{l} = \vec{K}dS.$$

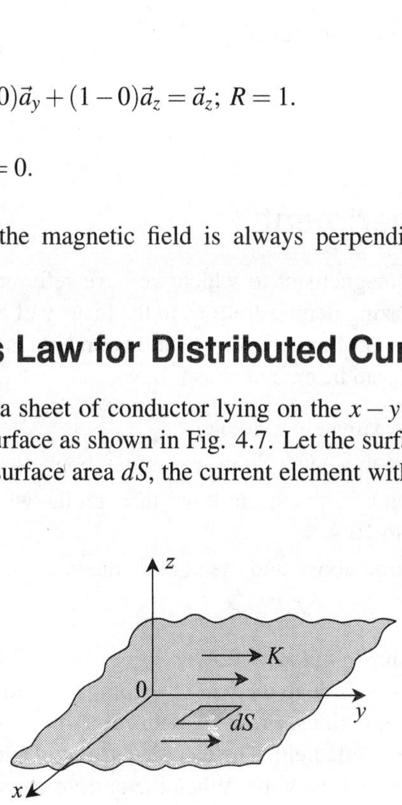

Fig. 4.7 *Sheet current*

Biot–Savart's law for sheet current is then

$$\vec{H} = \int_s \frac{\vec{K} dS \times \vec{a}_R}{4\pi R^2} \quad \text{A/m}. \tag{4.7}$$

Volume current Consider a solid conductor lying on the $x - y$ plane, carrying current I uniformly distributed over the volume as shown in Fig. 4.8. Let the current density be $\vec{J}$ A/m^2. If we consider a differential volume dv, the current element with differential length dl having current density is

$$I d\vec{l} = \vec{J}\, dv.$$

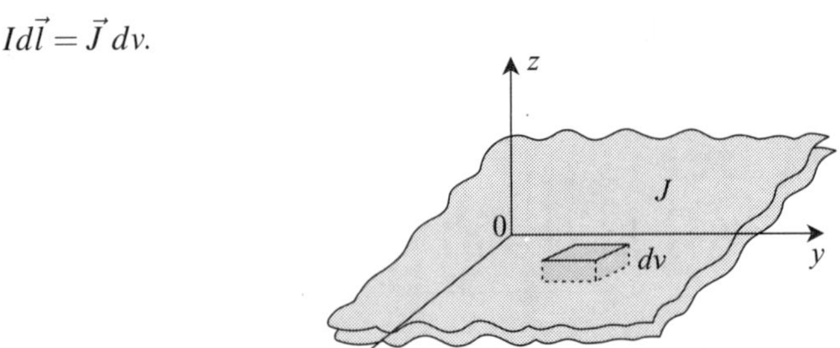

Fig. 4.8 *Volume current*

Biot–Savart's law for volume current is then

$$\vec{H} = \int_v \frac{\vec{J} \times \vec{a}_R\, dv}{4\pi R^2} \quad \text{A/m}. \tag{4.8}$$

4.6 Oersted's Experiment

Oersted's discovery in electromagnetism, to which we have referred to earlier, made on July 21 1820, was one of the most amazing demonstrations in the history of electromagnetism. His experiment proved that there is a relationship between electricity and magnetism.

His experimental procedure can be emulated as follows.

1. Take two copper wires approximately 1 m long.
2. Connect the ends of the wire to a battery power supply (about 1 to 2 volts).
3. Switch on the supply so that a large current flows through the wire. The power supply must be able to allow current of up to 10 A.
4. Hold a magnetic compass first above and then below the wire. Observe how it orients itself in each position.

The experimental setup is shown in Fig. 4.9.

When an electric current flows through the wire, the compass needle placed above and below the wire will point in the opposite directions. From the deflection of the compass, it can be inferred that the electric current causes a magnetic field. The electric current in a long straight wire will produce circular magnetic field lines around the wire. When the current-carrying conductor has a different configuration, the electric currents will cause a magnetic field of a different configuration.

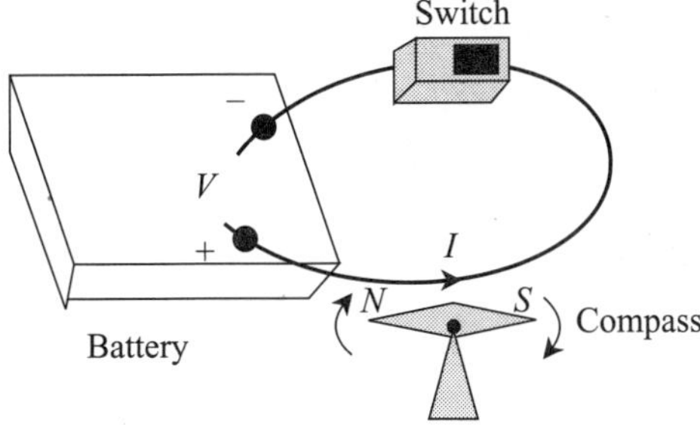

Fig. 4.9 *Oersted's experimental setup*

4.7 Magnetic Field Intensity due to an Infinitely Long Conductor

Consider an infinitely long conductor carrying current I along the z-axis. Let there be a current element $I\vec{dl}$ on the conductor at a distance z from the origin. Also let us consider a point P on the y-axis at a distance ρ from the origin, as shown in Fig. 4.10.

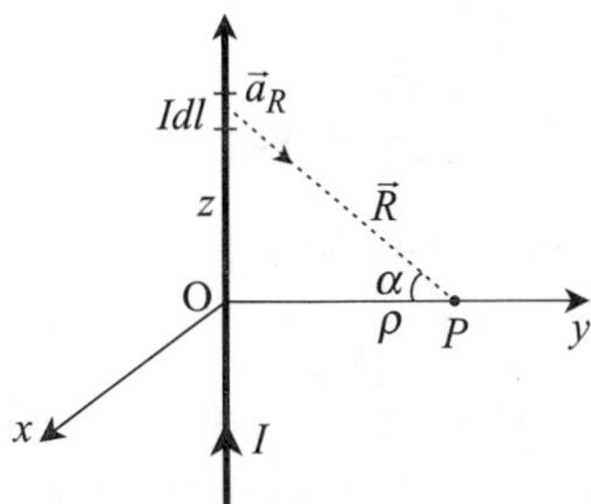

Fig. 4.10 *Magnetic field due to an infinitely long conductor*

In cylindrical coordinates, the position at the current element is $z\,\vec{a}_z$ and the position vector at the point P is $(\rho, 0, 0)$.

The distance vector is given by

$$\vec{R} = \rho\vec{a}_\rho - z\vec{a}_z.$$

The unit vector is $\vec{a}_R = \dfrac{\vec{R}}{R} = \dfrac{\rho\vec{a}_\rho - z\vec{a}_z}{\sqrt{\rho^2 + z^2}}$ and the current element is $I\vec{dl} = Idz\,\vec{a}_z$.

The magnetic field intensity at point P is

$$\vec{H} = \int_l \frac{I\vec{dl} \times \vec{a}_R}{4\pi R^2}.$$

Evaluating the cross product,

$$I d\vec{l} \times \vec{a}_R = I dz\, \vec{a}_z \times \frac{(\rho \vec{a}_\rho - z\vec{a}_z)}{\sqrt{\rho^2+z^2}} = \frac{I dz}{\sqrt{\rho^2+z^2}}(\rho \vec{a}_z \times \vec{a}_\rho - z\vec{a}_z \times \vec{a}_z).$$

Since $\vec{a}_z \times \vec{a}_\rho = \vec{a}_\phi$,

$$I d\vec{l} \times \vec{a}_R = \frac{I dz}{\sqrt{\rho^2+z^2}}\rho \vec{a}_\varphi.$$

$$\therefore \quad \vec{H} = \int_{-\infty}^{\infty} \frac{I\rho \vec{a}_\varphi\, dz}{4\pi(\rho^2+z^2)^{3/2}}.$$

To evaluate the integration, let $z = \rho \tan\alpha$. Then, $dz = \rho \sec^2 \alpha\, d\alpha$, where α is the angle at P due to the current element.

The limits of α are from $-\pi/2$ to $\pi/2$.

So, $$\frac{\rho dz}{(\rho^2+z^2)^{3/2}} = \frac{\rho\rho \sec^2 \alpha\, d\alpha}{(\rho^2+\rho^2 \tan^2 \alpha)^{3/2}} = \frac{\sec^2 \alpha\, d\alpha}{\rho \sec^3 \alpha} = \frac{\cos\alpha\, d\alpha}{\rho}.$$

$$\therefore \quad H = \int_{-\pi/2}^{\pi/2} \frac{I\, \vec{a}_\phi}{4\pi\rho} \cos\alpha\, d\alpha = \frac{I\, \vec{a}_\phi}{4\pi\rho}\int_{-\pi/2}^{\pi/2} \cos\alpha\, d\alpha$$

$$= \frac{I\, \vec{a}_\phi}{4\pi\rho}[\sin\alpha]_{-\pi/2}^{\pi/2} = \frac{I\, \vec{a}_\varphi}{4\pi\rho}(2) = \frac{I\, \vec{a}_\varphi}{2\pi\rho} \text{ A/m.} \tag{4.9}$$

If $l = 2\pi\rho$, the length of the magnetic path of radius ρ,

$$\vec{H} = \frac{I}{l}\vec{a}_\phi = \frac{I\vec{a}_\phi}{2\pi\rho} \text{ A/m.}$$

The magnetic flux density is

$$\vec{B} = \mu\vec{H} = \frac{\mu I \vec{a}_\phi}{2\pi\rho} \quad Wb/\text{m}^2 \tag{4.10}$$

Note:

1. The magnetic field intensity $\vec{H}$ at a point P due to an infinite length conductor is proportional to the current I and inversely proportional to the perpendicular distance ρ.
2. The direction of $\vec{H}$ is along ϕ. It is tangential to the concentric circles around the conductor as shown in Fig. 4.11 and follows the right-hand thumb rule.

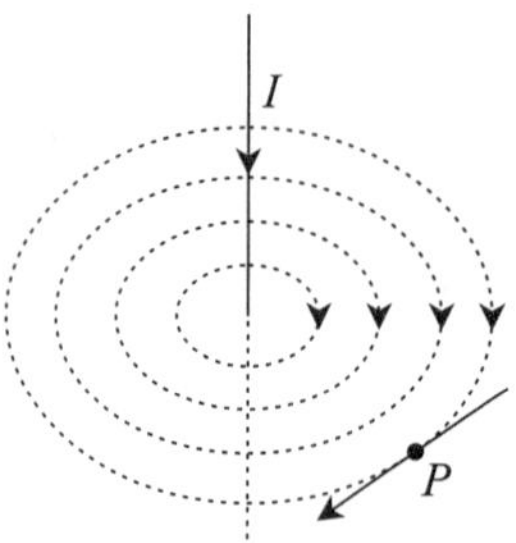

Fig. 4.11 *Direction of the magnetic field due to an infinitely long conductor*

For example, a long straight wire carries a current of $I = 1$ A. At a distance of 1 m around the conductor, the magnetic field produced is $H = \dfrac{1}{2\pi}$ A/m.

Example 4.4 A current filament, $I = 100$ A along the y-axis, is passing through the point $(4, 0, -5)$. Find the magnetic field intensity at the origin.

Solution Given $I = 100$ A and a point on the current element at $(4, 0, -5)$.
The magnetic field at the origin is given by

$$\vec{H} = \frac{I}{4\pi\rho}\,\vec{a}_\phi,$$

where ρ in terms of the radial distance is given by

$$\rho = \sqrt{4^2 + 5^2} = \sqrt{41} = 6.403.$$

$$\vec{H} = \frac{100}{4\pi \times 6.403}\vec{a}_\phi = 1.24\,\vec{a}_\phi.$$

4.8 Magnetic Field Intensity due to a Finite Length Conductor

Consider a straight conductor of finite length L carrying current I along the z-axis. Also consider a current element $Id\vec{l}$ on the conductor at a distance z from the origin. The end points of the conductor are at distances z_1 and z_2 on the z-axis as shown in Fig. 4.12.

Let a point P be on the y-axis at a distance ρ from the origin. Using cylindrical coordinates, the position vector at the current element is $(0, 0, z)$ and the position vector at point P is $(\rho, 0, 0)$.

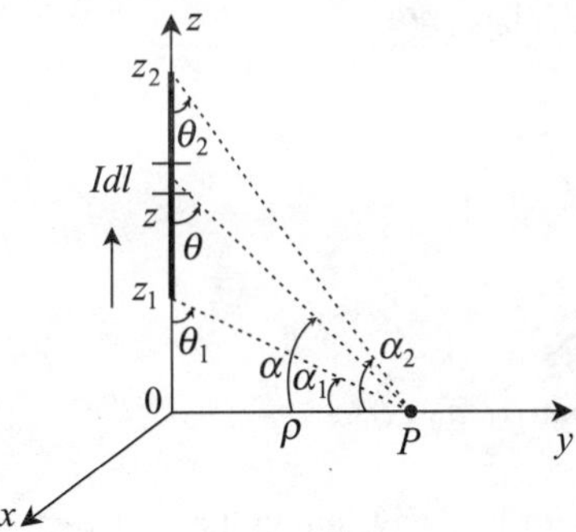

Fig. 4.12 *Magnetic field due to a finite length conductor*

The distance vector is $\vec{R} = \rho\vec{a}_\rho - z\vec{a}_z$. The unit vector is

$$\vec{a}_R = \frac{\vec{R}}{R} = \frac{\rho\vec{a}_\rho - z\vec{a}_z}{\sqrt{\rho^2 + z^2}}.$$

The current element is $Id\vec{l} = Idz\,\vec{a}_z$. The magnetic field intensity at point P is

$$\vec{H} = \int \frac{Id\vec{l} \times \vec{a}_R}{4\pi R^2}.$$

The cross product is

$$Id\vec{l} \times \vec{a}_R = Idz\,\vec{a}_z \times \frac{(\rho\vec{a}_\rho - z\vec{a}_z)}{\sqrt{\rho^2+z^2}} = \frac{Idz}{\sqrt{\rho^2+z^2}}(\rho\vec{a}_z \times \vec{a}_\rho - z\vec{a}_z \times \vec{a}_z).$$

Since $\vec{a}_z \times \vec{a}_\rho = \vec{a}_\phi$

$$Id\vec{l} \times \vec{a}_R = \frac{Idz}{\sqrt{\rho^2+z^2}}\rho\vec{a}_\phi.$$

$$\therefore \quad H = \int_{z_1}^{z_2} \frac{I\,\rho\vec{a}_\phi\,dz}{4\pi(\rho^2+z^2)^{3/2}}.$$

To evaluate the line integral, let α, α_1 and α_2 be angles at point P in the clockwise direction that the distance vectors from points z, z_1 and z_2 makes with the y-axis respectively.

Then, $\quad z = \rho\tan\alpha,\ dz = \rho\sec^2\alpha d\alpha.$

The limits are: $z_1 = \rho\tan\alpha_1$ or $\alpha_1 = \tan^{-1}(z_1/\rho)$.

$z_2 = \rho\tan\alpha_2$ or $\alpha_2 = \tan^{-1}(z_2/\rho)$.

Also, $\dfrac{\rho dz}{\rho^2+z^2} = \dfrac{\rho\,\rho\sec^2\alpha d\alpha}{(\rho^2+\rho^2\tan^2\alpha)^{3/2}} = \dfrac{\rho^2\sec^2\alpha\,d\alpha}{\rho^3\sec^3\alpha} = \dfrac{\cos\alpha\,d\alpha}{\rho}.$

$$\therefore \quad \vec{H} = \int_{\alpha_1}^{\alpha_2} \frac{I\vec{a}_\phi}{4\pi\rho}\cos\alpha\,d\alpha = \frac{I\vec{a}_\phi}{4\pi\rho}[\sin\alpha]_{\alpha_1}^{\alpha_2}$$

$$= \frac{I}{4\pi\rho}(\sin\alpha_2 - \sin\alpha_1)\vec{a}_\phi \text{ A/m}, \quad (4.11)$$

and the magnetic flux density is

$$\vec{B} = \mu\vec{H} = \frac{\mu I}{4\pi\rho}(\sin\alpha_2 - \sin\alpha_1)\vec{a}_\phi \text{ A/m}. \quad (4.12)$$

The direction of the magnetic field is always along the unit vector $\vec{a}_\phi$ irrespective of the length of the conductor or the position of the point P. The magnetic field forms concentric circles around the conductor.

Note:

1. If θ, θ_1 and θ_2 are the angles of the conductor at points z, z_1 and z_2 to the point P from the z-axis as shown in Fig. 4.13, then $\theta = 90 - \alpha$, $\theta_1 = 90 - \alpha_1$, and $\theta_2 = 90 - \alpha_2$.

$$\text{So,}\quad \vec{H} = \frac{I}{4\pi\rho}(\sin(90-\theta_2) - \sin(90-\theta_1))\vec{a}_\phi$$

$$= \frac{I}{4\pi\rho}(\cos\theta_2 - \cos\theta_1)\vec{a}_\rho. \tag{4.13}$$

2. For infinitely long conductors, $\theta_2 = 0$ and $\theta_1 = 180°$.

$$\therefore\quad \vec{H} = \frac{I}{4\pi\rho}2\vec{a}_\phi = \frac{I}{2\pi\rho}\vec{a}_\phi, \tag{4.14}$$

which is the same as Eq. (4.10).

3. For semi-infinitely long conductors at $z_1 = 0$ and $z_2 = \infty$, $\theta_2 = 0°$ and $\theta_1 = 90°$,

$$\text{then}\quad \vec{H} = \frac{I}{4\pi\rho}(\cos 0 - \cos 90°) = \frac{I}{4\pi\rho}\vec{a}_\rho.$$

4. If the point P makes angles θ_2 and θ_3 with the direction of the current, as shown in Fig. 4.13. If $\theta_3 = 180 - \theta_1$, and $\theta_1 > \frac{\pi}{2}$,

$$\text{then}\ \vec{H} = \frac{I}{4\pi\rho}(\cos\theta_2 - \cos(180-\theta_3))\vec{a}_\phi$$

$$= \frac{I}{4\pi\rho}(\cos\theta_2 + \cos\theta_3)\vec{a}_\rho. \tag{4.15}$$

5. If the point P is on the line perpendicular to the bisection of the current line, then

$$\theta_3 = \theta_2 = \theta \text{ and } \theta_1 = 180 - \theta.$$

$$\therefore\quad \vec{H} = \frac{I}{4\pi\rho}(\cos\theta - \cos(180-\theta))\vec{a}_\phi = \frac{I\cos\theta}{2\pi\rho}\vec{a}_\phi. \tag{4.16}$$

If L = length of the wire, then

$$\cos\theta = \frac{L/2}{\sqrt{(L/2)^2+\rho^2}},$$

$$\vec{H} = \frac{I}{2\pi\rho}\left(\frac{L/2}{\sqrt{(L/2)^2+\rho^2}}\right).$$

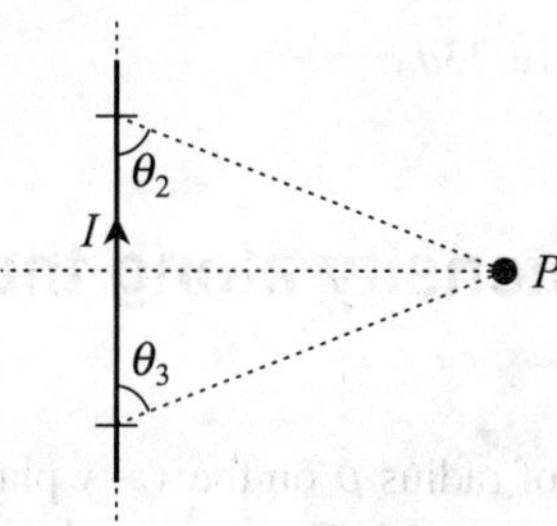

Fig. 4.13 *Point P on the triangle*

Example 4.5 Find the magnetic field intensity at point $P(2, 2, 3)$ caused by a current filament of 25 A in the $\vec{a}_z$ direction and extending from $z = 0$ to $z = 6$.

Solution Given a finite length current filament extending from $z = 0$ to $z = 6$, $I = 25\vec{a}_z$ A and point P at (2, 2, 3).

If θ_1 and θ_2 are the angles of the conductor points at $z = 0$ and $z = 6$ to the point P with the z-axis as shown in Fig. 4.14.

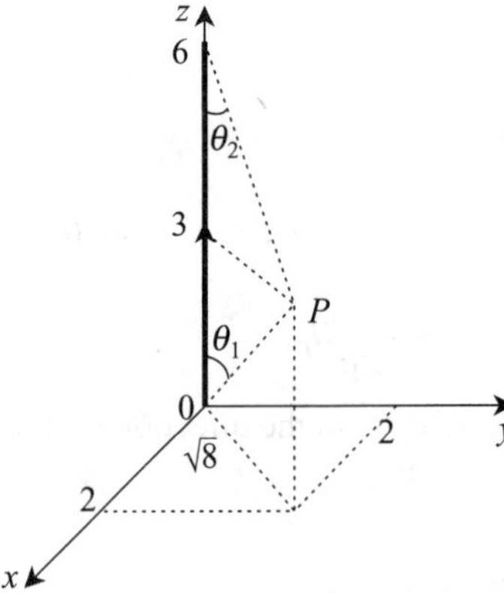

Fig. 4.14 *Field due to current element*

The magnetic field intensity due to a finite length conductor is

$$\vec{H} = \frac{I}{4\pi\rho}(\cos\theta_2 + \cos\theta_1)\vec{a}_\phi$$

Since the point P is on the line perpendicular to the bisection of the current line as shown in Fig. 4.14, the distance from point P to the line of current is

$$\rho = \sqrt{2^2 + 2^2} = \sqrt{8} \quad \text{and} \quad \theta_1 = \theta_2 = \theta.$$

$$\therefore \quad \vec{H} = \frac{I\cos\theta}{2\pi\rho}\vec{a}_\phi.$$

From Fig. 4.14, $\theta_1 = \theta_2 = \tan^{-1}\left(\frac{\sqrt{8}}{3}\right) = 43.3°.$

$$\therefore \quad \vec{H} = \frac{25}{2\pi \times \sqrt{8}}(\cos 43.3°)\vec{a}_\rho = 1.023\vec{a}_\phi.$$

4.9 Magnetic Field Intensity along the Axis of a Circular Loop

Consider a circular loop conductor of radius ρ on the $x - y$ plane centred at the origin as shown in Fig. 4.15. Let the conductor carry a current I. The current element with differential length $d\vec{l}$ is $Id\vec{l}$.

Let P be a point on the z-axis at a distance z from the origin.

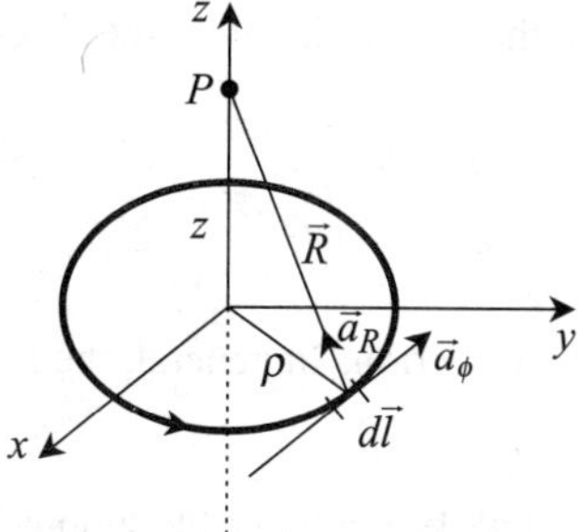

Fig. 4.15 *Circular loop conductor*

In cylindrical coordinates, the position vector at the current element is $(\rho,\ \phi,\ 0)$ and the position vector at point P is $(0,\ \phi,\ z)$.

$\therefore$ The distance vector is $\vec{R} = -\rho\vec{a}_\rho + z\vec{a}_z$.

The unit vector is $\vec{a}_R = \dfrac{-\rho\vec{a}_\rho + z\vec{a}_z}{\sqrt{\rho^2+z^2}}$.

The current element is $Id\vec{l} = I\ \rho d\phi\ \vec{a}_\phi$.

The magnetic field intensity at point P is $\vec{H} = \displaystyle\int_l \frac{Id\vec{l}\times\vec{a}_R}{4\pi R^2}$.

Evaluating the cross product,

$$Id\vec{l}\times\vec{a}_R = (I\ \rho d\phi\vec{a}_\phi)\times\frac{(-\rho\vec{a}_\rho + z\vec{a}_z)}{\sqrt{\rho^2+z^2}}$$

$$= \frac{I\rho d\phi}{\sqrt{\rho^2+z^2}}\vec{a}_\phi\times(-\rho\vec{a}_\rho + z\vec{a}_z)$$

$$= \frac{I}{\sqrt{\rho^2+z^2}}\begin{vmatrix}\vec{a}_\rho & \vec{a}_\phi & \vec{a}_z\\ 0 & \rho d\phi & 0\\ -\rho & 0 & z\end{vmatrix} = \frac{I}{\sqrt{\rho^2+z^2}}\left[z\rho d\phi\vec{a}_\rho + \rho^2 d\phi\vec{a}_z\right].$$

The magnetic field intensity is

$$\vec{H} = \int_l \frac{I}{4\pi}\frac{\left[z\rho d\phi\vec{a}_\rho + \rho^2 d\phi\vec{a}_z\right]}{(\rho^2+z^2)^{3/2}}.$$

For a circular loop, the limits are from 0 to 2π.

$$\therefore\quad \vec{H} = \frac{I}{4\pi}\int_0^{2\pi}\frac{z\rho\vec{a}_\rho}{(\rho^2+z^2)^{3/2}}d\phi + \frac{I}{4\pi}\int_0^{2\pi}\frac{\rho^2\vec{a}_z}{(\rho^2+z^2)^{3/2}}d\phi.$$

Since the loop is symmetrical along the radial direction, the $\vec{a}_\rho$ component is zero.

$$\therefore \quad \vec{H} = \frac{I}{4\pi}\int_0^{2\pi} \frac{\rho^2 \vec{a}_z}{(\rho^2+z^2)^{3/2}} d\phi = \frac{I\rho^2\vec{a}_z}{4\pi(\rho^2+z^2)^{3/2}} 2\pi = \frac{I\rho^2}{2(\rho^2+z^2)^{3/2}}\vec{a}_z. \tag{4.17}$$

Thus the field exists only along the z-direction. In general, the field intensity is perpendicular to the plane of the circular wire.

Note 1: If $z = 0$, the magnetic field at the centre of the loop is

$$\vec{H} = \frac{I\rho^2\vec{a}_z}{2(\rho^2+0)^{3/2}} = \frac{I}{2\rho}\vec{a}_z \text{ A/m} \tag{4.18}$$

and $$\vec{B} = \frac{\mu I}{2\rho}\vec{a}_z \text{ A/m.} \tag{4.19}$$

For example, a circular loop of radius 1 m carries a current of 10 A. It produces a magnetic field $H = 5$ A/m at the centre of the loop.

Note 2: If L is the length of the circular loop, i.e., $L = 2\pi\rho$, then the field at the centre of the loop is

$$\vec{H} = \frac{\pi I}{L}\vec{a}_z.$$

Note 3: At large distances from the loop, where $z \gg \rho$, $\rho^2 + z^2 \approx z^2$. The magnetic field is then given by $\vec{H} = \dfrac{I\rho^2}{2z^3}\vec{a}_z$.

Example 4.6 A thin ring of radius 10 cm is placed on plane $z = 5$ cm so that its centre is at (0, 0, 5). If the ring carries 100 mA, find $\vec{H}$ at (0, 0, −10).

Solution Given $\rho = 10$ cm $= 0.1$ m; $I = 100$ mA $= 0.1$ A.

The distance between the ring and the point where the field is to be determined is

$$z = 5 - (-10) = 15 \text{ cm} = 0.15 \text{ m}.$$

We know that the magnetic field intensity due to a circular ring is

$$\vec{H} = \frac{I\rho^2\vec{a}_z}{2(\rho^2+z^2)^{3/2}} = \frac{(0.1)(0.1)^2\vec{a}_z}{2(0.1^2+0.15^2)^{3/2}} = 85.33\vec{a}_z \text{ mA/m}.$$

4.10 Magnetic Field Intensity at the Centre of an Equilateral Triangle Formed by a Wire of Length L

Consider a wire of length L bent in the form of an equilateral triangle of side $L/3$ each and placed in the $x-y$ plane as shown in Fig. 4.16. Let I be the current in the conductor.

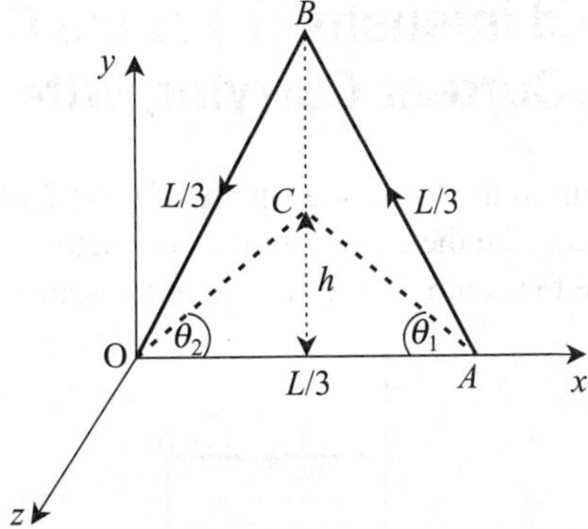

Fig. 4.16 Current-carrying wire in the form of an equilateral triangle

Consider the side OA of the triangle lying along the x-axis. Let C be the centre of the triangle such that it is at a distance h from the side OA.

We know that magnetic field intensity at height h due to finite line current is

$$\vec{H}_{OA} = \frac{I}{4\pi h}[\cos\theta_1 + \cos\theta_2]\vec{a}_z \text{ A/m.}$$

From Fig. 4.16, it can be seen that

$$OA = OB = AB = \frac{L}{3}\,;\ \theta = \theta_1 = \theta_2 = 30^\circ.$$

$$\tan 30^\circ = \frac{h}{(L/6)} \quad \text{or} \quad h = \frac{L}{6\sqrt{3}}.$$

$$\therefore \quad \vec{H}_{OA} = \frac{I}{2\pi h}\cos\theta\vec{a}_z$$

$$= \frac{I\vec{a}_z}{2\pi \times \left(\frac{L}{6\sqrt{3}}\right)}\cos 30^\circ = \frac{6\sqrt{3}\times I}{2\pi L}\left(\frac{\sqrt{3}}{2}\right)\vec{a}_z = \frac{9I}{2\pi L}\vec{a}_z \text{ A/m.} \tag{4.20}$$

Due to symmetrical geometry, at point C, the magnetic field intensity $(\vec{H})$ due to the three sides is given by

$$\vec{H} = 3 \times \vec{H}_{OA} = 3 \times \frac{9I}{2\pi L}\vec{a}_z = \frac{27I}{2\pi L}\vec{a}_z \text{ A/m.} \tag{4.21}$$

Note: If a be the length of the side of the triangle, then $L = 3a$.

$$\therefore \quad \vec{H} = \frac{27I}{2\pi(3a)}\vec{a}_z = \frac{9I}{2\pi a}\vec{a}_z. \tag{4.22}$$

4.11 Magnetic Field Intensity ($\vec{H}$) at the Centre of a Square Formed by a Current-Carrying Wire of Length L

Consider a wire of length L bent in the form of a square of side $(L/4)$ each and placed in the $x-y$ plane as shown in Fig. 4.17. Let I be the current in the conductor. Let OA, be a side of the square lying along the x-axis and Q be the centre of the square, at a distance of h metres from the side OA.

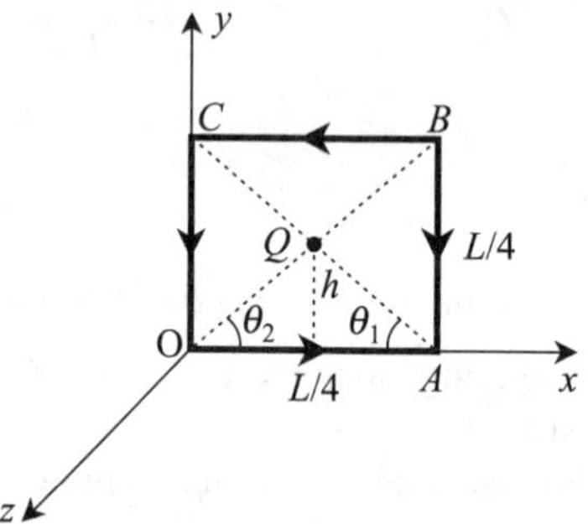

Fig. 4.17 *Current-carrying wire in the form of a square*

We know that the magnetic field intensity at height h due to finite current wire OA is given by

$$\vec{H}_{OA} = \frac{I}{4\pi h}[\cos\theta_1 + \cos\theta_2]\vec{a}_z \text{ A/m.}$$

From Fig. 4.17, $OA = AB = BA = CA = \frac{L}{4}$ and $\theta = \theta_1 = \theta_2 = 45°$.

So, $\tan 45° = \frac{h}{L/8}$ or $h = \frac{L}{8}$.

$$\therefore \quad \vec{H}_{OA} = \frac{I}{4\pi(L/8)}[\cos 45 + \cos 45]\vec{a}_z$$

$$\vec{H}_{OA} = \frac{2I}{\pi L} \times \frac{2}{\sqrt{2}}\vec{a}_z = \frac{2\sqrt{2}I}{\pi L}\vec{a}_z \text{ A/m.} \tag{4.23}$$

Due to symmetrical geometry at point Q, the magnetic field intensity $(\vec{H})$ due to the four sides of the square at the centre Q is given by

$$\vec{H} = 4 \times \vec{H}_{OA}$$

$$= \frac{4 \times 2\sqrt{2}I}{\pi L}\vec{a}_z = \frac{8\sqrt{2}I}{\pi L}\vec{a}_z \text{ A/m.} \tag{4.24}$$

Note: If a be the length of the side of a square, then $L = 4a$.

$$\text{Then,} \quad \vec{H} = \frac{8\sqrt{2}I}{\pi(4a)} = \frac{2\sqrt{2}I}{\pi a}\vec{a}_z \text{ A/m} \tag{4.25}$$

and $\vec{B} = \dfrac{2\sqrt{2}\mu I}{\pi a}\vec{a}_z$ A/m.

Example 4.7 Consider a wire carrying a current 1 A bent into a square of side 10 m. Find the magnetic field intensity at the centre of the square.

Solution Given side $a = 10$ m; $I = 1$ A.
We know that the magnetic field intensity at the centre of a square is

$$\vec{H} = \frac{2\sqrt{2}I}{\pi a}\vec{a}_z = \frac{2\sqrt{2}}{\pi} \times \frac{1}{10}\vec{a}_z = 0.09\vec{a}_z \text{ A/m}.$$

Example 4.8 For the current elements located in free space as shown in Fig. 4.18, if $I_1 = 1000$ A, $I_2 = 4000$ A, $I_3 = 20,000$ A and $I_4 = 8000$ A, find the magnetic field at the point P.

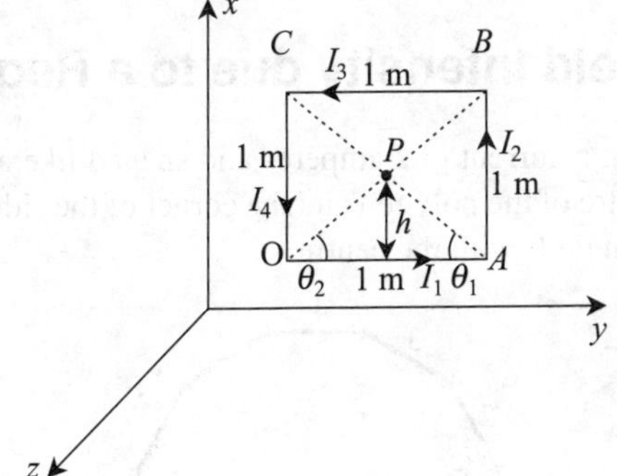

Fig. 4.18 *Field due to a square loop*

Solution The location of the current elements in the $x-y$ plane in free space is as shown in Fig. 4.18. Let H_1, H_2, H_3 and H_4 be the magnetic field intensities at point P due to the four current elements.

Given $I_1 = 1000$ A, $I_2 = 4000$ A, $I_3 = 20,000$ A and $I_4 = 8000$ A.

The total magnetic field intensity $\vec{H}$ at point P is

$$\vec{H} = \vec{H}_1 + \vec{H}_2 + \vec{H}_3 + \vec{H}_4.$$

We know that the magnetic field intensity due to the current element OA at point P is

$$\vec{H}_1 = \frac{I_1}{4\pi h}(\cos\theta_1 + \cos\theta_2)\vec{a}_z \text{ A/m}.$$

From Fig. 4.18, $\theta_1 = \theta_2 = 45°$; $a = 1$ m and $h = \dfrac{1}{2} = 0.5$ m.

Magnetic field intensity $\vec{H}_1 = \dfrac{I_1}{4\pi(0.5)}(\cos 45 + \cos 45)\vec{a}_z$

$$= \frac{2I_1}{2\pi\sqrt{2}} = \frac{I_1}{\sqrt{2}\pi}\vec{a}_z \text{ A/m}.$$

Due to symmetry, for other current elements,

$$\vec{H}_2 = \frac{I_2}{\sqrt{2}\pi}\vec{a}_z\,;\ \vec{H}_3 = \frac{I_3}{\sqrt{2}\pi}\vec{a}_z \text{ and } \vec{H}_4 = \frac{I_4}{\sqrt{2}\pi}\vec{a}_z.$$

The total magnetic field intensity at point P,

$$\vec{H} = \vec{H}_1 + \vec{H}_2 + \vec{H}_3 + \vec{H}_4$$

$$= \frac{1000\vec{a}_z}{\pi\sqrt{2}} + \frac{4000\vec{a}_z}{\pi\sqrt{2}} + \frac{20000\vec{a}_z}{\pi\sqrt{2}} + \frac{8000\vec{a}_z}{\pi\sqrt{2}} = 7427.6\,\vec{a}_z.$$

The magnetic flux density $\vec{B}$ at point P is given by

$$\vec{B} = \mu_0\vec{H} = 4\pi \times 10^{-7} \times 7427.6\vec{a}_z = 9.33 \times 10^{-3}\vec{a}_z \text{ tesla.}$$

4.12 Magnetic Field Intensity due to a Regular Polygon

Consider a loop wire carrying a current of I amperes and shaped like a regular polygon of n sides. The distance between the centre of the polygon and the corner of the sides is a as shown in Fig. 4.19.

Let the corners make an angle θ with the centre.

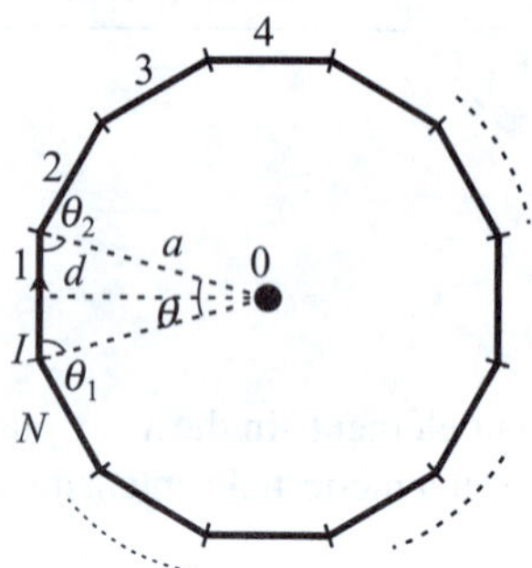

Fig. 4.19 *Field due to a polygon*

Let d be the distance of side 1 from the centre.

For an n sided polygon, $\theta = \dfrac{2\pi}{n}$.

We know that the magnetic field intensity due to the conductor of side 1 at the centre is

$$\vec{H} = \frac{I}{4\pi d}(\cos\theta_1 + \cos\theta_2)\vec{a}_\phi.$$

From Fig. 4.19, $\theta_1 = \theta_2 = 90 - \dfrac{\theta}{2}$.

$$\vec{H} = \frac{2I}{4\pi d}\left(\cos\left(90 - \frac{\theta}{2}\right)\right) \quad (4.26)$$

$$\vec{H} = \frac{I}{2\pi d}\sin\left(\frac{\theta}{2}\right). \quad (4.27)$$

Due to symmetry at point O, for a total of n sides,

$$\vec{H} = \frac{nI}{2\pi d}\sin\left(\frac{\pi}{n}\right). \tag{4.28}$$

Also, from Fig. 4.19, $d = a\cos\left(\frac{\pi}{n}\right)$.

$$\therefore \quad \vec{H} = \frac{nI\sin\left(\frac{\pi}{n}\right)}{2\pi a\cos\left(\frac{\pi}{n}\right)} = \frac{nI}{2\pi a}\tan\left(\frac{\pi}{n}\right)\vec{a}_\varphi \text{ A/m}. \tag{4.29}$$

Note: As n tends to infinity, the regular polygon becomes a circle. The magnetic field intensity is

$$H = \lim_{n\to\infty}\frac{nI}{2\pi a}\tan\left(\frac{\pi}{n}\right) = \frac{I}{2\pi a}\lim_{n\to\infty}\frac{n\sin(\pi/n)}{\cos(\pi/n)}.$$

Let $n = \dfrac{1}{m}$.

Then $H = \dfrac{I}{2\pi a}\lim\limits_{m\to\infty}\dfrac{\sin(m\pi)}{m\cos(m\pi)} = \dfrac{I\pi}{2\pi a} = \dfrac{I}{2a}$,

or $\vec{H} = \dfrac{I}{2a}\vec{a}_\phi$ and $\vec{B} = \dfrac{\mu I}{2a}\vec{a}_\phi$,

which is same as Eq. (4.19).

4.13 Magnetic Field Intensity due to a Solenoid

Consider a solenoid of length l and radius a consisting of N turns of wire carrying current I as shown in Fig. 4.20, lying along the z-axis.

Let P be a point along its axis. θ_1 and θ_2 are the angles at P between two end turns of the coil and the z-axis. Let the current element Idl make an angle θ with P.

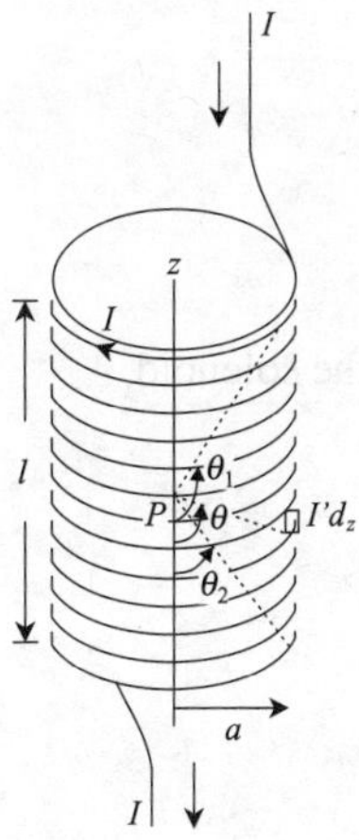

Fig. 4.20 *Field due to a solenoid*

Since the solenoid consists of N turns of circular loops, the current in the thickness dl is

$$I' = \frac{NI}{l} \text{ and } dl = dz.$$

Then $I'dl = \frac{IN}{l}dz.$ (4.30)

We know that the magnetic field intensity at point P is given by Eq. (4.17):

$$d\vec{H} = \frac{I'dl\, a^2}{2(a^2+z^2)^{3/2}}\vec{a}_z = \frac{NIa^2dz}{2l(a^2+z^2)^{3/2}}\vec{a}_z.$$

Now, $\vec{H} = \int_{\theta_1}^{\theta_2} \frac{NIa^2\, dz}{2l(a^2+z^2)^{3/2}}\vec{a}_z.$

To evaluate the integral, let $z = a\cot\theta$.
Then, $dz = -a\ \text{cosec}^2\theta d\theta$.

$$\therefore \quad \vec{H} = \frac{NI}{2l}\int_{\theta_1}^{\theta_2} \frac{a^2\vec{a}_z}{(a^2+a^2\cot^2\theta)^{3/2}}(-a\,\text{cosec}^2\theta)d\theta$$

$$= \frac{NI}{2l}\int_{\theta_1}^{\theta_2} \frac{a^2\vec{a}_z}{a^3\,\text{cosec}^3\theta}(-a\ \text{cosec}^2\theta)\, d\theta$$

$$= \frac{NI}{2l}\int_{\theta_1}^{\theta_2} \frac{-\vec{a}_z}{\text{cosec}\theta}d\theta = \frac{-NI}{2l}\int_{\theta_1}^{\theta_2}\sin\theta d\theta \vec{a}_z$$

$$= \frac{NI}{2l}(\cos\theta_2 - \cos\theta_1)\vec{a}_z. \quad (4.31)$$

Note 1: For a single loop, $N = 1$.

$$\therefore \quad \vec{H} = \frac{I}{2l}(\cos\theta_2 - \cos\theta_1)\vec{a}_z. \quad (4.32)$$

Note 2: When P lies in the middle of the solenoid,

$$\theta_1 = \pi - \theta_2\ ;\ \theta_2 = \tan^{-1}\left(\frac{a}{l/2}\right).$$

Then $\vec{H} = \frac{NI}{2l}(\cos\theta_2 - \cos(\pi-\theta_2))\vec{a}_z$

$$= \frac{NI}{2l}(\cos\theta_2 + \cos\theta_2)\vec{a}_z$$

$$= \frac{NI}{2l} 2\cos\theta_2 \vec{a}_z = \frac{NI\cos\theta_2}{l}\vec{a}_z. \tag{4.33}$$

Note 3: When the length of the solenoid $l >> a$, then

$$\theta_1 \approx \pi \text{ and } \theta_2 = 0.$$

$$\therefore \quad \vec{H} = \frac{NI}{2l}(\cos 0 - \cos\pi)\vec{a}_z = \frac{NI}{l}\vec{a}_z. \tag{4.34}$$

Note 4: When P lies at one end of the solenoid and ($l >> a$), then

$$\theta_1 = \pi/2 \text{ and } \theta_2 = 0 \text{ or } \theta_1 = \pi \text{ and } \theta_2 = \pi/2.$$

$$\therefore \quad \vec{H} = \frac{NI}{2l}(\cos 0 - \cos\pi/2)\,\vec{a}_z$$

$$= \frac{NI}{2l}(\cos\pi/2 - \cos\pi)\vec{a}_z = \frac{NI}{2l}\vec{a}_z. \tag{4.35}$$

That is, the magnetic field intensity at the end of a solenoid is equal to half the magnetic field intensity at the centre of the solenoid for $l >> a$.

Example 4.9 A solenoid has 3000 turns, a length of $l = 150$ cm, a radius of $a = 2$ cm and carries a current of 100 mA. Find H at (0, 0, 20) and (0, 0, 150).

Solution Given $N = 3000$, $I = 100$ mA $= 0.1$A, $a = 2$ cm $= 2 \times 10^{-2}$ m, $l = 150$ cm $= 1.5$ m.

We know that $\vec{H} = \frac{NI}{2l}(\cos\theta_2 - \cos\theta_1)\vec{a}_z$

$$= \frac{3000 \times 0.1}{2 \times 1.5}(\cos\theta_2 - \cos\theta_1)\vec{a}_z.$$

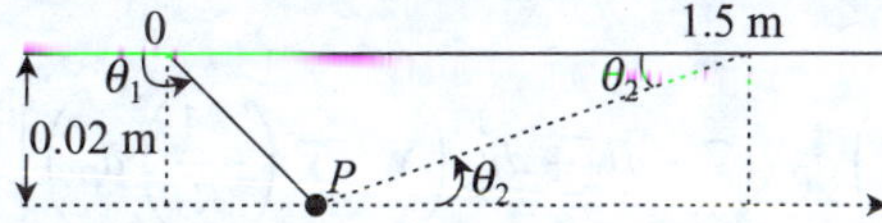

Fig. 4.21 *Field due to the solenoid*

(i) With point P at (0, 0, 20) cm, from Fig. 4.21,

$$\pi - \theta_1 = \tan^{-1}\left(\frac{2}{20}\right) \text{ or } \theta_1 = 174.28°.$$

$$\cos\theta_1 = \cos(174.28) = -0.995.$$

and $\quad \theta_2 = \tan^{-1}\left(\frac{2}{130}\right) = 0.88°.$

$$\cos\theta_2 = \cos(0.88°) = 0.999.$$

Then $\quad \vec{H} = 100(0.999 + 0.995)\vec{a}_z = 100 \times 1.995 = 199.4\vec{a}_z.$

(ii) With point P at (0, 0, 150),

$$\theta_1 = \pi - \tan^{-1}\left(\frac{2}{150}\right) = 179.23^\circ \text{ and } \quad \theta_2 = \pi/2.$$

Then $\quad \vec{H} = 100(\cos\pi/2 - \cos(179.23^\circ)) = 99.99\vec{a}_z.$

4.14 Maxwell's Second Equation for Magnetic Flux Density

We know from Biot–Savart's law that the magnetic flux density at distance R from a current-carrying conductor is

$$\vec{B} = \frac{\mu_0 I}{4\pi}\oint \frac{d\vec{l}\times\vec{a}_R}{R^2}.$$

Now, $\quad \nabla\cdot\vec{B} = \nabla\cdot\left[\frac{\mu_0 I}{4\pi}\oint\frac{d\vec{l}\times\vec{a}_R}{R^2}\right]$

$$= \frac{\mu_0 I}{4\pi}\oint\frac{\nabla\cdot(d\vec{l}\times\vec{a}_R)}{R^2}. \qquad (4.36)$$

We know that $\nabla\cdot(\mathrm{A}\times B) = B\cdot\nabla\times\ \mathrm{A} - \mathrm{A}.\ \nabla\times B$ (See Chapter 1).

$$\therefore \quad \nabla\cdot\vec{B} = \frac{\mu_0 I}{4\pi}\oint\left[\frac{\vec{a}_R}{R^2}\cdot(\nabla\times d\vec{l}) - d\vec{l}\cdot\left(\nabla\times\frac{\vec{a}_R}{R^2}\right)\right].$$

Since R is only in the radial direction,

$$\frac{\vec{a}_R}{R^2} = \frac{\vec{a}_\rho}{R^2} = -\nabla\left(\frac{1}{R}\right)\vec{a}_\rho.$$

$$\therefore \quad \nabla\cdot\vec{B} = \frac{\mu_0 I}{4\pi}\oint\left[-\nabla\left(\frac{1}{R}\right)\vec{a}_\rho\cdot(\nabla\times\vec{dl}) + d\vec{l}\cdot\left(\nabla\times\nabla\left(\frac{-1}{R}\right)\vec{a}_\rho\right)\right].$$

Let $\quad \frac{1}{R}\vec{a}_\rho = C$ and $d\vec{l} = D$.

Then $\quad \nabla\cdot\vec{B} = \frac{\mu_0 I}{4\pi}\oint[-\nabla C\cdot(\nabla\times D) + D\cdot(\nabla\times\nabla C)].$

From vector identity, $\nabla\times\nabla C = 0$ and $\nabla\cdot\nabla\times D = 0$,

$$\nabla\cdot\vec{B} = \frac{\mu_0 I}{4\pi}\oint(-0+0) = 0$$

$$\therefore \quad \nabla\cdot\vec{B} = 0. \qquad \text{(Proved). (4.37)}$$

Since div $\vec{B} = 0$, the magnetic flux lines are continuous. This equation is known as Maxwell's equation for magnetic flux density.

4.15 Ampere's Circuital Law or Ampere's Work Law

Ampere's circuital law can be stated as follows: *The line integral of magnetic field intensity $\vec{H}$ around a closed path is equal to the net current I_{encl} enclosed by that path.*

Mathematically,

$$\oint \vec{H} \cdot d\vec{l} = I_{\text{encl}}, \tag{4.38}$$

where I_{encl} is the current enclosed by the path called Amperian path and $d\vec{l}$, the differential length.

Proof Consider a long conductor carrying current I along the z-axis as shown in Fig. 4.22. Take a closed circular path of radius ρ, called an Amperian path, around the conductor and a current element $Id\vec{l}$.

Using Biot–Savart's law, $\vec{H}$ at a point P due to an infinite length current is

$$\vec{H} = \frac{I}{2\pi\rho}\vec{a}_\phi \quad \text{and} \quad d\vec{l} = \rho d\phi \vec{a}_\phi.$$

Then, $\vec{H} \cdot d\vec{l} = \dfrac{I}{2\pi\rho}\vec{a}_\phi\ \rho d\phi\ \vec{a}_\phi = \dfrac{I}{2\pi}d\phi.$

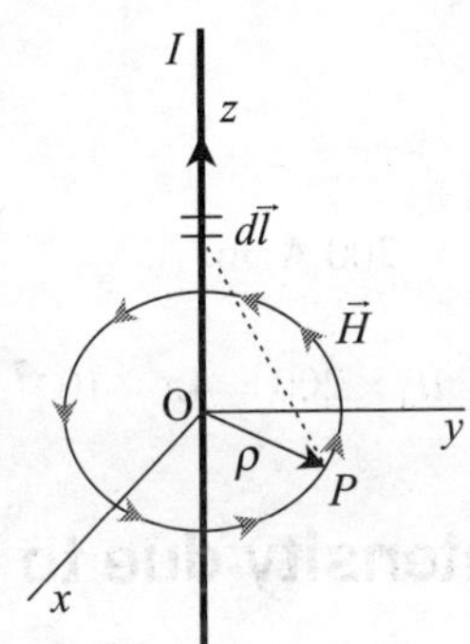

Fig. 4.22 *Current-carrying conductor*

The line integral over the closed path is

$$\oint \vec{H} \cdot d\vec{l} = \int_0^{2\pi} \frac{I}{2\pi} d\phi = \frac{I}{2\pi} \times 2\pi.$$

If $I = I_{\text{encl}}$, the current enclosed by the path,

$$\oint \vec{H} \cdot d\vec{l} = I_{\text{encl}} \qquad \text{(Proved).}$$

Note 1: Ampere's circuital law is analogous to Gauss' law in electrostatics. It can be applied to determine the magnetic field when the current distribution is symmetrical. The closed path (Amperian path) considered need not be circular. It may be of any shape.

Note 2: The closed path on which the Ampere's circuital law is to be applied is called Amperian path. It is analogous to the Gaussian surface for electric fields. Generally, a Gaussian surface can be used for both electric and magnetic fields.

Note 3: For a conductor of cross-section S m^2 with current density J A/m^2, Ampere's law can be expressed as

$$I_{\text{encl}} = \oint \vec{H} \cdot d\vec{l} = \int_{S} \vec{J} \cdot d\vec{S}.$$

Example 4.10 Using Ampere's circuital law, find $\vec{H}$ and $\vec{B}$ inside a long straight non-magnetic conductor radius 8 mm, carrying a current density of 50 kA/ m^2.

Solution Given $\vec{J} = 50$ kA/ m$^2 = 50 \times 10^3$ A/m^2, $\rho = 8$ mm $= 8 \times 10^{-3}$ m.

(i) Let us assume that the current is flowing in the $\vec{a}_z$ direction. Applying Ampere's circuital law,

$$\oint \vec{H} \cdot d\vec{l} = I.$$

$$\vec{H} \oint dl = I.$$

$$H(2\pi\rho) = |J||S| = J\pi\rho^2.$$

$$H = \frac{J\rho}{2} = \frac{50 \times 10^3 \times 8 \times 10^{-3}}{2} = 200 \text{ A/m}.$$

(ii) Magnetic flux density $B = \mu_0 H = \mu_0 \times 200 = 4\pi \times 10^{-7} \times 200 = 0.25$ mWb/m^2.

4.16 Magnetic Field Intensity due to a Solid Conductor

Consider a solid conductor of radius a of infinite length carrying a current of I amperes.

Let a Gaussian surface of radius ρ (Amperian path) be considered in the conductor as shown in Fig. 4.23. The magnetic field intensity $\vec{H}$ at various values of radius ρ are given below.

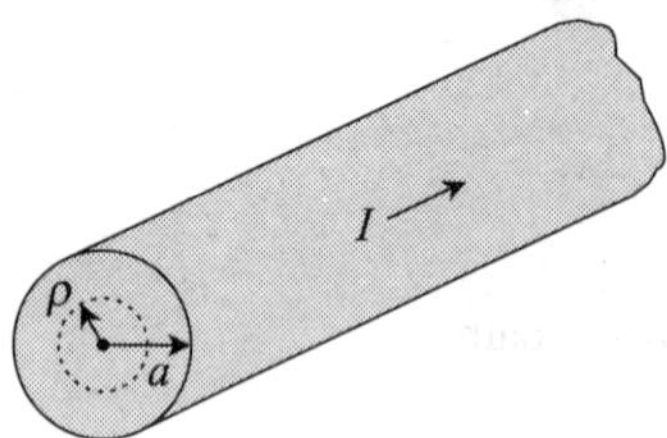

Fig. 4.23 *Solid conductor of infinite length*

Case 1 $\rho < a$

The current enclosed by the circular path of radius ρ is given by

$$I_{\text{encl}} = \frac{\pi\rho^2 I}{\pi a^2} = \frac{\rho^2 I}{a^2}.$$

According to Ampere's circuital law,

$$\oint \vec{H} \cdot d\vec{l} = I_{\text{encl}}.$$

$$\int_0^{2\pi} H_\phi \, \rho d\phi = I_{\text{encl}}.$$

$$H_\phi 2\pi\rho = \frac{\rho^2}{a^2} I.$$

$$H_\phi = \frac{\rho I}{2\pi a^2}$$

or $$\vec{H} = \frac{\rho I}{2\pi a^2} \vec{a}_\phi. \tag{4.39}$$

Case 2 $\rho = a$

$$\vec{H} = \frac{aI}{2\pi a^2} \vec{a}_\rho = \frac{I}{2\pi a} \vec{a}_\rho. \tag{4.40}$$

Case 3 $\rho > a$

The current enclosed by the circular path is

$$I_{\text{encl}} = I.$$

$$\therefore \quad H_\phi 2\pi\rho = I.$$

$$H_\phi = \frac{I}{2\pi\rho}$$

or $$\vec{H} = \frac{I}{2\pi\rho} \vec{a}_\phi. \tag{4.41}$$

Hence, $$\vec{H} = \begin{cases} \dfrac{\rho I}{2\pi a^2} \vec{a}_\phi & \rho < a \\ \dfrac{I}{2\pi a} \vec{a}_\phi & \rho = a \\ \dfrac{I}{2\pi\rho} \vec{a}_\phi & \rho > a \end{cases} \tag{4.42}$$

The variation of magnetic field intensity $\vec{H}$ with ρ is shown in Fig. 4.24.

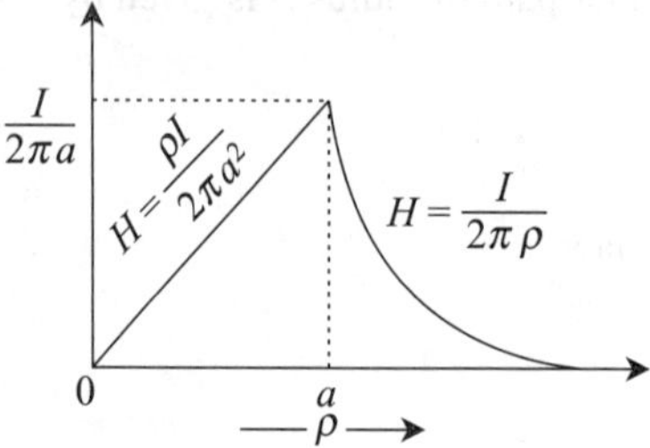

Fig. 4.24 *Variation of $\vec{H}$ along the radius in a solid conductor*

Magnetic field between two-wire transmission line Consider the system as two solid infinite conductors separated by a distance d. If the conductors are carrying currents I_1 and I_2 in opposite direction, then the magnetic field intensity at point P half way between the two wires is equal to the sum of the fields produced by the two lines at P.

The total magnetic field intensity is

$$\vec{H} = \left(\frac{I_1}{2\pi\rho} + \frac{I_2}{2\pi\rho}\right)\vec{a}_\phi.$$

If $I_1 = I_2$, $\rho = \frac{d}{2}$ then

$$\vec{H} = \frac{I}{\pi\rho}\vec{a}_\phi = \vec{H} = \frac{2I}{\pi d}\vec{a}_\varphi$$

If point P makes an angle α with the conductors, then

$$\cos\alpha = \frac{d}{2\rho},$$

$$\vec{H} = \frac{2I\cos\alpha}{\pi d}\vec{a}_\varphi.$$

4.17 Magnetic Field Intensity due to a Coaxial Cable

Consider a coaxial cable as shown in Fig. 4.25. The inner conductor has a radius a. The outer conductor has a thickness defined by the inner radius b and outer radius c. Let the inner conductor carry a current of I A and the outer conductor carry a current of $-I$ amperes along the z-axis. The space between the inner and outer conductor is filled with air.

Consider a closed surface of radius ρ as the Gaussian surface. The magnetic field intensity $\vec{H}$ inside the coaxial cable at different regions is given below.

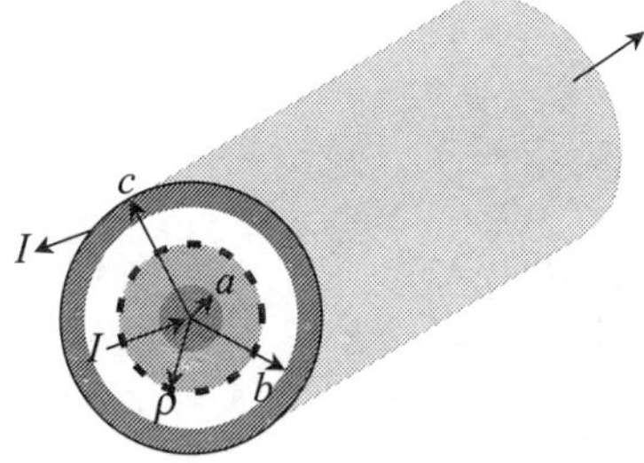

Fig. 4.25 *Field due to a coaxial cable*

Region 1 $\rho < a$

The current enclosed by the circular path of radius ρ (Amperian path) is given by

$$I_{\text{encl}} = \frac{\pi\rho^2 I}{\pi a^2} = \frac{\rho^2}{a^2} I.$$

According to Ampere's circuital law,

$$\oint_L \vec{H}.d\vec{l} = I_{\text{encl}}. \tag{4.43}$$

Since only H_ϕ component exists,

$$\int_0^{2\pi} H_\varphi \vec{\varphi}_\rho \cdot \rho d\phi \vec{a}_\phi = I_{\text{encl}}.$$

$$H_\phi 2\pi\rho = I_{\text{encl}} = \frac{\rho^2}{a^2} I.$$

$$H_\phi = \frac{\rho^2}{2\pi\rho a^2} I = \frac{I\rho}{2\pi a^2}.$$

$$\vec{H} = \frac{I\rho}{2\pi a^2} \vec{a}_\phi \text{ A/m}. \tag{4.44}$$

Region 2 $a < \rho < b$

If the circular path of radius ρ is within $a < \rho < b$, the current enclosed by the path is I amperes.

$$\therefore \quad H_\phi 2\pi\rho = I_{\text{encl}} = I.$$

$$H_\phi = \frac{I}{2\pi\rho}$$

or $$\vec{H} = \frac{I}{2\pi\rho} \vec{a}_\phi. \tag{4.45}$$

Region 3 $b < \rho < c$

If the circular path of radius ρ is within $b < \rho < c$, then, the current enclosed by the path is

$$I_{\text{encl}} = I + I'.$$

$$I' = -I\left(\frac{\pi(\rho^2 - b^2)}{\pi(c^2 - b^2)}\right).$$

$$\therefore \quad I_{\text{encl}} = I - I\left(\frac{\rho^2 - b^2}{c^2 - b^2}\right) = I\left[\frac{c^2 - \rho^2}{c^2 - b^2}\right]. \qquad (4.46)$$

According to Ampere's circuital law,

$$H_\phi 2\pi\rho = I_{\text{encl}} = I\left[\frac{c^2 - \rho^2}{c^2 - b^2}\right].$$

$$\therefore \quad H_\phi = \frac{I}{2\pi\rho}\left[\frac{c^2 - \rho^2}{c^2 - b^2}\right]\vec{a}_\phi \text{ A/m}. \qquad (4.47)$$

Region 4 $\rho > c$

If the circular path has radius $\rho > c$, the current enclosed by the path is

$$I_{\text{encl}} = I - I = 0.$$

$$\vec{H} = 0.$$

Therefore the field outside the cable is zero.

$$\text{Hence,} \quad \vec{H} = \begin{cases} \dfrac{I\rho}{2\pi a^2}\vec{a}_\phi & 0 \le \rho \le a \\ \dfrac{I}{2\pi\rho}\vec{a}_\phi & a \le \rho \le b \\ \dfrac{I}{2\pi\rho}\left[\dfrac{c^2 - \rho^2}{c^2 - b^2}\right]\vec{a}_\phi & b \le \rho \le c \\ 0 & \rho \ge c \end{cases} \qquad (4.48)$$

The variation in the magnetic field intensity with radius ρ of the coaxial cable is shown in Fig. 4.26.

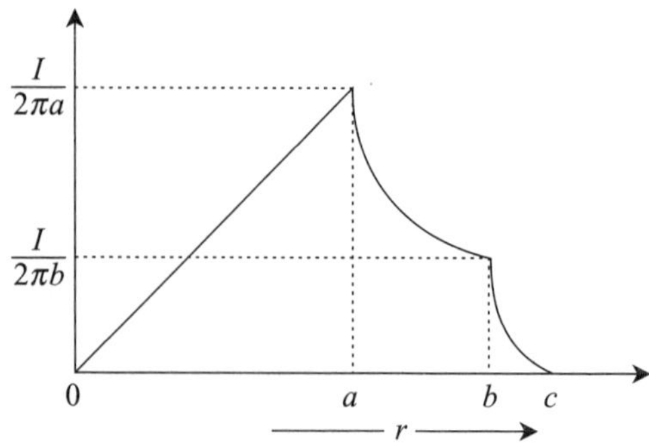

Fig. 4.26 *Variation of magnetic field with radius*

Example 4.11 A steady current of 10 A is established in a long straight hollow aluminum conductor which has an inner and outer radius of 1.5 cm and 3 cm respectively. Find the value of $\vec{B}$ as a function of radius.

Solution Given inner radius $\rho_1 = 1.5$ cm $= 1.5 \times 10^{-2}$ m; outer radius $\rho_2 = 3$ cm $= 3 \times 10^{-2}$ m, $I = 10$ A.

Assume the conductor's axis is along the z-axis. Let ρ be the radius of the Gaussian surface.

Case 1 $\rho_1 < \rho < \rho_2$

Amperian path $l = 2\pi\rho$.

Let J be the uniform current density.

$$I = JA, \quad \text{where } A \text{ is the cross-sectional area.}$$

$$\therefore \quad I_{\text{encl}} = \frac{I}{\pi(\rho_2^2 - \rho_1^2)} \pi(\rho^2 - \rho_1^2).$$

We know from Ampere's circuital law that

$$\oint \vec{H} \cdot d\vec{l} = I_{\text{encl}}.$$

$$H(2\pi\rho) = \frac{I(\rho^2 - \rho_1^2)}{(\rho_2^2 - \rho_1^2)} = \frac{I(\rho^2 - \rho_1^2)}{2\pi\rho(\rho_2^2 - \rho_1^2)} \vec{a}_\phi.$$

The magnetic field density is

$$\vec{B} = \mu_0 \vec{H} = \frac{\mu_0 I(\rho^2 - \rho_1^2)}{2\pi\rho(\rho_2^2 - \rho_1^2)} \vec{a}_\phi$$

$$= \frac{10 \times 10^{-7}}{36\pi \times 2\pi\rho} \frac{(\rho^2 - 1.5^2 \times 10^{-4})}{(3^2 \times 10^{-4} - 1.5^2 - 10^{-4})} \vec{a}_\phi.$$

$$\vec{B} = 2.96 \times 10^3 \frac{(\rho^2 - 2.25 \times 10^{-4})}{\rho} \vec{a}_\phi.$$

Case 2 $\rho < \rho_1 \quad I_{\text{encl}} = 0.$

$$\therefore \quad H = 0 \text{ and } B = 0.$$

Case 3 $\rho > \rho_2 \quad I_{\text{encl}} = I.$

$$\therefore \quad H 2\pi\rho = I.$$

$$H = \frac{10}{2\pi\rho} \quad \text{and} \quad B = \frac{\mu_0 10}{2\pi\rho} = \frac{2 \times 10^{-6}}{\rho} \text{ Wb/m}^2.$$

4.18 Magnetic Field Intensity due to an Infinite Sheet of Current

Consider an infinite sheet of conductor in the $x-y$ plane at $z=0$. Let the conductor has a surface current density $\vec{K}$ in the positive y direction as shown in Fig. 4.27.

Then, $\vec{K}=K_y\vec{a}_y$. (4.49)

Consider a closed path (Amperian path) of width b and length a perpendicular to the $x-y$ plane. The current flowing across width b is given by

$$I_{\text{encl}}=K_y b.$$

From Ampere's circuital law,

$$\oint \vec{H}\cdot d\vec{l}=I_{\text{encl}}. \tag{4.50}$$

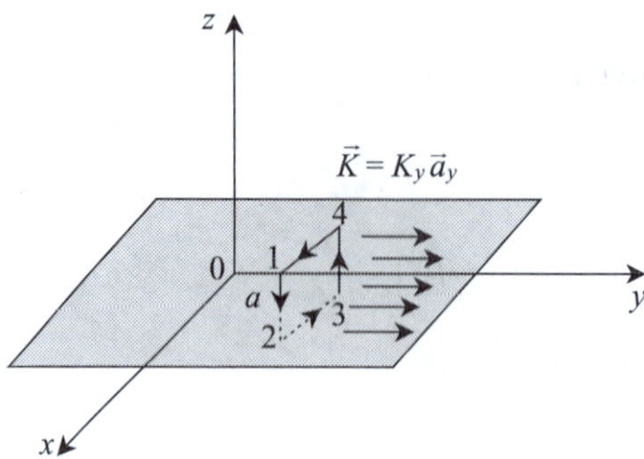

Fig. 4.27 *A uniform sheet of surface current on the x–y plane*

To evaluate the integral, the symmetrical conditions should be considered. Since the sheet extends to infinity in the $x-y$ plane, due to symmetry, the field along the z direction is zero.

As current is flowing in the y direction, the component H_y is also zero. Therefore, only H_x component exists.

The closed path integration is $\oint \vec{H}\cdot d\vec{l}=\int_1^2 \vec{H}\cdot d\vec{l}+\int_2^3 \vec{H}\cdot d\vec{l}+\int_3^4 \vec{H}\cdot d\vec{l}+\int_4^1 \vec{H}\cdot d\vec{l}=I_{\text{encl}}.$

For $z>0$,

$$\oint \vec{H}\cdot d\vec{l}=0(-a)+(-H_x)(-b)+0(a)+H_x b=I_{\text{encl}}.$$

$$2bH_x=I_{\text{encl}}=K_y b$$

$$H_x=\tfrac{1}{2}K_y \quad \text{for } z>0$$

or $$\vec{H}=\frac{1}{2}K_y\,\vec{a}_x \quad \text{for } z>0. \tag{4.51}$$

Similarly, $\vec{H} = \frac{-1}{2} K_y\, \vec{a}_x \quad \text{for } z < 0$ (4.52)

In general, $\vec{H} = \frac{1}{2} \vec{K} \times \vec{a}_n$ (4.53)

where $\vec{a}_n$ is the unit vector normal to the direction of the current flow on the sheet conductor.

4.19 Magnetic Field Intensity at Any Point in between Two Infinite Parallel Surface Current Sheets

Consider two infinite parallel surface current sheets in the x-y plane located along the z-axis some distance apart. Assume that a uniform sheet current density $\vec{K}$ is flowing in opposite directions along the y-axis on sheet surfaces as shown in Fig. 4.28. That is, the current density on the first sheet is $K\vec{a}_y$ and that on the second sheet is $-K\vec{a}_y$. Let P be a point in between the two current sheets.

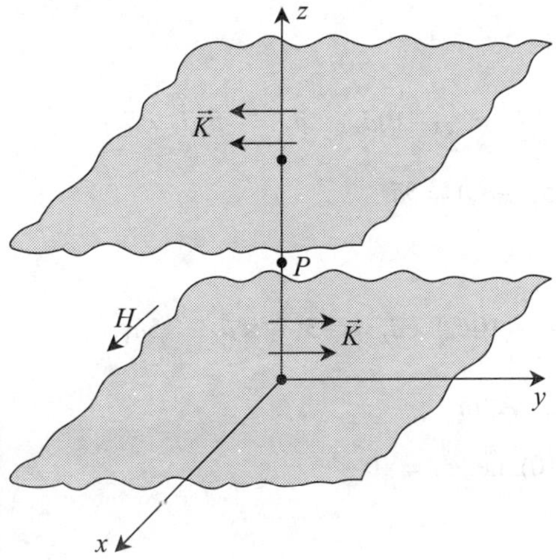

Fig. 4.28 *Field between two infinite current sheets*

From Eq. (4.53), the magnetic field intensity due to infinite sheet current is

$$\vec{H} = \frac{1}{2} \vec{K} \times \vec{a}_n.$$

The magnetic field intensity at point P due to the first current sheet is

$$\vec{H}_1 = \frac{1}{2} K\vec{a}_x. \tag{4.54}$$

Since the sheet current is in direction $\vec{a}_y$, the field is in $\vec{a}_x$ direction.

The magnetic field intensity at point P due to the second current sheet is

$$\vec{H}_2 = \frac{1}{2}(-K)(-\vec{a}_x) = \frac{K\vec{a}_x}{2}. \tag{4.55}$$

Therefore the total magnetic field intensity is

$$\vec{H} = \vec{H}_1 + \vec{H}_2 = \frac{K}{2}\vec{a}_x + \frac{K}{2}\vec{a}_x = K\vec{a}_x. \tag{4.56}$$

Hence the magnetic field intensity at any point in between two infinite parallel surface current sheets carrying equal current density in opposite directions is equal to the sheet current density. The field direction is normal to the current direction on the surfaces. The field is zero outside the current sheets.

Example 4.12 Planes $z = 0$ and $z = 4$ carry current $\vec{K}_1 = -10\vec{a}_x$ A/m and $\vec{K}_2 = 10\vec{a}_x$ A/m, respectively. Determine $\vec{H}$ at (i) (1, 1, 1) and (ii) (0, −3, 10).

Solution Given two parallel current sheets at $z = 0$ and $z = 4$.
At $z = 0$, $\vec{K}_1 = -10\vec{a}_x$ A/m, and at $z = 4$, $\vec{K}_2 = 10\vec{a}_x$.

Let $\vec{H}_1$ be the field due to the current sheet at $z = 0$ and $\vec{H}_2$ that due to the current sheet at $z = 4$. Total magnetic field intensity is $\vec{H} = \vec{H}_1 + \vec{H}_2$.

(i) Field at point (1, 1, 1), i.e., $z_1 = 1$:

Since $z_1 > 0$, $\vec{H}_1 = \frac{1}{2}\vec{K}_1 \times \vec{a}_z = \frac{1}{2}(-10\vec{a}_x) \times \vec{a}_z = -5(\vec{a}_x \times \vec{a}_z)$

$$= (-5)(-\vec{a}_y) = 5\vec{a}_y.$$

Since $z_1 < 4$,

$$\vec{H}_2 = \frac{-1}{2}\vec{K}_2 \times \vec{a}_z = \frac{-1}{2}(10\vec{a}_x) \times \vec{a}_z = -5(\vec{a}_x \times \vec{a}_z) = 5\vec{a}_y.$$

$$\vec{H} = 5\vec{a}_y + 5\vec{a}_y = 10\vec{a}_y \text{ A/m}.$$

(ii) Field at point (0, −3, 10), i.e., $z_1 = 10$

Since $z_1 > 0$ and $z_1 > 4$,

$$\vec{H} = \frac{1}{2}\vec{K}_1 \times \vec{a}_z + \frac{1}{2}\vec{K}_2 \times \vec{a}_z$$

$$= \frac{1}{2}(-10\vec{a}_x) \times \vec{a}_z + \frac{1}{2}(10\vec{a}_x) \times \vec{a}_z = 5\vec{a}_y - 5\vec{a}_y = 0.$$

Therefore the field is zero outside the current sheets.

4.20 Differential or Point Form of Ampere's Circuital Law (Maxwell's Third Equation)

The differential form of Ampere's circuital law states that *the curl of the magnetic field intensity $\vec{H}$ is equal to the conduction current density $\vec{J}$. It is also called Maxwell s third equation.*

Mathematically, $\nabla \times \vec{H} = \vec{J}$. (4.57)

Proof Consider a differential rectangular path of lengths Δx, Δy in the x-y plane centred at point P as shown in Fig. 4.29.

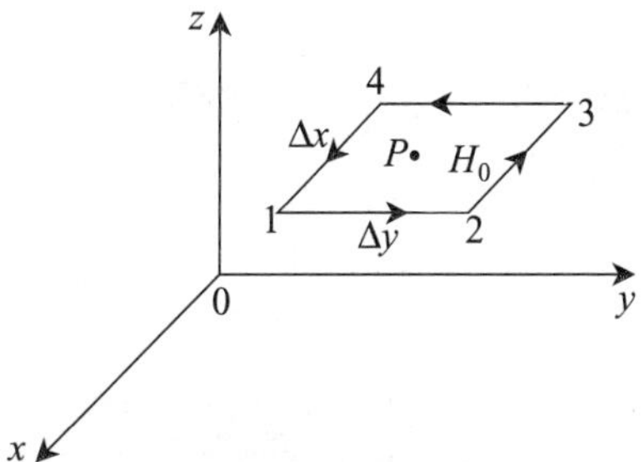

Fig. 4.29 *An incremental rectangular closed path*

Let current I with current density $\vec{J}$ A/m^2 flow through the enclosed path along the z- direction. Consider rectangular coordinates.

Let the magnetic field intensity at point P be

$$H_0 = H_{x0}\vec{a}_x + H_{y0}\vec{a}_y + H_{z0}\vec{a}_z \tag{4.58}$$

and the total current density be

$$\vec{J} = J_x\vec{a}_x + J_y\vec{a}_y + \vec{J}_z\vec{a}_z. \tag{4.59}$$

According to Ampere's circuital law,

$$\oint \vec{H}\cdot d\vec{l} = I_{\text{encl}}.$$

The closed path is along 1 to 2, 2 to 3, 3 to 4 and 4 to 1 as shown in Fig. 4.29.
The field along path 1 to 2 is

$$(\vec{H}\cdot d\vec{l})_{1,2} = H_y\Delta y \approx \left(H_{y0} + \frac{\Delta x}{2}\frac{\partial H_y}{\partial x}\right)\Delta y. \tag{4.60}$$

The field along path 2 to 3 is

$$(\vec{H}.\,d\vec{l})_{2,3} = -H_x\Delta x \approx -\left[H_{x_o} + \frac{\Delta y}{2}\frac{\partial H_x}{\partial y}\right]\Delta x. \tag{4.61}$$

The field along path 3 to 4 is

$$(\vec{H}.\,d\vec{l})_{3,4} = -H_y\Delta y \approx -\left[H_{y0} - \frac{\Delta x}{x}\frac{\partial H_y}{\partial x}\right]\Delta y. \tag{4.62}$$

The field along path 4 to 1 is

$$(\vec{H}.\,d\vec{l})_{4,1} = H_x\Delta x \approx \left[H_{x0} - \frac{\Delta y}{2}\frac{\partial H_x}{\partial y}\right]\Delta x. \tag{4.63}$$

The total field intensity along the closed path is obtained by combining the above equations

$$\oint \vec{H}.\, d\vec{l} = H_{y0}\Delta y + \frac{\Delta x \Delta y}{2}\frac{\partial H_y}{\partial x} - H_{x0}\Delta x - \frac{\Delta x \Delta y}{2}\frac{\partial H_x}{\partial y}$$

$$-H_{y0}\Delta y + \frac{\Delta x \Delta y}{2}\frac{\partial H_y}{\partial x} + H_{x0}\Delta x - \frac{\Delta x \Delta y}{2}\frac{\partial H_x}{\partial y}$$

$$= \left(\frac{\partial H_y}{\partial x} - \frac{\partial H_x}{\partial y}\right)\Delta x \Delta y = I_{\text{encl}}. \tag{4.64}$$

The current enclosed through the surface area $\Delta x \Delta y$ is given by

$$I_{\text{encl}} = J_z \Delta x \Delta y. \tag{4.65}$$

From Eqs (4.64) and (4.65),

$$\frac{\partial H_y}{\partial x} - \frac{\partial H_x}{\partial y} = J_z. \tag{4.66}$$

Similarly, for x- and y- components, the current densities are:

$$\frac{\partial H_x}{\partial z} - \frac{\partial H_z}{\partial x} = J_y \tag{4.67}$$

and $$\frac{\partial H_z}{\partial y} - \frac{\partial H_y}{\partial z} = J_x. \tag{4.68}$$

Substituting the above in Eq. (4.59),

$$\left(\frac{\partial H_z}{\partial y} - \frac{\partial H_y}{\partial z}\right)\vec{a}_x + \left(\frac{\partial H_x}{\partial z} - \frac{\partial H_z}{\partial x}\right)\vec{a}_y + \left(\frac{\partial H_y}{\partial x} - \frac{\partial H_x}{\partial y}\right)\vec{a}_z = \vec{J}. \tag{4.69}$$

Writing in matrix form,

$$\begin{vmatrix} \vec{a}_x & \vec{a}_y & \vec{a}_z \\ \dfrac{\partial}{\partial x} & \dfrac{\partial}{\partial y} & \dfrac{\partial}{\partial z} \\ H_x & H_y & H_z \end{vmatrix} = \vec{J}. \tag{4.70}$$

The above determinant is known as curl $\vec{H}$, given by $\nabla \times \vec{H}$, the cross product of the del operator and $\vec{H}$.

That is, $\nabla \times \vec{H} = \vec{J}$.

Note Curl in the cylindrical coordinate system is $$\nabla \times \vec{H} = \frac{1}{\rho}\begin{vmatrix} \vec{a}_\rho & \rho\vec{a}_y & \vec{a}_z \\ \dfrac{\partial}{\partial \rho} & \dfrac{\partial}{\partial \phi} & \dfrac{\partial}{\partial z} \\ H_\rho & \rho H\phi & H_z \end{vmatrix} \tag{4.71}$$

Curl in the spherical coordinate system is $$\nabla \times \vec{H} = \frac{1}{r^2 \sin\theta} \begin{vmatrix} \vec{a}_r & r\vec{a}\theta & r\sin\theta\vec{a}_\phi \\ \dfrac{\partial}{\partial r} & \dfrac{\partial}{\partial \theta} & \dfrac{\partial}{\partial \phi} \\ H_r & rH_\theta & r\sin\theta H_\phi \end{vmatrix} \tag{4.72}$$

Physical interpretation of curl When a current is flowing through a conductor, the magnetic field lines seem to swirl in concentric circles around the conductor with radius ρ. The direction of the curl of the magnetic field intensity vector, $\nabla \times \vec{H}$, is along the axis about which rotation of the vector field exists, i. e., along the current direction. Its direction can be obtained from the right-hand thumb rule . Curl of a vector exists for rotational fields. If curl is zero, then there is no rotation.

Example 4.13 The magnetic field intensity, $\vec{H}$, due to current source is given by $\vec{H} = (y\cos\ ax)\vec{a}_x + (ye^x)\vec{a}_z$. Describe the current density over the $y-z$ plane.

Solution Given $\vec{H} = (y\cos\ ax)\vec{a}_x + (ye^x)\vec{a}_z$.
From Ampere's circuital law in differential form,

$$\vec{J} = \nabla \times \vec{H}.$$

$$\nabla \times \vec{H} = \begin{vmatrix} \vec{a}_x & \vec{a}_y & \vec{a}_z \\ \dfrac{\partial}{\partial x} & \dfrac{\partial}{\partial y} & \dfrac{\partial}{\partial z} \\ H_x & H_y & H_z \end{vmatrix}$$

Since $H_x = y\cos ax,\ H_z = ye^x$,

$$\nabla \times \vec{H} = \vec{a}_x \left[\frac{\partial}{\partial y}(ye^x) - \frac{\partial}{\partial z}(H_y)\right] - \vec{a}_y \left[\frac{\partial}{\partial x}(ye^x) - \frac{\partial}{\partial z}(y\cos\ ax)\right]$$

$$+\vec{a}_z \left[\frac{\partial}{\partial x}(H_y) - \frac{\partial}{\partial y}(y\cos\ ax)\right]$$

$$= \vec{a}_x[e^x - 0] - \vec{a}_y[ye^x - 0] + \vec{a}_z[0 - \cos\ ax]$$

$$= (e^x)\vec{a}_x - (ye^x)\vec{a}_y - (\cos\ ax)\vec{a}_z.$$

Thus the current density $\vec{J} = \nabla \times \vec{H} = e^x\vec{a}_x - ye^x\vec{a}_y - \cos ax\vec{a}_z$

On the $y-z$ plane, $x = 0$.

Therefore, $\vec{J} = \nabla \times \vec{H} = (e^0)\vec{a}_x - (ye^0)\vec{a}_y - (\cos a(0))\vec{a}_z$

$$= (1)\vec{a}_x - (y)\vec{a}_y - (1)\vec{a}_z$$

$$\vec{J} = \vec{a}_x - y\vec{a}_y - \vec{a}_z.$$

4.21 Stokes' Theorem

Stokes' theorem for magnetic fields states that *the line integral of magnetic field intensity over a closed path is equal to the surface integral of the curl of magnetic field intensity over any surface enclosed by the path.*

Mathematically,

$$\oint \vec{H} \cdot d\vec{l} = \int_S (\nabla \times \vec{H}) \cdot d\vec{S} \tag{4.73}$$

The theorem relates line integral of a closed path to the surface integral.

Proof Consider a current-carrying conductor with current density J A/m^2. From Ampere's circuital law we know that the magnetic field intensity around the conductor is

$$\oint H \cdot d\vec{l} = I_{\text{encl}}.$$

Considering the current crossing a differential surface $d\vec{S}$ normal to the direction of current flow,

$$I = I_{\text{encl}} = \int_S \vec{J} \cdot d\vec{S}. \tag{4.74}$$

Also, we know that the curl of $\vec{H}$

$$\nabla \times \vec{H} = \vec{J}.$$

$$\therefore \quad I_{\text{encl}} = \int_S (\nabla \times \vec{H}) \cdot d\vec{S}. \tag{4.75}$$

Thus, $\oint \vec{H} \cdot d\vec{l} = \int_S (\nabla \times \vec{H}) d\vec{S}.$ (Proved).

Note: If the surface is a closed area, then the integral on the RHS is equal to zero.

Example 4.14 In a cylindrical region $0 < \rho < 0.8$ m, $\vec{J} = 3e^{-5\rho}\,\vec{a}_z$ A/m^2. Determine $H = H_\phi \vec{a}_\phi$ everywhere.

Solution Given, $\vec{J} = 3e^{-5\rho}\,\vec{a}_z$ A/m^2, for $0 < \rho < 0.8$m.
From Stokes' theorem, we have

$$\oint \vec{H} \cdot d\vec{l} = \int_S \vec{J} \cdot d\vec{S}.$$

In cylindrical coordinates,

$$d\vec{S} = \rho d\phi d\rho\, \vec{a}_z.$$

$$\int \vec{J}\cdot d\vec{S} = \int_{0}^{\rho}\int_{\phi=0}^{2\pi} (3e^{-5\rho}\vec{a}_z)\cdot(\rho d\phi d\rho\,\vec{a}_z) = \int_{0}^{\rho}\int_{\phi=0}^{2\pi} 3\rho e^{-5\rho} d\phi d\rho$$

$$= \int_{\phi=0}^{2\pi} d\phi \int_{0}^{\rho} 3\rho e^{-5\rho}\, d\rho = 6\pi \int_{0}^{\rho} \rho e^{-5\rho} d\rho.$$

Integrating by parts, we get

$$= 6\pi\left[\left[\frac{\rho e^{-5\rho}}{(-5)}\right]_0^{\rho} - \int_0^{\rho}\left[\frac{e^{-5\rho}}{(-5)}\right] d\rho\right] = 6\pi\left[\frac{\rho e^{-5\rho}}{(-5)} - \frac{e^{-5\rho}}{25}\right]_0^{\rho}$$

$$\therefore \int_s \vec{J}\cdot d\vec{S} = \frac{6\pi}{5}\left(-\rho e^{-5\rho} - \frac{e^{-5\rho}}{5}\right) + \frac{6\pi}{25} = I_{\text{encl}}.$$

From the cross-section of the cylinder, the line integral is,

$$\oint \vec{H}\cdot d\vec{l} = H_\phi(2\pi\rho).$$

Now $\oint \vec{H}\cdot d\vec{l} = \int_s \vec{J}\cdot d\vec{S}.$

$$H_\phi(2\pi\rho) = I_{\text{encl}}.$$

For $0 \le \rho \le 0.8$,

$$H_\psi - \frac{1}{2\pi\rho}\left[\frac{6\pi}{5}\left(-\rho e^{-5\rho} - \frac{e^{-5\rho}}{5}\right) + \frac{6\pi}{25}\right]$$

$$= \frac{3}{5\rho}\left(-\rho e^{-5\rho} - \frac{e^{-5\rho}}{5}\right) + \frac{3}{25\rho}\ \text{A/m}.$$

At $\rho = 0.8$,

$$H_\phi = 0.75\left(-0.8e^{-4} - \frac{e^{-4}}{5}\right) + 0.15$$

$$= -0.75e^{-4} + 0.15 = 0.136\ \text{A/m}.$$

For $\rho > 0.8$,

$$I_{\text{encl}} = \frac{6\pi}{5}\left(-0.8e^{-4} - \frac{e^{-4}}{5}\right) + \frac{6\pi}{25}$$

$$= -0.069 + 0.754 = 0.685\ \text{A}.$$

$$H_\phi(2\pi\rho) = I_{\text{encl}} = 0.685.$$

$$H_\phi = \frac{0.685}{2\pi\rho} = \frac{0.109}{\rho} \text{ A/m.}$$

$$\therefore \quad H_\phi = \begin{cases} \dfrac{3}{5\rho}\left(-\rho e^{-5\rho} - \dfrac{e^{-5\rho}}{5}\right) + \dfrac{3}{25\rho} & 0 < \rho \leq 0.8 \\ 0.136 & \rho = 0.8 \\ \dfrac{0.109}{\rho} & \rho > 0.8 \end{cases}$$

4.23 Point Form of Magnetic Flux Density

Consider a magnetic field with magnetic flux density $\vec{B}$ Tesla. The flux crossing area S is given by

$$\varphi = \int_S \vec{B} \cdot d\vec{S} \text{ Wb.} \tag{4.76}$$

Consider a closed surface. Since magnetic flux is continuous and exists in the form of a closed loop, the amount flux entering the surface is equal to the amount of flux leaving it.

Thus for a closed surface, the magnetic flux is zero, or the integral of the magnetic flux density over a closed surface is zero.

$$\int_S \vec{B} \cdot d\vec{S} = 0. \tag{4.77}$$

This is called Gauss' law for magnetic fields. The statement also means that magnetic monopoles do not exist and that magnetic field lines form closed loops that never begin or end.

Applying divergence theorem,

$$\int_S \vec{B} \cdot d\vec{S} = \int_V (\nabla \cdot \vec{B}) dv = 0, \tag{4.78}$$

where dv is the differential volume enclosed by the closed surface dS.

Since $dv \neq 0$,

$$\nabla \cdot \vec{B} = 0. \tag{4.79}$$

That is, the divergence magnetic flux density is always zero. This equation is also called the point form of Maxwell's equation for magnetic flux density.

4.24 Magnetic Field Intensity due to a Solenoid using Ampere's Circuital Law

Consider a solenoid of length l carrying a current I amperes. Let the solenoid have N number of turns wound on a magnetic material as shown in Fig. 4.30. Since the current in each turn is I, for N turns, the total current enclosed around the solenoid is $I_{\text{encl}} = NI$.

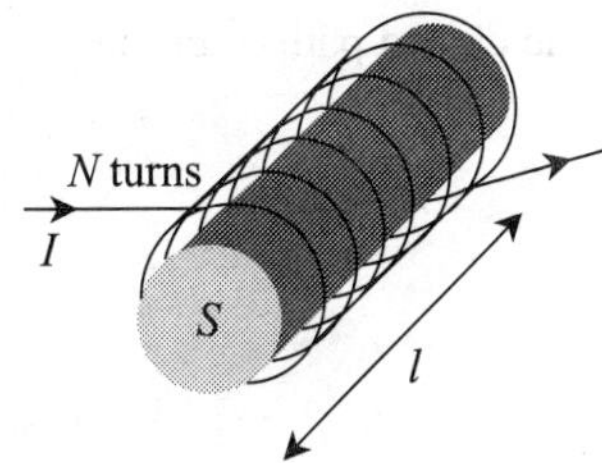

Fig. 4.30 *Solenoid*

We know from Ampere's circuital law that

$$\oint H \cdot dl = I_{\text{encl}} \qquad \text{or} \qquad \int_0^l H \cdot dl = I_{\text{encl}}. \tag{4.80}$$

For a uniform field, the magnetic field intensity at points within the solenoid is

$$Hl = NI \quad \text{or} \quad H = \frac{NI}{l} \text{ A/m} \quad \text{along the axis.}$$

If the solenoid is lying along the z-axis, then

$$\vec{H} = \frac{NI}{l}\vec{a}_z \text{ A/m} \tag{4.81}$$

and $\vec{B} = \dfrac{\mu NI}{l}\vec{a}_z \text{ Wb/m}^2$

where μ = permeability of the magnetic material.

4.25 Magnetic Field Intensity due to a Toroid using Ampere's Circuital Law

Consider a toroid having N turns wound on a magnetic material and carrying current I. Let the toroid have a mean, minimum and maximum radii of ρ, ρ_1 and ρ_2 respectively, as shown in Fig. 4.31. Since the current per each turn is I, the net current enclosed by the circle path is NI,

i.e., $I_{\text{encl}} = NI$.

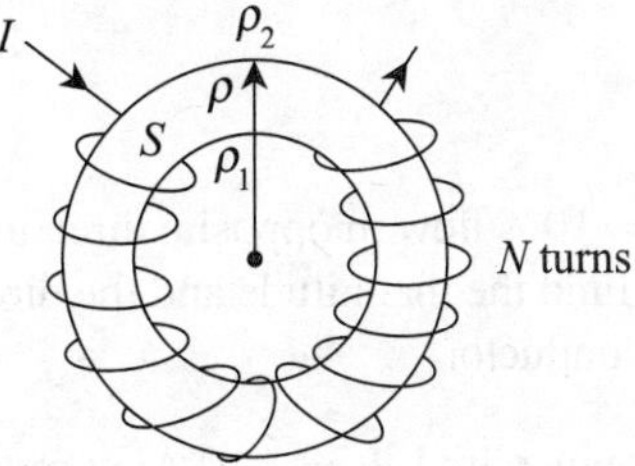

Fig. 4.31 *Toroid*

Apply Ampere's circuital law along the closed path of circle.

$$\oint H \cdot dl = NI.$$

$$\int_0^{2\pi} H \cdot \rho d\phi = NI.$$

The magnetic field intensity inside the toroid coil is then,

$$H = \frac{NI}{2\pi\rho} \text{ A/m}$$

$$\vec{H} = \frac{NI}{2\pi\rho}\vec{a}_\phi, \quad \rho_1 < \rho < \rho_2. \tag{4.82}$$

Outside the toroid, the field is zero.

The flux density inside the toroid coil is

$$\vec{B} = \frac{\mu NI}{2\pi\rho}\vec{a}_\phi \text{ Wb/m}^2$$

where μ = permeability of the magnetic material.

Additional Solved Problems

Example 4.15 Given magnetic flux density $\vec{B} = \rho \sin\phi \vec{a}_\phi$, find the total flux crossing the surface defined by $\phi = \pi/4$, $1 \le \rho \le 2$ and $0 \le z \le 5$.

Solution Given $\vec{B} = \rho \sin\phi \vec{a}_\phi$,

the flux $\varphi = \int_s B \cdot dS$ where $d\vec{S} = d\rho \, dz \vec{a}_\phi$.

$$\varphi = \int_{z=0}^{5}\int_{1}^{2} \rho \sin\phi \vec{a}_\phi d\,\rho\, dz\, \vec{a}_\phi = [z]_0^5 \left[\frac{r^2}{2}\right]_1^2 \sin\phi.$$

At $\phi = \pi/4$, $\varphi = 5\sin\frac{\pi}{4}\left(\frac{3}{2}\right)$.

$$\varphi = \frac{5}{\sqrt{2}} \times \frac{3}{2} = 5.303 \text{ Wb}.$$

Example 4.16 Currents $I_1 = I_2 = 10$ A flow in opposite directions through two long parallel wires separated by a distance of 20 cm. Find the magnitude and the direction of the magnetic flux density at a point 20 cm away from each conductor.

Solution Given that current through wire l, $I_1 = +10$ A; current through wire 2, $I_2 = -10$ A; and the distance between two wires, $d = 20$ cm; $\rho_1 = \rho_2 = 20$ cm $= 0.2$ m.

The magnetic field produced by wire 1 is given by

$$B_1 = \frac{\mu I_1}{2\pi\rho_1}.$$

Similarly, the magnetic field produced by wire 2 is

$$B_2 = \frac{\mu I_2}{2\pi\rho_2}.$$

Therefore, the required magnitude of the magnetic field is,

$$B = B_1 + B_2$$

$$= \frac{\mu}{2\pi}\left(\frac{I_1}{0.2} + \frac{I_2}{0.2}\right).$$

$$B = \frac{\mu}{2\pi}\left(\frac{100}{0.2} - \frac{100}{0.2}\right) = 0.$$

Example 4.17 A single-phase circuit comprises two parallel conductors, A and B, each 1 cm in diameter and spaced 1 m apart. The conductors carry current of +100 A and −100 A respectively. Determine the field intensity at the surface of each conductor and also in the space exactly midway between A and B.

Solution Given diameter of conductors, $a = 1$ cm $= 1 \times 10^{-2}$ m ; distance between the conductors, $b = 1$ m.

Current in conductor $A = +100$ A; current in conductor $B = -100$ A.

We know that the magnetic field intensity at the surface of the current-carrying conductor is given by

$$H = \frac{I}{2\pi\rho} \text{ A/m}$$

where, $\rho =$ radius of the conductor.

The field intensity due to the two conductors is

$$H_1 = H_A + H_B.$$

$$H_A = \frac{I}{2\pi\rho_1}, \text{ field intensity due to conductor } A.$$

$$\rho_1 = a/2 = 0.5 \times 10^{-2} \text{ m}.$$

$$H_B = \frac{I}{2\pi\rho_2}, \text{ field intensity due to conductor } B \text{ on the surface of the conductor } A.$$

$$\rho_2 = a/2 + b = -0.5 \times 10^{-2} + 1.0 = 1 - 0.05 = 0.95 \text{ m}.$$

$$H_1 = \frac{I}{2\pi\rho_1} + \frac{I}{2\pi\rho_2} = \frac{I}{2\pi}\left[\frac{1}{\rho_1} + \frac{1}{\rho_2}\right]$$

$$= \frac{100}{2\pi}\left[\frac{1}{0.5\times 10^{-2}} + \frac{1}{0.95}\right] = 3200 \ \text{A/m}.$$

The magnetic field (H) at any point P midway between A and B is given by the sum of fields due to A and B, i.e.,

$$H_2 = H_A + H_B.$$

$$= \frac{I}{2\pi\rho_1} + \frac{I}{2\pi\rho_2} = \frac{I}{2\pi}\left(\frac{1}{\rho_1} + \frac{1}{\rho_2}\right).$$

In this case, $\rho_1 = \rho_2 = \dfrac{b}{2} = 0.5$ m.

$$\therefore \quad H_2 = \frac{100}{2\pi}\left(\frac{1}{0.5} + \frac{1}{0.5}\right) = \frac{100}{2\pi} \times 4 = 63.67 \ \text{A/m}.$$

Example 4.18 Find $\vec{H}$ at $(-5,6,0)$ due to a current filament carrying 15 A current extending from 0 to infinity along the x-axis.

Solution Given $I = 15$ A and the filament extends from 0 to infinity along the x-axis.
At point $(-5,6,0)$, the distance along the y-axis, $\rho = 6$ m.

At infinity, $\theta_2 = 0$,

$$\theta_1 = \frac{\pi}{2} + \cos^{-1}\left(\frac{6}{\sqrt{5^2+6^2}}\right) = 129.8^\circ$$

The field intensity is

$$\vec{H} = \frac{I}{4\pi\rho}(\cos\theta_2 + \cos\theta_1)\vec{a}_\phi, \quad \text{here } \vec{a}_\phi = \vec{a}_z$$

$$\vec{H} = \frac{15}{4\pi(6)}(\cos 0 + \cos 129.8^\circ) = 71.6\vec{a}_z \ \text{mA/m}.$$

Example 4.19 A filament carrying current of 10 A is directed inward from infinity to the origin on the positive x-axis and then outward to infinity along the y-axis. Find the magnetic field intensity at point P (5, 8, 0).

Solution Given $I = 10$ A, and current filament as shown in Fig. 4.32.
The point on the $x-y$ plane where magnetic field intensity is to be found is P (5, 8, 0).

First consider the current on the x-axis extending from infinity to zero. We know that the magnetic field intensity due to a finite length wire is

$$\vec{H}_1 = \frac{I}{4\pi\rho_1}(\cos\theta_2 + \cos\theta_1)\vec{a}_\phi.$$

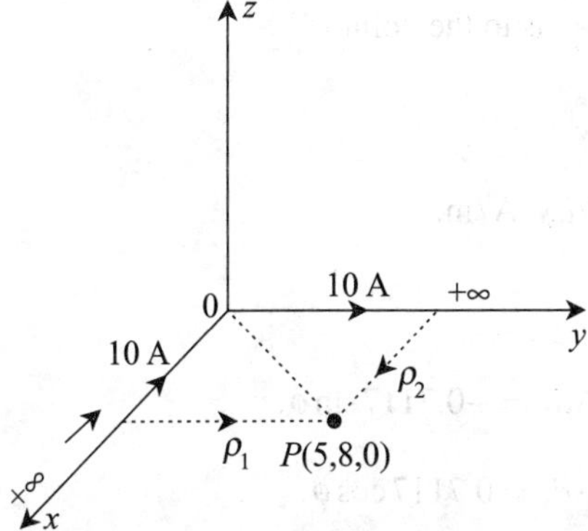

Fig. 4.32 *Field due to semi-infinite length currents*

Since one end of the line is infinity, $\theta_1 = 0°$, $\cos\theta_1 = 1$ and $\cos\theta_2 = \dfrac{5}{\sqrt{5^2+8^2}} = 0.53$.

$$\rho_1 = 8.$$

$$\therefore \quad \vec{H}_1 = \frac{10}{4\pi(8)}(0.53+1)\vec{a}_\phi = 0.152\vec{a}_\rho \text{ A/m} = 0.152\vec{a}_\phi.$$

Now consider another semi-infinite current on the y-axis extending from zero to infinity.

Since one end of the line is infinity, $\theta_2 = 0$, $\cos\theta_2 = 1$ and $\cos\theta_1 = \dfrac{8}{\sqrt{5^2+8^2}} = 0.848$.

$$\rho_2 = 5.$$

$$\therefore \quad \vec{H}_2 = \frac{I}{4\pi\rho_2}(\cos\theta_2 + \cos\theta_1)\vec{a}_\phi = \frac{10}{4\pi(5)}(1+0.848)\vec{a}_\phi = 0.294\vec{a}_\phi.$$

The total magnetic field intensity at point P is

$$\vec{H} = \vec{H}_1 + \vec{H}_2 = 0.152\vec{a}_\rho + 0.294\vec{a}_\rho = 0.446\vec{a}_\phi.$$

In rectangular coordinates, since the current is flowing in the clockwise direction in the $x-y$ plane,

$$\vec{a}_\phi = -\vec{a}_z.$$

$$\therefore \quad \vec{H} = -0.446\vec{a}_z.$$

Example 4.20 A current of 10 A is flowing in the $\vec{a}_z$ direction on the z-axis. Find the magnetic field intensity $\vec{H}$ at point $P(1, 2, 3)$ due to this filament if it extends from (a) $z = -\infty$ to ∞, (b) $z = 0$ to 5 m. Express your answers in Cartesian coordinates.

Solution Given $I = 10$ A in $\vec{a}_z$ direction as shown in Fig. 4.33.

(a) For current element extending from $z = -\infty$ to $+\infty$, the magnetic field intensity is

$$\vec{H} = \frac{I}{2\pi\rho}\vec{a}_\phi.$$

The radial distance from the z-axis to the point is

$$\rho = \sqrt{1+2^2} = \sqrt{5}.$$

$$\therefore \quad \vec{H} = \frac{10}{2\pi(\sqrt{5})}\vec{a}_\phi = 0.7117\vec{a}_\phi \ \text{A/m}.$$

In Cartesian coordinates,

$$H_x = \vec{H}\cdot\vec{a}_x = 0.7117\vec{a}_\phi\cdot\vec{a}_x = -0.7117\sin\phi,$$

$$H_y = \vec{H}\cdot\vec{a}_y = 0.7117\vec{a}_\phi\cdot\vec{a}_y = 0.7117\cos\phi,$$

$$H_z = \vec{H}\cdot\vec{a}_z = 0.7117\vec{a}_\phi\cdot\vec{a}_z = 0,$$

where, $\sin\phi = \dfrac{2}{\sqrt{5}} = 0.8944$; $\cos\phi = \dfrac{1}{\sqrt{5}} = 0.4472.$

$$\therefore \quad H_x = -0.6365\vec{a}_x,$$

$$H_y = 0.3183\vec{a}_y,$$

and $\vec{H} = -6365\vec{a}_x + 0.3183\vec{a}_y$ A/m.

(b) For current element extending from 0 to 5 m,

$$\vec{H} = \frac{I}{4\pi\rho}(\cos\theta_2 + \cos\theta_1)\vec{a}_\phi.$$

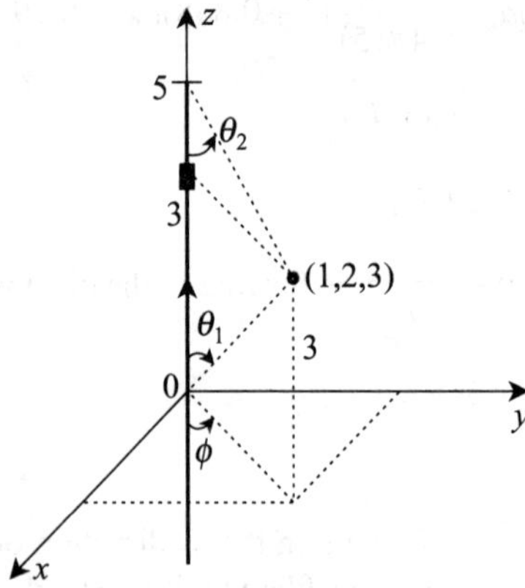

Fig. 4.33 *Field due to finite current element*

From Fig. 4.33,

$$\theta_2 = \tan^{-1}\left(\frac{\sqrt{5}}{2}\right) = 48.19^\circ.$$

$$\theta_1 = \tan^{-1}\left(\frac{\sqrt{5}}{3}\right) = 36.7°.$$

The field intensity is

$$\vec{H} = \frac{I}{4\pi\rho}(\cos\theta_2 + \cos\theta_1)\vec{a}_\phi,$$

$$\vec{H} = \frac{10}{4\pi(\sqrt{5})}(\cos 48.19° + \cos 36.7°) = 0.522\vec{a}_\phi \text{ mA/m.}$$

$$H_x = \vec{H}\cdot\vec{a}_x = 0.53(\vec{a}_\phi\cdot\vec{a}_x) = 0.53(-\sin\phi),$$

$$H_x = -0.522\times 0.8944 = -0.467$$

$$H_y = \vec{H}\cdot\vec{a}_y = 0.53(\vec{a}_\phi\cdot\vec{a}_y) = 0.53(\cos\phi)$$

$$H_y = -0.522\times 0.4472 = 0.233.$$

$$\vec{H} = -0.474\vec{a}_x + 0.233\vec{a}_y \text{ A/m.}$$

Example 4.21 A circular loop of radius $\rho = 0.2$ m causes a current of 10 A. Find the magnetic field intensity $\vec{H}$ at point (0, 0, 0.5).

Solution Given a loop of radius $\rho = 0.2$ m, $I = 10$ A, $z = 0.5$ m.
We know that the magnetic field intensity due to a circular loop is

$$\vec{H} = \frac{I\rho^2\,\vec{a}_z}{2(\rho^2+z^2)^{3/2}}.$$

$$\vec{H} = \frac{100(0.2)^2}{2(0.2^2+0.5^2)^{3/2}}\vec{a}_z = 12.8\vec{a}_z.$$

Example 4.22 Determine the magnetic field intensity H at the centre of a square current element. The length of each side is 2 m and the current $I = 2$ A.

Solution Given $a = 2$ m, $I = 2$ A.
We know that the magnetic field intensity at the centre of a square is

$$\vec{H} = \frac{2\sqrt{2}I}{\pi a}\vec{a}_z$$

$$= \frac{2\sqrt{2}\times 2}{2\pi}\vec{a}_z = \frac{2\sqrt{2}}{\pi}\vec{a}_z = 0.9\vec{a}_z \text{ A/m.}$$

Example 4.23 Two narrow circular coils, *A* and *B*, have a common axis and are placed 10 cm apart. Coil has 10 turns radius 5 cm, with a current of 1 A passing through it. Coil *B* has a single turn radius 7.5 cm. If the magnetic field at the centre of coil *A* is to be zero, what current should be passed through coil *B*?

Solution For coil A:

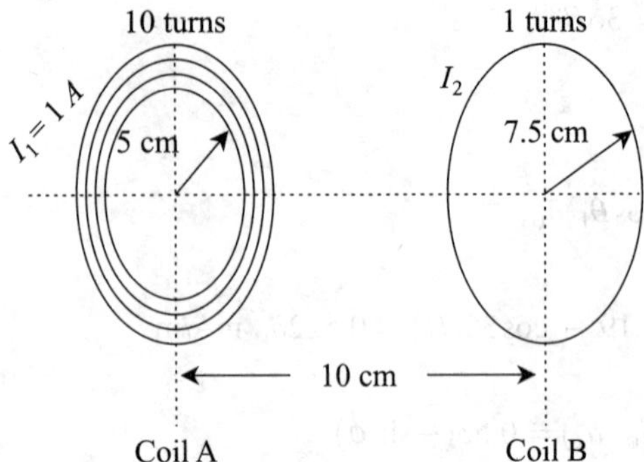

Fig. 4.34 *Field between two loops*

Current $I_1 = 1$ A; radius $r_A = 5$ cm $= 0.05$ m, turns $N_1 = 10$.
For coil $B : r_B = 7.5$ cm $= 0.075$ m, $N_2 = 1$.
Distance between coils $d = 10$ cm $= 0.1$ m.

Let the field due to coil A act in the opposite direction to the field in coil B.

It is given that the magnetic field at the centre of coil A is zero, which means that at the centre of coil A, the magnetic field due to coil A = the magnetic field due to coil B at the centre of coil A.

$$|H_A| = |H_B|.$$

We know that the magnetic field at the centre of the circular loop is

$$|H_A| = \frac{N_1 I}{2r_A} = \frac{10 \times 1}{2 \times 0.05} = 100 \text{ A/m}.$$

Let the axis be along z-direction. Then the magnetic field due to the circular loop (coil B) at a height $z = 0.1$ m is

$$|H_B| = \frac{N_2 I_B r_B^2}{2(r_B^2 + z^2)^{3/2}}$$

$$= \frac{I_B \times (7.5 \times 10^{-2})^2}{2(0.075^2 + 0.1^2)^{3/2}} = 1.44 I_B.$$

Since $|H_A| = |H_B|$,

$$100 = 1.44 I_B.$$

$$I_B = \frac{100}{144} = 69.44 \text{ A}.$$

Example 4.24 A solid non-magnetic conductor of cross-section $\rho = 2$ cm, carries a total current of 60 A, in the $\vec{a}_z$ direction. The conductor is inhomogeneous, having a conductivity that varies with ρ as $\sigma = 10^5(1 + 2.5 \times 10^5\ \rho^2)$ Siemens/m. Find the total flux crossing the radial plane defined by $\phi = 0$, $0 < z < 1$ m and $0 < \rho < 1$ cm.

Solution Given $\rho = 2$ cm $= 2 \times 10^{-2}$ m, $I = 60$ A along the $\vec{a}_z$ direction, conductivity, and $\sigma = 10^5(1 + 2.5 \times 10^5 \rho^2)$ S/m.

The plane is defined as $\phi = 0,\ 0 < z < 1$ and $1 < \rho < 0.01$, we know that current density $\vec{J} = \sigma\vec{E} = \sigma E\ \vec{a}_z\ \text{A/m}^2$

$$= 10^5(1 + 2.5 \times 10^5\ \rho^2)E\ \vec{a}_z\ \text{A/m}^2.$$

The differential surface is $d\vec{S} = \rho d\rho\ d\phi\ \vec{a}_z$.

$$\therefore \quad \text{Total current} \quad I = \int \vec{J}.d\vec{S} = \int \vec{J}(\rho d\rho\ d\phi\ \vec{a}_z)$$

$$= \int_0^{\rho}\int_0^{2\pi} 10^5(1 + 2.5 \times 10^5\ \rho^2)E\ \rho d\rho\ d\phi \vec{a}_z \cdot \vec{a}_z$$

$$= \int_0^{0.02}\int_0^{2\pi} 10^5(1 + 2.5 \times 10^5\ \rho^2)E\ \rho d\rho\ d\phi$$

$$= 10^5 E \int_0^{0.02}\int_0^{2\pi} (\rho + 2.5 \times 10^5\ \rho^3)d\rho\ d\phi.$$

Since, $I = 60$ A,

$$60 = 10^5 E\left[\frac{\rho^2}{2} + 2.5 \times 10^5 \frac{\rho^4}{4}\right]_0^{0.02} [\phi]_0^{2\pi}$$

$$= 10^5 E\left[\frac{(0.02)^2}{2} + 2.5 \times 10^5 \times \frac{(0.02)^4}{4}\right] \times (2\pi)$$

$$= 10^5 E[0.0002 + 0.01] \times 2\pi.$$

$$\therefore \quad E = \frac{60}{10^5 \times 0.0102 \times 2\pi} = 9.36\ \text{mV/m}.$$

The current density is $\vec{J} = \sigma\vec{E}$.

$$\vec{J} = 936(1 + 2.5 \times 10^5\ \rho^2)\vec{a}_z\ \ \text{A/m}^2.$$

From Ampere's circuital law, we know that

$$\therefore \quad \oint \vec{H} \cdot d\vec{l} = I_{\text{encl}} = 2\pi\ \rho\ H_\phi = \int_s \vec{J} \cdot d\vec{S}.$$

$$2\pi\rho\ H_\phi = \int_0^{\rho}\int_0^{2\pi} 936[1 + 2.5 \times 10^5\ \rho^2]\rho d\rho\ d\phi.$$

$$= \int_0^{\rho}\int_0^{2\pi} 936[\rho d\rho + 2.5 \times 10^5 \rho^3 d\rho]d\varphi = 936\left[\frac{\rho^2}{2} + \frac{2.5 \times 10^5 \rho^4}{4}\right]_0^{\rho} [\phi]_0^{2\pi}$$

$$H_\varphi = \frac{936}{2\pi}\left[\frac{\rho^2}{2} + \frac{2.5\times 10^5\,\rho^4}{4}\right]\times(2\pi).$$

$$\therefore \quad \vec{H} = 936\left[\frac{\rho}{2} + \frac{2.5\times 10^5\,\rho^3}{4}\right]\vec{a}_\phi \text{ A/m}.$$

We know that flux density $\vec{B} = \mu_0\vec{H} = 936\mu_0\left[\frac{\rho}{2} + \frac{2.5\times 10^5\,\rho^3}{4}\right]\vec{a}_\phi$ Wb/m^2.

Total flux φ in the limits $0 < z < 1$, $0 < \rho < 0.01$ is given by

$$\varphi = \int_s \vec{B}\cdot d\vec{S}.$$

$$d\vec{S} = d\rho\; dz\vec{a}_\phi.$$

$$\therefore \quad \varphi = \int_0^\rho\int_0^z 936\mu_0\left[\frac{\rho}{2} + \frac{2.5\times 10^5\,\rho^3}{4}\right] d\rho\; dz$$

$$= \int_0^{0.01}\int_0^1 936\mu_0\left[\frac{\rho}{2} + \frac{2.5\times 10^5\,\rho^3}{4}\right] d\rho\; dz = 936\mu_0\int_0^{0.01}\left[\frac{\rho}{2} + \frac{2.5\times 10^5\,\rho^3}{4}\right] d\rho\left[z\right]_0^1$$

$$= 936\mu_0\left[\frac{\rho^2}{4} + \frac{2.5\times 10^5\,\rho^4}{16}\right]_0^{0.01}$$

$$= 936\mu_0\left[\frac{(0.01)^2}{4} + \frac{2.5\times 10^5 (0.01)^4}{16}\right] = 0.169\mu_0 \text{ Wb}.$$

Example 4.25 The magnetic field intensity $\vec{H}$ is given by $\vec{H} = -y(x^2+y^2)\vec{a}_x + x(x^2+y^2)\vec{a}_y$ A/m in the $z = 0$ plane. Calculate the current passing through the $z = 0$ plane and through the region $-1 < x < 1$ and $-2 < y < 2$.

Solution Given $\vec{H} = -y(x^2+y^2)\vec{a}_x + x(x^2+y^2)\vec{a}_y$ A/m.
Consider the rectangular region $PQRS$ in the $x - y$ plane at $z = 0$ for $-1 < x < 1$ and $-2 < y < 2$, as shown in Fig. 4.35.

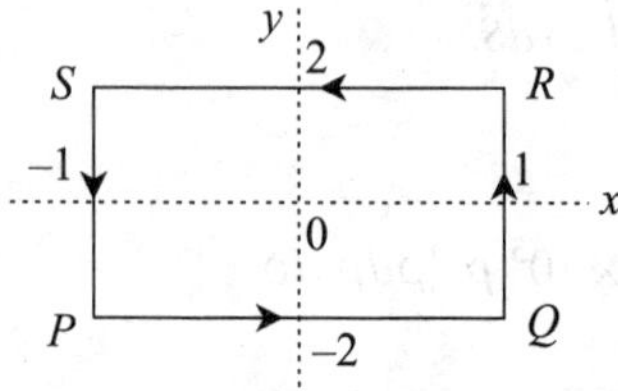

Fig. 4.35 *Rectangular region at $z = 0$ plane*

The total current passing through the region *PQRS* is given as,

$$I = \oint \vec{H}\cdot d\vec{l} = \int_P^Q \vec{H}\cdot d\vec{l} + \int_Q^R \vec{H}\cdot d\vec{l} + \int_R^S \vec{H}\cdot d\vec{l} + \int_S^P \vec{H}\cdot d\vec{l}$$

$$= \int_P^Q H_x dx + \int_Q^R H_y dy + \int_R^S H_x dx + \int_S^Q H_y dy$$

$$= \int_{-1}^{1} -(y)(x^2+y^2)dx + \int_{-2}^{2} x(x^2+y^2)dy + \int_{1}^{-1} -y(x^2+y^2)dx + \int_{2}^{-2} x(x^2+y^2)dy$$

$$= \int_{-1}^{1} -(-2)(x^2+(-2^2))dx + \int_{-2}^{2} 1(y^2+(1^2))dy + \int_{1}^{-1} -(2)(x^2+(2^2))dx$$

$$+ \int_{2}^{-2} (-1)(y^2+(-1^2))dy$$

$$= \int_{-1}^{1} 2(x^2+4)dx + \int_{-2}^{2} (y^2+1)dy + \int_{1}^{-1} -2(x^2+4)dx + \int_{2}^{-2} -(y^2+1)dy$$

$$= 2\left[\frac{x^3}{3}+4x\right]_{-1}^{1} + \left[\frac{y^3}{3}+y\right]_{-2}^{2} - 2\left[\frac{x^3}{3}+4x\right]_{1}^{-1} - \left[\frac{y^3}{3}+y\right]_{2}^{-2}$$

$$= 2\left[\left(\frac{1}{3}+4\right) - \left(\frac{-1}{3}-4\right)\right] + \left[\left(\frac{8}{3}+2\right) - \left(\frac{-8}{3}-2\right)\right] - 2\left[\left(\frac{-1}{3}-4\right) - \left(\frac{1}{3}+4\right)\right]$$

$$- \left[\left(\frac{-8}{3}-2\right) - \left(\frac{8}{3}+2\right)\right]$$

$$= 2[(4.33)-(-4.33)] + [4.66) - (-4.66)] - 2(-4.33) - (4.33)] - [(-4.66) - (4.66)]$$

$$= 17.32 + 9.32 + 17.32 + 9.32 = 53.28 \text{ A.}$$

Another method We know that the total current passing through the surface or region $z = 0$ is

$$I = \oint_s \vec{J}\cdot d\vec{S}.$$

From Ampere's circuital law, $\vec{J} = \nabla \times \vec{H}$.

$$\nabla \times \vec{H} = \begin{vmatrix} \vec{a}_x & \vec{a}_y & \vec{a}_z \\ \dfrac{\partial}{\partial x} & \dfrac{\partial}{\partial y} & \dfrac{\partial}{\partial z} \\ -y(x^2+y^2) & x(x^2+y^2) & 0 \end{vmatrix}$$

$$= \vec{a}_x\left[-\frac{\partial}{\partial z}x(x^2+y^2)\right] + \vec{a}_y\left[\frac{\partial}{\partial z}(-y(x^2+y^2))\right] + \vec{a}_z\left[\frac{\partial}{\partial x}x(x^2+y^2) - \frac{\partial}{\partial y}(-y(x^2+y^2))\right]$$

$$= 0 + 0 + \vec{a}_z\left[\frac{\partial}{\partial x}(x^3+xy^2) - \frac{\partial}{\partial y}(-yx^2-y^3).\right]$$

$$\nabla \times \vec{H} = \vec{a}_z[3x^2+y^2+x^2+3y^2] = 4(x^2+y^2)\vec{a}_z.$$

From Stokes' theorem,

$$I = \int_s (\nabla \times \vec{H})\cdot d\vec{S} = \int_{-2}^{2}\int_{-1}^{1} 4(x^2+y^2)dx\,dy$$

$$= 4\int_{-2}^{2}\left[\frac{x^2}{3}+xy^2\right]_{-1}^{1} dy = 4\int_{-2}^{2}\left[\left(\frac{1}{3}+y^2\right) - \left(\frac{-1}{3}-y^2\right)\right]dy$$

$$= 4\int_{-2}^{2}\left[\left(\frac{1}{3}+y^2\right) + \left(\frac{1}{3}+y^2\right)\right]dy = 8\int_{-2}^{2}\left[\frac{1}{3}+y^2\right]dy$$

$$= 8\left[\frac{y}{3}+\frac{y^3}{3}\right]_{-2}^{2} = 8\left[\left(\frac{2}{3}+\frac{8}{3}\right) - \left(\frac{-2}{3}-\frac{8}{3}\right)\right]$$

$$= 8\left[\frac{20}{3}\right] = \frac{160}{3} = 53.33 \text{ A}.$$

Example 4.26 Three infinite conductors are carrying currents of 1, 2, 3 amperes respectively in the same direction. The conductors are arranged in a straight line at a distance of 1 m. The conductor carrying current 1 A is leftmost and that carrying 3 A is rightmost. Find $\vec{H}$ at a point 1 m exactly above the conductor carrying 1 A current.

Solution Given three conductors carrying currents 1 A, 2 A, 3 A respectively in the $x = 0$ plane, as shown in Fig. 4.36.

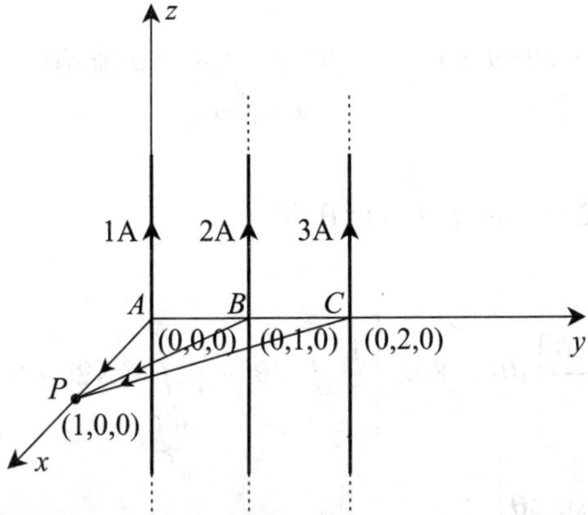

Fig. 4.36 *Field due to three infinite conductors*

The magnetic field at point $P(1, 0, 0)$ due to the three conductors is the sum of magnetic fields of individual conductors, i.e.,

$$\vec{H} = \vec{H}_1 + \vec{H}_2 + \vec{H}_3$$

where, H_1 H_2 and H_3 are magnetic fields due to the conductors carrying 1, 2 and 3 amperes respectively.

We known that the magnetic field intensity H produced at point P due to infinite length current is

$$\vec{H} = \frac{I}{2\pi\rho}\vec{a}_\phi = (\frac{I_1}{2\pi\rho_1} + \frac{I_2}{2\pi\rho_2} + \frac{I_3}{2\pi\rho_3})\vec{a}_\phi$$

where $\vec{\rho}_1 = (1, 0, 0) - (0, 0, 0) = \vec{a}_x$ and $\rho_1 = 1$;

$\vec{\rho}_2 = (1, 0, 0) - (0, 1, 0) = \vec{a}_x - \vec{a}_y$ and $\rho_2 = \sqrt{1+1} = \sqrt{2}$;

$\vec{\rho}_3 = (1, 0, 0) - (0, 2, 0) = \vec{a}_x - 2\vec{a}_y$ and $\rho_3 = \sqrt{1+2^2} - \sqrt{5}$.

$$\therefore \quad H = \left(\frac{1}{2\pi} + \frac{2}{2\pi\sqrt{2}} + \frac{3}{2\pi\sqrt{5}}\right)\vec{a}_\phi = \frac{1}{2\pi}\left(1 + \sqrt{2} + \frac{3}{\sqrt{5}}\right)\vec{a}_\phi = 0.6\vec{a}_\phi.$$

Example 4.27 Given $\vec{J} = 10^3 \sin\theta \vec{a}_r$ A/m^2, find the current passing through a spherical shell of $r = 0.2$ m.

Solution Given $\vec{J} = 10^3 \sin\theta \vec{a}_r$ A/m^2 and $r = 0.2$ m.

We know that current $I = \int_s \vec{J} \cdot d\vec{S}$.

In spherical coordinates, $d\vec{S} = r^2 \sin\theta d\varphi d\theta \vec{a}_r$.

$$\therefore \quad I = \int_{\theta=0}^{\pi} \int_{\phi=0}^{2\pi} 10^3 \sin\theta r^2 \sin\theta d\phi d\theta \; \vec{a}_r \cdot \vec{a}_r$$

$$= r^2 \int_{\theta=0}^{\pi} \int_{r=0}^{2\pi} 10^3 \sin^2\theta d\theta d\theta = 0.2^2 \times 10^3 \int_{\theta=0}^{\pi} \int_{\phi=0}^{2\pi} \sin^2\theta d\phi \, d\theta$$

$$= 40 \int_{\theta=0}^{\pi} \sin^2\theta \, d\theta (2\pi) = 80\pi \int_{\theta=0}^{\pi} \sin^2\theta \, d\theta$$

$$= 80\pi \int_{\theta=0}^{\pi} \left[\frac{1-\cos 2\theta}{2}\right] d\theta = 80\pi \left[\frac{1}{2}\int_{\theta=0}^{\pi} d\theta - \frac{1}{2}\int_{\theta=0}^{\pi} \cos 2\theta \, d\theta\right].$$

$$I = 80\pi \left[\frac{1}{2} \times \pi - \frac{1}{4}\left[\sin 2\theta\right]_0^{\pi}\right]$$

$$= 80\pi \left[\frac{\pi}{2} - 0\right] = 40\pi^2 = 394.78 \text{ A}.$$

Example 4.28 If the magnetic field intensity is $\vec{H} = x^2\vec{a}_x + 2yz\vec{a}_y + (-x^2)\vec{a}_z$ A/m, find the current density at point (a) (2, 3, 4); (b) $\rho = 6$, $\phi = 45°$, $z = 3$; (c) $r = 3.6$, $\theta = 60°$, $\phi = 90°$.

Solution

(a) Current density at point (2, 3, 4):

Given, $\vec{H} = x^2\vec{a}_x + 2yz\vec{a}_y + (-x^2)\vec{a}_z$ A/m.

Ampere's law in differential form, $\nabla \times \vec{H} = \vec{J}$, in Cartesian coordinate system

$$\vec{J} = \nabla \times \vec{H} = \begin{vmatrix} \vec{a}_x & \vec{a}_y & \vec{a}_z \\ \frac{\partial}{\partial x} & \frac{\partial}{\partial y} & \frac{\partial}{\partial z} \\ H_x & H_y & H_z \end{vmatrix} = \begin{vmatrix} \vec{a}_x & \vec{a}_y & \vec{a}_z \\ \frac{\partial}{\partial x} & \frac{\partial}{\partial y} & \frac{\partial}{\partial z} \\ x^2 & 2yz & -x^2 \end{vmatrix}.$$

$$\vec{J} = \left(\frac{\partial}{\partial y}(-x^2) - \frac{\partial}{\partial z}(2yz)\right)\vec{a}_x - \left(\frac{\partial}{\partial x}(-x^2) - \frac{\partial}{\partial z}(x^2)\right)\vec{a}_y + \left(\frac{\partial}{\partial x}(2yz) - \frac{\partial}{\partial y}(x^2)\right)\vec{a}_z$$

$$= (0 - (2y))\vec{a}_x - ((-2x) - 0)\vec{a}_y + (0 - 0)\vec{a}_z$$

$$= 2x\vec{a}_y - 2y\vec{a}_x = -2y\vec{a}_x + 2x\vec{a}_y.$$

At $P(2, 3, 4)$,

$$\vec{J} = \nabla \times \vec{H} = -6\vec{a}_x + 4\vec{a}_y.$$

(b) Current density at point (6, 45°, 3) in cylindrical coordinates:

Transforming the cylindrical coordinates (ρ, ϕ, z) into rectangular coordinates (x, y, z) using the relation,

$$x = \rho \cos\phi = 6 \times \cos 45° = 4.242,$$

$$y = \rho \sin\phi = 6 \times \sin 45^\circ = 4.242,$$

$$z = z = 3,$$

The point (6, 45°, 3) is converted into (4.242, 4.242, 3).

Now, $\nabla \times \vec{H} = -2y\vec{a}_x + 2x\vec{a}_y.$

At point (4.242, 4.242, 3),

$$\vec{J} = \nabla \times \vec{H} = -8.484\vec{a}_x + 8.484\vec{a}_y.$$

(c) Current density at point (3.6, 60°, 90°) in spherical coordinates
Transforming the spherical coordinates $(r,\ \theta,\ \phi)$ into rectangular coordinates $(x,\ y,\ z)$ using the relations

$$x = r\sin\theta\sin\phi = 3.6\sin 60 \times \cos 90^\circ = 0,$$

$$y = r\sin\theta\sin\phi = 3.6 \times \sin 60 \times \sin 90^\circ = 3.117,$$

$$z = r\cos\theta = 3.6 \times \cos 60^\circ = 1.8.$$

The point (3.6, 60°, 90°) is converted into (0, 3.117, 1.8).

Now, $\vec{J} = \nabla \times \vec{H} = -2y\vec{a}_x + 2x\,\vec{a}_y.$

At point (0, 3.117, 1.8),

$$\vec{J} = \nabla \times \vec{H} = -2 \times 3.117\ a_x + 0 = -6.234\vec{a}_z.$$

Example 4.29 A circular loop located at $x^2 + y^2 = 9$, $z = 0$ carries a direct current of 10 A along the $\vec{a}_\phi$ direction. Determine $\vec{H}$ at (0, 0, 5) and (0, 0, –5).

Solution Given $x^2 + y^2 = 9$, $\rho = \sqrt{9} = 3$ m, $I = 10$ A, and $\rho = (0,\ 0,\ 5)$ and (0, 0, 5).
We know that the magnetic field due to a circular loop at height h along its axis is

$$\vec{H} = \frac{I\rho^2\vec{a}_z}{2(\rho^2 + h^2)^{3/2}}.$$

(i) At (0, 0, 5), $\vec{H}_1 = \dfrac{10(3^2)\vec{a}_z}{2(3^2 + 5^2)^{3/2}} = 0.226\,\vec{a}_z.$

(ii) At (0, 0, –5), $\vec{H}_2 = \dfrac{10(3)^2\vec{a}_z}{2[3^2 + (-5)^2]^{3/2}} = 0.226\,\vec{a}_z.$

Example 4.30 Evaluate the closed line integral of $\vec{H}$ from (5, 4, 1) to (5, 6, 1) to (0, 6, 1) to (0, 4, 1) to (5, 4, 1) using straight line segments if $\vec{H} = 0.1y^3\,\vec{a}_x + 0.4x\vec{a}_z$ A/m.

Solution Given thc line segments as shown in Fig. 4.37.

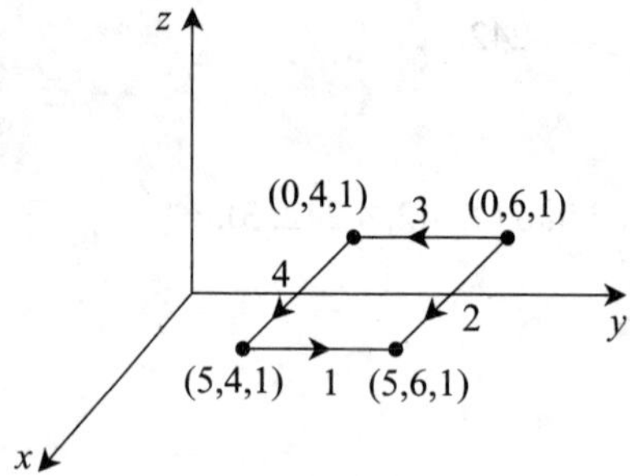

Fig. 4.37 *Field due to four lines segments*

$$\vec{H} = 0.1y^3\,\vec{a}_x + 0.4x\,\vec{a}_z.$$

The closed line integral is

$$\oint_c \vec{H}\cdot d\vec{l} = \int_1 \vec{H}\cdot d\vec{l}_1 + \int_2 \vec{H}\cdot d\vec{l}_2 + \int_3 \vec{H}\cdot d\vec{l}_3 + \int_4 \vec{H}\cdot d\vec{l}_4.$$

Now, dl_1 along the y-axis, $dl_1 = dy\vec{a}_y$.

dl_2 along the x-axis, $dl_2 = dx\vec{a}_x$.

dl_3 along the y-axis, $dl_3 = dy\vec{a}_y$.

dl_4 along the x-axis, $dl_4 = dx\vec{a}_x$.

$$\int_1 \vec{H}\cdot d\vec{l} = \int_4^6 (0.1y^3\,\vec{a}_x + 0.4x\vec{a}_z)\cdot dy\,\vec{a}_y = \int_4^6 (0.1y^3\,\vec{a}_x\cdot\,\vec{a}_y + 0.4x\vec{a}_z\cdot\vec{a}_y)dy = 0.$$

Similarly,

$$\int_3 \vec{H}\cdot d\vec{l} = \int_6^4 (0.1y^3\,\vec{a}_x + 0.4x\vec{a}_z)\cdot(-dy\,\vec{a}_y) = 0.$$

$$\int_2 \vec{H}\cdot d\vec{l} = \int_5^0 (0.1y^3\,\vec{a}_x + 0.4x\vec{a}_z)\cdot(dx)\,\vec{a}_x = -\int_5^0 0.1y^3dx = -0.1y^3x\Big|_5^0 = -0.1y^3x.$$

At $y = 6$, $\displaystyle\int_2 \vec{H}\cdot d\vec{l} = -0.1\times 6^3(5) = -108.$

$$\int_4 \vec{H}\cdot d\vec{l} = +0.1y^3x\Big|_0^5 = 0.5y^3.$$

And at $y = 4$, $\displaystyle\int_4 \vec{H}\cdot d\vec{l} = +0.5(4^3) = 32.$

$$\therefore \quad \int \vec{H}\cdot d\vec{l} = -108 + 32 = 76\text{ A}.$$

Example 4.31 Two infinitely long and parallel wires carry currents $I_1 = I_2 = I$ A in opposite directions. They are separated by a distance 17.32 cm. If the magnetic field intensity at a point 10 cm from one wire and 20 cm from the other is 13.783 A/m, determine the value of current I.

Solution Given $I_1 = I_2 = I$ A, separation between the conductors = 17.32 cm.
Let H_1 and H_2 be the field at point P located 10 cm and 20 cm away from the conductors, respectively.

$$H = H_1 + H_2 = 13.783 \text{ A/m}.$$

Let the conductors A and B be normal to the $x-y$ plane as shown in Fig. 4.38.

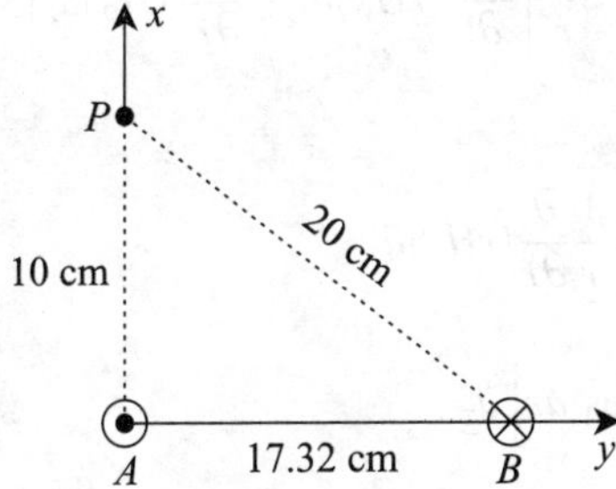

Fig. 4.38 *Position of the conductor and point*

The distance vector between the point $P(10, 0, 0)$ and the first conductor at $(0, 0, 0)$ is

$$\vec{R}_1 = 10\vec{a}_x \times 10^{-2} = 0.1\vec{a}_x.$$

The distance vector between the point $(10, 0, 0)$ and the second conductor at $(0, 17.32, 0)$ is

$$\vec{R}_2 = (10\vec{a}_x - 17.32\vec{a}_y)10^{-2} \text{ m} = 0.1\vec{a}_x - 0.1732\vec{a}_y.$$

We know that $\vec{H}_1 = \dfrac{I}{2\pi R_1}\vec{a}_{R_1} = \dfrac{I}{2\pi(0.1)}(-\vec{a}_x) = -1.5915I\vec{a}_x$

and $\vec{H}_2 = \dfrac{I}{2\pi R_2}\vec{a}_{R_2} = \dfrac{I}{2\pi R_2}(0.5\vec{a}_x + 0.866\vec{a}_y)$

$$= I(0.3978\vec{a}_x + 0.689\vec{a}_y).$$

$$\vec{H} = \vec{H}_1 + \vec{H}_2$$

$$= -1.5915I\vec{a}_x + I(0.3978\vec{a}_x + 0.689\vec{a}_y) = (-1.19\vec{a}_x + 0.689\vec{a}_y)I.$$

$$|H| = \sqrt{119^2 + 0689^2} = 1.378\, I$$

$$13.78 = 1.378\, I \text{ or } I = 10 \text{ A}.$$

Example 4.32 A circular conductor of 1 cm radius has an internal magnetic field of $H = \dfrac{1}{r}\left(\dfrac{1}{a^2}\sin ar - \dfrac{r}{a}\cos ar\right)\vec{a}_\phi$ A/m, where $a = \dfrac{\pi}{2r_0}$, r_0 being the radius of the conductor. Calculate the total current enclosed by the conductor.

Solution Given $\vec{H} = \frac{1}{r}\left(\frac{1}{a^2}\sin ar - \frac{r}{a}\cos ar\right)\vec{a}_\phi$ A/m.

$$H_\phi = \frac{1}{r}\left(\frac{1}{a^2}\sin ar - \frac{r}{a}\cos ar\right) = \left(\frac{1}{ra^2}\sin ar - \frac{1}{a}\cos ar\right) \text{ A/m.}$$

The field is in cylindrical coordinates. Consider $\rho = r$.

We know that $\vec{J} = \nabla \times \vec{H}$, in cylindrical coordinates

$$\vec{J} = \frac{1}{r}\begin{vmatrix} \vec{a}_r & r\vec{a}_\phi & \vec{a}_z \\ \frac{\partial}{\partial r} & \frac{\partial}{\partial \phi} & \frac{\partial}{\partial z} \\ 0 & rH_\phi & 0 \end{vmatrix} = \frac{1}{r}\left[\frac{-\partial}{\partial z}(rH_\phi)\vec{a}_r + \frac{\partial}{\partial r}(rH_\phi)\vec{a}_z\right]$$

$$= \frac{1}{r}\left[0 + \frac{\partial}{\partial r}(rH_\phi)\vec{a}_z\right] = \frac{1}{r}\frac{\partial}{\partial r}(rH_\phi)\vec{a}_z$$

$$= \frac{1}{r}\frac{\partial}{\partial r}\left[\frac{r}{ra^2}\sin ar - \frac{r}{a}\cos ar\right]\vec{a}_z$$

$$= \left(\frac{1}{ra^2}\frac{\partial}{\partial r}(\sin ar) - \frac{1}{ar}\left(\frac{\partial}{\partial r}r\cos ar\right)\right)\vec{a}_z$$

$$\vec{J} = \left(\frac{1}{ra}\cos ar - \frac{1}{ar}[\cos ar - ar\sin ar]\right)\vec{a}_z$$

$$\vec{J} = \sin ar\vec{a}_z$$

The total current I enclosed by the conductor is

$$I_{\text{encl}} = \int \vec{J}\cdot d\vec{S}$$

$$d\vec{S} = rdr\, d\phi\vec{a}_z$$

$$\therefore \quad I_{\text{encl}} = \int_0^1 \int_0^{2\pi} (\sin ar) rdr\, r\phi$$

$$= \int_0^1 r\sin ar \times dr \times 2\pi = 2pi\int_0^1 r\sin ar\, dr$$

$$= 2\pi\left[\frac{r\cos ar}{a} + \frac{\sin ar}{ar^2}\right] = 2\pi\left[\frac{\cos a}{a^2} + \frac{\sin a}{a^2}\right]$$

Given $a = \frac{\pi}{2r_0}$ and $r_0 = 10^{-2}$ m, then $a = 50\pi$

$$I_{\text{encl}} = 2\pi\left[\frac{\cos 50\pi}{(50\pi)} + \frac{\sin 50\pi}{(50\pi)^2}\right] = \frac{2\pi}{50\pi} = 40 \text{ mA.}$$

Example 4.33 In the cylindrical region $0 < \rho < 0.5$ m, the current density is $\vec{J} = 4.5e^{-2\rho}\,\vec{a}_z$ A/m^2 and 0 elsewhere. Use Ampere's law, find $\vec{H}$.

Solution Given $\vec{J} = \begin{cases} 4.5e^{-2\rho}\,\vec{a}_z \text{ A/m}^2 & 0 < \rho < 0.5 \\ 0 & \text{elsewhere} \end{cases}$

From Ampere's law, $\oint \vec{H}.d\vec{l} = \int_S \vec{J}\cdot d\vec{S}.$

In cylindrical coordinates, $d\vec{S} = \rho d\phi\, d\rho\, \vec{a}_z.$

Now, $$\int_S \vec{J}\cdot d\vec{S} = \int_0^{\rho}\int_{\phi=0}^{2\pi} 4.5e^{-2\rho}\,\vec{a}_z\cdot \rho d\phi\, d\rho\, \vec{a}_z$$

$$= \int_0^{\rho}\int_0^{2\pi} 4.5\rho e^{-2\rho}\, d\phi d\rho = \int_0^{2\pi} 4.5\left(\int_0^{\rho} \rho e^{-2\rho} d\rho\right) d\phi$$

$$= 4.5\left(\frac{\rho e^{-2\rho}}{(-2)} - \frac{e^{-2\rho}}{(4)}\right)_0^{\rho} 2\pi = 9\pi\left(-\frac{\rho e^{-2\rho}}{2} - \frac{e^{-2\rho}}{4} + \frac{1}{4}\right).$$

$$\therefore \quad \int_S \vec{J}\cdot d\vec{S} = 2.25\pi(-2\rho e^{-2\rho} - e^{-2\rho} + 1) \text{ A}.$$

From the cross-section of the cylindrical region, the line integral of the path is $l = 2\pi\rho$.

Then, $$\oint \vec{H}\cdot d\vec{l} = H_\phi(2\pi\rho) = \int_S \vec{J}\cdot d\vec{S}.$$

$$H_\phi(2\pi\rho) = 2.25\pi(-2\rho e^{-2\rho} - e^{-2\rho} + 1)$$

or $$H_\phi = \frac{1.125}{\rho}(-2\rho e^{-2\rho} - e^{-2\rho} + 1) \quad \text{for} \quad 0 < \rho \le 0.5.$$

At $\rho = 0.5$, $$\int_S \vec{J}\cdot d\vec{S} = 2.25\pi(-2(0.5)e^{-1} - e^{-1} + 1) = 1.867 \text{ A}.$$

For $\rho > 0.5$, $H_\phi(2\pi\rho) = 1.867$.

$$H_\phi = \frac{1.867}{2\pi\rho} = \frac{0.297}{\rho}.$$

Hence, $$\vec{H} = \begin{cases} \dfrac{1.125}{\rho}(-2\rho e^{-2\rho} - e^{-2\rho} + 1)\vec{a}_\phi & 0 < \rho \le 0.5 \\ 1.867\,\vec{a}_\phi & \rho = 0.5 \\ \dfrac{0.297}{\rho}\vec{a}_\phi & \rho > 0.5 \end{cases}$$

Example 4.34 Given $\vec{H} = \frac{y^2 z}{x}\vec{a}_x + \frac{0.5y^2z^2}{x^2}\vec{a}_z$ A/m, find the current crossing the square surface at $y = 2$ bounded by $x = z = 1$ and $x = z = 2$.

Solution Given $\vec{H} = \frac{y^2 z}{x}\vec{a}_x + \frac{0.5y^2z^2}{x^2}\vec{a}_z$ A/m, $d\vec{S} = dxdz\,\vec{a}_y$.

The current according to Stokes' theorem is

$$I = \int_s (\nabla \times \vec{H}) \cdot d\vec{S}.$$

Now $$\nabla \times \vec{H} = \begin{vmatrix} \vec{a}_x & \vec{a}_y & \vec{a}_z \\ \frac{\partial}{\partial x} & \frac{\partial}{\partial y} & \frac{\partial}{\partial z} \\ \frac{y^2 z}{x} & 0 & \frac{y^2z^2}{2x^2} \end{vmatrix}.$$

Since $d\vec{S}$ is in the $x - z$ plane, taking only the y component,

$$(\nabla \times \vec{H})_y = \left[\frac{-\partial}{\partial x}\left[\frac{y^2z^2}{2x^2}\right] + \frac{\partial}{\partial z}\left(\frac{y^2 z}{x}\right)\right]\vec{a}_y = \left(\frac{-y^2z^2}{2}\left(\frac{-2}{x^3}\right) + \frac{y^2}{x}\right)\vec{a}_y$$

$$= \left(\frac{y^2z^2}{x^3} + \frac{y^2}{x}\right)\vec{a}_y.$$

Now $$I = \int_1^2\int_1^2 \left(\frac{y^2z^2}{x^3} + \frac{y^2}{x}\right)\vec{a}_y \cdot dx\,dz\,\vec{a}_y = \int_1^2\int_1^2 \left(\frac{y^2z^2}{x^3} + \frac{y^2}{x}\right)dx\,dz.$$

At $y = 2$,

$$I = \int_1^2\int_1^2 \left(\frac{4z^2}{x^3} + \frac{4}{x}\right)dx\,dz = 4\int_1^2 \left[\frac{1}{x^3}\cdot\frac{z^3}{3} + \frac{z}{x}\right]_1^2 dx$$

$$= 4\int_1^2 \left(\frac{1}{3x^3}\left[2^3 - 1^3\right] + \frac{1}{x}(2-1)\right)dx = 4\int_1^2 \left(\frac{7}{3}\frac{1}{x^3} + \frac{1}{x}\right)dx$$

$$= 4\left[\frac{7}{3}\left[\frac{-1}{2x^2}\right]_1^2 + [\ln x]_1^2\right] = 4\left[\frac{-7}{3}\left(\frac{1}{8} - \frac{1}{2}\right) + \ln 2 - \ln 1\right]$$

$$= 4\left[\frac{7}{8} + \ln 2\right] = 6.27 \text{ A}.$$

Example 4.35 Consider the portion of a sphere specified by $r = 4$, $0 \le \theta \le 0.1\pi$, $0 \le \phi \le 0.3\pi$. Given $H = 6r\sin\phi\,\vec{a}_r + 18r\sin\theta\cos\phi\,\vec{a}_\phi$, find the current flowing through the surface.

Solution Given $\vec{H} = 6r\sin\phi\vec{a}_r + 18r\sin\theta\cos\phi\vec{a}_\phi$ and the surface under consideration at $r = 4$ defined by $0 \le \theta \le 0.1\pi$ and $0 \le \phi \le 0.3\pi$.

Now, $d\vec{S} = r^2 \sin\theta \, d\theta \, d\phi \vec{a}_r$.

Using Stokes' theorem,

$$I = \int_s (\nabla \times \vec{H}) \cdot d\vec{S}.$$

$$\nabla \times \vec{H} = \frac{1}{r^2 \sin\theta} \begin{vmatrix} \vec{a}_r & r\vec{a}_\theta & r\sin\theta\vec{a}_\phi \\ \dfrac{\partial}{\partial r} & \dfrac{\partial}{\partial \theta} & \dfrac{\partial}{\partial \phi} \\ H_r & rH_\theta & r\sin\theta H_\phi \end{vmatrix}$$

Since the surface is directed along $\vec{a}_r$, the radial component of $\nabla \times H$ is

$$\begin{aligned}
\nabla \times \vec{H} &= \frac{1}{r^2 \sin\theta}\left[\frac{\partial}{\partial \theta}(r\sin\theta H_\phi) - \frac{\partial}{\partial \phi}(r(0)\right] \vec{a}_r \\
&= \frac{1}{r^2 \sin\theta}\left[\frac{\partial}{\partial \theta}(r\sin\theta(18r\sin\theta\cos\phi))\right] \vec{a}_r \\
&= \frac{18r^2\cos\phi}{r^2 \sin\theta}\left[\frac{\partial}{\partial \theta}(\sin^2\theta)\right] \vec{a}_r \\
&= \frac{18\cos\phi}{\sin\theta}[2\sin\theta\cos\theta]\vec{a}_r = 36\cos\theta\cos\phi\,\vec{a}_r.
\end{aligned}$$

Now, $I = \int\limits_{\theta=0}^{0.1\pi} \int\limits_{\phi=0}^{0.3\pi} 36\cos\theta\cos\phi\vec{a}_r \cdot r^2 \sin\theta \, d\theta \, d\phi \, \vec{a}_r.$

At $r = 4$,

$$\begin{aligned}
I &= 576 \int\limits_{\theta=0}^{0.1\pi} \int\limits_{\phi=0}^{0.3\pi} \cos\theta\sin\theta\cos\phi \, d\theta \, d\phi \\
&= \frac{576}{2}\left[\sin^2\theta\right]_0^{0.1\pi}\left[\sin\phi\right]_0^{0.3\pi} = 288\sin^2(0.1\pi)\sin(0.3\pi). \\
I &= 22.2 \text{ A}.
\end{aligned}$$

Summary

1. The magnetic flux crossing an unit area normal to the direction of flux is called magnetic flux density, $\vec{B} = \dfrac{d\varphi}{dS}\vec{a}_n$ Wb/m^2 or tesla and the magnetic flux is given by

$$\varphi = \int_s \vec{B} \cdot d\vec{S} \text{ Wb}.$$

2. The magnetic flux density $\vec{B}$ and magnetic field intensity $(\vec{H})$ through the material are related by the equation

$$\vec{B} = \mu\vec{H}.$$

3. Biot–Savart's law can be expressed as

$$\vec{H} = \oint \frac{Id\vec{l} \times \vec{a}_R}{4\pi R^2} \text{ A/m}.$$

The direction of magnetic field can be determined by the right-handed thumb rule or right-handed screw rule.

4. Biot–Savart's law for distributed currents

$$\vec{H} = \int_s \frac{\vec{K}dS \times \vec{a}_R}{4\pi R^2} = \int_V \frac{\vec{J} \times \vec{a}_R dv}{4\pi R^2} \text{ A/m}.$$

5. The net magnetic flux over a closed surface is zero.

$$\varphi = \oint_s \vec{B} \cdot d\vec{S} = 0.$$

That is, the magnetic flux lines are continuous, $\text{div}\,\vec{B} = \nabla \cdot \vec{B} = 0$. This equation is known as Maxwell's second equation for magnetic fields.

6. Ampere's law can be expressed as

$$I_{\text{encl}} = \oint \vec{H} \cdot d\vec{l} = \int_s \vec{J} \cdot d\vec{S} \text{ A}$$

and $\vec{H}$ at point P due to infinite length current is $\vec{H} = \dfrac{I}{2\pi\rho}\vec{a}_\phi$.
Ampere's circuital law is analogous to Gauss' law.

7. The magnetic field intensity at any point in between two infinite parallel surface current sheets carrying equal current density in opposite directions is equal to the sheet current density. The field direction is normal to the current direction on the surface.

$$\vec{H} = \frac{K}{2}\vec{a}_x + \frac{K}{2}\vec{a}_x = K\vec{a}_x$$

8. The differential form of Ampere's circuital law can be expressed as $\nabla \times \vec{H} = \vec{J}$. It is also called Maxwell's third equation.

9. Stokes' theorem for magnetic field can be expressed as

$$\oint \vec{H} \cdot d\vec{l} = \int_s (\nabla \times \vec{H}) \cdot d\vec{S}.$$

10. The magnetic field intensity due to solenoid is $\vec{H} = \dfrac{NI}{l}$ A/m along the axis and due to toroid is $\vec{H} = \dfrac{NI}{2\pi\rho}\vec{a}_\rho$ A/m.

Review Questions

1. Using Biot–Savart's law, find $\vec{H}$ inside a long solenoid carrying a current of I A and show that $\vec{H}$ at the ends of such a solenoid is half of that in the middle. *(May 2010, 2013)*
2. A circular loop of wire radius a, lying in the $x-y$ plane with its centre at the origin, carries a current I in the $\vec{a}_\phi$ direction. Using Biot–Savart's law, find H (0, 0, z) and H (0, 0, 0). *(May 2010, 2011)*
3. State and explain Biot–Savart's law. *(May 2010, 2011, 2012, Aug 2014)*
4. Derive the expression for Maxwell's equation, $\nabla.\vec{B} = 0$. *(Nov 2009, May 2011, 2012)*
5. Using Biot–Savart's law, find $\vec{H}$ on the axis of a current loop. *(May 2010, 2011, 2013)*
6. A steady current of I amperes flow in a conductor bent in the form of square loop of side a. Find the magnetic field intensity at the centre of the loop. *(May 2010, Nov 2011)*
7. Explain Oersted's experiment with the help of a neat diagram. *(May 2011)*
8. Using Biot–Savart's law, find $\vec{H}$ due to an infinitely long straight conductor. *(Nov 2011, Jan 2014)*
9. Derive an expression for the magnetic flux density at a point due to a current in an infinitely long straight conductor. *(May 2011, 2012, Aug 2014)*
10. Using Biot–Savart's law, find an expression for the magnetic field intensity in the vicinity of a straight current-carrying conductor of finite length. *(May 2012, Aug 2014)*
11. Show that the field strength at the end of a long solenoid is one-half of that at the centre. *(Nov 2008, 2011)*
12. Find an expression for field intensity at the centre of a circular wire carrying a current I in the anti-clockwise direction. The radius of circle is r and the wire is in the $x-y$ plane. *(May 2008, 2009, 2010)*
13. Using Ampere's circuital law, obtain an expression for the magnetic field intensity at any point due to concentric cylindrical conductors, where the inner and outer conductors carry equal and opposite currents. *(Nov 2008, 2011, May 2011)*
14. Describe the application of Ampere's circuital law to an unsymmetrical field. *(May 2008)*
15. Define and establish Ampere's circuital law for an electromagnetic field. *(May 2008, 2012, 2013)*
16. Using Ampere's circuital law, find $\vec{H}$ due to an infinitely long straight conductor. *(Nov 2010, May 2012)*
17. Describe any two applications of Ampere's circuital law. *(May 2011, Nov 2013)*
18. Using Ampere's circuital law, find $\vec{H}$ due to an infinite sheet of current. *(May 2010, 2011)*
19. Derive Maxwell's third equation. *(May 2009, 2010, 2011, Nov 2013)*
20. Obtain the expression for $\vec{H}$ in all the regions if a cylindrical conductor carries a direct current I, its radius being R. Plot the variation of $\vec{H}$ against the distance r from the centre of the conductor. *(May 2010, 2011)*
21. Show that $\nabla \times \vec{H} = \vec{J}$. *(May 2010, 2011, 2012)*
22. Derive Maxwell's equation connected with Ampere's circuital law. *(May 2011, Aug 2014)*

Problems

1. Find $\vec{H}$ in Cartesian coordinates at P (2, 3, 4) if there is a current filament on the z-axis carrying 8 mA in the $\vec{a}_z$ direction.
2. A current filament of $3\vec{a}_x$ lies along the x-axis. Find $\vec{H}$ in Cartesian coordinates at $P(-1, 3, 2)$.
3. A filament is formed into a circle of radius a, centred at the origin in the plane $z = 0$. It carries a current I in the $\vec{a}_\phi$ direction. Find $\vec{H}$ at the origin.
4. Given points C (5, −2, 3) and P (4, −1, 2), a current element $Idl = 10^{-4}(4,\ -3,\ 1)$ A-m at C produces a field $d\vec{H}$ at P. Find $d\vec{H}$.
5. A current sheet of surface density $K = 8\ \vec{a}_x$ A/m flows in the region $-2 < y < 2$ in the plane $z = 0$. Calculate $\vec{H}$ at $P\ (0,\ 0,\ 3)$.
6. A current filament of 20 A in the $\vec{a}_z$ direction is of infinite length. Find $\vec{H}$ at (3, 4, 0).
7. A current sheet with surface current density $\vec{K} = 50\vec{a}_z$ A/m lies in the $x = 2$ m plane and a second sheet $50\vec{a}_z$ A/m is situated at the $x = -2$ m plane. Find $\vec{H}$ at all points.
8. A thin cylindrical shell conductor of radius 20 cm and infinite length carries a current of 10 A. Find $\vec{H}$ at a radius of (a) $r > 20$ cm, (b) $r < 20$ cm.
9. Given $\vec{A} = 10\rho \sin\ \phi\vec{a}_z$ in cylindrical coordinates, find $\vec{B}$ at (2, π, 0).
10. An air-core toroid having a cross-sectional area of 6 cm^2 and a mean radius of 15 cm is wound uniformly with 500 turns. Determine the magnetic flux density and the field intensity when a current of 100 A is applied.
11. Find the field intensity at a point on the axis 3 m from the centre of a circular coil of area 150 cm^2 and carrying a current of 65 A.
12. Find the magnetic field intensity at the origin due to a current element $(8\pi a_z)\ \mu$ A-m at the point (4, −5, 1) in free space.
13. A long solenoid has a radius of 2 mm and length of 1.2 cm. If the number of turns per unit length is 200 and the current is 12 A, calculate the magnetic flux density at (i) the centre and (ii) the ends of the solenoid.
14. Given the field $\vec{H} = 20p^2\vec{a}_\phi$ A/m, determine the current density $\vec{J}$.
15. The magnetic field intensity in a square region defined by $x = 0, 0.5 < y < 1, 1 < z < 1.5$ is given by $H = z^2\vec{a}_x + x^3\vec{a}_y + y^4\vec{a}_z$ A/m. Find $\nabla \times H$.
16. Given the field $\vec{H} = 20\rho^2\vec{a}_\phi$ A/m, determine the current density J.
17. A 3-cm-long solenoid carries a current of 400 mA. If the solenoid is to produce a magnetic flux density of 5 mWb/m, how many turns of wire are needed?
18. A solenoid of radius 4 mm and length 2 cm has 150 turns/m and carries a current of 500 mA. Find: (a) $\vec{H}$ at the centre, (b) $|H|$ at the ends of the solenoid.

18\. Plane $x = 10$ carries current 100 mA/m along $\vec{a}_z$ while line $x = 1,\ y = -2$ carries filamentary current of 20 πmA along $\vec{a}_r$. Determine $\vec{H}$ at (4, 3, 2).

20. The positive y-axis (semi-infinite line with respect to the origin) carries a filamentary current of 2 A in the $-\vec{a}_y$ direction. Assume it is part of a large circuit. Find $\vec{H}$ at (a) A (2, 3, 0) and (b) B (3, 12, −4).
21. If a solenoid has 2,000 turns, a length of 75 cm, a radius of 5 cm, and carries a current of 50 mA along $\vec{a}_\phi$, find $\vec{H}$ at (a) (0, 0, 0), (b) (0, 0, 75 cm) and (c) (0, 0, 50 cm).

22. Plane $y = 1$ carries a current of $\vec{K} = 50\vec{a}_z$ mA/m. Find $\vec{H}$ at (a) (0, 0, 0) (b) (1, 5, −3).
23. A toroid of circular cross-section whose centre is at the origin and whose axis is the same as z-axis, has 1000 turns with a mean radius of 10 cm; $a = 1$ cm. If the toroid carries a 100 mA current, find $|H|$ at (a) (3 cm, −4 cm, 0) and (b) (6 cm, 9 cm, 0).
24. Find the total current in a circular conductor of radius 2 mm if the current density varies with r according to $J = 10^3$ A/m^2.
25. In cylindrical coordinates, $\vec{J} = 10^5(\cos^2 2r)\vec{a}_z$ in a certain region. Obtain $\vec{H}$ from this current density, then take the curl of $\vec{H}$ and compare with $\vec{J}$.
26. The planes $z = 0$ and $z = 6$ carry current $\vec{K} = -40\vec{a}_x$ A/m and $\vec{K} = 20\vec{a}_x$ A/m respectively. Find the magnetic field at (a) P (1, 5, 7) and (b) (0, −4, 15).
27. A circular loop located on $x^2 + y^2 = 9$, $z = 0$ carries a direct current of 10 A along the a_ϕ direction. Determine $\vec{H}$ at (0, 0, 5) and (0, 0, −5).
28. Find the magnetic field intensity at the centre of a square of side 5 m carrying a current equal to 10 A.
29. A square loop of 10 cm side has 500 turns that are closely and tightly wound and carries a current of 120 A. Determine the magnetic flux density at the centre of the loop.

Answers

1. $-294\vec{a}_x + 196\vec{a}_y$ μA/m
2. $0.110\vec{a}_z - 0.073\vec{a}_y$ A/m
3. $\dfrac{0.5I}{a}\vec{a}_z$ A/m
4. 5.73 μA/m
5. $-1.50\vec{a}_y$ A/m
6. $0.127(-4\vec{a}_x + \vec{a}_y)$ A/m
7. Field outside the sheets = 0; for $-2 < x < 2$, $\vec{H} = -50\vec{a}_y$ A/m
8. $\dfrac{10}{2\pi r}$; 0
9. $-10\vec{a}_\rho$ Wb/m^2
10. $H = 530\vec{a}_\phi$ A/m, $B = \mu_0 530\vec{a}_\phi$ Wb/m^2
11. $H = 5.737$ mA/m
12. $0.617\vec{a}_\phi$ μA/m
13. $B = \mu_0 2277$ Wb/m^2, $B = \mu_0 1183.67$ Wb/m^2
14. $60\rho\vec{a}_z$ A/m^2
15. $4y^3\vec{a}_x + 2z\vec{a}_y + 4x^2\vec{a}_z$
16. $60\rho\vec{a}_z$ A/m^2
17. 299
18. (a) 69. 63 A/m (b) 36.77 A/m
19. $-1.923\vec{a}_x - 0.3346\vec{a}_y$ A/m
20. (a) 145.8 $\vec{a}_z$ mA/m, (b) 48.97 $\vec{a}_x + 36.73\vec{a}_z$ mA/m.
21. (a) 66.52 $\vec{a}_z$ A/m, (b) 66.52 $\vec{a}_z$ A/m, (c) 131.7 $\vec{a}_z$ A/m.
22. (a) 25 $\vec{a}_x$ mA/m, (b) $-25\vec{a}_x$ mA/m.
23. (*a*) 0, (*b*) 147.1 A/m.
24. 4π *A*
25. $H = 10^3\left(\dfrac{r}{4} + \dfrac{\sin 4r}{8} + \dfrac{\cos 4r}{32r} - \dfrac{1}{32r}\right)\vec{a}_\rho$
26. (a) $-10\vec{a}_y$, (b) $-10\vec{a}_y$
27. At both points, $\vec{H} = 0.227\vec{a}_z$ A/m
28. 1.8 A/m
29. 0.679 Wb/m^2

Multiple-choice Questions

1. If a conductor is lying on the z-axis, to find the magnetic field in cylindrical coordinates around the conductor, the magnetic field should be along

   (a) $\vec{a}_z$ (b) $\vec{a}_r$ (c) $\vec{a}_\phi$ (d) $\vec{a}_x$

2. Biot–Savart's law is also known as

   (a) Ampere's work law (b) Ampere's force law
   (c) Ampere's law for current element (d) Ampere's law

3. Stokes' theorem can be used to convert the

   (a) surface into volume integral (b) line into surface integral
   (c) line into volume integral (d) surface to line integral

4. Magnetic field in terms of magnetic flux density is given by

   (a) $H = B/\mu$ (b) $H = B\mu$ (c) $B = H/\mu$ (d) all of the above

5. One of the following cannot produce magnetic fields:

   (a) permanent magnets (b) direct currents
   (c) moving charges (d) static charges

6. A thin infinite current element along the z-axis is carrying a current of 1 A. The value of $\vec{H}$ at $(0, d, 0)$ in A/m is

   (a) $\dfrac{I}{2\pi d}$ (b) $\dfrac{I}{2\pi d}\vec{a}_\phi$ (c) $\dfrac{I}{2\pi d}\sin\phi\vec{a}_\phi$ (d) $\dfrac{I}{d}\vec{a}_\phi$

7. A wire carrying a current of 1 A subtends an angle 2α at a point P midway between the wire. If the distance of the point from the wire is d, then the value of $\vec{H}$ at P is

   (a) $\dfrac{I}{2\pi d}$ (b) $\dfrac{I}{\pi d}$ (c) $\dfrac{I\sin\alpha}{\pi d}$ (d) $\dfrac{I\sin\alpha}{2\pi d}$

8. A circular wire of radius R is carrying a current of 1 A. Then $\vec{H}$ at the centre of the wire is

   (a) $\dfrac{I}{R}$ (b) $\dfrac{I}{2R}$ (c) $\dfrac{I}{2\pi R}$ (d) $\dfrac{I}{\pi R}$

9. The flux crossing the portion of the plane $\phi = \pi/4$ defined by $0.02 < \rho < 0.06$ m and $0 < z < 4$ m due to a infinite current filament of 5A in $\vec{a}_z$ direction is

   (a) 1.60 μ Wb (b) 3.22 μ Wb (c) 1μ Wb (d) 2 μ Wb

10. A current sheet is situated in the $y = 0$ plane carrying a current of $K = 10\vec{a}_z$ A/m. The magnetic field intensity at (1, 5, 1) is

    (a) $-5\vec{a}_x$ A/m (b) $5\vec{a}_x$ A/m (c) $10\vec{a}_y$ A/m (d) $-10\vec{a}_y$ A/m

11. Which of the following statements is not true for static magnetic fields.

    (a) It is a solenoidal (b) It has no sinks or sources
    (c) It is conservative (d) Magnetic flux lines are closed

12. The field intensity due to an infinite line current is inversely proportional to

    (a) flux (b) distance (c) charge (d) square of distance

13. The field intensity at the centre of a loop carrying current of 20 A and having a radius of 20 m is

    (a) 20 A/m (b) 0.5 A/m (c) 2 A/m (d) 1 A/m

14. The right-hand cork-screw rule applied to a current-carrying conductor gives the direction of the rotation of the screw as the direction of

    (a) current (b) electric field intensity
    (c) vector potential (d) magnetic field intensity

15. A current sheet $\vec{K} = 10\vec{a}_y$ A/m is in the $z = 0$ plane. The value of $\vec{H}$ in the region $z > 0$ is

    (a) $-5\vec{a}_x$ (b) $5\vec{a}_x$ (c) $10\vec{a}_z$ (d) $10\vec{a}_x$

16. The magnetic field intensity at the circumference of a circular long conductor of radius b and carrying a current of I A is

    (a) $\dfrac{I}{\pi b}$ (b) $\dfrac{I}{2\pi b}$ (c) $\dfrac{I}{b}$ (d) $\dfrac{I}{4\pi b}$

17. The magnetic field intensity due to a long circular conductor of radius b and carrying a current of I A at radius $x\,(x < b)$ is

    (a) $\dfrac{I}{2\pi x}$ (b) $\dfrac{Ix}{4\pi b^2}$ (c) $\dfrac{Ix}{2\pi b^2}$ (d) $\dfrac{I}{4\pi x}$

18. Which of the following expressions indicate that the magnetic flux lines are continuous?

    (a) $\nabla \cdot \vec{D} = \rho$ (b) $\nabla \cdot \vec{B} = 0$ (c) $\nabla \cdot \vec{E} = 0$ (d) $\nabla \cdot \vec{J} = 0$

19. The magnetic field intensity on the axis of a long solenoid of n turns/metre and carrying a current of I A is

    (a) $2\,nI$ (b) nI (c) $\dfrac{nI}{2}$ (d) $3\,nI$

20. The direction of magnetic field around a conductor follows

    (a) right-hand thumb rule (b) right-hand screw rule
    (c) both of these (d) none of these

21. The unit magnetic flux density is

    (a) Wb/m (b) Tesla (c) Tesla/m^2 (d) Wb−m

22. The maximum flux density is always normal to the

    (a) surface area (b) surface length (c) enclosed volume (d) field intensity

23. The value absolute permeability is

    (a) $8\pi \times 10^{-7}$ H/m (b) $4\pi \times 10^{-7}$ H/m
    (c) $9\pi \times 10^{-9}$ H/m (d) none of these

24. The magnetic field strength around a conductor is independent of

    (a) current (b) distance (c) permeability (d) direction

25. The magnetic field strength at the centre of an equilateral triangle of side a carrying current I is

    (a) $\dfrac{9I}{4\pi a}$ (b) $\dfrac{I}{2\pi a}$ (c) $\dfrac{I}{\pi a}$ (d) $\dfrac{5I}{2\pi a}$

26. Which of the following is analogous to Ampere's law in electrostatics?
(a) Coulomb's law (b) Gauss' law (c) Ohm's law (d) Lenz law

27. A cylindrical conductor carries a current that produces $\vec{H} = 6y\,\vec{a}_x$ A/m. What is the current density?
(a) $6\,\vec{a}_z$ (b) $12\,\vec{a}_z$ (c) $15\,\vec{a}_z$ (d) $10\,\vec{a}_z$

28. Two current-carrying conductors of infinite length and equal radius r are situated at (0, 0, 0) and (0, 1, 0). The conductor at (0, 0, 0) is carrying 10 A along $\vec{a}_z$, and the other at (0, 1, 0) is carrying a current of 5 A along $-\vec{a}_z$. The value of $\oint \vec{H}\cdot d\vec{l}$ integrated over a circle of radius 2 m with centre at the origin is
(a) 5 A (b) −5 A (c) 15 A (d) 10 A

29. One of the following is not a Maxwell's equation for steady fields.
(a) $\nabla\cdot\vec{D} = \rho$ (b) $\nabla\times\vec{H} = \vec{J}$ (c) $\nabla\cdot\vec{B} = 0$ (d) $\vec{B} = \nabla\times\vec{A}$

30. Magnetic field intensity at the centre of an equilateral triangle of side b carrying a current of I A is
(a) $\dfrac{\mu_0 I}{2\pi b}$ (b) $\dfrac{\mu_0 I}{4\pi b}$ (c) $\dfrac{\mu_0 3I}{2\pi b}$ (d) $\dfrac{\mu_0 9I}{2\pi b}$

31. The expression $\oint H.d\vec{l} = I_{\text{encl}}$ is called
(a) Ampere's work law (b) both (a) & (d)
(c) Ampere's law for current element (d) Ampere's law

32. Magnetic flux in terms of the magnetic flux density is given by
(a) $\phi = \int (\vec{B}\,/\,\mu).d\vec{S}$ (b) $\phi = \int \vec{B}\cdot d\vec{S}$ (c) $\phi = \int \vec{H}\cdot d\vec{S}$ (d) $\phi = \int \vec{B}\cdot d\vec{l}$

33. The magnetic field intensity for a circular current-carrying loop at the centre is given by
(a) $\mu I/2a$ (b) $I/2a$ (c) $\mu I/2a^2$ (d) $I/2a^2$

34. The magnetic flux intensity in a long solenoid at the extreme right is given by
(a) $\mu NI/2l$ (b) $\mu N/2l$ (c) $\mu NI/2l^2$ (d) $NI/2l$

35. Magnetic flux density for a toroid is given by
(a) $\dfrac{\mu N}{2RI}$ (b) $\dfrac{\mu NI}{2\pi R^2}$ (c) $\dfrac{\mu N}{2\pi R}$ (d) $\dfrac{\mu NI}{2\pi R}$

36. The line integral magnetic field intensity around a closed path is equal to
(a) current flowing through the conductor (b) current enclosed
(c) differential current (d) total current

37. The magnetic field intensity due to a long conductor of radius ρ m carrying current J A/m^2 is
(a) $J\rho$ (b) $\dfrac{J\rho}{4}$ (c) $\dfrac{J\rho}{2}$ (d) $2J\rho$

38. If a conductor of radius a carries a current I, then the current enclosed by a radius $\rho(\rho < a)$ is
(a) $\dfrac{\rho^2 I}{a}$ (b) $\dfrac{\rho I}{a^2}$ (c) $\dfrac{\rho I}{a}$ (d) $\dfrac{\rho^2 I}{a^2}$

39. Magnetic field intensity due to an infinite sheet current of current density K A/m is

    (a) K (b) $K/2$ (c) $\sqrt{K}$ (d) K^2

40. Magnetic field intensity outside a current-carrying solid conductor is proportional to

    (a) ρ (b) $\dfrac{1}{\rho^2}$ (c) $\dfrac{1}{\rho}$ (d) ρ^2

41. The curl of the magnetic field intensity $\vec{H}$ is equal to

    (a) conduction current density (b) conduction current
    (c) magnetic potential (d) magnetic flux density

42 For a stationary magnetic field, the curl of the magnetic field intensity is

    (a) maximum (b) minimum (c) zero (d) one

43. The statement that the line integral of $\vec{H}$ over a closed path is equal to the surface integral of the curl of $\vec{H}$ over any surface enclosed by the path is

    (a) Stokes' theorem (b) divergence theorem
    (c) Faraday's theorem (d) Maxwell's theorem

44. An infinite sheet carries a current density of 10 A/m. The magnetic field intensity $\vec{H}$ is

    (a) 5 $\vec{a}_n$ (b) 10 $\vec{a}_n$ (c) 15 $\vec{a}_n$ (d) 20 $\vec{a}_n$

45. If sheet current $\vec{K}$ flows in the same direction on two parallel surfaces, the magnetic field intensity between the surfaces is

    (a) $\vec{K}$ (b) $2\vec{K}$ (c) $\dfrac{\vec{K}}{2}$ (d) zero

Answers

1. *c*	2. *c*	3. *b*	4. *a*	5. *d*	6. *b*	7. *d*	8. *b*	9. *b*	10. *c*
11. *c*	12. *b*	13. *b*	14. *d*	15. *b*	16. *b*	17. *c*	18. *b*	19. *b*	20. *c*
21. *b*	22. *a*	23. *b*	24. *c*	25. *a*	26. *b*	27. *a*	28. *a*	29. *d*	30. *d*
31. *b*	32. *b*	33. *a*	34. *a*	35. *d*	36. *b*	37. *c*	38. *d*	39. *b*	40. *c*
41. *a*	42. *c*	43. *a*	44. *a*	45. *d*					

Magnetic Force and Inductance | 5

Learning Objectives

The main objectives of this chapter are to understand thoroughly:

- Magnetic scalar and vector potentials
- Magnetic potentials due to conductors
- Magnetic force on moving charges and on current elements
- Torque and Lorentz's force equation
- Forces exerted on line, surface and volume current elements
- Ampere's force law and its applications
- Magnetic torque on current loops
- Magnetic dipole and dipole moment, permeability
- Boundary conditions for magnetic fields
- Inductor and inductances of different current elements
- Mutual inductance and Nuemann's formula
- Magnetic energy and energy stored in magnetic fields
- Magnetic circuits, magnetic materials

5.1 Introduction

In the previous chapters, we looked at the basic laws and techniques to calculate magnetic field quantities such as magnetic field intensity, flux density and flux using Biot–Savart's law and Ampere's law. These quantities are important in field measurements and in electromagnetic instruments. This chapter starts with the concepts of scalar and vector magnetic potentials in electromagnetic fields. The magnetic field parameters can be easily obtained from magnetic potentials by using Laplace's equation. We will explain the forces and torques exerted by a magnetic field on other current distributions. We will also explain boundary conditions, inductance, magnetic energy and magnetic circuits. Finally, the classification and characteristics of magnetic materials will also be discussed.

5.2 Scalar Magnetic Potential (V_m)

The scalar magnetic potential is defined *as the line integral of the magnetic field intensity along a given path, where the current density is zero.*
Mathematically,

$$V_m = -\int \vec{H} \cdot d\vec{l}. \tag{5.1}$$

The scalar magnetic potential between two points A and B in a magnetic field is

$$V_{m\,A,B} = -\int_B^A \vec{H} \cdot d\vec{l}. \tag{5.2}$$

The scalar magnetic potential is also called magnetic current. Its units are amperes. The magnetic field intensity can be expressed as

$$\vec{H} = -\nabla V_m \text{ if } \vec{J} = 0.$$

That is, the negative gradient of the scalar magnetic potential is equal to the magnetic field intensity at which the current density $\vec{J} = 0$.

This can be proved by taking curl on both sides.

$$\nabla \times \vec{H} = -\nabla \times \nabla V_m.$$

But we know that the curl of the gradient of any scalar is always zero.

That is, $$\nabla \times \vec{H} = -\nabla \times \nabla V_m = 0. \tag{5.4}$$

From Ampere's circuital law, we know that

$$\nabla \times \vec{H} = \vec{J}. \tag{5.5}$$

Equating Eq. (5.4) and Eq. (5.5),

$$\vec{J} = 0.$$

Therefore the scalar magnetic potential V_m exists in the region where the current density $\vec{J}$ is zero.

Laplace equation for scalar magnetic potential

We know that $\nabla \cdot \vec{B} = 0$, or $\nabla \cdot \mu\vec{H} = 0$ since $\mu \neq 0$.

That is, $\nabla \cdot \vec{H} = 0$.

But $\vec{H} = -\nabla V_m$.

So, $\nabla \cdot (-\nabla V_m) = 0$.

$$\therefore \quad \nabla^2 V_m = 0 \quad \text{for} \quad \vec{J} = 0. \tag{5.6}$$

That is, scalar magnetic potential satisfies Laplace equation.

Note: Scalar magnetic potential is used to find the magnetic field intensity and flux in a region where there is no current. This happens in the case of fields due to permanent magnets.

Example 5.1 A current sheet $\vec{K} = 2.4\vec{a}_z$ A/m is present on a circular surface of radius $\rho = 1.2$ in free space. Find $\vec{H}$ for $\rho > 1.2$. Find the scalar magnetic potential at $(1.5,\ 0.6\pi,\ 1)$ if the scalar magnetic potential is zero at $\phi = 0$, and the barrier is at $\phi = \pi$.

Solution Given $\vec{K} = 2.4\vec{a}_z$ A/m, $\rho = 1.2$ m.

$$d\vec{S} = 2\pi\rho\ \vec{a}_z.$$

$$I_{\text{encl}} = \vec{K} \cdot d\vec{S}.$$

$$I = (2.4)2\pi(1.2)\vec{a}_z \cdot \vec{a}_z = 18.09 \text{ A along the } z \text{ direction.}$$

Now, from Ampere's law, $\oint \vec{H} \cdot d\vec{l} = I_{\text{encl}}$.

For $\rho > 1.2$,

$$\vec{H} \cdot 2\pi\rho = 18.09.$$

$$\vec{H} = \frac{18.09}{2\pi\rho} = \frac{2.88}{\rho}\vec{a}_\phi.$$

To obtain scalar magnetic potential, we know that $\vec{H} = -\nabla V_m$.

In cylindrical coordinates,

$$\vec{H} = -\nabla V_m = \frac{\partial V_m}{\partial \rho}\vec{a}_\rho + \frac{1}{\rho}\frac{\partial V_m}{\partial \phi}\vec{a}_\phi + \frac{\partial V_m}{\partial z}\vec{a}_z.$$

Since H has only $\vec{a}_\phi$ component,

$$H_\phi = \frac{2.88}{\rho}\vec{a}_\phi = \frac{1}{\rho}\frac{\partial V_m}{\partial \phi}\vec{a}_\phi.$$

$$\frac{\partial V_m}{\partial \phi} = 2.88.$$

$$V_m = \int 2.88 d\phi = 2.88\ \phi.$$

At $\phi = 0.6\pi$, $V_m = 2.88(0.6\pi) = 5.42$ A.

5.3 Magnetic Potential at the Centre of a Square Loop

Let a square loop of side L metres lying on the $y - z$ plane with centre at the origin (as shown in Fig. 5.1) carry a current of I amperes.

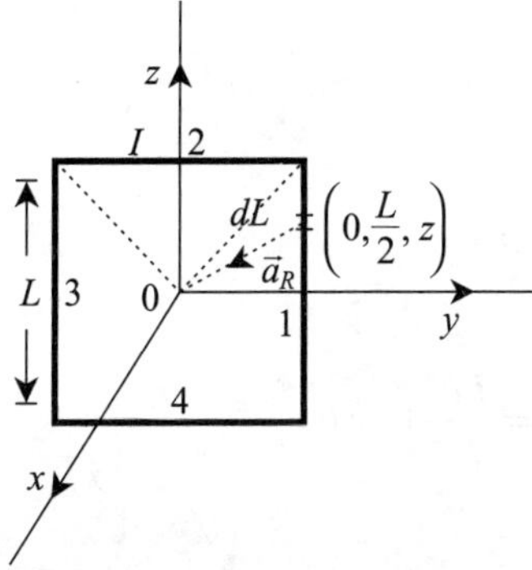

Fig. 5.1 *Scalar magnetic potential due to square loop*

Consider side 1 of length L and the differential current element Idl on it at $(0, \frac{L}{2}, z)$.

The distance vector is $R = (0,\ 0,\ 0) - \left(0,\ \frac{L}{2},\ z\right)$,

$$\vec{R} = -\frac{L}{2}\vec{a}_y - z\vec{a}_z.$$

$$\vec{a}_R = \frac{-(L/2)\vec{a}_y - z\vec{a}_z}{\sqrt{(L/2)^2 + z^2}},$$

and $d\vec{l}_1 = dz\vec{a}_z$.

Using Biot–Savart's law, $d\vec{H} = \dfrac{Id\vec{l}_1 \times \vec{a}_R}{4\pi R^2}$.

The differential magnetic field intensity at the origin due to side 1 is

$$d\vec{H}_1 = \frac{Idz\vec{a}_z}{4\pi\left((L/2)^2 + z^2\right)} \times \frac{\left(-(L/2)\vec{a}_y - z\vec{a}_z\right)}{\sqrt{(L/2)^2 + z^2}}$$

$$= \frac{(L/2)Idz}{4\pi\left((L/2)^2 + z^2\right)^{3/2}}\vec{a}_x.$$

$$\vec{H}_1 = \frac{LI\vec{a}_x}{8\pi}\int_{-L/2}^{L/2}\frac{dz}{\left((L/2)^2 + z^2\right)^{3/2}}.$$

By symmetry, $\vec{H}_1 = \dfrac{LI\,\vec{a}_x}{4\pi}\displaystyle\int_{0}^{L/2}\frac{dz}{\left((L/2)^2 + z^2\right)^{3/2}}.$

To evaluate the integration, let $z = \frac{L}{2}\tan\theta$. Then $dz = \frac{L}{2}\sec^2\theta d\theta$.

At the limit $z = \frac{L}{2}$, the angle is $\theta = \frac{\pi}{4}$,

$$\vec{H}_1 = \frac{LI}{4\pi}\vec{a}_x \int_0^{\pi/4} \frac{(L/2)\sec^2\theta d\theta}{(L/2)^3 \sec^3\theta}$$

$$= \frac{LI}{4\pi}\vec{a}_x \int_0^{\pi/4} \frac{4}{L^2}\cos\theta d\theta = \frac{I\,\vec{a}_x}{L\pi}\Big[\sin\theta\Big]_0^{\pi/4}$$

$$= \frac{I\,\vec{a}_x}{L\pi}\sin\frac{\pi}{4} = \frac{I}{\sqrt{2}L\pi}\vec{a}_x. \tag{5.7}$$

Since the loop is symmetric about the origin, the total magnetic field intensity due to four sides is

$$\vec{H} = 4\vec{H}_1 = \frac{4I}{\sqrt{2}L\pi}\vec{a}_x = \frac{2\sqrt{2}I}{L\pi}\vec{a}_x, \tag{5.8}$$

which is the same as Eq. (4.25). The direction of the magnetic field is normal to the plane of the loop.

We know that the scalar magnetic potential is

$$V_m = \oint \vec{H}\cdot d\vec{l} = H \times \text{length of the loop}$$

$$V_m = \frac{2\sqrt{2}I}{L\pi} \times 4L = \frac{8\sqrt{2}I}{\pi} \quad \text{Wb/m}. \tag{5.9}$$

We can see that the magnetic potential is independent of the loop dimensions.

5.4 Vector Magnetic Potential ($\vec{A}$)

In a given magnetic field, the curl of the vector magnetic potential $\vec{A}$ is equal to the magnetic flux density.

$$\vec{B} = \nabla \times \vec{A}, \tag{5.10}$$

where $\vec{A}$ is the vector magnetic potential. Its units are Wb/m.

Vector magnetic potential for line current element

The differential vector magnetic potential at a point P in the magnetic field is *proportional to the differential current element $Id\vec{l}$ flowing through the conductor and inversely proportional to the distance from the current element to the point P. It is denoted by $d\vec{A}$ and expressed as*

$$d\vec{A} = \frac{\mu}{4\pi}\left(\frac{Id\vec{l}}{R}\right)$$

$$\vec{A} = \int_l \frac{\mu I d\vec{l}}{4\pi R} \tag{5.11}$$

where $\mu/4\pi$ is a proportional constant.

Proof From Biot–Savart's law, we know that for a differential line current element, $Id\vec{l}$, the magnetic field intensity is

$$\vec{H} = \int_l \frac{Id\vec{l} \times \vec{a}_R}{4\pi R^2},$$

where $\vec{R}$ is the distance vector from the current element to the point P.

The magnetic flux density is

$$\vec{B} = \mu_0 \vec{H} = \frac{\mu_0}{4\pi} \int_l \frac{Id\vec{l} \times \vec{a}_R}{R^2}. \tag{5.12}$$

To bring the cross product out of the integral, let us do the following.

We know that $-\nabla\left(\frac{1}{R}\right) = \frac{\vec{a}_R}{R^2}$. Substituting this in Eq. (5.12),

$$\vec{B} = \frac{-\mu_0}{4\pi} \int_l \left(Id\vec{l} \times \nabla\left(\frac{1}{R}\right)\right). \tag{5.13}$$

From the vector identity $\nabla \times (f\vec{C}) = f\nabla \times \vec{C} + \nabla f \times \vec{C}$ (see Chapter 1), and letting

$$f = \frac{1}{R} \quad \text{and} \quad \vec{C} = Id\vec{l},$$

$$\nabla \times \left(\frac{1}{R}(Id\vec{l})\right) = \frac{1}{R}(\nabla \times Id\vec{l}) + \nabla\left(\frac{1}{R}\right) \times Id\vec{l}$$

or $$\nabla\left(\frac{1}{R}\right) \times Id\vec{l} = \nabla \times \left(\frac{Id\vec{l}}{R}\right) - \frac{1}{R}\left(\nabla \times Id\vec{l}\right).$$

Since $\nabla \times Id\vec{l} = 0$,

$$\nabla\left(\frac{1}{R}\right) \times Id\vec{l} = \nabla \times \left(\frac{Id\vec{l}}{R}\right)$$

or $$Id\vec{l} \times \nabla\left(\frac{1}{R}\right) = -\nabla \times \left(\frac{Id\vec{l}}{R}\right). \tag{5.14}$$

Substituting in Eq. (5.13),

$$\vec{B} = \frac{-\mu}{4\pi}\int_l \left(Id\vec{l} \times \nabla\left(\frac{1}{R}\right)\right)$$

$$= \frac{\mu}{4\pi}\int_l \left(\nabla \times \frac{Id\vec{l}}{R}\right) = \nabla \times \int_l \frac{\mu Id\vec{l}}{4\pi R}.$$

In terms of vector magnetic potential,

$$\vec{B} = \nabla \times \vec{A}, \quad \text{where,}$$

$$\vec{A} = \int_l \frac{\mu Id\vec{l}}{4\pi R}. \qquad \text{(Proved).}$$

Therefore, for a differential current element $Id\vec{l}$, the vector magnetic potential at a distance R from the current element is

$$\vec{A} = \int_l \frac{\mu Id\vec{l}}{4\pi R} \quad \text{Wb/m.} \tag{5.15}$$

The vector magnetic potential for other current distributions can be written as

(i) For differential surface current element $\vec{K}dS$, $\vec{A} = \int_s \frac{\mu \vec{K}dS}{4\pi R}$ Wb/m. (5.16)

(ii) For differential volume current element $\vec{J}dv$, $\vec{A} = \int_v \frac{\mu \vec{J}dv}{4\pi R}$ Wb/m. (5.17)

Poisson's equation for vector magnetic potential

From Ampere's circuital law,

$$\nabla \times \vec{H} = \vec{J}.$$

$$\nabla \times \frac{\vec{B}}{\mu} = \vec{J} \text{ or } \nabla \times \vec{B} = \mu\vec{J}.$$

But, $\vec{B} = \nabla \times \vec{A}$.

$$\therefore \quad \nabla \times \nabla \times \vec{A} = \mu\vec{J}. \tag{5.18}$$

Now, $\nabla \times \nabla \times \vec{A} = \nabla(\nabla \cdot A) - \nabla^2\vec{A}$.

Since $\nabla \cdot \vec{B} = 0$, $\nabla \cdot \nabla \times \vec{A} = 0$ (refer to Chapter 1)

or $\nabla \cdot A = 0$.

That is, the divergence of vector magnetic potential is zero.

Substituting in Eq. (5.18),

$$\nabla \times \nabla \times \vec{A} = -\nabla^2 \vec{A} = \mu \vec{J} \quad \text{or} \quad \nabla^2 \vec{A} = -\mu \vec{J}. \tag{5.19}$$

This equation is known as Poisson's equation for magnetic fields.

If $\vec{J} = 0$,

then $\nabla^2 \vec{A} = 0$ (becomes Laplace's equation). (5.20)

Properties of vector magnetic potential

1 Vector magnetic potential is proportional to the current density flowing through the conductor.
2 The curl of vector magnetic potential gives the magnetic flux density.
3 It satisfies Poisson's equation.

Note: Vector magnetic potential is used to obtain radiation characteristics of antennas.

Example 5.2 If $\vec{A} = 10\rho^{1.5}\vec{a}_z$ Wb/m in free space, find $\vec{H}, \vec{J}$.

Solution Given, $\vec{A} = 10\rho^{1.5}\vec{a}_z$ Wb/m.

We know that $\vec{B} = \nabla \times \vec{A}$ and $\vec{H} = \dfrac{\vec{B}}{\mu_0}$.

$$\nabla \times \vec{A} = \frac{1}{\rho} \begin{vmatrix} \vec{a}_\rho & \rho\vec{a}_\phi & \vec{a}_z \\ \dfrac{\partial}{\partial \rho} & \dfrac{\partial}{\partial \phi} & \dfrac{\partial}{\partial z} \\ 0 & 0 & 10\rho^{1.5} \end{vmatrix}$$

$$\vec{B} = \frac{\vec{a}_\rho}{\rho}\left[\frac{\partial}{\partial \phi}(10\rho^{1.5}) - \frac{\partial}{\partial z}(0)\right] - \vec{a}_\phi\left[\frac{\partial}{\partial \rho}(10\rho^{1.5}) - \frac{\partial}{\partial z}(0)\right] + \frac{\vec{a}_z}{\rho}\left[\frac{\partial}{\partial \rho}(0) - \frac{\partial}{\partial \phi}(0)\right]$$

$$= \vec{a}_\rho[0-0] - \vec{a}_\phi\left[\frac{\partial}{\partial \rho}(10\rho^{1.5}) - 0\right] + \vec{a}_z[0] = -\vec{a}_\phi\left[\frac{\partial}{\partial \rho}(10\rho^{1.5})\right]$$

$$= -\vec{a}_\phi(10 \times 1.5\rho^{1.5-1}) = -15\rho^{0.5}\vec{a}_\phi \ \text{Tesla.}$$

$$\therefore \quad \vec{H} = \frac{\vec{B}}{\mu_0} = \frac{-15\rho^{0.5}}{4\pi \times 10^{-7}}\vec{a}_\phi = -1.19 \times 10^7 \sqrt{\rho}\, \vec{a}_\phi \ \text{A/m.}$$

Current density $\vec{J}$ is given by

$$\vec{J} = \nabla \times \vec{H}.$$

$$\nabla \times \vec{H} = \frac{1}{\rho} \begin{vmatrix} \vec{a}_\rho & \rho\vec{a}_\phi & \vec{a}_z \\ \dfrac{\partial}{\partial \rho} & \dfrac{\partial}{\partial \phi} & \dfrac{\partial}{\partial z} \\ 0 & -1.19 \times 10^7 \rho\sqrt{\rho} & 0 \end{vmatrix}.$$

$$\vec{J} = \nabla \times \vec{H} = \frac{\vec{a}_\rho}{\rho}\left[\frac{\partial}{\partial \phi}(0) - \frac{\partial}{\partial z}(-1.19 \times 10^7 \rho\sqrt{\rho})\right] - \vec{a}_\phi\left[\frac{\partial}{\partial \rho}(0) - \frac{\partial}{\partial z}(0)\right]$$

$$+\frac{\vec{a}_z}{\rho}\left[\frac{\partial}{\partial \rho}(-1.19 \times 10^7 \rho\sqrt{\rho}) - \frac{\partial}{\partial \phi}(0)\right]$$

$$= \frac{\vec{a}_\rho}{\rho}[0-0] - \vec{a}_\phi[0-0] + \frac{\vec{a}_z}{\rho}\left[\frac{\partial}{\partial \rho}(-1.19 \times 10^7 \rho\sqrt{\rho}) - 0\right]$$

$$= \frac{\vec{a}_z}{\rho}\left[\frac{\partial}{\partial \rho}(-1.19 \times 10^7 \rho\sqrt{\rho})\right] = \frac{-1.19 \times 10^7 \vec{a}_z}{\rho}\frac{\partial}{\partial \rho}(\rho^{3/2}).$$

$$\vec{J} = \frac{-1.19 \times 3 \times 10^7}{2\sqrt{\rho}}\vec{a}_z = \frac{-1.875 \times 10^7}{\sqrt{\rho}}\vec{a}_z \text{ A/m}^2.$$

5.5 Vector Magnetic Potential in the Field due to an Infinite Length Conductor

Consider an infinitely long current-carrying conductor along the z-axis as shown in Fig. 5.2.

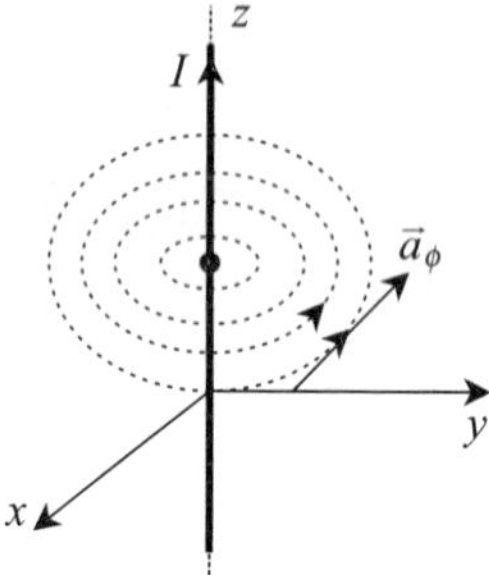

Fig. 5.2 *Magnetic field around a conductor*

The magnetic field intensity at a distance ρ from the conductor is given by

$$\vec{H} = \frac{I}{2\pi\rho}\vec{a}_\phi.$$

We know that $\vec{B} = \mu H = \dfrac{\mu I}{2\pi\rho}\vec{a}_\phi$ and $\vec{B} = \nabla \times \vec{A}$

or $\quad \vec{B} = \nabla \times A = \dfrac{\mu I}{2\pi\rho}\vec{a}_\phi.$

Consider cylindrical coordinates since $\vec{B}$ has only the ϕ component.

$$\nabla \times A = \frac{1}{\rho}\begin{vmatrix} \vec{a}_\rho & \rho\vec{a}_\phi & \vec{a}_z \\ \frac{\partial}{\partial \rho} & \frac{\partial}{\partial \phi} & \frac{\partial}{\partial z} \\ A_\rho & \rho A_\phi & A_z \end{vmatrix} = \frac{\mu I}{2\pi\rho}\vec{a}_\phi \tag{5.21}$$

Equating the $\vec{a}_\phi$ components,

$$\frac{\mu I}{2\pi\rho}\vec{a}_\phi = \left(\frac{\partial A_\rho}{\partial z} - \frac{\partial A_z}{\partial \rho}\right)\vec{a}_\phi.$$

$$\frac{\partial A_\rho}{\partial z} - \frac{\partial A_z}{\partial \rho} = \frac{\mu I}{2\pi\rho}.$$

Since $\vec{B}$ and $\vec{A}$ are constants along the z direction, $\frac{\partial A_\rho}{\partial z} = 0$.

$$\therefore \quad \frac{-\partial A_z}{\partial \rho} = \frac{\mu I}{2\pi\rho}. \tag{5.22}$$

Taking integration on both sides,

$$A_z = \int \frac{-\mu I}{2\pi\rho} d\rho = \frac{-\mu I}{2\pi}\ln(\rho) + A_1, \tag{5.23}$$

where A_1 is a constant. If A_z is zero at a distance ρ_0, then

$$A_1 = \frac{\mu I}{2\pi}\ln(\rho_0).$$

$$\therefore \quad A_z = \frac{\mu I}{2\pi}\ln\left(\frac{\rho_0}{\rho}\right)$$

$$\text{or,} \quad \vec{A} = \frac{\mu I}{2\pi}\ln\left(\frac{\rho_0}{\rho}\right)\vec{a}_z \quad \text{Wb/m.} \tag{5.24}$$

5.6 Vector Magnetic Potential in the Field due to Finite Length Conductor

Consider a line carrying current I along the z-axis placed at $z = a$ and $z = b$ and a current element $Id\vec{l} = Idz\,\vec{a}_z$ on it at (0, 0, z). Let P be a point at (d, 0, 0), as shown in Fig. 5.3.
The distance vector is $\vec{R} = d\vec{a}_\rho - z\vec{a}_z$ and $R = \sqrt{d^2 + z^2}$.

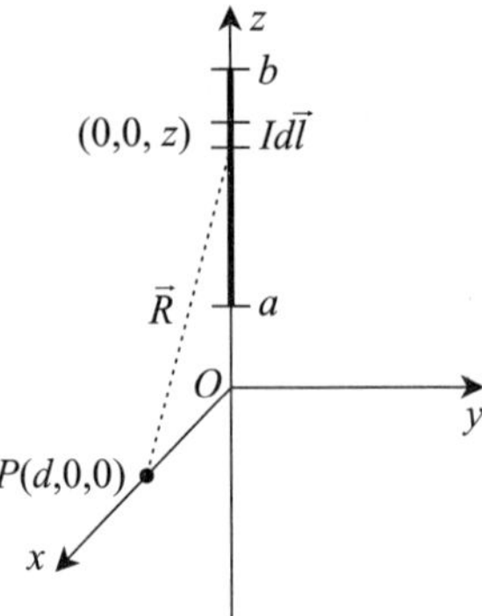

Fig. 5.3 *Vector magnetic potential due to finite conductor*

We know that the differential vector magnetic potential at P is

$$d\vec{A} = \frac{\mu_0 I dz}{4\pi R}\vec{a}_z. \tag{5.25}$$

The vector magnetic potential is

$$\vec{A} = \frac{\mu_0 I}{4\pi}\int_a^b \frac{dz}{\sqrt{d^2+z^2}}\vec{a}_z = \frac{\mu_0 I}{4\pi}\ln\left[\frac{z+\sqrt{z^2+d^2}}{d}\right]_a^b \vec{a}_z \qquad \text{(Refer to Chapter 1)}$$

$$= \frac{\mu_0 I}{4\pi}\ln\left[\frac{b+\sqrt{b^2+d^2}}{a+\sqrt{a^2+d^2}}\right]\vec{a}_z. \tag{5.26}$$

Note: If middle point of the finite conductor is at the origin. Let $2l$ be the conductor length, i.e., $a = -l$ and $b = l$, then

$$\vec{A} = \frac{\mu_0 I}{4\pi}\ln\left[\frac{l+\sqrt{l^2+d^2}}{-l+\sqrt{l^2+d^2}}\right]\vec{a}_z.$$

To simplify it,

$$\frac{l+\sqrt{l^2+d^2}}{-l+\sqrt{l^2+d^2}} = \frac{(l+\sqrt{l^2+d^2})(l+\sqrt{l^2+d^2})}{(-l+\sqrt{l^2+d^2})(l+\sqrt{l^2+d^2})}$$

$$= \frac{l^2+l^2+d^2+2l\sqrt{l^2+d^2}}{-l^2+l^2+d^2} \approx \frac{2l^2+d^2+2l\sqrt{l^2+d^2}}{d^2}.$$

If l is very large, i.e., $l >> d$,

$$\approx \frac{2l^2+d^2+2l\sqrt{l^2}}{d^2}$$

$$\approx \frac{2l^2+2l^2+d^2}{d^2} = \frac{4l^2+d^2}{d^2} = 1+4\left(\frac{l}{d}\right)^2.$$

$$\therefore \quad \vec{A} = \frac{\mu_0 I}{4\pi} \ln\left(1 + 4\left(\frac{l}{d}\right)^2\right)\vec{a}_z. \tag{5.27}$$

Example 5.3 Determine the magnetic potential at a distant point due to a small circular loop (called magnetic dipole) of radius a carrying a current of I amps.

Solution Consider a circular loop conductor of radius a on the x-y plane centred at the origin as shown in Fig. 5.4. Let the conductor carry a current of I amperes. Since the loop is circular, for convenience, consider spherical coordinates.

Let $P(r,\ \theta,\ \phi)$ be an arbitrary point at a distance r from the origin. Since the loop is circular, the field is symmetrical about ϕ.

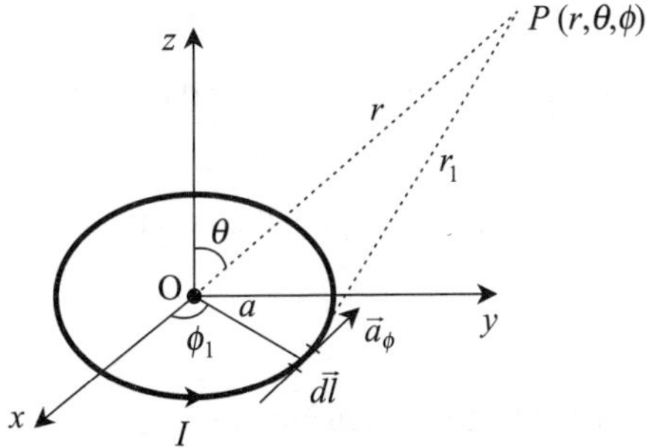

Fig. 5.4 *Field due to a magnetic dipole*

Let the differential length at $(a,\ \pi/2,\ \phi_1)$ be

$$d\vec{l} = a\sin\phi_1 d\phi_1 \vec{a}\phi.$$

The magnetic vector potential at point P due to the current is

$$\vec{A} = \frac{\mu_0 I}{4\pi}\oint \frac{d\vec{l}}{r_1} = \frac{\mu_0 I}{4\pi}\int_0^{2\pi} \frac{a\sin\phi_1}{r_1} d\phi_1 \vec{a}_\phi.$$

From Fig. 5.4 geometry,

$$r_1^2 = r^2 + a^2 - 2ar\sin\theta\sin\phi_1$$

$$r_1^2 = r^2\left(1 + \frac{a^2}{r^2} - \frac{2a}{r}\sin\theta\sin\phi_1\right).$$

$$\frac{1}{r_1} = \frac{1}{r}\left(1 + \frac{a^2}{r^2} - \frac{2a}{r}\sin\theta\sin\phi_1\right)^{-1/2}.$$

Since $r >> a$, using series expansion

$$\frac{1}{r_1} = \frac{1}{r}\left(1 + \frac{a}{r}\sin\theta\sin\phi_1 + \text{higher order terms}\right).$$

Neglecting higher order terms,

$$\frac{1}{r_1} = \frac{1}{r}\left(1 + \frac{a}{r}\sin\theta\sin\phi_1\right).$$

Substituting it in the expression for magnetic vector potential,

$$\vec{A} = \frac{\mu_0 I a\vec{a}_\phi}{4\pi r}\int_0^{2\pi}\left(1 + \frac{a}{r}\sin\theta\sin\phi_1\right)\sin\phi_1 d\phi_1$$

$$= \frac{\mu_0 I a\vec{a}_\phi}{4\pi r}\left(\int_0^{2\pi}\sin\phi_1 d\phi_1 + \frac{a\sin\theta}{r}\int_0^{2\pi}\sin^2\phi_1 d\phi_1\right)$$

$$= \frac{\mu_0 I a\vec{a}_\phi}{4\pi r}\left(0 + \frac{a\sin\theta}{r}\left(\frac{2\pi}{2}\right) + 1\right) = \frac{\mu_0 I a^2\sin\theta\vec{a}_\phi}{4r^2}$$

Therefore the magnetic vector potential due to a small current loop (magnetic dipole) is

$$\vec{A} = \frac{\mu_0 I a^2\sin\theta\vec{a}_\phi}{4r^2}$$

Note: The magnetic flux density due to a magnetic dipole is

$$\vec{B} = \Delta\times\vec{A} = \frac{\mu_0 I a^2}{4r^3}(2\cos\theta\vec{a}_r + \sin\theta\vec{a}_\theta)$$

5.7 Force and Torque on a Moving Charge

Force exerted by a magnetic field can be considered in three different situations.

- Force on a moving charge
- Force on a current element
- Force between two current elements

Lorentz's force equation

Consider a positive electric charge Q placed in static electric and static magnetic fields. According to Coulomb's law, an electric field exerts a force on both stationary and moving electric charges. The electric force exerted on the positive charge by the electric field intensity $\vec{E}$ is given by

$$\vec{F}_e = Q\vec{E} \quad \text{newtons (or N).} \tag{5.28}$$

The direction of the force is the same as that of the electric field intensity. No magnetic force is exerted on a stationary charge. Magnetic fields exert force only on moving charges.

If the charge is moving with velocity $\vec{u}$ in a magnetic field of flux density $\vec{B}$, then the magnetic force on the charge is

$$\vec{F}_m = Q\vec{u}\times\vec{B} \quad \text{N.} \tag{5.29}$$

The direction of magnetic force is along the normal to the plane of $\vec{u}$ and $\vec{B}$ as shown in Fig. 5.5. The total force on the moving charge in the presence of both static magnetic and static electric fields is obtained from Eq. (5.28) and Eq. (5.29) as

$$\vec{F} = \vec{F}_e + \vec{F}_m = Q(\vec{E} + \vec{u} \times \vec{B}) \text{ N.} \tag{5.30}$$

This equation is known as Lorentz's force equation for a moving charge. It relates mechanical force to electrical force. If m is the mass of the charged particle moving in both electric and magnetic fields, then

$$\vec{F} = m\frac{d\vec{u}}{dt} = Q(\vec{E} + \vec{u} \times \vec{B}) \text{ N.} \tag{5.31}$$

The solution of this equation is important in determining the locus of the charge particles in electromagnetic fields. It should be noted that $\vec{F}_m$ is small compared to $\vec{F}_e$ except at high velocities.

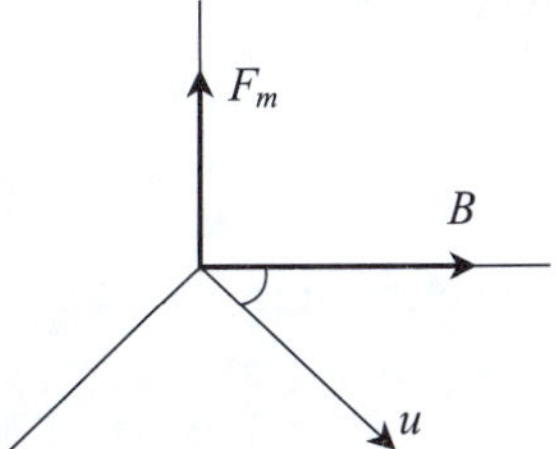

Fig. 5.5 *Direction of the magnetic force*

Note: From Eq. (5.28) it is found that the force due to the electric field is independent of the velocity of the charge. It can perform work on the charged particle and change its kinetic energy. The force due to the magnetic field, on the other hand, depends on the particle velocity and is perpendicular to the direction of both $\vec{u}$ and $\vec{B}$ (see Eq. (5.29)). It cannot perform work because it is normal to the direction of velocity and hence the work done is always zero. Since the kinetic energy is constant in a magnetic field, the energy transfer can be made only due to the electric field.

5.8 Force on a Differential Current Element

Consider a differential charge dQ moving with a velocity $\vec{u}$ in a steady magnetic field of flux density $\vec{B}$. It experiences an acceleration perpendicular to the direction of its motion.

The differential magnetic force exerted on the differential charge is

$$d\vec{F} = dQ(\vec{u} \times \vec{B}) \quad \text{newtons.} \tag{5.32}$$

If ρ_v is the volume charge density and $\vec{J}$ is the current density, then

$$\vec{J} = \frac{I}{A} = \frac{dQ}{dt} \times \frac{1}{dydz} = \frac{dx}{dt} \times \frac{dQ}{dxdydz} = \rho_v\, \vec{u} \tag{5.33}$$

where $u = \dfrac{dx}{dt}$, the velocity along the x direction.

$dQ = \rho_v\, dv$, $dv = dxdydz$ being the differential volume.

$\therefore \quad d\vec{F} = \rho_v\, dv(\vec{u}\times\vec{B}) = \rho_v\, \vec{u}\, dv\times\vec{B}.$

From Eq. (5.33), $d\vec{F} = (\vec{J}dv)\times\vec{B} = (\vec{J}\times\vec{B})dv.$ (5.34)

The force exerted on a volume current element is

$$\vec{F} = \int_v (\vec{J}\times\vec{B})dv. \tag{5.35}$$

Similarly, the force exerted on other current elements can be obtained from the relations for differential current elements,

$$\vec{J}dv = \vec{K}\, d\vec{S} = Id\vec{l}. \tag{5.36}$$

The force exerted on the surface current element is

$$\vec{F} = \int_S (\vec{K}\times\vec{B})\cdot d\vec{S} \tag{5.37}$$

and the force exerted on the line current element is

$$\vec{F} = \int_l Id\vec{l}\times\vec{B}. \tag{5.38}$$

Note: For a straight conductor of length l in a uniform magnetic field, the force is

$$\vec{F} = I(\vec{l}\times\vec{B}) = I\, l\, B\sin\theta\vec{a}_n, \tag{5.39}$$

where θ is the angle between the direction of current flow and the direction of the magnetic flux density.

Torque The torque T due to a force on the conductor is *equal to the cross product of the magnetic dipole moment and the magnetic flux density.*

$$\vec{T} = \vec{m}\times\vec{B} \quad \text{newton-metres (or N-m)} \tag{5.40}$$

Torque T is also defined as the cross product of the lever arm vector $\vec{r}$ and the force $\vec{F}$, where r is the width of the conductor.

So, $\vec{T} = \vec{r}\times\vec{F} = \vec{r}\times I\vec{l}\times\vec{B} = ISB\sin\theta\vec{a}_n = \vec{m}\times\vec{B},$ (5.41)

where $\vec{m} = I\,\vec{S}$ A-m^2 is called the *magnetic dipole moment*, $S = lr$ being the cross-sectional area of the conductor and $\vec{a}_n$, the unit vector normal to the cross-section.

Example 5.4 A magnetic field $\vec{B} = 3.5\times10^{-2}\vec{a}_z$ Wb/m^2 exerts a force on a 0.30 m long conductor along the x-axis. If the conductor current is 5 A in the $-\vec{a}_x$ direction, what force must be applied to hold the conductor in position?

Solution Given $I = 5$ A, $\vec{B} = 3.5 \times 10^{-2}\vec{a}_z$ Wb/m^2, $l = 0.3(-\vec{a}_x) = -0.3\vec{a}_x$.
We know that the force $\vec{F}$ on the conductor is given by

$$\vec{F} = I(\vec{l} \times \vec{B}) = 5(-0.3\vec{a}_x \times 3.5 \times 10^{-2}\vec{a}_z) = 5.25 \times 10^{-2}\vec{a}_y \text{ N}.$$

Since an equal and opposite force must be applied in order to hold the conductor in position, force to be applied, $\vec{F}_a = -5.25 \times 10^{-2}\vec{a}_y$ N.

5.9 Ampere's Force Law: Force between Two Current Elements

Ampere's force law states that *the force between two current-carrying elements $I_1 dl_1$ and $I_2 dl_2$ placed at a distance R_{12} is given by*

$$\vec{F} = \frac{\mu}{4\pi} \oint_{L_1} \oint_{L_2} \frac{I_2 d\vec{l}_2 \times (I_1 d\vec{l}_1 \times \vec{a}_{R_{12}})}{R_{12}^2}, \tag{5.42}$$

where, $I_1 d\vec{l}_1$ is the current element of the first conductor,
$I_2 d\vec{l}_2$ is the current element of the second conductor, and
R_{12} is the distance between the current elements of the two conductors.

Ampere's law is used to express the force on one current carrying conductor directly in terms of a second current-carrying conductor without needing to find the magnetic field between them.

Proof Consider two conductors, in the form of loop L_1 and loop L_2, carrying currents I_1 and I_2 respectively as shown in Fig. 5.6.

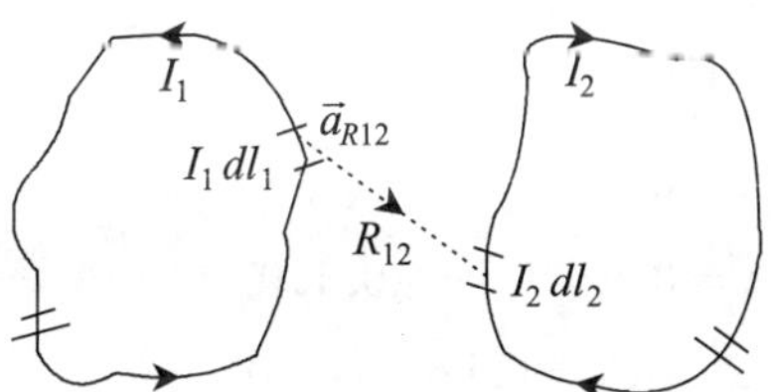

Fig. 5.6 *Force between two current loops*

Let the current element of the first conductor be $I_1 dl_1$ and that of the second conductor be $I_2 d\vec{l}_2$. The current elements are placed at a distance R_{12} from each other; the unit vector $\vec{a}_{R_{12}}$ is as shown in Fig. 5.6.

According to Biot–Savart's law, the magnetic field intensity at $I_2 d\vec{l}_2$ due to the current element $I_1 d\vec{l}_1$ is

$$d\vec{H}_2 = \frac{I_1 d\vec{l}_1 \times \vec{a}_{R_{12}}}{4\pi R_{12}^2},$$

or $$d\vec{B}_2 = \frac{\mu I_1 d\vec{l}_1 \times \vec{a}_{R_{12}}}{4\pi R_{12}^2}. \tag{5.43}$$

Due to the magnetic field B_2, the differential force on the current element $I_2 d\vec{l}_2$ is

$$d\vec{F}_2 = I_2 d\vec{l}_2 \times \vec{B}_2.$$

Taking differential magnetic flux density,

$$d(d\vec{F}_2) = I_2 d\vec{l}_2 \times d\vec{B}_2. \tag{5.44}$$

Substituting Eq. (5.43) in Eq. (5.44), we get

$$d(d\vec{F}_2) = \frac{\mu I_2 d\vec{l}_2 \times I_1 d\vec{l}_1 \times \vec{a}_{R_{12}}}{4\pi R_{12}^2}.$$

Taking double integration on closed paths, the total force $\vec{F}_2$ on current loop L_2 due to current loop L_1 is

$$\vec{F}_2 = \frac{I_1 I_2 \mu}{4\pi} \oint_{L_1} \oint_{L_2} \frac{d\vec{l}_2 \times d\vec{l}_1 \times \vec{a}_{R_{12}}}{R_{12}^2} \quad \text{(Proved).}$$

Note: The force $\vec{F}_1$ on current loop L_1 due to current loop L_2 is

$$\vec{F}_1 = -\vec{F}_2. \tag{5.45}$$

Example 5.5 Filamentary currents of $-25\vec{a}_z$ A and $25\vec{a}_z$ A are located in the $x = 0$ plane in free space at $y = -1$ m and $y = 1$ m respectively. A third filamentary current of $10^3\vec{a}_z$ A is located at $x = 1$ m, $y = 0$. Find the vector force on a 1 m length of 1 mA filament.

Solution Given filamentary currents:

$$I_1 d\vec{l}_1 = -25\,\vec{a}_z \text{ A} \quad \text{at} \quad x = 0, y = -1, \text{i.e., } (0, -1, 0)$$
$$I_2 d\vec{l}_2 = 25\,\vec{a}_z \text{ A} \quad \text{at} \quad x = 0, y = 1, \text{i.e., } (0, 1, 0)$$
$$I_3 d\vec{l}_3 = 10^{-3}\,\vec{a}_z \text{ A} \quad \text{at} \quad x = 1, y = 0, \text{i.e., } (1, 0, 0)$$

Force on 1 mA current element (i.e., I_3) due to other filamentary currents is

$$d\vec{F} = d\vec{F}_1 + d\vec{F}_2$$

where $d\vec{F}_1$ is the force due to current element $I_1 d\vec{l}_1$ and $d\vec{F}_2$ is the force due to current element $I_2 d\vec{l}_2$.

From Ampere's force law, $d\vec{F} = I_3 d\vec{l}_3 \times \vec{B}_1 + I_3 d\vec{l}_3 \times \vec{B}_2$.

$$\vec{F} = I_3 d\vec{l}_3 \times (\vec{B}_1 + \vec{B}_2),$$

where $\vec{B}_1$ is the field due to $I_1 d\vec{l}_1$ and B_2, the field due to $I_2 d\vec{l}_2$.

We know that $\vec{B} = \dfrac{\mu_0 I d\vec{l} \times \vec{a}_R}{2\pi R^2}$.

The distance vectors are

$$\vec{R}_{13} = (1, 0, 0)-(0, -1, 0) = \vec{a}_x + \vec{a}_y \quad \text{and} \quad \vec{R}_{23} = (1, 0, 0)-(0, 1, 0) = \vec{a}_x - \vec{a}_y.$$

$$|R_{13}| = \sqrt{1+1} = \sqrt{2} \quad \text{and} \quad |R_{23}| = \sqrt{1+1} = \sqrt{2}.$$

$$\vec{B}_1 = \frac{\mu_0 I_1 d\vec{l}_1 \times \vec{a}_{R_{13}}}{2\pi R_{13}^2} = \frac{\mu_0(-25\vec{a}_z) \times (\vec{a}_x + \vec{a}_y)}{4\pi(\sqrt{2})^2} = \frac{\mu_0(-25)(\vec{a}_y - \vec{a}_x)}{8\pi}$$

and $$\vec{B}_2 = \frac{\mu_0 I_2 d\vec{l}_2 \times \vec{a}_{R_{23}}}{2\pi R_{23}^2} = \frac{\mu_0(25\vec{a}_z) \times (\vec{a}_x - \vec{a}_y)}{4\pi(\sqrt{2})^2} = \frac{\mu_0(25)(\vec{a}_y + \vec{a}_x)}{8\pi}.$$

$$\therefore \quad d\vec{F}_1 = I_3 d\vec{l}_3 \times (\vec{B}_1 + \vec{B}_2).$$

$$\vec{F} = I_3 d\vec{l}_3 \times \left[\frac{\mu_0(-25)(\vec{a}_y - \vec{a}_x)}{8\pi} + \frac{\mu_0(25)(\vec{a}_y + \vec{a}_x)}{8\pi}\right] = 10^{-3}\vec{a}_z \times \frac{\mu_0 25}{4\pi}\vec{a}_x.$$

$$= \frac{25 \times 10^{-3}\mu_0}{4\pi}\vec{a}_y = 25 \times 10^{-10}\vec{a}_y \text{ N.}$$

5.10 Force between Two Straight, Infinitely Long Parallel Current Carrying Conductors

Consider two infinitely long parallel conductors carrying currents I_1 and I_2 respectively in the same direction. Let the distance between the conductors be d metres. Since I_1 and I_2 are in the same direction, an attractive force exists between the two conductors.

We know that the magnetic field intensity due to a current-carrying conductor at a distance is given by

$$\vec{H} = \frac{I}{2\pi\rho}\vec{a}_\phi. \tag{5.46}$$

Therefore the magnetic field intensity due to I_1 on the second conductor at a distance d is

$$\vec{H}_1 = \frac{I_1}{2\pi d}\vec{a}_\rho \quad \text{A/m}$$

and $$\vec{B}_1 = \frac{\mu I_1}{2\pi d}\vec{a}_\phi \quad \text{Wb/m}^2. \tag{5.47}$$

Let a current element on the second conductor be $I_2 d\vec{l}$. The differential force on the second conductor is

$$d\vec{F}_2 = I_2 d\vec{l} \times \vec{B}_1.$$

Since $d\vec{l}$ is normal to B_1,

$$d\vec{F_2} = I_2 dl\, B_1\, \vec{a}_n = I_2 dl \cdot \frac{\mu I_1}{2\pi d}\vec{a}_n. \tag{5.48}$$

Integrating, $\vec{F_2} = \dfrac{\mu I_1 I_2}{2\pi d}\vec{a}_n \displaystyle\int dl = \dfrac{\mu I_1 I_2}{2\pi d} l\vec{a}_n.$

Then force per unit length $\vec{F_2} = \dfrac{\mu I_1 I_2}{2\pi d}\vec{a}_n.$ (5.49)

Similarly, the force per unit length due to current I_2 on the first conductor is

$$\vec{F_1} = \frac{\mu I_1 I_2}{2\pi d}\vec{a}_n \text{ N/m.}$$

Then, $\vec{F_1} = \vec{F_2}.$ (5.50)

Therefore, the force between the two conductors is attractive.

Note: If currents I_1 and I_2 flow in opposite directions through the conductors,

$$\vec{F_1} = -\vec{F_2}, \tag{5.51}$$

i.e., the forces act in opposite directions and are repulsive.

Example 5.6 The force between two long parallel conductors placed 10 cm from each other is 15 kg/m. If one conductor carries twice the current as the other, calculate the current in each conductor.

Solution Given $F = 15$ kg/m $= 15 \times 10^{-5}$ N/m and $d = 10$ cm$= 0.1$ m.
Let I be the current in one conductor. The current in the second conductor is $2I$. We know that the force per unit length between two current-carrying wires is

$$F = \frac{\mu_0 I_1 I_2}{2\pi d} \text{ N/m.}$$

$$15 \times 10^{-5} = \frac{4\pi \times 10^{-7} I(2I)}{(2\pi)(0.1)} = 40I^2 \times 10^{-7}.$$

$$I = \sqrt{\frac{15}{40} \times 10^2} = 6.123 \text{ A.}$$

5.11 Magnetic Torque due to a Rectangular Differential Current Loop in a Magnetic Field

Consider a rectangular differential current loop with dimensions dx and dy in a magnetic field $\vec{B}$ as shown in Fig. 5.7. Let the loop lie in the $x-y$ plane with centre at the origin. The magnetic field at the centre of the loop is B_0. Since the loop is very small in size, the field B_0 is constant. In the equilibrium position, the total force on the loop is zero.

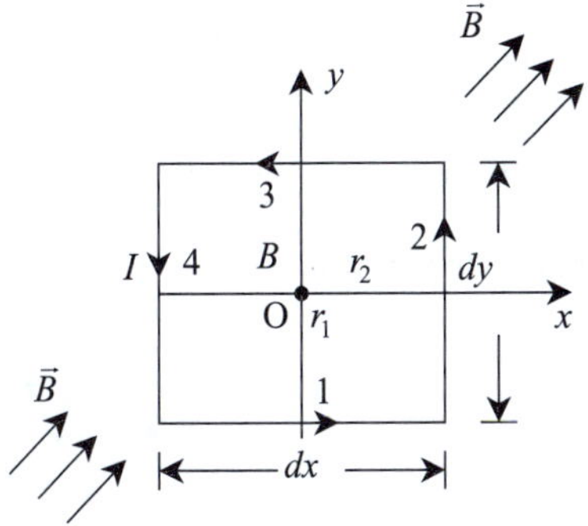

Fig. 5.7 *Rectangular loop in the field*

The differential force on side 1 is

$$d\vec{F}_1 = Id\vec{l} \times \vec{B} = Idx\vec{a}_x \times \vec{B}.$$

The torque at the centre of the loop is

$$d\vec{T}_1 = \vec{r}_1 \times d\vec{F}_1,$$

where, r_1 from Fig. 5.7 is, $r_1 = -\frac{1}{2}dy\,\vec{a}_y$.

$$d\vec{T}_1 = -\frac{1}{2}dy\,\vec{a}_y \times I\,dx\,\vec{a}_x \times \vec{B} = -\frac{I}{2}dx\,dy[\vec{a}_y \times (\vec{a}_x \times \vec{B})]. \tag{5.52}$$

Let $\vec{B} = B_{0x}\,\vec{a}_x + B_{0y}\,\vec{a}_y + B_{0z}\,\vec{a}_z$.

Then $d\vec{T}_1 = -\frac{I}{2}dx\,dy[\vec{a}_y \times \vec{a}_x \times (B_{0x}\,\vec{a}_x + B_{0y}\,\vec{a}_y + B_{0z}\,\vec{a}_z)]$

$$= \frac{-I}{2}dx\,dy[\vec{a}_y \times (B_{0y}\,\vec{a}_z - B_{0z}\,\vec{a}_y)] = \frac{-I}{2}dx\,dy\,B_{0y}\,\vec{a}_x. \tag{5.53}$$

Similarly, the torque at the centre due to side 3 is

$$d\vec{T}_3 = \frac{-I}{2}dxdy\,B_{0y}\vec{a}_x. \tag{5.54}$$

The differential force on side 2 is

$$d\vec{F}_2 = Id\vec{l} \times \vec{B} = I\,dy\,\vec{a}_y \times \vec{B}.$$

The torque at the centre of the loop is

$$d\vec{T}_2 = r_2 \times d\vec{F}_2, \quad \text{where} \quad \vec{r}_2 = \frac{1}{2}dx\,\vec{a}_x.$$

$$d\vec{T}_2 = \frac{I}{2}dx\,\vec{a}_x \times (Idy\,\vec{a}_y \times \vec{B})$$

$$= \frac{I}{2} dx\, dy(\vec{a}_x \times \vec{a}_y \times (B_{0x}dx + B_{0y}dy + B_{0z}dz))$$

$$= \frac{I}{2} dx\, dy[\, \vec{a}_x \times (-B_{0x}\vec{a}_z + B_{0z}\vec{a}_x)] = \frac{I}{2} dxdy\, B_{0x}\, \vec{a}_y. \tag{5.55}$$

Similarly, the torque at the centre due to side 4 is

$$d\vec{T}_4 = \frac{I}{2} dxdy\, B_{0x}\, \vec{a}_y. \tag{5.56}$$

The total torque at the centre is

$$d\vec{T} = d\vec{T}_1 + d\vec{T}_2 + d\vec{T}_3 + d\vec{T}_4$$

$$= \frac{-I}{2} dxdy\, B_{0y}\, \vec{a}_x - \frac{I}{2} dxdy\, B_{0y}\, \vec{a}_x + \frac{I}{2} dxdy\, B_{0x}\, \vec{a}_y + \frac{I}{2} dxdy\, B_{0x}\, \vec{a}_y$$

$$= Idxdy\, B_{0x}\, \vec{a}_y - Idxdy\, B_{0y}\, \vec{a}_x = Idxdy\, (B_{0x}\, \vec{a}_y - B_{0y}\, \vec{a}_x) = Idxdy\vec{a}_z \times \vec{B}$$

$$d\vec{T} = I\, d\vec{S} \times \vec{B},$$

where $d\vec{S} = dx\, dy\, \vec{a}_z$ = cross-sectional area of the current loop.

Therefore, the total torque is

$$\vec{T} = I\vec{S} \times \vec{B}. \tag{5.57}$$

We know that the differential magnetic dipole moment

$$d\vec{m} = I\, d\vec{S}.$$

$$\vec{m} = I\vec{S}.$$

$$\therefore \quad \vec{T} = \vec{m} \times \vec{B}. \tag{5.58}$$

If θ be the angle between the normal to the plane of the loop and the direction of the field, then

$$\vec{T} = mB \sin\theta \vec{a}_x. \tag{5.59}$$

Note: The torque on the current loop always tends to turn the loop so as to align the magnetic field produced by the loop along the applied magnetic field.

Example 5.7 (i) What is the maximum torque on a square loop of 100 turns in a field of uniform flux density 1 Wb/m^2? The loop has a 10 cm side and carries a current of 3 A. (ii) What is the magnetic moment of the loop?

Solution Given $N = 100$; $B = 1$ Wb/ m^2; $I = 3$ A; Area of the loop $= 10 \times 10 = 10^{-2}$ m^2.

(i) Max torque $\quad T = BIS$

For single loop, $\quad T_{\max} = 1 \times 3 \times 10^{-2}$ N-m/turn

For N turns, $\quad T_{\max} = 3 \times 10^{-2} \times 100 = 3$ N-m

(ii) Magnetic moment $\quad m = IS = 3 \times 10^{-2}$ A-m^2.

5.12 Magnetic Dipole and Dipole Moment

A small filamentary current loop carrying a current I or a permanent bar magnet of length l is called a magnetic dipole. If Q_m is the pole strength of the magnetic bar, then the magnetic dipole moment is

$$\vec{m} = Q_m\vec{l}. \tag{5.60}$$

When the bar magnet is in a uniform magnetic field $\vec{B}$, the force on the bar is

$$\vec{F} = Q_m\vec{B}$$

and the torque is $\vec{T} = \vec{m} \times \vec{B} = Q_m\vec{l} \times \vec{B}$. (5.61)

The torque tends to align the bar magnet with the external magnetic field.
For a small current loop under the magnetic field $\vec{B}$, the torque is given by

$$\vec{T} = I\vec{S} \times \vec{B}.$$

Since $\vec{T} = \vec{m} \times \vec{B}$, the dipole moment is $\vec{m} = I\vec{S}$.

If both the bar and the current loop have the same torque, then

$$Q_m\vec{l} = I\vec{S}. \tag{5.62}$$

5.13 Boundary Conditions for Magnetic Field

Tangential magnetic field intensity *The tangential component of magnetic field $\vec{H}$ is continuous across the boundary except at the surface of a perfect conductor, i.e.,*

$$H_{t1} - H_{t2} = K, \tag{5.63}$$

where K is the surface current density.

For a non-conducting boundary, $K = 0$, and therefore,

$$H_{t1} = H_{t2}. \tag{5.64}$$

Proof Consider the boundary between two isotropic homogeneous conductors with different permeabilities, μ_1 and μ_2, as shown in Fig. 5.8. Let Δx, Δy be the dimensions of a closed path of differential rectangular loop situated across the boundary. Also let the magnetic field H move from medium 2 and medium 1.

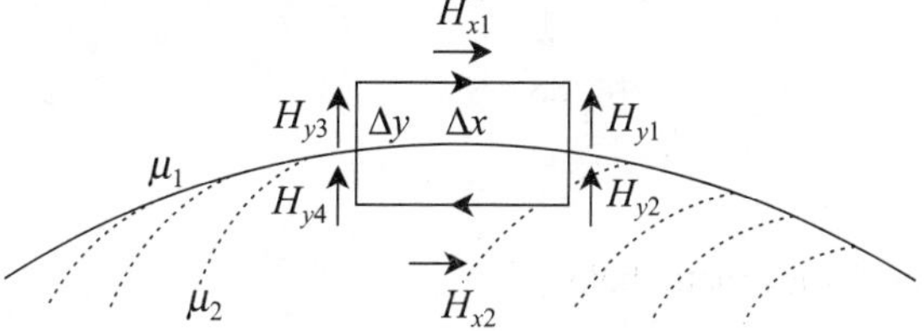

Fig. 5.8 *Boundary conditions for $\vec{H}$*

According to Ampere's circuital law, we know that

$$\oint \vec{H} \cdot d\vec{l} = I.$$

For the loop, taken in the clockwise direction,

$$H_{x1}\Delta x - H_{y1}\frac{\Delta y}{2} - H_{y2}\frac{\Delta y}{2} - H_{x2}\Delta x + H_{y4}\frac{\Delta y}{2} + H_{y3}\frac{\Delta y}{2} = I. \tag{5.65}$$

At the boundary, as $\Delta y \to 0$, we get

$$H_{x1}\Delta x - H_{x2}\Delta x = I.$$

$$H_{x1} - H_{x2} = \frac{I}{\Delta x} = K.$$

The fields $\vec{H}_{x1}$ and $\vec{H}_{x2}$ are tangential to the boundary.

$\therefore \quad H_{t1} - H_{t2} = K.$

If the surface current density $K = 0$,

then $H_{t1} = H_{t2}$ (Proved).

Also, $$\frac{B_{t1}}{\mu_1} = \frac{B_{t2}}{\mu_2}. \tag{5.66}$$

Normal component of magnetic flux density *The normal component of the flux density $\vec{B}$ is continuous across the boundary*, i.e.,

$$B_{n1} = B_{n2}.$$

Proof Consider a Gaussian surface in the form of a cylinder as shown in Fig. 5.9. The cylinder is placed at the centre of the boundary between the two media. Let Δh be the height and dS, the cross-sectional area of the cylinder.

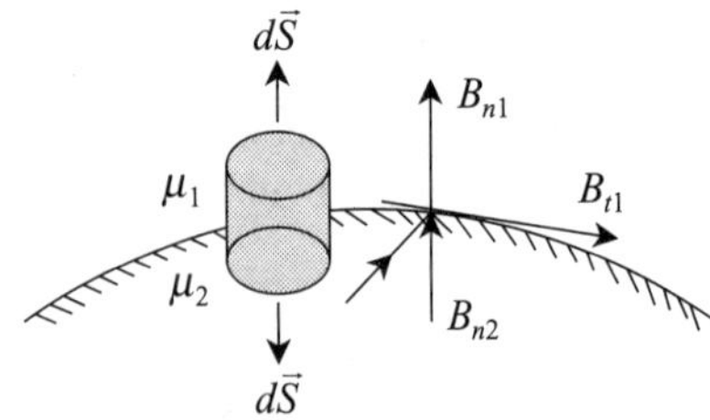

Fig. 5.9 *Boundary conditions for B*

According to Gauss' law for magnetic fields,

$$\oint \vec{B} \cdot d\vec{S} = 0$$

$$\text{or} \quad \oint_{\text{top}} \vec{B}\cdot d\vec{S} + \oint_{\text{bottom}} \vec{B}\cdot d\vec{S} + \oint_{\text{side}} \vec{B}\cdot d\vec{S}_1 = 0. \tag{5.67}$$

As $\Delta h \to 0$ at the boundary, the side surface area $d\vec{S}_1$ becomes zero.

$$\therefore \quad \oint_{\text{top}} \vec{B}\cdot d\vec{S} + \oint_{\text{bottom}} \vec{B}\cdot d\vec{S} = 0.$$

$$B_{n1}\Delta S - B_{n2}\Delta S = 0$$

$$\text{or} \quad B_{n1} = B_{n2}, \tag{5.68}$$

where B_{n1} and B_{n2} are normal components of $\vec{B}$. That is, the normal component of the flux density is continuous at the boundary.

$$\text{Also,} \quad \mu_1 H_{n1} = \mu_2 H_{n2}. \tag{5.69}$$

Note: Let the fields make angles θ_1 and θ_2 with the normal to the interface line as shown in Fig. 5.10.

$$\therefore \quad \tan\theta_1 = \frac{B_{t1}}{B_{n1}} \quad \text{and} \quad \tan\theta_2 = \frac{B_{t2}}{B_{n2}}.$$

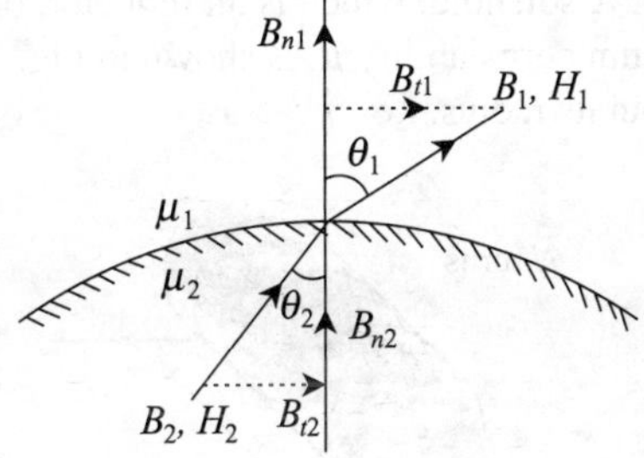

Fig. 5.10 *Field components at the boundary*

We know that $B_{n1} = B_{n2}$.

$$\therefore \quad \frac{B_{t1}}{B_{t2}} = \frac{B_{n1}\tan\theta_1}{B_{n2}\tan\theta_2} = \frac{\tan\theta_1}{\tan\theta_2} = \frac{\mu_1}{\mu_2}. \tag{4.70}$$

5.14 Inductor and Inductance

An insulated conducting wire wound on a magnetic material with N turns, called a coil or an inductor, is shown in Fig. 5.11. The medium inside the coil may be a magnetic material or free space. When a current I passes through the coil, it produces a magnetic field $\vec{B}$ which causes a flux φ in each turn of the coil. Inductors store energy in magnetic fields.

The total flux linkage of the coil is $N\varphi$ Wb-turns.

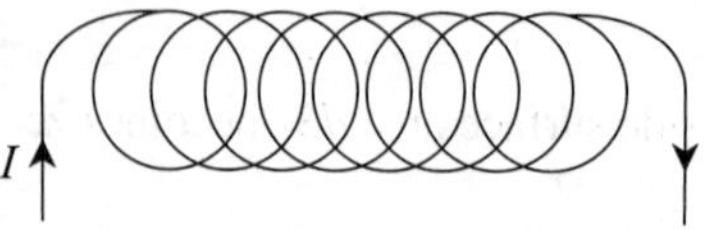

Fig. 5.11 *Inductor*

If the medium of the coil is linear, then the flux linkage of the coil is proportional to the current flowing through it.

$$N\varphi \propto I. \tag{5.71}$$

The ratio of the total flux linkage to the current flowing through the coil is called inductance (L). Its unit is Henry (H).

Since $N\varphi \propto I$, $N\varphi = LI$

$$\text{or} \quad L = \frac{N\varphi}{I} \quad \text{Wb-turn/A or H (henry)} \tag{5.72}$$

This is also called the *self-inductance* of the coil since the flux is produced by the coil itself.

Inductance of a solenoid A solenoid, which is an inductor, of length l, area of cross-section S, number of turns N, and medium permeability μ, is shown in Fig. 5.12. Assume that the length of the solenoid is much greater than its radius, i.e., $l >> r$.

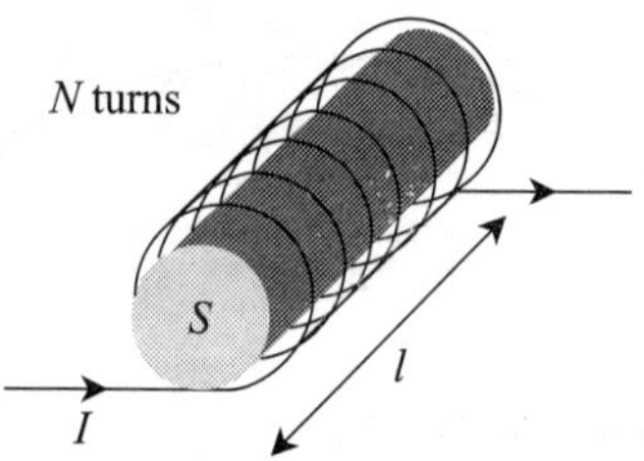

Fig. 5.12 *Solenoid*

Let I be the current flowing through the solenoid. The magnetic field intensity in the solenoid can be obtained from

$$\oint H \cdot dl = I_{\text{encl}}.$$

$$H \cdot l = NI \quad \text{or} \quad H = \frac{NI}{l},$$

where NI is called the *magnetomotive force* (mmf). The total flux linkage is

$$N\varphi = NBS = N\mu HS,$$

where μ is the permeability of the magnetic material and $S = \pi r^2$.

From the definition, inductance is

$$L = \frac{N\varphi}{I} = \frac{N\mu HS}{I}.$$

Thr inductance of the solenoid is $L = \frac{N\mu NIS}{lI} = \frac{N^2\mu S}{l}$ H. (5.73)

Inductance per unit length $\frac{L}{l} = \left(\frac{N}{l}\right)^2 \mu S = n^2\mu S$ H/m,

where $n = \frac{N}{l}$ = turns/unit length.

Note: Consider the solenoid is converted into a single flat sheet of conductor wound on a medium of permeability μ. If the sheet current density is K, then the total current flowing through the sheet is

$$NI = Kl \quad \text{or} \quad K = \frac{NI}{l} = H.$$

Thus, it may be noted that the field inside the solenoid is equal to the sheet current density.

Example 5.8 An air core solenoid of length 50 cm long has an inductance of 10 mH. The radius this solenoid is 2 cm. Find the number of turns.

Solution Given $l = 50$ cm $= 0.5$ m; $L = 10$ mH; radius $r = 2$ cm $= 0.02$ m.
For air core, $\mu = \mu_0$, $S = \pi(0.02)^2$ m^2.

We know that the inductance of a solenoid is

$$L = \frac{N^2\mu_0 S}{l}.$$

$$N^2 = \frac{Ll}{\mu_0 S} = \frac{10 \times 10^{-3} \times 0.5}{4\pi \times 10^{-7} \times \pi \times 0.02^2} = 3166 \times 10^3.$$

No. of turns, $N = 1780$.

Inductance of a toroid A toroid is a solid ring of magnetic material with a coil of N turns wound on it as shown in Fig. 5.13. Let R be the radius of the toroid and let I be the current through the coil. Let S be the cross-sectional area of the ring.

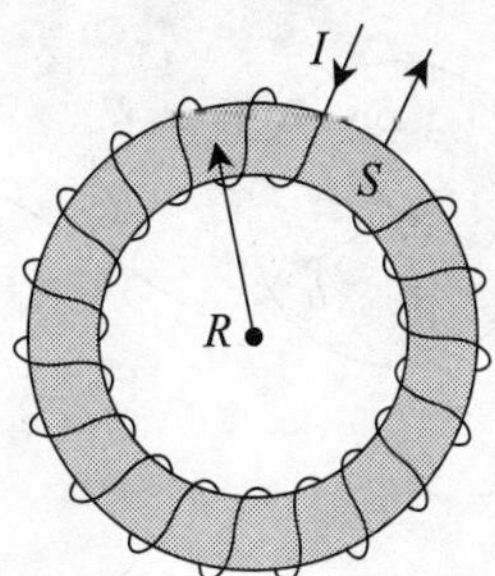

Fig. 5.13 *Toroid*

Applying Ampere's circuital law,

$$\oint \vec{H}.d\vec{l} = I_{\text{encl}}.$$

$$H\,2\pi R = NI \text{ or } H = \frac{NI}{2\pi R}.$$

Magnetic flux density $B = \mu H = \dfrac{\mu NI}{2\pi R}$. (5.74)

Total flux linkage $= N\varphi = NBS$

From the definition of inductance, $L = \dfrac{N\varphi}{I} = \dfrac{NBS}{I}$.

$$\therefore \quad L = \frac{N\mu NIS}{2\pi RI} = \frac{N^2\mu S}{2\pi R} \text{ henrys.} \tag{5.75}$$

The above equation gives the inductance of a toroid.

Example 5.9 A toroid has 600 turns of coil, a circular cross-section of 6 cm^2 and a mean diameter of 38 cm. The permeability of the toroid is 1000. Calculate the inductance of the coil.

Solution Given $N = 600$; $S = 6 \times 10^{-4}$ m^2; $R = 0.38/2 = 0.19$ m; $\mu_r = 1000$.
For a toroid, the inductance is

$$L = \frac{\mu N^2 S}{2\pi R}$$

$$= \frac{1000 \times 4\pi \times 10^{-7} \times 600^2 \times 6 \times 10^{-4}}{2\pi \times 0.19} = 0.227 \text{ henrys.}$$

Inductance of a coaxial cable Consider a coaxial cable as shown in Fig. 5.14. Let a be the solid inner conductor radius, b, the outer conductor radius and l, the length of the cable. Also, let the current through the inner conductor be I and that through the outer conductor be $-I$.

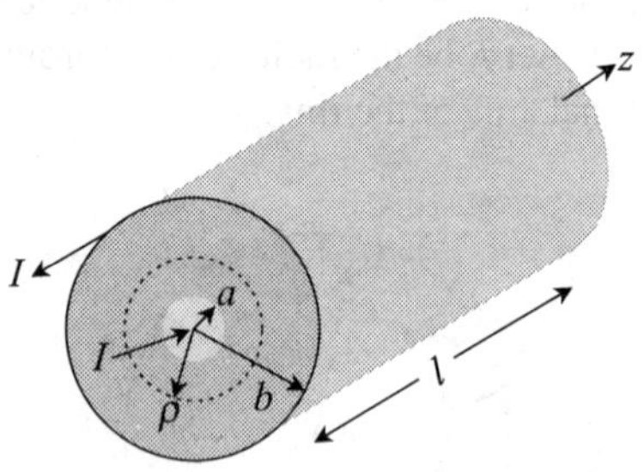

Fig. 5.14 *Coaxial cable*

Let ρ be the radius of a Gaussian surface inside the cable.

For $0 < \rho < a$, the current enclosed is

$$I_{\text{encl}} = I\frac{\rho^2}{a^2}.$$

From Ampere's circuital law, we know that the magnetic field intensity is $\oint \vec{H}.d\vec{l} = I_{\text{encl}}$.

$$\text{Then,} \quad H(2\pi\rho) = I_{\text{encl}} = I\frac{\rho^2}{a^2}.$$

$$\therefore \quad H = \frac{I\rho}{2\pi a^2} \quad \text{or} \quad \vec{B}_1 = \frac{\mu I\rho}{2\pi a^2}\vec{a}_\phi.$$

For the region $a < \rho < b$,

$$H(2\pi\rho) = I \quad \text{or} \quad \vec{B}_2 = \frac{\mu I}{2\pi\rho}\vec{a}_\phi.$$

The flux linkage due to the inner conductor is

$$\varphi_1 = \int_s \vec{B}_1 \cdot d\vec{S}_1.$$

In the region $0 < \rho < a$, the fraction of the area through which the flux leaves is,

$$d\vec{S}_1 = \left(\frac{\pi\rho^2}{\pi a^2}\right) d\rho dz\vec{a}_\phi = \frac{\rho^2}{a^2} d\rho dz\vec{a}_\phi.$$

$$\therefore \quad \varphi_1 = \int_0^l \int_0^a \frac{\mu I\rho}{2\pi a^2}\vec{a}_\rho \cdot \frac{\rho^2}{a^2} d\rho dz\vec{a}_\phi$$

$$= \int_0^l \int_0^a \frac{\mu I}{2\pi a^4}\rho^3 d\rho dz = \frac{\mu Il}{2\pi a^4}\left[\frac{\rho^4}{4}\right]_0^a$$

$$= \frac{\mu Il}{8\pi} \text{ Wb.} \tag{5.76}$$

Let L_m be the internal inductance of the inner conductor of the cable in the region $0 < \rho < a$.

$$\text{Then} \quad L_{\text{in}} = \frac{\varphi_1}{I} = \frac{\mu l}{8\pi} \text{ henrys.} \tag{5.77}$$

Thus, when a current I flows through a long conductor, a magnetic flux is produced around the conductor which leads to internal inductance. The internal inductance is independent of the wire radius.

The external inductance of the cable can be obtained by using the flux linkage in the region $a < \rho < b$.

The flux in the region $a < \rho < b$ is

$$\varphi_2 = \int_s \vec{B}_2 \cdot d\vec{S}_2 = \int_0^l \int_a^b \frac{\mu I}{2\pi\rho} \vec{a}_\phi \cdot d\rho dz\, \vec{a}_\phi$$

$$= \frac{\mu Il}{2\pi} [\ln \rho]_a^b = \frac{\mu Il}{2\pi} \ln\left(\frac{b}{a}\right) \text{ Wb.} \tag{5.78}$$

Let L_{ext} be the external inductance of the cable.

$$\text{Then} \quad L_{ext} = \frac{\varphi_2}{I} = \frac{\mu l}{2\pi} \ln\left(\frac{b}{a}\right). \tag{5.79}$$

The total inductance is the sum of the internal inductance and external inductance.
Therefore the inductance of the cable is

$$L = L_{in} + L_{ext} = \frac{\mu l}{8\pi} + \frac{\mu l}{2\pi} \ln\left(\frac{b}{a}\right)$$

$$= \frac{\mu l}{2\pi}\left(\frac{1}{4} + \ln\left(\frac{b}{a}\right)\right) \text{ henrys.} \tag{5.80}$$

Note: At low frequencies, the internal inductance is significant, while at high frequencies, as the current in the conductor will concentrate near the surface, I_{encl} becomes very small in the region $0 < \rho < a$. Thus the internal inductance is negligible.

Inductance of a two-wire transmission line Consider a two-wire transmission line with a separation distance of d. Each wire has radius a, as shown in Fig. 5.15.

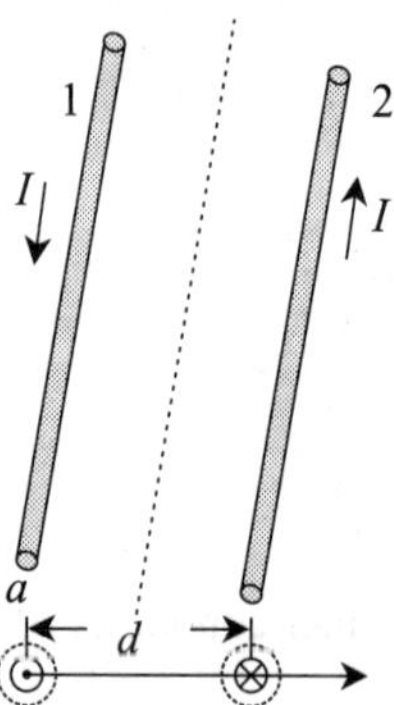

Fig. 5.15 *Two-wire transmission line*

Let I be the current flowing through wire 1 and $-I$ that flowing through wire 2.
The magnetic field intensity inside wire 1 is

$$\oint \vec{H} \cdot d\vec{l} = I_{encl}.$$

If ρ be the radius of a Gaussian surface in the wire $(\rho < a)$, the current enclosed is

$$I_{\text{encl}} = \frac{I\rho^2}{a^2}.$$

Then $H(2\pi\rho) = I_{\text{encl}}$.

$$\therefore \quad H = \frac{I_{\text{encl}}}{2\pi\rho} = \frac{I\rho}{2\pi a^2}$$

$$\text{or} \quad \vec{B}_1 = \frac{\mu I\rho}{2\pi a^2}\vec{a}_\phi.$$

Then flux linkage in wire 1 is

$$\varphi_1 = \int_s \vec{B}_1 \cdot d\vec{S}_1.$$

Inside conductor wire 1, in the region $0 < \rho < a$, the fraction of the area through which the flux leaves is

$$d\vec{S}_1 = \frac{\pi\rho^2}{\pi a^2} d\rho dz\vec{a}_\phi = \frac{\rho^2}{a^2} d\rho dz\vec{a}_\phi.$$

$$\therefore \quad \varphi_1 = \int_0^l \int_0^a \frac{\mu I\rho}{2\pi a^2}\vec{a}_\rho \cdot \frac{\rho^2}{a^2} d\rho dz\vec{a}_\phi$$

$$= \int_0^l \int_0^a \frac{\mu I}{2\pi a^4}\rho^3 d\rho dz = \frac{\mu Il}{2\pi a^4}\left[\frac{\rho^4}{4}\right]_0^a$$

$$= \frac{\mu Il}{8\pi} \text{ Wb.} \tag{5.81}$$

The inductance due to wire 1 is

$$L_1 = \frac{\varphi_1}{I} = \frac{\mu l}{8\pi} \text{ henrys} \tag{5.82}$$

Since wire 2 is kept at a distance 'd' from wire 1 in free space, the flux linkage between the wires due to the flux density in the region $a \leq \rho \leq d - a$ is

$$\varphi_2 = \int_s \vec{B}_2 \cdot d\vec{S}_2.$$

$$\text{Since} \quad \vec{B}_2 = \frac{\mu_0 I}{2\pi\rho}\vec{a}_\phi,$$

$$\varphi_2 = \int_0^l \int_a^{d-a} \frac{\mu_0 I}{2\pi\rho}\vec{a}_\phi \cdot d\rho dz\vec{a}_\phi$$

$$= \frac{\mu_0 Il}{2\pi} [\ln\rho]_a^{d-a} = \frac{\mu_0 Il}{2\pi} \ln\left(\frac{d-a}{a}\right). \tag{5.83}$$

The total flux linkage between the wires produced by wire 1 is

$$\varphi_1 + \varphi_2 = \frac{\mu Il}{8\pi} + \frac{\mu_0 Il}{2\pi} \ln\left(\frac{d-a}{a}\right). \tag{5.84}$$

By symmetry, the same amount of flux will be produced by wire 2.

Thus the total flux linkage produced between the two-wire line is

$$\varphi = 2(\varphi_1 + \varphi_2) = \frac{\mu Il}{4\pi} + \frac{\mu_0 Il}{\pi} \ln\left(\frac{d-a}{a}\right) \tag{5.85}$$

Therefore, the self inductance of a two-wire transmission line is

$$L = \frac{\varphi}{I} = \frac{\mu_0 l}{\pi}\left[\frac{1}{4} + \ln\left(\frac{d-a}{a}\right)\right] \text{ henrys.} \tag{5.86}$$

5.15 Mutual Inductance

When two coils are coupled together, i.e., they are kept in a common magnetic field as shown in Fig. 5.16, the flux produced by one coil links with that of the other and as a result, voltage is generated in the second coil. The inductance due to the flux linkages between the coils is called *mutual inductance*.

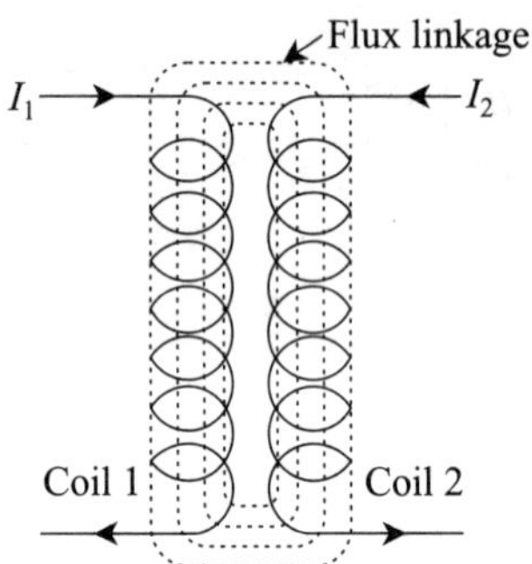

Fig. 5.16 *Mutual inductance between two coils*

This system has many practical benefits such as being the basis of the operation of transformers and generators. The interference and cross talk noise in communication systems is also a consequence of this phenomenon, and is an important issue to be tackled at the circuit design stage itself.

Let L_1 be the inductance of coil 1 of N_1 turns carrying current I_1 and producing a magnetic flux φ_{11}; and L_2, the inductance of coil 2 of N_2 turns carrying a current I_2 and producing a magnetic flux φ_{22}.

Let flux φ_{12} be passing through coil 1 due to current I_2 of coil 2 and flux φ_{21} be passing through coil 2 due to the current I_1 of coil 1.

The total flux linkage of coil 1 due to flux produced by current I_2 of coil 2 is $N_1\varphi_{12}$. Similarly, the total flux linkage of coil 2 due to the flux produced by current I_1 of coil 1 is $N_2\varphi_{21}$.

The mutual inductance (M) between two coils is defined as the ratio of *the total flux linkage in one coil to the current in the other coil.* It is expressed as

$$M_{12} = \frac{N_1 \varphi_{12}}{I_2} \quad \text{H}$$

$$\text{and} \quad M_{21} = \frac{N_2 \varphi_{21}}{I_1} \quad \text{H.}$$

If the surrounding medium is linear, then

$$M_{12} = M_{21}. \tag{5.87}$$

Mutual inductance can also defined as *the ratio of the voltage generated in one circuit and the rate of change of current in the other circuit. That is,*

$$M = \frac{V_2}{dI_1/dt},$$

where V_2 is the voltage induced in coil 2 due to current in coil 1. The units of mutual inductance are henrys (or H).

$$\therefore \quad V_2 = M_{21}\frac{dI_1}{dt}.$$

$$\text{Similarly,} \quad V_1 = M_{12}\frac{dI_2}{dt}.$$

Coefficient of coupling The coefficient of coupling between two coils is defined as *the ratio of the total flux linkage between two coils to the flux produced by one coil.*

Mathematically,

$$k = \frac{\text{Total flux linkage between coil 1 and coil 2}}{\text{Flux produced by coil 1 or coil 2}}$$

$$k = \frac{\varphi_{12}}{\varphi_1} = \frac{\varphi_{21}}{\varphi_2}. \tag{5.88}$$

$$\text{It can also be} \quad k = \frac{M}{\sqrt{L_1 L_2}}, \tag{5.89}$$

where M is the mutual inductance between the coils, $M = M_{12} = M_{21}$. The range of k is $0 \le k \le 1$.

$$k = \begin{cases} 0 & \text{No coupling} \\ 1 & \text{Perfect coupling} \end{cases} \tag{5.90}$$

A comparison between inductance and mutual inductance is summarised in Table 5.1.

Table 5.1 *A comparison between self and mutual inductances*

Self inductance	Mutual inductance
1. It is the flux linkage per unit current due to the current flowing through the same coil.	Mutual inductance of a coil is defined as the flux linkage per unit current due to the current in some other neighbouring coil.
2. It is associated with only one coil.	It is associated with more than one coil.
3. It is denoted by L.	It is denoted by M.
4. It is the actual value of an inductor element.	It is not the actual value of an inductance, but only a property associated with it.
5. It is given by $L = \frac{N\varphi}{I} H$, where, N = number of turns, φ = flux and I = current.	It is given by $M_{12} = \frac{\text{Flux linkage with coil}}{\text{current in coil 2}} = \frac{N_1\varphi_{21}}{I_2}$ or $M_{21} = \frac{N_1\varphi_{21}}{I_1}$.
6. The self inductance of a coil can be increased by increasing the number of turns (N), area of cross-section (A) and by selecting a material of high permeability (μ)	The mutual inductance of a coil can be increased by increasing the relative permeability (μ_r) , maintaining low spacing between the two coils.
7. It depends on the current I passing through its own coil.	Mutual inductance depends on current in its neighbouring coil.

5.16 Neumann's Formula for Mutual Inductance

If two coils are kept in a common magnetic field, the mutual inductance per turn between the coils is given by

$$M_{12} = \frac{\varphi_{12}}{I_2} = \frac{\mu_0}{4\pi} \oint_1 \oint_2 \frac{d\vec{l}_1 . d\vec{l}_2}{R_{12}}. \tag{5.91}$$

The above equation is known as Neumann's formula for mutual inductance.

Proof Consider two coils, 1 and 2, carrying currents I_1 and I_2 respectively, as shown in Fig. 5.17. Let φ_{12} denote the flux produced by current I_2 in circuit B linked with circuit A.

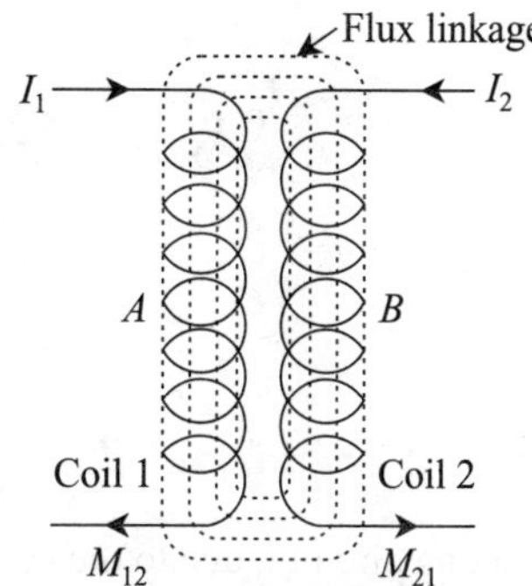

Fig. 5.17 *Mutual inductance between two coils*

If $\vec{B}_2$ denotes the magnetic flux density, then flux φ_{12} is given by

$$\varphi_{12} = \int_s \vec{B}_2 \cdot d\vec{S}. \tag{5.92}$$

But from vector magnetic potential, we have,

$$\vec{B}_2 = \nabla \times \vec{A}_2.$$

Substituting this in Eq. (5.92), we get,

$$\varphi_{12} = \int_s (\nabla \times \vec{A}_2) \cdot d\vec{S}. \tag{5.93}$$

From Stokes' theorem, we have,

$$\int_s (\nabla \times \vec{A}_2) \cdot d\vec{S} = \oint \vec{A}_2 \cdot d\vec{l}.$$

Applying this to Eq. (5.93), we get,

$$\varphi_{12} = \oint_s \vec{A}_2 \cdot d\vec{l}_1. \tag{5.94}$$

But the vector magnetic potential $\vec{A}_2$ due to current I_2 is given by

$$\vec{A}_2 = \frac{\mu_0 I_2}{4\pi} \oint_2 \frac{d\vec{l}_2}{r}, \tag{5.95}$$

where r is the radius of the coil. Substituting Eq. (5.95) in Eq. (5.94), we get

$$\begin{aligned} \varphi_{12} &= \oint_1 \left[\frac{\mu_0 I_2}{4\pi} \oint_2 \frac{d\vec{l}_2}{r} \right] d\vec{l}_1 \\ &= \frac{\mu_0 I_2}{4\pi} \oint_1 \oint_2 \frac{d\vec{l}_1 \cdot d\vec{l}_2}{r}. \end{aligned} \tag{5.96}$$

But the mutual inductance is given by,

$$M_{12} = \frac{\varphi_{12}}{I_2}. \tag{5.97}$$

From Eq. (5.96) and Eq. (5.97), we get,

$$M_{12} = \frac{\mu_0}{4\pi} \oint_1 \oint_2 \frac{d\vec{l}_1 \cdot d\vec{l}_2}{r}. \tag{5.98}$$

Similarly, we can find the flux φ_{12} linking circuit 2 due to the current in circuit 1, which is given by

$$\varphi_{21} = M_{21} I_1,$$

where, M_{21} = mutual inductance.

Example 5.10 The current in a coil is increased from zero to 15 amperes at a uniform rate in 6 seconds. It is found that this coil develops self induced emf of 150 V, whereas an emf of 25 V is produced in a neighbouring coil. Compute the self inductance of the first coil and the mutual inductance between the two coils.

Solution Given $i_1 = 0$ A; $i_2 = 15$ A; time elapsed = 6 seconds; self induced emf $V_s = 150$ V; mutual induced emf, $V_m = 25$ V.

We know that self induced emf is given by

$$V_s = L\frac{di}{dt}.$$

Now, $\dfrac{di}{dt} = \dfrac{15}{6} = 2.5.$

$$\therefore \quad V_s = L(2.5) \quad \text{or} \quad L = \frac{V_s}{2.5}.$$

$$L = \frac{150}{2.5} = 60 \text{ H}.$$

Mutual induced emf, $V_m = M\dfrac{di}{dt} = M(2.5).$

$$M = \frac{V_m}{2.5} = \frac{25}{2.5} = 10 \text{ H}.$$

5.17 Mutual Inductance between Two Solenoids

Consider two coaxial solenoids of radius a_1 and $a_2 (a_2 > a_1)$, carrying currents I_1 and I_2 along the z-axis respectively, as shown in Fig. 5.18. Let N_1 and N_2 be the number of turns and l_1 and l_2, the respective lengths of the solenoids.

We know that the magnetic fields inside the solenoids are

$$\vec{H}_1 = \frac{N_1 I_1}{l_1}\vec{a}_z \quad l_1 \gg a_1$$

and $$\vec{H}_2 = \frac{N_2 I_2}{l_2}\vec{a}_z \quad l_2 \gg a_2. \tag{5.99}$$

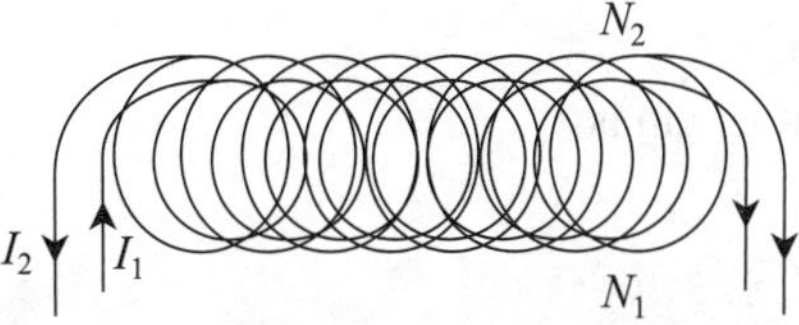

Fig. 5.18 *Inductance between two solenoids*

In the uniform field, the flux produced by the first solenoid is

$$\varphi_{11} = B_1 S_1$$

where $S_1 =$ cross-sectional area $= \pi a_1^2$.

$$\therefore \quad \varphi_{11} = \mu H_1 S_1 = \frac{\mu N_1 I_1 S_1}{l_1}. \tag{5.100}$$

Assume that there is no flux leakage. The flux flowing through the second solenoid (cross-sectional area $S_2 = \pi a_2^2$) due to first solenoid is

$$\varphi_{21} = \varphi_{11} \quad (\text{since } S_1 < S_2).$$

Therefore, the mutual inductance is given by

$$M_{21} = \frac{\varphi_{21} N_2}{I_1} = \frac{\mu N_1 N_2 S_1}{l_1} \text{ H}.$$

Mutual inductance per unit length is

$$M_{21} = \frac{\mu N_1 N_2 S_1}{l_1 l_2} \text{ H/m}.$$

If $n_1 = \dfrac{N_1}{l_1}$ turns /m and $n_2 = \dfrac{N_2}{l_2}$ turns /m, then the mutual inductance per unit length is

$$M_{21} = \mu n_1 n_2 S_1. \tag{5.101}$$

Similarly, the flux produced by the second solenoid is

$$\varphi_{22} = B_2 S_2,$$

$$\therefore \quad \varphi_{22} = \mu H_2 S_2 = \frac{\mu N_2 I_2 S_2}{l_2}. \tag{5.102}$$

The flux flowing through the first coil due to the second coil

$$\varphi_{12} = \varphi_{22} \times \frac{S_1}{S_2} = \frac{\mu N_2 I_2 S_1}{l_2} \quad (\text{since } S_2 > S_1). \tag{5.103}$$

The mutual inductance is given by

$$M_{12} = \frac{\varphi_{12} N_1}{I_2} = \frac{\mu N_1 N_2 S_1}{l_2} \text{ H.}$$

The mutual inductance per unit length is

$$M_{12} = \frac{\mu N_1 N_2 S_1}{l_1 l_2} = \mu n_1 n_2 S_1. \tag{5.104}$$

From Eqs (5.101) and (5.104), it is observed that $M_{12} = M_{21}$.

Example 5.11 An air core solenoid 60 cm long and 1 cm radius is wound with 2000 turns. Another coil of 200 turns of the same radius with length 30 cm is located near it. Find the mutual inductance between the coils.

Solution Given $l_1 = 60$ cm $= 0.6$ m; $N_1 = 2000$; $l_2 = 0.3$ m; $N_2 = 200$; $a = 0.01$ m;

$$n_1 = \frac{N_1}{l_1} = \frac{2000}{0.6} \quad \text{and} \quad n_2 = \frac{N_2}{l_2} = \frac{200}{0.3}.$$

Mutual inductance per length is

$$M = n_1 n_2 \mu S$$

$$= \frac{2000 \times 200 \times 4\pi \times 10^{-7} \times \pi(0.01)^2}{0.3 \times 0.6} = \frac{8}{9}\pi^2 \times 10^{-4} \text{ H.}$$

Example 5.12 Determine the mutual inductance between a conducting triangular loop and a very long straight wire.

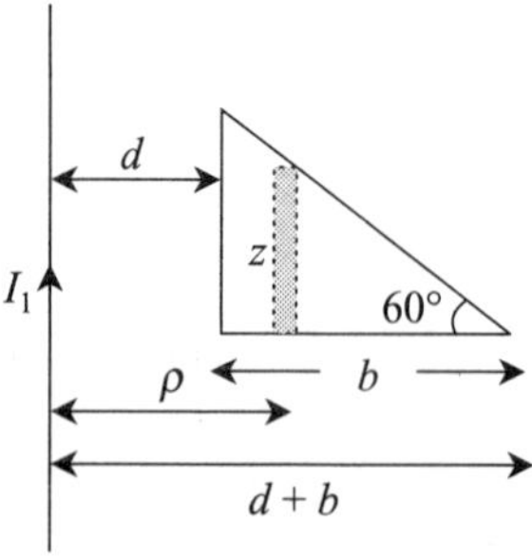

Fig. 5.19 *Mutual inductance between a conducting triangular loop and a very long straight wire*

Solution Consider a long straight conductor and triangular loop as shown in Fig. 5.19. Let I_1 be the current in the long straight line, d be the distance between the triangle and long conductor, and let us assume that the base angle of the triangle is 60° and the base is b.

We know that the magnetic flux density produced at a distance ρ by a long wire is

$$\vec{B}_1 = \frac{\mu_0 I_1}{2\pi\rho}\vec{a}_\varphi.$$

Since the triangular loop is kept at a distance $d(\rho > d)$, the flux linkage between the wire and the loop is

$$\varphi_{12} = \int B_1 \cdot d\vec{S}_2 = \int \frac{\mu_0 I_1 \vec{a}_\varphi}{2\pi\rho} \cdot d\vec{S}_2.$$

Let a differential strip of width $d\rho$ and height z be considered inside the triangular loop at a distance ρ from the straight conductor. From the given triangle's geometry, the differential strip is

$$d\vec{S}_2 = z\,d\rho\,\vec{a}_\phi \qquad \text{(shaded area in Fig. 5.19)}$$

and $z = [d+b-\rho]\tan 60^\circ = -\sqrt{3}(\rho - d - b).$

$$\therefore \quad \varphi_{12} = \frac{-\sqrt{3}\mu_0 I_1}{2\pi}\int_d^{d+b} \frac{(\rho - b - d)}{\rho} d\rho.$$

$$\text{Mutual inductance} \quad M_{12} = \frac{\varphi_{12}}{I_1} = \frac{\sqrt{3}\mu_0}{2\pi}\left[(d+b)\ln\left(1+\frac{d}{b}\right) - b\right] \text{ H.}$$

Example 5.13 Derive an expression for the mutual inductance between a straight long wire and a square loop wire in the same plane.

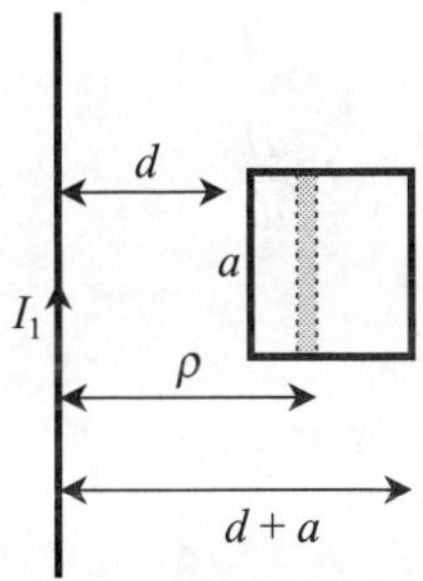

Fig. 5.20 *Mutual inductance between a straight long wire and a square loop wire*

Solution Consider a straight long wire and square loop of side a as shown in Fig. 5.20. Let I_1 be the current in the straight long wire, d be the distance between the long wire and the square loop.

We know that the magnetic flux density produced at a distance ρ by a long wire is

$$\vec{B}_1 = \frac{\mu_0 I_1}{2\pi\rho}\vec{a}_\phi.$$

Since the square loop is kept at a distance d, the flux linkage between the loop and the wire is

$$\varphi_{12} = \int \vec{B}_1 \cdot d\vec{S}_2 = \int \frac{\mu_0 I_1}{2\pi\rho}\vec{a}_\varphi \cdot d\vec{S}_2.$$

Let a differential strip of width $d\rho$ be considered inside the square loop at a distance ρ from the straight conductor. From Fig. 5.20, the area of the differential strip is $d\vec{S}_1 = ad\rho\vec{a}_\varphi$.

$$\therefore \quad \varphi_{12} = \int_d^{d+a} \frac{\mu_0 I_1 a d\rho}{2\pi\rho} = \frac{\mu_0 I_1 a}{2\pi}\Big[\ln\rho\Big]_d^{d+a}.$$

Mutual inductance $M_{12} = \dfrac{\varphi_{12}}{I_1} = \dfrac{\mu_0 a}{2\pi}\ln\left(\dfrac{d+a}{d}\right).$

5.18 Magnetic Energy

The energy stored in a magnetic field produced by an inductor is given by

$$W_m = \frac{1}{2}LI^2 \text{ joules (or J)} \tag{5.105}$$

where L is the inductance and I is the applied current.

Proof If a current I is applied on a coil having N turns, it produces a magnetic field with flux φ. The voltage induced across the coil is

$$V = \frac{-Nd\varphi}{dt} = L\frac{dI}{dt}.$$

Since $N\varphi = LI$

We know that power is $P = VI = NI\dfrac{d\varphi}{dt} = LI\dfrac{dI}{dt}$. (5.106)

Therefore the energy stored is

$$W_m = \int P\,dt = \int LI\frac{dI}{dt}dt = \int LI\,dI = \frac{1}{2}LI^2.$$

$$\therefore \quad W_m = \frac{1}{2}LI^2 \text{ joules} \qquad \text{(Proved).}$$

Note: The inductance of an inductor can also be defined as

$$L = \frac{2W_m}{I^2} \text{ H.} \tag{5.107}$$

5.19 Energy Density Stored in the Magnetic Field

Energy density is defined as *energy per unit volume*. The energy density stored in the magnetic field of an inductor is given by

$$w_m = \frac{1}{2}\mu\vec{H}^2 = \frac{1}{2}\vec{B}\cdot\vec{H} \text{ joules/m}^3 \text{ or } (J/m^3) \tag{5.108}$$

Proof The energy stored in an inductor is given by

$$W_m = \frac{1}{2}LI^2.$$

Consider the case of a solenoid whose self inductance is

$$L = \frac{\mu N^2 S}{l}.$$

Substituting the value in the equation for stored energy,

$$W_m = \frac{1}{2}\mu\frac{N^2 S}{l}I^2 = \frac{1}{2}\mu\left(\frac{NI}{l}\right)^2 lS. \tag{5.109}$$

We know that volume $v = lS$, and $H = \dfrac{NI}{l}$.

$$\therefore \quad W_m = \frac{1}{2}\mu H^2 v. \tag{5.110}$$

The energy density is

$$w_m = \frac{W_m}{v} = \frac{1}{2}\mu H^2 = \frac{1}{2}\vec{B}\cdot\vec{H} \text{ J/m}^3.$$

If dv is the differential volume, the energy stored is

$$W_m = \int_v w_m dv = \frac{1}{2}\int_v \vec{B}.\vec{H}dv \text{ joules.} \tag{5.113}$$

Example 5.14 A solenoid with 300 turns is 300 mm long and has a diameter of 30 mm. If the current flowing through it is 500 mA, find (i) inductance and (ii) energy stored. Assume $\mu_r = 1$.

Solution Given $N = 300$; $l = 0.3$ m; $d = 0.3$ m; $I = 500$ mA.

Then $S = \dfrac{\pi}{4}(0.03)^2$.

(i) Inductance $L = \dfrac{\mu N^2 S}{l} = \dfrac{4\pi \times 10^{-7}(300)^2(\pi)(0.03)^2}{(0.3)(4)} = 0.266$ mH.

(ii) Energy stored $W_m = \dfrac{1}{2}LI^2 = \dfrac{1}{2}(0.266 \times 10^{-3})(500)^2 \times 10^{-6} = 33\ \mu$J.

5.20 Energy Stored due to Mutual Inductance

The self inductances of coils 1 and 2 are given by

$$L_1 - \frac{N_1\varphi_1}{I_1} \quad \text{and} \quad L_2 = \frac{N_2\varphi_2}{I_2} \tag{5.114}$$

and the mutual inductances are

$$M_{12} = \frac{\varphi_{12} N_1}{I_2} \quad \text{and} \quad M_{21} = \frac{\varphi_{21} N_2}{I_1}. \tag{5.115}$$

The total flux in coil 1 due to the interaction between the coils is

$$\varphi_1 = \varphi_{11} \pm \varphi_{12}$$

and the total flux in coil 2 is

$$\varphi_2 = \varphi_{22} \pm \varphi_{21}.$$

Since $\varphi_{12} = \varphi_{21}$ (assuming that there is no flux leakage), the total flux between the two coils is

$$\varphi = \varphi_1 + \varphi_2 = \varphi_{11} + \varphi_{22} \pm \varphi_{12} \pm \varphi_{21} = \varphi_{11} + \varphi_{22} \pm 2\varphi_{12}. \tag{5.116}$$

The positive sign indicates that both currents strengthen the magnetic field and negative sign indicates that the current flow is producing magnetic fields that oppose each other.

The total inductance is

$$L = L_1 + L_2 \pm 2M_{12}. \tag{5.117}$$

The total energy stored is equal to the sum of the energies in inductances and energy stored due to mutual inductance.

Energy stored in L_1 is $W_1 = \frac{1}{2} L_1 I_1^2$.

Energy stored in L_2 is $W_2 = \frac{1}{2} L_2 I_2^2$.

Energy stored in mutual inductance is

$$W_{12} = \int V_{12} I_2 dt = \int M_{12} \frac{dI_1}{dt} I_2 dt = M_{12} I_2 \int dI_1 = M_{12} I_1 I_2.$$

The total energy in the magnetic field is

$$W_m = W_1 + W_2 \pm W_{12} = \frac{1}{2} L_1 I_1^2 + \frac{1}{2} L_2 I_2^2 \pm M_{12} I_1 I_2 \quad \text{joules}. \tag{5.118}$$

5.21 Magnetic Circuits

When a magnetic flux flows through a medium, reluctance is the property of the medium which opposes the flow of the magnetic flux. The medium through which the magnetic flux is flowing in a closed loop is called a magnetic circuit.

Reluctance is defined as *the ratio of magnetomotive force to the total flux.*

Mathematically, $\mathbb{R} = \frac{\text{mmf}}{\varphi} = \frac{NI}{\varphi}$ AT/Wb.

If a current I is flowing through a coil of length l and of N turns, then the magnetic field intensity

$$\vec{H} = \frac{NI}{l}.$$

But $\vec{B} = \mu\vec{H}$ or $\text{H} = \frac{\vec{B}}{\mu}$ or $\vec{H} = \frac{\varphi}{\mu S}$ (since $\varphi = BS$)

where S is the cross-sectional area of the circuit.

$$\therefore \quad \frac{\varphi}{\mu S} = \frac{NI}{l}$$

or reluctance $\mathbb{R} = \frac{NI}{\varphi} = \frac{l}{\mu S}.$

Reluctance is analogous to resistance in an electric circuit. Since reluctance is inversely proportional to permeability, its value is very high for air and non-magnetic materials. For ferromagnetic materials such as steel or iron, the reluctance is small.

Relationship between reluctance and inductance When a current I is passing through a coil of N turns, the flux produced is φ Wb. Then the self inductance of the magnetic path is

$$L = \frac{N\varphi}{I} \tag{5.119}$$

and the reluctance of the magnetic path is

$$\mathbb{R} = \frac{NI}{\varphi} \quad \text{or} \quad \varphi = \frac{NI}{\mathbb{R}}. \tag{5.120}$$

$$\therefore \quad L = \frac{N}{I} \times \frac{NI}{\mathbb{R}} = \frac{N^2}{\mathbb{R}}\ \text{H}.$$

Note: If there are two coils of turns N_1 and N_2, then the mutual inductance between the two coils is

$$M = \frac{N_1 N_2}{\mathbb{R}}\ \text{H}. \tag{5.121}$$

where $\mathbb{R}$ is the reluctance of the magnetic path between the two coils.

5.22 Magnetic Materials

Materials can be classified in terms of their magnetic behaviour as magnetic or non-magnetic materials. Magnetic materials have non-zero susceptibility $(\chi_m \neq 0)$ and their relative permeability μ_r is not equal to 1 $(\mu_r \neq 1)$. For non-magnetic materials, $\chi_m = 0$ and $\mu_r = 1$. Magnetic materials classified according to their material properties—magnetic susceptibility or relative permeability—are either:

diamagnetic materials,
paramagnetic materials or
ferromagnetic materials.

Diamagnetic materials Diamagnetic materials acquire magnetisation only in the presence of an external field, which acts opposite to the applied magnetic field intensity. In these materials, the magnetic field due to the orbiting of electrons and the magnetic field due to the spinning of electrons in the atoms are equal and cancel with each other.

Example : Copper, lead, silicon, diamond, and bismuth are diamagnetic at room temperature.

Paramagnetic materials Magnetic materials that exhibit a positive magnetisation with susceptibility slightly greater than zero in the presence of an external field are called paramagnetic materials. In the presence of the external field, these materials are magnetised parallel to the magnetic field intensity. Here, magnetic fields due to orbiting and spinning electrons do not cancel each other. So they have permanent magnetic moment greater than zero.

Example: Air, tungsten, platinum, potassium, etc.

Ferromagnetic materials The magnetisation in these materials depends on the magnetic properties of the material. It is not dependent on the applied field. Ferromagnetic materials produce strong magnetic fields. They can be used as permanent magnets or shielding to protect electrical devices from external flux.

Example: Iron, nickel, cobalt, etc.

A comparison of the above categories of materials is shown in Table 5.2.

Table 5.2 *Comparison between materials*

Diamagnetic	Paramagnetic	Ferromagnetic
1. Permanent magnetic moment of each atom is zero	Permanent magnetic moment is not zero	Very large permanent magnetic moment
2. Flux density $\vec{B} = 0$	$\vec{B} = \mu\vec{H}$	$\vec{B} \neq \mu\vec{H}$, since μ depends on B
3. Linear and non-magnetic characteristics	Linear and non-magnetic characteristics	Non-linear and magnetic characteristics
4. Weak magnetic properties	Slightly stronger magnetic properties	Very strong magnetic properties
5. Diamagnetism is independent of temperature. $\mu_r \leq 1$; $\chi_m < 0$ (very small $\chi_m \approx -10^{-5}$)	Para-magnetism depends on temperature $\mu_r \geq 1$; $\chi_m > 0$ $(10^{-5} \leq \chi_m \leq 10^{-3})$	Ferromagnetism depends on temperature. If temperature rises above curie temperature, magnets lose their magnetic property. $\mu_r >> 1$; $\chi_m >> 0$ (very large χ_m)

The differences between static electric field and static magnetic fields are shown in Table 5.3.

Table 5.3 *Differences between a static electric field and a static magnetic fields*

Static electric field	Static magnetic field
1. Static electric field is produced by a stationary charge.	Static magnetic field is produced by a permanent magnet or current carrying conductor.
2. Electric field exerts a force on stationary as well as on moving charges.	Magnetic field exerts force only on moving charges.
3. The force acting on a charge is proportional to the electric field intensity.	The force acting on a charge, in a magnetic field, is proportional to the cross product of velocity of the charge and magnitude flux density.
4. The force acting on a unit charge, in an electric field is called electric field intensity. It is denoted by $\vec{E}$ and expressed in volts per metre or newtons per coulombs (N/C).	The force experienced by a unit magnetic pole in a magnetic field is called the magnetic field intensity. It is denoted by $\vec{H}$ and expressed in ampere per metre (A/m) or newtons per weber (N/Wb)
5. The imaginary lines that visualize the electric field and show the direction of force on charges in the field are called electric field lines or electric flux.	The magnetic lines of force form concentric circles around the current carrying conductor, the direction of which can be determined by the right-hand screw rule. Magnetic lines are called magnetic flux.
6. Electric field lines always leave a positive charge and enter a negative charge. They do not form closed paths.	Magnetic lines of force always form closed paths.
7. Electric field lines do not intersect each other.	Magnetic lines do not intersect each other. They are always parallel to each other.
8. The force due to an electric field can perform work on the charged particle and change its kinetic energy	The force due to a magnetic field cannot perform work on the charged particle and the kinetic energy is constant.

5.23 Characteristics of Magnetic Materials

The orbiting and spinning of electrons in the atoms of a material produce internal magnetic field. The electric dipole moment as shown in Fig. 5.21 is given by

$$m = Q_m\vec{d}. \tag{5.122}$$

The dipole moments are randomly oriented in the material and hence the net magnetic dipole moment is zero. When an external magnetic field is applied, the dipole moments of the electrons align themselves in the direction of the applied field so that the net magnetic dipole moment is not zero.

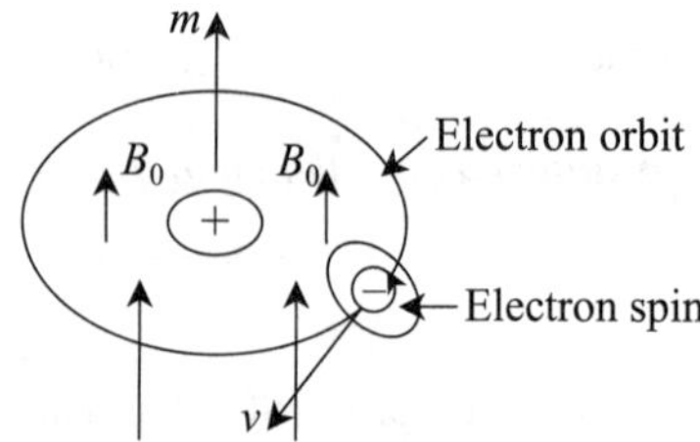

Fig. 5.21 *Dipole moment due to electron spin and electron orbiting around the nucleus*

The magnetic dipole moment is defined as

$$\vec{m} = I_b d\vec{S} \quad \text{A-m}^2 \tag{5.123}$$

where I_b is the bound current enclosed by differential area $d\vec{S}$.

Since magnetic dipole moment is equal to the electric dipole moment,

$$\vec{m} = I_b d\vec{S} = Q_m \vec{d}. \tag{5.124}$$

If N number of magnetic dipoles per unit volume (Δv) exist, then the total magnetic dipole moment is

$$m_t = \sum_{i=1}^{N} m_i.$$

Magnetisation is defined as the magnetic dipole per unit volume

$$\vec{M} = \lim_{\Delta v \to 0} \frac{1}{\Delta v} \sum_{i=1}^{N} \vec{m}_i \quad \text{A/m}. \tag{5.125}$$

For linear and isotropic materials, the magnetisation $\vec{M}$ depends linearly on magnetic field intensity $\vec{H}$.

Thus, magnetisation $\vec{M} = \chi_m \vec{H}$ (5.126)

where χ_m is called *magnetic susceptibility* of the medium.

For magnetic materials, the total flux density is

$$\vec{B} = \mu_0 \vec{H} + \vec{M}\mu_0 = \mu_0(\vec{H} + \vec{M}) = \mu_0 \vec{H} + \mu_0 \chi_m \vec{H}$$

$$= \mu_0(\chi_m + 1)\vec{H} = \mu_0 \mu_r \vec{H} \quad \text{where,} \quad \mu_r = \chi_m + 1.$$

$$\therefore \quad B = \mu \vec{H}. \tag{5.127}$$

Additional Solved Problems

Example 5.15 The magnetic field intensity is given by $\vec{H} = \frac{I}{2\pi\rho}(\rho^3 - a^3)\,\vec{a}_\phi$. Find the magnetic potential $\vec{A}$.

Solution Given $\vec{H} = \frac{I}{2\pi\rho}(\rho^3 - a^3)\,\vec{a}_\phi$; $\vec{B} = \mu_0\vec{H} = \frac{\mu_0 I}{2\pi\rho}(\rho^3 - a^3)\,\vec{a}_\phi$.

In cylindrical coordinates,
flux density $\vec{B} = \nabla \times \vec{A}$.

$$\vec{B} = \frac{\mu_0 I}{2\pi\rho}(\rho^3 - a^3)\vec{a}_\phi = \frac{1}{\rho}\begin{vmatrix} \vec{a}_\rho & \rho\vec{a}_\varphi & \vec{a}_z \\ \frac{\partial}{\partial\rho} & \frac{\partial}{\partial\phi} & \frac{\partial}{\partial z} \\ A_\rho & \rho A_\phi & A_z \end{vmatrix}.$$

Comparing the ϕ components,

$$\frac{\mu_0 I}{2\pi\rho}(\rho^3 - a^3)\vec{a}_\phi = -\left(\frac{\partial A_z}{\partial\rho} - \frac{\partial A_\rho}{\partial z}\right)\vec{a}_\phi.$$

Since the field is symmetrical about the z-axis, $\frac{\partial A_\rho}{\partial z} = 0$.

$$\therefore \quad \frac{-dA_z}{d\rho} = \frac{\mu_0 I}{2\pi\rho}(\rho^3 - a^3).$$

$$A_z = \frac{-\mu_0 I}{2\pi}\int\left(\frac{\rho^3 - a^3}{\rho}\right)d\rho$$

$$= \frac{-\mu_0 I}{2\pi}\int\left(\rho^2 - \frac{a^3}{\rho}\right)d\rho$$

$$= \frac{-\mu_0 I}{2\pi}\left[\frac{\rho^3}{3} - a^3\ln\rho\right] \text{ Wb/m.}$$

Example 5.16 A wire is bent into the form of a square coil. Each side of the coil has a length of 20 cm. The coil carries a current of 10 A. The medium is air. Find the magnetic potential at the centre of the coil.

Solution Given $L = 20$ cm $= 0.2$ m; current $I = 10$ A.

The magnetic potential at the centre of the square coil (see Eq. (5.9)) is

$$V_m = \frac{\mu_0 8\sqrt{2}I}{\pi}$$

$$= \frac{4\pi \times 10^{-7} \times 8\sqrt{2} \times 10}{\pi} = 45.25\ \mu \text{ Wb/m.}$$

Example 5.17 If the magnetic field is $\vec{H} = (0.01/\mu_0)\vec{a}_x$ A/m, what is the force on a charge of 1 pC moving with a velocity of $10^6\,\vec{a}_y$ m/s?

Solution Given $\vec{H} = \left(\dfrac{0.01}{\mu_0}\right)\vec{a}_x$ A/m; $Q = 1$ pC $= 10^{-12}$ C; $\vec{u} = 10^6\,\vec{a}_y$ m/s.

We know that the force experienced by a charge Q moving with a velocity $\vec{u}$ in a magnetic field $\vec{B}$ is,

$$\vec{F}_m = Q\vec{u} \times \vec{B}.$$

We also know that,

$$\vec{B} = \mu_0 \vec{H} = \mu_0 \left(\frac{0.01}{\mu_0}\right)\vec{a}_x \quad \text{or} \quad \vec{B} = 0.01\vec{a}_x.$$

Also, $\vec{u} \times \vec{B} = 10^6\,\vec{a}_y \times 0.01\,\vec{a}_x = -10^4\,\vec{a}_z$.

Now, $\vec{F}_m = 10^{-12}(\vec{u} \times \vec{B}) = 10^{-12}(-10^4\,\vec{a}_z) = -10^8\,\vec{a}_z$ N.

Example 5.18 A wire 1 metre long carries a current of 10 A and makes an angle of 30° with uniform magnetic field $B = 1.5$ Wb/m^2. Calculate the magnitude of the force on the wire.

Solution Given $I = 10$ A; $B = 1.5$ Wb/m^2; $l = 1$ m; $\theta = 30°$.
Magnitude of force $F = I(l \times B) = IlB\sin\theta$

$$= 10 \times 1.5 \times \sin 30° = 7.5 \text{ N}.$$

Example 5.19 If a point charge 4 C moves with velocity of $5\vec{a}_x + 6\vec{a}_y - 7\vec{a}_z$ m/s, find the force exerted. The flux density is $5\vec{a}_x + 7\vec{a}_y + 9\vec{a}_z$ Wb/m^2.

Solution Given $Q = 4$C; $\vec{u} = 5a_x + 6a_y - 7a_z$; $\vec{B} = 5a_x + 7a_y - 9a_z$.
The force acting on a point charge in the presence of a magnetic field is given by,

$$\vec{F} = Q(\vec{u} \times \vec{B}).$$

$$\vec{u} \times \vec{B} = \begin{vmatrix} \vec{a}_x & \vec{a}_y & \vec{a}_z \\ 5 & 6 & -7 \\ 5 & 7 & 9 \end{vmatrix}$$

$$= \vec{a}_x[(9 \times 6) - (-7 \times 7)] - \vec{a}_y[(5 \times 9) - (-7 \times 5)] + \vec{a}_z[(5 \times 7) - (5 \times 6)]$$

$$= \vec{a}_x(54 + 49) - \vec{a}_y(45 + 35) + \vec{a}_z(35 - 30).$$

$$\vec{u} \times \vec{B} = 103\vec{a}_x - 80\vec{a}_y + 5\vec{a}_z.$$

$$\vec{F} = Q(\vec{u} \times \vec{B}) = 4(103\vec{a}_x - 80\vec{a}_y + 5\vec{a}_z) = 412\vec{a}_x - 320\vec{a}_y + 20\vec{a}_z.$$

Magnitude of force is $|F| = \sqrt{(412)^2 + (-320)^2 + (20)^2} = 522.05$ newtons.

Example 5.20 Calculate the force on a straight conductor of length 30 cm carrying a current of 5 A along the z-axis. The magnetic field is $\vec{B} = 3.5 \times 10^{-3}(\vec{a}_x - \vec{a}_y)$ tesla, where $\vec{a}_x$ and $\vec{a}_y$ are unit vectors.

Solution Given $l = 0.3$ m; $I = 5\vec{a}_z$; $\vec{B} = 3.5 \times 10^{-3}\vec{a}_x - 3.5 \times 10^{-3}\vec{a}_y$.
The force on the conductor is given by

$$F = \vec{I}l \times \vec{B}.$$

$$\vec{I} \times \vec{B} = \begin{vmatrix} \vec{a}_x & \vec{a}_y & \vec{a}_z \\ 0 & 0 & 5 \\ 3.5 \times 10^{-3} & -3.5 \times 10^{-3} & 0 \end{vmatrix}$$

$$= a_x(3.5 \times 10^{-3} \times 5) + a_y(3.5 \times 10^{-3} \times 5) + a_z(0) = +0.0175\vec{a}_x + 0.0175\vec{a}_y$$

$$\vec{F} = 0.3(0.0175\vec{a}_x + 0.0175\vec{a}_y) = 5.25 \times 10^{-3}\vec{a}_x + 5.25 \times 10^{-3}\vec{a}_y.$$

$$\text{Magnitude of } F = \sqrt{(525 \times 10^{-3})^2 + (525 \times 10^{-3})^2} = 7.42 \times 10^{-3} \text{ N} = 7.42 \text{ mN}.$$

Example 5.21 A solenoid 25 cm long and with a 1 cm mean diameter of the coil-turns has a uniformly distributed winding of 2000 turns. If the solenoid is placed in a uniform field of 2 Wb/m^2 flux density and a current of 5 A is passed through the solenoid winding, what is the maximum (i) force on the solenoid and (ii) torque on the solenoid?

Solution Given $l = 25$ cm $= 0.25$ m; $d = 1$ cm.

$$S = \frac{\pi d^2}{4} = 0.25\pi \times 10^{-4} \text{ m}^2.$$

$$N = 2000; \; B = 2 \text{ Wb/m}^2; \; I = 5 \text{ A}.$$

(i) $F = I\vec{l} \times \vec{B}$.

Maximum force $F = IlB = 5 \times 0.25 \times 2 = 2.5$ newtons per turn.

For N turns, $F = 2.5 \times 2000 = 5000$ N.

(ii) Torque $T = BIS = 2 \times 5 \times 0.25\pi \times 10^{-4} = 2.5\pi \times 10^{-4}$ N-m.

For N turns, $T = 2.5\pi \times 10^{-4} \times 2000 = 0.5\pi$ N-m.

Example 5.22 A conductor of length 0.25 m lies along the y-axis and carries a current of 25 A in the $\vec{a}_y$ direction. Find the power needed for parallel translation of the conductor to $x = 5$ m at constant speed in 3.0 seconds if the uniform field is $\vec{B} = 0.06\,\vec{a}_z$ tesla.

Solution Given $l = 0.25$ m; $I = 25 \text{ A} = 25\vec{a}_y$; $\vec{B} = 0.06\vec{a}_z$ tesla.

We know that $\vec{F} = (\vec{I} \times \vec{B})l = 25(0.25)0.06\,\vec{a}_y \times \vec{a}_z = (0.015)25\,\vec{a}_x = 0.375\vec{a}_x.$

Now, the work done required for parallel translation is

$$W = \int \vec{F} \cdot d\vec{l} = \int_0^5 \vec{F} \cdot dx\vec{a}_x$$

$$= \int_0^5 0.375dx\vec{a}_x \cdot \vec{a}_x = 0.375 \int_0^5 dx = 0.375x\Big|_0^5 = 1.875 \text{ joules.}$$

Power needed $P = W/t = 1.875/3 = 0.625$ watts

Example 5.23 A distribution line consists of two straight parallel conductors supported on the cross arms of wooden poles spaced 100 m apart. The normal spacing between the two conductors is 20 cm. Suppose a current of 10,000 A flows down one conductor and back the other during a fault, determine the force on each 100 m section of the conductor.

Solution Given two straight parallel conductors with separation $d = 20$ cm $= 0.2$ m, and $I_1 = I_2 = 10,000 = 10^4$ A.

We know that the force per metre is $\vec{F} = \dfrac{\mu_0 I_1 I_2}{2\pi d} = \dfrac{4\pi \times 10^{-7}(10^4)^2}{2\pi \times 0.2} = 10 \times 10^{-7} \times 10^8 = 100$ N/m.

For 100 m section, $F = 100 \times 100 = 10000$ N.

Example 5.24 Two long parallel conductors carry 80 A. If the conductors are separated by 30 mm, find the force per metre of each conductor. The current flow through them is in the opposite directions.

Solution Given $I_1 = I_2 = 80$ A; $d = 30$ mm $= 30 \times 10^{-4}$ m.

Force per unit length $F = \dfrac{\mu_0 I_1 I_2}{2\pi d} = \dfrac{4\pi \times 10^{-7} \times 80 \times 80}{2\pi 30 \times 10^{-4}} = \dfrac{20}{3} \times 64 \times 10^{-3} = 0.426$ N/m.

Example 5.25 A triangular loop of wire in free space points A (1, 0, 1), B (3, 0, 1) and C (3, 0, 4). The wire carries a current of 6 mA flowing in the $\vec{a}_z$ direction from B to C. A filamentary current of 15 A flows along the entire z-axis in the $\vec{a}_z$ direction. Find the force on the loop.

Solution Given points A (1, 0, 1), B (3, 0, 1) and C (3, 0, 4) forming a triangle loop with sides AB, BC and CA. The total force on the loop is equal to the sum of the forces due to currents along sides AB, BC and CA.

$$\vec{F} = \vec{F}_{AB} + \vec{F}_{BC} + \vec{F}_{CA}.$$

Given the current flowing through the loop is $I = 6$ mA, and the current flowing through the z-axis is $I_1 = 15$ A.

We know that $\vec{F} = Id\vec{l} \times \vec{B}$ and $\vec{B} = \mu_0 H = \mu_0 \dfrac{I_1}{2\pi\rho} \vec{a}_\phi$.

In rectangular coordinates, along the x-axis, $\rho = x$.

$$\vec{B} = \mu_0 \frac{I_1}{2\pi x}(-\vec{a}_y) = \frac{-15\mu_0}{2\pi x}\,\vec{a}_y = \frac{-30\times 10^{-6}}{x}\vec{a}_y.$$

Force along the side AB:

$$d\vec{F}_{AB} = Id\vec{l}_{AB} \times \vec{B}.$$

Now, $\quad d\vec{l}_{AB} = \vec{a}_x dx.$

$$\therefore \quad d\vec{F}_{AB} = Idx\,\vec{a}_x \times \left(\frac{-3\times 10^{-6}}{x}\vec{a}_y\right)$$

$$= \frac{(6\times 10^{-3})(-3\times 10^{-6})}{x}(\vec{a}_x \times \vec{a}_y) = \frac{18dx}{x}\vec{a}_z \times 10^{-9}.$$

$$\vec{F}_{AB} = 18\int_1^3 \frac{dx}{x} \times 10^{-9}\vec{a}_z = 18[\ln(3) - \ln(1)] \times 10^{-9}\vec{a}_z = 19.77\vec{a}_z \ \text{nN}.$$

Force along the side BC:

at BC, $\ x = 3.$

$$d\vec{F}_{BC} = Id\vec{l}_{BC} \times \vec{B}.$$

$$d\vec{l}_{BC} = dz\vec{a}_z.$$

$$d\vec{F}_{BC} = 6\times 10^{-3} dz\vec{a}_z \times \left(\frac{-3\times 10^{-6}}{3}\right)\vec{a}_y = -6\times 10^{-9} dz\vec{a}_x.$$

$$\vec{F}_{BC} = -6\times 10^{-9}\int_1^4 dz\vec{a}_x = -18\times 10^{-9}\vec{a}_x.$$

Force along the side CA:

$$d\vec{F}_{CA} = Id\vec{l}_{CA} \times \vec{B}$$

$$d\vec{l}_{CA} = -dx\vec{a}_x - dz\vec{a}_z$$

$$d\vec{F}_{CA} = 6\times 10^{-3}(-dx\vec{a}_x - dz\vec{a}_z) \times \left(\frac{-3\times 10^{-6}}{x}\vec{a}_y\right)$$

$$= \frac{18\times 10^{-9}}{x}(dx\vec{a}_z + dz\vec{a}_x).$$

$$\vec{F}_{CA} = 18\times 10^{-9}\int_3^1 \frac{dx}{x}\vec{a}_z + \int_4^1 \frac{18\times 10^{-9}}{3} dz\,\vec{a}_x$$

$$= 18 \times 10^{-9} \Big[\ln x\Big]_3^1 \vec{a}_z - 6 \times 10^{-9} z \Big|_4^1 \vec{a}_x$$

$$= -19.77 \times 10^{-9} \vec{a}_z - 18 \times 10^{-9} \vec{a}_x.$$

The total force is $\vec{F} = \vec{F}_{AB} + \vec{F}_{BC} + \vec{F}_{CA}$.

$$\vec{F} = +19.77 \times 10^{-9} \vec{a}_z - 18 \times 10^{-9} \vec{a}_x - 19.77 \times 10^{-9} \vec{a}_z - 18 \times 10^{-9} \vec{a}_z = -36\vec{a}_x \text{ nN}.$$

The total force acting on the triangular loop is in the direction opposite to the positive x-axis.

Example 5.26 A current strip 2 cm wide carries a current of 15 A in the $\vec{a}_x$ direction. Find the force on a strip of unit length, if the uniform field is $\vec{B} = 0.20\vec{a}_z$ tesla.

Solution Given $d = 2$ cm $= 0.02$ m; $I = 15$ A in the z direction;

$$\vec{K} = \frac{I}{b} = \frac{15}{0.02} \vec{a}_x; \quad \vec{B} = 0.2\vec{a}_z \text{ tesla}.$$

Now, $dS = dxdy$.

We know that $d\vec{F} = \vec{K} dS \times \vec{B}$

$$= \frac{15}{0.02} dxdy \vec{a}_x \times (0.2\vec{a}_z) = (150) dxdy \vec{a}_y.$$

Integrating the above equation, we get,

$$\vec{F} = \int d\vec{F} = \int_{y=0}^{0.02} \int_0^l 150 dxdy \vec{a}_y = \int_0^{0.02} (150l) dy \cdot \vec{a}_y.$$

$$\vec{F}/l = 150 \times 0.02 \vec{a}_y = 3\vec{a}_y \text{ N/m}.$$

The force 3 newton per unit length acts on the current element in the direction of the y-axis.

Example 5.27 What is the maximum torque on a square loop of 1000 turns in a field of uniform flux density B tesla? The loop has a 10 cm side and carries a current of 3 A. What is the magnetic moment of the loop?

Solution Given $a = 10$ cm $= 0.1$ m; $I = 3$ A; $N = 1000$; $B = 1$ telsa.
The maximum torque on a single loop, $T_{\max} = B\,I\,S$.

$$S = a^2 = 0.1^2 = 0.01.$$

$$T_{\max} = 1 \times 3 \times 0.01 = 0.03 \text{ N-m/turns}.$$

For N turns,

$$T_{\max} = 0.03 \times 1000 = 30 \text{ N-m}.$$

Magnetic moment, $m = IS = 3 \times 0.01 = 0.03$ A-m^2

Example 5.28 Two long parallel wires 2 metres apart carry currents of 50 A and 100 A respectively in the same direction. Determine the magnitude and direction of the force between them per unit length.

Solution Given $I_1 = 50$ A; $I_2 = 100$ A; $d = 2$ m.
We know that the force between two parallel conductors is

$$\vec{F} = \frac{\mu_0 I_1 I_2}{2\pi d}\,\vec{a}_n$$

$$\vec{F} = \frac{4\pi \times 10^{-7}(50)(100)}{(2\pi)(2)} = 5 \times 10^{-4}\,\vec{a}_n \text{ N/m}.$$

Since the conductors are in the same direction, the force is attractive. The direction of the force is normal to the plane of the conductors.

Example 5.29 A conductor of length 4 m, with current 10 A, is lying along the y-axis between $y = \pm 2$. If the field is $\vec{B} = 0.05\vec{a}_x$ T, find the work done in moving the conductor at constant speed parallel to itself at $x = y = 2$ m.

Solution Given $l = 4$ m; $I = 10\vec{a}_y$; $\vec{B} = 0.05\vec{a}_x$ tesla.
Location of the conductor, $y = \pm 2$.

The work done by the force is given by $W = \int_{\text{initial}}^{\text{final}} \vec{F}\cdot d\vec{z}.$

We know that $\vec{F} = Id\vec{l} \times \vec{B}.$

Since $Id\vec{l} = 4(10\vec{a}_y) = 40\vec{a}_y,$

$$\vec{F} = 40\vec{a}_y \times 0.05\vec{a}_x = -2\vec{a}_z.$$

Work done $W = \int_{\text{inital}}^{\text{final}} (-2\vec{a}_z)\cdot dz\vec{a}_z.$

The conductor is moved parallel to itself at a constant speed to the final position (2, 2, 0), i.e., along the z-direction from 0 to 2.

$$\therefore \quad W = \int_0^2 (2)dz = 2z\Big|_0^2 = 4 \text{ joules}.$$

Example 5.30 Prove that the force on a closed filamentary circuit in a uniform magnetic field is zero.

Solution The force exerted on a closed current element in a magnetic field $\vec{B}$ is given by

$$\vec{F} = -I\oint \vec{B} \times d\vec{l}.$$

Assume that the magnetic flux intensity is uniform throughout the field. Then,

$$\vec{F} = -I\vec{B} \times \oint d\vec{l}.$$

Since the closed line integral in a uniform field is zero, $\oint d\vec{l} = 0$.

$$\therefore \quad \vec{F} = -I\vec{B} \times 0 = 0.$$

Therefore, the force on a closed current circuit in a uniform magnetic field is zero. For non-uniform field, the force need not be zero.

Example 5.31 A rectangular coil carrying a current of 5 A is placed in the magnetic field $\vec{B} = 0.3(\vec{a}_x + \vec{a}_y)$ Wb/m^2. The coil is lying in the $y-z$ plane and has dimensions 0.8 m $\times$ 0.4 m. Find the torque on the coil.

Solution Given $I = 5\,A$, $\vec{B} = 0.3(\vec{a}_x + \vec{a}_y)$; area $\vec{S} = 0.8 \times 0.4\,\vec{a}_x = 0.32\vec{a}_x$.
The torque on the coil is $\vec{T} = \vec{m} \times \vec{B} = I\vec{S} \times \vec{B}$

$$= 5(0.32)\vec{a}_x \times (0.3)(\vec{a}_x + \vec{a}_y)$$

$$= 5(0.35)(0.3)\vec{a}_z = 0.48\vec{a}_z \text{ N-m}.$$

Example 5.32 A rectangular current loop in the $z = 0$ plane has corners at (0, 0, 0), (1, 0, 0), (1, 2, 0) and (0, 2, 0). The loop carries a current of 5 A. Find the total force on the loop produced by the magnetic field $\vec{B} = 5\vec{a}_x + 2\vec{a}_y - 4\vec{a}_z$ Wb/m^2.

Solution Given a rectangular current loop as shown in Fig. 5.22.

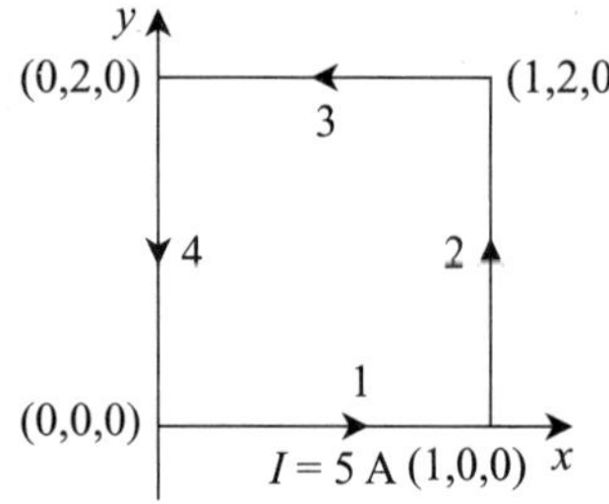

Fig. 5.22 *Rectangular current loop*

Flux density $\vec{B} = 5\vec{a}_x + 2\vec{a}_y - 4\vec{a}_z$; $I = 5$ A.

The force on side 1: $\vec{F}_1 = \int I d\vec{l} \times \vec{B} = \int_0^1 5dx\vec{a}_x \times (5\vec{a}_x + 2\vec{a}_y - 4\vec{a}_z)$

$$= 5\int_0^1 (2\vec{a}_z + 4\vec{a}_y)dx = 20\vec{a}_y + 10\vec{a}_z.$$

Similarly, force on side 3 is

$$\vec{F}_3 = -(20\vec{a}_y + 10\vec{a}_z) \text{ N}.$$

Force on side 2 is

$$\vec{F}_2 = \int_0^2 5dy\vec{a}_y \times (5\vec{a}_x + 2\vec{a}_y - 4\vec{a}_z) = 5\int_0^2 (-5\vec{a}_z - 4\vec{a}_x)dy = -(50\vec{a}_z + 40\vec{a}_x).$$

Similarly, force on side 4 is

$$\vec{F}_4 = 50\vec{a}_z + 40\vec{a}_x.$$

The total force $\vec{F} = \vec{F}_1 + \vec{F}_2 + \vec{F}_3 + \vec{F}_4 = 0.$

Example 5.33 In region 1, let $\mu_{r1} = 10$ and in region 2, let $\mu_{r2}, = 2$. Also, $B_1 = 10\vec{a}_x + 20\vec{a}_y + 30\vec{a}_x$ mT everywhere in region 1. Find (a) the angle θ_1 that B_1 makes with the normal to the boundary, (b) the angle θ_2 that B_2 makes with the normal to the boundary.

Solution Given $\vec{B}_1 = 10\vec{a}_x + 20\vec{a}_y + 30\vec{a}_z$ mT $= 0.01\vec{a}_x + 0.02\vec{a}_y + 0.03\vec{a}_z$ tesla

$$\mu_{r_1} = 10; \ \ \mu_{r_2} = 2.$$

$$\vec{H}_1 = \frac{\vec{B}_1}{\mu_1} = \frac{1}{10\mu_0}(0.01\vec{a}_x + 0.02\vec{a}_y + 0.03\vec{a}_z) \text{ A/m}$$

Let z-axis be normal to the boundary. If H_{n1} and H_{n2} are the normal components of H_1 and H_2, then

$$\frac{H_{n1}}{H_{n2}} = \frac{\mu_{r2}}{\mu_{r1}}.$$

From H_1, $H_{n1} = \dfrac{0.03}{10\mu_0}$.

$$\therefore \ \ H_{n2} = \frac{H_{n1}\mu_{r1}}{\mu_{r2}} = \frac{10}{2} \times \frac{0.03}{10\mu_0} = \frac{0.03}{2\mu_0} = \frac{0.015}{\mu_0}.$$

At the boundary, $\vec{H}_{t1} = \vec{H}_{t2} = \dfrac{1}{10\mu_0}(0.01\vec{a}_x + 0.02\vec{a}_y)$.

$$\vec{H}_2 = \vec{H}_{t2} + \vec{H}_{n2} = \frac{1}{\mu_0}(0.001\vec{a}_x + 0.002\vec{a}_y + 0.015\vec{a}_z).$$

$$\vec{B}_2 = \frac{\mu}{\mu_0}(0.001\vec{a}_x + 0.002\vec{a}_y + 0.015\vec{a}_z)$$

$$= 2(0.001\vec{a}_x + 0.002\vec{a}_y + 0.015\vec{a}_z)$$

$$= 0.002\vec{a}_x + 0.004\vec{a}_y + 0.03\vec{a}_z \text{ tesla}.$$

Also at the boundary, $B_1 \cdot \vec{a}_z = |B_1|\ |\vec{a}_z| \cos\theta_1$.

$$\cos\theta_1 = \frac{\vec{B}_1 \cdot \vec{a}_z}{B_1}$$

$$= \frac{0.03}{\sqrt{001^2 + 002^2 + 003^2}} = \frac{0.03}{0.0374} = 0.8021.$$

$$\theta_1 = \cos^{-1}(0.8021) = 36.66^\circ$$

and $$\cos\theta_2 = \frac{\vec{B}_2 \cdot \vec{a}_z}{B_2} = \frac{0.03}{\sqrt{0002^2 + 0004^2 + 003^2}} = \frac{0.03}{0.0303}$$

$$\theta = 8.1^\circ.$$

Example 5.34 A solenoid of 10 cm length consists of 1000 turns having a cross-section of radius 1 cm. Find the inductance of the solenoid. What is the value of current required to maintain a flux of 1 mWb in a toroid? Take $\mu_r = 1500$.

Solution Given $l = 10$ cm; $N = 1000$; $r = 1$ cm; $\varphi = 1$ mWb; $\mu_r = 1500$.
We know that the inductance of a solenoid is

$$L = \frac{\mu_0 \mu_r N^2\, S}{l}$$

$$= \frac{4\pi \times 10^{-7} \times 1500 \times (1000)^2 \times \pi \times (10^{-2})^2}{10 \times 10^{-2}} = 5.92 \text{ H}.$$

Inductance of a toroid can also be written as,

$$L = \frac{N\varphi}{I}.$$

Then, the current required to maintain a flux of φ in the toroid is

$$I = \frac{N\varphi}{L} = \frac{1000 \times 1 \times 10^{-3}}{5.92} = 0.17 \text{ A}.$$

Example 5.35 A solenoid of 500 turns has a length of 50 cm and a radius of 10 cm. A steel rod of circular cross-section is fitted in the solenoid coaxially. The relative permeability of steel is 3000. A dc current of 10 A is passed through the solenoid. Evaluate the inductance of the solenoid.

Solution Given $N = 500$; $l = 0.5$ m; $r = 0.1$m; $S = \pi r^2 = \pi(0.1)^2\text{m}^2$; $\mu_r = 3000$; $I = 10$ A.

Inductance $$L = \frac{\mu N^2 S}{l} = \frac{3000 \times 4\pi \times 10^{-7} \times 500^2 \times \pi(0.1)^2}{0.5} = 59.21 \text{ H}.$$

Example 5.36 A toroidal coil of 1200 turns has a mean radius of 30 cm and a radius of winding of 2 cm. What is the average self inductance with (i) an air core, (ii) an iron core of relative permeability, $\mu_r = 900$?

Solution Given $N = 1200$; $R = 0.3$ m; $r = 2$ cm $= 0.02$ m. $S = \pi r^2 = \pi(0.02)^2$.

(i) With air core: $\mu = \mu_0 = 4\pi \times 10^{-7}$.

$$L_0 = \frac{\mu N^2 S}{2\pi R} = \frac{4\pi \times 10^{-7} \times 1200^2 \times \pi(0.02)^2}{2\pi \times 0.3} = 1.206 \text{ mH}.$$

(ii) With iron core: $\mu = \mu_0\mu_r = 900\mu_0$.

$\therefore \quad L_i = 900\, L_0 = 1.086$ H.

Example 5.37 A solenoid with an air core has 1000 turns of wire. Its length is 800 mm and its core radius is 60 mm. What is the inductance?

Solution Given $N = 1000$; $l = 0.8$ m; $r = 0.06$ m, then $S = \pi(0.06)^2$ m^2.
We know that the inductance of the solenoid is

$$L = \frac{\mu N^2 S}{l} = \frac{4\pi \times 10^{-7} \times (1000)^2 \times \pi(0.06)^2}{0.8} = 17.7 \text{ mH}.$$

Example 5.38 A toroidal coil of 500 turns is wound on a steel ring of 0.3 m mean diameter and 2×10^{-2} m^2 cross-sectional area. An excitation of 4 A produces a flux density of 1 tesla. Find the inductance and reluctance of the coil.

Solution Given $N = 500$; $D = 0.3$ m; $R = \dfrac{0.3}{2} = 0.15$ m.

$$S = 2 \times 10^{-2} \text{ m}^2; \text{ excitation } I = 4000 \text{ A/m}; \text{ flux density} = 1 \text{ tesla}.$$

Assume that for a steel ring $\mu_r = 3000$.

$$L = \frac{\mu N^2 S}{2\pi R} = \frac{\mu_0\mu_r N^2 S}{2\pi R} = \frac{4\pi \times 10^{-7} \times 3000 \times (500)^2 2 \times 10^{-2}}{2\pi \times (0.15)} = 20 \text{ H}.$$

$$\text{Reluctance } \mathbb{R} = \frac{NI}{\varphi} = \frac{NI}{BS} = \frac{500 \times 4}{1 \times 2 \times 10^{-2}} = 100000 \text{ AT/ Wb}.$$

Example 5.39 A spiral upright spring consists of 1000 turns copper wire. The length of the spring is 10 cm. Find the inductance of the spring if the current flowing through the spring is 5 A and radius of the spring is 1 cm. If this spring is bent into a circle, find the new inductance of the spring. (Neglect elongation of spring.)

Solution Given $N = 1000$; $L = 0.1$ m; $I = 5$ A; $r = 0.01$, then $S = \pi(0.01)^2$ m^2.
Inductance of the spring (solenoid)

$$L = \frac{\mu_0 N^2 S}{l} = \frac{4\pi \times 10^{-7} \times 1000^2(\pi \times (0.01)^2)}{0.1} = 3.945 \text{ mH}.$$

When the spring is bent into a circle, it forms a toroid.

Here $l = 2\pi R$.

$$L = \frac{\mu_0 N^2 S}{2\pi R} = 3.945 \text{ mH}.$$

There is no change in the inductance.

Example 5.40 A two-conductor transmission line is made up of conductors separated by a distance of 2 m. The radius of each conductor is 1 cm. The medium is air. Compute the exact value of inductance of each conductor per km length.

Solution Given $d = 2$ m; $r = 1$ cm $= 10^{-2}$ m; $\mu_r = 1$.
We know that the inductance of each conductor is

$$L = \frac{\mu l}{\pi} \ln\left(\frac{d-a}{a}\right).$$

$$\frac{L}{l} = \frac{\mu}{\pi} \ln\left(\frac{d-a}{a}\right) = 4\times 10^{-7} \ln\left(\frac{2-0.01}{0.01}\right)$$

$$= 4\times 10^{-7} \ln(200) = 2.17 \mu\text{H/m} = 2.17 \text{ mH/km}.$$

Example 5.41 A toroid is made up of two semi-circular rings of iron and steel held together tightly. Cross-sectional area of each part is 5 cm^2 and the mean radius of the toroid is 20 cm. Relative permeabilities of steel and iron are respectively 2000 and 500. The exciting coil has 500 turns. Find the inductance of the system.

Solution Given $S = 5 \text{ cm}^2 = 5\times 10^{-4}$ m; $N = 500$; $\mu_{r1} = 2000$, $\mu_{r2} = 500$.
R = mean radius = 20 cm = 0.2 m.
For a toroid, the total inductance is

$$L = L_{\text{iron}} + L_{\text{steel}}$$

$$L = \frac{\mu_1 N^2 S}{l} + \frac{\mu_2 N^2 S}{l},$$

where $l = \dfrac{2\pi R}{2} = \pi R$, length of the semi-circular ring.

$$L = \frac{\mu_0 N^2 S}{\pi R}(\mu_{r1} + \mu_{r2})$$

$$= \frac{4\pi\times 10^{-7}\times 50.0^2\times 5\times 10^{-4}}{\pi(02)}(2000+500) = 625 \text{ mH}.$$

Example 5.42 A solenoid of 500 turns has a length of 50 cm and radius of 10 cm. A steel rod of circular cross-section is fitted in the solenoid coaxially and tightly. The relative permeability of steel is 3000. A dc current of 10 A is passed through the solenoid. Compute the inductance of the system, energy stored in the system and the mean flux density inside the solenoid.

Solution Given $N = 500;\ l = 0.5$ m; $R = 0.1$ m; $\mu_r = 3000;\ I = 10$ A.

We know that $L = \dfrac{\mu N^2 S}{l} = \dfrac{4\pi \times 10^{-7} \times 3000 \times 500^2 \times \pi(0.1)^2}{0.5} = 59.22$ H.

Energy stored in the system

$$W = \frac{1}{2}LI^2 = \frac{1}{2}(59.22)(10)2 = 2961 \text{ joules.}$$

Flux density $B = \dfrac{\mu NI}{l} = \dfrac{4\pi \times 10^{-7} \times 3000 \times 500 \times 10}{0.5} = 37.7\ \text{Wb/m}^2$

Example 5.43 Consider two coaxial solenoids of radius 2 cm and 3 cm having 50 turns/cm and 80 turns/cm respectively. Find the mutual inductance.

Solution Given $a_1 = 0.02$ m; $a_2 = 0.03$ m.

For $a_2 > a_1,\ S_1 = \pi(0.02)^2$,

$n_1 = 50$ turns/cm $= 5000$ turns/m; $n_2 = 80$ turns/cm $= 8000$ turns/m; $l_1 = l_2 = 1$ m.

The mutual inductance is

$$M_{12} = M_{21} = \frac{\mu N_1 N_2 S_1^2}{l_1 l_2} = \mu n_1 n_2 S_1^2$$

$$= 4\pi \times 10^{-7}(5000)(8000)\pi(0.02)^2 = 63.2\ \text{mH/m.}$$

Example 5.44 The current in a coil is increased from 0 to 10 A at a uniform rate in 5 seconds. It is found that this coil develops self induced emf of 100 V whereas an emf of 20 V is induced in a neighbouring coil. Compute self inductance of the first coil and mutual inductance between the two coils.

Solution Given $i_1 = 0$ A; $i_2 = 10$ A; $dt = 5$ s.
Self induced emf $V_s = 100$ V; mutual induced emf $V_m = 20$ V.

(i) Self induced emf is given by $V_s = L\dfrac{di}{dt}$,

where $di = 10 - 0 = 10$ A and $\dfrac{di}{dt} = \dfrac{10}{5} = 2$.

$$\therefore \quad V_s = L(2) \text{ or } L = \frac{V_s}{2} = \frac{100}{2} = 50 \text{ H.}$$

(ii) Mutual induced emf is given by

$$V_m = M\frac{di}{dt} = M(2).$$

$$\therefore \quad M = \frac{V_m}{2} = \frac{20}{2} = 10 \text{ H.}$$

Example 5.45 A magnetic circuit comprising a toroid of 500 turns, an area of 6 cm^2 and mean radius of 15 cm carries a current of 4 A. Find reluctance and flux. Assume $\mu_r = 1$.

Solution Given $N = 500$; $S = 6 \times 10^{-4}\ \text{m}^2$; $R = 15 \times 10^{-2}$ m; $I = 4$ A; $\mu_r = 1$.
We know that the reluctance of the magnetic circuit is

$$\mathbb{R} = \frac{l}{\mu \text{A}} = \frac{2\pi R}{\mu_0 \mu_r A}$$

$$= \frac{2\pi \times 15 \times 10^{-2}}{(4\pi \times 10^{-7}) \times 1 \times (6 \times 10^{-4})} = 1.25 \times 10^9 \text{ AT/ Wb.}$$

Flux through the magnetic circuit is given by

$$\varphi = \frac{NI}{\mathbb{R}} = \frac{500 \times 4}{1.25 \times 10^9} = 1.6 \times 10^{-6} = 1.6\mu \text{ Wb.}$$

Example 5.46 Derive the mutual inductance between two coaxial circular wires of radii a and b ($b > a$) separated by distance h.

Solution Given two coaxial circular wires of radii a and b ($b > a$), separated by distance h.
Let I_1 be the current flowing through the circular wire of radius a and P be any point on the circular wire of radius b (see Fig. 5.23).

The magnetic vector potential at point P due to current I_1 is

$$\vec{A} = \frac{\mu_0 I_1}{4\pi} \oint \frac{d\vec{l}}{r} = \frac{\mu_0 I_1}{4\pi} \int \frac{d\vec{l}_1 \times \vec{a}_r}{r^2}$$

$$= \frac{\mu_0 I_1}{4} \frac{a^2 \sin\theta}{r^2} \vec{a}_\phi \qquad \text{(Refer to Example 5.11)}$$

where r is the distance between point P and the centre of the small circular wire and θ the angle point P makes at the centre of the small circular wire.

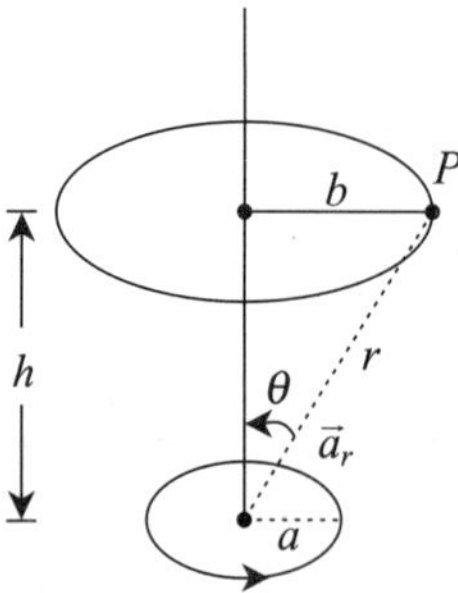

Fig. 5.23 *Coaxial circular conductors*

From geometry, $\sin\theta = \dfrac{b}{\sqrt{b^2 + h^2}}$ and $r = \sqrt{b^2 + h^2}$.

$$\therefore \quad \vec{A} = \frac{\mu_0 I}{4} \frac{a^2 b}{(b^2+h^2)^{3/2}} \vec{a}_\phi.$$

The flux linkage with the big circular wire is

$$\varphi_{12} = \int_0^l \vec{A} \cdot d\vec{l_2}$$

where $d\vec{l_2} = dl\vec{a}_\phi$.

$$\varphi_{12} = \int_0^l \frac{\mu \pi I_1 a^2 b \vec{a}_\phi}{4(b^2+h^2)^{3,2}} dl\, \vec{a}_\phi.$$

Since $l = 2\pi b$,

$$\varphi_{12} = \frac{\mu \pi l_1 a^2 b^2}{2(h^2+b^2)^{3/2}}.$$

The mutual inductance is $M_{12} = \dfrac{\varphi_{12}}{I_1} = \dfrac{\mu \pi a^2 b^2}{2(h^2+b^2)^{3/2}}.$

If $h >> b$, then $M_{12} = \dfrac{\mu \pi a^2 b^2}{2h^3}$ H.

Example 5.47 A toroid has a core of cross-section area 20 cm^2 and a mean diameter of 250 mm. The core material has a relative permeability of 1000. Calculate the number of turns to be wound on the core to obtain an inductance of 1 H.

Solution Given $S = 20$ cm$^2 = 20 \times 10^{-4}$ m^2; $d = 250$ mm $= 0.25$ m; $R = 0.125$ m; $\mu_r = 1000$; $L = 1$ H.

We know that $L = \dfrac{\mu N^2 S}{2\pi R}.$

$$1 = \frac{1000 \times 4\pi \times 10^{-7} \times N^2 \times 20 \times 10^{-4}}{2\pi \times 0.125}.$$

$\therefore \quad N = 560.$

We know that $\mathbb{R} = \dfrac{l}{\mu S} = \dfrac{0.1\pi}{500 \times 4\pi \times 10^{-7} \times 20 \times 10^{-4}} = 0.25 \times 10^6$ AT/Wb.

Flux $\varphi = BS = 1 \times 20 \times 10^{-4} = 2 \times 10^{-3}$ Wb.

$$NI = \varphi\, R = 2 \times 10^{-3} \times 0.25 \times 10^6.$$

$\therefore \quad I = 2.5$ A.

Inductance $L = \dfrac{\mu N^2 S}{l} = \dfrac{N^2}{\mathbb{R}} = 0.04\ \text{H}.$

Energy stored $W_m = \dfrac{1}{2}LI^2 = \dfrac{1}{2}(0.04)(2.5)^2 = 0.125$ joules

Example 5.48 The magnetic field intensity is $H = 1200$ A/m in a material when $B = 2$ T. When H is reduced to 400 A/m, $B = 1.4$ T. Calculate the change in magnetisation.

Solution Given $H_1 = 1200$ A/m; $B_1 = 2$ T; $H_2 = 400$ A/m; $B_2 = 1.4$ T.
We know that $B = \mu(H + M)$, where M is the magnetisation of the material.

$$B_1 = \mu(H_1 + M_1).$$

Let $\mu = \mu_0$.

$$M_1 = \frac{B_1}{\mu_0} - H_1 = \frac{2}{4\pi \times 10^{-7}} - 1200 = 1.59 \times 10^6 \ \text{A/m}.$$

Similarly, $M_2 = \dfrac{B_2}{\mu_0} - H_2 = \dfrac{1.4}{4\pi \times 10^{-7}} - 400 = 1.11 \times 10^6 \ \text{A/m}.$

Change in magnetisation

$$\Delta M = M_1 - M_2 = 0.48 \times 10^6 \ \text{A/m}.$$

Summary

1. The scalar magnetic potential between two points A and B in the magnetic field is

   $$V_{m\,A,B} = -\int_B^A \vec{H} \cdot d\vec{l} \text{ amperes if } \vec{J} = 0$$

   The negative gradient of the scalar magnetic potential is equal to the magnetic field intensity at which the current density $J = 0$.

   i.e., $\vec{H} = -\nabla V_m$ if $\vec{J} = 0$

2. Laplace equation for scalar magnetic potential is

   $$\nabla^2 V_m = 0 \text{ for } \vec{J} = 0$$

3. Vector magnetic potential at a point P can be expressed as

   $$\vec{A} = \int_L \frac{\mu I d\vec{l}}{4\pi R} \text{ Wb/m}$$

   and the curl of the vector magnetic potential $\vec{A}$ is equal to the magnetic flux density, i.e., $\vec{B} = \nabla \times \vec{A}$

4. The vector magnetic potential for other current distributions are,

$$\vec{A} = \int_s \frac{\mu \vec{K} dS}{4\pi R} = \int_v \frac{\mu \vec{J} dv}{4\pi R}.$$

5. Poisson's equation for vector magnetic potential is $\nabla^2 \vec{A} = -\mu \vec{J}$.
   If $\vec{J} = 0$, it becomes Laplace's equation, i.e., $\nabla^2 \vec{A} = 0$.

6. Lorentz's force equation for a moving charge can be expressed as

$$\vec{F} = m\frac{d\vec{u}}{dt} = Q(\vec{E} + \vec{u} \times \vec{B}) \text{ N}.$$

   The solution of this equation gives the locus of the charge particles in electromagnetic fields.

7. The magnetic force exerted on different current distributions is

$$\vec{F} = \int_v (\vec{J} \times \vec{B}) dv = \int_s (\vec{K} \times \vec{B}) \cdot d\vec{S}\, \vec{F} = \int_l I d\vec{l} \times \vec{B} \text{ N}.$$

8. The torque T due to a force on the conductor is

$$\vec{T} = m \times B = Q_m l \times B \text{ N-m, and the magnetic dipole moment is } \vec{m} = Q_m \vec{l}$$

9. Ampere's force law between two current-carrying elements can be expressed as

$$\vec{F} = \frac{\mu}{4\pi} \oint_{L_1} \oint_{L_2} \frac{I_2 d\vec{l}_2 \times (I_1 d\vec{l}_1 \times \vec{a}_{R_{12}})}{R_{12}^2} \text{ N}.$$

10. The tangential components of the magnetic field $\vec{H}$ across the boundary are

$$H_{\tan 1} - H_{\tan 2} = K.$$

    For a non-conducting boundary, $H_{\tan 1} = H_{\tan 2}$ and $\dfrac{\vec{B}_{\tan 1}}{\mu_1} = \dfrac{\vec{B}_{\tan 2}}{\mu_2}$

11. The normal component of the flux density $\vec{B}$ is continuous across the boundary, i.e.,

$$B_{N_1} = B_{N_2} \text{ and } \mu_1 H_{N_1} = \mu_2 H_{N_2}$$

12. If the fields make angles θ_1 and θ_2 with the normal to the interface

$$\frac{B_{\tan 1}}{B_{\tan 2}} = \frac{\tan\theta_1}{\tan\theta_2} = \frac{\mu_1}{\mu_2}$$

13. The ratio of the total flux linkage to the current flowing through the coil is called Inductance, $L = \dfrac{N\phi}{I}$ henrys.

14. The inductance of a solenoid is $L = \dfrac{N^2 \mu S}{l}$ henrys.

    The inductance of a toroid is $L = \dfrac{\mu N^2 S}{2\pi R}$ henrys.

    The inductance of a coaxial cable is $L = \dfrac{\mu l}{2\pi}\left(\dfrac{1}{4} + \ln(b/a)\right)$ henrys.

The inductance of a two-wire transmission line is

$$L = \frac{\mu l}{\pi}\left[\frac{1}{4}+\left(\frac{d-a}{a}\right)\right] \text{ henrys.}$$

15. Mutual inductance between two coils is defined as the ratio of the total flux linkage in one coil to the current in another coil. It is given by

$$M_{12} = \frac{N_1\varphi_{12}}{I_2} \text{ and } M_{21} = \frac{N_2\varphi_{21}}{I_1} \text{ henrys}$$

For a linear medium, $M_{12} = M_{21} = M = \dfrac{V_2}{dI_1/dt}$.

16. The coefficient of coupling between two coils $k = \dfrac{M}{\sqrt{L_1L_2}}$.

The range of k is $0 \leq k \leq 1$.

17. Neumann's formulae for mutual inductance is

$$M_{12} = \frac{\varphi_{12}}{I_2} = \frac{\mu_0}{4\pi}\oint_1\oint_2 \frac{dl_1 dl_2}{R_{12}}$$

18. Energy density is defined as the energy per unit volume. The energy density stored in the magnetic field of an inductor is given by

$$W = \frac{1}{2}\mu\vec{H}^2 = \frac{1}{2}\vec{B}\cdot\vec{H} \text{ J/m}^3$$

Energy stored is $W_m = \dfrac{1}{2}LI^2$ joules.

19. Reluctance is defined as the ratio of magnetomotive force to the total flux.

$$\mathbb{R} = \frac{\text{mmf}}{\phi} = \frac{NI}{\phi} \text{ AT/Wb; also, } L = \frac{N^2}{\mathbb{R}} \text{ and } M = \frac{N_1N_2}{\mathbb{R}}.$$

20. Magnetic materials are classified into diamagnetic materials, paramagnetic materials or ferromagnetic materials according to their material properties—magnetic susceptibility or relative permeability.

Review Questions

1 Explain the difference between magnetic vector potential and magnetic scalar potential. *(Nov 2011, 2013)*

2 Explain the concept of scalar and vector magnetic potentials. *(May 2011, 2012, Aug 2014)*

3 Derive the expression for vector magnetic potential $\vec{A}$ which satisfies Poisson's equation. *(May 2010, 2011, 2012)*

4 Using the concept of vector magnetic potential, find the magnetic field density at a point due to a long straight filamentary conductor carrying a current I in the z direction. *(May 2011)*

5 Give the properties of vector magnetic potential. *(Nov 2011, Aug 2014)*

6 Derive an expression for the force on a current element in a magnetic field. *(Nov 2008, 2011)*

7 Derive the expression for Lorentz's force equation. *(May 2011, 2012, Aug 2014)*

8 Derive the expression for force on a straight current-carrying conductor placed in a magnetic field. *(May 2010, 2011, 2012)*

9 Derive the expression for force between two long parallel current-carrying conductors placed in a magnetic field. *(May 2010, 2011)*

10 Find the torque that will be produced on a rectangular current loop if it is placed in a magnetic field $\vec{B}$. Show that $\vec{T} = \vec{m} \times \vec{B}$ also holds for the system. *(Nov 2008, May 2009, 2012)*

11 Show that $\vec{T} = \vec{m} \times \vec{B}$ also holds for the torque on a solenoid situated in a uniform magnetic field. *(May 2008, 2011)*

12 Derive an expression for the torque developed in a current loop placed in a magnetic field. *(Nov 2010)*

13 Obtain the expression for inductance of a toroid. *(May 2009, 2010, 2012)*

14 Obtain the expression for inductance of a solenoid. *(Jan 2014)*

15 Derive the boundary conditions for magnetic field intensity and flux density. *(Nov 2010, 2011)*

16 Derive the inductance of a two-wire transmission line. *(Nov 2010, May 2012)*

17 Explain the differences between self inductance and mutual inductance. *(May 2010, 2011, 2012)*

18 Explain the terms: (i) reluctance, (ii) inductance. *(May 2010, Nov 2011)*

19 Define inductance and mutual inductance. *(May 2010, 2011, 2012)*

20 Derive an expression for mutual inductance. Use Newmann's formula. *(Nov 2010, 2013)*

21 Explain the relationship between reluctance and inductance.

22 Explain the concept mutual inductance and derive the energy stored in a magnetic field due to mutual inductance. *(May 2011, 2012)*

23 Explain coefficient coupling between two coils. *(May 2011, Nov 2013)*

24 Derive the mutual inductance between two solenoids. *(May 2009, 2012, Nov 2011)*

25 What is a magnetic dipole? How does a magnetic dipole differ from an electric dipole? *(May 2012, 2013)*

26 Define magnetic dipole. What is magnetic moment? Discuss how a differential current loop behaves like a magnetic dipole. *(May 2009, 2010, 2011)*

27 Derive the expression for energy stored and energy density in a magnetic field. *(May 2008, 2011)*

Problems

1 Let $\vec{A} = (3y - z)\vec{a}_x + 2xz\vec{a}_y$ Wb/m in a certain region of free space. Show that $\nabla \cdot A = 0$.

2 Let $\vec{A} = (3y^2 - 2z)\vec{a}_x - 2x^2z\vec{a}_y + (x + 2y)\vec{a}_z$ Wb/m in free space. Find $\nabla \times \nabla \times \vec{A}$ at $(-2, 3, -1)$.

3 Assume that $\vec{A} = 50\rho^2\vec{a}_z$ Wb/m in a certain region of free space. Find $\vec{H}$ and $\vec{B}$.

4 Given the magnetic vector potential $\vec{A} = 5r \sin\phi\vec{a}_r$ in spherical coordinates, find the flux density at $(2, \pi/2, 0)$.

5 Given the magnetic vector $\vec{A} = e^{-2z}\left(\sin\frac{1}{2}\phi\right)\vec{a}_\phi$ in cylindrical coordinates, find the curl of $\vec{A}$ at $(\rho, \pi/3, 0.5)$.

6 Given that the vector magnetic potential within a cylindrical conductor of radius b is $\vec{A} = \frac{-\mu_0 I r^2}{4\pi b^2}\vec{a}_z$, find the corresponding $\vec{H}$.

7 A solenoid 5 cm in length consists of 2000 turns and has a cross-sectional radius of 1 cm. Find the inductance of the solenoid. What is the value of current required to maintain a flux of 2 mWb in the toroid? Take $\mu_r = 80$.

8 A toroidal core has a rectangular cross-section defined by the surfaces $\rho = 2$ cm, $\rho = 3$ cm, $z = 4$ cm, and $z = 4.5$ cm. The core material has a relative permeability of 80. If the core is wound with a coil containing 8000 turns of wire, find its inductance.

9 A rectangular coil is composed of 150 turns of a filamentary conductor. Find the mutual inductance in free space between this coil and an infinite straight filament on the z-axis if the four corners of the coil are located at (0, 1, 0), (0, 3, 0), (0, 3, 1), and (0, 1, 1).

10 A conducting current strip carrying current of density $\vec{K} = 12\vec{a}_z$ A/m lies in the $x = 0$ plane between $y = 0.5$ and $y = 1.5$ m. There is also a current filament of $I = 5$ A in the $\vec{a}_z$ direction along the z-axis. Find the force exerted on the filament by the current strip.

11 Two circular loops of radii 1 m carrying currents 10 A and 5 A are situated in the $z = 0$ and $z = 1$ m planes respectively and with their centres lying on the z-axis. Find the force experienced by current elements dI_1 and dI_2 located at (1, 0, 0) and (0, 1, 0) respectively.

12 A current element 2 m in length carries 5 A current along the y-axis, centred at the origin. If it experiences a force $1.5(\vec{a}_x + \vec{a}_z)/\sqrt{2}$N due to a uniform magnetic field, determine the magnetic flux density $\vec{B}$.

13 The rails of a 1.44 m gauge railways are insulated from each other, but connected through a milli-voltmeter. The vertical component of Earth's field is 45×10^{-6} Wb/m^2 Find the reading of the voltmeter when a train is travelling along a neighbouring rail at 100 km/h.

14 A long current filament carrying a current of 12 A lies along the z-axis. A rectangular loop 2 m $\times$ 0.5 m is approaching the filament at a velocity of 1000 m/s with its larger side parallel to the z-axis. Determine the induced current in the rectangular loop when the nearest side is at a distance 1 m from the current filament.

15 A single-phase circuit comprises two parallel conductors, A and B, each 1 cm diameter and spaced 1 metre apart. The conductors carry currents of $+100$ A and -100 A respectively. Determine the magnetic field intensity at the surface of each conductor and also exactly midway between A and B.

16 A differential current loop in the $z=0$ plane is placed in a magnetic field of density $\vec{B}$ Wb/ m^2. Find the total torque on the loop about the origin located at the centre of the loop.

17 A charge of value $Q=5\times10^{-18}$ C is moving though a uniform magnetic field $\vec{B}=-0.4\vec{a}_x+0.2\vec{a}_y-0.1\vec{a}_z$ T with a velocity $v=10(2\vec{a}_x-3\vec{a}_y+6\vec{a}_z)$ km/s at $t=0$. (a) What electric field is present at $t=0$ if the net force on the charge is zero? (b) If the electric field intensity is entirely in the $\vec{a}_x$ direction and $|\vec{F}|=2\times10^{-12}N$ at $t=0$, find $\vec{E}_x$.

18 A point charge of value −40 nC is moving with a velocity of 6000 km/s in a direction specified by the unit vector $\vec{a}_v=-0.48\vec{a}_x-0.6\vec{a}_y+0.64\vec{a}_z$. Using Lorentz' force equation, find the force $\vec{F}$ if (a) $\vec{B}=2\vec{a}_x-3\vec{a}_y+5\vec{a}_z$ mT, (b) $\vec{E}=2\vec{a}_x-3\vec{a}_y+5\vec{a}_z$ kV/m.

Answers

2. $-6\vec{a}_x-4\vec{a}_y$ Wb/ m^3
3. $\vec{H}=-100\rho\vec{a}_\varphi$ Wb/ m^2 $\vec{B}=-100p\vec{a}_\varphi/\mu_0$ A/m
4. 5 Wb/ m^2
5. $\nabla\times\text{A}=-0.368\vec{a}_\rho+0.16\vec{a}_z$
6. $\frac{1}{\mu}(\nabla\times\vec{A})$
7. $L=25.26$ H, $I=0.158$ A
8. 2.08 H
9. 33 μH
10. 13.2 $\vec{a}_y$ μN/m
13. 1.8 mV
14 1.6 mV
15. On conductor l, $\vec{H}=3199\vec{a}_\varphi$ A/m, on conductor 2, $\vec{H}=-3199\vec{a}_\varphi$ A/m, at mid point, $\vec{H}=63.03\vec{a}_\varphi$ A/m
17. (a) $9\vec{a}_x+22\vec{a}_y+8\vec{a}_z$ kV/m (b) 409 kV/m
18. (a) $-0.26\vec{a}_x+0.88\vec{a}_y+0.63\vec{a}_z$ N (b) $-80\vec{a}_x+12\vec{a}_y-200\vec{a}_z$ μN

Multiple-choice Questions

1 Magnetic potential is a
(a) scalar quantity (b) vector quantity
(c) both (a) & (b) are true in different situations

2 The expression for a magnetic vector potential for a volume is given by
(a) $\int(\mu JdV/4\pi R)$ (b) $\int(\mu JdV/4\pi R^2)$
(c) $\int(\mu JdS/4\pi R^2)$ (d) none of the above

3 Self inductance of a long solenoid is given by
(a) $\mu N^2/AI$ (b) $\mu N^2A/I$ (c) $\mu N^2IA/2\pi R$ (d) none of these

4 The ratio of magnetic flux linkages to the current through the coil is defined as
(a) reluctance (b) inductance (c) unit flux (d) susceptance

5 Inductance of a coil depends on
(a) cross-sectional area (b) current
(c) length (d) both (a) and (c)

6 The unit vector magnetic potential is

(a) Wb-m (b) Wb/m (c) Wb/m^2 (d) Wb-m^2

7 If $\vec{H}$ is the magnetic field intensity and $\vec{A}$ is the vector magnetic potential at any point due to a current-carrying element, then the phase difference between $\vec{A}$ and $\vec{H}$ is

(a) $\frac{\pi}{4}$ (b) π (c) $\frac{\pi}{2}$ (d) $\frac{\pi}{3}$

8 If $\vec{A}$ is the vector magnetic potential, then for steady magnetic fields $\nabla \cdot \vec{A}$ is equal to

(a) zero (b) $\vec{B}$ (c) 2 (d) $\vec{H}$

9 The inductance of a long solenoid of N turns and cross-sectional area A and length L is

(a) $\frac{\mu NA}{L}$ (b) $\frac{\mu N^2 A}{L}$ (c) $\frac{N^2 A}{L}$ (d) $\frac{\mu N^2}{L}$

10 Internal inductance of a round solid conductor of radius a is

(a) $\frac{\mu}{8\pi}$ (b) $\frac{\mu}{4\pi}$ (c) $\frac{\mu}{2\pi}\ln a$ (d) $\frac{\mu}{4\pi}\ln a$

11 The approximate self inductance of a toroid of N turns wound on a wooden material of rectangular cross-section $a \times b$ and inner radius R is

(a) $\frac{\mu_0 N^2 ab}{2\pi R}$ (b) $\frac{\mu_0 N^2 ab}{2R}$ (c) $\frac{\mu_0 Nab}{2\pi R}$ (d) $\frac{\mu_0 N^2}{2\pi R}$

12 A coil has 30 turns and is wound on an iron circuit of cross-section 1100 cm^2, and mean length of 150 cm. μ_r of iron is 2000. The self inductance of the coil is

(a) 0.03 H (b) 0.045 H (c) 0.010 H (d) 0.015 H

13 A coil of 600 turns is wound on an iron circuit of cross-section 1100 cm^2, and mean length of 150 cm. μ_r of iron is 2000. The self inductance of the coil is

(a) 6.032 H (b) 5.045 H (c) 4.010 H (d) 0.015 H

14 A coil with 2000 turns carries a current of 2 A and produces a flux of 0.2 mWb. The inductance of the coil is

(a) 300 mH (b) 200 mH (c) 500 mH (d) 150 mH

15 A coil with 250 turns carries a current of 2 A and produces a flux of 0.3 mWb. The self inductance of the coil is

(a) 30 mH (b) 42.5 mH (c) 50 mH (d) 37.5 mH

16 A non-magnetic material ring having a mean diameter of 100 m and a cross-sectional area of 1 m^2 is uniformly wound with a coil of 100 turns. The value of the reluctance of coil, if $\mu_r = 1000$, is

(a) 2,50,00 AT/Wb (b) 3,50,000 AT/Wb

(c) 2,50,000 AT/Wb (d) 2,50,00 AT/Wb

17 Coils A and B of 100 and 150 turns respectively, have reluctance each of 62500 AT/ Wb. The value of self inductance of coil B is then

(a) 0.125 H (b) 0.353 H (c) 0.25 H (d) 0.4 H

18 Coils *A* and *B* have turns 100 and 200 respectively. The reluctance of each coil is 25000 AT/Wb. The value of mutual inductance between the coils is

(a) 0.8 H (b) 0.4 H (c) 0.2 H (d) 1.6 H

19 The mutual inductance between coils *A* and *B* of turns 100 and 150 respectively, and with a reluctance each of 50000 AT/Wb is

(a) 0. 3H (b) 0.6 H (c) 0.9 H (d) 0.15 H

20 Two coils have self inductances of 10 mH and 40 mH and a mutual inductance between the coils of 2.5 mH. The value of coefficient coupling is

(a) 0.25 (b) 0.5 (c) 0.8 (d) 0.125

21 Two identical coils having 2000 turns each, lie in parallel planes such that 80% of the flux produced by coil *A* with a current of 4 A produces a flux of 0.04 mWb. The value of mutual inductance is

(a) 0.016 H (b) 0.008 H (c) 0.032 H (d) 0.162 H

22 A very long solenoid with cross-sectional area 0.4 m^2 has an iron core with 4000 turns/metres. If it carries a current of 500 mA, the self inductance per metre is

(a) 8.145 H (b) 8.042 H (c) 8.93 H (d) 8.62 H

23 If the inductances of two coils are 10 H and 250 H, the value of mutual inductance between them is

(a) 25 H (b) 50 H (c) 100 H (d) 10 H

24 The mutual inductance between two coils is 20 mH. If the current in the first coil changed from 5A to 2A in 3 milliseconds, the induced voltage in the second coil is

(a) 20 V (b) 40 V (c) 50 V (d) 10 V

25 In two coils situated nearby, the current in one coil changed its magnitude from 5 A to 2 A in 3 milliseconds, inducing a voltage in the other coil of 20 V. The mutual conductance between them is

(a) 30 mH (b) 10 mH (c) 20 mH (d) 40 mH

26 A charged particle moves with a uniform velocity 4 $\vec{a}_x$ m/s in a region where $\vec{E} = 20\vec{a}_y$ V/m. The maximum flux density such that the velocity of the particle remains the same is

(a) 5 Wb/ m^2 (b) 80 Wb/ m^2 (c) 25 Wb/ m^2 (d) 15 Wb/ m^2

27 The total force experienced by a charge moving with velocity $\vec{\mu}$ in the presence of both the magnetic field and the electric field is

(a) $\vec{\mu} \times (\vec{E} \times \vec{B})$ (b) $Q(\vec{E} \times \vec{B})$ (c) $Q(\vec{E} + \mu \times \vec{B})$ (d) $\vec{E} \times \vec{B}$

28 Force acting on a current element of *dl* in a magnetic field *B* is given by

(a) $Id\vec{l} \cdot \vec{B}$ (b) $Id\vec{l} \times \vec{B}$ (c) $Id\vec{l}/ B$ (d) $B^2 Id\vec{l}$

29 Force acting on a volume element $\vec{J}dv$ in a magnetic field $\vec{B}$ is given by

(a) $\vec{J}dv \times \vec{B}$ (b) $Jdv \cdot \vec{B}$ (c) $Jdv \,/\, B$ (d) $B^2 \vec{J}dv$

30 The force acting on a stationary charge particle in a magnetic field is

(a) one (b) zero (c) maximum (d) not known

31 The force acting on surface current in a magnetic field $\vec{B}$ is given by

(a) $\vec{K}dS \times \vec{B}$ (b) $Jdv \times \vec{B}$ (c) $\vec{I}dS \,/\, \vec{B}$ (d) $\vec{K}dS \cdot \vec{B}$

32 If a charge 3 C placed in a magnetic field $\vec{B} = 2\vec{a}_x$ is moving with a uniform velocity of 2 m/s, the force experienced by it is

(a) 3 N (b) 4 N (c) 12 N (d) 6 N

33 If the direction of the magnetic field is along $\vec{a}_x$ and the dipole moment is along $\vec{a}_y$, then the direction of force is along

(a) $\vec{a}_z$ (b) $-\vec{a}_z$ (c) $\vec{a}_y$ (d) $\vec{a}_x$

34 If the direction of the surface current element is along $\vec{a}_x$ and the magnetic field direction is along $\vec{a}_x$, then force experienced by it is

(a) maximum (b) one (c) zero (d) not known

35 Lorentz's force equation is very useful to find the

(a) position of charged particles (b) intensity of electric field

(c) direction of the electric field (d) velocity of charged particles

36 Two infinitely long straight conductors are parallel to each other but the direction of current flow is opposite to one another. Then the mutual force is

(a) attractive (b) repulsive (c) zero

37 The direction of force is along $\vec{a}_x$ on a charge particle of Q coulombs moving along $\vec{a}_y$. Then the work done on the particle is

(a) zero (b) maximum (c) minimum (d) not known

38 The direction of magnetic field is $\vec{a}_y$ and velocity of charge is $\vec{a}_x$. Then the direction of force will be along

(a) $\vec{a}_z$ (b) $-\vec{a}_z$ (c) $\vec{a}_x$ (d) $-\vec{a}_x$

39 A point charge $Q = 4\pi\ C$ is moving with a velocity of $100a_y$ m/s at (0, 1, 0), then the value of $\vec{H}$ in A/m at the point (1, 0, 0) is

(a) $-100\vec{a}_z$ (b) $\dfrac{-100\vec{a}_z}{\sqrt{2}}$ (c) $\dfrac{\vec{a}_z}{\sqrt{2}}$ (d) $\dfrac{-50\vec{a}_z}{\sqrt{2}}$

40 A circular wire of radius R is carrying a current of l A. The magnetic dipole moment of the loop is

(a) $I\pi R^2$ (b) $I2\pi R$ (c) $I/\pi R$ (d) $2I\pi H^2$

41 Two infinitely long wires of radius r are situated at $y = \pm d$ and carries a current I in the direction $\pm\vec{a}_x$ respectively. The magnetic field intensity at the origin is

(a) $\dfrac{I}{\pi d}\vec{a}_z$ (b) $\dfrac{-I}{\pi d}\vec{a}_z$ (c) $\dfrac{I}{2\pi d}\vec{a}_z$ (d) $\dfrac{-I\vec{a}_z}{2\pi d}$

42 Lorentz's force is given by

(a) $Q(\vec{u} + \vec{E} \times \vec{B})$ (b) $Q(\vec{B} + \vec{u} \times \vec{E})$ (c) $Q(\vec{E} + \vec{u} \times \vec{B})$ (d) $\vec{E}(\vec{Q} + \vec{u} \times \vec{B})$

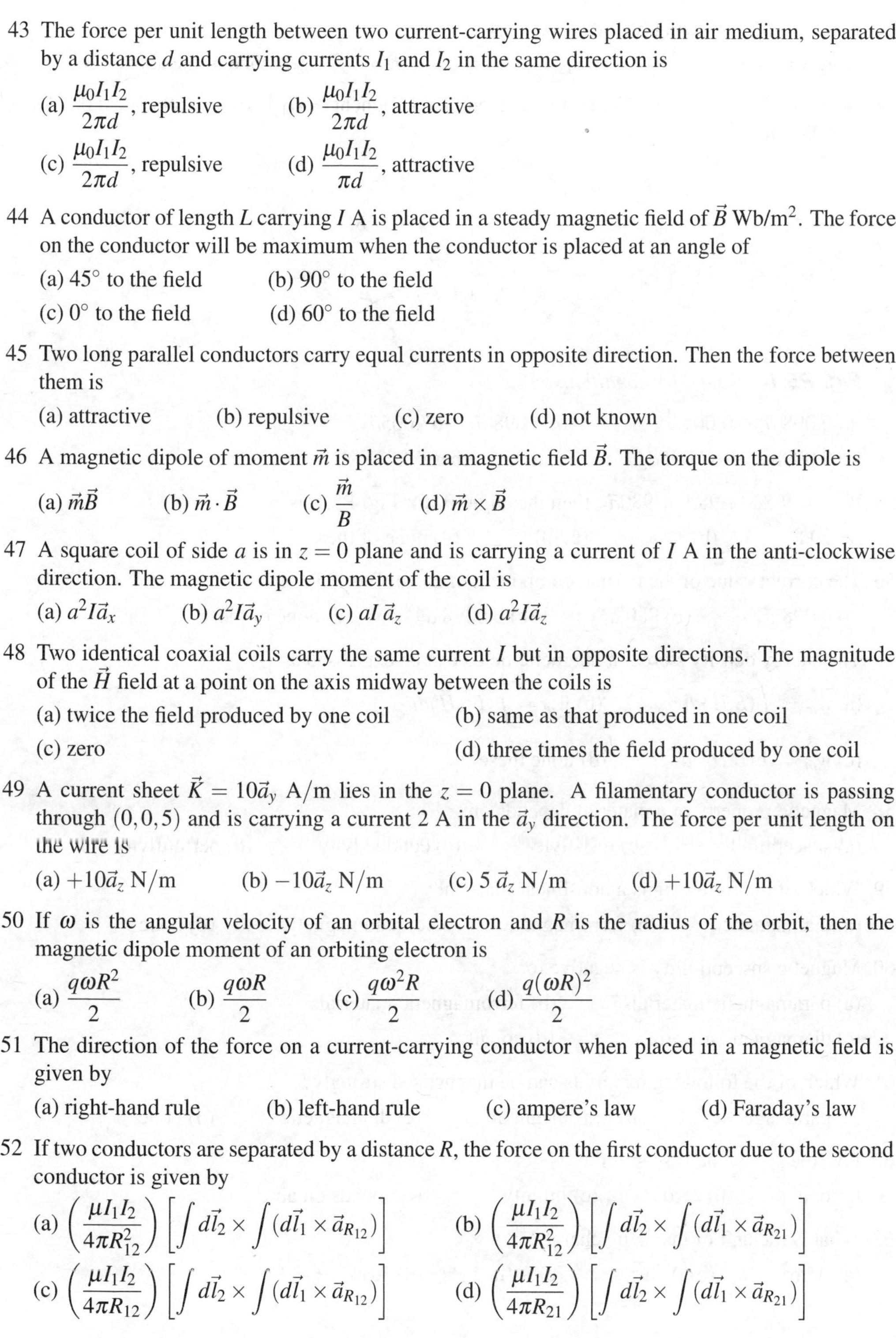

43 The force per unit length between two current-carrying wires placed in air medium, separated by a distance d and carrying currents I_1 and I_2 in the same direction is

(a) $\dfrac{\mu_0 I_1 I_2}{2\pi d}$, repulsive (b) $\dfrac{\mu_0 I_1 I_2}{2\pi d}$, attractive

(c) $\dfrac{\mu_0 I_1 I_2}{2\pi d}$, repulsive (d) $\dfrac{\mu_0 I_1 I_2}{\pi d}$, attractive

44 A conductor of length L carrying I A is placed in a steady magnetic field of $\vec{B}$ Wb/m^2. The force on the conductor will be maximum when the conductor is placed at an angle of

(a) 45° to the field (b) 90° to the field

(c) 0° to the field (d) 60° to the field

45 Two long parallel conductors carry equal currents in opposite direction. Then the force between them is

(a) attractive (b) repulsive (c) zero (d) not known

46 A magnetic dipole of moment $\vec{m}$ is placed in a magnetic field $\vec{B}$. The torque on the dipole is

(a) $\vec{m}\vec{B}$ (b) $\vec{m} \cdot \vec{B}$ (c) $\dfrac{\vec{m}}{B}$ (d) $\vec{m} \times \vec{B}$

47 A square coil of side a is in $z = 0$ plane and is carrying a current of I A in the anti-clockwise direction. The magnetic dipole moment of the coil is

(a) $a^2 I\vec{a}_x$ (b) $a^2 I\vec{a}_y$ (c) $aI\,\vec{a}_z$ (d) $a^2 I\vec{a}_z$

48 Two identical coaxial coils carry the same current I but in opposite directions. The magnitude of the $\vec{H}$ field at a point on the axis midway between the coils is

(a) twice the field produced by one coil (b) same as that produced in one coil

(c) zero (d) three times the field produced by one coil

49 A current sheet $\vec{K} = 10\vec{a}_y$ A/m lies in the $z = 0$ plane. A filamentary conductor is passing through $(0, 0, 5)$ and is carrying a current 2 A in the $\vec{a}_y$ direction. The force per unit length on the wire is

(a) $+10\vec{a}_z$ N/m (b) $-10\vec{a}_z$ N/m (c) $5\,\vec{a}_z$ N/m (d) $+10\vec{a}_z$ N/m

50 If ω is the angular velocity of an orbital electron and R is the radius of the orbit, then the magnetic dipole moment of an orbiting electron is

(a) $\dfrac{q\omega R^2}{2}$ (b) $\dfrac{q\omega R}{2}$ (c) $\dfrac{q\omega^2 R}{2}$ (d) $\dfrac{q(\omega R)^2}{2}$

51 The direction of the force on a current-carrying conductor when placed in a magnetic field is given by

(a) right-hand rule (b) left-hand rule (c) ampere's law (d) Faraday's law

52 If two conductors are separated by a distance R, the force on the first conductor due to the second conductor is given by

(a) $\left(\dfrac{\mu I_1 I_2}{4\pi R_{12}^2}\right)\left[\int d\vec{l}_2 \times \int (d\vec{l}_1 \times \vec{a}_{R_{12}})\right]$ (b) $\left(\dfrac{\mu I_1 I_2}{4\pi R_{12}^2}\right)\left[\int d\vec{l}_2 \times \int (d\vec{l}_1 \times \vec{a}_{R_{21}})\right]$

(c) $\left(\dfrac{\mu I_1 I_2}{4\pi R_{12}}\right)\left[\int d\vec{l}_2 \times \int (d\vec{l}_1 \times \vec{a}_{R_{12}})\right]$ (d) $\left(\dfrac{\mu I_1 I_2}{4\pi R_{21}}\right)\left[\int d\vec{l}_2 \times \int (d\vec{l}_1 \times \vec{a}_{R_{21}})\right]$

53 Which of the following boundary condition is correct?

(a) $H_{n1} = H_{n2}$ (b) $B_{t1} = B_{t2}$ (c) $B_{n1} = B_{n2}$ (d) none of the above

54 Which of the following is the correct tangential component of H_1 if $H_1 = 0.098\vec{a}_x + 0.0098\vec{a}_y + 67.005\vec{a}_z$ for Fig. P5.1?

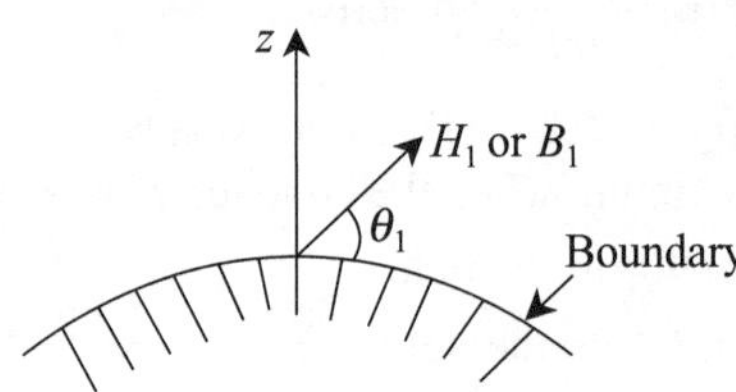

Fig. P5.1 *Boundary conditions*

(a) $0.098\ \vec{a}_x + 0.0098\vec{a}_y$ (b) $0.098\ \vec{a}_x + 67.005\vec{a}_z$

(c) $0.0098\ \vec{a}_x + 67.005\vec{a}_z$ (d) none of these

55 If $\vec{B}_1 = 9.8\vec{a}_x + 98\vec{a}_y + 980\vec{a}_z$, then the angle θ_1 for Fig. P5.1, is

(a) 20.8° (b) 10° (c) 40° (d) none of these

56 The correct value of the normal component of B_1 for Fig. P5.1 is

(a) $0.098\ \vec{a}_x$ (b) $980\ \vec{a}_z$ (c) $0.0098\ \vec{a}_y$ (d) none of these

57 The energy density stored in magnetic fields can be expressed as

(a) $w_m = \int (\vec{B}.\vec{H})dt$ (b) $w_m = \int (B^2.\vec{H})dt$

(c) $w_m = \int (\vec{B}.\vec{H})^2 dt$ (d) none these

58 Magnetic property of a material depends on

(a) susceptibility (b) resistivity (c) conductivity (d) permittivity

59 Which of the following is a non-linear material?

(a) diamagnetic (b) ferromagnetic (c) paramagnetic (d) none

60 Magnetic susceptibility is negative for

(a) paramagnetic materials (b) ferromagnetic materials

(c) diamagnetic materials (d) cobalt

61 Which of the following materials can be magnetised strongly?

(a) paramagnetic (b) ferromagnetic (c) diamagnetic (d) none

62 For free space, the susceptibility is

(a) one (b) zero (c) infinity (d) depends on air

63 What is the unit of magnetic charge?

(a) A-m^2 (b) A/m^2 (c) A/m (d) A-m

Answers

1. *b* 2. *b* 3. *c* 4. *b* 5. *d* 6. *a* 7. *c* 8. *d* 9. *a* 10. *a*

11. *a* 12. *d* 13. *a* 14. *b* 15. *d* 16. *c* 17. *b* 18. *a* 19. *a* 20. *d*

21. *a* 22. *b* 23. *a* 24. *a* 25. *c* 26. *a* 27. *c* 28. *b* 29. *a* 30. *b*

31. *a* 32. *c* 33. *a* 34. *c* 35. *d* 36. *b* 37. *a* 38. *b* 39. *d* 40. *a*

41. *b* 42. *c* 43. *b* 44. *b* 45. *b* 46. *d* 47. *d* 48. *c* 49. *b* 50. *a*

51. *b* 52. *a* 53. *c* 54. *a* 55. *c* 56. *b* 57. *a* 58. *a* 59. *b* 60. *c*

61. *b* 62. *b* 63. *d*

Time-varying Fields | 6

Learning Objects

The main objectives of this chapter are to understand thoroughly the following topics:

- Faraday's law, transformer and generator emfs
- Induced emf in AC generators
- Equation of continuity for time-varying fields
- Inconsistency of Ampere's law, modified Ampere's circuital law
- Displacement current density, differences between conduction, convection and displacement currents
- Maxwell's equations in different forms
- Boundary conditions: Dielectric–dielectric and dielectric–conductor interfaces.

6.1 Introduction

In the previous chapters on electrostatics and magnetostatics, we learnt about static electric fields produced by stationary charge distributions and static magnetic fields associated with charges in motion or steady currents. In this chapter, we will look at electric and magnetic fields that vary with time and study the dynamic interdependence of each on the other. We shall observe that time-varying or dynamic fields produce time-varying voltages and currents and vice versa. Finally, we will explore Faraday's law, Lenz's law and Maxwell's equations that describe the relationship among the field vectors in the time-varying case.

6.2 Faraday's Law

Michael Faraday in 1831 discovered experimentally that a current is induced in a conducting loop when there is a change in the magnetic flux linked with the loop. His experiment comprised two insulated coils wound on an iron ring as shown in Fig. 6.1. The first coil is connected to a voltage source through a switch, while the second is connected to a galvanometer. With the switch closed, he observed deflection in the meter in one direction; the deflection was in the opposite direction when the switch was opened. He also observed that if the switch remained closed for a long time, the meter showed no deflection. From this he inferred that when there is a change in the flux linkages with time, there is an emf induced in the circuit.

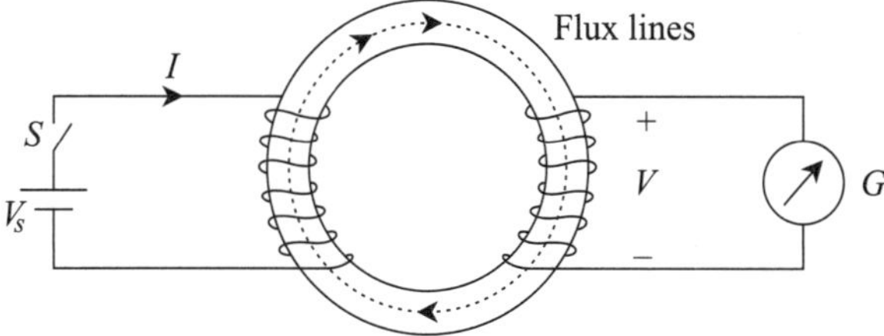

Fig. 6.1 *Faraday's experiment*

Faraday's law states that *in a magnetic field, if a closed circuit is in motion or if the magnetic field is time-varying, then the induced electromagnetic force (emf) or voltage across the terminals of the closed circuit is equal to the time rate of change of magnetic flux linkage by the circuit.*

It can be expressed as

$$V = -\frac{Nd\phi}{dt}, \tag{6.1}$$

where N is the number of turns of the circuit and φ is the flux through each turn. The negative sign is explained by Lenz's law given below.

Heinrich Lenz in 1834 stated a simple rule to find the direction of induced emf. The rule states that *the direction of the induced emf is always such that it tends to set up an induced current separately, which produces a flux opposing the flux linkages.* In other words, the direction of current flow in the closed circuit is such that the induced magnetic field produced by the current will oppose the original magnetic field.

The negative sign in Eq. (6.1) thus indicates that according to Lenz's law, the induced emf will produce a magnetic field which will oppose the original field. It can be observed that Faraday's law links dynamic electric and magnetic fields.

Induced emfs may be of the following kinds.

1. Transformer emf
2. Generator emf
3. Both transformer and generator emf

Transformer emf Consider a transformer (stationary circuit) as shown in Fig. 6.2. When a time-varying current is applied, it produces a time-varying magnetic flux in the primary coil that induces an emf in the secondary coil. This emf is called transformer emf, or, statically induced emf.

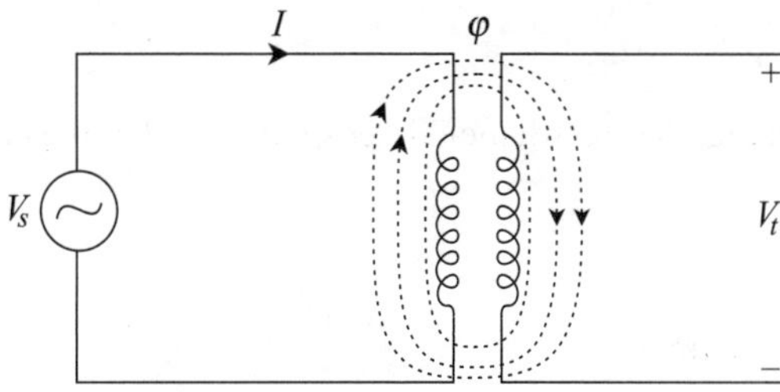

Fig. 6.2 *Transformer emf*

According to Faraday's law (for $N = 1$),

$$V_t = \frac{-d\varphi}{dt}.$$

We know that $\varphi = \int_S \vec{B} \cdot d\vec{S}$ and $V_t = \oint \vec{E} \cdot d\vec{l}$,

$$\text{i.e.,} \quad V_t = \oint \vec{E} \cdot d\vec{l} = \frac{-d}{dt} \int_S \vec{B} \cdot d\vec{S} \tag{6.2}$$

where S is the surface area of the circuit bounded by the closed circuit of length l. Equation (6.2) is called the transformer emf. The induced electromagnetic field is equal to the surface integral of the time rate of decrease of flux density for a given closed circuit.

Using Stokes' theorem,

$$\oint \vec{E} \cdot d\vec{l} = \oint_s (\nabla \times \vec{E}) \cdot d\vec{S} = -\int_S \frac{\partial \vec{B}}{\partial t} \cdot d\vec{S}.$$

By eliminating the surface integrals,

$$\nabla \times \vec{E} = \frac{-\partial \vec{B}}{\partial t}. \tag{6.3}$$

The curl of the time-varying electric field is equal to the time rate of decrease of flux density.

This is the fourth law of Maxwell's equations and shows that the time-varying field is nonconservative. We know, in contrast, that a static electric field is conservative as its curl is zero.

i.e., $\nabla \times \vec{E} = 0$

$$\text{and} \quad \oint \vec{E} \cdot d\vec{l} = 0. \tag{6.4}$$

Generator emf (Motional emf) Consider a stationary magnetic field is applied on a conducting circuit which is moving or revolving on its axis with respect to time. When the circuit cuts across the magnetic field, an electromagnetic field is induced across the circuit terminals. This emf is called generator emf and the action is called generator action.

If the circuit is moving with linear velocity $\vec{u}$ in the magnetic field of uniform flux density $\vec{B}$, from Lorentz's force equation, the force on the free charge in the circuit is

$$\vec{F} = Q\vec{u} \times \vec{B}.$$

When a force acts on a charge, the electric field intensity induced in the circuit is

$$\vec{E} = \frac{\vec{F}}{Q} = \vec{u} \times \vec{B}.$$

Then the induced emf in the circuit is

$$V_m = \oint_l \vec{E} \cdot d\vec{l} = \oint_l (\vec{u} \times \vec{B}) \cdot d\vec{l}. \tag{6.5}$$

This emf is called generator emf or motional emf, or dynamic emf. It arises due to a motional action.

Thus, in a magnetic field with flux density $\vec{B}$, if a conductor of length l is moving with a velocity $\vec{u}$, an electric voltage is produced across the conductor.

If $\vec{B}$ and $\vec{u}$ are perpendicular to each other, then the induced emf from Eq. (6.5) is $V = uBl$ volts.

Dynamic emf is found in electrical machines such as dc generators and alternators.

Applying Stokes' theorem,

$$\int_S (\nabla \times \vec{E}) \cdot d\vec{S} = \int_S \nabla \times (\vec{u} \times \vec{B}) \cdot d\vec{S}.$$

By eliminating the surface integrals, the differential form of the field is

$$\nabla \times \vec{E} = \nabla \times \vec{u} \times \vec{B}. \tag{6.6}$$

Moving circuit in a time-varying field In general, if a conducting circuit is moving in a time-varying magnetic field, both transformer emf and motional emf will be present in the conductor.

The total induced emf is

$$V = \oint_l \vec{E} \cdot d\vec{l} = -\int_S \frac{\partial \vec{B}}{\partial t} \cdot d\vec{S} + \oint_l (\vec{u} \times \vec{B}) \cdot d\vec{l}. \tag{6.7}$$

The first term of Eq. (6.7) gives the voltage induced by the time-varying magnetic flux and the second term gives the voltage induced by the circuit in motion.

Applying Stokes' theorem,

$$\int_S (\nabla \times \vec{E}) \cdot d\vec{S} = -\int_S \frac{\partial \vec{B}}{\partial t} \cdot d\vec{S} + \int_S \nabla \times (\vec{u} \times \vec{B}) \cdot d\vec{S} \tag{6.8}$$

and eliminating the surface integrals, the differential form of the total field is

$$\nabla \times \vec{E} = -\frac{\partial \vec{B}}{\partial t} + \nabla \times \vec{u} \times \vec{B}. \tag{6.9}$$

Example 6.1 A circular loop of 10 cm radius is located in the $x-y$ plane in a field given by $\vec{B} = 0.5\cos 377t\ (3\vec{a}_y + 4\vec{a}_z)$ tesla. Find the emf induced in the loop.

Solution Given $\vec{B} = 0.5(3\vec{a}_y + 4\vec{a}_z)\cos 377t$; $r = 10$ cm $= 0.1$m.

We know that the induced emf is

$$V = -\frac{d\varphi}{dt} \text{ and } \varphi = \int \vec{B} \cdot d\vec{S}.$$

But, $V = -\dfrac{d}{dt}\displaystyle\int \vec{B} \cdot d\vec{S} = -\int_S \frac{\partial \vec{B}}{\partial t} \cdot d\vec{S}$

$$= -\int_S 0.5 \times 377 \sin(377t)(3\vec{a}_y + 4\vec{a}_z) \cdot dS\vec{a}_z = -\int_S 0.5 \times 377 \sin(377t) \times 4\,dS$$

$$V = -188.5\sin(377t) \times 4 \times S.$$

Now $S = \pi r^2 = \pi(0.1)^2$.

$$V = -754\sin(377t)(0.01\pi)$$

$$= -23.687\sin(377t) \text{ volts.}$$

6.3 Induced EMF in an AC Generator

Consider an ac generator with a single rectangular wire loop of length a metres and breadth b metres mounted on a rotor. The loop is rotating in a steady magnetic field of B tesla at an angular velocity ω rad/s, as shown in Fig. 6.3. Assume that the axis of loop rotation is perpendicular to the magnetic field. Let the loop be placed in the $x-y$ plane and the field be in the z-direction. Since only two sides of the square loop affect the magnetic field, the electric length of the conductor is $l = 2a$ metres. The induced voltage appears at the terminals of the slip rings.

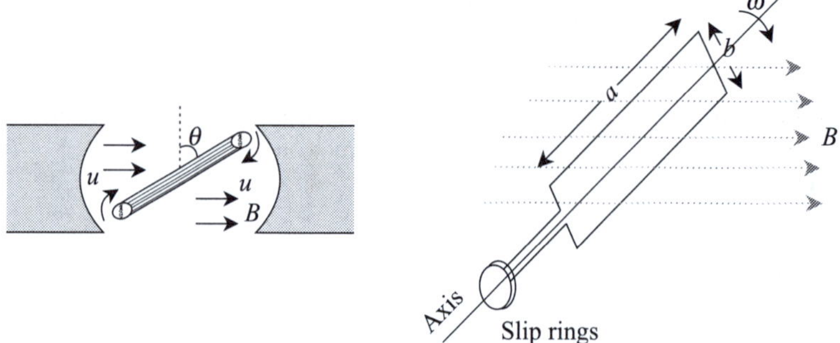

Fig. 6.3 *Loop rotating in a magnetic field*

We know that the linear velocity,

$$u = \frac{b\omega}{2} \text{ m/s.} \tag{6.10}$$

Let θ be the angle between $\vec{u}$ and $\vec{B}$. Then the angle of rotation is $\theta = \omega t$.

Area of the loop $S = ab$ m^2.

Case 1 Steady magnetic field

We know that the induced emf is

$$V = \oint (\vec{u} \times \vec{B}) \cdot d\vec{l}$$

$$= \oint uB\sin\theta dl = uB\sin\theta(l) = Bul\sin\theta$$

$$= \left(\frac{b\omega}{2}\right)(B)(2a)\sin\theta = ab\ \omega B\sin\theta \ \text{ volts}$$

or $\quad V = S\ \omega B\sin\omega t\ $ volts. (6.11)

Note: If the loop is a square with side a $(S = a^2)$, then the induced emf is

$$V = a^2\omega B\sin\omega t \text{ volts.} \tag{6.12}$$

Case 2 Varying magnetic field

When the loop is rotating in a time-varying magnetic field, two types of emfs will be induced in the loop: (1) transformer emf (V_t) and (2) motional emf (V_m). The total emf induced in the loop is the sum of these emfs, i.e.,

$$V = V_t + V_m. \tag{6.13}$$

Let, for the time-varying field,

$$B = B_0\sin\omega t.$$

Now, $\quad V_t = -\int \frac{\partial \vec{B}}{\partial t}\cdot d\vec{S} = -\int \frac{\partial B}{\partial t} dS\cos\theta,$

where $\theta = \omega t$ is the angle between $\vec{B}$ and the axis of the loop.

$$V_t = -\int \frac{\partial}{\partial t}(B_0\sin\omega t)dS\ \cos\theta \quad = -\int B_0 dS(\omega\cos\omega t)(\cos\omega t)$$

$$= -B_0\omega\cos^2\omega t\int dS = -B_m S\omega\cos^2\omega t. \tag{6.14}$$

We know that the motional emf is

$$V_m = S\ \omega B\sin\omega t$$

$$= S\ \omega B_0(\sin\omega t)(\sin\omega t) = S\ \omega B_0\sin^2\omega t. \tag{6.15}$$

$\therefore$ The total emf is

$$V = V_t + V_m \quad = S\ \omega B_0(\sin^2\omega t - \cos^2\omega t)V = -B_0\omega S\ \cos(2\omega t). \tag{6.16}$$

Thus the induced emf will have twice the frequency of the field.

Note: If the rotor is rotating with n rpm (revolutions per minute), then the angular velocity is

$$\omega = \frac{2\pi n}{60} \text{ rad/s.} \tag{6.17}$$

6.4 Induced EMF in a Coil

Consider a coil of N turns with radius a placed in a uniform time-varying magnetic field. (Fig. 6.4). Let θ be the angle between the coil axis and the direction of the magnetic field.

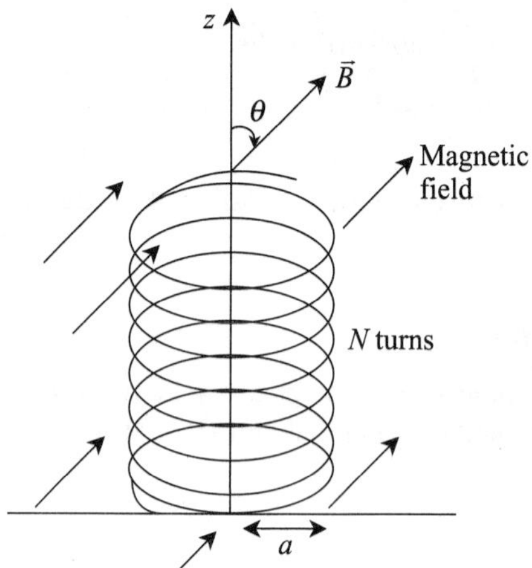

Fig. 6.4 *Coil of N turns in time-varying magnetic field*

Since the field is varying, $B = B_0 \sin \omega t$. The area of coil, $S = \pi a^2$, $\vec{B}.\vec{dS} = BdS\cos(\frac{\pi}{2} - \theta)$.

The emf induced across the terminals of the coil is

$$V = -\int \frac{\partial \vec{B}}{\partial t} \cdot d\vec{S} = -\frac{\partial B}{\partial t} \int dS \cos\left(\frac{\pi}{2} - \theta\right)$$

$$= -S\frac{\partial}{\partial t}(B_0 \sin \omega t) \sin \theta = -B_0 S(\omega \cos \omega t) \sin \theta$$

$$= -B_0\, S\, \omega \sin \theta \cos \omega t. \tag{6.18}$$

Let $V_{\max} = -B_0 \omega S \sin \theta$.

Then, $V = V_{\max} \cos \omega t$.

If the flux for N turns is $N\varphi = B_0\, S$,

then $V_{\max} = -\varphi\, \omega N \sin \theta$.

The magnitude of the rms voltage is $|V_{\text{rms}}| = \dfrac{|V_{\max}|}{\sqrt{2}}$.

$$\therefore \quad |V_{\text{rms}}| = \frac{1}{\sqrt{2}} \varphi\, 2\pi f\, N \sin \theta = 4.44 \varphi\, f\, N \sin \theta \text{ volts,} \tag{6.19}$$

where f is the frequency of rotation $(\omega = 2\pi f)$. Thus the maximum induced emf in the coil depends on the frequency and the magnetic flux of the time-varying magnetic field.

Example 6.2 A square coil side 60 cm is situated in a uniform magnetic field with its axis normal to the field. If the coil is rotated at 1500 rpm about its central axis, find the maximum value of the induced emf. Consider $B = 0.8\ \text{Wb/m}^2$.

Solution Given $a = 60$ cm $= 0.6$ m; rpm $= 1500$; $B = 0.8$ Wb/m^2; $\theta = \dfrac{\pi}{2}$; $N = 1$.
Area of the loop, $S = a^2 = 0.6^2$ m$^2 = 0.36$ m^2.

Maximum induced emf is $V_{\max} = \varphi\ \omega N\ \sin\theta$.

$$\varphi = BS = 0.8 \times 0.36 = 0.288 \text{ Wb.}$$

$$\omega = \text{angular velocity} = \frac{2\pi(\text{rpm})}{60} = \frac{2\pi(1500)}{60} = 157\text{rad/s.}$$

$$\therefore \quad |V_{\max}| = 0.288 \times 157 = 45.2 \text{ volts.}$$

Example 6.3 A square coil of area 0.02 m^2 and 100 turns is rotated at 1000 rpm about its axis at right angle to a uniform magnetic field, $B = 1$ tesla. Calculate the instantaneous value of emf induced in the coil when its plane is (a) at right angle to the field, (b) in the plane of the field and (c) when the coil is 45° to the field.

Solution Given area of the coil $S = 0.02$ m^2; $N = 100$, speed $= 1000$ rpm; $B = 1$ tesla.
Maximum emf inducted, $|V_{\max}| = \varphi\ \omega N\ \sin\theta$.
But $\varphi = BS = 1 \times 0.02 = 0.02$ Wb

$$\text{and} \quad \omega = \text{angle velocity} = \frac{2\pi \times \text{rpm}}{60} = \frac{2\pi \times 1000}{60} = 104.7 \text{ r/s.}$$

$$\therefore \quad |V_{\max}| = 100 \times 0.02 \times 104.7 \sin\theta \text{ volts} = 209.4 \sin\theta \text{ volts.}$$

(a) When the plane of the coil is at right angles to the field, $\theta = \pi/2$.

$$V = 209.4 \sin(\pi/2) = 209.4 \text{ volts.}$$

(b) When the plane of the coil is along the field, $\theta = 0$.

$$V = 209.4 \sin(0) = 0 \text{ volts.}$$

(c) When the plane of the coil is at angle 45° to the field, $\theta = 45$,

$$V = 209.4 \sin(45) = 148.06 \text{ volts.}$$

Example 6.4 A straight conductor of 2 m lies along the x-axis from the origin. When a field of $\vec{B} = 0.04\vec{a}_x$ tesla is applied with a velocity of $\vec{u} = 2.5 \sin 10^3 t\ \vec{a}_z$ m/s, calculate the motional emf.

Solution Given $l = 2$ m; $\vec{B} = 0.04\ \vec{a}_x$; $\vec{u} = 2.5\ \sin 10^3 t\ \vec{a}_z$.

We know that the motional emf, $V = \int (\vec{u} \times \vec{B}) \cdot d\vec{l}$.

Since $\vec{B}$ and $\vec{u}$ are perpendicular to each other,

$$V = \int_0^2 2.5(0.04) \sin 10^3 t \cdot dl$$

$$= 0.1(\sin\ 1000t)(l) = 0.2 \sin(1000t) \text{ volts.}$$

6.5 Equation of Continuity for Time-Varying Fields

Consider an alternating current is flowing through a conductor, due to conservation of charge, the total current flowing out from a volume should be equal to the rate of decrease of charge within the volume.

To get the equation for this process, we know that

$$I = \frac{-dQ}{dt} \text{ or } \oint \vec{J}\cdot d\vec{S} = \frac{-d}{dt}\int_v \rho_v \, dv,$$

where $\vec{J}$ is the current density and ρ_v is the volume charge density.

$$\oint \vec{J}\cdot d\vec{S} = -\int_v \frac{d\rho_v}{dt} dv.$$

From the divergence theorem,

$$\oint \vec{J}\cdot d\vec{S} = \int_v (\nabla\cdot\vec{J}) dv$$

$$\text{or} \quad \int_v (\nabla\cdot\vec{J}) dv = \int_v \frac{-\partial\rho_v}{\partial t}\cdot dv \tag{6.20}$$

Eliminating volume integral on both sides, for the same volume,

$$\nabla\cdot\vec{J} = \frac{-\partial\rho_v}{\partial t}. \tag{6.21}$$

This equation is called the equation of continuity for time-varying fields, which is same as Eq. (3.30).

6.6 Modified Ampere's Circuital Law for Time-Varying Fields

According to Ampere's circuital law, for static electromagnetic fields, the current density is

$$\nabla\times\vec{H} = \vec{J}.$$

Taking divergence of both sides,

$$\nabla\cdot(\nabla\times\vec{H}) = \nabla\cdot\vec{J} = 0, \tag{6.22}$$

since the divergence of the curl of any vector field is zero.

But the continuity equation for time-varying fields is given by from Eq. (6.21),

$$\nabla\cdot\vec{J} = \frac{-\partial\rho_v}{\partial t}. \tag{6.23}$$

As the quantity $\nabla \cdot \vec{J}$ does not vanish in time-varying fields, Eqs (6.22) and (6.23) are incompatible. Therefore, Ampere's circuital law is not consistent with time-varying fields.

James Clark Maxwell developed a modification to Ampere's law by substituting Gauss' law in the equation of continuity,

Now, $\nabla \cdot \vec{J} = \dfrac{-\partial \rho_v}{\partial t}$

From Gauss' law, $\nabla \cdot \vec{D} = \rho_v$.

$$\therefore \quad \nabla \cdot \vec{J} = \frac{-\partial}{\partial t}(\nabla \cdot D) = -\nabla \cdot \frac{\partial \vec{D}}{\partial t} \tag{6.24}$$

$$\text{or} \quad \nabla \cdot \vec{J} + \nabla \cdot \frac{\partial \vec{D}}{\partial t} = 0.$$

$$\nabla \cdot \left(\vec{J} + \frac{\partial \vec{D}}{\partial t} \right) = 0.$$

In integral form, using the divergence theorem,

$$\oint_S \left(\frac{\partial \vec{D}}{\partial t} + \vec{J} \right) \cdot d\vec{S} = 0 \tag{6.25}$$

Thus the total current flowing through a closed surface is zero.

Therefore the total current density for time-varying fields is

$$\vec{J}_t = \vec{J} + \frac{\partial \vec{D}}{\partial t}. \tag{6.26}$$

Since $\vec{D}$ is the displacement charge density, $(\partial \vec{D} \,/\, \partial t)$ is called the displacement current density. The additional $(\partial \vec{D} \,/\, \partial t)$ term makes Ampere's circuital law consistent with the principle of conservation of charge. Therefore, the modified Ampere's circuital law is

$$\nabla \times \vec{H} = \vec{J}_t \quad \text{or} \quad \nabla \times \vec{H} = \vec{J} + \frac{\partial \vec{D}}{\partial t}.$$

It is observed that a time-varying electric field will produce a magnetic field even when there is no current flow in the medium. This indicates that the displacement current density flowing through a dielectric material can also develop a magnetic field.

Using Stokes' theorem, the integral form is

$$\oint_l \vec{H} \cdot d\vec{l} = \int_s \left(\vec{J} + \frac{\partial \vec{D}}{\partial t} \right) d\vec{S} = I_{\text{encl}}. \tag{6.27}$$

This states that the magnetomotive force around a closed path is equal to the total current enclosed by the path.

6.7 Displacement Current

Consider a current flowing through a capacitor as shown in Fig. 6.5. If i_c is the current through the circuit, then the conduction current density is $\vec{J} = i_c/S = \sigma\vec{E}$, where S is the area of cross-section and E, the electric field across the capacitor.

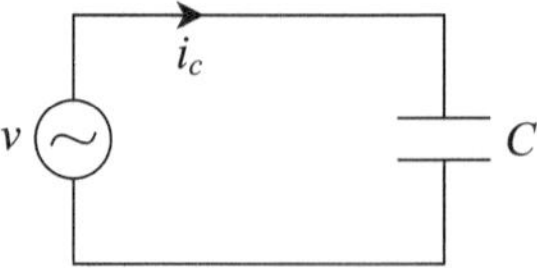

Fig. 6.5 *Capacitor circuit*

The current flowing through the capacitor is

$$i_d = C\frac{dV}{dt} = \frac{\varepsilon S}{d}\frac{dV}{dt},$$

where ε is the permittivity of the dielectric material in the capacitor and d, the separation between plates.

We know that $V = \vec{E}d = \dfrac{\vec{D}\,d}{\varepsilon}$.

$$\text{So} \quad i_d = \frac{\varepsilon S}{d}\cdot\frac{\partial}{\partial t}\left(\frac{\vec{D}\cdot d}{\varepsilon}\right) = S\cdot\frac{\partial \vec{D}}{\partial t}. \tag{6.28}$$

This current is called the *displacement current* flowing through the capacitor.

The displacement current in the dielectric must be equal to the conduction current in the circuit,

i.e., $i_c = i_d$.

The displacement current density is $J_d = \dfrac{i_d}{S} = \dfrac{\partial \vec{D}}{\partial t}$. (6.29)

Therefore, the total current density

$$J = J_c + J_d$$

$$\text{or} \quad J = \sigma\vec{E} + \frac{\partial \vec{D}}{\partial t}. \tag{6.30}$$

This shows that a time-varying current will flow through the capacitor circuit even though there is no flow of current between the capacitor plates.

Note: The conduction current i_c flows through the conductor while the displacement current i_d flows through the capacitor.

Example 6.5 Moist soil has a conductivity of 10^{-3} siemens/m and $\varepsilon_r = 2.5$. Find J_c and J_d where, $E = 6\times 10^{-6}\sin(9\times 10^9 t)$ V/m.

Solution Given conductivity, $\sigma = 10^{-3}$ S/m; relative permittivity $\varepsilon_r = 2.5$; and electric field intensity $E = 6 \times 10^{-6} \sin(9 \times 10^9 t)$ V/m.

We know that conduction current density is

$$J_c = \sigma E = 10^{-3} \times 6 \times 10^{-6} \sin(9 \times 10^9\, t)$$

$$= 6 \times 10^{-9} \sin(9 \times 10^9 t) \text{ A/m}^2$$

or $$= 6 \sin(9 \times 10^9 t) \text{ nA/m}^2$$

Displacement current density,

$$J_d = \frac{\partial D}{\partial t} = \varepsilon \frac{\partial E}{\partial t} = \varepsilon_o \varepsilon_r \frac{\partial E}{\partial t}$$

$$= 8.854 \times 10^{-12} \times 2.5 \times \frac{\partial}{\partial t}[6 \times 10^{-6} \sin(9 \times 10^9\, t)]$$

$$= 8.854 \times 10^{-12} \times 2.5 \times 6 \times 10^{-6} \cos(9 \times 10^9\, t) \times 9 \times 10^9$$

$$= 1.1952 \cos(9 \times 10^9\, t)\ \mu\text{A/m}^2.$$

6.8 Ratio between Conduction Current Density and Displacement Current Density

Let $\vec{E}$ be a time-varying field given by

$$\vec{E} = E\, e^{j\omega t}. \tag{6.31}$$

Then $\dfrac{\partial \vec{E}}{\partial t} = j\omega \vec{E}$.

The conduction current density is $J_c = \sigma\, \vec{E}$ and the displacement current density is

$$J_d = \frac{\partial \vec{D}}{\partial t} = \varepsilon \cdot \frac{\partial \vec{E}}{\partial t} = j\omega\varepsilon\vec{E}. \tag{6.32}$$

The ratio of conduction current density to displacement current density is

$$\frac{J_c}{J_d} = \frac{\sigma \vec{E}}{j\omega\varepsilon\vec{E}} = \frac{\sigma}{j\omega\varepsilon},$$

whose magnitude is $$\left|\frac{J_c}{J_d}\right| = \frac{\sigma}{\omega\varepsilon}. \tag{6.33}$$

At low frequencies, the displacement current density J_d is usually negligible compared to conduction current density J_c. But at radio frequencies (high frequencies), in wave propagation, the value of J_d becomes comparable with J_c.

Note: The current densities depend on the medium constants σ, ε and frequency ω. The medium is classified based on the current densities in the wave, i.e.,

1. If $\dfrac{\sigma}{\omega\varepsilon} >> 1$, the medium is a perfect conductor.
2. If $\dfrac{\sigma}{\omega\varepsilon} << 1$, the medium is a perfect dielectric.
3. If $\dfrac{\sigma}{\omega\varepsilon} = 0$, the medium is an insulator or free space.

It is also noted that a conducting material at very low frequency may become dielectric at very high frequencies.

Example 6.6 Find the frequency at which conduction current density and displacement current density are equal in a medium with $\sigma = 2 \times 10^{-4}$ mho/m and $\varepsilon_r = 81$.

Solution Given $\sigma = 2 \times 10^{-4}$ mhos/m, and $\varepsilon_r = 81$.
Let $E = E_m \sin \omega t$, where $\omega = 2\pi f$.

Then $\left|\vec{J_c}\right| = |\sigma E_m \sin \omega t| = \sigma E_m$

and $\left|\vec{J_d}\right| = \left|\dfrac{\partial \vec{D}}{\partial t}\right| = \varepsilon \left|\dfrac{\partial}{\partial t}(E_m \sin \omega t)\right| = \varepsilon |E_m \omega|$

or $\left|\vec{J_d}\right| = \varepsilon E_m \omega.$

When $|J_c| = |J_d|$, $\sigma E_m = \varepsilon E_m \omega$

i.e., $\omega = \dfrac{\sigma}{\varepsilon}$ or $2\pi f = \dfrac{\sigma}{\varepsilon}.$

$$f = \frac{\sigma}{2\pi\varepsilon_0\varepsilon_r} = \frac{2 \times 10^{-4}}{2\pi \times 8.854 \times 10^{-12} \times 81} = 44.38 \times 10^3 \text{ Hz}$$

6.9 Conduction, Convection and Displacement Currents

The summary of different types of currents and their current densities are given in Tables 6.1 and 6.2 respectively.

Table 6.1 *Differences between conduction, convection and displacement currents*

	Conduction current	**Convection current**	**Displacement current**
1.	Conduction current is the flow of electrons through any conducting medium.	Convection current is the flow of electrons through a non-conducting (insulating) medium.	Displacement current is the flow of charge in dielectrics which results due to a time-varying electric fields.
2.	Conduction current is the current passing through resistors and conductors.	It is the leakage current passing through the dielectric (insulating) medium of the capacitor.	It is the rate of flow of charge between the capacitor plates in a capacitor circuit.
3.	It is given by $I_c = \int_S \vec{J} \cdot d\vec{S} = \int_S \sigma \vec{E} \cdot d\vec{S}$ where σ = conductivity, $\vec{E}$ = electric field intensity and $\vec{S}$ =surface area	It is given by $I = \int\int \vec{J} \cdot d\vec{S} = \int\int \rho_v \vec{u} \cdot d\vec{S}$, where $\vec{J} = \rho_v \vec{u}$, the convection current density, and ρ_v = volume density and $\vec{u}$ = velocity.	It is given by $I_d = \varepsilon S \frac{\partial E}{\partial t}$ where, ε = permittivity of the dielectric material.
4.	It is independent of frequency.	Its value increases with frequency.	It is directly proportional to frequency.
5.	It obeys Ohm's law and hence has linear charge characteristics.	It does not obey Ohm's law and so has non-linear characteristics.	It also does not obey Ohm's law and so has non-linear characteristics.
6.	It exists in both time variant and invariant cases.	It also exists in both time variant and invariant cases.	It exists only in time variant case.
7.	Typical examples are current through conductors, resistors, etc.	Typical examples are electron beam moving through vaccum, CRT, liquids, etc.	Typical examples are the currents flowing through capacitor and all imperfect conductors carrying a time-varying conduction current.
8.	It flows only through closed paths in circuits.	It flows only through open paths in circuits.	It provides a closed path in circuits having capacitor elements or in open circuits where the conduction current cannot flow further.

Table 6.2 *Differences between displacement current density and conduction current density*

	Displacement current density	**Conduction current density**
1.	It is defined as the displacement current at a given point, passing through a unit surface area when the surface is normal to the direction of the displacement current. It is denoted by $\vec{J}_d$.	It is defined as the conduction current at a given point, passing through a unit surface area normal to the direction of the current. It is denoted by J_c .
2.	Displacement current results when a potential is applied across the dielectric medium.	Conduction current results in conductors according to Ohm's law.
3.	Displacement current density is given by $\vec{J}_d = \dfrac{\partial \vec{D}}{\partial t}$.	Conduction current density is given by $\vec{J}_C = \dfrac{i_C}{A} = \sigma \vec{E}$.
4.	Only displacement current density exists and conduction current density is negligible in a dielectric medium.	Only conduction current exists and displacement current density is negligible in conductors.
5.	Displacement current density exists only in time-varying fields.	Conduction current density exists both in steady and time-varying fields.

6.10 Maxwell's Equations for Static Fields

As stated earlier, in the previous chapters we had considered static electric field due to charges at rest and static magnetic field due to charges in motion or steady currents. The equations so derived are summarised in Table 6.3.

For time-varying fields, according to Faraday's law, an emf is induced either by a time-varying magnetic field or by a generator action.

Maxwell's equation for time-varying fields is a modification of Eq. (6.34).

$$\nabla \times \vec{E} = \frac{-\partial \vec{B}}{\partial t} \quad \text{and} \quad \oint \vec{E} \cdot d\vec{l} = -\int_S \frac{\partial \vec{B}}{\partial t} \cdot d\vec{S}. \tag{6.39}$$

James Clerk Maxwell, a professor in Cambride University, England, while performing experiments on electromagnetic fields in the year 1873, modified the laws of Faraday, Ampere and Gauss into a set of equations for time-varying fields. These equations are well known as Maxwell's equations. Maxwell developed mathematical analysis for electromagnetic waves (field theory) by defining the relationship between electric and magnetic fields using these equations.

Since Maxwell's equations are valid at every point in free space, to analyse electromagnetic fields for specific objects and boundaries, it is convenient to convert the point forms into integral forms.

The point or differential forms are obtained by using the del operator while the integral forms are obtained by using divergence or Stokes' theorem.

Table 6.3 *Maxwell's equations for static electric and magnetic fields*

Point form	Integral form	
$\nabla \times \vec{E} = 0$	$\oint \vec{E} \cdot d\vec{l} = 0$ (Potential around a closed path is zero)	(6.34)
$\nabla \times \vec{H} = \vec{J}$	$\oint \vec{H} \cdot d\vec{l} = \int_S \vec{J} \cdot d\vec{S} = I_{\text{encl}}$ (Ampere's circuital law)	(6.35)
$\nabla . \vec{D} = \rho_v$	$\oint_S \vec{D} \cdot d\vec{S} = \int_v \rho_v \, dv = Q$ (Gauss' law)	(6.36)
$\nabla \cdot \vec{B} = 0$	$\oint_S \vec{B} \cdot d\vec{S} = 0$ (Magnetic field is continuous, conservation of flux)	(6.37)
The continuity equation for steady current is		
$\nabla \cdot \vec{J} = 0$	$\oint_S \vec{J} \cdot d\vec{S} = 0$	(6.38)

The general expressions for Maxwell's equations valid for both static and time-varying fields are given in Table 6.4.

Table 6.4 *Maxwell's equations*

Point or differential form	Integral form	
$\nabla \times \vec{H} = \vec{J} + \dfrac{\partial \vec{D}}{\partial t}$	$\int_l \vec{H} \cdot d\vec{l} = \int_S \left(\dfrac{d\vec{D}}{dt} + \vec{J} \right) d\vec{S}$ (modified Ampere's circuital law)	(6.40)
$\nabla \times \vec{E} = \dfrac{-\partial \vec{B}}{\partial t}$	$\int_l \vec{E} \cdot d\vec{l} = -\int_S \dfrac{\partial \vec{B}}{\partial t} \cdot d\vec{S}$ (Faraday's law of induction)	(6.41)
$\nabla \cdot \vec{D} = \rho_v$	$\oint_s \vec{D} \cdot d\vec{S} = \int_v \rho_v dv$ (Gauss' law)	(6.42)
$\nabla \cdot \vec{B} = 0$	$\oint_s \vec{B} \cdot d\vec{S} = 0$ (conservation of flux)	(6.43)
The continuity equation is		
$\nabla \cdot \vec{J} = -\dfrac{\partial \rho_v}{\partial t}$	$\oint_s \vec{J} \cdot d\vec{S} = -\int_v \dfrac{\partial \rho_v}{\partial t} dv$	(6.44)

Eq. (6.40) is a modified form of Ampere's circuital law applicable to time-varying electric and magnetic fields. It is observed that current density $\vec{J}$ may comprise both conduction current density (σE) and convection current density $(\rho\mu)$ due to the motion of free charge. Eq. (6.41) is Faraday's law of electromagntic induction due to a time-varying field.

Eq. (6.42), Gauss' law, is valid for both static and time-varying fields and states that the divergence of charge density is equal to Q. Eq. (6.43) indicates that flux is conservative and that the total outward magnetic flux through any closed surface is zero. Also, that isolated magnetic charges do not exist.

Maxwell's equations stated in words

1 The first equation Eq. (6.40) states that the magnetomotive force (mmf) around a closed path is equal to the sum of the conduction current and the displacement current through any surface bounded by the path.

In other words:

The magnetic voltage around a closed path is equal to the electric current through the path.

2 The second equation Eq. (6.41) states that the electromotive force (emf) around a closed path is equal to the negative of the time derivative of the magnetic flux density (magnetic displacement) through any surface bounded by the path.

In other words:

The electric voltage around a closed path is equal to the magnetic current through the path.

3 The third equation, Eq. (6.42), states that the total electric flux density (electric displacement) through any surface enclosing a volume is equal to the total charge within that volume.

4 The fourth equation, Eq. (6.43), states that the net magnetic flux emerging through any closed surface is zero.

Proof of Maxwell's equations

First equation Consider that time-varying electric and magnetic fields exist in a medium.

Method 1 The continuity equation for time-varying fields is

$$\nabla\cdot\vec{J}=\frac{-\partial\rho_v}{\partial t}$$

or $$\nabla\cdot\vec{J}+\frac{\partial\rho_v}{\partial t}=0. \tag{6.45}$$

We know that

$$\nabla\cdot\nabla\times\vec{H}=0.$$

$$\therefore\quad \nabla\cdot\nabla\times\vec{H}=\nabla\cdot\vec{J}+\frac{\partial\rho_v}{\partial t}. \tag{6.46}$$

From Gauss' law,

$$\nabla\cdot\vec{D}=\rho_v.$$

Substituting in Eq. (6.46),

$$\nabla\cdot(\nabla\times\vec{H})=\nabla\cdot\vec{J}+\frac{\partial}{\partial t}(\nabla\cdot\vec{D}).$$

$$\nabla\cdot\nabla\times\vec{H}=\nabla\cdot\vec{J}+\nabla\cdot\frac{\partial\vec{D}}{\partial t}=\nabla\cdot\left(\vec{J}+\frac{\partial\vec{D}}{\partial t}\right) \tag{6.47}$$

Eliminating the del operator,

$$\nabla\times\vec{H}=\vec{J}+\frac{\partial\vec{D}}{\partial t}. \qquad \text{(Proved)}$$

Method 2 From Ampere's circuital law, we know that

$$\oint\vec{H}\cdot d\vec{l}=I_{\text{encl}}=\int_S\vec{J}\cdot d\vec{S}.$$

If $\vec{J}=\vec{J}_c+\vec{J}_d$ for time-varying fields.

$$\oint\vec{H}\cdot d\vec{l}=\int_S(\vec{J}_c+\vec{J}_d)\cdot d\vec{S}=\int_S\left(\vec{J}+\frac{\partial\vec{D}}{\partial t}\right)\cdot d\vec{S} \tag{6.48}$$

Applying Stokes' theorem,

$$\oint\vec{H}\cdot d\vec{l}=\int_S(\nabla\times\vec{H})\cdot d\vec{S}.$$

$$\therefore\quad\int_S(\nabla\times\vec{H})\cdot d\vec{S}=\int_S\left(\vec{J}+\frac{\partial\vec{D}}{\partial t}\right)\cdot d\vec{S}. \tag{6.49}$$

Eliminating surface integrals on both sides,

$$\nabla\times\vec{H}=\vec{J}+\frac{\partial\vec{D}}{\partial t} \qquad \text{(Proved).}$$

Second equation Consider Faraday's law of transformer action,

$$\oint_l\vec{E}\cdot d\vec{l}=-\int_S\frac{\partial\vec{B}}{\partial t}\cdot d\vec{S}. \tag{6.50}$$

Using Stokes' theorem,

$$\int_S(\nabla\times\vec{E})\cdot d\vec{S}=-\int_S\frac{\partial\vec{B}}{\partial t}\cdot d\vec{S}. \tag{6.51}$$

Eliminating surface integrals on both sides,

$$\nabla \times \vec{E} = -\frac{\partial \vec{B}}{\partial t}. \qquad \text{(Proved)}$$

Third equation According to Gauss' law, the total flux flowing out of a closed surface is equal to the charge enclosed by that surface.

$$\text{That is,} \quad \int_S \vec{D} \cdot d\vec{S} = Q_{\text{encl}}.$$

If ρ_v is the volume charge density and V is the volume enclosed by the surface, then

$$\int_S \vec{D} \cdot d\vec{S} = \int_v \rho_v \, dv. \tag{6.52}$$

Using divergence theorem,

$$\int_S \vec{D} \cdot d\vec{S} = \int_v (\nabla \cdot \vec{D}) dv$$

$$\therefore \quad \int_v (\nabla \cdot \vec{D}) dv = \int_v \rho_v \, dv. \tag{6.53}$$

Eliminating volume integrals on both sides

$$\nabla \cdot \vec{D} = \rho_v. \qquad \text{(Proved).}$$

Fourth equation In any magnetic field, due to conservation of flux, the total flux enclosed by a surface is always zero (non-existence of monopole in magnetic fields).

$$\int_S \vec{B} \cdot d\vec{S} = 0.$$

Using divergence theorem,

$$\int_S \vec{B} \cdot d\vec{S} = \int_v (\nabla \cdot \vec{B}) dv = 0. \tag{6.54}$$

Since the volume is finite, i.e., $v \neq 0$,

$$\nabla \cdot \vec{B} = 0.$$

Maxwell's equations for different cases are summerised in Table 6.5, Table 6.6 and Table 6.7.

Table 6.5 *Maxwell's equations for free space and time-varying fields* $(\sigma = 0,\ J = 0 \text{ and } \rho_v = 0)$

Point (differential) form	Integral form
1. $\nabla \times \vec{H} = \dfrac{\partial \vec{D}}{\partial t}$	$\int_l \vec{H} \cdot \vec{dl} = \int_S \dfrac{\partial \vec{D}}{\partial t} \cdot d\vec{S}$
2. $\nabla \times \vec{E} = -\dfrac{\partial \vec{B}}{\partial t}$	$\int_l \vec{E} \cdot \vec{dl} = -\int_S \dfrac{\partial \vec{B}}{\partial t} \cdot d\vec{S}$
3. $\nabla \cdot \vec{D} = 0$	$\oint_S \vec{D} \cdot d\vec{S} = 0$
4. $\nabla \cdot \vec{B} = 0$	$\oint_S \vec{B} \cdot d\vec{S} = 0$
Continuity equation $\nabla \cdot \vec{J} = 0$	

(6.55)

Table 6.6 *Maxwell's equations for free space and static fields*

Point form	Integral form
1. $\nabla \times \vec{H} = 0$	$\int_l \vec{H} \cdot \vec{dl} = 0$
2. $\nabla \times \vec{E} = 0$	$\int_l \vec{E} \cdot \vec{dl} = 0$
3. $\nabla \cdot \vec{D} = 0$	$\oint_S \vec{B} \cdot d\vec{S} = 0$
4. $\nabla \cdot \vec{B} = 0$	$\oint_S \vec{B} \cdot d\vec{S} = 0$
Continuity equation $\nabla \cdot \vec{J} = 0$	

(6.56)

Table 6.7 *Maxwell's equations for a good conductor* $(\sigma >> \omega\varepsilon,\ \rho_v = 0)$

Point form	Integral form
1. $\nabla \times \vec{H} = \vec{J}$	$\oint_l \vec{H} \cdot \vec{dl} = \int_S J \cdot d\vec{S} = I$
2. $\nabla \times \vec{E} = \dfrac{-\partial \vec{B}}{\partial t}$	$\oint_l \vec{E} \cdot \vec{dl} = -\int_S \dfrac{\partial \vec{B}}{dt} \cdot d\vec{S}$
3. $\nabla \times \vec{D} = 0$	$\oint_S \vec{D} \cdot d\vec{S} = 0$
4. $\nabla \cdot \vec{B} = 0$	$\oint_S \vec{B} \cdot d\vec{S} = 0$
Continuity equation $\nabla \cdot \vec{J} = 0$	

(6.57)

Example 6.7 Find the displacement current density next to your radio in air, where the local FM station provides a carrier having $\vec{H} = 0.2\cos[210(3\times10^8 t - x]\vec{a}_z$ A/m.

Solution Given $H = 0.2\cos[210(3\times10^8 t - x)]\vec{a}_z = 0.2\cos[6.3\times10^{10}t - 210x]\vec{a}_z$.

or $H_z = 0.2\cos(6.3\times10^{10}t - 210x)$.

We know that the displacement current density

$$J_d = \nabla\times\vec{H} = \begin{vmatrix} \vec{a}_x & \vec{a}_y & \vec{a}_z \\ \dfrac{\partial}{\partial x} & \dfrac{\partial}{\partial y} & \dfrac{\partial}{\partial z} \\ 0 & 0 & H_z \end{vmatrix}$$

$$J_d = \frac{\partial}{\partial x}[0.2\cos(6.3\times10^{10}t - 210x]\vec{a}_y$$

$$J_d = 42.03\sin(6.3\times10^{10}t - 210x]\vec{a}_y$$

6.11 Maxwell's Equation for Sinusoidal (Harmonic) Time-Varying Fields

Consider electric and magnetic fields varying harmonically with time. The fields can be expressed in terms of harmonic frequencies in exponential or phasor form.

The displacements with time can be expressed as

$$\vec{D} = De^{j\omega t} \tag{6.58}$$

and $$\vec{B} = B\,e^{j\omega t}, \tag{6.59}$$

where D and B are magnitudes of displacements and $\omega = 2\pi f$ rad/s is the angular frequency.

Then, $$\frac{\partial\vec{D}}{\partial t} = j\omega De^{j\omega t} = j\omega\vec{D} \tag{6.60}$$

and $$\frac{\partial\vec{B}}{\partial t} = j\omega\vec{B}. \tag{6.61}$$

Also, for conductors, $\vec{J} = \sigma\vec{E}$; for dielectrics, $\vec{D} = \varepsilon\vec{E}$; and for magnetic materials, $\vec{B} = \mu\vec{H}$.

Then, $$\nabla\times\vec{E} = -\frac{\partial\vec{B}}{\partial t} = -j\omega\vec{B} = -j\omega\mu\vec{H}.$$

$$\nabla\times\vec{H} = \vec{J} + \frac{\partial\vec{D}}{\partial t} = \sigma\vec{E} + j\omega\vec{D} = \sigma\vec{E} + j\omega\varepsilon\vec{E}$$

$$= (\sigma + j\omega\varepsilon)\vec{E} \tag{6.62}$$

and $\nabla\cdot\vec{J}=\dfrac{-\partial\rho_v}{\partial t}=j\omega\rho_v.$ (6.63)

Maxwell's equations for sinusoidal fields are summarised in Table 6.8.

Table 6.8 *Maxwell's equations for sinusoidal fields*

Point form	Integral form
$\nabla\times\vec{H}=(\sigma+j\omega\varepsilon)\vec{E}$	$\int_l \vec{H}\cdot d\vec{l}=(\sigma+j\omega\varepsilon)\int_S \vec{E}\cdot d\vec{S}$
$\nabla\times\vec{E}=-j\omega\mu\vec{H}$	$\int_l \vec{E}\cdot d\vec{l}=-j\omega\mu\int_S \vec{H}\cdot d\vec{S}$
$\nabla\cdot\vec{D}=\rho_v$	$\oint_S \vec{D}\cdot d\vec{S}=\int_v \rho_v dv$
$\nabla\cdot\vec{B}=0$	$\oint_S \vec{B}\cdot d\vec{S}=0$
The continuity equation $\nabla\cdot\vec{J}=-j\omega\rho_v$	$\oint_S \vec{J}\cdot d\vec{S}=-j\omega\int_v \rho_v dv$

(6.64)

Example 6.8 If $\vec{E}=200e^{(4y-\omega t)}\vec{a}_x$ V/ m in free space, use Maxwell's equation to find $\vec{H}$, knowing that all fields vary harmonically and for which $\nabla\times\vec{E}=-j\omega\mu\vec{H}$.

Solution Maxwell's equation for time-varying fields is given as

$$\nabla\times\vec{E}=-j\omega\mu\vec{H}.$$

In free space $\mu=\mu_0$; $\vec{H}=\dfrac{-1}{j\omega\mu_0}(\nabla\times\vec{E})$.

In rectangular coordinates, $\nabla\times\vec{E}=\begin{vmatrix}\vec{a}_x & \vec{a}_y & \vec{a}_z\\ \dfrac{\partial}{\partial x} & \dfrac{\partial}{\partial y} & \dfrac{\partial}{\partial_z}\\ E_x & E_y & E_z\end{vmatrix}$

Given $\vec{E}=20e^{(4y-\omega t)}\vec{a}_x$,

$$E_x=200e^{(4y-\omega t)},\quad E_y=E_z=0.$$

$$\therefore\quad \nabla\times\vec{E}=-\vec{a}_y\left(\frac{-\partial E_x}{\partial z}\right)+\vec{a}_z\left(\frac{-\partial E_x}{\partial y}\right)$$

$$=\vec{a}_y(0)+\vec{a}_z\left[\frac{-\partial}{\partial y}\left(200e^{(4y-\omega t)}\right)\right].$$

$$\nabla \times \vec{E} = -800e^{(4y-\omega t)}\vec{a}_z.$$

$$\therefore \quad \vec{H} = \frac{-1}{j\omega\mu_0}(-800e^{(4y-\omega t)})\vec{a}_z$$

$$= \frac{800}{j\omega\mu_0}e^{(4y-\omega t)}\vec{a}_z.$$

Example 6.9 The magnetic field intensity in free space is given as $\vec{H} = H_0 \sin\theta \vec{a}_y$ A/m, where $\theta = \omega t - \beta z$ and β is a constant quantity. Determine the displacement current density.

Solution

$$\vec{H} = H_0 \sin\theta \vec{a}_y = H_0 \sin(\omega t - \beta z)\vec{a}_y.$$

The displacement current density in free space is given by

$$\vec{J} = \nabla \times \vec{H}.$$

$$\text{Now,} \quad \nabla \times \vec{H} = \begin{vmatrix} \vec{a}_x & \vec{a}_y & \vec{a}_z \\ \frac{\partial}{\partial x} & \frac{\partial}{\partial y} & \frac{\partial}{\partial z} \\ 0 & H_0 \sin\theta & 0 \end{vmatrix}.$$

$$\nabla \times \vec{H} = \vec{a}_x\left[0 - \frac{\partial}{\partial z}(H_0 \sin\theta)\right] + 0 + \vec{a}_z\left[\frac{\partial}{\partial x}(H_0 \sin\theta) - 0\right]$$

$$= -\vec{a}_x\frac{\partial}{\partial z}(H_0 \sin(\omega t - \beta z)) + \vec{a}_z\frac{\partial}{\partial x}(H_0(\sin\omega t - \beta z))$$

$$= -\vec{a}_x H_0(-\beta)\cos(\omega t - \beta z) + \vec{a}_y H_0(0)$$

$$= H_0\beta\cos(\omega t - \beta z)\vec{a}_x.$$

$$\therefore \quad \vec{J} = \beta H_0 \cos(\omega t - \beta z)\vec{a}_x.$$

6.12 Boundary Conditions

Maxwell's equations give the relationship between the four field vectors, $\vec{H}, \vec{E}, \vec{D}$ and $\vec{B}$, at any point within a continuous medium. The conditions of the field vectors at the boundary surface between different media can be obtained by applying Maxwell's equation at the interface. The boundary conditions can be stated as

(1) The tangential component of $\vec{E}$ is continuous at the surface.

$$E_{\tan 1} = E_{\tan 2} \tag{6.65}$$

(2) The tangential component of $\vec{H}$ is continuous across the surface except at the surface of a perfect conductor.

$$H_{\tan 1} = H_{\tan 2}. \tag{6.66}$$

At the surface of a perfect conductor, the tangential component of $\vec{H}$ is discontinuous by an amount equal to the surface current density $K\ A/m$. Since the current will flow only on the surface of a perfect conductor,

$$H_{\tan 2} - H_{\tan 1} = K. \tag{6.67}$$

(3) The normal component of $\vec{D}$ is discontinuous across the surface of the boundary by the amount of the surface charge density, i.e.,

$$D_{n1} - D_{n2} = \rho_s. \tag{6.68}$$

(4) The normal components of $\vec{B}$ is always continuous across the boundary surface. That is,

$$B_{n1} = B_{n2}. \tag{6.69}$$

Proof of boundary condition for electric field intensity $(\vec{E})$ Consider the surface of discontinuity at $x = 0$ plane between the two dielectrics media as shown in Fig. 6.6.

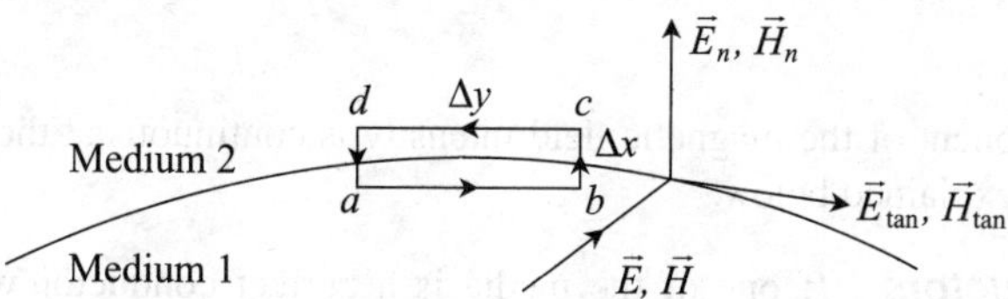

Fig. 6.6 *Boundary conditions between two media for E and H*

Consider a small rectangular surface of width Δx and length Δy on the boundary between two media. Maxwell's equation is given by

$$\oint \vec{E} \cdot d\vec{l} = -\int_s \frac{\partial \vec{B}}{\partial t} \cdot d\vec{S}.$$

For the rectangular path in the anti-clockwise direction,

$$E_{y2}\Delta y - E_{x2}\frac{\Delta x}{2} - E_{x1}\frac{\Delta x}{2} - E_{y1}\Delta y + E_{x3}\frac{\Delta x}{2} + E_{x4}\frac{\Delta x}{2} = -\frac{\partial B_z}{\partial t}\Delta x \Delta y, \tag{6.70}$$

where B_z is the magnetic flux density through the rectangular area $\Delta x \Delta y$. E_{y1} and E_{y2} are the tangential components of $\vec{E}$.

Now, at the boundary, $\Delta x \to 0$.

$$E_{y2}\Delta y - E_{y1}\Delta y = 0.$$

$$E_{y2} = E_{y1} \text{ (tangential components of } \vec{E}).$$

or $$E_{\tan 1} = E_{\tan 2}. \qquad \text{(Proved)} \tag{6.71}$$

Therefore, the tangential component electric field intensity is continuous at the boundary.

Proof of boundary condition for magnetic field intensity $(\vec{H})$

Consider Maxwell's equation,

$$\oint \vec{H} \cdot d\vec{l} = \int_s \left(\vec{J} + \frac{\partial \vec{D}}{\partial t}\right) \cdot d\vec{S}.$$

For the rectangular path in the anti-clockwise direction,

$$H_{y2}\Delta y - H_{x2}\frac{\Delta x}{2} - H_{x1}\frac{\Delta x}{2} - H_{y1}\Delta y + H_{x3}\frac{\Delta x}{2} + H_{x4}\frac{\Delta x}{2} = (J + \frac{\partial D}{\partial t})\Delta x \Delta y \tag{6.72}$$

At the boundary, $\Delta x \to 0$, $D \neq 0$ and $J \neq 0$. Then,

$$H_{y2}\Delta y - H_{y1}\Delta y = 0.$$

$$H_{y2} = H_{y1} \text{ (tangential components of } \vec{H}\text{)}.$$

i.e., $H_{\tan 1} = H_{\tan 2}$ (Proved). (6.73)

The tangential component of the magnetic field intensity is continuous at the boundary except for perfect conductor as explained below.

For perfect conductors If one of the media is a perfect conductor whose conductivity is infinity at the boundary, the current will flow only on the surface of the conductor. This current is called sheet current. In good conductors, at high frequency, the current will flow through the surface and the depth of current penetration becomes zero as the conductivity tends to infinity.

If K is the sheet current per unit width along the z direction on the conductor surface, the current density through a small depth Δx as $\Delta x \to 0$ is

$$\vec{J} = \lim_{\Delta x \to 0} \frac{K}{\Delta x} \text{ A/m}^2.$$

Therefore, Eq. (6.72) becomes

$$H_{y2}\Delta y - H_{y1}\Delta y = \lim_{\Delta x \to 0} \frac{K}{\Delta x}\Delta x \Delta y. \tag{6.74}$$

$$H_{y2} - H_{y1} = K.$$

$$H_{y1} = H_{y2} - K, \tag{6.75}$$

i.e., $H_{\tan 2} - H_{\tan 1} = K.$

Within a perfect conductor (say medium 2), the electric and magnetic fields are zero, i.e.,

$$H_{y2} = H_{\tan 2} = 0.$$

$$\therefore \quad H_{y1} = K. \tag{6.76}$$

That is, the current per unit width along surface of a perfect conductor is equal to the magnetic field strength $\vec{H}$ just outside the surface.

In vector notation,

$$\vec{K} = \vec{a}_n \times \vec{H}. \tag{6.77}$$

The magnetic field and surface current density are perpendicular to each other.

Proof of boundary conditions for electric flux density ($\vec{D}$) Consider a small cylindrical surface area dS with height Δx between the media at boundary $x = 0$ as shown in Fig. 6.7.

Maxwell's equation for charge density is

$$\oint_s \vec{D} \cdot d\vec{S} = \int_v \rho_v dv,$$

where dS = closed surface area of the cylinder and ρ_v = charge density enclosed by the volume.

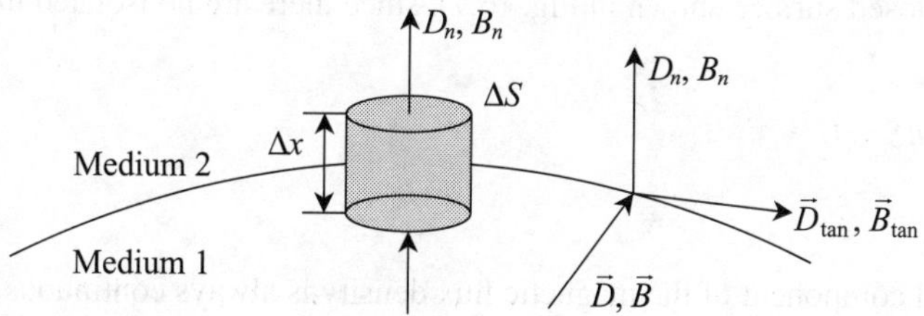

Fig. 6.7 *Boundary conditions between two media for D and B*

From the Fig. 6.7,

$$D_{n1}dS - D_{n2}dS + D_{\tan}\Delta x(2\pi r) = \rho_v \Delta x dS. \tag{6.78}$$

For zero surface charge density At the boundary, as $\Delta x \to 0$, if surface charge density ρ_v is zero, then

$$D_{n1}dS - D_{n2}dS = 0 \tag{6.79}$$

or $D_{n1} = D_{n2}$.

Therefore, if there is no surface charge, the normal component of flux density is continuous across the surface.

For metallic surface If there is a charge on the boundary of say, a metallic surface(medium 2), and ρ_s is a surface charge density, then $\rho_s = \lim_{\Delta x \to 0} \rho_v \Delta x$.

At the boundary as $\Delta x \to 0$, Eq. (6.78) becomes

$$D_{n1}dS - D_{n2}dS = \lim_{\Delta x \to 0} \rho_v \Delta x dS$$

or $D_{n1}dS - D_{n2}dS = \rho_s dS$.

$$\therefore \quad D_{n1} - D_{n2} = \rho_s \qquad \text{(Proved).} \tag{6.80}$$

That is, if there is a surface charge density ρ_s, then the normal component of flux density is discontinuous across the surface by an amount equal to the surface charge density.

Since for metallic conductor $\vec{E} = 0$, therefore $D_{n2} = 0$.

$$\text{Then} \quad D_{n1} = \rho_s. \tag{6.81}$$

That is, at the boundary between conductor and a dielectric, the normal component of flux density in the dielectric is equal to the surface charge density on the conductor.

Proof of boundary conditions for magnetic flux density ($\vec{B}$)

Similarly, Maxwell's equation for flux density is

$$\oint \vec{B} \cdot d\vec{S} = 0.$$

For a cylindrical closed surface shown in Fig. (6.7), since there are no isolated magnetic charges at the boundary,

$$B_{n1}dS - B_{n2}dS = 0$$

$$\text{or} \quad B_{n1} = B_{n2} \qquad \text{(Proved).} \tag{6.82}$$

That is, the normal component of the magnetic flux density is always continuous across a boundary surface.

Example 6.10 Given $\vec{E} = E_0(\omega t - \beta z)\vec{a}_y$ A/m, where $\beta = \omega\sqrt{\mu\varepsilon}$, find $\vec{H}$ and $|H|$.

Solution From Faraday's law we know that

$$\nabla \times \vec{E} = -\mu \frac{\partial \vec{H}}{\partial t}$$

$$\text{or} \quad \vec{H} = \frac{-1}{\mu} \int (\nabla \times \vec{E}) dt.$$

$$\text{Now,} \quad \nabla \times \vec{E} = \begin{vmatrix} \vec{a}_x & \vec{a}_y & \vec{a}_z \\ \frac{\partial}{\partial x} & \frac{\partial}{\partial y} & \frac{\partial}{\partial z} \\ 0 & E_y & 0 \end{vmatrix} = \left(0 - \frac{\partial E_y}{\partial z}\right)\vec{a}_x + \left(\frac{\partial E_y}{\partial x} - 0\right)\vec{a}_z$$

$$= \frac{-\partial E_y}{\partial z}\vec{a}_x = \frac{-\partial}{\partial z}(E_0 \sin(\omega t - \beta z)\vec{a}_x).$$

$$\text{Then} \quad \nabla \times \vec{E} = E_0 \beta \cos(\omega t - \beta z)\vec{a}_x.$$

$$\vec{H} = \frac{-1}{\mu} \int (\nabla \times \vec{E}) dt$$

$$\vec{H} = \frac{-E_0\beta}{\mu} \int \cos(\omega t - \beta z)\vec{a}_x = \frac{-E_0\beta}{\mu\omega} \sin(\omega t - \beta z)\vec{a}_x$$

$$\therefore \quad \vec{H} = \frac{-\beta}{\omega\mu} E_0 \sin(\omega t - \beta z)\vec{a}_x$$

$$\vec{H} = \frac{-\beta}{\mu\omega}\vec{E}$$

The magnitude is $|\vec{H}| = \dfrac{\beta}{\mu\omega}|\vec{E}|$.

Since given $\beta = \omega\sqrt{\mu\varepsilon}$,

$$|\vec{H}| = \frac{\omega\sqrt{\mu\varepsilon}|\vec{E}|}{\mu\omega} = \sqrt{\frac{\varepsilon}{\mu}}E.$$

Therefore the magnitude is $H_0 = \sqrt{\dfrac{\varepsilon}{\mu}}E_0$.

Additional Solved Problems

Example 6.11 A conductor of length 100 cm moves at right angles to a uniform field of strength 10,000 lines/cm^2, with a velocity of 50 m/s. Calculate the emf induced in it. Find also the value of the induced emf when the conductor moves at an angle of 30° to the direction of the field.

Solution Given $l = 100$ cm $= 1$ m; B=10,000 lines /cm^2 $= 1$ Wb/m^2;
velocity of the conductor $u = 50$m/ s; angle of the conductor $\theta = 30°$.
We know that the induced emf $V = Blu\sin\theta = 1 \times 1 \times 50 \times \sin 30° = 25$ V.

Example 6.12 A stationary 10 turns square coil of 1 m side is situated with its lower left corner coincident with the origin and with sides x_1 and y_1 along the x-axis and y-axis. If the field $\vec{B}$ is normal to the plane of the coil and has its amplitude given by $B = \sin(\pi x/x_1)\sin(\pi y/y_1)$ tesla, determine the rms value of emf induced in the coil if B varies harmonically at a frequency of 1 kHz.

Solution Given $N = 10$; $B_0 = \sin\left(\dfrac{\pi x}{x_1}\right)\sin\left(\dfrac{\pi y}{y_1}\right)$;

side of the square coil $a = 1$ m; $x_1 = y_1 = 1$.

$$f = 1 \text{ kHz}; \ \omega = 2\pi \times 10^3 \text{ rad/s}.$$

We know that the induced emf is $V = -N\displaystyle\int_S \frac{\partial B}{\partial t}\cdot dS$.

Given that B varies with time as $B = B_0 \cos\omega_0 t$,

$$\frac{\partial B}{\partial t} = \omega B_0 \sin\omega t.$$

The maximum value is at $\omega t = \pi/2$.

$$\frac{\partial B}{\partial t}\Big|_{\max} = \omega B_0.$$

$$\therefore \quad V_{\max} = -N\omega \int B_0 dS$$

$$= -N\omega \int_0^1 \int_0^1 \sin(\pi x)\sin(\pi y)dxdy.$$

$$V_{\max} = N\omega \int_0^1 \int_0^1 \sin(\pi x)\sin(\pi y)dxdy = N\omega \left.\frac{\cos \pi x}{\pi}\right|_0^1 \left.\frac{\cos \pi y}{\pi}\right|_0^1$$

$$= \frac{N\omega}{\pi^2}(\cos\pi - \cos 0)(\cos\pi - \cos 0)$$

$$= \frac{N\omega}{\pi^2}(2\times 2) = \frac{4N\omega}{\pi^2}.$$

$$V_{\max} = \frac{4\times 10\times 2\pi \times 10^3}{\pi^2}$$

$$V_{\max} = 25.465 \text{ kV}.$$

$$V_{\text{rms}} = \frac{V_{\max}}{\sqrt{2}} = 18 \text{ kV}.$$

Example 6.13 If a square loop of side 4 cm with a resistor of 10 Ω on one side is placed in a uniform magnetic field of 50 m tesla in the direction of the x-axis, calculate (i) induced current at time $t = 1$ ms, and (ii)induced emf at time $t = 3$ ms. It is given that the square loop cuts the flux lines at a frequency of 10 Hz and the axis of loop rotation is perpendicular to the field. Also, the loop lies in the $y-z$ plane at time $t = 0$.

Solution Given side of a square loop $a = 4$ cm $= 0.04$ m; $B = 50\times 10^{-3}$ tesla;

$$f = 10 \text{ Hz}, \; \omega = 2\pi f = 20\pi \text{ rad/s}; \; \theta = 90°; \; R = 10\Omega.$$

We know that the induced emf from Eq. (6.12) is

$$V = a^2\omega B \sin \omega t$$

$$= (0.04)^2 \times 20\pi \times 50\times 10^{-3} \sin(20\pi t)$$

$$= 5.03 \sin(20\pi t) \text{ mV}.$$

(i) Induced emf at $t = 1$ ms:

$$V = 5.03\sin(20\pi \times 10^{-3}) \text{ mV} = 0.3158 \text{ mV}.$$

Induced current $i = \frac{V}{R} = \frac{0.3158}{10} \times 10^{-3} = 31.58\ \mu\text{A}.$

(ii) Induced emf at $t = 3$ ms:

$$V = 5.03 \sin(20\pi \times 3 \times 10^{-3}) = 0.9425 \text{ mV}.$$

Example 6.14 In a given lossy dielectric medium, the conduction current density $J_c = 0.02\sin 10^9 t$ A/m^2. Find the displacement current density if $\sigma = 10^3$ S/m and $\varepsilon_r = 6.5$.

Solution Given $J_c = 0.02 \sin 10^9 t$ A/m^2; $\sigma = 10^3$ S/m; $\varepsilon_r = 6.5$.

We know that $J_c = \sigma E$, $J_d = \frac{\partial D}{\partial t}$ and $D = \varepsilon E$.

Displacement current density is

$$J_d = \frac{\partial D}{\partial t} = \frac{\partial \varepsilon E}{\partial t} = \frac{\partial}{\partial t}\left(\varepsilon \frac{J_c}{\sigma}\right)$$

$$= \frac{\varepsilon}{\sigma}\frac{dJ_c}{dt} = \frac{\varepsilon_0 \varepsilon_r}{\sigma}\frac{d}{d\varepsilon}(0.02 \sin 10^9 t)$$

$$= \frac{8.854 \times 10^{-12} \times 6.5 \times 0.02 \times 10^9 \times \cos(10^9 t)}{10^3}$$

$$= 1.15 \cos(10^9 t)\ \mu\text{A/m}^2.$$

Example 6.15 Find the displacement current density within a parallel plate capacitor having a dielectric with $\varepsilon_r = 10$, area plates $S = 0.01$ m^2, distance of separation, $d = 0.05$ mm, applied voltage is $V = 200 \sin 200t$.

Solution Given $S = 0.01\text{m}^2$; $d = 0.05 \times 10^{-3}$ m; $\varepsilon_r = 10$; $V = 200 \sin 200t$.
We know that displacement current density is

$$\vec{J}_d = \frac{\partial \vec{D}}{\partial t} = \frac{\partial \varepsilon \vec{E}}{\partial t}$$

$$= \frac{d}{dt}\left(\frac{\varepsilon V}{d}\right) = \frac{\varepsilon}{d}\frac{dV}{dt} = \frac{\varepsilon_0 \varepsilon_r}{d}\frac{dV}{dt}$$

$$= \frac{8.854 \times 10^{-12} \times 10}{0.05 \times 10^{-3}}\frac{d}{dt}(200 \sin 200t)$$

$$= \frac{8.854 \times 10^{-8}}{5} \times 200^2 \cos 200t = 0.7 \cos 200t \text{ mA/m}^2.$$

Example 6.16 A copper wire carries a conduction current of 1 A. Determine the displacement current in the wire at 1 MHz. For copper, $\varepsilon = \varepsilon_0$ and $\sigma = 5.8 \times 10^7$S/m.

Solution Given $I_c = 1$ A; $f = 1$ MHz $= 10^6$ Hz; $\omega = 2\pi f = 2\pi \times 10^6$ rad/s; $\varepsilon = \varepsilon_0$; $\sigma = 5.8 \times 10^7$ S/m.

We know that conduction current density, $J_c = \sigma E$.

$$\text{Conduction current} \quad I_c = \sigma E \times S \quad \text{or} \quad E = \frac{I_c}{\sigma S}.$$

The displacement current density is

$$J_d = \omega \varepsilon E = \omega \varepsilon \frac{I_c}{\sigma S}.$$

and displacement current $I_d = J_d S = \dfrac{\omega \varepsilon_0 I_c}{\sigma}$.

$$I_d = \frac{6.28 \times 10^6 \times 8.854 \times 10^{-12}}{5.8 \times 10^7} = 9.59 \times 10^{-13} \text{A}.$$

Thus, for perfect conductors the displacement current is negligible.

Example 6.17 Find the displacement current density within a parallel plate capacitor, where $\varepsilon = 100\varepsilon_0$, $S = 0.01$ m^2, $d = 0.05$ mm and the capacitor voltage is $100 \sin 200\pi t$ volts?

Solution Given $\varepsilon = 100\varepsilon_0$; $S = 0.01$ m^2; $d = 0.05$ mm and $V = 100 \sin 200\pi t$.

We know that capacitance $C = \dfrac{\varepsilon S}{d}$

$$= \frac{100 \times 8.854 \times 10^{-12} \times 0.01}{0.05 \times 10^{-3}} = 0.1771\ \mu\text{F}.$$

Displacement current $I_d = C\dfrac{dV}{dt}$

$$= 0.1771 \times 10^{-6} \times \frac{d}{dt}(100 \sin 200\ \pi t)$$

$$= 0.1771 \times 10^{-6} \times 100 \times (\cos 200\ \pi t)(200\pi)$$

$$= 11.127 \cos(200\pi t) \text{ mA}.$$

Displacement current density $J_d = \dfrac{I_d}{S} = \dfrac{11.127 \times 10^{-3} \cos 200\ \pi t}{0.01}$

$$= 1.1127 \cos(200\pi t) \text{ A/m}^2.$$

Alternative method

We know that,

$$J_d = \frac{\partial D}{\partial t} = \frac{\partial}{\partial t}(\varepsilon E)$$

$$= \varepsilon \frac{\partial E}{\partial t} = \varepsilon \frac{\partial}{\partial t}\left(\frac{V}{d}\right)$$

$$= \frac{\varepsilon}{d}\frac{\partial V}{\partial t} = \frac{100\varepsilon_0}{0.05 \times 10^{-3}} \times \frac{\partial}{\partial t}(\ 100 \sin 200\ \pi t)$$

$$= \frac{100 \times 8.854 \times 10^{-12}}{0.05 \times 10^{-3}} \times 100 \cos(200\ \pi t) \times 200\pi$$

$$= 1.1127 \cos(200\ \pi t)\ \text{A/m}^2.$$

Example 6.18 A parallel plate capacitor with plate area of 5 cm^2 and plate separation of 3 mm has a voltage of 50 sin $10^3\ t$ volts applied to its plates. Calculate the displacement current assuming $\varepsilon = 2\varepsilon_0$.

Solution Given $S = 5\ \text{cm}^2 = 5 \times 10^{-4}\ \text{m}^2$; $d = 3 \times 10^{-3}$ m and $V = 50 \sin 10^3\ t$.

We know that displacement current $I_d = C\dfrac{dV}{dt}$, where capacitance,

$$C = \frac{\varepsilon S}{d} = \frac{2\varepsilon_0 \times 5 \times 10^{-4}}{3 \times 10^{-3}}$$

$$= \frac{2 \times 8.85 \times 10^{-12} \times 5 \times 10^{-4}}{3 \times 10^{-3}} = 2.95 \times \text{pF}.$$

Displacement current $I_d = C\dfrac{dV}{dt}$

$$= 2.95 \times 10^{-12} \times \frac{d}{dt}(50 \sin 10^3\ t)$$

$$= 2.95 \times 10^{-12} \times 50 \cos(10^3 t) \times 10^3 = 0.147 \times 10^{-6} \cos(10^3 t)$$

$$= 0.147 \cos(10^3 t)\ \mu\text{A}.$$

Example 6.19 Find the conduction and displacement current densities in a material having conductivity of 10^{-3} S/m and $\varepsilon_r = 2.5$, if the electric field in the material is $E = 5.8 \times 10^{-6} \sin(9 \times 10^9 t)$V/m.

Solution Given $\sigma = 10^{-3}$S/m; $\varepsilon_r = 2.5$; $E = 5.8 \times 10^{-6} \sin(9 \times 10^9 t)$ V/m.

The conduction current density is given by

$$J_c = \sigma E = 10^{-3} \times 5.8 \times 10^{-6} \sin(9 \times 10^9 t)$$

$$= 5.8 \sin(9 \times 10^9 t)\ \text{nA/m}^2.$$

The displacement current density is given by

$$J_d = \frac{\partial D}{\partial t} = \frac{\partial}{\partial t}(\varepsilon E)$$

$$= \varepsilon \frac{\partial E}{\partial t} = \varepsilon_0 \varepsilon_r \frac{\partial}{\partial t}(5.8 \times 10^{-9} \sin(9 \times 10^9 t))$$

$$= 8.85 \times 10^{-12} \times 2.5 \times 5.8 \times 10^{-6} \times \cos(9 \times 10^9 t)(9 \times 10^9)$$

$$= 1.155 \times 10^{-9} \cos(9 \times 10^9 t) = 1.155 \cos(9 \times 10^9 t) \text{ nA/m}^2.$$

Example 6.20 Find $\vec{J}_d$ in air within a large power distribution transformer where $\vec{B} = 1.1 \cos (1.257 \times 10^{-6}(3 \times 10^8\, t - y))\vec{a}_x$ tesla.

Solution Given $\vec{B} = 1.1 \cos(1.257 \times 10^{-6}(3 \times 10^8 t - y))\vec{a}_x$

$$\vec{B} = 1.1 \cos(377t - 1.257 \times 10^{-6} y)\vec{a}_x.$$

$$B_x = 1 \cdot 1 \cos(377t - 1.257 \times 10^{-6} y)$$

We know that $\vec{J}_d = \nabla \times \vec{H} = \dfrac{1}{\mu_0}(\nabla \times \vec{B})$

$$\vec{J}_d = \frac{1}{\mu_0} \begin{vmatrix} \vec{a}_x & \vec{a}_y & \vec{a}_z \\ \dfrac{\partial}{\partial x} & \dfrac{\partial}{\partial y} & \dfrac{\partial}{\partial z} \\ B_x & 0 & 0 \end{vmatrix} = \frac{1}{\mu_0}\left(\frac{\partial B_x}{\partial y}\vec{a}_z\right)$$

$$= \frac{1}{\mu_0}\left[\left(\frac{\partial}{\partial y}\left(1.1 \cos(377t - 1.257 \times 10^{-6} y)\right)\right)\right]\vec{a}_y$$

$$= \frac{1.1}{\mu_0}(1.257 \times 10^{-6}) \sin(377t - 1.257 \times 10^{-6} y)\vec{a}_y.$$

$$\therefore \quad \vec{J}_d = 1.1 \sin(377t - 1.257 \times 10^{-6} y)\vec{a}_y.$$

Example 6.21 Find the emf developed around a circular path with radius $r = 0.5$ m in the plane $z = 0$ at $t = 0$ if (i) $B = 0.1 \sin(377t) \cdot \vec{a}_z$, (ii) $B = 0.1 \sin(377t/r)\vec{a}_r$.

Solution For circular path, consider cylindrical coordinates.

Given radius $r = \rho = 0.5$ m in the $z = 0$ plane.

Area $d\vec{S} = \rho d\rho d\phi\, \vec{a}_z$.

Induced emf $V = -\displaystyle\int \frac{\partial \vec{B}}{\partial t} \cdot d\vec{S}$.

(i) Given $B = 0.1 \sin (377t)\, \vec{a}_z$

$$V = -\int_S 0.1 \times 377 \cos(377t)(\rho d\rho d\phi)\vec{a}_z \cdot \vec{a}_z$$

$$= -37.7\cos(377t)(\pi\rho^2)$$

$$= -37.7(\pi 0.5^2)\cos(377t) = -29.6\cos(377t)$$

At $t = 0$, $V = -29.6$ V.

(ii) Given $B = 0.1\sin\left(\dfrac{377t}{\rho}\right)\vec{a}_\rho$.

$$V = -\int \frac{\partial \vec{B}}{\partial t}\cdot d\vec{S} = -\int \frac{\partial}{\partial t}(0.1\sin\frac{377t}{\rho})\rho d\rho d\phi(\vec{a}_\rho\cdot\vec{a}_z)$$

Since $\vec{a}_\rho\cdot\vec{a}_z = 0$, $V = 0$.

Example 6.22 A coaxial capacitor has the parameters $a = 5$ mm, $b = 30$ mm, $l = 20$ cm, $\varepsilon_r = 8$ and $\sigma = 10^{-6}$ siemens/m. If the conduction current density in the capacitor is $(2/\rho)\sin(10^6 t)\vec{a}_\rho$ A/m^2, find

(a) the total conduction current through the capacitor;

(b) the maximum value of the displacement current density;

(c) the total displacement current.

Solution Given $a = 5\times 10^{-3}$ m; $b = 0.03$ m; $l = 0.2$ m; $\varepsilon_r = 8$; $\sigma = 10^{-6}$ S/m;

and $\vec{J}_c = \dfrac{2}{\rho}\sin(10^6 t)\vec{a}_\rho$ in cylindrical coordinates.

(a) The total conduction current $I_c = \displaystyle\int \vec{J}_c\cdot d\vec{S}$.

$$d\vec{S} = \rho d\phi dz\vec{a}_\rho.$$

$$I_c = \int_S \left(\frac{2}{\rho}\sin(10^6 t)\vec{a}_\rho \rho d\phi dz\cdot\vec{a}_\rho\right)$$

$$= \int_0^{2\pi}\int_0^{0.2} 2\sin(10^6 t)d\phi dz = 2(2\pi)(0.2)\sin(10^6 t) \quad = 2.512\sin 10^6 t \text{ A.}$$

(b) Displacement current density

$$\vec{J}_d = \frac{\partial \vec{D}}{\partial t} = \frac{\partial \varepsilon\vec{E}}{\partial t}$$

$$= \frac{\partial}{\partial t}\left(\frac{\varepsilon\vec{J}_c}{\sigma}\right) = \frac{\varepsilon_r\varepsilon_0}{\sigma}\frac{\partial \vec{J}_c}{\partial t}$$

$$= \frac{8\varepsilon_0}{\sigma}\frac{d}{dt}\left(\frac{2}{\rho}\sin(10^6 t)\right)\vec{a}_\rho = \frac{8\varepsilon_0}{\sigma}\left(\frac{2}{\rho}\right)10^6\cos(10^6 t)\vec{a}_\rho$$

$$= \frac{16 \times 8.854 \times 10^{-12} \times 10^{6}}{10^{-6} \times \rho} \cos(10^6 t)\vec{a}_\rho$$

$$= \frac{141.664}{\rho} \cos(10^6 t)\vec{a}_\rho.$$

The maximum value of displacement current density at $\rho = 5$ mm is

$$J_{d_{\max}} = \frac{141.664}{5 \times 10^{-3}} = 28332.8 \ \text{A/m}^2.$$

(c) Total displacement current I_d

$$I_d = \int \vec{J}_d \cdot d\vec{S} = \int \frac{141.664}{\rho} \cos(10^6 t)\vec{a}_\rho \cdot \rho d\phi \, dz \, \vec{a}_\rho$$

$$= 141.664 \int_0^{0.2} \int_0^{2\pi} \cos(10^6 t) d\phi \, dz = 141.664 \cos(10^6 t)(0.2)(2\pi)$$

$$I_d = 178.02 \cos(10^6 t) \ \text{A/m}^2.$$

Example 6.23 In a material for which $\sigma = 5.0$ S/m and $\varepsilon_r = 1$, the electric field intensity is $E = 250 \sin 10^{10} t \text{V/m}$. Find the conductor and displacement current densities and the frequency at which they have equal magnitudes.

Solution Given $\sigma = 5$ S/m; $\varepsilon_r = 1$; $E = 250 \sin(10^{10} t)$ V/m.

(i) Conduction current density $J_c = \sigma E$

$$= 5 \times 250 \sin(10^{10} t) = 1250 \sin(10^{10} t).$$

(ii) Displacement current density $J_d = \dfrac{\partial D}{\partial t} = \dfrac{\partial \varepsilon E}{\partial t}$

$$= \varepsilon \frac{\partial E}{\partial t} = 8.854 \times 10^{-12} \times 250 \times 10^{10} \cos(10^{10} t)$$

$$= 22.135 \cos(10^{10} t) \ \text{A/m}^2.$$

(iii) When $|J_c| = |J_d|$,

$$|\sigma E| = \left| \varepsilon \frac{\partial E}{\partial d} \right|.$$

Let $E = E_0 \sin \omega t$. Then,

$$\left| \frac{\partial E}{\partial t} \right| = \omega E_0$$

$$|\sigma E| = \sigma E_0 = \varepsilon \omega E_0$$

$$\sigma = \varepsilon\omega$$

$$\omega = \frac{\sigma}{\varepsilon} \text{ or } f = \frac{\sigma}{2\pi\varepsilon}.$$

$$f = \frac{5}{2\pi \times 8.854 \times 10^{-12}} = 89.87 \times 10^9 \text{ Hz}.$$

Example 6.24 Find the displacement current within a parallel plate capacitor where $\varepsilon = 100\varepsilon_0$, $S = 0.1 \text{ m}^2$, $d = 0.05$ mm and the capacitor voltage is $100 \sin 2000\pi t$ volts.

Solution Given $\varepsilon = 100\varepsilon_0$; $S = 0.1 \text{ m}^2$; $d = 0.05 \times 10^{-3}$ m; and $V = 100 \sin(2000\pi t)$.

We know that $I_d = C\dfrac{dV}{dt}$,

where $C = \dfrac{\varepsilon S}{d} = \dfrac{100 \times 8.854 \times 10^{-12} \times 0.1}{0.05 \times 10^{-3}} = 1.77\ \mu\text{F}$.

$$I_d = 1.77 \times 10^{-6} \frac{d}{dt}(100 \sin(2000\pi t))$$

$$= 1.77 \times 10^{-6} \times 100 \times 2000\pi \cos(2000\pi t)$$

$$= 1.11 \cos(2000\pi t) \text{ A}.$$

Example 6.25 In a material for which $\sigma = 5.0$ S/m, $\varepsilon_r = 1$ and $E = 250 \sin(10^{10}t)$V/m, find the conduction and displacement current densities.

Solution Given $\sigma = 5$ S/m; $\varepsilon_r = 1$; $E = 250 \sin(10^{10}t)$ V/m.

Conduction current density $\vec{J_c} = \sigma\vec{E} = 1250 \sin(10^{10}t) \text{ V/m}^2$.

Displacement current density $\vec{J_d} = \dfrac{\partial \vec{D}}{\partial t} = \varepsilon\dfrac{\partial \vec{E}}{\partial t}$

$$= 8.854 \times 10^{-12} \times 10^{10} \times 250 \cos(10^{10}t) = 22.1 \cos(10^{10}t) \text{ V/m}^2.$$

Example 6.26 Find the total current in a circular conductor of radius 4 mm if the current density varies according to $J = 10^4/\rho \text{ A/m}^2$.

Solution Given $J = \dfrac{10^4}{\rho}$.

Current $I = \displaystyle\int J \cdot dS$

$$= \int_{\phi=0}^{2\pi} \int_{\rho=0}^{0.004} \frac{10^4}{\rho} \cdot \rho \, d\rho \, d\phi = 10^4 \times 0.004 \times 2\pi = 8\pi \text{ A}.$$

Example 6.27 In a cylindrical conductor of radius 2 mm, the current density varies with the distance from the axis according to $J = 10^3 e^{-400\rho}$ A/m^2. Find the total current I.

Solution Given $J = 10^3 e^{-400\rho}$ A/m^2; $\rho = 0.002$ m.

Total current $I = \int_s J \cdot dS = \int_0^{2\pi} \int_0^{0.002} 10^3 e^{-400\rho} \rho \, d\rho \, d\phi$

$$= 2\pi \times 10^3 \int_0^{0.002} e^{-400\rho} \rho \, d\rho = 2\pi \times 10^3 \left[\frac{-\rho}{400} e^{-400\rho} - \frac{e^{-400\rho}}{400^2} \right]_0^{0.002}$$

$$= 2\pi \times 10^3 \left[\frac{e^{-400\rho}}{(-400)^2} (-400\rho - 1) \right]_0^{0.002}$$

$$= \frac{2\pi \times 10^3}{16 \times 10^4} \left[-e^{-400\rho} (400\rho + 1) \right]_0^{0.002}$$

$$= 39.26 \times 10^{-3} \left[-e^{-0.8} (0.8 + 1) + 1 \right].$$

$I = 39.26 \times 10^{-3} (-0.8088 + 1) = 7.51$ mA.

Example 6.28 A parallel plate capacitor with a plate area of 5 cm^2 and plate separation of 3 mm has a voltage of $50 \sin 10^3 t$ V applied to its plates. Calculate the displacement current assuming $\varepsilon = 2\varepsilon_r$.

Solution Given $S = 5 \times 10^{-4}$ m; $d = 0.003$ m; $V = 50 \sin(10^3 t)$,

$\varepsilon = 2\varepsilon_0$.

$$I_d = C \frac{dV}{dt} = \frac{\varepsilon S}{d} \frac{d}{dt} (50 \sin 10^3 t)$$

Displacement current $I_d = \frac{\varepsilon S}{d} \times 50 \times 10^3 \cos(10^3 t)$

$$= \frac{5 \times 10^{-4} \times 2 \times 8.854 \times 10^{-12} \times 10^3 \times 50 \cos(10^3 t)}{0.003}$$

$$= 0.147 \cos(10^3 t) \ \mu\text{A}.$$

Example 6.29 A circular loop conductor of radius 0.1 m lies in the $z = 0$ plane and has a resistance of 5 Ω. Given $B = 0.2 \sin 10^3 t \vec{a}_z$ T, determine the current.

Solution Given $B = 0.2 \sin(10^3 t) \vec{a}_z$ tesla; $r = 0.1$ m; $R = 5\ \Omega$.

Area $S = \pi r^2 = \pi \times (0.1)^2$.

Flux $\varphi = BS$

$$= \pi \times 0.01 \times 0.2\ \sin(10^3 t) = 2\pi \times 10^{-3} \sin(10^3 t)\ \text{Wb}.$$

Voltage $V = \dfrac{d\varphi}{dt} = 2\pi \times 10^{-3} \times 10^3 \cos(10^3 t)$

$$= 2\pi \cos(10^3 t)\ \ \text{volts}.$$

Current $I = \dfrac{V}{R} = \dfrac{2\pi}{5}\cos(10^3 t) = 0.4\pi \cos(10^3 t)\ \text{A}.$

Summary

1. Faraday's law states that for time-varying magnetic fields, the induced emf is given by $V = -N\dfrac{d\varphi}{dt}$ volts. The negative sign indicates that, according to Lenz's law, the induced emf will produce a magnetic field which will oppose the original field. Faraday's law links electric and magnetic fields.
2. The induced emf in time-varying fields can be obtained from

$$V = \oint_l \vec{E} \cdot \vec{l} = -\int_S \frac{\partial \vec{B}}{\partial t} \cdot d\vec{S} + \oint_l (\vec{u} \times \vec{B}) \cdot \vec{l}$$

and $\nabla \times \vec{E} = -\dfrac{\partial \vec{B}}{\partial t} + \nabla \times \vec{u} \times \vec{B},$

where the transformer emf is $V_t = -\int_S \dfrac{\partial \vec{B}}{\partial t} \cdot d\vec{S}$ and the generator emf or motional emf is

$$V_m = \oint_l (\vec{u} \times \vec{B}) \cdot d\vec{l}.$$

3. The induced emf in an AC generator is $V = -B_0 \omega S \cos 2\omega t$.
4. The maximum induced emf in a coil is $V_{\max} = \varphi \omega N \sin\theta$.
5. The continuity equation for time-varying fields is $\nabla \cdot \vec{J} = \dfrac{-\partial \rho_v}{\partial t}$.
6. The modified Ampere's circuital law is $\nabla \times \vec{H} = \vec{J} + \dfrac{\partial \vec{D}}{\partial t}$, where D is the displacement charge density.
7. The displacement current which flows through the capacitor is $i_d = S\dfrac{\partial \vec{D}}{\partial t}$.
8. Ratio between conduction current density and displacement current density is

$$\left|\frac{J_c}{J_d}\right| = \frac{\sigma}{\omega\varepsilon}.$$

9. The medium is classified depending on the current densities of the wave, i.e., for a perfect conductor $\dfrac{\sigma}{\omega\varepsilon} \gg 1$, for a perfect dielectric $\dfrac{\sigma}{\omega\varepsilon} \ll 1$, and for an insulator or free space $\dfrac{\sigma}{\omega\varepsilon} = 0$.

10. Maxwell's equations for dynamic fields are given as

Point or differential form	Integral form
$\nabla \times \vec{H} = \vec{J} + \frac{\partial \vec{D}}{\partial t}$	$\int_l \vec{H} \cdot d\vec{l} = \int_S \left(\frac{d\vec{D}}{dt} + \vec{J} \right) \cdot d\vec{S}$ (Modified Ampere's circuital law)
$\nabla \times \vec{E} = \frac{-\partial \vec{B}}{\partial t}$	$\int_l \vec{E} \cdot d\vec{l} = -\int_s \frac{\partial \vec{B}}{\partial t} \cdot d\vec{S}$ (Faraday's law of induction)
$\nabla \cdot \vec{D} = \rho_v$	$\oint_S \vec{D} \cdot d\vec{S} = \int_v \rho_v dv$ (Gauss's law)
$\nabla \cdot \vec{B} = 0$	$\oint_S \vec{B} \cdot d\vec{S} = 0$ (Conservation of flux)
$\nabla \cdot \vec{J} = \frac{\partial \rho_v}{\partial t}$	$\oint_S \vec{J} \cdot d\vec{S} = -\int_v \frac{\partial \rho_v}{\partial t} dv$ (Continuity equation)

11. Maxwell's equations for harmonically time-varying fields are

Point form	Integral form
(I) $\nabla \times \vec{H} = (\sigma + j\omega\varepsilon)\vec{E}$	$\int_l \vec{H} \cdot d\vec{l} = (\sigma + j\omega\varepsilon) \int_S \vec{E} \cdot d\vec{S}$
(II) $\nabla \times \vec{E} = -j\omega\mu\vec{H}$	$\int_l \vec{E} \cdot d\vec{l} = -j\omega\mu \int_S \vec{H} \cdot d\vec{S}$
(III) $\nabla \cdot \vec{D} = \rho_v$	$\oint_S \vec{D} \cdot d\vec{S} = \int_v \rho_v dv$
(IV) $\nabla \cdot \vec{B} = 0$	$\oint_S \vec{B} \cdot d\vec{S} = 0$
$\nabla \cdot \vec{J} = -j\omega\rho_v$	$\oint_S \vec{J} \cdot d\vec{S} = -j\omega \int_v \rho_v dv$ (Continuity equation)

12. The boundary conditions of the field vectors are:

    The tangential component of $\vec{E}$ is continuous at the surface

    $E_{\tan 1} = E_{\tan 2}$.

    The tangential component of $\vec{H}$ is $H_{\tan 2} - H_{\tan 1} = K$.

    If $K = 0$, $H_{\tan 1} = H_{\tan 2}$.

    The normal component of $\vec{D}$ is $D_{n1} - D_{n2} = \rho_s$. The normal components of $\vec{B}$ is $B_{n1} = B_{n2}$.

13. If the interfacing surface is a conductor $H_{y2} = H_{\tan 2} = 0$, then $H_{y1} = K$ and $D_{n1} = \rho_s$.

Review Questions

1 State and explain Maxwell's equations in integral and point form for general time- varying fields. (*May 2010, 2012*)

2 State and explain Maxwell's equation in integral and point form in free space and write down their word statements. (*May 2010, 2011, Jan 2014*)

3 Write Maxwell's equation in point form using phasor notation and derive the wave equation from them for free space condition. (*May 2009, 2011*)

4 Show that the normal components of $\vec{B}$ is continuous for conductors. (*Nov 2010, 2011*)

5 Show that the tangential component of $\vec{H}$ is discontinuous for conductors. (*Nov 2009, 2011*)

6 State and explain Faraday's law electromagnetic induction. (*May 2010, 2012, Jan 2014*)

7 Discuss the physical interpretation of Maxwell's equations. (*May 2009, Nov 2011*)

8 Explain Faraday's law of electromagnetic induction and derive the expression for induced emf. (*May 2008, 2010, 2012*)

9 Derive the expression for one of Maxwell's equations, $\nabla \times E$, for time-varying fields. (*Nov 2009, May 2012*)

10 Explain induced emf and derive the expressions for statically and dynamically induced emfs. (*May 2010, 2012*)

11 State and explain the differential and integral forms of Maxwell's equations for fields varying harmonically with time. (*May 2010, 2011, 2012*)

12 Show that for a capacitor the conduction current in the wire equals the displacement current in the dielectric if subjected to a time-varying field. (*May 2011*)

13 State Maxwell's equations for static fields. Explain how they are modified for time-varying electric and magnetic fields. (*May 2011*, 2012)

14 Explain the concept of displacement current and obtain an expression for the displacement current density. (*Nov 2010, 2011, Aug 2014*)

15 Explain the terms (i) motional emf and (ii) static emf. (*May 2012*)

16 Differentiate between conduction and displacement current densities. (*Jan 2014*)

17 Derive the equation of continuity for time-varying fields. (*Jan 2014*)

18 Verify that the displacement current in the parallel plate capacitor is the same as the conduction current in conducting wires. (*Aug 2014*)

Problems

1 A circular loop of 10 cm radius located in the xy-plane is in a magnetic field. Find out the induced emf if $\vec{B} = (3\vec{a}_y + 4\vec{a}_z)\cos 377t$ tesla.

2 In free space, $\vec{E} = 0.5\cos(10^3 t - 50x)\vec{a}_y$ V/m. Calculate $\vec{J}_d$ and $\vec{H}$.

3 A medium is characterised by $\sigma = 0$, $\mu = 2\mu_0$ and $\varepsilon = 5\varepsilon_0$. If $\vec{H} = 2\cos(\omega t - 3y)\vec{a}_z$ A/m, calculate $\vec{E}$.

4 For a medium with $\sigma = 1 \times 10^{-3}$ S/m and $\varepsilon_r = 20$, find the frequency at which the conduction current density is equal to the displacement current density.

5 Find the displacement current within a parallel plate capacitor, where $\varepsilon_r = 100$, $A = 0.1$ m^2, $d =$ 0.05 mm and the capacitor voltage is $200 \sin 2000\pi t$ volts.

6 In a conducting medium, $\vec{H} = y^2 z\vec{a}_x + 2(x+1)yz\vec{a}_y - (x+1)z^2\vec{a}_z$ A/m. Find the current density at (1, 0, −3) and calculate the current passing through the $y = 1$ plane, $0 \le x \le 1, 0 \le z \le 1$.

7 In free space, $\vec{D} = D_m \sin(\omega t + \beta z)\vec{a}_x$. Determine $\vec{B}$ and the displacement current density.

8 A moist coil has a conductivity of 10^{-3} S/m and $\varepsilon_r = 2.5$. Find J_c and J_d, where $E = 6 \times 10^{-6} \sin(9 \times 10^9 t)$ V/m.

9 A magnetic field, $\vec{H} = 3\cos x\vec{a}_x + z\cos x\vec{a}_y$ A/m, for $z \ge 0 = 0$ for $z = 0$ is applied to a perfectly conducting surface in the $x - y$ plane. Find the current density on the conductor surface.

10 An electric field in a medium which is source free is given by $\vec{E} = 1.5\cos(10^8 t - \beta z)\vec{a}_x$ V/m. Find $\vec{B}, \vec{H}$ and $\vec{D}$. Assume $\varepsilon_r = 1, \mu_r = 1, \sigma = 0$.

11 A conduction current in a copper conductor is 10 A. Find the displacement current in the conductor at 100 MHz. For copper, $\varepsilon_r = 1, \mu_r = 1$ and $\sigma = 5.8 \times 10^7$ S/m.

Answers

1. 47.4 $\sin 377t$ volts
2. (a) $-500\varepsilon_0 \sin(10^3 t - 50x)\vec{a}_y$ A/m^2
   (b) $10\varepsilon_0 \cos(10^3 t - 50x)\vec{a}_z$ A/m
3. $-476.8\cos(\omega t - 3y)\vec{a}_x$ V/m
4. 2.219 kHz
5. $I_d = 2.22\cos(2000\pi t)$
6. $-3\vec{a}_x + 0.67\vec{a}_y$
7. $\dfrac{j\beta D_m}{\omega\varepsilon}\cos(\omega t + \beta z)\vec{a}_y$
8. (a) $6 \times 10^{-9} \sin(9 \times 10^9 t)$
   (b) $1.2 \times 10^{-6} \cos(9 \times 10^9 t)$
9. $\vec{J} = -\cos x\vec{a}_x$ A/m^2 for $z \ge 0$
10. (a) $3.15 \times 10^8 \cos(10^8 t - 2.1z)\vec{a}_z$ Wb/m^2
    (b) $24.9\cos(10^8 t - 2.1z)\vec{a}_y$ mA/m
    (c) 13.26 $\cos(10^8 t - 2.1z)\vec{a}_x$ pC/m^2
11. 0.96 nA

Multiple-choice Questions

1 A coil is placed in a uniform magnetic field that decreases at a constant rate. If the plane of the coil is parallel to the field line, which of the following is true?

(a) induced emf is zero (b) induced emf is maximum
(c) induced emf is minimum (d) none

2 A conductor length 150 cm moves at right angles to a uniform magnetic field of flux density 1.5 Wb/m^2 with a velocity 60 m/s. What is the induced emf in it?

(a) 230 volts (b) 135 volts (c) 67.5 volts (d) 270 volts

3 A conductor of length 1 m moves at right angles to a uniform magnetic field of flux density 1Wb/m^2 with a velocity of 50 m/s. What is the induced emf in it?

(a) 50 V (b) 135 V (c) 67.5 V (d) 270 V

4 A conductor of length 1 m moves parallel to the direction of magnetic field of flux density 1 Wb/m^2 with a velocity of 10 m/s. What is the induced emf in it?

(a) 20 V (b) 5 V (c) 10 V (d) 0 V

5 A conductor length 1 m makes an angle of 30° to the direction of magnetic field of flux density 10 Wb/m^2 with a velocity 10 m/s. What is the induced emf in it?

(a) 230 V (b) 135 V (c) 67.5 V (d) 50 V

6 A loop is rotating about the y-axis in a magnetic field $B = B_0 \sin \omega t \vec{a}_x$ Wb/m^2. The voltage induced in the loop is due to

(a) Combination of motional and transformer emf (b) Transformer emf

(c) Motional emf (d) None of the above

7 Which one of the following illustrates the principle of operation of a dc motor?

(a) $\nabla \times \vec{B} = 0$ (b) $\nabla \times \vec{H} = \vec{J}$ (c) $\vec{F} = \int I\vec{dl} \times \vec{B}$ (d) $\vec{B} = \mu\vec{H}$

8 A lossy dielectric has $\varepsilon_r = 10$, $\sigma = 40$ nS/ m. An electric field $E = 100 \sin \omega t \vec{a}_y$ V/m exists in the dielectric. At what frequency will the conduction and displacement current densities have equal magnitude?

(a) 144 Hz (b) 100 Hz (c) 67.5 Hz (d) 72 Hz

9 A lossy dielectric has $\varepsilon_r = 10$, $\sigma = 20$ nS/ m. An electric field $E = 100 \sin \omega t \vec{a}_z$ V/m exists in the dielectric. At what frequency will the conduction and displacement current density have equal magnitude?

(a) 36 Hz (b) 135 Hz (c) 72 Hz (d) 25 Hz

10 In a capacitor, conduction current is equal to

(a) Convection current (b) zero

(c) Displacement current (d) Infinity

11 A square coil of 10 m side having 100 turns is rotated at 1000 rpm about its axis at 0° to the direction of the magnetic field having a flux density of 1.5 Wb/m^2. The total induced emf is

(a) 6 $\vec{a}_z$ (b) 12 $\vec{a}_z$ (c) 15 $\vec{a}_z$ (d) 0

12 A square coil of 0.1 m side having 400 turns is rotated at 1000 rpm about its axis at 90° to the direction of the magnetic field having a flux density of 1.5 Wb /m^2. The total induced emf is

(a) 628.4 $\vec{a}_z$ (b) 114.24 $\vec{a}_z$ (c) 5 $\vec{a}_z$ (d) 10 $\vec{a}_z$

13 A square coil of 0.1 m side having 200 turns is rotated at 1000 rpm about its axis at 90° to the direction of the magnetic field having a flux density of 1.5 Wb /m^2. The total induced emf is

(a) 314.2 V (b) 206 V (c) 356.6 V (d) 300 V

14 A capacitor has a capacitance $C = 1$ pF. The displacement current if a voltage of $5\sin(10^6\text{t})$ is applied to it is

(a) 5 μA (b) 1 μA (c) 10 μA (d) 0.5 μA

15 A circular loop of area 2 m^2 is kept in the xz-plane in a field given by $\vec{B} = 5 \sin \omega t(3\vec{a}_x + 5\vec{a}_z)$. The voltage induced in the loop is

(a) 3 V (b) 5 V (c) 6 V (d) zero V

16 A circular loop of area of cross-section 1 m^2 is located in the *xy*-plane in a magnetic field given by $\vec{B} = \sin\omega t(2\vec{a}_x + 4\vec{a}_z)$. The induced voltage is

(a) zero (b) $4\omega\sin\omega t$ (c) $4\omega\cos\omega t$ (d) $2\omega\cos\omega t$

17 The induced emf in a dc generator is due to

(a) Convection emf (b) Motional emf

(c) Transformer emf (d) Both (b) and (c)

18 The induced emf in an induction motor is due to

(a) Convection emf (b) Motional emf

(c) Transformer emf (d) Both (b) and (c)

19 For an insulator, the relation between $\vec{J}_c$ and $\vec{J}_d$ is

(a) $\vec{J}_c = \varepsilon\vec{J}_d$ (b) $\vec{J}_c = \sigma\vec{J}_d$ (c) $\vec{J}_c = \sigma/\omega\varepsilon$ (d) $\omega\varepsilon\vec{J}_c = \sigma\vec{J}_d$

20 If a moving loop is kept in a static *B* field, the emf induced is

(a) Convection (b) Motional (c) Transformer (d) Both (b) and (c)

21 If a stationary loop is kept in a time-varying *B* field, the emf induced is

(a) Motional (b) Transformer (c) Conduction (d) None

22 In which of the following cases is the induced emf zero?

(a) Moving loop in static field (b) Stationary loop in static field

(c) Moving loop in time-varying field (d) Stationary loop in time-varying field

23 The direction of induced emf is given by

(a) Lenz' law (b) Ampere's law (c) Coulomb's law (d) Ohm's law

24 The direction of induced emf is opposite to the direction of its cause. This statement is given by

(a) Lenz' law (b) Ampere's law (c) Coulomb's law (d) Ohm's law

25 Time-varying magnetic fields can be produced by

(a) permanent magnets (b) direct current

(c) static charges (d) charges under acceleration

26 One of the following produce time-varying electric fields

(a) moving charges (b) direct current

(c) rotating magnet (d) time-varying magnetic field

27 Faraday's law of electromagnetic induction can be expressed mathematically as

(a) $-\frac{\partial}{\partial t}\int \vec{B}\cdot d\vec{S} = \text{emf}$ (b) $-\int \frac{\partial \vec{B}}{\partial S}\cdot d\vec{S} = \text{emf}$

(c) $\oint (\vec{u}\times\vec{B})\cdot d\vec{l} = \text{emf}$ (d) $\nabla\cdot\vec{B} = 0$

28 The equation $\nabla \times \vec{H} = \vec{J}_c$ should be modified for time-varying magnetic fields as

(a) $\nabla \times \vec{H} = 0$ (b) $\nabla \times \vec{H} = \vec{J}_d$

(c) $\nabla \times \vec{H} = \vec{J}_c + \frac{\partial \vec{D}}{dt}$ (d) $\nabla \times \vec{H} = \vec{J}_c + \varepsilon \frac{\partial \vec{D}}{dt} \vec{a}_r$

29 A square coil of N turns is rotating at an angular speed of ω rad/s in the magnetic field B tesla. Then the emf induced in the coil when the plane of the coil is normal to the field is

(a) zero (b) $2NBb^2\omega$ (c) $NBb\omega$ (d) $4NBb^2\omega$

30 Which one of the following is the necessary condition at the boundary of composite magnetic materials?

(a) $H_{t_1} = H_{t_2}$ (b) $B_{t_1} = B_{t_2}$ (c) $B_{n_1} = B_{n_2}$ (d) (a) and (c)

31 Which of the following is not a Maxwell's equation?

(a) $\nabla \times \vec{H} = -\frac{\partial \vec{B}}{\partial t}$ (b) $\nabla \times \vec{E} = \frac{-\partial \vec{B}}{\partial t}$ (c) $\nabla \cdot \vec{E} = \frac{\rho_v}{\varepsilon}$ (d) $\nabla \cdot \vec{D} = 0$

32 Which of the following is the correct Maxwell equation in free space?

(a) $\nabla \times \vec{H} = \vec{J} + \frac{\partial \vec{E}}{\partial t}$ (b) $\nabla \cdot \vec{D} = \rho_v$ (c) $\nabla \cdot \vec{H} = \frac{\rho_v}{\varepsilon}$ (d) $\nabla \times \vec{E} = -\frac{\partial \vec{B}}{\partial t}$

33 Which of the following is the correct Maxwell equation in time-varying fields?

(a) $\nabla \times \vec{H} = (\sigma + j\omega\varepsilon)\vec{E}$ (b) $\nabla \times \vec{E} = \partial \vec{B}/\partial t$ (c) $\nabla \cdot \vec{H} = \mu$ (d) $\nabla \cdot \vec{D} = 0$

34 Which of the following is not a Maxwell equation in time-varying fields?

(a) $\nabla \times \vec{H} = (\sigma + j\omega\varepsilon)\vec{E}$ (b) $\nabla \times \vec{E} = -\mu \frac{\partial^2 \vec{H}}{\partial t^2}$ (c) $\nabla \cdot \vec{H} = 0$ (d) $\nabla \cdot \vec{D} = \rho$

35 Continuity of the current equation can be written as

(a) $\nabla \cdot \vec{J} = \frac{\partial \rho_v}{\partial t}$ (b) $\nabla \cdot \vec{J} = -\frac{\partial \rho_v}{\partial t}$

(c) $\nabla \cdot \vec{I} = -\frac{\partial \rho_v}{\partial t}$ (d) None of the above

36 The physical significance of the Maxwell equation stated as "the electromotive force around a closed path is equal to the time derivative of the magnetic flux density or magnetic displacement B through the surface bounded by the path" can be expressed as

(a) $\nabla \times \vec{H} = (\sigma + j\omega\varepsilon)\vec{E}$ (b) $\nabla \times \vec{E} = -\frac{\partial \vec{B}}{\partial t}$ (c) $\nabla \cdot \vec{H} = 0$ (d) $\nabla \cdot \vec{D} = \rho_v$

37 $\nabla \times \vec{H} = \vec{J} + \frac{\partial \vec{D}}{\partial t}$ in integral form can be expressed as

(a) $\int (\vec{H} \cdot d\vec{l}) = (\sigma + j\omega\varepsilon) \int (\vec{E} \cdot d\vec{S})$ (b) $\int (\vec{H} \cdot d\vec{l}) = (\sigma + j\omega) \int (\vec{E} \cdot d\vec{S})$

(c) $\int (\vec{H} \cdot d\vec{S}) = (\sigma + j\omega\varepsilon) \int (\vec{E} \cdot d\vec{l})$ (d) $\int (\vec{H} \cdot d\vec{l}) = \int \left(\vec{J} + \sigma \frac{\partial \vec{D}}{\partial t}\right) dV$

38 Faraday's law is given by

(a) $\int(\vec{E}\cdot d\vec{l}) = -d\varphi/dt$ (b) $\int(\vec{E}\cdot d\vec{l}) = -\frac{d}{dt}\int(\vec{B}\cdot d\vec{S})$

(c) $\int(\vec{E}\cdot d\vec{l}) = -V(d\varphi/dt)$ (d) All of the above.

39 Which one of the following expression gives the time-varying electric fields due to time-varying magnetic fields?

(a) $\nabla\times\vec{E} = -\nabla V$ (b) $\vec{E} = -\frac{\partial\vec{A}}{dt}$ (c) $\nabla\times\vec{E} = \frac{-d\vec{B}}{dt}$ (d) (b) or (c)

40 Displacement current density is given by

(a) $\vec{J}_c = \sigma\vec{E}$ (b) $\vec{J}_d = \frac{d\vec{D}}{dt}$ (c) $\nabla\cdot\vec{J} = \frac{-\partial\rho}{dt}$ (d) $\nabla\times\vec{H} = \vec{J}_c + \vec{J}_d$

41 A loop is rotating about the *y*-axis in a magnetic field $B = B_0 \sin\omega t\vec{a}_x$ T. The voltage induced in the loop is due to

(a) transformer emf (b) motional emf

(c) both motional and transformer emf (d) all of the above

42 The equation for displacement current was introduced by

(a) Lenz (b) Lorentz (c) Faraday (d) Maxwell

43 The modified expression for $\nabla\times\vec{H} = \vec{J}_c$ for time-varying magnetic fields is

(a) $\nabla\times\vec{H} = 0$ (b) $\nabla\times\vec{H} = \vec{J}_c + \frac{\partial\vec{D}}{dt}$

(c) $\nabla\times\vec{H} = \frac{d\vec{D}}{dt}$ (d) $\oint\vec{H}\cdot d\vec{l} = \vec{J}_c$

44 A metal ring of radius *a* is falling vertically through a horizontal magnetic field of *B* tesla. The velocity of the ring is *u*. The instantaneous emf induced in the ring is

(a) $uB\pi a$ (b) uBa (c) uBa (d) $uB2\pi a$

Answers

1. *a*	2. *b*	3. *a*	4. *d*	5. *d*	6. *a*	7. *c*	8. *d*	9. *a*	10. *c*
11. *d*	12. *d*	13. *a*	14. *a*	15. *d*	16. *c*	17. *d*	18. *d*	19. *d*	20. *b*
21. *b*	22. *b*	23. *a*	24. *a*	25. *d*	26. *d*	27. *a*	28. *c*	29. *a*	30. *d*
31. *a*	32. *d*	33. *a*	34. *b*	35. *b*	36. *a*	37. *a*	38. *a*	39. *d*	40. *b*
41. *c*	42. *d*	43. *b*	44. *d*						

Electromagnetic Waves | 7

Learning Objects

The main objectives of this chapter are to understand thoroughly the following topics:

- Wave equations for conducting and perfect dielectric media
- Uniform plane waves, plane wave equations in different media such as free space, conductors, good conductors and good dielectrics
- Derivation of wave equations in phasor and time harmonic domains
- Propagation characteristics: wave velocity, wavelength, propagation constant, attenuation constant, phase shift constant, intrinsic impedances, depth of penetration
- Wave propagation in good conductors, good dielectrics, lossy and lossless ones
- Polarization and its types, direction cosines

7.1 Introduction

Waves such as sound, tides, earthquake signals, sun radiation, radio signals etc., transport energy to distant points. A wave can propagate either in longitudinal or in transverse mode. It is a function of both space and time. As the wave propagates away from the source, it loses energy and attenuates more at larger distances. Wave speed, wave mode, energy, attenuation, direction and distance of travel etc. are some of the important characteristics of a wave. Wave characteristics depend on the material through which the wave propagates. For example, a sound wave travels at a speed of about 340 m/s in air, while it travels at about 1500 m/s in water. Light propagates at the highest speed of about 3×10^8 m/s in air, whereas tidal waves created by tsunami propagate with a minimum speed of 70 m/s.

James Clark Maxwell (1831–1879) predicted the existence of electromagnetic waves (EM waves). The existence was experimentally verified by Heinrich Hertz in 1880. It was proved that the propagation of electromagnetic waves can be described with Maxwell's equations. The energy generated from a source is stored in time-varying electric and magnetic fields and propagates as electromagnetic waves or radiation. Electromagnetic waves, also called Hertzian waves propagate in transverse mode in which the direction of propagation is perpendicular to both electric and magnetic fields. Some examples of electromagnetic waves are radio waves, microwave signals, infrared radiation, light rays and cosmic rays. The analysis of Maxwell's equations under certain conditions provide solutions to electromagnetic wave equations. An electromagnetic wave can be represented by a second-order differential equation in terms of a single variable in three dimensions with time and special domains which can be obtained from Maxwell's equations.

The wave equation can be either source free, called a homogeneous wave equation, or with source terms, called a non-homogeneous wave equation. It should be noted that wave homogeneity is not in any way related to medium homogeneity. In this chapter, we will assume that source generators (applied external sources) do not exist in the medium (i.e., $\rho_v = 0$). So we will consider only homogeneous wave equations throughout our discussion.

We will first develop electromagnetic wave equations in time domain using Maxwell's equations. Later, we will consider plane wave propagation in different media such as conductors and dielectrics and derive wave equations in phasor and time harmonic domains. We will also discuss wave velocity, propagation constant, wave impedance, depth of penetration and polarization. Wave radiation will be explained in the next chapter.

7.2 Wave Equations in a Homogeneous Medium

Consider a medium with constants ε, μ and σ. Let the medium be homogeneous, linear, isotropic and source free. A homogeneous medium is one for which the quantities ε, μ and σ are constant throughout the medium.

A medium is linear with respect to the field if the flux density is proportional to the field intensity. Otherwise, the medium is non-linear. A homogeneous medium is always linear.

A medium is isotropic if the permittivity ε is a scalar quantity so that $\vec{E}$ and $\vec{D}$ are always in the same direction.

A medium is source free if there is no source generator in it.

The field relations are

$$\vec{D} = \varepsilon \vec{E}, \ \vec{B} = \mu \vec{H}, \ \text{and} \ \vec{J} = \sigma \vec{E}.$$

7.2.1 Wave Equations for a Perfect Dielectric (Free Space)

Consider a wave propagating in a free space medium.

Free space is a perfect dielectric with $\rho_v = 0,\ J = 0,\ \sigma = 0,\ \varepsilon = \varepsilon_0$ and $\mu = \mu_0$. Maxwell's equations for free space are

$$\nabla \times \vec{E} = \frac{-\partial \vec{B}}{\partial t}. \tag{7.1}$$

$$\nabla \times \vec{H} = \frac{\partial \vec{D}}{\partial t}. \tag{7.2}$$

$$\nabla \cdot \vec{D} = 0. \tag{7.3}$$

$$\nabla \cdot \vec{B} = 0. \tag{7.4}$$

Differentiating Eq. (7.1) and Eq. (7.2),

$$\nabla \times \frac{\partial \vec{E}}{\partial t} = \frac{-\partial^2 \vec{B}}{\partial t^2}$$

and $$\nabla \times \frac{\partial \vec{H}}{\partial t} = \frac{\partial^2 \vec{D}}{\partial t^2}.$$

But $$\vec{D} = \varepsilon_0 \vec{E}, \text{ and } \vec{B} = \mu_0 \vec{H}.$$

$$\therefore \quad \nabla \times \frac{\partial \vec{E}}{\partial t} = -\mu_0 \frac{\partial^2 \vec{H}}{\partial t^2} \tag{7.5}$$

and $$\nabla \times \frac{\partial \vec{H}}{\partial t} = \varepsilon_0 \frac{\partial^2 \vec{E}}{\partial t^2}. \tag{7.6}$$

Taking curl on both sides of Eq. (7.1),

$$\nabla \times \nabla \times \vec{E} = -\nabla \times \frac{\partial \vec{B}}{\partial t} = -\mu_0 \left(\nabla \times \frac{\partial \vec{H}}{\partial t} \right).$$

From Eq. (7.6),

$$\nabla \times \nabla \times \vec{E} = -\mu_0 \varepsilon_0 \frac{\partial^2 \vec{E}}{\partial t^2}.$$

From vector identity, we know that

$$\nabla \times \nabla \times \vec{E} = \nabla(\nabla \cdot \vec{E}) - \nabla^2 \vec{E}.$$

$$\nabla(\nabla \cdot \vec{E}) - \nabla^2 \vec{E} = -\mu_0 \varepsilon_0 \frac{\partial^2 \vec{E}}{\partial t^2}.$$

Since $\nabla \cdot \vec{E} = \frac{1}{\varepsilon_0} \nabla \cdot \vec{D} = 0,$

$$\therefore \quad \nabla^2 \vec{E} = \mu_0 \varepsilon_0 \frac{\partial^2 \vec{E}}{\partial t^2}.$$

Similarly, $$\nabla \times \nabla \times \vec{H} = \varepsilon_0 \left(\nabla \times \frac{\partial \vec{E}}{\partial t} \right).$$

$$-\nabla^2 \vec{H} = -\mu_0 \varepsilon_0 \frac{\partial^2 \vec{H}}{\partial t^2}$$

or $$\nabla^2 \vec{H} = \mu_0 \varepsilon_0 \frac{\partial^2 \vec{H}}{\partial t^2}.$$

$\therefore$ The partial differential equations

$$\nabla^2 \vec{E} = \mu_0 \varepsilon_0 \frac{\partial^2 \vec{E}}{\partial t^2} \tag{7.7}$$

$$\text{and}\quad \nabla^2 \vec{H} = \mu_0 \varepsilon_0 \frac{\partial^2 \vec{H}}{\partial t^2} \tag{7.8}$$

are called wave equations for free space.

The wave velocity in free space is $c = \dfrac{1}{\sqrt{\mu_0 \varepsilon_0}} = 3 \times 10^8$ m/s.

7.2.2 Wave Equations for a Conducting Medium

Consider a wave propagating in a conducting medium ($J \neq 0,\ \rho_v = 0$).

For a conducting medium, Maxwell's equations are

$$\nabla \times \vec{E} = \frac{-\partial \vec{B}}{\partial t} = -\mu \frac{\partial \vec{H}}{\partial t},$$

$$\text{and}\quad \nabla \times \vec{H} = \vec{J} + \frac{\partial \vec{D}}{\partial t} = \sigma \vec{E} + \varepsilon \frac{\partial \vec{E}}{\partial t}. \tag{7.9}$$

Taking partial differentiation,

$$\nabla \times \frac{\partial \vec{E}}{\partial t} = -\mu \frac{\partial^2 \vec{H}}{\partial t^2}$$

$$\text{and}\quad \nabla \times \frac{\partial \vec{H}}{\partial t} = \sigma \frac{\partial \vec{E}}{\partial t} + \varepsilon \frac{\partial^2 \vec{E}}{\partial t^2}. \tag{7.10}$$

Taking curl on both sides of Eq. (7.9),

$$\nabla \times \nabla \times \vec{E} = -\mu \left(\nabla \times \frac{\partial \vec{H}}{\partial t} \right).$$

Substituting Eq. (7.10) in the above equation,

$$\nabla \times \nabla \times \vec{E} = -\mu \left(\sigma \frac{\partial \vec{E}}{\partial t} + \varepsilon \frac{\partial^2 \vec{E}}{\partial t^2} \right). \tag{7.11}$$

Since the medium is homogenous,

$$\nabla \cdot \vec{D} = 0, \quad \nabla \cdot \vec{E} = 0 \quad \text{and} \quad \nabla \times \nabla \times \vec{E} = -\nabla^2 \vec{E}.$$

$$\therefore \quad \nabla^2 \vec{E} = \mu \sigma \frac{\partial \vec{E}}{\partial t} + \mu \varepsilon \frac{\partial^2 \vec{E}}{\partial t^2}$$

$$\text{or}\quad \nabla^2 \vec{E} - \mu \varepsilon \frac{\partial^2 \vec{E}}{\partial t^2} - \mu \sigma \frac{\partial \vec{E}}{\partial t} = 0. \tag{7.12}$$

Similarly, for magnetic field intensity, taking curl on both sides of Eq. (7.9),

$$\nabla \times \nabla \times \vec{H} = \sigma(\nabla \times \vec{E}) + \varepsilon\left(\nabla \times \frac{\partial \vec{E}}{\partial t}\right).$$

$$\text{or} \quad \nabla^2\vec{H} - \mu\,\varepsilon\frac{\partial^2 \vec{H}}{\partial t^2} - \mu\,\sigma\frac{\partial \vec{H}}{\partial t} = 0. \tag{7.13}$$

Equations Eq. (7.12) and Eq. (7.13) are called wave equations for a conducting medium.

Note: For free space, the wave equations can be obtained by substituting $\sigma = 0$ in Eq. (7.12) and Eq. (7.13).

7.3 Time Harmonic Wave Equations (Phasor Notation)

To obtain wave equations in phasor or time harmonic notation, we start with electric and magnetic fields in phasor form. Assume that the medium is homogeneous, linear, isotropic and source free.

The fields can be written in phasor form as,

$$\vec{E} = |E|e^{j\omega t} \text{ and } \vec{H} = |H|e^{j\omega t}.$$

$$\therefore \quad \frac{\partial \vec{E}}{\partial t} = j\omega\vec{E} \text{ and } \frac{\partial^2 \vec{E}}{\partial t^2} = -\omega^2\vec{E}. \tag{7.14}$$

$$\text{Also,} \quad \frac{\partial \vec{H}}{\partial t} = j\omega\vec{H} \text{ and } \frac{\partial^2 \vec{H}}{\partial t^2} = -\omega^2\vec{E}. \tag{7.15}$$

where ω rad/s is the wave angular frequency.

We know from Eq. (7.12) that the wave equation for an electric field in the time domain for a conducting medium is

$$\nabla^2\vec{E} - \mu\,\varepsilon\frac{\partial^2 \vec{E}}{\partial t^2} - \mu\,\sigma\frac{\partial \vec{E}}{\partial t} = 0.$$

Substituting Eq. (7.14) in this equation,

$$\nabla^2\vec{E} - \mu\,\varepsilon(-\omega^2)\vec{E} - \mu\,\sigma(j\omega)\vec{E} = 0.$$

$$\nabla^2\vec{E} + \omega^2\mu\,\varepsilon\vec{E} - \mu\,\sigma\,j\omega\vec{E} = 0.$$

$$\nabla^2\vec{E} - (j\omega\mu\,\sigma - \omega^2\mu\,\varepsilon)\vec{E} = 0.$$

Similarly, for a magnetic field,

$$\nabla^2\vec{H} - (j\omega\mu\,\sigma - \omega^2\mu\,\varepsilon)\vec{H} = 0.$$

Therefore, the time harmonic wave equations for a conducting medium are

$$\nabla^2\vec{E} - (j\omega\mu\,\sigma - \omega^2\mu\,\varepsilon)\vec{E} = 0$$

$$\text{and} \quad \nabla^2\vec{H} - (j\omega\mu\,\sigma - \omega^2\mu\,\varepsilon)\vec{H} = 0. \tag{7.16}$$

Note: For a lossless medium ($\sigma = 0$), the wave equations become

$$\nabla^2 \vec{E} + \omega^2 \mu \varepsilon \vec{E} = 0$$

and $$\nabla^2 \vec{H} + \omega^2 \mu \varepsilon \vec{H} = 0. \tag{7.17}$$

These equations are called Helmholtz equations.

7.4 Uniform Plan Wave Propagation

It is very difficult to solve the wave equations without considering the conditions of the medium and wave properties. To simplify the solution, the following set of conditions can be assumed

(1) The fields have a single dimension in space and varies perpendicular to the direction propagation.
(2) Fields are time harmonic, i.e., there are sinusoidal time variations.
(3) The medium is homogeneous, source free and lossless ($\rho_v = 0,\ \sigma = 0$).

These assumptions can be used in many practical cases. A wave which satisfies the above assumptions is called a uniform plane wave.

A uniform plane wave is defined as an electromagnetic wave in which electric and magnetic field intensities are directed in fixed directions in space and constant on infinite planes perpendicular to the direction of propagation.

That is

(1) the plane wave has no electric and magnetic field components along the direction propagation.
(2) E and H fields have constant amplitude and phase on infinite planes perpendicular to the direction propagation.

We cannot generate a uniform plane wave in practice, since it is not possible to keep constant E and H fields on infinite planes with infinite energy. However many practical waves can be approximated as uniform plane waves. For example, the radiation received by a small antenna at far distances can be approximated as a plane wave, since the wave front becomes almost spherical and a very small portion of the sphere acts like a plane at the receiving antenna.

Uniform plane waves have great importance in practical cases. They can also be considered for propagation through an isotropic medium.

7.5 Solution for the Uniform Plane Wave Equation

Consider a uniform plane wave travelling in the z direction in free space. Assume that in Cartesian coordinate system, the electric field vector is oriented along the x direction and the magnetic field vector is oriented along the y direction. The fields are independent in the x and y directions and depend only on z and t.

For the electric field, we know that

$$\nabla^2 \vec{E} = \frac{\partial^2 \vec{E}}{\partial x^2} + \frac{\partial^2 \vec{E}}{\partial y^2} + \frac{\partial^2 \vec{E}}{\partial z^2}.$$

Since for a plane wave, $\frac{\partial^2 \vec{E}}{\partial x^2} = \frac{\partial^2 \vec{E}}{\partial y^2} = 0,$

$$\nabla^2 \vec{E} = \frac{\partial^2 \vec{E}}{\partial z^2}.$$

$\therefore$ The plane wave equation for a conducting medium is

$$\nabla^2 \vec{E} - (j\omega\mu\sigma - \omega^2 \mu\varepsilon)\vec{E} = 0.$$

or
$$\frac{d^2 \vec{E}}{dz^2} - (j\omega\mu\sigma - \omega^2 \mu\varepsilon)\vec{E} = 0. \tag{7.18}$$

For free space or lossless medium ($\sigma = 0$ $\mu = \mu_0$ and $\varepsilon = \varepsilon_0$), the plane wave equation becomes

$$\frac{d^2 \vec{E}}{dz^2} + \omega^2 \mu_0 \varepsilon_0 \vec{E} = 0.$$

or
$$\frac{d^2 \vec{E}}{dz^2} + \beta^2 \vec{E} = 0, \tag{7.19}$$

where $\beta = \omega\sqrt{\mu\varepsilon}$ rad/m is called the phase shift constant.

When the uniform plane wave is propagating in the z direction, the electric field and magnetic field may have components in the x and y directions but not in the z direction.

i.e., $E_z = 0$, and $H_z = 0$.

Let the electric field be in the x direction with component E_x, then the wave equation becomes

$$\frac{d^2 \vec{E}_x}{dz^2} + \beta^2 \vec{E}_x = 0. \tag{7.20}$$

This is a second-order differential equation which represents the uniform plane wave propagating in the z direction and the electric field in the x direction.

The general solution for this equation is

$$\vec{E}_x(z) = E^+ e^{-j\beta z} + E^- e^{j\beta z}, \tag{7.21}$$

where E^+ and E^- are complex arbitrary constants which can be determined from the boundary conditions of the media. E^+ represents the electric field component when the wave is propagating in the positive z direction and E^- represents the electric field component when the wave is propagating in the negative z direction. Assume that the initial phase angle of the wave is zero.

In time domain, the wave can be expressed as

$$E_x(z, t) = \Re e[\vec{E}_x(z) e^{j\omega t}], \tag{7.22}$$

where $\Re e$ represents the real value of the function.

$$E_x(z,t) = \Re e\left[\left(E^+ e^{-j\beta z} + E^- e^{j\beta z}\right) e^{j\omega t}\right]$$

$$= \Re e\left[E^+ e^{j(\omega t - \beta z)} + E^- e^{j(\omega t + \beta z)}\right].$$

or $E_x = E_x(z,t) = E^+ \cos(\omega t - \beta z) + E^- \cos(\omega t + \beta z)$ V/m. (7.23)

If the wave propagates in free space without bounds, it travels only in the forward direction. Then the electric field component is

$$\vec{E}_x(z) = E^+ e^{-j\beta z}$$

or $E_x = E^+ \cos(\omega t - \beta z)$. (7.24)

Similarly, for the magnetic field in the y direction

$$H_y = H_y(z,t) = H^+ \cos(\omega t - \beta z) + H^- \cos(\omega t + \beta z) \text{ A/m}. \quad (7.25)$$

If the wave travels only in the forward direction,

$$\vec{H}_y(z) = H^+ e^{-j\beta z}$$

or $H_y = H^+ \cos(\omega t - \beta z)$. (7.26)

Note: Eq. (7.23) and Eq. (7.25) represent a solution for the uniform plane wave equation when the wave is travelling in free space along both the +ve and −ve directions of z. Eq. (7.24) and Eq. (7.26) represent the same but the wave is travelling only in the forward z direction.

Phase velocity The velocity at which the constant phase of the wave travels is called phase velocity.

From Eq. (7.26), the phase of the wave is

$$\omega t - \beta z = \text{constant}$$

or $z = \dfrac{\omega t}{\beta} - \dfrac{\text{constant}}{\beta}$.

Differentiating, $\dfrac{dz}{dt} = \dfrac{\omega}{\beta}$.

The phase velocity of the wave is

$$v_p = \frac{dz}{dt} = \frac{\omega}{\beta}$$

or $v_p = \dfrac{\omega}{\beta} = \dfrac{1}{\sqrt{\mu\varepsilon}}$ m/s. (7.27)

Therefore, the phase velocity is material dependent. It depends on permittivity and permeability of the medium.

For free space,

$$v_p = \frac{1}{\sqrt{\mu_0 \varepsilon_0}} = 3 \times 10^8 \text{ m/s}. \quad (7.28)$$

It is equal to the velocity of light, $c = v_p$.

In materials, the phase velocity is, from Eq. (7.27), $v_p = \dfrac{c}{\sqrt{\mu_r \varepsilon_r}}$.

Thus the phase velocity in materials is lower than the velocity of light.

Wave velocity The velocity at which the energy of a wave propagates is called wave velocity.
It is the real speed of the wave whereas phase velocity is not real. In unbounded lossless space, the wave velocity is equal to the phase velocity. For bounded medium, they may be different.

Wave length and frequency Wavelength λ is defined as the distance travelled by the sinusoidal wave through a full cycle of 2π radians.

i.e., $\beta\lambda = 2\pi$

$$\text{or} \qquad \lambda = \frac{2\pi}{\beta}. \tag{7.29}$$

If T is the time period of one full cycle, then the frequency is $f = \frac{1}{T}$ Hz and angular frequency is $\omega = 2\pi f$.

Therefore the velocity of wave propagation is $v_p = f\,\lambda = \frac{\omega}{\beta}$ m/s.

Example 7.1 The electric field in free space is given by $\vec{E} = 50\cos(10^8 t + \beta x)\vec{a}_y$ V/m.

(i) Find the direction of propagation
(ii) Calculate β and the time it takes to travel a distance of $\lambda/2$.
(iii) Sketch the wave at $t = 0$, $T/4$ and $T/2$.

Solution Given:

$$\vec{E} = 50\cos(10^8 t + \beta x)\vec{a}_y \text{ V/m};$$

or $E_y = 50\cos(10^8 t + \beta x)$

$$\omega = 10^8 \text{ rad/s}.$$

(i) The positive sign for x in the βx term represents that the direction of propagation of wave is along the negative x-axis or backward x direction.
(ii) In free space,

$$\beta = \frac{\omega}{c} = \frac{10^8}{3\times 10^8}.$$

$$\therefore \quad \beta = 0.33 \text{ rad/m}.$$

Let T be the period of the wave. That is, the time taken by the wave to travel λ m distance.

$\therefore$ The time taken to travel a distance of $\frac{\lambda}{2}$ is

$$t = \frac{T}{2} = \frac{1}{2}\cdot\frac{2\pi}{\omega}$$

or $$t = \frac{\pi}{\omega} = \frac{\pi}{10^8} = 3.14 \text{ ns}.$$

(iii) At $t = 0$, the electric field is $E_y = 50\cos(\beta\, x)$.

At $t = \dfrac{T}{2}$ or $t = \dfrac{\pi}{\omega}$, the electric field is

$$E_y = 50\cos\left(\omega\left(\frac{\pi}{\omega}\right) + \beta x\right) = 50\cos(\pi + \beta x).$$

$$\therefore \quad E_y = -50\cos(\beta x).$$

At $t = \dfrac{T}{4}$ or $t = \dfrac{1}{4}\cdot\dfrac{2\pi}{\omega} = \dfrac{\pi}{2\omega}$, the electric field is

$$E_y = 50\cos\left(\omega\frac{\pi}{2\omega} + \beta x\right) = 50\cos\left(\beta x + \frac{\pi}{2}\right).$$

$$E_y = -50\ \sin\ \beta x.$$

Figure 7.1 represents the electric field at $t = 0,\ T/4$ and $T/2$.

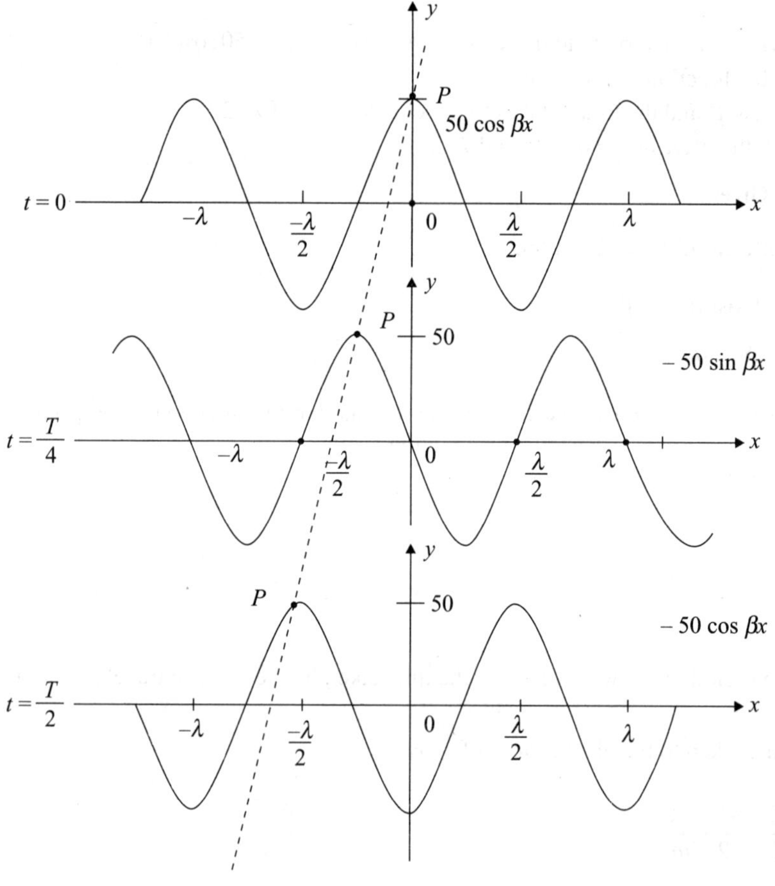

Fig. 7.1 *Electric field of the wave travelling along the* $-\vec{a}_x$ *direction*

7.6 Characteristic Impedance

When an EM wave is travelling in a medium, the ratio between the electric field and the magnetic field is a constant. This ratio is called the characteristic impedance or wave impedance or intrinsic impedance of the medium.

It is denoted as

$$\eta = \frac{\vec{E}}{\vec{H}}\ \Omega. \tag{7.30}$$

Relationship between $\vec{E}$ and $\vec{H}$ in a lossless medium Consider a plane wave propagating in the forward z direction in a lossless medium, i.e.,

$$J = 0.$$

For a uniform plane wave along the z direction, the conditions are

$$E_z = 0,\ \ H_z = 0,\ \ \frac{\partial E_x}{\partial y} = 0,\ \ \frac{\partial E_y}{\partial x} = 0,$$

and $$\frac{\partial H_x}{\partial y} = 0,\ \ \frac{\partial H_y}{\partial x} = 0. \tag{7.31}$$

We know that the Maxwell's equations for a lossless medium are

$$\nabla \times \vec{H} = \varepsilon \frac{\partial \vec{E}}{\partial t} \tag{7.32}$$

and $$\nabla \times \vec{E} = -\mu \frac{\partial \vec{H}}{\partial t}. \tag{7.33}$$

Expanding Eq. (7.31) in Cartesian coordinates,

$$\begin{vmatrix} \vec{a}_x & \vec{a}_y & \vec{a}_z \\ \dfrac{\partial}{\partial x} & \dfrac{\partial}{\partial y} & \dfrac{\partial}{\partial z} \\ H_x & H_y & H_z \end{vmatrix} = \varepsilon \left[\frac{\partial E_x}{\partial t}\vec{a}_x + \frac{\partial E_y}{\partial t}\vec{a}_y + \frac{\partial E_z}{\partial t}\vec{a}_z \right].$$

Substituting the conditions from Eq. (7.30),

$$\vec{a}_x \left(\frac{-\partial H_y}{\partial z} \right) - \vec{a}_y \left(\frac{-\partial H_x}{\partial z} \right) = \varepsilon \frac{\partial E_x}{\partial t}\vec{a}_x + \varepsilon \frac{\partial E_y}{\partial t}\vec{a}_y.$$

$$\frac{-\partial H_y}{\partial z}\vec{a}_x + \frac{\partial H_x}{\partial z}\vec{a}_y = \varepsilon \frac{\partial E_x}{\partial t}\vec{a}_x + \varepsilon \frac{\partial E_y}{\partial t}\vec{a}_y. \tag{7.34}$$

Similarly, expanding Eq. (7.32) in Cartesian coordinates,

$$\frac{-\partial E_y}{\partial z}\vec{a}_x + \frac{\partial E_x}{\partial z}\vec{a}_y = -\mu \frac{\partial H_x}{\partial t}\vec{a}_x - \mu \frac{\partial H_y}{\partial t}\vec{a}_y.$$

Equating the respective unit vector components with phasor notation, we get

$$\frac{-\partial H_y}{\partial z} = \varepsilon \frac{\partial E_x}{\partial t} = j\omega\varepsilon\, E_x,$$

$$\frac{\partial H_x}{\partial z} = \varepsilon \frac{\partial E_y}{\partial t} = j\omega\varepsilon E_y,$$

$$\frac{\partial E_y}{\partial z} = \mu \frac{\partial H_x}{\partial t} = j\omega\mu H_x,$$

and $$\frac{\partial E_x}{\partial z} = -\mu \frac{\partial H_y}{\partial t} = -j\omega\mu H_y. \tag{7.35}$$

Since the wave is propagating in the forward z direction, the electric field component is

$$E_x = E_x(z) = E^+ e^{-j\beta z}$$

or $$\frac{\partial E_x}{\partial z} = -j\beta E_x.$$

From Eq. (7.35), $\dfrac{\partial E_x}{\partial z} = -j\beta E_x = -j\omega\mu H_y.$

Then $$\frac{E_x}{H_y} = \frac{\omega\mu}{\beta} = \frac{\omega\mu}{\omega\sqrt{\varepsilon\mu}}$$

or $$\frac{E_x}{H_y} = \sqrt{\frac{\mu}{\varepsilon}}.$$

Similarly, we can show that $$\frac{E_y}{H_x} = -\sqrt{\frac{\mu}{\varepsilon}}. \tag{7.36}$$

The total electric and magnetic fields of the wave propagating in the z direction can be written as

$$E = \sqrt{E_x^2 + E_y^2}.$$

$$H = \sqrt{H_x^2 + H_y^2}.$$

The ratio is $$\frac{E}{H} = \frac{\sqrt{E_x^2 + E_y^2}}{\sqrt{H_x^2 + H_y^2}} = \sqrt{\frac{\mu}{\varepsilon}}.$$

Therefore, the magnitude of the intrinsic impedance of the lossless medium is

$$\eta = \frac{E}{H} = \sqrt{\frac{\mu}{\varepsilon}} \text{ ohms}. \tag{7.37}$$

Thus the relationship between E and H depends only on medium properties.

$$\text{Also,}\quad \frac{E_x}{H_y} = \eta \quad \text{and} \quad \frac{-E_y}{H_x} = \eta. \tag{7.38}$$

For free space:
The intrinsic impedance for free space is

$$\eta_0 = \sqrt{\frac{\mu_0}{\varepsilon_0}}$$

$$= \sqrt{4\pi \times 10^7 \times 36\pi 10^9} = 377\Omega = 120\pi\Omega.$$

$$\therefore \quad \eta_0 = 377\Omega. \tag{7.39}$$

Note 1: The dot product of E and H is always zero.

$$\text{That is,}\quad \vec{E}\cdot\vec{H} = (E_x\vec{a}_x + E_y\vec{a}_y)(H_x\vec{a}_x + H_y\vec{a}_y)$$

$$= E_xH_x + E_yH_y.$$

From Eq. (7.37),

$$\vec{E}\cdot\vec{H} = \eta H_y H_x - \eta H_x H_y = 0. \tag{7.40}$$

Thus, E and H are always perpendicular to each other.

Note 2: The cross product of E and H fields represents signal power.

$$\text{It is given by}\quad \vec{E}\times\vec{H} = (E_x\vec{a}_x + E_y\vec{a}_y)\times(H_x\vec{a}_x + H_y\vec{a}_y)$$

$$= \begin{vmatrix} \vec{a}_x & \vec{a}_y & \vec{a}_z \\ E_x & E_y & 0 \\ H_x & H_y & 0 \end{vmatrix} = (E_xH_y - H_xE_y)\vec{a}_z$$

$$= [\eta H_y H_y - H_x(-\eta H_x)]\vec{a}_z$$

$$\vec{E}\times\vec{H} = (\eta H_y^2 + \eta H_x^2)\vec{a}_z \tag{7.41}$$

$$= \eta(H_y^2 + \eta H_x^2)\vec{a}_z.$$

$$\therefore \quad \vec{E}\times\vec{H} = \eta H^2\vec{a}_z$$

$$\text{or}\quad \vec{E}\times\vec{H} = \frac{E^2}{\eta}\vec{a}_z$$

Thus the signal power $\vec{E}\times\vec{H}$ flows along the direction of propagation.

7.7 Wave Propagation in a Conducting Medium

Consider a uniform plane wave propagating in a conducting medium. The wave equations are given by (From Eq. (7.16)),

$$\nabla^2 E - (j\omega\mu\sigma - \omega^2\mu\varepsilon)E = 0,$$

$$\nabla^2 H - (j\omega\mu\sigma - \omega^2\mu\varepsilon)H = 0,$$

or $\nabla^2 E - \gamma^2 E = 0$, and $\nabla^2 H - \gamma^2 H = 0$, (7.42)

where $\gamma^2 = j\omega\mu(\sigma + j\omega\varepsilon)$ and γ is called the propagation constant of a medium at a given frequency of the wave. It is a complex quantity given by

$$\gamma = \alpha + j\beta, \tag{7.43}$$

where α is called attenuation constant and β is called phase shift constant.

When a wave is propagating in a conducting medium, its power reduces due to losses. The amount of attenuation in the wave can be represented by the attenuation constant (α). Its units are nepers (Np) or decibels (dB).

1 neper = 8.686 dB.

For a lossless medium, $\alpha = 0$.

A change in phase shift of the wave can be represented by the phase shift constant (β). Its units are radians (rad).

So $\gamma = \alpha + j\beta = \sqrt{j\omega\mu(\sigma + j\omega\varepsilon)}$. (7.44)

Therefore, the plane wave equation for the electric field component can be written as

$$\frac{d^2\vec{E}_x}{dz^2} - \gamma^2\vec{E}_x = 0. \tag{7.45}$$

This is a second-order differential equation which represents the uniform plane wave propagating in the z direction and the electric field vector in the x direction.

The general solution for this wave equation is

$$\vec{E}_x(z) = E^+ e^{-\gamma z} + E^- e^{\gamma z}, \tag{7.46}$$

where E^+ and E^- are complex arbitrary constants which can be determined from the boundary conditions of the media. E^+ represents the electric field component when the wave is propagating along the positive z direction and E^- represents the electric field component when the wave is propagating in the negative z direction. Assume that the initial phase angle of the wave is zero.

In time domain, the wave can be expressed as

$$E_x(z,t) = \Re e\left[\vec{E}_x(z)e^{j\omega t}\right], \tag{7.47}$$

where $\Re e$ represents the real value of the function.

$$E_x\ (z,t) = \Re e\left[(E^+e^{-\gamma z} + E^-e^{\gamma z})e^{j\omega t}\right].$$

$$E_x\ (z,t) = \Re e\left[E^+e^{-\alpha z}e^{j(\omega t-\beta z)} + E^-e^{\alpha z}e^{j(\omega t+\beta z)}\right].$$

$$E_x = E_x(z,t) = E^+e^{-\alpha z}\cos(\omega t - \beta z) + E^-e^{\alpha z}\cos(\omega t + \beta z)\ \text{V/m}. \tag{7.48}$$

If the wave propagates without bounds, it travels only in the forward direction. The electric field component is

$$\vec{E}_x(z) = \Re e\left[E^+e^{-\alpha z}e^{j(\omega t-\beta z)}\right]$$

or $$E_x = E^+e^{-\alpha z}\cos(\omega t - \beta z). \tag{7.49}$$

Similarly, for the magnetic field in the y direction,

$$H_y = H_y(z,\,t) = H^+e^{-\alpha z}\cos(\omega t - \beta z) + H^-e^{\alpha z}\cos(\omega t + \beta z)\ \ \text{A/m}. \tag{7.50}$$

If the wave travels only in the forward direction,

$$\vec{H}_y(z) = \Re e\left[H^+e^{-\alpha z}e^{j(\omega t-\beta z)}\right]$$

or $$H_y = H^+e^{-\alpha t}\cos(\omega t - \beta z). \tag{7.51}$$

Note: Eq. (7.48) and Eq. (7.50) represent a solution for the uniform plane wave equation when the wave is travelling in a conducting medium along both +ve and −ve directions of z. Eq. (7.49) and Eq. (7.51) represent the same but the wave travels only in the forward z direction.

7.8 Characteristic Impedance in a Lossy Dielectric or Conducting Medium

Consider a plane wave propagating in the forward z direction in a lossy conducting medium, i.e., $J \neq 0$.

For a uniform plane wave, the conditions are

$$E_z = 0 \quad H_z = 0, \quad \frac{\partial E_x}{\partial y} = 0, \quad \frac{\partial E_y}{\partial x} = 0,$$

and $$\frac{\partial H_x}{\partial y} = 0, \quad \frac{\partial H_y}{\partial x} = 0. \tag{7.52}$$

We know that Maxwell's equations for a lossy medium are

$$\nabla \times \vec{H} = \sigma\vec{E} + \varepsilon\frac{\partial \vec{E}}{\partial t}, \tag{7.53}$$

and $$\nabla \times \vec{E} = -\mu\frac{\partial \vec{H}}{\partial t}. \tag{7.54}$$

Also from Eq. (7.35),

$$\frac{\partial E_x}{\partial z} = -\mu\frac{\partial H_y}{\partial t} = -j\omega\mu H_y. \tag{7.55}$$

Since the wave is propagating in the forward z direction in a lossy medium, the electric field component is

$$E_x = E_x(z) = E^+ e^{-j\gamma z}. \tag{7.56}$$

$$\frac{\partial E_x}{\partial z} = -\gamma E_x.$$

From Eq. (7.35), $\dfrac{\partial E_x}{\partial z} = -\gamma E_x = -j\omega\mu H_y.$

Then $\dfrac{E_x}{H_y} = \dfrac{j\omega\mu}{\gamma} = \dfrac{j\omega\mu}{\sqrt{j\omega\mu(\sigma + j\omega\varepsilon)}}$

or $$\frac{E_x}{H_y} = \sqrt{\frac{j\omega\mu}{\sigma + j\omega\varepsilon}}. \tag{7.57}$$

Similarly, we can show that $$\frac{E_y}{H_x} = \frac{-j\omega\mu}{\gamma} = \frac{-j\omega\mu}{\sqrt{j\omega\mu(\sigma + j\omega\varepsilon)}} = -\sqrt{\frac{j\omega\mu}{\sigma + j\omega\varepsilon}}. \tag{7.58}$$

The total electric and magnetic fields of the wave propagating in the z direction can be written as

$$E = \sqrt{E_x^2 + E_y^2}.$$

$$H = \sqrt{H_x^2 + H_y^2}.$$

The ratio is $\dfrac{E}{H} = \dfrac{\sqrt{E_x^2 + E_y^2}}{\sqrt{H_x^2 + H_y^2}} = \sqrt{\dfrac{j\omega\mu}{\sigma + j\omega\varepsilon}}.$

Therefore, the intrinsic impedance of the lossy medium is

$$\eta = \frac{E}{H} = \sqrt{\frac{j\omega\mu}{\sigma + j\omega\varepsilon}} \text{ ohms.} \tag{7.59}$$

For a non-conducting medium or lossless medium, $\sigma = 0$.

$$\therefore \quad \eta = \sqrt{\frac{\mu}{\varepsilon}} \text{ ohms}$$

which is same as Eq. (7.37).

Complex permittivity and loss tangent for lossy medium

A lossy medium (conductor or dielectric) is a poor insulator. It has both polarization of charges for field establishment and free charges for conduction. Both displacement current and conduction current will flow through the lossy medium.

The wave equation for a lossy medium is

$$\nabla^2 \vec{E} = j\omega\mu(\sigma + j\omega\varepsilon)\vec{E},$$

and that for a lossless material ($\sigma = 0$) is

$$\nabla^2 \vec{E} = (j\omega\mu)(j\omega\varepsilon_c)\vec{E}.$$

Comparing the above equations for the same field,

$$j\omega\varepsilon_c = \sigma + j\omega\varepsilon$$

or $$\varepsilon_c = \varepsilon - j\frac{\sigma}{\omega} = \varepsilon - j\varepsilon'' \text{ F/m},$$

where ε_c is called complex permittivity. The additional term $\varepsilon'' = \dfrac{\sigma}{\omega}$ results due to power losses in the material.

Loss tangent: The ratio between the imaginary and real parts of the complex permittivity ε_c is called loss tangent of the material. It can be expressed as

$$\tan\theta_{\text{loss}} = \frac{\varepsilon''}{\varepsilon} = \frac{\sigma}{\omega\varepsilon}$$

and the loss angle is $\theta_{\text{loss}} = \tan^{-1}\left(\dfrac{\sigma}{\omega\varepsilon}\right)$.

Losses tangent indicates both damping and ohmic losses in the material. It is a measure of the power loss in the medium. If ε_c is real, the material is lossless.

7.9 Wave Propagation in Perfect Dielectric Medium

Consider a plane wave propagating in a lossless or perfect dielectric medium, i.e., $\alpha = 0$.

The wave equation is

$$\nabla^2 \vec{E} + \omega^2\mu\varepsilon\vec{E} = 0,$$

where $\gamma^2 = -\omega^2\mu\varepsilon$

or $$\gamma = \sqrt{-\omega^2\mu\varepsilon} = j\omega\sqrt{\mu\varepsilon} = j\beta. \tag{7.60}$$

Therefore, for perfect dielectrics, the phase shift constant is

$$\beta = \omega\sqrt{\mu\varepsilon}$$

and the velocity of propagation is

$$v_p = \frac{\omega}{\beta}.$$

For free space,

$$\beta = \omega\sqrt{\mu_0\,\varepsilon_0}.$$

$$\therefore \quad v_0 = \frac{\omega}{\beta} = \frac{1}{\sqrt{\mu_0\,\varepsilon_0}} = 3\times 10^8 \text{ m/s.} \tag{7.61}$$

Wavelength is $\lambda = \dfrac{2\pi}{\beta}$

and intrinsic impedance is

$$\eta = \frac{E}{H} = \sqrt{\frac{\mu_0}{\varepsilon_0}} = 120\pi\ \Omega. \tag{7.62}$$

Example 7.2 A uniform plane wave at a frequency of 1 GHz is travelling in a large block of Teflon ($\varepsilon_r = 2.1$, $\mu_r = 1$ and $\sigma = 0$). Determine γ, η, β and λ.

Solution Given:

$$\varepsilon_r = 2.1,\ \ \mu_r = 1,\ \ \sigma = 0 \ \text{ and } \ f = 1\times 10^9 \text{ Hz.}$$

Since $\sigma = 0$, for a lossless dielectric medium, $\alpha = 0$.

We know that

$$\beta = \omega\sqrt{\mu\varepsilon} = \omega\sqrt{\mu_0\,\mu_r\,\varepsilon_0\,\varepsilon_r}$$

$$\beta = \frac{\omega}{c}\sqrt{\mu_r\,\varepsilon_r} = \frac{2\pi\times 10^9}{3\times 10^8}\sqrt{1\times 2.1} = 30.35 \text{ rad/m.}$$

Phase velocity, $v_p = \dfrac{\omega}{\beta} = \dfrac{2\pi f}{\beta} = \dfrac{2\pi\times 10^9}{30.35} = 2\times 10^8$ m/s.

Impedance $\eta = \sqrt{\dfrac{\mu}{\varepsilon}} = \sqrt{\dfrac{\mu_0\,\mu_r}{\varepsilon_0\varepsilon_r}} = \dfrac{120\pi}{\sqrt{2.1}}$

$$= 260\ \Omega.$$

Wavelength $\lambda = \dfrac{2\pi}{\beta} = \dfrac{2\pi}{30.35} = 0.2$ m.

7.10 Expressions for α and β for a Conducting Medium

Consider a wave propagating in a lossy conducting medium. The propagation constant is

$$\gamma = \alpha + j\beta$$

$$= \sqrt{j\omega\mu(\sigma + j\omega\varepsilon)}.$$

$$\gamma^2 = (\alpha + j\beta)^2 = j\omega\mu(\sigma + j\omega\varepsilon).$$

$$\alpha^2 - \beta^2 + 2j\alpha\beta = -\omega^2\mu\varepsilon + j\omega\mu\sigma. \tag{7.63}$$

Equating real and imaginary terms,

$$\alpha^2 - \beta^2 = -\omega^2\mu\varepsilon \tag{7.64}$$

and $2\alpha\beta = \omega\mu\sigma.$

Since $(a+b)^2 = (a-b)^2 + 4ab.$

$$(\alpha^2 + \beta^2)^2 = (\alpha^2 - \beta^2)^2 + 4\alpha^2\beta^2.$$

or $(\alpha^2 + \beta^2)^2 = (-\omega^2\mu\varepsilon)^2 + (\omega\mu\sigma)^2.$

$$\therefore \quad \alpha^2 + \beta^2 = \sqrt{\omega^4\mu^2\varepsilon^2 + \omega^2\mu^2\sigma^2} = \omega\mu\sqrt{\omega^2\varepsilon^2 + \sigma^2}. \tag{7.65}$$

Solving Eq. (7.64) and Eq. (7.65),

$$2\alpha^2 = \omega\mu\sqrt{\omega^2\varepsilon^2 + \sigma^2} - \omega^2\mu\varepsilon.$$

$$= \omega^2\mu\varepsilon\sqrt{1 + \frac{\sigma^2}{\omega^2\varepsilon^2}} - \omega^2\mu\varepsilon$$

or

$$\alpha = \omega\sqrt{\frac{\mu\varepsilon}{2}\left[\sqrt{1 + \left(\frac{\sigma}{\omega\varepsilon}\right)^2} - 1\right]} \tag{7.66}$$

and $2\beta^2 = \omega\mu\sqrt{\omega^2\varepsilon^2 + \sigma^2} + \omega^2\mu\varepsilon.$

$$\beta = \omega\sqrt{\frac{\mu\varepsilon}{2}\left[\sqrt{1 + \left(\frac{\sigma}{\omega\varepsilon}\right)^2} + 1\right]}. \tag{7.67}$$

Thus the propagation constant of the wave depends on material properties at a given frequency.

Example 7.3 A conductor (σ = 10 S/m, $\varepsilon_r = \mu_r = 1$) support a uniform plane wave at 60 GHz. Compute the attenuation constant, propagation constant, intrinsic impedance, wavelength and phase velocity of propagation.

Solution Given: $\sigma = 10$ S/m; $\varepsilon_r = \mu_r = 1$; wave frequency, $f = 60$ GHz;

then, $\omega\varepsilon = 2\pi \times 60 \times 10^9 \times \varepsilon_0 = 3.338.$

Attenuation constant: $\alpha = \omega\sqrt{\frac{\mu\varepsilon}{2}\left[\sqrt{1 + \left(\frac{\sigma}{\omega\varepsilon}\right)^2} - 1\right]}$

$$= 120\pi \times 10^9 \times \sqrt{\frac{\mu_0 \varepsilon_0}{2}\left[\sqrt{1+\left(\frac{10}{3.338}\right)^2} - 1\right]}.$$

$$= 1306.32 \text{ Np/m}.$$

Phase shift constant:

$$\beta = \omega\sqrt{\frac{\mu\varepsilon}{2}\left[\sqrt{1+\left(\frac{\sigma}{\omega\varepsilon}\right)^2} + 1\right]}$$

$$= 120\pi \times 10^9 \times \sqrt{\frac{\mu_0 \varepsilon_0}{2}\left[\sqrt{1+\left(\frac{10}{3.338}\right)^2} + 1\right]}.$$

$$\beta = 1813.23 \text{ rad/m}.$$

Propagation constant:

$$\gamma = \alpha + j\beta = 1306.32 + j1813.23 \text{ per metre}.$$

Intrinsic impedance:

$$\eta = \sqrt{\frac{j\omega\mu}{\sigma + j\omega\varepsilon}}$$

$$= \sqrt{\frac{j120\pi \times 10^9 \mu_0}{10 + j3.338}} = \sqrt{14228.11 + j42624.6}$$

$$= \sqrt{44936.6\angle 71.54} = 212\angle 35.77\ \Omega.$$

Wavelength:

$$\lambda = \frac{2\pi}{\beta} = \frac{2\pi}{1813.23} = 3.465 \text{ mm}.$$

Phase velocity of propagation:

$$v_p = \frac{\omega}{\beta} = \frac{2\pi f}{\beta} = \frac{120\pi \times 10^9}{1813.23} = 2.08 \times 10^8 \text{ m/s}.$$

7.11 Wave Propagation in Good Conductors

Consider a wave propagating in a good conductor. For good conductors, the conductivity is very high compared to the wave frequency, i.e., $\sigma >> \omega\varepsilon$.

Now, $\gamma = \sqrt{j\omega\mu(\sigma + j\omega\varepsilon)}$.

Since $\sigma >> \omega\varepsilon$, we can neglect the imaginary part.

Hence, $\gamma = \sqrt{j\omega\mu\sigma} = \sqrt{\omega\mu\sigma}\angle 45°$.

$$= \sqrt{\omega\mu\sigma}\left[\frac{1}{\sqrt{2}} + j\frac{1}{\sqrt{2}}\right] = \sqrt{\frac{\omega\mu\sigma}{2}} + j\sqrt{\frac{\omega\mu\sigma}{2}}.$$

$$\therefore \quad \alpha = \sqrt{\frac{\omega\mu\sigma}{2}} \tag{7.68}$$

and

$$\beta = \sqrt{\frac{\omega\mu\sigma}{2}}. \tag{7.69}$$

Thus, for good conductors, α and β are equal.

Note: In a highly conducting medium, since α is proportional to $\sqrt{\sigma}$, the losses are very high and the wave attenuates rapidly.

Intrinsic impedance We know that the intrinsic impedance for a conducting medium is

$$\eta = \sqrt{\frac{j\omega\mu}{\sigma + j\omega\varepsilon}}.$$

For a good conductor, $\sigma >> \omega\varepsilon$.

$$\eta = \sqrt{\frac{j\omega\mu}{\sigma}}$$

or

$$\eta = \sqrt{\frac{\omega\mu}{\sigma}}\angle 45°. \tag{7.70}$$

Phase velocity We know that the velocity of the wave is

$$v_p = \frac{\omega}{\beta} = \frac{\omega}{\sqrt{\frac{\omega\mu\sigma}{2}}}.$$

Phase velocity in good conductors is

$$v_p = \sqrt{\frac{2\omega}{\mu\sigma}}. \tag{7.71}$$

Note: The speed of the wave decreases as the conductivity of the medium increases at a given frequency.

Example 7.4 Determine the phase velocity of propagation, attenuation constant, phase constant and intrinsic impedance for a forward travelling wave in a large block of copper at 1 MHz ($\sigma = 5.8 \times 10^7$, $\varepsilon_r = \mu_r = 1$).

Solution Given for a block of copper, $\sigma = 5.8 \times 10^7$, $\varepsilon_r = \mu_r = 1$.
Wave frequency, $f = 1$ MHz $= 10^6$ Hz, then $\omega\varepsilon = 2\pi \times 10^6 \times 8.854 \times 10^{-12} = 17.708\pi \times 10^{-6}$.

Since $\sigma \gg \omega\varepsilon$, for a good conductor phase velocity is

$$v_p = \frac{\omega}{\beta} = \sqrt{\frac{2\omega}{\mu\sigma}} = \sqrt{\frac{2 \times 2\pi \times 10^6}{4\pi \times 10^{-7} \times 1 \times 5.8 \times 10^7}}.$$

$$v_p = 415.22 \text{ m/s}.$$

The attenuation and phase shift constants are

$$\alpha = \beta = \sqrt{\frac{\omega\mu\sigma}{2}} = \sqrt{\frac{2\pi \times 10^6 \times 4\pi \times 10^{-7} \times 5.8 \times 10^7}{2}}.$$

$$\alpha = 15132 \text{ Np/m}; \ \beta = 15132 \text{ rad/m}.$$

Intrinsic impedance is $\eta = \sqrt{\dfrac{j\omega\mu}{\sigma}}$

$$= \sqrt{\frac{(2\pi \times 10^6) \times (4\pi \times 10^{-7})}{5.8 \times 10^7}} \angle 45^\circ$$

$$= 3.68 \times 10^{-4} \angle 45^\circ \text{ ohms}.$$

7.12 Wave Propagation in Good Dielectrics

Consider a wave propagating in a good dielectric medium. For good dielectrics, the conductivity is very low compared to $\omega\varepsilon$, i.e., $\sigma << \omega\varepsilon$.

The expression for the propagation constant is

$$\gamma = \sqrt{j\omega\mu(\sigma + j\omega\varepsilon)}$$

$$= (\sqrt{j\omega\mu})(\sqrt{j\omega\varepsilon})\sqrt{1 + \frac{\sigma}{j\omega\varepsilon}}.$$

$$\gamma = j\omega\sqrt{\mu\varepsilon}\sqrt{1 + \frac{\sigma}{j\omega\varepsilon}}. \tag{7.72}$$

Now, using binomial expansion,

$$\sqrt{1 + \frac{\sigma}{j\omega\varepsilon}} \approx 1 + \frac{\sigma}{2j\omega\varepsilon} + \left(\frac{-1}{4}\right)\frac{\sigma^2}{2j^2\omega^2\varepsilon^2} + \ldots$$

Since $\dfrac{\sigma}{\omega\varepsilon} \ll 1$, higher order terms can be neglected

Therefore a good approximation is

$$\sqrt{1+\frac{\sigma}{j\omega\varepsilon}} \approx 1+\frac{\sigma}{2j\omega\varepsilon}+\frac{\sigma^2}{8\omega^2\varepsilon^2}. \tag{7.73}$$

Substituting this value in Eq. (7.72),

$$\gamma = j\omega\sqrt{\mu\varepsilon}\left[1+\frac{\sigma}{2j\omega\varepsilon}+\frac{\sigma^2}{8\omega^2\varepsilon^2}\right].$$

$$\alpha + j\beta = j\omega\sqrt{\mu\varepsilon}+\frac{j\omega\sqrt{\mu\varepsilon}\sigma}{2j\omega\varepsilon}+\frac{j\omega\sqrt{\mu\varepsilon}\sigma^2}{8\omega^2\varepsilon^2}.$$

or $$\alpha + j\beta = j\omega\sqrt{\mu\varepsilon}+\frac{\sigma}{2}\sqrt{\frac{\mu}{\varepsilon}}+\frac{j\sigma^2\sqrt{\mu\varepsilon}}{8\omega^2\varepsilon^2}. \tag{7.74}$$

Equating real and imaginary terms, the attenuation contant is

$$\alpha = \frac{\sigma}{2}\sqrt{\frac{\mu}{\varepsilon}} \tag{7.75}$$

and the phase shift constant is

$$\beta = \omega\sqrt{\mu\varepsilon}\left(1+\frac{\sigma^2}{8\omega^2\varepsilon^2}\right), \tag{7.76}$$

Note: For a perfect or lossless dielectric, $\sigma = 0$, and $\beta = \omega\sqrt{\mu\varepsilon}$.

Intrinsic impedance We know that the intrinsic impedance for a lossy medium is

$$\eta = \sqrt{\frac{j\omega\mu}{\sigma + j\omega\varepsilon}}$$

$$= \sqrt{\frac{\mu}{\varepsilon}}\sqrt{\frac{1}{\left(1+\frac{\sigma}{j\omega\varepsilon}\right)}} = \sqrt{\frac{\mu}{\varepsilon}}\left(1+\frac{\sigma}{j\omega\varepsilon}\right)^{-1/2}.$$

Using the binomial theorem expansion,

$$\eta = \sqrt{\frac{\mu}{\varepsilon}}\left(1-\frac{\sigma}{2j\omega\varepsilon}+.....\right).$$

Neglecting higher order terms, the intrinsic impedance for good dielectrics is

$$\eta = \sqrt{\frac{\mu}{\varepsilon}}\left(1+j\frac{\sigma}{2\omega\varepsilon}\right). \tag{7.77}$$

Thus, when a wave is travelling in a lossy dielectric medium, due to conductivity a reactive component is added in the impedance.

Phase velocity We know that the velocity of the wave is

$$v_p = \frac{\omega}{\beta} = \frac{\omega}{\omega\sqrt{\mu\varepsilon}\left(1+\dfrac{\sigma^2}{8\omega^2\varepsilon^2}\right)}.$$

$$v_p = \frac{1}{\sqrt{\mu\varepsilon}}\left(1+\frac{\sigma^2}{8\omega^2\varepsilon^2}\right)^{-1}.$$

Now, using binomial expansion and neglecting higher order terms, phase velocity in good dielectrics is

$$v_p = \frac{v_0}{\sqrt{\mu_r\varepsilon_r}}\left(1-\frac{\sigma^2}{8\omega^2\varepsilon^2}\right). \tag{7.78}$$

For practical dielectrics, since the conductivity σ is very small, the velocity of propagation reduces slighty compared to free space velocity v_0. As conductivity σ of the medium increases, the loss of the energy in the wave increases and the velocity decreases. For example, the phase velocity in sea water is less than the phase velocity in rain water.

Example 7.5 A uniform plane wave at a frequency of 1 GHz is travelling in a large block of dielectric ($\varepsilon_r = 55$, $\mu_r = 1$ and $\sigma = 0.05$ S/m). Determine γ, η, β and λ.

Solution Given data, $\varepsilon_r = 55, \ \mu_r = 1, \ \sigma = 0.05$ and $f = 1\times 10^9$ Hz.

$$\omega = 2\pi f = 2\pi\times 1\times 10^9 = 2\pi\times 10^9 \text{ Hz}$$

$$\omega\varepsilon = 2\pi\times 10^9\times 55\varepsilon_0 = 3.06$$

Since $\sigma << \omega\varepsilon$, the material is a good dielectric.

That attenuation contant is

$$\alpha = \frac{\sigma}{2}\sqrt{\frac{\mu}{\varepsilon}}.$$

$$\alpha = \frac{0.05}{2}\times 120\pi\sqrt{\frac{1}{55}} = 6.354 \text{ Np/m},$$

and the phase shift constant is

$$\beta = \omega\sqrt{\mu\varepsilon}\left(1+\frac{\sigma^2}{8\omega^2\varepsilon^2}\right)$$

$$= \frac{2\pi\times 10^9}{3\times 10^8}\times\sqrt{55}\left(1+\frac{1}{8}\left(\frac{0.05}{3.06}\right)^2\right)$$

$$\beta = 155.32 \text{ rad/m}.$$

Propagation constant $\gamma = \alpha + j\beta = 6.354 + j155.32$.

Phase velocity, $v_p = \frac{\omega}{\beta} = \frac{2\pi f}{\beta} = \frac{2\pi \times 10^9}{155.32} = 0.405 \times 10^8$ m/s.

The intrinsic impedence for good dielectrics is

$$\eta = \sqrt{\frac{\mu}{\varepsilon}}\left(1 + j\frac{\sigma}{2\omega\varepsilon}\right) = 120\pi \times \frac{1}{\sqrt{55}}\left(1 + j\frac{1}{2} \times \frac{0.05}{3.06}\right)$$

$$= 50.83\angle 0.468^\circ.$$

Wavelength $\lambda = \frac{2\pi}{\beta} = \frac{2\pi}{155.32} = 0.0404$ m.

Example 7.6 A 100 V/m plane wave of frequency 300 MHz travels in an infinite, lossless medium having $\mu_r = 1,\ \varepsilon_r = 9, \sigma = 0$. Write the complete time domain equations for E and H field vectors.

Solution Given:

$$E_0 = 100 \text{ V/m}, \quad f = 300 \text{ MHz}, \quad \mu_r = 1, \quad \varepsilon_r = 9 \quad \text{and} \quad \sigma = 0,$$
$$\omega = 2\pi f = 2\pi \times 300 \times 10^6 = 1.8 \times 10^9 \text{ Hz}.$$

For a lossless medium, $\alpha = 0$

$$\beta = \omega\sqrt{\mu\varepsilon} = \omega\sqrt{\mu_0\,\mu_r\,\varepsilon_0\,\varepsilon_r} = \frac{\omega}{c}\sqrt{\mu_r\,\varepsilon_r}$$

$$= \frac{1.8\pi \times 10^9}{3 \times 10^8}\sqrt{1 \times 9} = 18 \text{ rad/m}.$$

Assume that the plane wave travelling in the z direction with E field has a component in the x direction and H field has a component in the y direction.

Then the electric field intensity in the time domain is

$$\vec{E} = E_0 \cos(\omega t - \beta z) a_x \text{ V/m}.$$

$$\therefore \quad \vec{E} = 100\cos[(1.8 \times 10^9 t) - 18z]\,\vec{a}_x \text{ V/m}.$$

The impedance is

$$\eta = \sqrt{\frac{\mu}{\varepsilon}} = \sqrt{\frac{\mu_r\,\mu_0}{\varepsilon_r\,\varepsilon_0}} = 120\pi\frac{1}{\sqrt{9}} = 125.6 \text{ ohms}.$$

The magnetic field intensity in the time domain is

$$\vec{H} = \frac{E_0}{\eta}\cos(\omega t - \beta z)\,\vec{a}_y \text{ A/m} = \frac{100}{125}\cos[(1.8 \times 10^9 t) - 18z]\,\vec{a}_y.$$

$$\vec{H} = 0.8\cos(1.8 \times 10^9 t - 18z)\,\vec{a}_y \text{ A/m}.$$

7.13 Depth of Penetration (Skin Depth δ)

When a wave propagates through a conducting medium, it is attenuated. As frequency ω increases, the rate of attenuation increases. The wave penetrates only a very small depth before it *attenuates* to a negligible value. The depth of penetration is defined as *the distance at which the wave attenuates to $1/e$ or approximately 37% of its maximum amplitude.* It is also defined as the reciprocal of *the attenuation constant*, i.e;

$$\delta = \frac{1}{\alpha}, \tag{7.79}$$

where α is the attenuation constant.

Proof Consider a wave propagating through a conducting medium in the z direction. We know that the amplitude of the electric field vector attenuates exponentially with the attenuation constant α. That is,

$$E = E_0 e^{-\alpha z},$$

where E_0 is the maximum amplitude.

At a depth equal to the skin depth, $z = \delta$, E attenuates to $1/e$ times.

i.e., $$E = \frac{E_0}{e} = E_0 e^{-\alpha\delta}.$$

$$\therefore \quad \frac{E}{E_0} = \frac{1}{e} = e^{-\alpha\delta}.$$

$$e^{-1} = e^{-\alpha\delta} \quad \text{or} \quad \alpha\delta = 1.$$

$\therefore$ The skin depth is $\delta = \dfrac{1}{\alpha}$ (Proved).

Note:

1. For a lossy medium,

$$\text{skin depth is } \delta = \frac{1}{\alpha} = \frac{1}{\omega\sqrt{\frac{\mu\varepsilon}{2}\left[\sqrt{1+\left(\frac{\sigma}{\omega\varepsilon}\right)^2} - 1\right]}}. \tag{7.80}$$

2. For good conductors,

$$\text{skin depth is } \quad \delta = \frac{1}{\alpha} = \sqrt{\frac{2}{\omega\mu\sigma}}. \tag{7.81}$$

3. For good dielectrics,

$$\text{skin depth is } \delta = \frac{1}{\alpha} = \frac{2}{\sigma}\sqrt{\frac{\varepsilon}{\mu}}$$

4. For perfect conductors ($\sigma = \infty$), skin depth is $\delta = 0$.
5. For lossless dielectrics ($\alpha = 0$), skin depth is $\delta = \infty$.
6. At microwave frequencies, the skin depth is very small (of the order of μm) in good conductors. The field strength reduces to almost zero at 2δ to 3δ depth. All the fields and currents are confined to only a very thin surface layer of the good conductor. No fields and currents exist inside the conductor.

Example 7.7 Find the skin depth at a frequency of 1.6 MHz in aluminium, where $\sigma = 38.2$ MS/m and $\mu_r = 1$. Also the find propagation constant and wave velocity in the medium.

Solution Given: frequency $f = 1.6\text{ MHz} = 1.6 \times 10^6$ Hz, $\mu_r = 1$,

$$\sigma = 38.2\text{ MS/m} = 38.2 \times 10^6\text{ S/m}.$$

$$\omega\varepsilon = 2\pi \times 1.6 \times 10^6 \times 8.854 \times 10^{-12}$$

Since $\sigma \gg \omega\varepsilon$, for a good conductor:

Skin depth is given by $\delta = \sqrt{\dfrac{2}{\omega\mu\sigma}}$.

Now, $\omega\mu\sigma = 2\pi \times 1.6 \times 10^6 \times 4\pi \times 10^{-7} \times 38.2 \times 10^6 = 4825.84 \times 10^5$.

$$\delta = \sqrt{\frac{2}{4825.84 \times 10^5}} = 64.37\mu\text{ m}.$$

Propagation constant

$$\gamma = \alpha + j\beta.$$

For good conductors,

$$\alpha = \beta = \frac{1}{\delta} = \frac{1}{64.37 \times 10^{-6}} = 15.533 \times 10^3.$$

$$\therefore \quad \gamma = (15.533 + j15.533) \times 10^3 = 21.97 \times 10^3\ \angle 45^\circ.$$

Velocity of propagation $v_p = \dfrac{\omega}{\beta} = \dfrac{2\pi \times 1.6 \times 10^6}{15.533 \times 10^3} = 648$ m/s.

Example 7.8 A 3 GHz uniform plane wave propagates through a material (ε_r = 2.54, σ = 0.8 S/m) in the positive z direction. The E field at $z = 0$ is 100 V/m.

(a) Calculate propagation constant and attenuation in dB.
(b) Calculate electric field at a distance of 2 cm.

Solution Given:
Wave frequency $f = 3\text{ GHz} = 3 \times 10^9$ Hz, $E_0 = 100$ V at $z = 0$, $\varepsilon_r = 2.54, \sigma = 0.8$ S/m.

(a) The propagation constant is

$$\gamma = \alpha + j\beta.$$

$$\gamma = \sqrt{j\omega\mu(\sigma + j\omega\varepsilon)}.$$

Now $\omega\mu = 6\pi \times 10^9 \times 4\pi \times 10^{-7} = 23.687 \times 10^3$.

$$\sigma + j\omega\varepsilon = 0.8 + j6\pi \times 10^9\,\varepsilon_0 = 0.8 + j0.1669.$$

$$\therefore \quad \gamma = \sqrt{j(23687 \times 10^3)(0.8 + j0.1669\ )}$$

$$= \sqrt{19.357 \times 10^3 \angle 101.78^\circ)} = 139\angle 50.89^\circ.$$

$$\gamma = \alpha + j\beta = 87.68 + j107.85 \text{ per metre.}$$

$\therefore$ Attenuation constant $\alpha = 87.68$ Np/m and phase shift constant $\beta = 107.85$ rad/m.

Attenuation in dB:

$$\alpha \text{ (in dB)} = 87.68 \times 8.686. = 761.58 \text{ dB/m.}$$

(b) We know that the electric field is

$$E = E_0 e^{-\alpha z}.$$

At $z = 2$ cm, we get, $= 100 \times e^{-87.68 \times 0.02} = 47.06$ V.

7.14 Polarization of a Uniform Plane Wave

Polarization of a uniform plane wave is defined as *the time-varying behaviour of the electric field strength $\vec{E}$ at a given fixed point in space.*

The fields of the uniform plane wave have directions in the medium. Depending on the direction or orientation of the electric field strength, there are three types of polarization.

1. Linear polarization
2. Elliptical polarization
3. Circular polarization

7.14.1 Linear Polarization

A wave is said to be linearly polarized, if the electric field strength remains along a straight line at a given point in space.

Linear polarization is divided into two types:

(a) horizontal polarization and

(b) vertical polarization.

Consider a uniform plane wave travelling in the z direction with electric and magnetic fields laying on the $x - y$ plane.

The electric field vector in time domain for linear polarization is

$$\vec{E}(z,\, t) = E_1 e^{-\alpha t} \cos(\omega t - \beta z)\, \vec{a}_x + E_2 e^{-\alpha t} \cos(\omega t - \beta z)\, \vec{a}_y \tag{7.82}$$

$$= E_x \vec{a}_x + E_y \vec{a}_y.$$

Thus the electric field strength $\vec{E}$ has two components E_x and E_y. If E_x is present and E_y is zero, the wave is said to be *polarized* in the x direction—this phenomena is called horizontal polarization.

Similarly, if E_y is present and E_x is zero, the wave is said to be *polarized* in the y direction—this phenomena is called vertical polarization.

If both E_x and E_y present and are in-phase, the resultant vector makes an angle $\theta = \tan^{-1}(E_y/E_x)$ with the x-axis. This angle is constant with time. Therefore, in linear polarization, the direction of the resultant vector is constant with time as shown in Fig. 7.2.

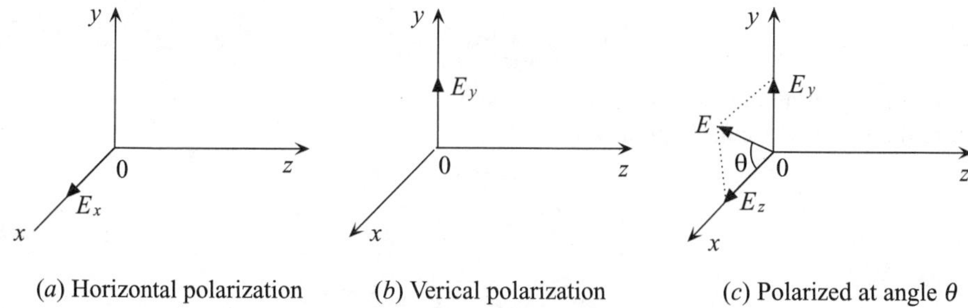

Fig. 7.2 *Linear polarization*

7.14.2 Circular Polarization

Consider a uniform plane wave travelling in the z-direction with electric field vector components $\vec{E}_x$ and $\vec{E}_y$ not in-phase, that is, they reach their maximum values at different instants of time. Then the direction of the resultant electric vector $\vec{E}$ will vary with time.

If $\vec{E}_x$ and $\vec{E}_y$ have equal magnitude with a 90° phase difference, then the locus of the resultant vector $\vec{E}$ is a circle. This wave is called a circularly polarized wave.

Proof Consider a uniform plane wave travelling in the z direction in a lossless medium $\alpha = 0$. The electric field vector in the time domain is

$$\vec{E}(z,\, t) = E_1 \cos(\omega t - \beta z)\vec{a}_x + E_2 \cos(\omega t - \beta z + \phi)\vec{a}_y, \tag{7.83}$$

where the $\vec{E}_y$ component leads the $\vec{E}_x$ component by an angle ϕ.

The electric field vector lies in the $x-y$ plane. The vector $\vec{E}$ has two components, $\vec{E}_x$ and $\vec{E}_y$, along the x- and y-axes respectively.

$$\therefore \quad \vec{E}(z,\, t) = E_x \vec{a}_x + E_y \vec{a}_y. \tag{7.84}$$

For circular polarization, $\vec{E}_x$ and $\vec{E}_y$ have equal magnitudes and $\phi = 90°$ phase difference

i.e., $E_1 = E_2 = E_a$ (say).

$$\therefore \quad \vec{E}(z,\, t) = E_a \cos(\omega t - \beta z)\vec{a}_x + E_a \cos(\omega t - \beta z + 90°)\vec{a}_y$$

at a fixed point $z = 0$ (say).

$$\vec{E}(0,\, t) = E_a \cos(\omega t)\vec{a}_x - E_a \sin(\omega t)\,\vec{a}_y.$$

From Eq. (7.84), $E_x = E_a \cos(\omega t)$ and $E_y = -E_a \sin(\omega t)$.

So, $\dfrac{E_x}{E_a} = \cos \omega t, \quad \dfrac{E_y}{E_a} = -\sin \omega t.$

Squaring and adding the components,

$$E_x^2 + E_y^2 = E_a^2 \cos^2(\omega t) + E_a^2 \sin^2(\omega t)$$

or $E_x^2 + E_y^2 = E_a^2.$ (7.85)

This equation represents the locus of a circle as shown in Fig. 7.3.
Therefore the wave is circularly polarized. (Proved)

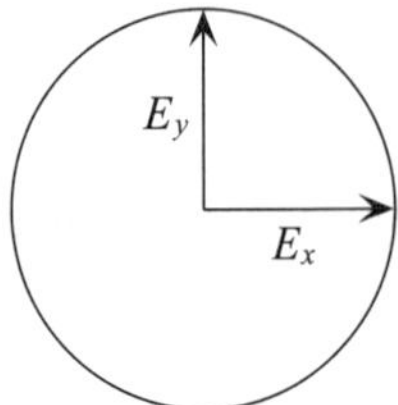

Fig. 7.3 *Locus of circular polarization*

Direction of rotation In circular polarization, if $\vec{E}_y$ leads $\vec{E}_x$ by $90°(\phi = 90°)$, the direction of rotation is a left-handed screw advancing in the z direction as shown in Fig. 7.4(a). Hence the wave is called a left-circularly polarized wave.

Similarly, if $\vec{E}_y$ lags behind $\vec{E}_x$ by $90°(\phi = -90°)$, the direction of rotation is a right-handed screw advancing in the z direction as shown in Fig. 7.4(b). Then the wave is called a right-circularly polarized wave.

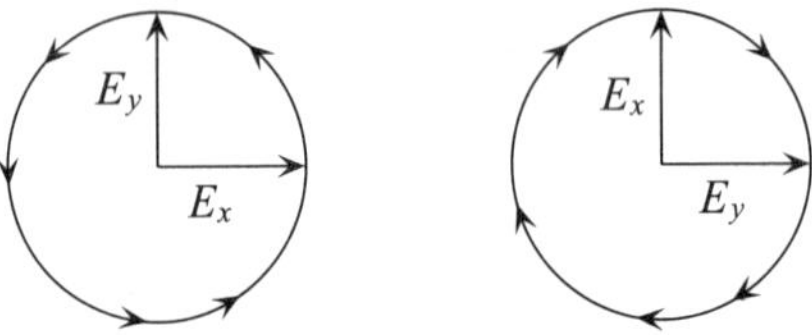

Fig. 7.4 *Left-circularly polarization, (b) right-circularly polarization*

7.14.3 Elliptical Polarization

Consider a uniform plane wave travelling in the z-direction with electric field vector components $\vec{E}_x$ and $\vec{E}_y$ not in-phase, that is, they reach their maximum values at different instants of time. Then the direction of the resultant electric vector $\vec{E}$ will vary with time.

If $\vec{E}_x$ and $\vec{E}_y$ have different magnitudes with a 90° phase difference, then the locus of the resultant vector $\vec{E}$ is an ellipse. The wave is called an elliptical polarized wave.

Proof Consider a uniform plane wave travelling in the z direction in a lossless medium $\alpha = 0$. The electric field vector in time domain is

$$\vec{E}(z,\,t) = E_1\cos(\omega t - \beta z)\vec{a}_x + E_2\cos(\omega t - \beta z + \phi)\vec{a}_y,$$

where the $\vec{E}_y$ component leads the $\vec{E}_x$ component by an angle ϕ.

The electric field vector lies in the $x-y$ plane. The vector $\vec{E}$ has two components, $\vec{E}_x$ and $\vec{E}_y$, along the x- and y-axes respectively.

$$\therefore \quad \vec{E}(z,\,t) = E_x\vec{a}_x + E_y\vec{a}_y. \tag{7.86}$$

For elliptical polarization, $\vec{E}_x$ and $\vec{E}_y$ have different magnitudes and $\phi = 90^\circ$ phase.

$$\therefore \quad \vec{E}(z,\,t) = E_1\cos(\omega t - \beta z)\vec{a}_\chi + E_2\cos(\omega t - \beta z\ + 90^\circ)\vec{a}_y.$$

At a fixed point $z = 0$ (say),

$$\vec{E}(0,\,t) = E_1\cos(\omega t)\vec{a}_x - E_2\sin(\omega t)\vec{a}_y.$$

From Eq. (7.86), $\quad E_x = E_1\cos(\omega t),$

$$E_y = -E_2\sin(\omega t).$$

So, $\quad \dfrac{E_x}{E_1} = \cos\omega t, \quad \dfrac{E_y}{E_2} = -\sin\omega t.$

Squaring and adding the components, since $\sin^2\omega t + \cos^2\omega t = 1$,

$$\frac{E_x^2}{E_1^2} + \frac{E_y^2}{E_2^2} = 1. \tag{7.87}$$

This equation represents the locus of an ellipse as shown in Fig. 7.5.

The wave is elliptically polarized. (Proved)

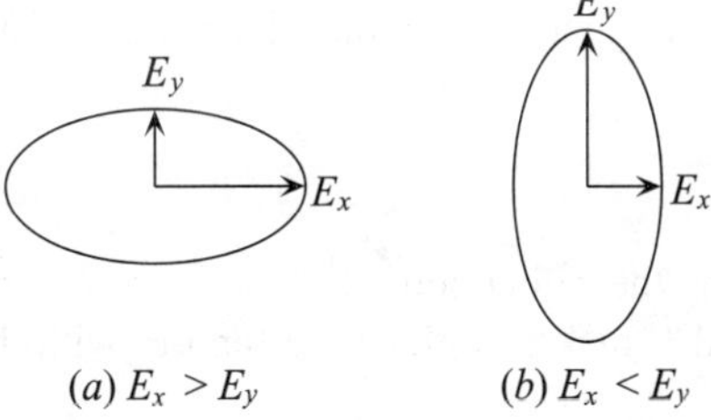

Fig. 7.5 *Locus of elliptical polarization*

Direction of rotation In elliptical polarization, if $\vec{E}_y$ leads $\vec{E}_x$ by $90^\circ (\phi = 90^\circ)$, the direction of rotation is a left-handed screw advancing in the z direction as shown in Fig. 7.6(a). Hence the wave is called a left-elliptically polarized wave.

Similarly, if $\vec{E}_y$ lags behind $\vec{E}_x$ by $90^\circ (\phi = -90^\circ)$, the direction of rotation is a right-handed screw advancing in the z direction as shown in Fig. 7.6(b). Then the wave is called a right-elliptically polarized wave.

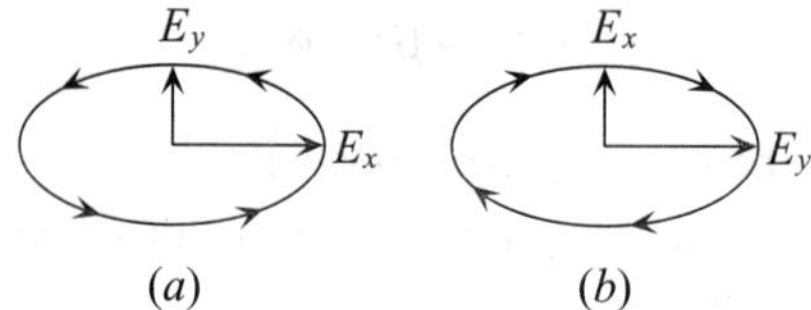

Fig. 7.6 *Left-elliptical polarization, (b) right-elliptical polarization*

7.15 Direction Cosines of a Vector Field

Consider a plane wave travelling in some arbitary direction say along $\vec{r}$. Since the electric field vector is expressed in terms of a fixed set of axes, x, y and z, it is necessary to write the wave field directions and direction of propagation in terms of x, y and z axes.

The cosines of the angles made by the field vectors with the fixed coordinate axes are called direction cosines of vector fields.

Consider the field vector $\vec{E}$ oriented arbitarly with respect to the Cartesian coordinates as shown in Fig. 7.7.

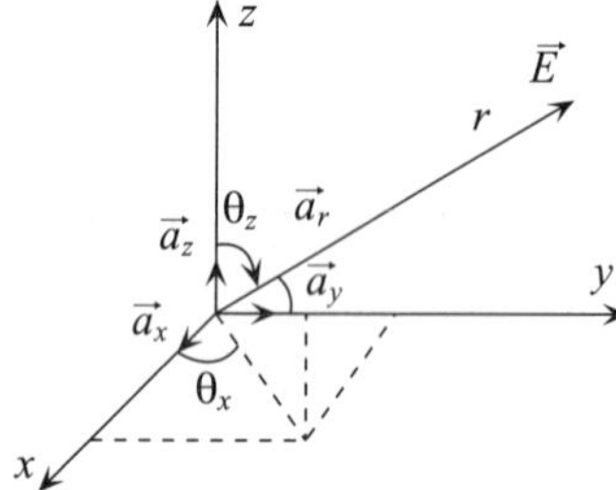

Fig. 7.7 *Direction cosines for $\vec{E}$*

Let a plane wave be travelling along the $\vec{r}$ direction. The electric vector of the plane wave in phasor notation is

$$\vec{E} = E_0 e^{-j\beta(\vec{r}\cdot\vec{a}_r)}, \tag{7.88}$$

where $\vec{a}_r$ is the unit vector along the r direction.

Assume that the vector field $\vec{E}$ makes angles θ_x, θ_y and θ_z with the x, y and z axes respectively. Then the direction cosines are

$$\vec{r}\cdot\vec{a}_r = x\cos\theta_x + y\cos\theta_y + z\cos\theta_z. \tag{7.89}$$

$$\therefore \qquad \vec{E} = E_0 e^{-j\beta(x\cos\theta_x + y\cos\theta_y + z\cos\theta_z)}. \tag{7.90}$$

The wavelength and velocity of the wave measured along the z-axis are greater than when measured along the direction r of wave propagation.

They are given by $\lambda_z = \dfrac{\lambda}{\cos\theta_z}$ and $v_z = \dfrac{v}{\cos\theta_z}$. (7.91)

Additional Solved Problems

Example 7.9 The electric field intensity associated with a plane wave travelling in a perfect dielectric medium is given by $E_x(z, t) = 10\cos(2\pi \times 10^7\, t - 0.1\pi z)$ V/m.

(i) What is the velocity of propagation?

(ii) Write down the magnetic field intensity associated with the wave if $\mu = \mu_0$.

Solution Given: Electric field intensity associated with the plane wave is,

$$E_x(z, t) = 10\cos(2\pi \times 10^7 t - 0.1\pi z) \text{ V/m}.$$

For a perfect dielectric medium, $\mu = \mu_0$, $\sigma = 0$. The general equation for electric field intensity is

$$E_x(z, t) = E_0 e^{-\alpha z} \cos(\omega t - \beta z).$$

Then $E_0 = 10$, $\alpha = 0$, $\omega = 2\pi \times 10^7$ and $\beta = 0.1\pi$.

Velocity of propagation:

$$v_p = \frac{\omega}{\beta} = \frac{2\pi \times 10^7}{0.1\pi} = 2 \times 10^8 \text{ m/s}.$$

Magnetic field intensity:
The magnetic field associated with the plane wave is given as,

$$\vec{H} = \frac{E_0}{|\eta|} e^{-\alpha z} \cos(\omega t - \beta z - \theta)\, \vec{a}_y.$$

We have, $v_p = \dfrac{1}{\sqrt{\mu\varepsilon}}$.

$$\varepsilon = \frac{1}{v_p^2 \mu_0} = \frac{1}{\mu_0 (2 \times 10^8)^2} = 1.989 \times 10^{-11}.$$

The intrinsic impedance for a perfect dielectric is

$$|\eta| = \sqrt{\mu/\varepsilon} = \sqrt{\mu_0 / 1.989 \times 10^{-11}} = 251.32 \text{ ohms}$$

and given $\theta = 0$.

Substituting the values of θ, ω, β, E_0, α,

$$H = \frac{10}{251.32} e^{-\alpha z} \cos(2\pi \times 10^7 t - 0.1\pi z).$$

Therefore the expression for the magnetic field intensity associated with the wave is,

$$\vec{H} = 0.0398 \cos(2\pi \times 10^7 t - 0.1\pi z)\vec{a}_y \text{ A/m}.$$

Example 7.10 Find μ_r, ε_r and σ for a material in which at 100 MHz, the uniform planc wave has $\alpha = 2$ Np/m, $\lambda = 1$ m and $|\eta| = 200$ ohms.

Solution Given for a uniform plane wave, $\alpha = 2$ Np/m, $\lambda = 1$ m, $\eta = 200\,\Omega$ and $f = 100\times10^6$ Hz. We know that the velocity of propagation

$$v_p = f\,\lambda = 100\times10^6\times1 = 10^8 \text{ m/s}.$$

Also, $$v_p = \frac{1}{\sqrt{\mu\,\varepsilon}}.$$

$$\frac{c}{\sqrt{\mu_r\,\varepsilon_r}} = 10^8.$$

$$\sqrt{\mu_r\,\varepsilon_r} = \frac{3\times10^8}{10^8} = 3.$$

The intrinsic impedance is

$$\eta = \sqrt{\frac{\mu}{\varepsilon}} = \sqrt{\frac{\mu_0\,\mu_r}{\varepsilon_0\,\varepsilon_r}}.$$

$$200 = 120\pi\sqrt{\frac{\mu_r}{\varepsilon_r}}$$

or $$\sqrt{\frac{\mu_r}{\varepsilon_r}} = 0.53.$$

Now, $$\sqrt{\frac{\mu_r}{\varepsilon_r}}\cdot\sqrt{\mu_r\,\varepsilon_r} = 0.53\times3.$$

$$\therefore\quad \mu_r = 1.59 \quad\text{and}\quad \sqrt{1.59\times\varepsilon_r} = 3.$$

$$\therefore\quad \varepsilon_r = 9/1.59 = 5.66.$$

To find the conductivity, we know that $\alpha = \sqrt{\dfrac{\omega\mu\sigma}{2}}.$

$$\therefore\quad \sigma = \frac{2\alpha^2}{\mu\omega} = \frac{2\alpha^2}{\mu_0\,\mu_r\,\omega} = \frac{2\times4}{2\times10^{-6}\times2\times\pi\times10^8} = 6.36\times10^{-3} \text{ S/m}.$$

Example 7.11 Find α, β, γ and η for ferrite at 10 GHz when $\varepsilon_r = 9$, $\mu_r = 4$, $\sigma = 10$ mS/m.

Solution Given: $\varepsilon_r = 9$, $\mu_r = 4$, $f = 10\times10^9$ Hz and $\sigma = 10\times10^{-3}$ mS/m.

Since $\sigma \ll \omega\varepsilon$, for a dielectric we know that,

$$\alpha = \frac{\sigma}{2}\sqrt{\frac{\mu}{\varepsilon}} = \frac{10\times10^{-3}}{2}\sqrt{\frac{4\mu_0}{9\varepsilon_0}} = 1.256 \text{ Np/m}.$$

$$\beta = \omega\sqrt{\mu\,\varepsilon}\left(1+\frac{\sigma^2}{8\omega^2\varepsilon^2}\right)$$

$$= 2\pi \times 10^{10}\sqrt{4\mu_0 \times 9\varepsilon_0}\left(1 + \frac{10^{-4}}{8 \times 0.556^2}\right) = 1256.6 \text{ rad/m}.$$

Velocity of propagation, $v_p = \dfrac{\omega}{\beta} = \dfrac{2\pi f}{\beta} = \dfrac{2\pi \times 10^{10}}{1256.6} = 0.5 \times 10^8$ m/s.

Intrinsic impedance $\eta = \sqrt{\dfrac{\mu}{\varepsilon}}\left(1 + j\dfrac{\sigma}{2\omega\varepsilon}\right)$

$$= \sqrt{\frac{4\mu_0}{9\varepsilon_0}}\left(1 + j\frac{10^{-2}}{2 \times 0.556}\right) = 251.32\left(1 + j9 \times 10^{-3}\right).$$

$\therefore \quad \eta = 251.33\angle 0.52^\circ\ \Omega.$

Example 7.12 Determine the phase velocity of propagation, attenuation constant, phase constant and intrinsic impedance for a forward travelling wave in a large block of copper at 1 GHz ($\sigma = 5.8 \times 10^7$, $\varepsilon_r = \mu_r = 1$). Determine the distance that the wave travelled in copper (skin depth).

Solution Given for a block of copper, $\sigma = 5.8 \times 10^7$, $\varepsilon_r = \mu_r = 1$; wave frequency $f = 1$ GHz $= 10^9$ Hz, $\omega\varepsilon = 0.0556$.

Since $\sigma \gg \omega\varepsilon$ for a good conductor, phase velocity is

$$v_p = \frac{\omega}{\beta} = \sqrt{\frac{2\omega}{\mu\sigma}}$$

$$= \sqrt{\frac{2 \times 2\pi \times 10^9}{4\pi \times 10^{-7} \times 1 \times 5.8 \times 10^7}} = 13,130 \text{ m/s}.$$

The attenuation and phase shift constants are

$$\alpha = \beta = \sqrt{\frac{\omega\mu\sigma}{2}}$$

$$= \sqrt{\frac{2\pi \times 10^9 \times 4\pi \times 10^{-7} \times 5.8 \times 10^7}{2}}.$$

$$\alpha = 4.78 \times 10^5 \text{ Np/m};$$

$$\beta = 4.78 \times 10^5 \text{ rad/m}.$$

Intrinsic impedance $\eta = \sqrt{\dfrac{j\omega\mu}{\sigma}} = \sqrt{\dfrac{(2\pi \times 10^9) \times (4\pi \times 10^{-7})}{5.8 \times 10^7}}\ \angle 45^\circ = 0.01167\angle 45^\circ.$

The distance travelled in copper is

$$\text{skin depth,} \quad \delta = \sqrt{\frac{2}{\omega\mu\sigma}} = \frac{1}{\alpha} = 2 \times 10^{-6} \text{ m}.$$

Example 7.13 A travelling wave has two linearly polarized components $E_x = 2\cos\omega t$ and $E_y = 3\cos(\omega t + \pi/2)$.

(i) What is the axial ratio?
(ii) What is the tilt angle of the major axis of the polarization ellipse?
(iii) What is the sense of rotation?

Solution Given: $E_x = 2\cos\omega t$;

$$E_y = 3\cos(\omega t + \pi/2).$$

We know that $E_x = E_1\cos(\omega t)$ and $E_y = E_2\cos(\omega t + \phi)$.

(i) Axial ratio $= \dfrac{E_2}{E_1} = \left(\dfrac{3}{2}\right) = 1.5.$

(ii) Tilt angle is the phase difference between E_x and E_y.

$$\text{Tilt angle} = \phi = 90^\circ.$$

(iii) $\dfrac{E_x}{2} = \cos(\omega t),\ \dfrac{E_y}{3} = \cos\left(\omega t + \dfrac{\pi}{2}\right) = -\sin(\omega t).$

$$\left(\frac{E_x}{2}\right)^2 + \left(\frac{E_y}{3}\right)^2 = \cos^2(\omega t) + \sin^2(\omega t) = 1.$$

This is the equation of an ellipse. Since E_y leads E_x by 90°, the sense of rotation is left-handed.

Example 7.14 Calculate the intrinsic impedance η, the propagation constant γ and the wave velocity v for a conducting medium in which $\sigma = 58$ MS/m, $\mu_r = 1$ at frequency $f = 100$ MHz.

Solution Given: $\sigma = 58$ M S/m $= 58\times10^6$ S/m, $\mu_r = 1$ and $f = 100$ MHz.
Since $\sigma \gg \omega\varepsilon$, for a good conductor:

$$\text{Intrinsic impedance,}\quad \eta = \sqrt{\frac{\omega\mu}{\sigma}}\ \angle 45^\circ$$

$$= \sqrt{\frac{2\times\pi\times100\times10^6\times4\pi\times10^{-7}}{58\times10^6}}\ \angle 45^\circ$$

$$= 3.69\times10^{-3}\ \angle 45^\circ\Omega.$$

$$\text{Propagation constant,}\quad \gamma = \sqrt{\omega\mu\sigma}\angle 45^\circ$$

$$= \sqrt{2\times\pi\times100\times10^6\times4\pi\times10^{-7}\times58\times10^6}\ \angle 45^\circ$$

$$= 214\times10^3\ \angle 45^\circ.$$

$$\gamma = \alpha + j\beta.$$

$$\alpha + j\beta = 214\times 10^3 \angle 45^\circ = (151.32 + j151.32)\times 10^3.$$

$$\alpha = \beta = 151.32\times 10^3.$$

Depth penetration $\delta = \dfrac{1}{\alpha} = \dfrac{1}{151.32\times 10^3} = 6.6\ \mu\text{m}.$

Wave velocity $v = \dfrac{\omega}{\beta} = \dfrac{2\pi\times 100\times 10^6}{151.32\times 10^3} = 4.15\times 10^3$ m/s.

Example 7.15 Find the skin depth at a frequency of 2 MHz in aluminium where $\sigma = 38.2$ MS/m and $\mu_r = 1$. Also find the distance at which the amplitude of the field becomes 50% of its value at $z = 0$.

Solution Given: $\sigma = 38.2\times 10^6$ S/m, $\mu_r = 1$, $\omega = 2\pi f = 2\pi\times 2\times 10^6 = 4\pi\times 10^6$.
Since $\sigma \gg \omega\varepsilon$, for a good conductor:

Skin depth $\delta = \sqrt{\dfrac{2}{\omega\mu\sigma}}.$

$$\delta = \sqrt{\frac{2}{4\pi\times 10^6\times 4\pi\times 10^{-7}\times 38.2\times 10^6}} = 5.758\times 10^{-5}\ \text{m}.$$

To find out the distance, the field inside the material is

$$E = E_0\, e^{-\alpha z}.$$

If the field becomes 50%. i.e., $\dfrac{E}{E_0} = 0.5,$

then $e^{-\alpha z} = 0.5$

or $z = \dfrac{1}{\alpha}\ln\left(\dfrac{1}{0.5}\right).$

For a good conductor, $\alpha = \dfrac{1}{\delta} = \dfrac{1}{5.758\times 10^{-5}}.$

Therefore the distance $z = 5.758\times 10^{-5}\ln(2) = 4\times 10^{-5}$ m.

Example 7.16 If $\varepsilon_r = 9$, $\mu = \mu_0$, for the medium in which a wave with a frequency of $f = 0.3$ GHz is propagating, determine the propagation constant and intrinsic impedance of the medium when $\sigma = 0$.

Solution Given: $\varepsilon_r = 9$, $\mu = \mu_0$, $\sigma = 0$, and frequency $f = 0.3$ GHz.
Since $\sigma = 0$ for lossless dielectric medium, $\alpha = 0$, we know that

$$\beta = \omega\sqrt{\mu\varepsilon} = \omega\sqrt{\mu_0\,\mu_r\,\varepsilon_0\,\varepsilon_r} = \frac{\omega}{c}\sqrt{\mu_r\,\varepsilon_r}$$

$$= \frac{2\pi \times 0.3 \times 10^9}{3 \times 10^8}\sqrt{1 \times 9} = 18.85 \text{ rad/m}.$$

Propagation constant $\gamma = j18.85$ per metre.

Intrinsic impedance $\eta = \sqrt{\frac{\mu}{\varepsilon}} = \sqrt{\frac{\mu_0 \mu_r}{\varepsilon_0 \varepsilon_r}} = 120\pi \frac{1}{\sqrt{9}} = 125.67\Omega.$

Example 7.17 In a lossless medium for which $\eta = 60\pi$, $\mu_r = 1$ and $\vec{H} = 0.2\cos(\omega t - z)\,\vec{a}_x + 0.3\sin(\omega t - z)\,\vec{a}_y$ A/m, calculate ε, γ, ω.

Solution Given, $\eta = 60\pi$, $\mu_r = 1$, for lossless medium $\sigma = 0$ and

$$\vec{H} = 0.2\cos(\omega t - z)\,\vec{a}_x + 0.3\sin(\omega t - z)\,\vec{a}_y \text{A/m}.$$

From the above equation, $\beta = 1$ rad/m.

Characteristic impedance $\eta = \sqrt{\frac{\mu}{\varepsilon}} = \sqrt{\frac{\mu_0 \mu_r}{\varepsilon_0 \varepsilon_r}}.$

$$\eta = \eta_0 \sqrt{\frac{\mu_r}{\varepsilon_r}} = 120\pi \sqrt{\frac{1}{\varepsilon_r}} = 60\pi.$$

$$\varepsilon_r = 4.$$

Propagation constant $\gamma = \alpha + j\beta = 0 + j1.$

$$\beta = \omega\sqrt{\mu\varepsilon} = 1.$$

$$\omega = \frac{1}{\sqrt{\mu\varepsilon}} = \frac{3 \times 10^8}{\sqrt{4}} = 150 \times 10^6 \text{ rad/s}.$$

Example 7.18 A plane wave propagation through a medium with $\varepsilon_r = 8$, $\mu_r = 2$ has $\vec{E} = 0.5e^{-0.33z}\sin(10^8 t - \beta z)\vec{a}_x \text{V/m}$. Determine wave velocity, wave impedance and the magnetic field intensity.

Solution Given, $\varepsilon_r = 8$, $\mu_r = 2$, and $\vec{E} = 0.5e^{-0.33z}\sin(10^8 t - \beta z)\vec{a}_x$ V/m.
From the above equation, $\alpha = 0.33$ Np/m and frequency $\omega = 10^8$ rad/m.

We know that the attenuation constant $\alpha = \omega\sqrt{\frac{\mu\varepsilon}{2}\left[\sqrt{1 + \left(\frac{\sigma}{\omega\varepsilon}\right)^2} - 1\right]}.$

$$0.33 = \frac{10^8}{3 \times 10^8} \times \sqrt{\frac{8 \times 2}{2}\left[\sqrt{1 + \left(\frac{\sigma}{10^8 \times 8\varepsilon_0}\right)^2} - 1\right]} = 0.943 \times \sqrt{\left[\sqrt{1 + \left(\frac{\sigma}{10^8 \times 8\varepsilon_0}\right)^2} - 1\right]}.$$

$$1 + \left(\frac{\sigma}{7.083 \times 10^{-3}}\right)^2 = 1.26.$$

$\therefore \quad \sigma = 3.61 \times 10^{-3}\ \text{S/m}$

and $\quad \beta = \omega\sqrt{\frac{\mu\varepsilon}{2}\left[\sqrt{1+\left(\frac{\sigma}{\omega\varepsilon}\right)^2}+1\right]}.$

$$\beta = \frac{10^8}{3\times 10^8} \times \sqrt{\frac{8\times 2}{2}\left[\sqrt{1+\left(\frac{3.61\times 10^{-3}}{7.083\times 10^{-3}}\right)^2}+1\right]}.$$

$$\beta = 1.3736\ \text{rad/m}.$$

Phase velocity of propagation

$$v_p = \frac{\omega}{\beta} = \frac{2\pi f}{\beta} = \frac{1\times 10^8}{1.3736} = 0.728\times 10^8\ \text{m/s}.$$

$$v_p = 0.728\times 10^8\ \text{m/s}.$$

Wave impedance

$$\eta = \sqrt{\frac{j\omega\mu}{\sigma + j\omega\varepsilon}} = \sqrt{\frac{j10^8\times 2\mu_0}{(3.612+j7.083)\times 10^{-3}}}$$

$$= \sqrt{31610\angle 27^\circ} = 177.8\angle 13.5\ \Omega.$$

$\therefore \quad \vec{E} = 0.5e^{-0.33z}\sin(10^8 t - 1.3736z)\,\vec{a}_x.$

The magnetic field intensity

$$\vec{H} = \frac{E_x}{\eta}\vec{a}_y = \frac{0.5e^{-0.33z}}{177.8\ \angle 13.5}\sin(10^8 t - \beta z)\,\vec{a}_y.$$

$$\vec{H} = 2.812e^{-0.33z}\sin(10^8 t - 1.373z - 13.5^\circ)\,\vec{a}_y\ \text{A/m}.$$

Example 7.19 Calculate the depth penetration for silver and brass at 1000 Hz given $\sigma_{\text{silver}} = 6.17\times 10^7$ mhos/m, $\sigma_{\text{brass}} = 1.5\times 10^7$ mhos/m, $\mu = \mu_r = 4\pi\times 10^7$.

Solution Given: $\mu_r = 1,\ f = 1000$ Hz; for silver, $\sigma_{\text{silver}} = 6.17\times 10^7$ mho/m; for brass,

$$\sigma_{\text{brass}} = 1.5\times 10^7\ \text{mho/m}$$

$$\omega = 2\pi f = 2\pi\times 1\times 10^3 = 2000\pi$$

Since $\sigma \gg \omega\varepsilon$, for a good conductor depth of penetration is $\delta = \sqrt{\frac{2}{\omega\mu\sigma}}$.

For silver, depth of penetration is

$$\delta = \sqrt{\frac{2}{2000\pi \times \mu_0 \times 6.17 \times 10^7}} = 2.026 \times 10^{-3} \text{ m.}$$

For brass, depth of penetration is

$$\delta = \sqrt{\frac{2}{2000\pi \times \mu_0 \times 1.5 \times 10^7}} = 4.11 \times 10^{-3} \text{ m.}$$

Example 7.20 A plane wave is propagating in a medium having properties $\mu_r = 4$, $\varepsilon_r = 36$, $\sigma = 1$ S/m and $\vec{E} = 100e^{-\alpha z}\cos(10^9 t - \beta z)\,\vec{a}_x$ V/m. Determine associated magnetic field.

Solution Given, $\vec{E} = 100e^{-\alpha z}\cos(10^9 t - \beta z)\,\vec{a}_x$ V/m, frequency $\omega = 10^9$ rad/s, $\mu_r = 4$, $\varepsilon_r = 36$, and $\sigma = 1$ S/m.

For lossy dielectric,
propagation constant $\gamma = \sqrt{j\omega\mu(\sigma + j\omega\varepsilon)}$.

$$\gamma = \sqrt{j10^9 \times 4\mu_0 \times (1 + j10^9 \times 36\varepsilon_0)}$$

$$= \sqrt{j5026.55 \times (1 + j0.3187)}$$

$$= \sqrt{5275.65\angle 107.67} = 72.63\angle 53.83.$$

$$\gamma = \alpha + j\beta = 42.865 + j58.632 \text{ per metre.}$$

$$\therefore \quad \vec{E} = 100e^{-42.865z}\cos(10^9 t - 58.632z)\,\vec{a}_x \text{ V/m.}$$

Intrinsic impedance

$$\eta = \sqrt{\frac{j\omega\mu}{\sigma + j\omega\varepsilon}} = \sqrt{\frac{j5026.55}{(1 + j0.3187)}} = \sqrt{1454.2 + j4563}$$

$$= \sqrt{4789.1\angle 72.32^\circ} = 69.2\angle 36.16\ \Omega.$$

The magnetic field intensity

$$\vec{H} = \frac{E_x}{\eta}\vec{a}_y = \frac{100e^{-42.865z}}{69.2\ \angle 36.16}\sin(10^9 t - 58.632z)\,\vec{a}_y.$$

$$H_y = 1.445e^{-42.865z}\sin(10^9 t - 58.632z - 36.16^\circ) \text{ A/m.}$$

Example 7.21 A uniform plane wave propagating in a medium has $E = 2e^{-\alpha z}\sin(10^8 t - \beta z)\,\vec{a}_y$ V/m. If medium is characterized by $\mu_r = 20$, $\varepsilon_r = 1$, $\sigma = 3$ mhos/m and $\omega = 10^8$ rad/s. Find α, β.

Solution Given, $E = 2e^{-\alpha z}\cos(10^8 t - \beta z)\,\vec{a}_y$V/m, frequency $\omega = 10^8$ rad/s, $\mu_r = 20$, $\varepsilon_r = 1$, and $\sigma = 3$ S/m.

Now, $\omega\varepsilon = 10^8\ \varepsilon_0 = 8.854 \times 10^{-4}$.

Since $\sigma \gg \omega\varepsilon$, for a good conductor, the attenuation and phase shift constants are

$$\alpha = \beta = \sqrt{\frac{\omega\mu\sigma}{2}} = \sqrt{\frac{10^8 \times 20\mu_0 \times 3}{2}} = 61.4.$$

$$\alpha = 61.4 \text{ Np/m}; \beta = 61.4 \text{ rad/m}.$$

$$\therefore \quad \vec{E} = 2e^{-61.4z}\cos(10^8 t - 61.4z)\,\vec{a}_y \text{ V/m}.$$

Example 7.22 For a uniform plane wave in space $\lambda = 12$ cm. In a lossless material of unknown characteristics $\lambda = 8$ cm. In this material, $E = 50$ V/m, H = 0.1 A/m. Find f, μ_r and ε_r.

Solution Given, $E = 50$ V/m, H = 0.1 A/m; wavelength in space $\lambda_0 = 12$ cm; wavelength in material $\lambda_m = 8$ cm.

$$\text{Frequency } f = \frac{c}{\lambda_0} = \frac{3 \times 10^8}{0.12} = 25 \times 10^8 \text{ Hz}.$$

For lossless material, $\sigma = 0$.

We know that the intrinsic impedance $\eta = \dfrac{E}{H} = \sqrt{\dfrac{\mu}{\varepsilon}} = \dfrac{50}{0.1}$.

$$\sqrt{\frac{\mu_0\,\mu_r}{\varepsilon_0\,\varepsilon_r}} = 500.$$

$$120\pi\sqrt{\frac{\mu_r}{\varepsilon_r}} = 500.$$

$$\sqrt{\frac{\mu_r}{\varepsilon_r}} = 1.3263, \quad \mu_r = 1.76\varepsilon_r.$$

$$\text{Also,} \quad \lambda_m = \frac{2\pi}{\beta} = \frac{2\pi}{\omega\sqrt{\mu\varepsilon}}\lambda_m = \frac{2\pi}{\beta_0\sqrt{\mu_r\,\varepsilon_r}} = \frac{\lambda_0}{\sqrt{\mu_r\,\varepsilon_r}}.$$

$$\sqrt{\mu_r\,\varepsilon_r} = \frac{\lambda_0}{\lambda_m} = \frac{12}{8} = 1.5.$$

$$\sqrt{1.76\varepsilon_r \times \varepsilon_r} = 1.5, \quad \varepsilon_r = 1.13 \text{ and } \quad \mu_r = 1.99.$$

Example 7.23 A 300 MHz uniform plane wave propagates through fresh water for which $\sigma = 0$, $\mu_r = 1$ and $\varepsilon_r = 78$. Calculate (i) the attenuation constant, (ii) the phase constant (iii) the wavelength and (iv) intrinsic impedance.

Solution Given, $\varepsilon_r = 78$, $\mu = \mu_0$, lossless medium $\sigma = 0$, and frequency $f = 0.3$ GHz.

(i) Since $\sigma = 0$, for lossless dielectric medium, attenuation constant $\alpha = 0$.

(ii) We know that

$$\beta = \omega\sqrt{\mu\varepsilon} = \omega\sqrt{\mu_0\,\mu_r\,\varepsilon_0\,\varepsilon_r} = \frac{\omega}{c}\sqrt{\mu_r\,\varepsilon_r} = \frac{2\pi\times 03\times 10^9}{3\times 10^8}\sqrt{1\times 78}.$$

Phase constant $\beta = 55.5$ rad/m.

$\therefore$ Propagation constant $= j55.5$ per metre.

(iii) Wavelength $\lambda = \dfrac{2\pi}{\beta} = \dfrac{2\pi}{55.5} = 0.1113$ m.

Intrinsic impedance $\eta = \sqrt{\dfrac{\mu}{\varepsilon}} = \sqrt{\dfrac{\mu_0\,\mu_r}{\varepsilon_0\,\varepsilon_r}} = 120\pi\dfrac{1}{\sqrt{78}} = 42.68\ \Omega.$

Example 7.24 Find the depth of penetration of an EM wave in copper at $f = 60$ Hz and $f = 100$ MHz. For copper $\sigma = 5.8\times 10^7$ S/m, $\varepsilon_r = 1$, and $\mu_r = 1$.

Solution Given $\varepsilon_r = 1$, $\mu_r = 1$.

For copper, $\sigma = 5.8\times 10^7$ mhos/m.

At $f = 60$ Hz, $\omega = 2\pi f = 2\pi\times 60 = 120\pi$ rad/m, $\omega\varepsilon = 120\pi\varepsilon_0$.

Since $\sigma \gg \omega\varepsilon$, the depth of penetration is $\delta = \sqrt{\dfrac{2}{\omega\mu\sigma}}$.

$$\delta = \sqrt{\frac{2}{120\pi}\times\mu_0\times 5.8\times 10^7} = 8.53\times 10^{-3}\text{ m}.$$

At $f = 100$ MHz, $\omega = 2\pi f = 2\pi\times 100\times 10^6 = 2\pi\times 10^8$ rad/s, $\omega\varepsilon = 2\pi\times 10^8\varepsilon_0$.

Since $\sigma \gg \omega\varepsilon$, the depth of penetration is

$$\delta = \sqrt{\frac{2}{\omega\mu\sigma}} = \sqrt{\frac{2}{2\pi}\times 10^6\times\mu_0\times 5.8\times 10^7} = 6.6\times 10^{-6}\text{ m}.$$

Example 7.25 At what frequency can earth be considered a perfect dielectric, if $\sigma = 6\times 10^{-3}$ S/m, $\mu_r = 1$ and $\varepsilon_r = 10$?

Solution Given: $\varepsilon_r = 10$, $\mu_r = 1$ and $\sigma = 6\times 10^{-3}$ S/m.

Now, $\dfrac{\sigma}{\omega\varepsilon} = \dfrac{6\times 10^{-3}}{\omega\times 10\varepsilon_0} = \dfrac{6\times 10^{-3}}{2\pi f\times 10\varepsilon_0} = \dfrac{10.78\times 10^6}{f}$.

For a perfect dielectric, $\dfrac{10.78\times 10^6}{f} \ll 1$.

$\therefore$ $f \gg 10.78$ MHz.

If frequency is much greater than 10.78 MHz, it acts as a perfect dielectric.

Example 7.26 Find the conducting behaviour of ground at 1 kHz, 10 MHz and 10 GHz. Given $\varepsilon_r = 10$ and $\sigma = 5 \times 10^{-3}$ mho/m.

Solution Given: $\varepsilon_r = 10,\ \sigma = 5 \times 10^{-3}$ S/m.

The ratio of conduction current to displacement current is $\dfrac{\sigma}{\omega\varepsilon}$ which determines conducting behaviour.

(a) For 1 kHz

$$\frac{\sigma}{\omega\varepsilon} = \frac{5 \times 10^{-3}}{2\pi \times 1 \times 10^3 \times 10\varepsilon_0} = 9 \times 10^3.$$

$$\frac{\sigma}{\omega\varepsilon} \gg 1, \text{ it is a condition for a good conductor.}$$

(b) For 10 MHz

$$\frac{\sigma}{\omega\varepsilon} = \frac{5 \times 10^{-3}}{2\pi \times 10 \times 10^6 \times 10\varepsilon_0} = 0.9.$$

$$\frac{\sigma}{\omega\varepsilon} \simeq 1, \text{ it is a condition for a lossy or quasi conductor.}$$

(c) For 10 GHz

$$\frac{\sigma}{\omega\varepsilon} = \frac{5 \times 10^{-3}}{2\pi \times 10 \times 10^9 \times 10\varepsilon_0} = 9 \times 10^{-4}.$$

$$\frac{\sigma}{\omega\varepsilon} \ll 1, \text{ it is a condition for an insulator.}$$

Thus, at 1 kHz, the ground acts like a good conductor and at 10 GHz, it acts like an insulator. At 10 MHz the ground acts like quasi-conductor.

Example 7.27 A lossy dielectric is characterised with $\varepsilon_r = 2.5$, $\mu_r = 4$ and $\sigma = 10^{-3}$ mhos/m at 10 MHz. Let $\vec{E} = 10e^{-\alpha z}\vec{a}_x$ V/m. Find (i) α (ii) β (iii) λ (iv) v_p (v) η.

Solution Given: $\sigma = 10^{-3}$ mhos/m, $\varepsilon_r = 2.5$, $\mu_r = 4$, $f = 10$ MHz and $E = 10e^{-\alpha z}\,\vec{a}_x$.

Attenuation constant $\alpha = \omega\sqrt{\dfrac{\mu\varepsilon}{2}\left(\sqrt{1+\left(\dfrac{\sigma}{\omega\varepsilon}\right)^2} - 1\right)}.$

Now, $\dfrac{\sigma}{\omega\varepsilon} = \dfrac{10^{-3}}{2\pi \times 10 \times 10^6 \times 2.5\varepsilon_0} = 0.72$

and $\mu\varepsilon = 4\pi \times 10^{-7} \times 4 \times 2.5 \times 8.854 \times 10^{-12} = 1112.62 \times 10^{-19}.$

$$\therefore \quad \alpha = 2\pi \times 10^7\sqrt{\frac{111.2.62}{2} \times 10^{-19}(\sqrt{1+0.72^2} - 1)} = 0.228 \text{ Np/m.}$$

Phase shift constant $\beta = \omega\sqrt{\frac{\mu\varepsilon}{2}\left(\sqrt{1+\left(\frac{\sigma}{\omega\varepsilon}\right)^2}+1\right)}.$

$$= 2\pi \times 10^7\sqrt{\frac{1112.62}{2} \times 10^{-19}(\sqrt{1+0.72^2}+1)} = 0.7 \text{ rad/m.}$$

Wavelength $\lambda = \frac{2\pi}{\beta} = \frac{2\pi}{0.7} = 8.97$ m.

Phase velocity $v_p = f\lambda = 10 \times 10^6 \times 8.97 = 8.97 \times 10^7$ m/s.

Intrinsic impedance $\eta = \sqrt{\frac{\mu}{\varepsilon}} = \eta_0\sqrt{\frac{\mu_r}{\varepsilon_r}} = 120\pi\sqrt{\frac{4}{25}} = 476.86\ \Omega.$

Example 7.28 The complex electric field vector of a uniform plane wave propagating in free space is given by $\vec{E} = (-\vec{a}_x - \sqrt{3}\vec{a}_y + \sqrt{3}\vec{a}_z)\, e^{-j0.04\pi(\sqrt{3}x-2y-3z)}$ V/m.

(a) What is the direction of propagation?

(b) Find the wavelength in the direction of propagation.

(c) Find the frequency of wave.

(d) Find the apparent wavelengths and the apparent velocities along the x, y and z axes.

Solution Given, the complex electric field vector,

$$\vec{E} = (-\vec{a}_x - \sqrt{3}\vec{a}_y + \sqrt{3}\vec{a}_z)\, e^{-j0.04\pi(\sqrt{3}x-2y-3z)}.$$

We know that the phasor form of the electric field is

$$E = E_0\, e^{j\vec{k}\cdot\vec{r}},$$

where $\vec{k} = k_x\vec{a}_x + k_y\vec{a}_y + k_z\vec{a}_z$

and $\vec{r} = x\vec{a}_x + y\vec{a}_y + z\vec{a}_z.$

$\therefore\quad \vec{k}.\vec{r} = k_x x + k_y y + k_z z.$

Given, $\vec{k}.\vec{r} = -0.04\pi(\sqrt{3}x - 2y - 3z).$

$$\beta = |k| = \sqrt{k_x^2 + k_y^2 + k_z^2} = 0.04\pi\sqrt{3 + 2^2 + 3^2} = 0.16\pi \text{ rad/m.}$$

Assume that the vector field $\vec{E}$ makes angles θ_x, θ_y and θ_z with the x, y and z axes respectively. The direction cosines are

$$k_x = \beta\cos\theta_x = 0.04\pi \times \sqrt{3}$$

$$\cos\theta_x = \frac{0.04\pi \times \sqrt{3}}{0.16\pi} = 0.433$$

$$k_y = \beta \cos\theta_y = 0.04\pi \times 2$$

$$\cos\theta_y = \frac{0.04\pi \times 2}{0.16\pi} = 0.5$$

and $k_z = \beta \cos\theta_z = 0.04\pi \times 3$

$$\cos\theta_z = \frac{0.0.4\pi \times 3}{0.16\pi} = 0.75.$$

In free space, we have, $\lambda = \frac{2\pi}{\beta} = \frac{2\pi}{0.16\pi} = 12.5$ m.

The frequency of the wave is $f = \frac{c}{\lambda} = \frac{3 \times 10^8}{12.5} = 24$ MHz.

The apparent wavelength along the x, y and z axes are:

$$\lambda_x = \frac{\lambda}{\cos\theta_x} = \frac{12.5}{0.433} = 28.86 \text{ m.}$$

$$\lambda_y = \frac{\lambda}{\cos\theta_y} = \frac{12.5}{0.5} = 25 \text{ m.}$$

$$\lambda_z = \frac{\lambda}{\cos\theta_z} = \frac{12.5}{0.75} = 16.66 \text{ m.}$$

The apparent velocity along the x, y and z axes are:

$$v_x = \frac{c}{\cos\theta_x} = \frac{3 \times 10^8}{0.433} = 6.93 \times 10^8 \text{ m/s.}$$

$$v_y = \frac{c}{\cos\theta_y} = \frac{3 \times 10^8}{0.5} = 6 \times 10^8 \text{ m/s.}$$

$$v_z = \frac{c}{\cos\theta_z} = \frac{3 \times .10^8}{0.75} = 4 \times 10^8 \text{ m/s.}$$

Summary

1. A homogeneous medium is always linear, i.e., the flux density is proportional to the field intensity. For an isotropic medium, $\vec{E}$ and $\vec{D}$ are always in the same direction. If there is no source generator in the medium, then the medium is source free.
2. For a homogeneous and isotropic medium, the field relations are

$$\vec{D} = \varepsilon\vec{E},\ \vec{B} = \mu\vec{H} \text{ and } \vec{J} = \sigma\vec{E}.$$

3. The wave equations for a conducting medium are

$$\nabla^2\vec{E} - \mu\varepsilon\frac{\partial^2\vec{E}}{\partial t^2} - \mu\sigma\frac{\partial\vec{E}}{\partial t} = 0$$

$$\nabla^2\vec{H} - \mu\varepsilon\frac{\partial^2\vec{H}}{\partial t^2} - \mu\sigma\frac{\partial\vec{H}}{\partial t} = 0$$

For free space, the wave equations can be obtained by substituting $\sigma = 0$

4. Helmholtz wave equations for conducting medium are

$$\nabla^2\vec{E} - (j\omega\mu\sigma - \omega^2\mu\varepsilon)\vec{E} = 0,\ \nabla^2 H - (j\omega\mu\sigma - \omega^2\mu\varepsilon)H = 0.$$

For a free space lossless medium ($\sigma = 0$), the wave equations became

$$\nabla^2\vec{E} + \omega^2\mu\varepsilon\vec{E} = 0, \text{ and } \nabla^2\vec{H} + \omega^2\mu\varepsilon\vec{H} = 0.$$

5. An electromagnetic wave in which electric and magnetic field intensities are directed in fixed directions in space and constant on infinite planes perpendicular to the direction propagation is called a uniform plane wave.
6. The radiation received by a small antenna at far distances can be approximated as a plane wave.
7. The wave equation for a uniform plane wave travelling in the z-direction in free space is $\dfrac{d^2\vec{E}}{dz^2} + \beta^2\vec{E} = 0$, where $\beta = \omega\sqrt{\mu\varepsilon}$ rad/m is called phase shift constant.

   The general solution for a wave equation in the forward direction is

$$E_x = E\cos(\omega t - \beta z),\quad H_y = H\cos(\omega t - \beta z).$$

8. The velocity at which the constant phase of the wave travels is called phase velocity.

   Phase velocity $v_p = \dfrac{\omega}{\beta} = \dfrac{1}{\sqrt{\mu\varepsilon}}$ m/s

9. The intrinsic impedance of the lossless medium is

$$\eta = \frac{E}{H} = \sqrt{\frac{\mu}{\varepsilon}} \text{ ohms.}$$

For a lossy medium,

$$\eta = \frac{E}{H} = \sqrt{\frac{j\omega\mu}{\sigma + j\omega\varepsilon}} \text{ ohms.}$$

10. The plane wave equation for the electric field component propagating in a conducting medium along z-direction is

$$\frac{d^2\vec{E}_x}{dz^2} + \gamma^2\vec{E}_x = 0.$$

The general solution for the wave equation in the forward direction is

$$E_x = Ee^{-\alpha z}\cos(\omega t - \beta z),\quad H_y = He^{-\alpha z}\cos(\omega t - \beta z),$$

where α is called attenuation constant, and

$$\gamma = \alpha + j\beta = \sqrt{j\omega\mu(\sigma + j\omega\varepsilon)}$$

12. Attenuation constant and phase shift constant can be expressed as

$$\alpha = \omega\sqrt{\frac{\mu\varepsilon}{2}\left[\sqrt{1+\left(\frac{\sigma}{\omega\varepsilon}\right)^2}-1\right]},\quad \beta = \sqrt{\frac{\mu\varepsilon}{2}\left[\sqrt{1+\left(\frac{\sigma}{\omega\varepsilon}\right)^2}+1\right]}.$$

13. For a lossless perfect dielectric, i.e, $\alpha = 0$.

$$\beta = \omega\sqrt{\mu\varepsilon},\ v_p = \frac{\omega}{\beta},\ \eta = \sqrt{\frac{\mu}{\varepsilon}} = \sqrt{\frac{\mu_0\mu_r}{\varepsilon_0\varepsilon r}}$$

For free space, $\beta = \omega\sqrt{\mu_0\varepsilon_0}$, $v_0 = \dfrac{\omega}{\beta} = \dfrac{1}{\sqrt{\mu_0\varepsilon_0}} = 3\times 10^8$ m/s.

$$\eta = \frac{E}{H} = \sqrt{\frac{\mu_0}{\varepsilon_0}} = 120\pi\Omega.$$

14. For good conductors,

$$\alpha = \sqrt{\frac{\omega\mu\sigma}{2}},\ \beta = \sqrt{\frac{\omega\mu\sigma}{2}},\ \eta = \sqrt{\frac{\omega\mu}{\sigma}}\angle 45^\circ \text{ and } v_p = \sqrt{\frac{2\omega}{\mu\sigma}}.$$

15. For good dielectrics,

$$\alpha = \frac{\sigma}{2}\sqrt{\frac{\mu}{\varepsilon}},\quad \beta = \omega\sqrt{\mu\varepsilon}\left(1+\frac{\sigma^2}{8\omega^2\varepsilon^2}\right),\quad \eta = \sqrt{\frac{\mu}{\varepsilon}}\left(1+j\frac{\sigma}{2\omega\varepsilon}\right)$$

and $v_p = v_0\left(1-\dfrac{\sigma^2}{8\omega^2\varepsilon^2}\right)$.

16. Skin depth for a lossy conducting medium is

$$\delta = \frac{1}{\alpha} = \frac{1}{\omega\sqrt{\frac{\mu\varepsilon}{2}\left[\sqrt{1+\left(\frac{\sigma}{\omega\varepsilon}\right)^2}-1\right]}}.$$

For good conductors,

Skin depth is $\delta = \dfrac{1}{\alpha} = \sqrt{\dfrac{2}{\omega\mu\sigma}}$.

For good dielectrics, $\delta = \dfrac{2}{\sigma}\sqrt{\dfrac{\varepsilon}{\mu}}$.

For perfect conductors ($\sigma = \infty$), skin depth $\delta = 0$.

For lossless dielectrics ($\alpha = 0$), skin depth $\delta = \infty$.

17. Polarization of a uniform plane wave is defined as the time-varying behaviour of the electric field strength $\vec{E}$ at a given fixed point in space.

18. There are three types of polarizations:
    (i) Linear polarization:
        (a) Vertical polarization, (b) horizontal polarization
    (ii) Circular polarization
    (iii) Elliptical polarization

Review Questions

1. Write Maxwell's equation in point form using phasor notation and derive the wave equation from them for free space condition. *(Nov 2013)*
2. State and explain wave equation in point form for time varying fields in a conducting medium. *(Aug 2014)*
3. Obtain the wave equations for sinusoidally varying fields.
4. Find all the relations between E and H in a uniform plane wave. Find the value of the intrinsic impedance of a free space. *(May 2013)*
5. Prove that a uniform plane wave propagating in the x direction has no longitudinal components of electric and magnetic fields. *(Aug 2014)*
6. For a wave propagating in good dielectrics, derive the expression for intrinsic impedance. *(May 2011, 2013)*
7. Explain wave propagation in a conducting medium. *(Nov 2013)*
8. Define uniform plane wave. Prove that a uniform plane wave does not have field components in the direction of propagation. *(May 2011)*
9. Discuss uniform plane wave in lossless dielectrics and establish the relationship between the electric and magnetic fields in the medium. *(May 2011)*
10. Derive the expression for the phase shift constant and attenuation constant of a plane wave propagation in a lossy dielectric medium. *(May 2011, Aug 2014)*
11. For good dielectrics, derive the expression for α, β, v and η. *(May 2013)*
12. Show that in a good conductor, the skin depth δ is always much shorter than the wavelength.
13. Derive Helmholtz wave equation from the fundamentals and determine loss tangent. What is the significance of the loss tangent? *(Jan 2014)*
14. What is polarization of an EM wave? Explain different types of polarizations. *(May 2011)*
15. What is polarization? Explain linear polarization. *(May 2011, Aug 2014)*
16. Explain circular and elliptical polarizations. *(May 2008, Aug 2014)*
17. Explain horizontal and vertical polarizations *(May 2010)*
18. Explain the terms:
    (i) Linear polarization, (ii) circular polarization and (iii) elliptical polarization *(May 2011)*

Exercise Problems

1. Given $E(z,t) = 10^3 \sin(6\times 10^8 t - \beta z)\vec{a}_y$ V/m in free space, sketch the wave at $t = 0$ and at time t_1 when it has travelled $T/4$ along the z axis. Find t_1, β and λ.
2. For a lossy dielectric material having $\mu_r = 1,\ \varepsilon_r = 48,\ \sigma = 20$ S/m. Calculate the propagation constant at a frequency of 16 GHz.
3. Determine the propagation constant γ for a material having $\mu_r = 1, \varepsilon_r = 8$ and $\sigma = 0.25$ pS/m, if the wave frequency is 1.6 MH_z.
4. In a medium, $\vec{H}(z,\ t) = H_0 e^{j(\omega t+\beta z)}\vec{a}_x$ A/m. Obtain on expression for $E(z,\ t)$.
5. In a conducting medium $\varepsilon_r = 9,\ \mu_r = 800$ and $\sigma = 4\times 10^6$ S/m. Find $\alpha,\ \beta,\ \eta$ and the velocity for a frequency of 1 MHz.
6. Find the propagation constant at 400 MHz for a medium in which $\varepsilon_r = 16,\ \mu = 4.5$ and $\sigma = 0.6$ S/m. Also find the ratio of the velocity v to free space.
7. For a medium in which $\mu_r = 1, \varepsilon_r = 15, \sigma = 0$, determine γ at 500 KHz. At what velocity will an electromagnetic wave travel in this medium?
8. The electric flux intensity of a plane wave travelling in a perfect dielectric medium is $E_x = 10\cos(2\pi\times 10^7 t - 0.1\pi z)$ V/m. Find the expression for magnetic field intensity if μ_r is unity.
9. A lossy dielectric ($\sigma = 60$ S/m, $\varepsilon_r = \mu_r = 1$) support, a uniform plane wave at 15.9 MHz. Compute the attenuation constant, phase constant, intrinsic impedance and phase velocity of propagation.
10. A uniform plane wave is propagating in a semi-infinite region $z > 0$. The electric field intensity is expressed as 100 $\cos(10^9(\pi t)\vec{a}_x)$ V/m at $z = 0$. The region has $\sigma = 0.025$ S/m, $\varepsilon_r = 9$ and $\mu_r = 10$. Determine the propagation, attenuation constants and phase constants.
11. A sinusoidal plane wave is transmitted through a medium whose electric field strength is 10 kV/m and relative permittivity of the medium is 4. Determine the magnetic field strength.
12. Determine the intrinsic impedance, attenuation constant and phase constant for sea water at 1 GHz frequency whose parameters are $\mu_r = 80,\ \varepsilon = 1$ and $\sigma = 4$ mho/m. Also calculate the distance at which the field amplitude reduces to 1% of its value at $z = 0$.
13. Find the characteristics impedance of the medium whose relative permittivity is 80, $\sigma = 0$ and relative permeability is 1.
14. Determine the intrinsic impedance, phase velocity and skin depth of a wave travelling at 5 MHz frequency in sea water whose parameters are $\mu_r = 1,\ \varepsilon_r = 72$, and $\sigma = 4$ mho/m.
15. Find the velocity of a plane wave in a lossless medium having a relative permittivity of 40 and a relative permeability of 1.5.
16. The velocity of a uniform plane wave in a lossless dielectric is 1×10^8 m/s. Find the dielectric constant if relative permeability is 1.5.

Answers

1. $t_1 = 2.62$ ns, $\beta = 2$ rad/m $\lambda = \pi$ m.
2. 1590.3 $\angle 46.27^\circ$ per metre
3. $94.8\times 10^{-3}\angle 90^\circ$ per metre
4. $\dfrac{\beta H_0}{\omega\varepsilon} e^{j(\omega t+\beta z)}\vec{a}_y$ V/m

5. 112.4×10^3 Np/m, 112.4×10^3 rad/m, $0.04\angle45^\circ$ Ω, 55.9 m/s.

6. $99.58\angle50.34$ m^{-1}, 0.097

7. $j4.06\times10^{-2}$ m^{-1}, 7.74×10^7 m/s

8. $H_y = 39.8\cos(2\pi\times10^7 t - 0.1\pi z)$ mA/m

9. 61.34 Np/m, 61.38 rad/m, $1.445\angle44.98^\circ\Omega$, 1.627×10^6 m/s.

10. $70.23\angle67.5^\circ$ 26.785 Np/m, 64.88 rad/m.

11. 53.05 A/m

12. $36.55\angle21^\circ$ Ω 77.8485 Np/m, 202.8 rad/m, 0.059 m

13. 42.15 ohms

14. $3.1416\angle45^\circ$, 3.53×10^6 m/s, 0.112 m

15. 38.73×10^6 m/s

16. 6

Multiple-choice Questions

1. Which of the following wave equation in a a conducting medium is correct?
   (a) $\nabla^2\vec{H} = \mu\varepsilon\frac{\partial^2\vec{H}}{\partial t^2} + \mu\sigma\frac{\partial\vec{H}}{\partial t}$ (b) $\nabla^2\vec{H} = \mu\sigma\frac{\partial^2\vec{H}}{\partial t^2} + \mu\varepsilon\frac{\partial\vec{H}}{\partial t}$
   (c) $\nabla^2\vec{H} = \mu\varepsilon\frac{\partial^2\vec{H}}{\partial t^2} + \mu\sigma\frac{\partial^2\vec{H}}{\partial t^2}$ (d) $\nabla^2\vec{H} = \mu\sigma\frac{\partial^2\vec{H}}{\partial t^2} + \mu\varepsilon\frac{\partial^2\vec{H}}{\partial t^2}$

2. Which of the following wave equation in a conducting medium is correct?
   (a) $\nabla^2\vec{E} = (j\omega\mu\sigma + \omega^2\mu\varepsilon)\vec{E}$ (b) $\nabla^2\vec{E} = (j\omega\mu\sigma - \omega^2\mu\varepsilon)\vec{E}$
   (c) $\nabla^2\vec{E} = (-j\omega\mu\sigma - \omega^2\mu\varepsilon)\vec{E}$ (d) $\nabla^2\vec{E} = (-j\omega\mu\sigma + \omega^2\mu\varepsilon)\vec{E}$

3. Propagation constant γ^2 in time-varying fields is equal to
   (a) $(\sigma + j\omega\mu)(j\omega\varepsilon)$ (b) $(\sigma - j\omega\varepsilon)(j\omega\mu)$ (c) $(\sigma + j\omega\varepsilon)(j\omega\mu)$ (d) $(\sigma - j\omega\mu)(j\omega\varepsilon)$

4. The expression for attenuation constant α expression is given by
   (a) $\omega\sqrt{\frac{\mu\varepsilon}{2}\left[\sqrt{1-\left(\frac{\sigma}{\omega\varepsilon}\right)^2}-1\right]}$ (b) $\omega\sqrt{\frac{\mu\varepsilon}{2}\left[\sqrt{1+\left(\frac{\sigma}{\omega\varepsilon}\right)^2}+1\right]}$
   (c) $\omega\sqrt{\frac{\mu\varepsilon}{2}\left[\sqrt{1-\left(\frac{\sigma}{\omega\varepsilon}\right)^2}+1\right]}$ (d) $\omega\sqrt{\frac{\mu\varepsilon}{2}\left[\sqrt{1+\left(\frac{\sigma}{\omega\varepsilon}\right)^2}-1\right]}$

5. The expression for phase constant β is given by
   (a) $\omega\sqrt{\frac{\mu\varepsilon}{2}\left[\sqrt{1-\left(\frac{\sigma}{\omega\varepsilon}\right)^2}-1\right]}$ (b) $\omega\sqrt{\frac{\mu\varepsilon}{2}\left[\sqrt{1+\left(\frac{\sigma}{\omega\varepsilon}\right)^2}+1\right]}$
   (c) $\omega\sqrt{\frac{\mu\varepsilon}{2}\left[\sqrt{1-\left(\frac{\sigma}{\omega\varepsilon}\right)^2}+1\right]}$ (d) $\omega\sqrt{\frac{\mu\varepsilon}{2}\left[\sqrt{1+\left(\frac{\sigma}{\omega\varepsilon}\right)^2}-1\right]}$

6. Which of the following is the correct wave equation in free space?

   (a) $\nabla^2\vec{E} = \mu_0\varepsilon_0\dfrac{\partial^2\vec{E}}{\partial t^2}$ (b) $\nabla^2\vec{E} = \sqrt{\mu_0\varepsilon_0}\dfrac{\partial^2\vec{E}}{\partial t^2}$ (c) $\nabla^2\vec{E} = \mu_0\varepsilon_0\dfrac{\partial\vec{E}}{\partial t}$ (d) $\nabla^2\vec{E} = \dfrac{\mu_0}{\varepsilon_0}\dfrac{\partial^2\vec{E}}{\partial t^2}$

7. In a uniform plane wave,

   (a) E and H fields may not be perpendicular to the direction of propagation.

   (b) E and H fields are always parallel to the direction of propagation.

   (c) E and H fields are always perpendicular to the direction of propagation.

   (d) E and H fields are independent of the direction of propagation.

8. Which of the following wave equation is called Helmholtz equation?

   (a) $\nabla^2\vec{H} - \omega\mu\varepsilon\vec{H} = 0$ (b) $\nabla^2\vec{H} + \omega^2\mu\varepsilon\vec{H} = 0$

   (c) $\nabla^2\vec{E} + \omega^2\mu\varepsilon\vec{H} = 0$ (d) $\nabla^2\vec{H} - \omega^2\mu\varepsilon\vec{H} = 0$

9. If the plane wave is propagating in the z-direction with phase constant β and electric field component E_x in x-direction, then the wave equation becomes

   (a) $\dfrac{d^2\vec{E}_x}{dz^2} - \beta^2\vec{E}_x = 0$ (b) $\dfrac{d^2\vec{E}_x}{dz^2} + \beta\vec{E}_x = 0$ (c) $\dfrac{d^2\vec{E}_x}{dz^2} + \beta^2\vec{E}_x = 0$ (d) $\dfrac{d^2\vec{E}_x}{dz^2} - \beta\vec{E}_x = 0$

10. If the wave propagates in free space along the z-direction, the electric field component is

    (a) $E_x = E_0\cos(\omega z - \beta t)$ (b) $E_x = E_0\cos(\omega t - \beta z)$

    (c) $E_x = E_0\cos(\omega - \beta z)$ (d) $E_x = E_0\cos(\omega t - \beta)$

11. If the wave travels at a frequency of 10 MHz with a phase constant π rad/m, the phase velocity is

    (a) 10^7 m/s (b) 2×10^7 m/s (c) 1×10^6 m/s (d) 2×10^6 m/s

12. The velocity of a plane wave in a lossless medium having a relative permittivity of 81 and a relative permeability of 1 is

    (a) 3.3×10^7 m/s (b) 0.33×10^7 m/s (c) 3×10^8 m/s (d) 3.3×10^5 m/s

13. The electric field of the wave is given by $E_x = 10\cos(2\pi\times10^9 t - 0.5\pi z)$ V/m. The phase constant (rad/m) and wavelength (m) of the wave are

    (a) 3.14, 2 (b) 1.57, 2 (c) 3.14, 4 (d) 1.57, 4

14. The intrinsic impedance for perfect dielectrics is

    (a) $\sqrt{\mu/\varepsilon}$ (b) $\sqrt{\mu\varepsilon}$ (c) $\sqrt{\varepsilon/\mu}$ (d) μ/ε

15. The time-varying behaviour of the electric field strength $\vec{E}$ at a given fixed point in space is called

    (a) magnetisation (b) neutralisation (c) polarization (d) electrification

16. The propagation constant for a good conductor is proportional to

    (a) $\sqrt{\sigma/f}$ (b) $\sqrt{\sigma/f}$ (c) $\sqrt{f/\sigma}$ (d) σ/f

17. If the wave amplitude decreases to 36.8% as it travels to a distance of 1 m, the attenuation is

    (a) 1 Np/m (b) 0.869 dB/m (c) 0.8 dB/m (d) π Np/m

18. The intrinsic impedance for a good conductor is
    (a) $(1+j)\frac{\alpha}{\sigma}$ (b) $(1-j)\frac{\alpha}{\sigma}$ (c) $(1+j)\frac{\sigma}{\alpha}$ (d) $(1-j)\frac{\sigma}{\alpha}$

19. The waves radiated by AM broadcast stations from transmitting antenna are oriented in
    (a) horizontal polarization (b) circular polarization
    (c) vertical polarization (d) elliptical polarization

20. The waves radiated by FM broadcast stations from transmitting antenna are oriented in
    (a) horizontal polarization (b) circular polarization
    (c) vertical polarization (d) elliptical polarization

21. The intrinsic impedance for free space is
    (a) 377 ohms (b) 187 ohms (c) 75 ohms (d) 300 ohms

22. In circular polarization if $\vec{E}_y$ leads $\vec{E}_x$ by $90^\circ (\phi = 90^\circ)$, the wave is called
    (a) right circularly polarized wave (b) left circularly polarized wave
    (c) right elliptically polarized wave (d) left elliptically polarized wave

23. In circular polarization if $\vec{E}_y$ lags $\vec{E}_x$ by $90^\circ (\phi = 90^\circ)$, the wave is called
    (a) right circularly polarized wave (b) left circularly polarized wave
    (c) right elliptically polarized wave (d) left elliptically polarized wave

24. If $\vec{E}_x$ and $\vec{E}_y$ have different magnitudes with a 90° phase difference, then the wave is called
    (a) circularly polarized wave (b) linearly polarized wave
    (c) right circularly polarized wave (d) elliptically polarized wave

25. The distance at which the wave attenuates to $1/e$ times of its original amplitude is
    (a) end depth (b) wave depth (c) skin depth (d) attenuated depth

26. the loss tangent of the dielectric material can be expressed as
    (a) $\frac{\sigma\omega}{\varepsilon}$ (b) $\frac{\omega}{\sigma\varepsilon}$ (c) $\frac{\sigma}{\omega\varepsilon}$ (d) $\frac{\sigma\varepsilon}{\omega}$

Answers

1. *a*	2. *b*	3. *c*	4. *a*	5. *b*	6. *a*	7. *c*	8. *b*	9. *c*	10. *b*
11. *b*	12. *a*	13. *d*	14. *a*	15. *c*	16. *b*	17. *a*	18. *a*	19. *c*	20. *b*
21. *a*	22. *b*	23. *a*	24. *d*	25. *c*	26. *c*				

Electromagnetic Wave Characteristics | 8

Learning objectives

The main objectives of this chapter are to understand thoroughly the following topics:

- Normal and oblique incidences of waves in different media
- Reflection and transmission coefficients of a uniform plane wave
- Types of polarization: Horizontal and vertical polarization
- Snell's law, total internal reflection, Brewster angle, surface impedance
- Poynting vector and Poynting theorem,
- Average power, power loss in conductors

8.1 Introduction

In the previous chapter, we have discussed the propagation of plane waves through unbounded media. The wave properties in free space, in conductors and in dielectrics were explained. In this chapter, we will explore wave properties when the wave travels from one medium to another. Properties like direction of propagation, polarization, reflection and refraction of the wave at the interface depend on the materials involved. We will also discuss the normal and oblique incidences of the wave, Snell's law, Brewster angle, types of polarization, surface impedance, Poynting vector and Poynting theorem.

8.2 Reflection and Refraction of a Uniform Plane Wave

The propagation of an electromagnetic wave through different materials (media) having different constant parameters such as ε, μ, σ, η, etc., undergoes partial reflections and refractions at the boundaries of the media. Some part of the incident wave is transmitted and the other part is reflected back at the boundary depending on the angle of incidence.

For example, when a plane wave in air is incident normally on the surface of a perfect conductor, the wave completely reflects back. The combination of incident (forward) wave and reflected (backward) wave result in standing waves.

Depending upon the direction of incident wave, two cases can be considered.

(1) Normal incidence: The incident wave is normal to the plane of the boundary surface.

(2) Oblique incidence: The incident wave makes an angle θ with the normal to the plane of the boundary surface.

When the wave is incident obliquely at the boundary between two materials, the following terms are considered (Fig. 8.1).

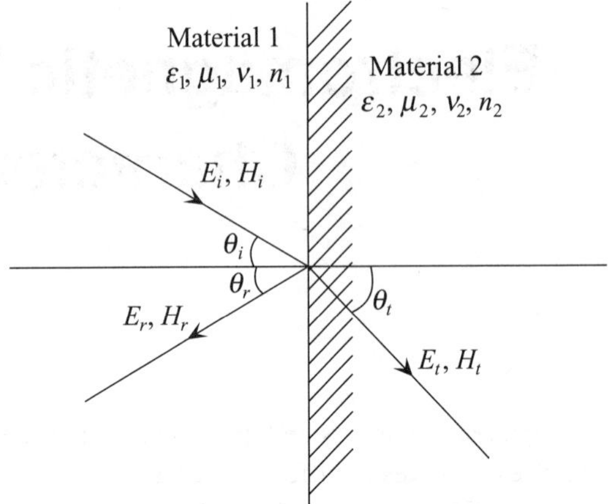

Fig. 8.1 *Reflection and refraction of a plane wave*

Incidence angle θ_i The angle at which the incident wave makes with the normal to the interface (boundary) is called the incidence angle θ_i.

Reflection angle θ_r The angle at which the reflected wave makes with the normal to the interface (boundary) is called the reflected angle θ_r.

Transmission angle θ_r The angle at which the transmitted (refracted) wave makes with the normal to the interface (boundary) is called the transmitted angle or the refracted angle θ_t.

Snell's law (law of refraction) The law of refraction is also known as Snell's law. It is used to determine the direction of waves travelling through refractive media with varying indices of refraction.

Snell's law states that the ratio of the sines of the angles of incidence and refraction is equivalent to the ratio of phase velocities in the two media, or equivalent to the reciprocal of the ratio of the refractive indices. That is

$$\frac{\sin \theta_i}{\sin \theta_t} = \frac{v_1}{v_2} = \frac{n_2}{n_1}, \tag{8.1}$$

where v_1 and v_2 are the velocities of the wave in medium 1 and medium 2, respectively and n_1 and n_2 are the refractive indices of medium 1 and medium 2, respectively.

For dielectric media,

$$\mu_1 = \mu_2 = \mu_0.$$

$$\therefore \quad v_1 = \frac{1}{\sqrt{\varepsilon_1 \mu_0}} \quad \text{and} \quad v_2 = \frac{1}{\sqrt{\varepsilon_2 \mu_0}}.$$

$$\therefore \quad \frac{\sin\theta_i}{\sin\theta_t} = \frac{\dfrac{1}{\sqrt{\varepsilon_1 \mu_0}}}{\dfrac{1}{\sqrt{\varepsilon_2 \mu_0}}} = \sqrt{\frac{\varepsilon_2}{\varepsilon_1}} = \frac{n_2}{n_1}$$

or $\quad \sqrt{\varepsilon_1}\sin\theta_i = \sqrt{\varepsilon_2}\sin\theta_t$

$$\text{and} \quad \frac{n_2}{n_1} = \sqrt{\frac{\varepsilon_2}{\varepsilon_1}}. \tag{8.2}$$

Law of reflection The law of reflection states that a wave incident upon a reflective surface will be reflected at an angle equal to the incident angle

i.e., incidence angle = reflected angle, $\theta_i = \theta_r$.

Reflection coefficient Γ It is the ratio of the amplitudes of reflected and incident waves. It gives the fraction of the incident wave that is reflected back from the interface. It is a dimensionless quantity and exists in the range of –1 to 1.

It can be expressed as $\Gamma = \dfrac{E_r}{E_i}$ or $\Gamma = \dfrac{H_r}{H_i}$,

where E_i and H_i are incident wave fields and E_r and H_r are reflected wave fields, respectively.

Transmission coefficient T It is the ratio of the amplitudes of refracted and incident waves. It gives the fraction of the incident wave that is transmitted into the second material from the interface. It is a dimensionless quantity and exists in the range of 0 to 2. It can be expressed as $T = \dfrac{E_t}{E_i}$ or $T = \dfrac{H_t}{H_i}$, where E_t, H_t are refracted fields in the second material.

The study of reflection and transmission coefficients of a uniform plane wave is important in various applications where signal transmission is made through transmission lines such as optical fibre systems, radar systems, remote sensing, missile tracking and guidance, stealth operations and medical imaging. For example, in optical fibre communications, light wave propagates due to total internal reflection and in satellite communications, the wave is refracted through the ionosphere.

8.3 Reflection by a Perfect Conductor with Normal Incidence

Consider a plane wave in a homogeneous medium incident normally on a perfect conductor at $z = 0$ plane as shown in Fig. 8.2. Assume that the electric field is in the x direction and the magnetic field is in the y direction. Since no EM fields exist within a perfect conductor, none of the energy is transmitted and absorbed. The incident wave is completely reflected back.

The field vectors of the incidence wave in phasor notation are

$$\vec{E}_i = E_i e^{-j\beta z}\vec{a}_x; \quad \vec{H}_i = H_i e^{-j\beta z}\vec{a}_y$$

and that of the reflected wave,

$$\vec{E}_r = E_r e^{j\beta z}\vec{a}_x; \ \vec{H}_r = H_r e^{j\beta z}\vec{a}_y, \tag{8.3}$$

where E_i, H_i and E_r, H_r are the field amplitudes of the incident and reflected waves at $z = 0$ respectively.

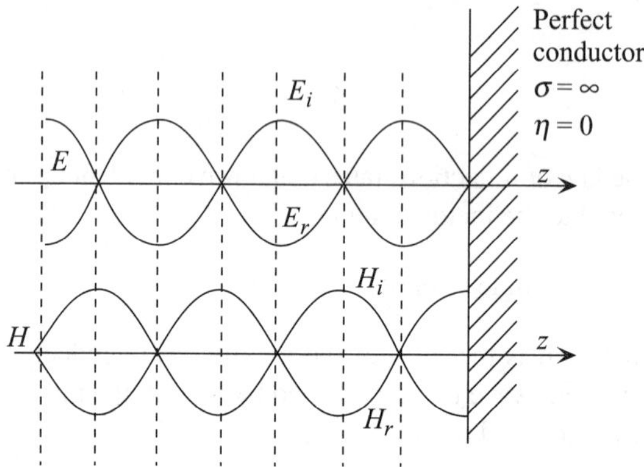

Fig. 8.2 *Standing waves of $\vec{E}$ and $\vec{H}$*

We know that the tangential component of the electric field vector is continuous at the boundary surface. Therefore the fields at the perfect conductor are

$$E_i + E_r = E_t = 0$$

or $$E_i = -E_r. \tag{8.4}$$

Thus the electric field vector reflects with 180° phase shift at the boundary surface. The superposition of incident and reflected waves travelling in opposite directions produces a standing wave.

The electric field strength of the standing wave is

$$E(z) = E_i e^{-j\beta z} + E_r e^{+j\beta z}$$

$$= E_i(e^{-j\beta z} - e^{+j\beta z}) = -2j\frac{(e^{j\beta z} - e^{-j\beta z})}{2j}$$

or $$\vec{E}(z) = -2jE_i \sin(\beta z)\vec{a}_x. \tag{8.5}$$

The field in time domain is

$$\vec{E}(z,t) = \Re e[-2jE_i \sin(\beta z)e^{j\omega t}\,\vec{a}_x]. \tag{8.6}$$

Since E_i is a real quantity,

$$\vec{E}(z,t) = 2E_i \sin\,(\beta z)\ \sin(\omega t)\vec{a}_x. \tag{8.7}$$

Thus the magnitude of the electric field vector of a standing wave varies sinusoidaly with distance from the boundary. It is zero at the boundary surface. The maximum amplitude 2 E_i occurs at odd multiples of $\lambda/4$ from the surface as shown in Fig. 8.2.

Similarly, the magnetic field vector, since it is normal to the electric field, reflects without phase reversal at the boundary surface. So at the boundary,

$$H_r = H_i.$$

The magnetic field vector of the standing wave is

$$H(z) = H_i e^{-j\beta z} + H_r e^{j\beta z} = 2H_i \frac{(e^{j\beta z} + e^{-j\beta z})}{2}$$

$$\text{or} \quad \vec{H}(z) = 2H_i \cos(\beta z)\,\vec{a}_y. \tag{8.8}$$

The field in time domain is

$$\vec{H}(z,t) = \Re e\{2H_i \cos(\beta z)\, e^{j\omega t}\}\,\vec{a}_y. \tag{8.9}$$

Since H_i is a real quantity,

$$\vec{H}(z,t) = 2H_i \cos(\beta z)\cos(\omega t)\,\vec{a}_y.$$

$$\vec{H}(z,t) = \frac{2E_i}{\eta} \cos(\beta z)\cos(\omega t)\,\vec{a}_y. \tag{8.10}$$

Thus the magnitude of the magnetic field vector of a standing wave varies sinusoidaly with distance from the boundary. The standing wave distribution of the magnetic field has maximum amplitude, $2H_i$ at the surface of the boundary and at even multiples of $\lambda/4$ from the surface.

The reflection coefficient for the electric field is

$$\Gamma = \frac{E_r}{E_i} = -1.$$

The transmission coefficient is $T = 0$.

Note: From the boundary conditions, the magnetic field strength $\vec{H}$ is equal to the surface current density $\vec{K}$ on the surface.

$$\vec{H} = \vec{K}.$$

8.4 Reflection by a Perfect Dielectric—Normal Incidence

Consider a uniform plane wave travelling along the z direction and incident normally on the surface of a perfect dielectric media. At the interface, the wave is partly reflected and partly transmitted. Assume that the electric field is in the x direction and magnetic field is in the y direction.

Let medium 1 and 2 have the properties ε_1, μ_1, v_1, η_1 and ε_2, μ_2, v_2, η_2 respectively. A perfect dielectric (lossless) has zero conductivity so that there is no loss of power in propagation.

Let E_i and H_i be the incident wave field strengths, E_r and H_r, the reflected wave field strengths, and E_t and H_t, the transmitted wave field strengths, respectively.

If the electric field vector is oriented upwards parallel to the plane of the paper, then the magnetic field vectors for incident and transmitted waves are oriented outward to the plane of the paper, indicated as symbol "$\odot H$", and for the reflected wave, it is oriented into the plane of the paper indicated as symbol "$\otimes H$" as shown in Fig. 8.3.

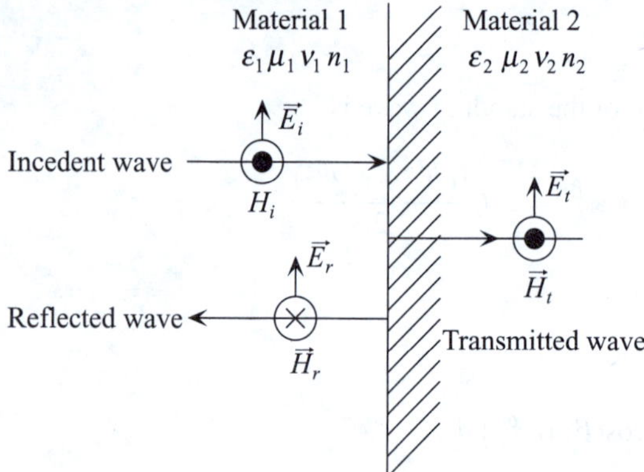

Fig. 8.3 *Normal incidence on a perfect dielectric*

The field vectors for the incident wave in phasor notation are

$$\vec{E}_i = E_i e^{-j\beta z}\vec{a}_x, \quad \vec{H}_i = H_i e^{-j\beta z}\vec{a}_y,$$

and for the reflected wave travelling in the $-\vec{a}_z$ direction,

$$\vec{E}_r = E_r e^{j\beta z}\vec{a}_x, \quad \vec{H}_r = -H_r e^{j\beta z}\vec{a}_y$$

and for the transmitted wave

$$\vec{E}_t = E_t e^{-j\beta z}\vec{a}_x, \quad \vec{H}_t = H_t e^{-j\beta z}\vec{a}_y, \tag{8.11}$$

where E_i, H_i, E_r, H_r and E_t, H_t are the field amplitudes of incident, reflected and transmitted waves at $z = 0$ respectively.

We know that the relationships between electric and magnetic fields are

$$E_i = \eta_1 H_i; \quad E_r = -\eta_1 H_r; \quad E_t = \eta_2 H_t, \tag{8.12}$$

where η_1 and η_2 are the intrinsic impedances of medium 1 and medium 2 respectively.

At the boundary, the tangential components of the field are continuous, i.e.,

$$E_i + E_r = E_t;$$

$$H_i + H_r = H_t. \tag{8.13}$$

Substituting Eq. (8.12) for electric fields,

$$\frac{E_i}{\eta_1} - \frac{E_r}{\eta_1} = \frac{E_t}{\eta_2}.$$

$$\frac{E_i}{\eta_1} - \frac{E_r}{\eta_1} = \frac{(E_i + E_r)}{\eta_2}.$$

$$\eta_2 E_i - \eta_2 E_r = \eta_1 E_i + \eta_1 E_r.$$

$$E_i(\eta_2 - \eta_1) = E_i(\eta_1 + \eta_2).$$

$\therefore$ The reflection coefficient for the electric field is

$$\Gamma = \frac{E_r}{E_i} = \frac{\eta_2 - \eta_1}{\eta_1 + \eta_2}. \tag{8.14}$$

The transmission coefficient for the electric field is

$$T = \frac{E_t}{E_i} = \frac{E_i + E_r}{E_i} = 1 + \frac{E_r}{E_i} = 1 + \Gamma.$$

$$T = 1 + \frac{\eta_2 - \eta_1}{\eta_1 + \eta_2} = \frac{2\eta_2}{\eta_1 + \eta_2}.$$

$$T = \frac{E_t}{E_i} = \frac{2\eta_2}{\eta_1 + \eta_2}. \tag{8.15}$$

The reflection coefficient for the magnetic field is

$$\frac{H_r}{H_i} = \frac{-E_r/\eta_1}{E_i/\eta_1} = \frac{-E_r}{E_i} = \frac{\eta_1 - \eta_2}{\eta_1 + \eta_2} \tag{8.16}$$

and the transmission coefficient for the magnetic field is

$$\frac{H_t}{H_i} = \frac{E_t/\eta_2}{E_i/\eta_1} = \frac{\eta_1}{\eta_2} \times \frac{E_t}{E_i} = \frac{2\eta_1}{\eta_1 + \eta_2}. \tag{8.17}$$

Reflection and transmission coefficients in terms of dielectric constants

For free space and dielectrics, $\mu_1 = \mu_2 = \mu_0$.
Also, the intrinsic impedances are

$$\eta_1 = \sqrt{\frac{\mu_0}{\varepsilon_1}} \text{ and } \eta_2 = \sqrt{\frac{\mu_0}{\varepsilon_2}}.$$

Substituting in Eq. (8.14), the reflection coefficient is

$$\frac{E_r}{E_i} = \frac{\sqrt{\frac{\mu_0}{\varepsilon_2}} - \sqrt{\frac{\mu_0}{\varepsilon_1}}}{\sqrt{\frac{\mu_0}{\varepsilon_2}} + \sqrt{\frac{\mu_0}{\varepsilon_1}}} = \frac{\frac{1}{\sqrt{\varepsilon_2}} - \frac{1}{\sqrt{\varepsilon_1}}}{\frac{1}{\sqrt{\varepsilon_2}} + \frac{1}{\sqrt{\varepsilon_1}}}.$$

$$\Gamma = \frac{E_r}{E_i} = \frac{\sqrt{\varepsilon_1} - \sqrt{\varepsilon_2}}{\sqrt{\varepsilon_1} + \sqrt{\varepsilon_2}} \tag{8.18}$$

and $$\frac{H_r}{H_i} = \frac{-E_r}{E_i} = \frac{\sqrt{\varepsilon_2} - \sqrt{\varepsilon_1}}{\sqrt{\varepsilon_1} + \sqrt{\varepsilon_2}}. \tag{8.19}$$

The transmission coefficient is

$$T = \frac{E_t}{E_i} = \frac{2\sqrt{\dfrac{\mu_0}{\varepsilon_2}}}{\sqrt{\dfrac{\mu_0}{\varepsilon_1}}+\sqrt{\dfrac{\mu_0}{\varepsilon_2}}} = \frac{2\sqrt{\varepsilon_1}}{\sqrt{\varepsilon_1}+\sqrt{\varepsilon_2}} \tag{8.20}$$

and $$\frac{H_t}{H_i} = \frac{2\sqrt{\varepsilon_2}}{\sqrt{\varepsilon_1}+\sqrt{\varepsilon_2}}. \tag{8.21}$$

Example 8.1 A plane wave travelling in air is normally incident on a block of paraffin with $\varepsilon_r = 2.2$. Find the reflection coefficient.

Solution Given: For air, $\varepsilon_{r_1} = 1,\ \mu_{r_1} = \mu_{r_2} = 1,\ \varepsilon_{r_2} = 2.2.$

Intrinsic impedance of air is $\eta_1 = 120\pi\ \Omega$.

Intrinsic impedance of paraffin is

$$\eta_2 = \sqrt{\frac{\mu_0\,\mu_r}{\varepsilon_0\,\varepsilon_r}} = 120\pi\sqrt{\frac{\mu_r}{\varepsilon_r}}.$$

$$= 120\pi\sqrt{\frac{1}{2.2}} = \frac{120\pi}{\sqrt{2.2}} = 254\ \Omega.$$

$$\text{Reflection coefficient} = \frac{\eta_1-\eta_2}{\eta_1+\eta_2} = \frac{377-254}{377+254} = 0.195.$$

8.5 Reflection of a Plane Wave—Oblique Incidence

Consider a plane wave travelling in a lossless homogeneous medium. It is incident obliquely on the surface of another medium. At the interface, some part of the wave is reflected and the other part is refracted (transmitted) through the medium. The fields and direction of propagation follow the right-hand screw law. There are two situations which occur depending on the orientation of the electric field vector.

(1) Horizontal polarization (or perpendicular polarization)

(2) Vertical polarization (or parallel polarization)

Horizontal polarization In horizontal polarization the electric field vector of the wave is perpendicular to the plane incidence. It is also called perpendicular polarization. If the interfacing surface is normal to the paper surface, the plane of incidence is parallel to the plane of the paper. So the electric vector is normal to the plane of the paper. As shown in Fig. 8.4, if the electric vector is oriented outward to the plane of the paper, indicated as '$\odot E$', the magnetic vector is perpendicular to E_i and parallel to the plane of the paper.

Horizontally polarized waves can be generated from a horizontal antenna.

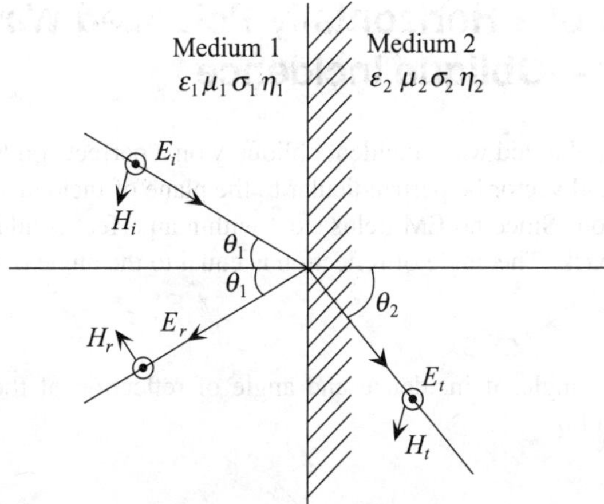

Fig. 8.4 *Reflection and refraction of a horizontally polarized wave*

Vertical polarization In vertical polarization, the electric field vector is parallel to the plane of incidence. It is also called parallel polarization. If the interfacing surface is normal to the paper surface, the plane of incidence is parallel to the plane of the paper. So the electric vector is parallel to the plane of the paper.

As shown in Fig. 8.5, if the electric vector is oriented parallel to the plane of the paper surface, the magnetic vector is oriented outward to the plane of the paper, indicated as "$\odot H$".

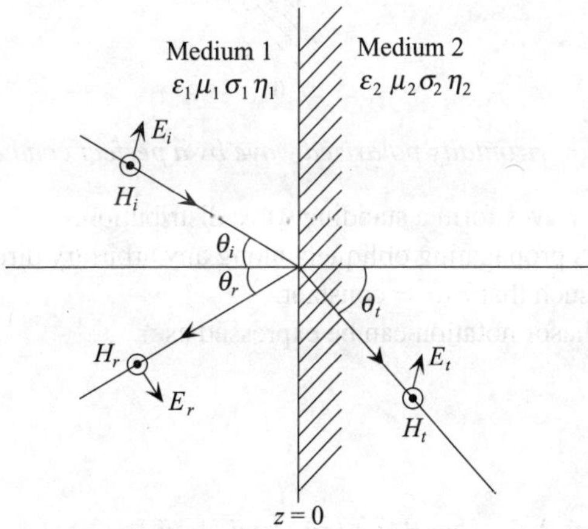

Fig. 8.5 *Reflection and refraction of a vertically polarized wave*

Vertically polarized waves can be generated from a vertical antenna. It is to be noted that the electric field vector in vertical polarization is not completely vertical—it may have some horizontal components.

8.6 Reflection of a Horizontally Polarized Wave by a Perfect Conductor—Oblique Incidence

Consider a horizontally polarized wave incident obliquely on a perfect conductor lying on the $z = 0$ plane. Let the electric field vector be perpendicular to the plane of incidence ($x - z$ plane) and H_i is in the downward direction. Since no EM fields exist within a perfect conductor, the incident wave is completely reflected back. The angle of reflection is equal to the angle of incidence,

i.e., $\theta_r = \theta_i = \theta$,

where θ_i and θ_r are the angle of incidence and angle of reflection of the wave with the z-axis respectively, as shown in Fig. 8.6.

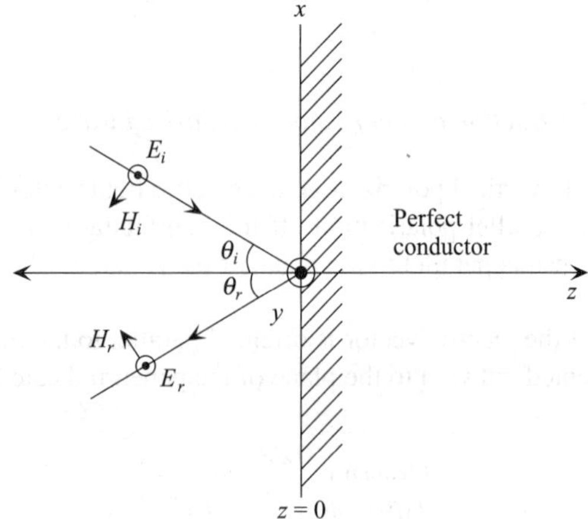

Fig. 8.6 *Reflection of a horizontally polarized wave by a perfect conductor*

The incident and reflected waves form a standing wave distribution.

Assume that the wave is propagating obliquely along any arbitrary direction with unit vector $\vec{a}_r$. Let a position vector $\vec{r}$ be such that $\vec{r}.\vec{a}_r$ = constant.

The incident wave in phasor notation can be expressed as

$$\vec{E}_i = E_i e^{-j\beta(\vec{r}.\vec{a}_r)}\,\vec{a}_r. \tag{8.22}$$

Using direction cosines,

$$\vec{r}.\vec{a}_r = (\cos\theta_x \vec{a}_x + \cos\theta_y \vec{a}_y + \cos\theta_z \vec{a}_z).(x\vec{a}_x + y\vec{a}_y + z\vec{a}_z).$$

$$\vec{r}.\vec{a}_r = x\cos\theta_x + y\cos\theta_y + z\cos\theta_z, \tag{8.23}$$

where θ_x, θ_y and θ_z are the incident angles between the direction of propagation and respective axes.

For horizontal polarization, the electric vector is along the y direction, i.e., normal to the plane of incidence ($x - z$ plane) and direction of propagation.

If θ is the incident angle with the z-axis, then

$$\theta_z = \theta,\ \ \theta_y = \pi/2,\ \ \text{and}\ \ \theta_x = \theta - \pi/2.$$

Hence, $\vec{r}.\vec{a}_r = x\cos(\theta - \pi/2) + y\cos(\pi/2) + z\cos\theta.$

$$\vec{r}\cdot\vec{a}_r = x\sin\theta + z\cos\theta.$$

$$\therefore\quad \vec{E}_i = E_i e^{-j\beta(x\sin\theta + z\cos\theta)}\,\vec{a}_y. \tag{8.24}$$

Similarly, for a reflected wave in the $(-z)$ direction,

$$\vec{r}.\vec{a}_r = x\sin\theta - z\cos\theta.$$

$$\therefore\quad \vec{E}_r = E_r e^{-j\beta(x\sin\theta - z\cos\theta)}\,\vec{a}_y. \tag{8.25}$$

The electric field vector for a standing wave is

$$E = E_i + E_r = E_i e^{-j\beta(x\sin\theta + z\cos\theta)} + E_r e^{-j\beta(x\sin\theta - z\cos\theta)}.$$

Since $E_i = -E_r$,

$$E = E_i[e^{-j\beta(x\sin\theta + z\cos\theta)} - e^{-j\beta(x\sin\theta - z\cos\theta)}]$$
$$= E_i e^{-j\beta x\sin\theta}[e^{-j\beta(z\cos\theta)} - e^{j\beta(z\cos\theta)}]$$
$$= -2jE_i e^{-j\beta x\sin\theta}\left(\frac{e^{j\beta z\cos\theta} - e^{-j\beta z\cos\theta}}{2j}\right).$$

$$E = -2jE_i\sin(z\beta\cos\theta)e^{-jx\beta\sin\theta}. \tag{8.26}$$

If $\beta = \dfrac{2\pi}{\lambda}$, the phase shift constant along the direction of propagation, $\beta_z = \beta\cos\theta$ is the phase shift constant along the z direction and $\beta_x = \beta\sin\theta$ is the phase shift constant along the x direction.

The wavelengths and velocities of the wave along the axes are:

wavelengths: $\lambda_z = \dfrac{2\pi}{\beta_z} = \dfrac{2\pi}{\beta\cos\theta} = \dfrac{\lambda}{\cos\theta}$,

$$\lambda_x = \frac{2\pi}{\beta_x} = \frac{2\pi}{\beta\sin\theta} = \frac{\lambda}{\sin\theta}$$

velocities: $v_z = \dfrac{v}{\cos\theta},\quad v_x = \dfrac{v}{\sin\theta}.$ (8.27)

Therefore the electric field vector for a standing wave distribution is given by

$$\vec{E} = -2jE_i\sin(z\beta_z)e^{-jx\beta_x}\,\vec{a}_y. \tag{8.28}$$

The magnetic field vector is

$$\vec{H} = \frac{1}{\eta}\left(\vec{a}_r \times \vec{E}\right).$$

Since the magnetic field vector is parallel to the $x-z$ plane, only x- and z- field components exist.

$$\vec{H} = \frac{1}{\eta}(H_x\vec{a}_x + \vec{E}_z\vec{a}_z)$$

$$\therefore \quad \vec{H} = H_x\vec{a}_x + H_z\vec{a}_z.$$

The magnetic field along the x-axis is

$$H_x = -2H_i \sin\theta \sin(z\beta_z)e^{-jx\beta_x}$$

and the magnetic field along the z-axis is

$$H_z = -2jH_i \cos\theta \cos(z\beta_z)e^{-jx\beta_x},$$

where $H_i = E_i/\eta$.

$$\therefore \quad \vec{H} = \frac{-2jE_i}{\eta}\left[\sin\theta \sin(z\beta_z)\vec{a}_x + \cos\theta \cos(z\beta_z)\vec{a}_z\right]e^{-jx\beta_x} \tag{8.29}$$

Note: In horizontal polarization, since the electric field vector is normal to the plane of incidence, only E_y, H_x and H_z components of the standing wave exist.

8.7 Reflection of a Vertically Polarized Wave by a Perfect Conductor—Oblique Incidence

Consider a vertically polarized wave incident obliquely on a perfect conductor lying on the $z = 0$ plane. Let the electric field vector is parallel to the plane of incidence ($x - z$ plane) and E_i is in the upward direction. Since no EM fields exist within a perfect conductor, the incident wave is completely reflected back. The angle of reflection is equal to the angle of incidence.

i.e., $\theta_r = \theta_i = \theta$,

where θ_i and θ_r are the angle of incidence and angle of reflection of the wave with the z-axis respectively as shown in Fig. 8.7.

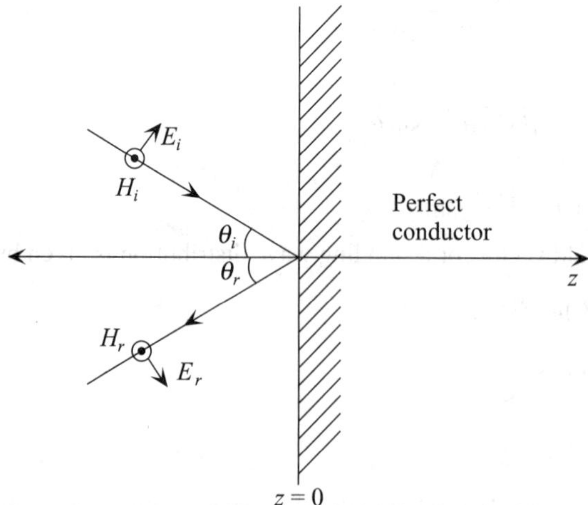

Fig. 8.7 *Reflection of a vertically polarized wave by a perfect conductor*

The incident and reflected waves form a standing wave distribution.

Since the magnetic field vector is parallel to the boundary, it will be reflected without phase reversal.

That is, $H_i = H_r$ and $\dfrac{E_i}{H_i} = \dfrac{E_r}{H_r} = \eta.$

Assume that the wave is propagating obliquely along any arbitrary direction with unit vector $\vec{a}_r$. Let a position vector $\vec{r}$ be such that $\vec{r}.\vec{a}_r =$ constant.

The incident wave in phasor notation can be expressed as

$$\vec{H}_i = H_i e^{-j\beta(\vec{r}.\vec{a}_r)} \vec{a}_r. \tag{8.30}$$

Using direction cosines,

$$\vec{r}.\vec{a}_r = (\cos\theta_x \vec{a}_x + \cos\theta_y \vec{a}_y + \cos\theta_z \vec{a}_z).(x\vec{a}_x + y\vec{a}_y + z\vec{a}_z)$$

$$= x\cos\theta_x + y\cos\theta_y + z\cos\theta_z, \tag{8.31}$$

where θ_x, θ_y and θ_z are the incident angles between direction of propagation and respective axes.

For vertical polarization, the magnetic vector is along the y direction, i.e., normal to the plane of incidence $(x - z$ plane) and direction of propagation. If θ is the incident angle with the z-axis, then

$$\theta_z = \theta,\ \theta_y = \pi/2, \text{ and } \theta_x = \theta - \pi/2.$$

$$\vec{r}.\vec{a}_r = x\cos(\theta - \pi/2) + y\cos(\pi/2) + z\cos\theta$$

$$= x\sin\theta + z\cos\theta.$$

$$\therefore\quad \vec{H}_i = H_i e^{-j\beta(x\sin\theta + z\cos\theta)} \vec{a}_y. \tag{8.32}$$

Similarly, for the reflected wave in the $(-z)$ direction,

$$\vec{r}\cdot\vec{a}_r = x\sin\theta - z\cos\theta.$$

$$\vec{H}_r = H_r e^{-j\beta(x\sin\theta - z\cos\theta)} \vec{a}_y. \tag{8.33}$$

The total magnetic field vector of the standing wave is

$$\vec{H} = \vec{H}_i + \vec{H}_r$$

$$= H_i e^{-j\beta(x\sin\theta + z\cos\theta)} \vec{a}_y + H_r e^{-j\beta(x\cos\theta - z\cos\theta)} \vec{a}_y.$$

Since $H_i = H_r$,

$$\vec{H} = H_i e^{-jx\beta\sin\theta}(e^{-jz\beta\cos\theta} + e^{jz\beta\cos\theta})\vec{a}_y.$$

$$\vec{H} = 2H_i\cos(z\beta\cos\theta)e^{-jx\beta\sin\theta} \vec{a}_y. \tag{8.34}$$

Since $\beta = \dfrac{2\pi}{\lambda}$, the phase shift constant along the direction of propagation, $\beta_z = \beta\cos\theta$ and $\beta_x = \beta\sin\theta$ are phase shift constants along the z and x directions, respectively. (8.29)

The velocity and wavelength of the wave along the x-axis is

$$v_x = \frac{v}{\sin\theta} \quad \text{and} \quad \lambda_x = \frac{\lambda}{\sin\theta}. \tag{8.35}$$

Therefore the magnetic field vector along the y-axis for a standing wave distribution is given by

$$\vec{H} = 2H_i \cos(z\beta_z)e^{-jx\beta_x}\,\vec{a}_y. \tag{8.36}$$

Since the electric field vector is parallel to the $x-z$ plane, only the x and z field components exist.

$$\vec{E} = E_x\vec{a}_x + E_z\vec{a}_z.$$

The electric field vector along the x-axis is

$$E_x = -2jE_i \cos\theta \sin(z\beta_z)e^{-jx\beta_x}.$$

The electric field vector along the z-axis is

$$E_z = -2E_i \sin\theta \cos(z\beta_z)e^{-jx\beta_x}$$

where $E_i = \eta H_i$

$$\vec{E} = -2j\eta H_i[\cos\theta \sin(z\beta_z)\vec{a}_x + \sin\theta\cos(z\beta_z)\vec{a}_z]e^{-jx\beta_x} \tag{8.37}$$

Note: In vertical polarization, since the electric field vector is parallel to the plane of incidence, only H_y, E_x and E_z components of the standing wave exist.

8.8 Reflection by a Dielectric or Perfect Insulator—Oblique Incidence

Consider a plane wave incident obliquely on the surface of a dielectric medium or a perfect insulator. Some part of the wave is reflected and the other part is refracted through the medium at the boundary as shown in Fig. 8.8.

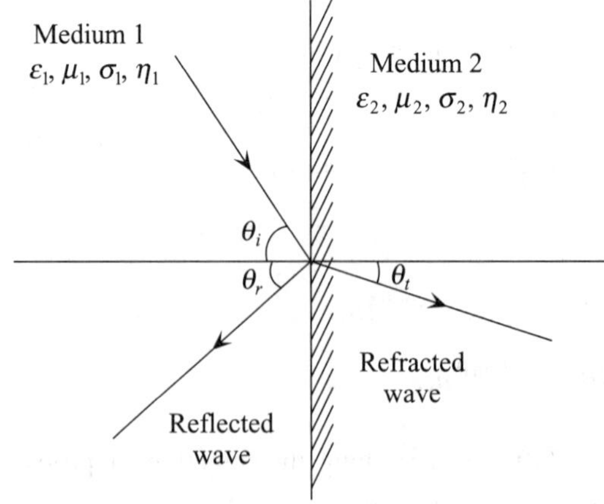

Fig. 8.8 *Oblique incidence of a plane wave on a dielectric medium*

Let θ_i = incident angle, θ_r = reflected angle and θ_t = refracted angle. $\theta_i = \theta_r = \theta_1,\ \theta_t = \theta_2$..

Also let medium 1 and 2 have the properties $\varepsilon_1,\ \eta_1,\ \beta_1$ and $\varepsilon_2,\ \eta_2,\ \beta_2$ respectively.

From Snell's law for dielectric media,

$$\sqrt{\varepsilon_1}\sin\theta_1 = \sqrt{\varepsilon_2}\sin\theta_2. \tag{8.38}$$

The field vectors can be obtained from the distribution of powers at the boundary surface.

Assume that the power absorbed by the material is neglected. We know that the wave power density is

$$P = |\vec{E}\times\vec{H}| = \frac{E^2}{\eta}. \tag{8.39}$$

Incident wave power density is

$$P_i = \frac{E_i^2}{\eta_1}\cos\theta_1. \tag{8.40}$$

Similarly, the reflected power density is

$$P_r = \frac{E_r^2}{\eta_1}\cos\theta_1 \tag{8.41}$$

and the transmitted wave power density is

$$P_t = \frac{E_t^2}{\eta_2}\cos\theta_2. \tag{8.42}$$

From the principle of conservation of energy, we know that

Incident power density = reflected power density + refracted power density

i.e., $P_i = P_r + P_t$.

$$\frac{E_i^2}{\eta_1}\cos\theta_1 = \frac{E_r^2}{\eta_1}\cos\theta_1 + \frac{E_t^2}{\eta_2}\cos\theta_2.$$

$$1 = \frac{E_r^2}{E_i^2} + \frac{E_t^2}{E_i^2}\left(\frac{\eta_1}{\eta_2}\right)\frac{\cos\theta_2}{\cos\theta_1}.$$

We know that $\dfrac{\eta_1}{\eta_2} = \dfrac{\sqrt{\mu_0/\varepsilon_1}}{\sqrt{\mu_0/\varepsilon_2}} = \sqrt{\dfrac{\varepsilon_2}{\varepsilon_1}}.$

$$\therefore\quad \frac{E_r^2}{E_i^2} = 1 - \frac{E_t^2}{E_i^2}\sqrt{\frac{\varepsilon_2}{\varepsilon_1}}\left(\frac{\cos\theta_2}{\cos\theta_1}\right), \tag{8.43}$$

where $\dfrac{E_r}{E_i} = \Gamma$, is the reflection coefficient and $\dfrac{E_r}{E_i} = T$, is the transmission coefficient.

To find the reflection and transmission coefficients, the wave polarizations should be considered.

Case 1: Horizontal (perpendicular) polarization

Assume that the electric vector is perpendicular to the plane of incidence. Let the electric field vector E_i of the obliquely incident wave be oriented outward (positive y direction) from the plane paper as shown in Fig. 8.9.

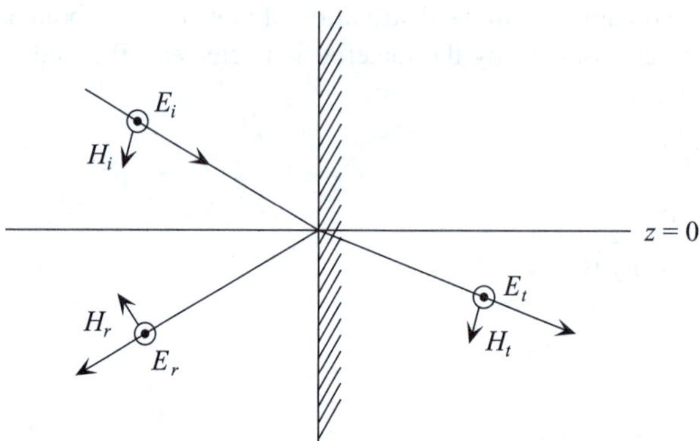

Fig. 8.9 *Reflection and refraction of a perpendicularly polarized wave*

E_i, E_r and E_t are in the positive y direction. Applying boundary conditions, the tangential components are continuous.

$$\text{i.e.,}\quad E_i + E_r = E_t \quad \text{or} \quad \frac{E_t}{E_i} = 1 + \frac{E_r}{E_i}. \tag{8.44}$$

Substituting this in Eq. (8.43),

$$\frac{E_r^2}{E_i^2} = 1 - \sqrt{\frac{\varepsilon_2}{\varepsilon_1}}\left(1 + \frac{E_r}{E_i}\right)^2 \frac{\cos\theta_2}{\cos\theta_1}.$$

$$1 - \left(\frac{E_r}{E_i}\right)^2 = \sqrt{\frac{\varepsilon_2}{\varepsilon_1}}\left(1 + \frac{E_r}{E_i}\right)^2 \frac{\cos\theta_2}{\cos\theta_1}.$$

$$\left(1 + \frac{E_r}{E_i}\right)\left(1 - \frac{E_r}{E_i}\right) = \sqrt{\frac{\varepsilon_2}{\varepsilon_1}}\left(1 + \frac{E_r}{E_i}\right)^2 \frac{\cos\theta_2}{\cos\theta_1}.$$

$$1 - \frac{E_r}{E_i} = \sqrt{\frac{\varepsilon_2}{\varepsilon_1}}\left(1 + \frac{E_r}{E_i}\right) \frac{\cos\theta_2}{\cos\theta_1}.$$

$$1 - \frac{E_r}{E_i} = \frac{\sqrt{\varepsilon_2}\cos\theta_2}{\sqrt{\varepsilon_1}\cos\theta_1} + \frac{E_r}{E_i}\frac{\sqrt{\varepsilon_2}\cos\theta_2}{\sqrt{\varepsilon_1}\cos\theta_1}.$$

$$\frac{E_r}{E_i}\left(\frac{\sqrt{\varepsilon_2}\cos\theta_2}{\sqrt{\varepsilon_1}\cos\theta_1} + 1\right) = 1 - \frac{\sqrt{\varepsilon_2}\cos\theta_2}{\sqrt{\varepsilon_1}\cos\theta_1}.$$

$$\frac{E_r}{E_i}\frac{[\sqrt{\varepsilon_2}\cos\theta_2+\sqrt{\varepsilon_1}\cos\theta_1]}{\sqrt{\varepsilon_1}\cos\theta_1}=\frac{\sqrt{\varepsilon_1}\cos\theta_1-\sqrt{\varepsilon_2}\cos\theta_2}{\sqrt{\varepsilon_1}\cos\theta_1}.$$

The reflection coefficient $\Gamma = \dfrac{E_r}{E_i}=\dfrac{\sqrt{\varepsilon_1}\cos\theta_1-\sqrt{\varepsilon_2}\cos\theta_2}{\sqrt{\varepsilon_1}\cos\theta_1+\sqrt{\varepsilon_2}\cos\theta_2}.$ (8.45)

Similarly, the transmission coefficient is

$$T=\frac{E_t}{E_i}=1+\frac{E_r}{E_i}=1+\frac{\sqrt{\varepsilon_1}\cos\theta_1-\sqrt{\varepsilon_2}\cos\theta_2}{\sqrt{\varepsilon_1}\cos\theta_1+\sqrt{\varepsilon_2}\cos\theta_2} \tag{8.46}$$

$$=\frac{2\sqrt{\varepsilon_1}\cos\theta_1}{\sqrt{\varepsilon_1}\cos\theta_1+\sqrt{\varepsilon_2}\cos\theta_2}.$$

θ_2 can be obtained from Snell's law,

$$\sin\theta_2=\frac{\sqrt{\varepsilon_1}\sin\theta_1}{\sqrt{\varepsilon_2}}. \tag{8.47}$$

The fields can be expressed in phasor notation as

Incident wave:

$$\vec{E}_i=E_ie^{-j\beta_1(x\sin\theta_1+z\cos\theta_1)}\,\vec{a}_y;$$

$$\vec{H}_i=H_i(-\cos\theta_1\,\vec{a}_x+\sin\theta_1\,\vec{a}_z)e^{-j\beta_1(x\sin\theta_1+z\cos\theta_1)}. \tag{8.48}$$

Reflected wave:

$$\vec{E}_r=E_re^{-j\beta_1(x\sin\theta_1-z\cos\theta_1)}\,\vec{a}_y;$$

$$\vec{H}_r=H_r(\cos\theta_1\,\vec{a}_x+\sin\theta_1\,\vec{a}_z)e^{-j\beta_1(x\sin\theta_1-z\cos\theta_1)}. \tag{8.49}$$

Transmitted wave:

$$\vec{E}_t=H_te^{-j\beta_2(x\sin\theta_2+z\cos\theta_2)}\vec{a}_y;$$

$$\vec{H}_t=H_t(-\cos\theta_2\,\vec{a}_x+\sin\theta_2\,\vec{a}_z)e^{-j\beta_2(x\sin\theta_2+z\cos\theta_2)}. \tag{8.50}$$

Case 2: Vertical (parallel) polarization

Assume that the electric vector is parallel to the plane of incidence. Let the electric field vector E_i of the obliquely incident wave be upward and the magnetic vector be outward (positive y direction) from the plane paper as shown is Fig. 8.10.

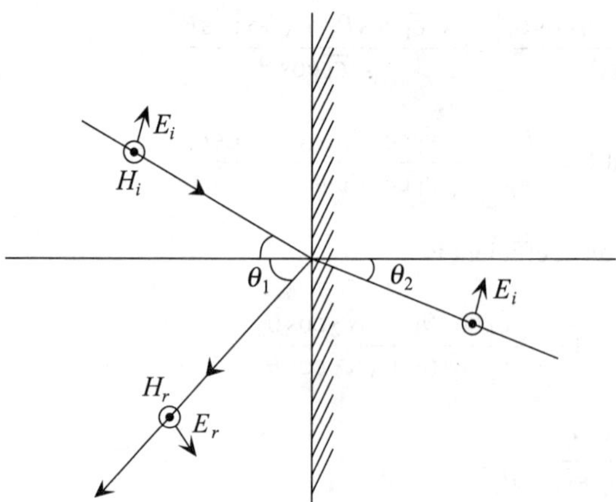

Fig. 8.10 *Obliquely incidence of a vertically polarized wave*

Applying boundary conditions, the tangential component of E is continuous at the boundary.

Since E_i and E_r make an angle θ_1 with the axis and E_t makes an angle θ_2 with the axis,

$$E_i \cos\theta_1 - E_r \cos\theta_1 = E_t \cos\theta_2.$$

$$1 - \frac{E_r}{E_i} = \frac{E_t \cos\theta_2}{E_i \cos\theta_1}.$$

$$\therefore \quad \frac{E_t}{E_i} = \left(1 - \frac{E_r}{E_i}\right)\frac{\cos\theta_1}{\cos\theta_2}. \tag{8.51}$$

Substituting this in Eq. (8.43),

$$\left(\frac{E_r}{E_i}\right)^2 = 1 - \sqrt{\frac{\varepsilon_2}{\varepsilon_1}}\left(\frac{E_t}{E_i}\right)^2 \frac{\cos\theta_2}{\cos\theta_1}$$

or $$\left(\frac{E_r}{E_i}\right)^2 = 1 - \sqrt{\frac{\varepsilon_2}{\varepsilon_1}}\left(1 - \frac{E_r}{E_i}\right)^2 \left(\frac{\cos^2\theta_1}{\cos^2\theta_2}\right)\frac{\cos\theta_2}{\cos\theta_1}.$$

$$1 - \left(\frac{E_r}{E_i}\right)^2 = \sqrt{\frac{\varepsilon_2}{\varepsilon_1}}\left(1 - \frac{E_r}{E_i}\right)^2 \frac{\cos\theta_1}{\cos\theta_2}.$$

$$\left(1 - \frac{E_r}{E_i}\right)\left(1 + \frac{E_r}{E_i}\right) = \sqrt{\frac{\varepsilon_2}{\varepsilon_1}}\left(1 - \frac{E_r}{E_i}\right)^2 \frac{\cos\theta_1}{\cos\theta_2}.$$

$$1 + \frac{E_r}{E_i} = \sqrt{\frac{\varepsilon_2}{\varepsilon_1}}\left(1 - \frac{E_r}{E_i}\right)\frac{\cos\theta_1}{\cos\theta_2}.$$

$$1 + \frac{E_r}{E_i} = \frac{\sqrt{\varepsilon_2}\cos\theta_1}{\sqrt{\varepsilon_1}\cos\theta_2} - \left(\frac{E_r}{E_i}\right)\frac{\sqrt{\varepsilon_2}\cos\theta_1}{\sqrt{\varepsilon_1}\cos\theta_2}.$$

$$\frac{E_r}{E_i}\left(1+\frac{\sqrt{\varepsilon_2}\cos\theta_1}{\sqrt{\varepsilon_1}\cos\theta_2}\right)=\frac{\sqrt{\varepsilon_2}\cos\theta_1}{\sqrt{\varepsilon_1}\cos\theta_2}-1.$$

The reflection coefficient $\Gamma=\frac{E_r}{E_i}=\frac{\sqrt{\varepsilon_2}\cos\theta_1-\sqrt{\varepsilon_1}\cos\theta_2}{\sqrt{\varepsilon_2}\cos\theta_1+\sqrt{\varepsilon_1}\cos\theta_2}$ (8.52)

and the transmission coefficient is

$$T=\frac{E_t}{E_i}=1-\left(\frac{E_r}{E_i}\right)\frac{\cos\theta_1}{\cos\theta_2}$$

$$=1-\left(\frac{\sqrt{\varepsilon_2}\cos\theta_1-\sqrt{\varepsilon_1}\cos\theta_2}{\sqrt{\varepsilon_2}\cos\theta_1+\sqrt{\varepsilon_1}\cos\theta_2}\right)\left(\frac{\cos\theta_1}{\cos\theta_2}\right).$$

$$\frac{E_t}{E_i}=\frac{\sqrt{\varepsilon_2}\cos\theta_1\cos\theta_2+\sqrt{\varepsilon_1}\cos^2\theta_2-\sqrt{\varepsilon_2}\cos^2\theta_1+\sqrt{\varepsilon_1}\cos\theta_1\cos\theta_2}{\sqrt{\varepsilon_2}\cos\theta_1\cos\theta_2+\sqrt{\varepsilon_1}\cos^2\theta_2}.$$

$$T=\frac{E_t}{E_i}=\frac{\sqrt{\varepsilon_2}\cos\theta_1(\cos\theta_2-\cos\theta_1)+\sqrt{\varepsilon_1}\cos\theta_2(\cos\theta_1+\cos\theta_2)}{(\sqrt{\varepsilon_2}\cos\theta_1+\sqrt{\varepsilon_1}\cos\theta_2)\cos\theta_2}. \quad (8.53)$$

The refracted angle can be obtained from Snell's law

$$\sin\theta_2=\sqrt{\frac{\varepsilon_1}{\varepsilon_2}}\sin\theta_1.$$

The fields can be expressed in phasor notation as

Incident wave:

$$\vec{E}_i=E_i(\cos\theta_1\vec{a}_x-\sin\theta_1\vec{a}_z)e^{-j\beta_1(x\sin\theta_1+z\cos\theta_1)};$$

$$\vec{H}_i=H_ie^{-j\beta_1(x\sin\theta_1+z\cos\theta_1)}\,\vec{a}_y. \quad (8.54)$$

Reflected wave:

$$\vec{E}_r=E_r(\cos\theta_1\vec{a}_x+\sin\theta_1\vec{a}_z)\,e^{-j\beta_1(x\sin\theta_1-z\cos\theta_1)};$$

$$\vec{H}_r=-H_re^{-j\beta_1(x\sin\theta_1-z\cos\theta_1)}\,\vec{a}_y. \quad (8.55)$$

Transmitted wave:

$$\vec{E}_t=E_t(\cos\theta_2\,\vec{a}_x-\sin\theta_2\vec{a}_z)e^{-j\beta_2(x\sin\theta_2+z\cos\theta_2)};$$

$$\vec{H}_t=H_te^{-j\beta_2(x\sin\theta_2+z\cos\theta_2)}\vec{a}_y. \quad (8.56)$$

Example 8.2 Consider two dielectric media, where medium 1 is free space and medium 2 has $\varepsilon_2=4\varepsilon_0$ and $\mu_2=\mu_0$. Determine the reflection coefficient for a wave obliquely incident at $\theta_1=30°$ for (a) perpendicular polarization and (b) parallel polarization.

Solution Given: $\varepsilon_2=4\varepsilon_0,\ \mu_2=\mu_0$, and $\theta_1=30°$.

(a) *Perpendicular polarization*:
From Snell's law,

$$\frac{\sin\theta_1}{\sin\theta_2} = \sqrt{\frac{\varepsilon_2}{\varepsilon_1}} = \sqrt{\frac{4\varepsilon_0}{\varepsilon_0}} = 2.$$

$$\sin\theta_2 = \frac{\sin 30^\circ}{2}.$$

$$\theta_2 = 14.478^\circ.$$

$$\text{Reflection coefficient, } \Gamma = \frac{E_r}{E_i} = \frac{\sqrt{\varepsilon_1}\cos\theta_1 - \sqrt{\varepsilon_2}\cos\theta_2}{\sqrt{\varepsilon_1}\cos\theta_1 + \sqrt{\varepsilon_2}\cos\theta_2}.$$

$$\Gamma = \frac{\sqrt{\varepsilon_0}\cos 30^\circ - 2\sqrt{\varepsilon_0}\cos 14.478^\circ}{\sqrt{\varepsilon_0}\cos 30^\circ + 2\sqrt{\varepsilon_0}\cos 14.478^\circ} = -0.381.$$

(b) *Parallel polarization* :

$$\text{Reflection coefficient} = \Gamma = \frac{E_r}{E_i} = \frac{\sqrt{\varepsilon_2}\cos\theta_1 - \sqrt{\varepsilon_1}\cos\theta_2}{\sqrt{\varepsilon_2}\cos\theta_1 + \sqrt{\varepsilon_1}\cos\theta_2}$$

$$\Gamma = \frac{2\sqrt{\varepsilon_0}\cos 30^\circ - \sqrt{\varepsilon_0}\cos 14.478^\circ}{2\sqrt{\varepsilon_0}\cos 30^\circ + \sqrt{\varepsilon_0}\cos 14.478^\circ} = 0.283.$$

8.9 Brewster Angle

When a wave is incident obliquely on the surface of a dielectric medium, the angle of incidence at which no reflection takes place is called the Brewster angle θ_B. It exists only for vertically polarized waves.

It can be expressed as $\theta_B = \tan^{-1}\sqrt{\frac{\varepsilon_2}{\varepsilon_1}}$. (8.57)

Proof Consider a vertically polarized wave incident obliquely on the surface of a dielectric medium.

The Brewster angle can be obtained from Eq. (8.52) by substituting the reflection coefficient as zero,

$$\text{i.e.,}\quad \frac{E_r}{E_i} = \frac{\sqrt{\varepsilon_2}\cos\theta_1 - \sqrt{\varepsilon_1}\cos\theta_2}{\sqrt{\varepsilon_2}\cos\theta_1 + \sqrt{\varepsilon_1}\cos\theta_2}.$$

At $\theta_1 = \theta_B$, reflection coefficient is zero,

$$\frac{\sqrt{\varepsilon_2}\cos\theta_B - \sqrt{\varepsilon_1}\cos\theta_2}{\sqrt{\varepsilon_2}\cos\theta_B + \sqrt{\varepsilon_1}\cos\theta_2} = 0$$

or $\sqrt{\varepsilon_2}\cos\theta_B - \sqrt{\varepsilon_1}\cos\theta_2 = 0.$

$$\sqrt{\varepsilon_2}\cos\theta_B = \sqrt{\varepsilon_1}\cos\theta_2$$

or $\varepsilon_2 \cos^2 \theta_B = \varepsilon_1 \cos^2 \theta_2.$ (8.58)

We know from Snell's law that

$$\sin \theta_2 = \sqrt{\frac{\varepsilon_1}{\varepsilon_2}} \sin \theta_B.$$

Since $\cos^2 \theta_2 = 1 - \sin^2 \theta_2,$

$$\cos^2 \theta_2 = 1 - \frac{\varepsilon_1}{\varepsilon_2} \sin^2 \theta_B.$$

Substituting the value in Eq. (8.58),

$$\varepsilon_2 \cos^2 \theta_B = \varepsilon_1 \left(1 - \frac{\varepsilon_1}{\varepsilon_2} \sin^2 \theta_B\right).$$

$$\frac{\varepsilon_2}{\varepsilon_1} \cos^2 \theta_B = 1 - \frac{\varepsilon_1}{\varepsilon_2} \sin^2 \theta_B.$$

$$\frac{\varepsilon_2}{\varepsilon_1} \cos^2 \theta_B = \cos^2 \theta_B + \sin^2 \theta_B - \frac{\varepsilon_1}{\varepsilon_2} \sin^2 \theta_B.$$

$$\cos^2 \theta_B \left(\frac{\varepsilon_2}{\varepsilon_1} - 1\right) = \sin^2 \theta_B \left(1 - \frac{\varepsilon_1}{\varepsilon_2}\right).$$

$$\cos^2 \theta_B \left(\frac{\varepsilon_2 - \varepsilon_1}{\varepsilon_1}\right) = \sin^2 \theta_B \left(\frac{\varepsilon_2 - \varepsilon_1}{\varepsilon_2}\right).$$

$$\varepsilon_2 \cos^2 \theta_B = \varepsilon_1 \sin^2 \theta_B.$$

$$\frac{\sin^2 \theta_B}{\cos^2 \theta_B} = \frac{\varepsilon_2}{\varepsilon_1}.$$

$$\tan^2 \theta_B = \frac{\varepsilon_2}{\varepsilon_1}.$$

$$\tan \theta_B = \sqrt{\frac{\varepsilon_2}{\varepsilon_1}}.$$

Brewster angle is $\theta_B = \tan^{-1} \sqrt{\frac{\varepsilon_2}{\varepsilon_1}}$ (Proved).

Therefore, when the wave is incident at the Brewster angle θ_B with parallel polarization, no reflections take place. The wave is completely transmitted through the media. Thus the Brewster angle is also called the angle of total transmission.

This phenomenon can be used when we want to transmit a signal from one medium to another medium without loss of power.

For example, when a laser beam is launched into an optical fibre, the incidence angle at the fibre surface is kept at the Brewster angle in order to minimise reflection losses.

When the wave is perpendicularly polarized, no Brewster angle exists. Therefore, when an unpolarized wave is incident obliquely on the dielectric surface at the Brewster angle θ_B, only the field components with perpendicular polarization will be reflected; those with parallel polarization will be transmitted. This method is used to separate the polarization components from the unpolarized wave.

8.10 Total Internal Reflection

Consider a plane wave incident obliquely on the surface between two media. Let ε_1 of medium 1 be greater than ε_2 of medium 2. If θ_1 and θ_2 are the incident and refracted angles, then from Snell's law, we know that

$$\frac{\sin\theta_1}{\sin\theta_2} = \sqrt{\frac{\varepsilon_2}{\varepsilon_1}}. \tag{8.59}$$

When $\theta_2 = \pi/2$, the transmitted wave travels along the boundary surface as shown in Fig. 8.11.

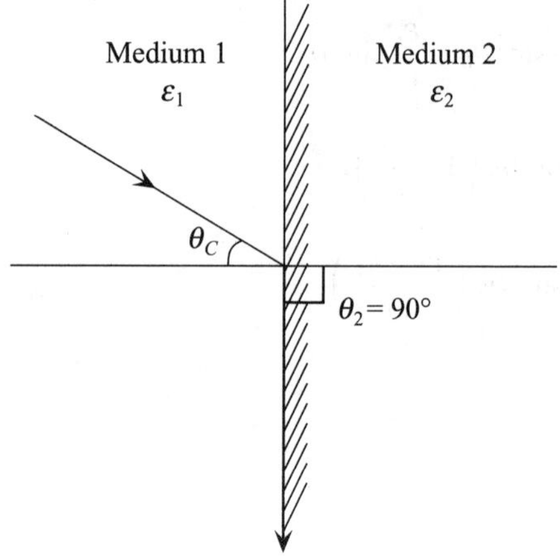

Fig. 8.11 *Total internal reflection*

The angle of incidence at which the refracted angle becomes $\pi/2$ is called the critical angle θ_c, i.e., if $\theta_1 = \theta_c$, then $\theta_2 = \pi/2$.

From Snell's law,

$$\frac{\sin\theta_c}{\sin\pi/2} = \sqrt{\frac{\varepsilon_2}{\varepsilon_1}}$$

or the critical angle is $$\theta_c = \sin^{-1}\left(\sqrt{\frac{\varepsilon_2}{\varepsilon_1}}\right). \tag{8.60}$$

If the incidence angle $\theta_1 > \theta_c$, no transmission takes place. The wave completely reflects back. This phenomenon is called total internal reflection.

Let n_1 be the refractive index of medium 1, and n_2, the refractive index of medium 2.

For lossless media, we know that $n_1 = \sqrt{\varepsilon_{r1}\mu_{r1}}$ and $n_2 = \sqrt{\varepsilon_{r2}\mu_{r2}}$. Also since $\mu_{r1} = \mu_{r2} = 1$.

$$\sin\theta_c = \sqrt{\frac{\varepsilon_2}{\varepsilon_1}} = \frac{n_2}{n_1}$$

$$\theta_c = \sin^{-1}\left(\frac{n_2}{n_1}\right). \tag{8.61}$$

Total internal reflection takes place if $n_1 > n_2$, i.e., the wave travels from denser medium to a lighter medium.

For example, if the wave travels from fresh water ($\varepsilon_{r_1} = 81$) to air ($\varepsilon_{r_2} = 1$), the critical angle is

$$\theta_c = \sin^{-1}\left(\sqrt{\frac{\varepsilon_0\varepsilon_{r2}}{\varepsilon_0\varepsilon_{r1}}}\right) = \sin^{-1}\sqrt{\frac{1}{81}} = 6.38^\circ.$$

That is, if the incidence angle of the wave is greater than 6.38°, the entire wave is reflected back into the water.

Example 8.3 A wave propagates from a dielectric medium to the interface with free space. If the angle of incidence is the critical angle of $\theta_c = 40^\circ$, find the relative permittivity.

Solution Given: $\theta_c = 40^\circ$
From Snell's law,

$$\sin\theta_c = \sqrt{\frac{\varepsilon_2}{\varepsilon_1}} = \sqrt{\frac{\varepsilon_{r_2}}{\varepsilon_{r_1}}}.$$

Given $\varepsilon_{r_2} = 1$ for free space.

$$\sin\theta_c = \frac{1}{\sqrt{\varepsilon_{r_1}}} \quad \text{or} \quad \sqrt{\varepsilon_{r_1}} = \frac{1}{\sin 40^\circ} = 1.5557$$

or $\quad \varepsilon_{r_1} = 2.42.$

8.11 Surface Impedance

When an electromagnetic wave travels along the conductor at a very high frequency, because of the tangential electric field, the total field current is confined only to the thin sheet of the surface.

Surface impedance is defined as *the ratio of the tangential component of the electric field to the surface current density at the surface.* It can be expressed as

$$Z_s = \frac{E_{\tan}}{K} = \frac{E_{\tan}}{J_s} = \frac{\gamma}{\sigma}\ \Omega, \tag{8.62}$$

where $E_{\tan}$ = tangential component of the electric field vector V/m,

$K = J_s$ = total surface current density A/m.

γ = propagation constant and σ = conductivity of the sheet.

Proof Consider an electromagnetic wave travelling along a conducting plate of thickness t lying at $z = 0$ plane as shown in Fig. 8.12.

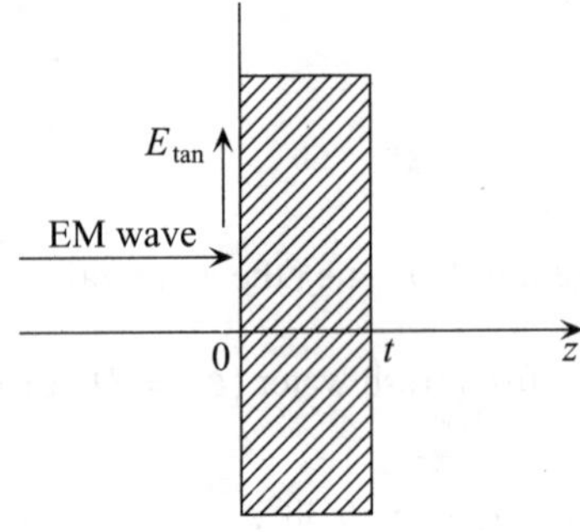

Fig. 8.12 *Conducting plate*

The current flows along the surface of the conductor due to the tangential component of the electric field vector.

Assume that the depth of penetration of the current into the plate along the z direction is less than the plate thickness t.

Let J be the current density. It can be expressed as

$$J = J_0 e^{-\gamma z} \tag{8.63}$$

where J_0 = current density in A/m^2 at the surface at $z = 0$ and
γ = propagation constant.

The variation of current density inside the plate is shown in Fig. 8.13.

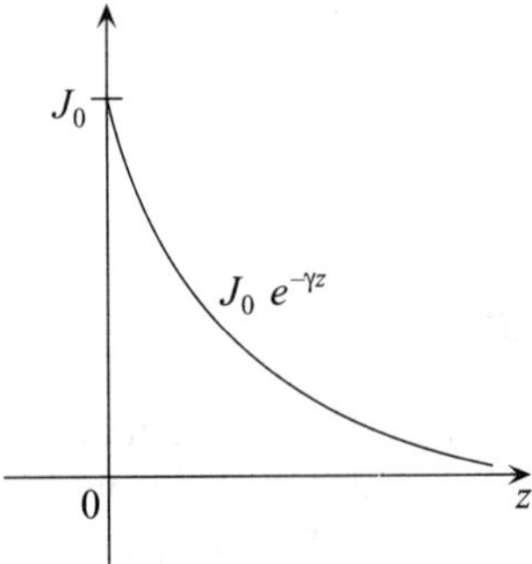

Fig. 8.13 *Variation of current density along the thickness of a conductor.*

If the current is distributed along the conductor surface, the linear current density or surface current density is

$$K = J_s = \int_0^\infty J dz, \quad \text{or} \quad J_s = \int_0^\infty J_0 e^{-\gamma z} dz.$$

$$J_s = J_0 \left[\frac{e^{-\gamma z}}{-\gamma} \right]_0^\infty \quad \text{or} \quad J_s = \frac{J_0}{\gamma} \ \text{A/m}. \tag{8.64}$$

If σ is the conductivity of the conductor, the current density on the surface is

$$J_0 = \sigma E_{\tan}.$$

$$\therefore \quad J_s = \frac{\sigma E_{\tan}}{\gamma}.$$

The surface impedance is

$$Z_s = \frac{E_{\tan}}{J_s} = \frac{\gamma}{\sigma} \ \Omega. \qquad \text{(Proved)}$$

We know that the propagation constant is $\gamma = \sqrt{j\omega\mu(\sigma + j\omega\varepsilon)}$.

$$\therefore \quad Z_s = \frac{\sqrt{j\omega\mu(\sigma + j\omega\varepsilon)}}{\sigma} \ \Omega. \tag{8.65}$$

Thus the surface impedance depends on the conductor properties, ε, μ, σ at a given frequency.

Note: For a good conductor, $\sigma >> \omega\varepsilon$.

$$\therefore \quad \gamma = \sqrt{j\omega\mu\sigma}.$$

Surface impedance is $Z_s = \dfrac{\sqrt{j\omega\mu\sigma}}{\sigma} = \sqrt{\dfrac{\omega\mu}{\sigma}} \angle 45^\circ$.

We know that the characteristic impedance $\eta = \sqrt{\dfrac{\omega\mu}{\sigma}} \angle 45^\circ$. (8.66)

Therefore, for a good conductor, the surface impedance is equal to the characteristic impedance.

8.12 Surface Resistance and Surface Reactance

The surface impedance of a good conductor at high frequency is from Eq. (8.66),

$$Z_s = \sqrt{\frac{\omega\mu}{\sigma}} \angle 45^\circ = \sqrt{\frac{\omega\mu}{\sigma}} \left(\frac{1}{\sqrt{2}} + j \frac{1}{\sqrt{2}} \right).$$

$$Z_s = \sqrt{\frac{\omega\mu}{2\sigma}} + j\sqrt{\frac{\omega\mu}{2\sigma}}. \tag{8.67}$$

The real component of surface impedance is called surface resistance R_s.

$$R_s = \sqrt{\frac{\omega\mu}{2\sigma}} \ \Omega. \tag{8.68}$$

The imaginary component of the surface impedance is called surface reactance X_s.

$$X_s = \sqrt{\frac{\omega\mu}{2\sigma}}\ \Omega. \tag{8.69}$$

The surface resistance R_s is defined as *the skin effect resistance per unit length of a conductor of unit width.*

Surface resistance in terms of depth of penetration The surface resistance R_s can also be expressed in terms of the depth of penetration in a conductor.

We know that the depth penetration of current in a conductor is

$$\delta = \sqrt{\frac{2}{\omega\mu\sigma}}.$$

∴ The surface resistance is

$$R_s = \sqrt{\frac{\omega\mu}{2\sigma}} = \frac{1}{\sigma}\sqrt{\frac{\omega\mu\sigma}{2}} = \frac{1}{\sigma\delta}\ \Omega. \tag{8.70}$$

Therefore, for a given conductor, the surface resistance is inversely proportional to the depth of penetration.

Due to surface resistance, the power loss per unit area of the plane conductor is

$$P_{\text{loss}} = J_{\text{seff}}^2 R_s, \tag{8.71}$$

where J_{seff} = effective current per metre width flowing in the conductor.

8.13 Poynting Vector and Poynting Theorem

Poynting vector The Poynting vector is defined as *the cross product of the electric field vector $\vec{E}$ and the magnetic field vector of an electromagnetic wave.*

Thus $\vec{P} = \vec{E} \times \vec{H}$ watts/m², (8.72)

where $\vec{P}$ = Poynting vector in watts/m²;
$\vec{E}$ = electric field vector in V/m;
$\vec{H}$ = magnetic field vector in A/m.

The Poynting vector gives instantaneous power density of the wave. The direction of the Poynting vector is perpendicular to the plane of the field and follows the right-hand screw law.

Consider an EM wave which propagates in free space along the z direction. The electric and magnetic fields are

$$\vec{E} = E_0 \cos(\omega t - \beta z)\,\vec{a}_x \quad \text{and} \quad \vec{H} = H_0 \cos(\omega t - \beta z)\,\vec{a}_y. \tag{8.73}$$

In free space, the intrinsic impedance is $\eta_0 = \dfrac{E_0}{H_0} = 377\Omega.$

The Poynting vector is

$$\vec{P} = \vec{E} \times \vec{H}$$

$$= E_0 \cos(\omega t - \beta z)\vec{a}_x \times H_0 \cos(\omega t - \beta z)\vec{a}_y$$

$$= E_0 H_0 \cos^2(\omega t - \beta z)\vec{a}_z.$$

$$\vec{P} = \frac{E_0^2}{\eta_0} \cos^2(\omega t - \beta z)\vec{a}_z \ \text{W/m}^2. \tag{8.74}$$

The Poynting vector is along the direction of wave propagation.

If S is the area of cross-section of power flow, then the instantaneous power is given by

$$\vec{p} = \int_S \vec{P} \cdot d\vec{S} \text{ watts.} \tag{8.75}$$

Average Poynting vector or average power density The average Poynting vector is the average of the instantaneous Poynting vector over one period.

$$\vec{P}_{av} = \frac{1}{T}\int_0^T \vec{P}\, dt.$$

$$\therefore \quad P_{av} = \frac{1}{T}\int_0^T E_0 H_0 \cos^2(\omega t - \beta z)dt.$$

$$= \frac{E_0 H_0}{T}\int_0^T \left(\frac{1}{2} + \frac{\cos 2(\omega t - \beta z)}{2}\right) dt.$$

$$= \frac{E_0 H_0}{T}\int_0^T \frac{1}{2} dt + 0.$$

$$\vec{P}_{av} = \frac{E_0 H_0}{2T}(T) = \frac{E_0 H_0}{2}\vec{a}_z \ \text{watts/m}^2. \tag{8.76}$$

In free space, $\dfrac{E_0}{H_0} = \eta_0$. The average power density is

$$\vec{P}_{av} = \frac{E_0^2}{2\eta_0}\vec{a}_z = \frac{1}{2}\eta_0 H_0^2 a_z \ \text{watts/m}^2. \tag{8.77}$$

In a medium, if the impedance is $\eta = |\eta|\angle\theta$, the electric and magnetic fields are

$$\vec{E} = E_0 e^{-\alpha z}\cos(\omega t - \beta z)\vec{a}_x,$$

$$\vec{H} = H_0 e^{-\alpha z}\cos(\omega t - \beta z)\vec{a}_y = \frac{E_0}{|\eta|}e^{-\alpha z}\cos(\omega t - \beta z - \theta)\vec{a}_y$$

and the Poynting vector is

$$\vec{P} = \frac{E_0^2}{|\eta|} e^{-2\alpha z} \cos(\omega t - \beta z) \cos(\omega t - \beta z - \theta) \vec{a}_z$$

$$\vec{P} = \frac{E_0^2}{|\eta|} e^{-2\alpha z} [\cos\theta + \cos(2\omega t - 2\beta z - \theta)] \vec{a}_z.$$

Then the avarage power density is

$$\vec{P}_{\text{av}} = \frac{1}{T} \int_0^T \vec{P} dt = \frac{E_0^2}{2|\eta|} e^{-2\alpha z} \cos\theta \vec{a}_z.$$

Also, $\vec{P}_{\text{av}} = \frac{1}{2} H_0^2 |\eta| e^{-2\alpha z} \cos\theta \, \vec{a}_z$ watts/m^2.

Average power $p_{av} = |\vec{P}_{av}| S$ watts. (8.78)

Complex Poynting vector The complex Poynting vector is defined as

$$\vec{P}_{\text{com}} = \frac{1}{2} (\vec{E} \times \vec{H}^*).$$

$$\vec{P}_{\text{com}} = \vec{P}_{\text{av}} + \vec{P}_{\text{react}}, \tag{8.79}$$

where $\vec{P}_{\text{av}}$ = average Poynting vector;

$\vec{P}_{\text{react}}$ = reactive Poynting vector.

In complex notation, average Poynting vector is

$$\vec{P}_{\text{av}} = \frac{1}{2} \Re e [E \times H^*]$$

and reactive Poynting vector is

$$\vec{P}_{\text{react}} = \frac{1}{2} \text{Im} [E \times H^*]. \tag{8.80}$$

8.14 Poynting Theorem

The Poynting theorem is one of the most important results in EM theory. It is a statement of energy conservation for electric and magnetic fields and tells us about the power flowing in an EM field. The Poynting theorem states that *the vector product $\vec{P} = \vec{E} \times \vec{H}$ at any point is a measure of the rate of energy flow per unit area at that point. The direction of energy flow is perpendicular to the E and H planes.*

It can be expressed as

$$\int_v \vec{E} . \vec{J} dv = \frac{-\partial}{\partial t} \int_v \left(\frac{\mu H^2}{2} + \frac{\varepsilon E^2}{2} \right) dv - \oint_S (\vec{E} \times \vec{H}) dS. \tag{8.81}$$

From the perspective of conservation of energy, it can be stated that *the rate of energy dissipation in the volume v must be equal to the rate at which the stored energy in volume v is decreasing plus the rate at which energy is entering the volume v from outside.*

Proof Consider an EM wave travelling through a medium. From Maxwell's equation, we know that

$$\nabla \times \vec{H} = \vec{J} + \frac{\partial \vec{D}}{\partial t}$$

or $$\vec{J} = \nabla \times \vec{H} - \varepsilon \frac{\partial \vec{E}}{\partial t}.$$

Taking dot product with $\vec{E}$,

$$\vec{E} \cdot \vec{J} = \vec{E} \cdot (\nabla \times \vec{H}) - \varepsilon \vec{E} \cdot \frac{\partial \vec{E}}{\partial t}. \tag{8.82}$$

We know that the vector identity

$$\nabla \cdot (\vec{E} \times \vec{H}) = \vec{H} \cdot (\nabla \times \vec{E}) - \vec{E}(\nabla \times \vec{H})$$

or $$\vec{E} \cdot (\nabla \times \vec{H}) = \vec{H} \cdot (\nabla \times \vec{E}) - \nabla \cdot (\vec{E} \times \vec{H}).$$

Substituting in Eq. (8.83),

$$\vec{E} \cdot \vec{J} = \vec{H} \cdot (\nabla \times \vec{E}) - \nabla \cdot (\vec{E} \times \vec{H}) - \varepsilon \vec{E}.\frac{\partial \vec{E}}{\partial t}. \tag{8.83}$$

From Maxwell's equation, we know that

$$\nabla \times \vec{E} = \frac{-\partial \vec{B}}{\partial t} = -\frac{\mu \, \partial \vec{H}}{\partial t}.$$

Substituting this in Eq. (8.83),

$$\vec{E} \cdot \vec{J} = \vec{H} \cdot \left(-\mu \frac{\partial \vec{H}}{\partial t} \right) - \nabla \cdot (\vec{E} \times \vec{H}) - \varepsilon \vec{E} \cdot \frac{\partial \vec{E}}{\partial t}$$

or $$\vec{E} \cdot \vec{J} = -\mu \vec{H} \cdot \frac{\partial \vec{H}}{\partial t} - \varepsilon \vec{E} \cdot \frac{\partial \vec{E}}{\partial t} - \nabla \cdot (\vec{E} \times \vec{H}).$$

We know that,

$$\vec{H} \cdot \frac{\partial \vec{H}}{\partial t} = \frac{1}{2} \frac{\partial H^2}{\partial t} \quad \text{and} \quad \vec{E} \cdot \frac{\partial \vec{E}}{\partial t} = \frac{1}{2} \frac{\partial E^2}{\partial t}.$$

$$\therefore \quad \vec{E} \cdot \vec{J} = \frac{-\mu}{2} \frac{\partial H^2}{\partial t} - \frac{\varepsilon}{2} \frac{\partial E^2}{\partial t} - \nabla \cdot (\vec{E} \times \vec{H}). \tag{8.84}$$

Integrating over the volume,

$$\int_v (\vec{E}\cdot\vec{J})\,dv = \frac{-\partial}{\partial t}\int_v \left(\frac{\mu H^2}{2}+\frac{\varepsilon E^2}{2}\right)dv - \int_v \nabla\cdot(\vec{E}\times\vec{H})\,dv. \tag{8.85}$$

Using divergence theorem, we know that,

$$\int_v \nabla\cdot(\vec{E}\times\vec{H})dv = \oint_S (\vec{E}\times\vec{H})d\vec{S}.$$

Substituting this in Eq. (8.85), the final expression is

$$\int_v (\vec{E}\cdot\vec{J})dv = \frac{-\partial}{\partial t}\int_v \left(\frac{\mu H^2}{2}+\frac{\varepsilon E^2}{2}\right)dv - \oint_S (\vec{E}\times\vec{H})\cdot d\vec{S} \qquad \text{(Proved).}$$

The physical interpretation of poynting theorem is given below.

The first term $\int_v (\vec{E}\cdot\vec{J})dv =$ Total power dissipated in a volume when the electric field vector produces a current density J in the conducting medium. It represents the power dissipated due to ohmic (I^2R) loss of the conductor.

$\vec{E}\cdot\vec{J}$ represents power density in watts /m^3.

The second term $\dfrac{-\partial}{\partial t}\displaystyle\int_v \left(\frac{\mu H^2}{2}+\frac{\varepsilon E^2}{2}\right)dv =$ rate at which the stored energy in the volume is decreasing.

where, $\dfrac{\mu H^2}{2}$ = energy density stored in the magnetic field.

and $\dfrac{\varepsilon E^2}{2}$ = energy density stored in the electric field.

The third term $-\oint_S (\vec{E}\times\vec{H})\cdot d\vec{S} =$ The rate of flow of energy inward through the surface of the volume.

where $\vec{E}\times\vec{H} = \vec{P}$ is a Poynting vector in watts/m^2.

Example 8.4 Show that the power flow for a plane wave is $\vec{P} = \vec{E}\times\vec{H}$.

Solution The rate of energy flow per unit area can be obtained by considering a uniform plane travelling with velocity,

$$v = \frac{1}{\sqrt{\mu\varepsilon}}.$$

The total energy density stored due to the electromagnetic field is

$$E_{\text{stored}} = \frac{1}{2}\varepsilon E^2 + \frac{1}{2}\mu H^2.$$

The rate of flow of energy per unit area along the direction of propagation,

$$\vec{P} = E_{\text{stored}} \cdot \vec{v} = \frac{1}{2}(\varepsilon E^2 + \mu H^2) v \vec{a}_z.$$

We know that for lossless medium, $\dfrac{\vec{E}}{\vec{H}} = \eta = \sqrt{\dfrac{\mu}{\varepsilon}}.$

$$\therefore \quad \vec{P} = \frac{1}{2}\left(\varepsilon \vec{E}\sqrt{\frac{\mu}{\varepsilon}}\vec{H} + \mu \vec{H}\sqrt{\frac{\varepsilon}{\mu}}\vec{E}\right) v \vec{a}_z$$

$$= \frac{1}{2}(\sqrt{\mu\varepsilon}\vec{E}\cdot\vec{H} + \sqrt{\mu\varepsilon}\vec{E}\cdot\vec{H}) v \vec{a}_z.$$

$$\vec{P} = EH \cdot \sqrt{\mu\varepsilon} \cdot \frac{1}{\sqrt{\mu\varepsilon}} \vec{a}_z = EH \vec{a}_z.$$

Since E and H are perpendicular to each other,

$$\vec{P} = \vec{E} \times \vec{H}. \qquad \text{(Proved).}$$

Example 8.5 Using Poynting vector $\vec{P} = \vec{E} \times \vec{H}$, show that the total power flow in a concentric cylindrical cable is $P = VI$ watts.

Solution Consider a concentric cylinder with inner conductor of radius a and outer conductor of radius b. Let V = voltage between the conductors and I = current flowing through the inner conductor. From Amprere's law, the magnetic field between the conductors is

$$\oint \vec{H} \cdot d\vec{l} = I_{\text{encl}}.$$

Consider an arbitrary circular path of radius r, $a \le r \le b$.

Then $H_\phi \cdot 2\pi r = I$

$$\vec{H} = \frac{I}{2\pi r}\vec{a}_\phi.$$

We know that the electric field in a cable is

$$\vec{E} = \frac{V}{r\ln(b/a)}\vec{a}_r.$$

The power flow, directed along the axis of the cable is

$$\vec{P} = \vec{E} \times \vec{H} = \frac{VI}{2\pi r^2 \ln(b/a)} \vec{a}_\phi \times \vec{a}_r.$$

$$\therefore \quad \vec{P} = \frac{VI}{2\pi r^2 \ln(b/a)} \vec{a}_z.$$

The total power flow along the cable is

$$p = \oint_s \vec{P} \cdot d\vec{S},$$

where $dS = r\,dr\,d\phi$.

$$p = \int_0^{2\pi}\int_a^b \frac{VI}{2\pi r^2 \ln(b/a)} .rdr\,d\phi = \frac{VI(2\pi)}{2\pi \ln(b/a)} \int_a^b \frac{1}{r} dr$$

$$= \frac{VI}{\ln(b/a)} \ln(b/a) = VI$$

$\therefore \quad p = VI$ watts.

8.15 Power Loss in a Plane Conductor

The average power flow per unit area normal to the surface is given by

$$\vec{P}_{av} = \frac{1}{2} \Re e[\vec{E} \times \vec{H}^*]. \tag{8.86}$$

For a good conductor, the tangential components are at 45°.

$$\therefore \quad \vec{P}_{av} = \frac{1}{2} |E_{\tan}|\, |H_{\tan}| \angle 45^\circ. \tag{8.87}$$

The real average power is

$$P_{av} = \frac{1}{2} |E_{\tan}|\, |H_{\tan}| \times \frac{1}{\sqrt{2}}.$$

$$P_{av} = \frac{1}{2\sqrt{2}} \frac{1}{\eta} |E_{\tan}|^2 = \frac{1}{2\sqrt{2}} \eta\, |H_{\tan}|^2. \tag{8.88}$$

We know that for good conductors,

$$Z_s = \eta = \frac{E_{\tan}}{J_s}.$$

Therefore the effective value of the average power is

$$P_{\text{eff}} = \frac{1}{\sqrt{2}} |Z_s| \left| \frac{J_s}{\sqrt{2}} \right|^2.$$

$$P_{\text{eff}} = \frac{1}{\sqrt{2}} |Z_s|\, |J_{s\ \text{eff}}|^2 \text{ watts/m}^2. \tag{8.89}$$

Example 8.6 A linearly polarized plane wave propagates through free space at an angle θ with respect to the normal to a plane circular surface on the $x-y$ plane. If the peak amplitude of E is 50 mV/m, calculate the average power that passes through a 1.2 m^2 surface.

Solution Given: $E_0 = 50 \times 10^{-3}$ V/m, area $5 = 1.2$ m^2, and impedance in free space $\eta_0 = 120\pi\ \Omega$.

The power flow is

$$P = \vec{E} \times \vec{H}$$

or $$P = \frac{E_0^2}{\eta_0} = \frac{(50 \times 10^{-3})^2}{120\pi} = 6.63 \times 10^{-6} \text{ watts/m}^2.$$

Average power

$$p = \oint_s \vec{P} \cdot d\vec{S} = P\cos\theta = 6.63 \times 10^{-6} \times 1.2$$

$$= 2.65 \times 10^{-7} \cos\theta \text{ watts.}$$

Example 8.7 A plane wave having a frequency of 10 MHz has an average Poynting vector of 1 W/m^2. If the medium is lossless with relative permeability 4 and relative permittivity 2, find the velocity propagation, wavelength, impedance of the medium and rms value of the electric field.

Solution Given: $P_{\text{av}} = 1$ watts/m^2, $f = 10$ MHz $= 10^7$ Hz, $\mu_r = 2$, $\varepsilon_r = 3$.

Velocity of propagation $v_p = \sqrt{\dfrac{1}{\mu\,\varepsilon}} = \dfrac{v_0}{\sqrt{\mu_r\,\varepsilon_r}}$.

$$v_p = \frac{3 \times 10^8}{\sqrt{2 \times 3}} = 1.06 \times 10^8 \text{ m/s.}$$

Wavelength $\lambda = \dfrac{v_p}{f} = \dfrac{1.06 \times 10^8}{10^7} = 10.6$ m.

Impedance $\eta = \sqrt{\dfrac{\mu}{\varepsilon}} = 120\pi\sqrt{\dfrac{\mu_r}{\varepsilon_r}}\ \eta = 120\pi\sqrt{\dfrac{4}{2}} = 533\ \Omega$.

Average Poynting vector $P = \dfrac{E^2}{2\eta}$

$$E^2 = 2P\eta = 2 \times 1 \times 533, \quad E^2 = 1066.3.$$

$$E = \sqrt{1.66.3} = 32.65 \text{ V/m.}$$

The rms value of the electric field

$$E_{\text{rms}} = \frac{32.65}{\sqrt{2}} \text{ V/m.}$$

Additional Solved Problems

Example 8.8 A perpendicularly polarized wave is incident at an angle of $\theta_i = 15°$. It is propagating from medium 1 to medium 2. Medium 1 is defined by $\varepsilon_{r_1} = 8.5$, $\mu_{r_1} = 1$ and $\sigma_1 = 0$; and medium 2 is free space. If $E_i = 1.0$ mV/m, determine, E_r, H_i, H_r, E_t and H_t.

Solution Given: A perpendicularly polarized wave, incident angle $\theta_i = 15°$; for medium 1, $\varepsilon_{r_1} = 8.5$, $\mu_{r_1} = 1$; for medium 2 (free space) $\varepsilon_{r_2} = \mu_{r_2} = 1$.

$$E_i = 1 \text{ mV/m} = 1 \times 10^{-3} \text{ V/m}.$$

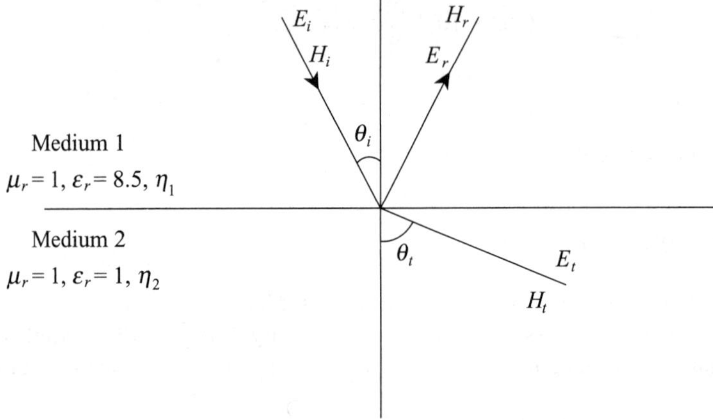

Fig. 8.14 *Plane wave reflection and refraction at the boundary*

From Snell's law, we know that,

$$\frac{\sin\theta_i}{\sin\theta_t} = \sqrt{\frac{\varepsilon_2}{\varepsilon_1}}.$$

$$\sin\theta_t = \sin\theta_i\sqrt{\frac{\varepsilon_{r_1}}{\varepsilon_{r_2}}} = \sqrt{8.5}\sin(15°) = 0.75.$$

$$\therefore \quad \theta_t = \sin^{-1}(0.75), \quad \theta_t = 49°.$$

We know that the reflection coefficient for perpendicularly polarized wave is

$$\Gamma = \frac{E_r}{E_i} = \frac{\sqrt{\varepsilon_1}\cos\theta_1 - \sqrt{\varepsilon_2}\cos\theta_2}{\sqrt{\varepsilon_1}\cos\theta_1 + \sqrt{\varepsilon_2}\cos\theta_2} = \frac{\sqrt{\varepsilon_{r_1}}\cos\theta_i - \sqrt{\varepsilon_{r_2}}\cos\theta_r}{\sqrt{\varepsilon_{r_1}}\cos\theta_i + \sqrt{\varepsilon_{r_2}}\cos\theta_r}.$$

$$\theta_1 = \theta_i = 15°, \quad \theta_2 = \theta_t = 49°.$$

$$E_r = \left(\frac{\sqrt{8.5}\cos 15° - \sqrt{1}\cos 49°}{\sqrt{8.5}\cos 15° + \sqrt{1}\cos 49°}\right) E_i = \left(\frac{2.816 - 0.656}{2.816 + 0.656}\right) \times 10^{-3}.$$

$$E_r = 0.622 \text{ mV/m}.$$

Also, we know that

$$E_t = E_i + E_r$$

or $E_t = 1 + 0.622 = 1.622$ mV.

For medium 1 We know that the impedance of the medium is

$$\eta_1 = \frac{E_i}{H_i} = \sqrt{\frac{\mu}{\varepsilon}}.$$

$$\frac{E_i}{H_i} = \sqrt{\frac{\mu_{r_1}\mu_0}{\varepsilon_{r_1}\varepsilon_0}} = \sqrt{\frac{4\pi \times 10^{-7}}{8.5 \times 8.85 \times 10^{-12}}}.$$

$$\therefore \quad \eta_1 = \frac{E_i}{H_i} = 129\ \Omega.$$

$$\therefore \quad H_i = \frac{E_i}{\eta_1} = \frac{1 \times 10^{-3}}{129} = 7.75\ \mu\text{A/m}.$$

Also, $H_r = \dfrac{E_r}{\eta_1} = \dfrac{0.622 \times 10^{-3}}{129} = 4.82\ \mu\text{A/m}.$

For medium 2:

$$\eta_2 = \frac{E_t}{H_t} = \sqrt{\frac{\mu}{\varepsilon}} = \sqrt{\frac{\mu_{r_2}\mu_0}{\varepsilon_{r_2}\varepsilon_0}}$$

$$= \sqrt{\frac{1 \times 1.25 \times 10^{-6}}{1 \times 8.85 \times 10^{-12}}} = 375.8\ \Omega.$$

$$H_t = \frac{E_t}{\eta_2} = \frac{1.622 \times 10^{-3}}{375.8} = 4.31\ \mu\text{A/ m}.$$

Example 8.9 A y-polarized uniform plane wave with fields (E_i, H_i) and a frequency of 100 MHz propagates in air in the $+x$ direction and impinges normally on a perfectly conducting plane at $x = 0$, assuming the amplitudes of E_i to be 6 mV/m. Write the phasor expressions for

(a) E_i and H_i of the incident wave

(b) E_r and H_r of the reflected wave

(c) E_T and H_T of the total wave in air

(d) Determine the location nearest to the conducting plane where E_T and H_T are zero.

Solution Given a y-polarized uniform plane wave,

$f = 100$ MHz, $= 10^8$ Hz

$E_0 = 6$ mV/m $= 6 \times 10^{-3}$ V/m

For air, $\lambda = \frac{c}{f} = \frac{3 \times 10^8}{10^8} = 3$ m, $\beta = \frac{2\pi}{\lambda} = \frac{2\pi}{3} = 2.09$ rad/m and $\eta_0 = 120\pi\ \Omega$.

For a uniform plane wave travelling in the $+x$ direction,

(a) $E_i = E_0\, e^{-j\beta x}$.

$$= 6 \times 10^{-3}\, e^{-j(2\pi/3)x} = 6\, e^{-j(2\pi/3)x} \text{ mV/m}$$

and $H_i = H_0\, e^{-j(2\pi/3)x} = \frac{E_0}{\eta_0} e^{-j(2\pi/3)x}$.

$$= \frac{6 \times 10^{-3}}{120\pi} e^{-j(2\pi/3)x} \text{ A/m} = 15.9\, e^{-j(2\pi/3)x}\ \mu\text{A/m}.$$

(b) For a perfectly conducting plane with normal incidence, the reflected fields are

$E_r = -E_i$ and $H_r = H_i$.

$\therefore\quad E_r = -6\, e^{-j(2\pi/3)x}$ mV/m.

$H_r = 15.9\, e^{-j(2\pi/3)x}\ \mu$A/m.

(c) The fields for the total wave (standing wave) are

$$E_T(x, t) = 2E_i \sin(\beta x)\sin(\omega t)$$

$$= 12 \sin\left(\frac{2\pi x}{3}\right) \sin(2\pi \times 10^8 t) \text{ mA/m}$$

and $H_T(x, t) = 2H_i \cos(\beta x)\cos(\omega t)$

$$= 32 \cos\left(\frac{2\pi x}{3}\right) \cos(2\pi \times 10^8 t)\ \mu\text{A/m}.$$

(d) $E_T = 0$ at $x = 0$ and is maximum at $x = \lambda/4$.

$H_T = 0$ at $x = \lambda/4$ and is maximum at $x = 0$.

Example 8.10 A plane wave has E = 2 V/m and a frequency of 300 MHz. It is moving in free space impinging on a thick copper sheet located perpendicular to the direction of propagation. Find

(a) E and H at the plate surface

(b) depth of penetration and

(c) surface impedance.

Solution Given: Incident electric field intensity, $E_i = 2$ V/m, frequency, $f = 300 \times 10^6$ Hz; for copper, $\sigma = 5.8 \times 10^7$ mhos/m, for free space $\sigma = 0$. Assume, $\mu = \mu_o$, $\varepsilon = \varepsilon$.

(a) *E and H at the plate surface*

We know that the transmission coefficient is

$$T = \frac{E_t}{E_i} = \frac{2\eta_2}{\eta_1 + \eta_2}$$

or $$E_t = \frac{2\eta_2}{\eta_1 + \eta_2} E_i,$$

where η_1 and η_2 are intrinsic impedances of medium 1 and 2 respectively.

For free space, $\eta_1 = 377\ \Omega$.

For medium 2, copper sheet, the intrinsic impedance is

$$\eta_2 = \sqrt{\frac{j\omega\mu}{\sigma + j\omega\varepsilon_0}}.$$

$$= \sqrt{\frac{j(2\pi \times 300 \times 10^6) \times 4\pi \times 10^{-7}}{5.8 \times 10^7 + j(2\pi \times 300 \times 10^6)(8.85 \times 10^{-12})}}$$

$$= \sqrt{\left(\frac{j2368.7}{5.8 \times 10^7 + j167 \times 10^{-4}}\right)}$$

$$\eta_2 = 6.39 \times 10^{-3} \angle 44.99°.$$

From the values of η_1 and η_2, we observe that $\eta_1 >> \eta_2$.

$$\therefore \quad \eta_1 + \eta_2 \cong \eta_1.$$

Now,

$$E_t = \frac{2\eta_2}{\eta_1} E_i.$$

Substituting the values of E_i, η_1 and η_2 in the above equation, we get,

$$E_t = \frac{2 \times 6.39 \times 10^{-3} \angle 44.99°}{377} \times 2$$

$$E_t = 6.78 \times 10^{-5}\ \angle 44.99°\ \text{V/m}.$$

For a plane wave with tangential components E_t and H_t, we have,

$$H_t = \frac{E_t}{\eta}.$$

Here,

$$\eta = \eta_2 = 6.39 \times 10^{-3}\ \angle\, 44.99°.$$

$$H_t = \frac{E_t}{6.39 \times 10^{-3} \angle 44.99^\circ} = \frac{6.78 \times 10^{-5} \angle 44.99^\circ}{6.39 \times 10^{-3} \angle 44.99^\circ}.$$

$$H_t = 10.6 \times 10^{-3} \text{ A/m}.$$

Therefore, the components E and H at the plate surface are

$$E = 6.78 \times 10^{-5} \text{ V/m and } H = 10.6 \times 10^{-3} \text{ A/m}.$$

(b) *Depth of penetration* (δ)
Depth of penetration, also known as skin depth, is given by,

$$\delta = \frac{1}{\sqrt{\pi f \mu \sigma}}$$

$$= \frac{1}{\sqrt{\pi \times 300 \times 10^6 \times 4\pi \times 10^7 \times 5.8 \times 10^7}} = 3.81 \times 10^{-3} \text{ mm}.$$

(c) *Surface impedance* Z_s
The surface impedance is given by

$$Z_s = \frac{\gamma}{\sigma}.$$

For good conductors,

$$\gamma = \sqrt{j\omega\mu(\sigma + j\omega\varepsilon)} \approx \sqrt{j\omega\mu\sigma}$$

$$= (j2\pi \times 300 \times 10^6 \times 4\pi \times 10^{-7} \times 5.8 \times 10^7)^{1/2}.$$

$$\gamma = (13.73 \times 10^{10})^{1/2} \angle 45^\circ = 3.7 \times 10^5 \angle 45^\circ.$$

$$Z_s = \frac{3.7 \times 10^5 \times \angle 45^\circ}{5.8 \times 10^7} = 6.39 \times 10^{-3} \angle 45^\circ.$$

which is the same as η_2.

Example 8.11 A free space conductor interface has $H_i = 1$ A/m on the free space side. The frequency is 31.8 MHz and the conductor constants are $\varepsilon_r = \mu_r = 1$ and $\sigma = 1.26$ MS/m. Determine H_r, H_t and the depth of penetration of H_i.

Solution Given: $f = 31.8$ MHz, $H_i = 1$ A/m, $\sigma = 1.26 \times 10^6$ S/m, $\varepsilon_r = \mu_r = 1$.

For normal incidence on free space-conductor interface, we know that, $\Gamma = \dfrac{H_r}{H_i} = \dfrac{\eta_1 - \eta_2}{\eta_1 + \eta_2}.$

For free space, $\eta_1 = 120\pi$,

For a conductor

$$\eta_2 = \sqrt{\frac{j\omega\mu}{\sigma + j\omega\varepsilon}}.$$

Now $\omega\mu = 2\pi \times 31.8 \times 10^6 \times 4\pi \times 10^{-7} = 251.08$

and $\omega\varepsilon = 2\pi \times 31.8 \times 10^6 \times 8.854 \times 10^{-12} = 1.769 \times 10^{-3}$.

Since $\sigma \gg \omega\varepsilon,\ \sigma + j\omega\varepsilon \approx \sigma,$

$$\eta_2 = \sqrt{\frac{j251.083}{1.26 \times 10^6}} = \sqrt{j2 \times 10^{-4}} = 1.414 \times 10^{-2}\ \angle 45^\circ \text{ ohms.}$$

$$\therefore \quad \frac{H_r}{H_i} = \frac{120\pi - 1.414 \times 10^{-2}\ \angle 45^o}{120\pi + 1.414 \times 10^{-2}\ \angle 45^o}$$

$$= \frac{120\pi - (0.01 + j0.01)}{120\pi + (0.01 + j0.01)} = \frac{376.98 - j0.01}{377 + j0.01}.$$

$$\frac{H_r}{H_i} = 0.99995.$$

$$H_r = 0.99995 \times H_i = 0.99995\text{A/m}.$$

Also, $H_t = \dfrac{2\eta_1}{\eta_1 + \eta_2} = \dfrac{2 \times 120\pi}{377\ \angle 1.52 \times 10^{-3^\circ}}.$

$$H_t = 1.99995\ \angle -0.00153^\circ \approx 2 \text{ A/m}.$$

Depth of penetration $\delta = \dfrac{1}{\alpha} = \sqrt{\dfrac{2}{\omega\mu\sigma}}$

$$= \sqrt{\frac{2}{251.08 \times 1.26 \times 10^6}} = 7.95 \times 10^{-5} \text{ m}$$

Note: Since the interfacing surface is a good conductor,

$$H_r \approx H_i$$

$$H_t \approx H_r + H_i = 2H_i.$$

Example 8.12 A parallel polarized wave propagates from air into a dielectric at a Brewster angle of 75°. Find ε_r.

Solution Given a parallel polarized wave,

Brewster angle $\theta_B = 75^\circ$

For medium 1, air, $\varepsilon_{r_1} = 1$,

For medium 2, let ε_{r_2} be the dielectric constant

We know that the Brewster angle is $\theta_B = \tan^{-1}\sqrt{\dfrac{\varepsilon_2}{\varepsilon_1}}$.

$$\tan 75^\circ = \sqrt{\frac{\varepsilon_2}{\varepsilon_1}} = \sqrt{\frac{\varepsilon_{r_2}}{\varepsilon_{r_1}}}.$$

$$\varepsilon_{r_2} = 13.93.$$

Example 8.13 A parallel polarized wave propagates from a dielectric medium 1, $\varepsilon_{r_1} = 78$ to the interface with dielectric medium 2, $\varepsilon_{r_2} = 2.5$. Find the critical angle and the Brewster angle.

Solution Given: $\varepsilon_{r_2} = 78, \quad \varepsilon_{r_2} = 2.5.$
From Snell's law

$$\text{Critical angle} \quad \theta_c = \sin^{-1}\sqrt{\frac{\varepsilon_{r2}}{\varepsilon_{r_1}}} = \sin^{-1}\sqrt{\frac{2.5}{78}} = 10.31^\circ.$$

$$\text{Brewster angle is} \quad \theta_B = \tan^{-1}\sqrt{\frac{\varepsilon_{r2}}{\varepsilon_{r_1}}} = \tan^{-1}\sqrt{\frac{2.5}{78}} = 10.15^\circ.$$

At angle 10.15°, the wave is completely transmitted and at angle 10.31°, the wave is completely reflected.

Example 8.14 Prove that under the condition of no reflection at an interface, the sum of the Brewster angle and the angle of refraction is 90 degrees for parallel polarization for the case of reflection by a conductor under oblique incidence.

Solution Given a parallel (vertically) polarized wave incident obliquely on a surface of a conducting medium.

Let ε_1 be the dielectric constant of medium 1, and ε_2 be the dielectric constant of medium 2.

$$\text{Brewster angle is} \quad \theta_B = \tan^{-1}\sqrt{\frac{\varepsilon_2}{\varepsilon_1}} \quad \text{or} \quad \tan\theta_B = \sqrt{\frac{\varepsilon_2}{\varepsilon_1}}.$$

Let ε_1 be greater than ε_2. If θ_1 and θ_2 are the incident and refracted angles, then from Snell's law, we know that,

$$\frac{\sin\theta_1}{\sin\theta_2} = \sqrt{\frac{\varepsilon_2}{\varepsilon_1}}.$$

At Brewster angle, $\theta_B = \theta_1$. Let the refracted (transmitted) angle be $\theta_2 = \theta_{tb}$.

$$\therefore \quad \frac{\sin\theta_B}{\sin\theta_{tb}} = \sqrt{\frac{\varepsilon_2}{\varepsilon_1}} = \tan\theta_B.$$

$$\sin\theta_{tb} = \frac{\sin\theta_B}{\tan\theta_B} = \cos\theta_B.$$

$$\sin\theta_{tb} = \cos\theta_B = \sin(90 - \theta_B).$$

$$\theta_{tb} = (90 - \theta_B).$$

$$\theta_B + \theta_{tb} = 90^\circ$$ (Proved)

Example 8.15 An EM wave travelling in air is incident normally on a boundary between air and a dielectric having permeability same as free space and permittivity as 4. Prove that one-ninth of the incident power is reflected and eight-ninths of it is transmitted into the second medium.

Solution Given $\varepsilon_{r_2} = 1, \ \ \varepsilon_{r_2} = 4$
We know that the reflection coefficient is

$$\Gamma = \frac{E_r}{E_i} = \frac{\sqrt{\varepsilon_1} - \sqrt{\varepsilon_2}}{\sqrt{\varepsilon_1} + \sqrt{\varepsilon_2}} = \frac{\sqrt{1} - \sqrt{4}}{\sqrt{1} + \sqrt{4}} = \frac{-1}{3}.$$

Let P_i, P_r and P_t be the incident, reflected and refracted powers respectively. Then

$$\frac{P_r}{P_i} = \frac{E_r^2/\eta_1}{E_i^2/\eta_1} = \frac{E_r^2}{E_i^2}.$$

$$\frac{P_r}{P_i} = \Gamma^2 = \frac{1}{9}.$$

Also $P_i = P_t + P_r$ or $P_t = P_i - P_r$.

$$\frac{P_t}{P_i} = 1 - \frac{P_r}{P_i} = 1 - \frac{1}{9} = \frac{8}{9} \qquad \text{(Proved)}$$

Example 8.16 Evaluate the reflection and transmission coefficients of an EM wave in air incident normally upon a copper sheet of frequency 1 MHz. Given $\mu_0 = \mu_1 = \mu_2$, $\varepsilon_0 = \varepsilon_1 = \varepsilon_2$ and $\sigma_1 = 0$, $\sigma_2 = 3.8 \times 10^7$ S/m.

Solution Given: $\mu_0 = \mu_1 = \mu_2$, $\varepsilon_0 = \varepsilon_1 = \varepsilon_2$ and $\sigma_1 = 0$, $\sigma_2 = 3.8 \times 10^7$ S/m,
frequency = 1 MHz.

For air $\eta_1 = 120\pi\ \Omega$, and for copper sheet, $\eta_2 = \sqrt{\frac{\omega\mu}{\sigma}}\angle 45^\circ = \sqrt{\frac{2\pi \times 10^6 \times \mu_0}{3.8 \times 10^7}}\angle 45^\circ$
$= 3.689 \times 10^{-4}\angle 45^\circ.$

We know that the reflection coefficient is

$$\Gamma = \frac{E_r}{E_i} = \frac{\eta_2 - \eta_1}{\eta_1 - \eta_2} = \frac{3.689 \times 10^{-4}\angle 45 - 120\pi}{3.689 \times 10^{-4}\angle 45 + 120\pi} = 0.99\angle 179.9^\circ.$$

$$\Gamma = 0.99\angle 179.9^\circ.$$

Transmission coefficient $\mathrm{T} = \Gamma + 1 = 0.99\angle 179.9^\circ + 1 = 0.01\angle 9.8^\circ$.

Example 8.17 An EM wave in free space is incident normally on a dielectric whose $\varepsilon_r = 5.0$. Find the reflection and transmission coefficients

Solution Given $\varepsilon_{r_1} = 1, \ \ \varepsilon_{r_2} = 5.$
We know that the reflection coefficient is

$$\Gamma = \frac{E_r}{E_i} = \frac{\sqrt{\varepsilon_1} - \sqrt{\varepsilon_2}}{\sqrt{\varepsilon_1} + \sqrt{\varepsilon_2}}. = \frac{\sqrt{1} - \sqrt{5}}{\sqrt{1} + \sqrt{5}} = -0.381.$$

Transmission coefficient $\mathrm{T} = \Gamma + 1 = -0.381 + 1 = 0.619$.

Example 8.18 A polarized wave is incident from air to poly-starin, $\varepsilon_2 = 2.6\ \varepsilon_0$ with $\mu = \mu_0$, $\varepsilon = 2.6\ \varepsilon_0$ at Brewster angle. Determine the transmission angle.

Solution Assume that the wave is a parallel polarized wave.

For medium 1, air, $\varepsilon_1 = \varepsilon_0,\ \mu_1 = \mu_0$.

For medium 2, poly-starin, $\mu_2 = \mu_0,\ \varepsilon_2 = 2.6\varepsilon_0$.

We know that the Brewster angle is $\theta_B = \tan^{-1}\sqrt{\dfrac{\varepsilon_2}{\varepsilon_1}} = \tan^{-1}\left(\sqrt{\dfrac{2.6}{1}}\right) = \tan^{-1}(1.612) = 58.2°$.

If θ_1 and θ_2 are the incident and refracted angles, then from Snell's law, we know that,

$$\frac{\sin\theta_1}{\sin\theta_2} = \sqrt{\frac{\varepsilon_2}{\varepsilon_1}}.$$

At Brewster angle, $\theta_B = \theta_1$. Let the refracted (transmitted) angle be $\theta_2 = \theta_{tb}$.

$$\therefore \quad \frac{\sin\theta_B}{\sin\theta_{tb}} = \sqrt{\frac{\varepsilon_2}{\varepsilon_1}} = \tan\theta_B.$$

$$\sin\theta_{tb} = \frac{\sin\theta_B}{\tan\theta_B} = \cos\theta_B.$$

$$\sin\theta_{tb} = \cos 58.2° = 0.527.$$

$$\theta_{tb} = \sin^{-1} 0.527 = 31.8°.$$

Note: The sum of the Brewster angle and the transmission angle is 90°.

Example 8.19 A plane wave travelling in a medium of $\varepsilon_r = 1,\ \mu_r = 1$ has an electric field intensity of $100 \times \sqrt{\pi}$ V/m. Determine the magnetic field intensity and energy density in the magnetic field and also the total energy density.

Solution Given: $\varepsilon_r = 1,\ \mu_r = 1$ and electric field intensity, $E = 100 \times \sqrt{\pi}$ V/m.
We know that,

$$\frac{E}{H} = \sqrt{\frac{\mu}{\varepsilon}}.$$

$$H = \sqrt{\frac{\varepsilon_0\, E_r}{\mu_0\, \mu_r}} \times E = \sqrt{\frac{885 \times 10^{12} \times 1}{4\pi \times 10^7 \times 1}} \times 100 \times \sqrt{\pi} = 0.47\ \text{A/m}.$$

Energy density in the magnetic field,

$$w_m = \frac{1}{2}\mu H^2 = \frac{1}{2}\mu_0\, \mu_r H^2$$

$$= \frac{1}{2} \times 4\pi \times 10^{-7} \times 1 \times (0.47)^2.$$

$$w_m = 1.38 \times 10^{-7}\ \text{J/m}^3.$$

For a plane wave, energy density stored in the electric field = energy density stored in the magnetic field.

$\therefore \quad w_e = w_m = 1.38 \times 10^{-7}$ J/m^3.

Total energy density $= w_e + w_m = 2w_e = 2 \times 1.38 \times 10^{-7} = 2.77 \times 10^{-7}$ J/m^3.

Example 8.20 In free space, $\vec{E}(z,t) = 50\cos(\omega t - \beta z)\,\vec{a}_x$ V/m. Find the average power crossing a circular area of radius 2.5 m in the plane $z =$ constant.

Solution Given, $\vec{E}(z,t) = 50\cos(\omega t - \beta z)\vec{a}_x$ V/m, radius, $r = 2.5$ m.
For free space, $\eta_1 = 120\pi\ \Omega$; Area $S = \pi \times 2.5^2 = 19.635$ m^2
The power flow or Poynting vector,

$$\vec{P} = \vec{E} \times \vec{H} = \frac{\vec{E}^2}{\eta} = \frac{50^2}{120\pi}\cos^2(\omega t - \beta z)\,\vec{a}_x.$$

Average power flow is $\vec{P}_{\text{av}} = \dfrac{1}{T}\displaystyle\int \vec{P}dt = \dfrac{E_0^2}{2\eta_0}\vec{a}_x$

$$= \frac{50^2}{2 \times 120\pi}\vec{a}_x = 3.315\vec{a}_x \quad \text{watts/m}^2.$$

Average power

$$p_{\text{av}} = \int_S \vec{P}_{\text{av}} \cdot d\vec{S}.$$

Given the plane $z =$ constant, i.e., the surface is normal to the power flow.

$$p_{\text{av}} = P_{\text{av}}\ S.$$

The average power is $\text{P}_{\text{av}} = 3.315 \times 19.635 = 65$ watts.

Example 8.21 A sinusoidal plane wave is transmitted through a medium whose electric field strength is 10 kV/m and relative permittivity of the medium is 4. Determine the mean rms power flow/unit area.

Solution Given: $E = 10 \times 10^3$ V/m, $\varepsilon_r = 4$, $\mu_r = 1$.

$$\frac{E}{H} = \sqrt{\frac{\mu}{\varepsilon}}.$$

$$\frac{E}{H} = \eta_0\sqrt{\frac{1}{\varepsilon_r}} = \frac{120\pi}{\sqrt{4}} = 188.5 \text{ ohms.}$$

$$\therefore \quad H = \frac{E}{188.5} = \frac{10^4}{188.5}.$$

We know that power flow/unit area is

$$P = E \times H = EH = E\frac{E}{188.5} = \frac{E^2}{188.5}$$

$$= \frac{(10 \times 10^3)^2}{188.5} = 530.503 \text{ kW/m}^2.$$

rms power flow/unit area $P_{\text{rms}} = \dfrac{530.503}{2} = 265.25 \text{ kW/m}^2.$

Example 8.22 A plane wave travelling in free space has an average Poynting vector of 5 watts/m^2. Find the average energy density.

Solution Given: $P = 5$ watts/m^2.
We know that $P = EH = 5$ watts/m^2 and the average energy density or total energy density in free space

$$W = \mu_0 H^2 = \varepsilon_0 E^2.$$

$$W = \frac{\mu_0 HP}{E}$$

In free space, $\dfrac{H}{E} = \sqrt{\dfrac{\varepsilon_0}{\mu_0}}.$

$$W = \mu_0 \sqrt{\frac{\varepsilon_0}{\mu_0}}\, P = \sqrt{\mu_0\, \varepsilon_0}\, P = \frac{P}{v_0},$$

where free space velocity $v_0 = \dfrac{1}{\sqrt{\mu_0\, \varepsilon_0}} = 3 \times 10^8$ m/s.

$$\therefore \quad W = \frac{5}{3 \times 10^8} = 16.67 \times 10^{-9} \text{ J/m}^3.$$

Example 8.23 A plane sinusoidal electromagnetic wave travelling in space has $E_{\max} = 1500$ mV/m. (i) Find the accompanying $H_{\max}$ and (ii) the average power transmitted.

Solution Given: $E_{\max} = 1500 \text{ mV/m} = 1500 \times 10^{-3}$ V/m
The electromagnetic wave is travelling in free space. So, $\sigma = 0,\ \varepsilon = \varepsilon_0,\ \mu = \mu_0,\ \eta_0 = 377\ \Omega.$

(i) Magnetic field intensity is

$$H_{\max} = \frac{E_{\max}}{\eta_0} = \frac{1500 \times 10^{-3}}{377}, = 3.98 \text{ mA/m}.$$

(ii) Average power transmitted

$$P_{\text{av}} = \frac{E_{\max}^2}{2|\eta_0|} = \frac{(1500 \times 10^{-3})^2}{2 \times 377} = 2.984 \text{ mW/m}^2.$$

Example 8.24 A plane wave of frequency 2 MHz is incident on a copper conductor normally. The wave has an electric field amplitude of $E = 2$ mV/m. Copper has $\mu_r = 1$, $\varepsilon_r = 1$ and $\sigma = 5.8 \times 10^7$ mho/m. Find the average power density absorbed by the copper.

Solution Given: A plane wave of frequency $f = 2$ MHz $= 2 \times 10^6$ Hz, $\mu_r = 1$, $\varepsilon_r = 1$ and $\sigma = 5.8 \times 10^7$ mho/m.

Electric field $E = 2 \text{ mV/m} = 2 \times 10^{-3}$ V/m.

The average power density or the average power flow is $P_{av} = \dfrac{E_0^2}{2|\eta|}$,

$$\text{where } \eta = \sqrt{\frac{\omega\mu}{\sigma}} = \sqrt{\frac{2 \times \pi \times 2 \times 10^6 \times 4 \times \pi \times 10^{-7}}{5.8 \times 10^7}} = 5.218 \times 10^{-4}\ \Omega.$$

$$P_{av} = \frac{(2 \times 10^{-3})^2}{2 \times 5.218 \times 10^{-4}} = 3.83 \text{ mW/m}^2.$$

Example 8.25 Find skin depth and surface resistance of a copper conductor at 100 MHz having $\sigma = 5.8 \times 10^7$ S/m, $\mu_r = 100$.

Solution Given for copper, $f = 100 \text{ MHz} = 10^8$ Hz; conductivity $\sigma = 5.8 \times 10^7$ S/m, $\mu_r = 100$.
We know that the skin depth is

$$\delta = \sqrt{\frac{2}{\omega\mu\sigma}} = \frac{1}{\sqrt{\pi f \mu \sigma}} = \frac{1}{\sqrt{\pi \times 10^8 \times 100 \times 4\pi \times 10^{-7} \times 5.8 \times 10^7}} = 6.6 \times 10^{-7} \text{ m}.$$

Surface resistance is

$$R_s = \frac{1}{\sigma\delta} = \frac{1}{6.6 \times 10^{-7} \times 5.8 \times 10^7} = 0.026 = 26 \text{ m}\Omega.$$

Summary

1. Snell's law can be expressed as

$$\frac{\sin\theta_i}{\sin\theta_t} = \frac{v_1}{v_2} = \frac{n_2}{n_1}$$

For dielectric media,

$$\mu_1 = \mu_2 = \mu_0 \text{ or } \sqrt{\varepsilon_1}\sin\theta_i = \sqrt{\varepsilon_2}\sin\theta_t \text{ and } \frac{n_2}{n_1} = \sqrt{\frac{\varepsilon_2}{\varepsilon_1}}.$$

2. The law reflection is given by
Incidence angle = reflected angle, i.e., $\theta_i = \theta_r$.

3. The reflection coefficient can be expressed as $\Gamma = \dfrac{E_r}{E_i}$ or $\Gamma = \dfrac{H_r}{H_i}$

4. The transmission coefficient can be expressed as $T = \dfrac{E_t}{E_i}$ or $T = \dfrac{H_t}{H_i}$.
5. If a uniform plane wave is incident normally on a perfect conductor, the reflection coefficient is $\Gamma = 1$, and the transmission coefficient is $T = 0$.
6. If a uniform plane wave is incident normally on the surface of a perfect dielectric media, the reflection coefficient for electric field is $\Gamma = \dfrac{E_r}{E_i} = \dfrac{\eta_2 - \eta_1}{\eta_1 + \eta_2}$.

   The transmission coefficient for an electric field is $T = 1 + \Gamma = \dfrac{E_t}{E_i} = \dfrac{2\eta_2}{\eta_1 + \eta_2}$.

7. When a plane wave is incident obliquely on the surface of another medium, depending on the orientation of the electric field vector, there are two types of polarizations.
   (i) Horizontal polarization (or perpendicular polarization)
   (ii) Vertical polarization (or parallel polarization)

   (i) For horizontal polarization, the reflection coefficient is

$$\Gamma = \frac{E_r}{E_i} = \frac{\sqrt{\varepsilon_1}\cos\theta_1 - \sqrt{\varepsilon_2}\cos\theta_2}{\sqrt{\varepsilon_1}\cos\theta_1 + \sqrt{\varepsilon_2}\cos\theta_2}.$$

   The transmission coefficient is $T = \dfrac{2\sqrt{\varepsilon_1}\cos\theta_1}{\sqrt{\varepsilon_1}\cos\theta_1 + \sqrt{\varepsilon_2}\cos\theta_2}$.

   (ii) For vertical polarization, the reflection coefficient is

$$\Gamma = \frac{E_r}{E_i} = \frac{\sqrt{\varepsilon_2}\cos\theta_1 - \sqrt{\varepsilon_1}\cos\theta_2}{\sqrt{\varepsilon_1}\cos\theta_1 + \sqrt{\varepsilon_1}\cos\theta_2}.$$

   The transmission coefficient is

$$T = \frac{E_t}{E_i} = \frac{\sqrt{\varepsilon_2}\cos\theta_1(\cos\theta_2 - \cos\theta_1) + \sqrt{\varepsilon_1}\cos\theta_2(\cos\theta_1 + \cos\theta_2)}{(\sqrt{\varepsilon_2}\cos\theta_1 + \sqrt{\varepsilon_1}\cos\theta_2)\cos\theta_2}.$$

8. The angle of incidence at which no reflection takes place is called Brewster angle θ_B. It exists only for a vertically polarized wave. It can be expressed as

$$\theta_B = \tan^{-1}\sqrt{\frac{\varepsilon_2}{\varepsilon_1}}.$$

9. The angle of incidence at which there is no transmission or the refracted angle becomes $\pi/2$ is called critical angle and is given by $\theta_c = \sin^{-1}\left(\sqrt{\dfrac{\varepsilon_2}{\varepsilon_1}}\right)$ or $\theta_c = \sin^{-1}\left(\dfrac{n_2}{n_1}\right)$.
10. Total internal reflection occurs, if the incidence angle $\theta_1 > \theta_c$, and no transmission takes place. The wave should travel from a denser medium to a lighter medium, i.e., $n_1 > n_2$.
11. The surface impedance of the conductor can be expressed as

$$Z_s = \frac{E_{\tan}}{K} = \frac{E_{\tan}}{J_s}\,\Omega.$$

12. For a good conductor, surface impedance is equal to the characteristic impedance,

$$Z_s = \frac{\sqrt{j\omega\mu\sigma}}{\sigma} = \sqrt{\frac{\omega\mu}{\sigma}}\underline{|45^\circ}.$$

13. The Poynting vector is defined as $\vec{P} = \vec{E} \times \vec{H}$ watts/m^2. It gives the instantaneous power density of the wave. The instantaneous power is given by

$$P = \int_S \vec{P} \cdot d\vec{S} \text{ watts}$$

14. The complex Poynting vector is defined as

$$\vec{P}_{\text{comp}} = \frac{1}{2}(\vec{E} \times \vec{H}^*).$$

15. Poynting theorem can be expressed as

$$\int \vec{E} \cdot \vec{J} dv = \frac{-\partial}{\partial t}\int_v \left(\frac{\mu H^2}{2} + \frac{\upsilon E^2}{2}\right) - \oint_s (\vec{E} \times \vec{H}) d\vec{S}.$$

16. The effective value of the average power is

$$P_{\text{eff}} = \frac{1}{\sqrt{2}}|Z_s||J_{s\ \text{eff}}|^2 \text{ watts/m}^2.$$

Review Questions

1. What is Snell's law?
2. State and explain the Poynting vector and where it will be useful.
3. Define surface impedance.
4. Derive the expression for power loss in a plane conductor. *(May 2011)*
5. What is the difference between normal incidence and oblique incidence? *(May 2012)*
6. Obtain the expression for surface impedance of conductors in terms of skin depth. *(May 2012, Aug 2014)*
7. Distinguish between a travelling wave and a standing wave. *(May 2009)*
8. Obtain an expression for reflection coefficient when a wave is incident on a dielectric obliquely with perpendicular polarization. *(Aug 2014)*
9. Derive the expression for skin depth of a good conductor. *(Nov 2010)*
10. Derive the expression for the resultant electric field and resultant magnetic field when a wave is incident normally on a perfect conductor. *(May 2011, 2012)*
11. Discuss Brewster angle, critical angle, total internal reflection and surface impedance. *(May 2012)*
12. Define instantaneous, average and complex Poynting vectors. *(May 2012)*
13. Explain the significance of the Poynting theorem and the Poynting vector. *(Jan 2014)*
14. State and prove Poynting theorem. *(May 2010, 2011)*
15. Show that the vector product $\vec{E} \times \vec{H} = \vec{P}$ represents the rate of energy flow per unit area at a point. *(May 2011, 2012)*

16. Show that the power flow in a concentric cylindrical cable is $w = VI$ watts using Poynting vector. *(May 2013)*
17. From the Maxwell's equations, derive the expression for Poynting vector. Also, explain the application of the Poynting vector. *(Nov 2012)*
18. Prove that under the condition of no reflection at an interface, the sum of the Brewster angle and the angle of reflection is $\pi/2$ for parallel polarization for the case of reflection by a perfect conductor under oblique incidance with the helps of neat sketches. *(Jan 2014)*
19. Define Brewster angle and derive an expression for Brewster angle when a wave is is parallelly polarized. *(Aug 2014)*

Problems

1. In spherical coordinates, the spherical wave $\vec{E} = \dfrac{100}{r}\sin\theta\cos(\omega t - \beta r)\vec{a}_\theta$ V/m, $\vec{H} = \dfrac{0.265}{r}\sin\theta(\cos\omega t - \beta r)\vec{a}_\phi$ A/m represents the electromagnetic field at a large distance r from a certain dipole antenna in free space. Find the average power crossing the hemispherical shell $r = 1$ km, $0 \le \theta \le \dfrac{\pi}{2}$.
2. A wave propagates from a dielectric medium to the interface with free space. If the angle incidence is the critical angle of 20°, find the relative permittivity.
3. Compute the ratios $\dfrac{E_r}{E_i}$ and $\dfrac{E_t}{E_i}$ for normal incidence and oblique incidence at $\theta_1 = 10^\circ$. For region, 1, $\varepsilon_{r_1} = 8.5$, $\mu_{r_1} = 1$, $\sigma_1 = 0$; region 2 is free space.
4. Determine the reflection coefficient for an EM wave at frequency 1 MHz incident normally on a sheet of iron. Assume $\sigma = 1\times10^6$ mhos/m, $\mu = 1000\mu_0$ for iron.
5. A polarized wave is incident from air to polystyrene with $\mu = \mu_0$, $\varepsilon = 2.6\,\varepsilon_0$ at Brewster angle. Determine the transmission angle.
6. Determine the reflection coefficient for an EM wave at frequency 1 GHz incident normally on a sea water. Assume $\sigma = 4$ mhos/m, $\mu = 1$, $\varepsilon_r = 80$.
7. An EM wave in free space is incident normally on a dielectric whose $\varepsilon_r = 3$. Find the reflection and transmission coefficients.
8. A plane wave is incident from air to polystyrene at an angle $\theta_1 = 30^0$. Calculate the angle of transmission if with $\mu = \mu_0$, $\varepsilon = 2.7\,\varepsilon_0$.
9. A perpendicularly polarized wave is incident at an angle of $\theta_1 = 15^0$ from medium 1 to free space. Medium 1 has $\varepsilon_{r1} = 9$, $\sigma_1 = 0$. If $E_i = 10$ mV/m, determine θ_2, E_r, H_i, H_r.
10. A plane wave travelling in a medium of $\varepsilon_r = 5.5$, $\mu_r = 1$ has an electric field intensity of $100\sqrt{\pi}$. Determine the energy density in the magnetic field and also the total energy density.
11. A plane wave travelling in glass has $H = 25$ mA/m. Compute the average energy density and also the velocity of this wave in glass; the relative permittivity of glass of glass is 5.5.
12. A 10 MHz plane wave travelling in free space has an amplitude $E_0 = 50$ V/m. What is the average and peak energy density of the wave?

Answers

1. 55.5 W   2. 8.55

3. For normal incidence, $\frac{E_r}{E_i} = 0.5$; $\frac{E_t}{E_i} = 1.5$; for oblique incidence, $\frac{E_r}{E_i} = 0.54$; $\frac{E_t}{E_i} = 1.54$.

4. $0.999 \angle 179.98^0$ 5. 31.8^0 6. $0.8836 \angle 176.08^0$

7. $E_r/E_i = -0.268, E_t/E_i = 0.732$ 8. 17.71^0

9. 51^0, 5 mV/m, 39.78 μA/m, 19.9 μA/m   10. 0.765 μJ/m^3, 1.53 μJ/m^3

11. 0.39 nJ/m^3, 1.28×10^8 m/s   12. 11.09 μJ/m^3

Multiple-choice Questions

1. In the case of reflection by a perfect conductor with normal incidence, the sum of the electric field strengths of the incident and reflected waves at any point is given by
   (a) $E_T(x,t) = 2E_1 \sin\beta x \sin\omega t$ (b) $H_T(x,t) = 2H_1 \cos\beta x \cos\omega t$
   (c) $E_T(x,t) = 2E_1 \sin\beta x \cos\omega t$ (d) $E_T(x,t) = 2E_1 \cos\beta x \sin\omega t$

2. In the case of reflection by a perfect conductor with normal incidence, sum of the magnetic field strengths of the incidence and reflected waves at any point is given by
   (a) $H_T(x,t) = 2H_1 \sin\beta x \cos\omega t$ (b) $H_T(x,t) = 2H_1 \cos\beta x \cos\omega t$
   (c) $E_T(x,t) = 2H_1 \sin\beta x \cos\omega t$ (d) $H_T(x,t) = 2H_1 \cos\beta x \sin\omega t$

3. In case of reflection by a perfect conductor with normal incidence, the value of the magnetic field strength at the surface of the conductor and at multiples of half wavelengths from the surface is
   (a) minimum (b) maximum (c) zero (d) none

4. In the case of reflection by a perfect conductor with normal incidence, the value of the electric field strength of quarter wavelengths from the surface is
   (a) minimum (b) maximum (c) zero (d) none

5. Parallel polarization is also known as
   (a) horizontal polarization, (b) vertical polarization
   (c) both (*a*) and (*b*) (d) none

6. Perpendicular polarization is also known as
   (a) horizontal polarization (b) vertical polarization
   (c) both (*a*) and (*b*) (d) none

7. The electric field vector parallel to the boundary surface or perpendicular to the plane of incidence is called.
   (a) horizontal polarization (b) vertical polarization
   (c) both (*a*) and (*b*) (d) none

8. When the magnetic field vector is parallel to the boundary surface, the electric field vector is parallel to the plane of incidences and is called

   (a) horizontal polarization (b) vertical polarization

   (c) both (*a*) and (*b*) (d) none

9. When a wave is reflected by a perfect dielectric, normal incidence, the ratio of the transmitted electric field to the incident electric field is given by

   (a) $\dfrac{2\eta_1}{\eta_1+\eta_2}$ (b) $\dfrac{(\eta_1-\eta_2)}{(\eta_1+\eta_2)}$ (c) $\dfrac{(\eta_2-\eta_1)}{(\eta_1+\eta_2)}$ (d) $\dfrac{2\eta_2}{\eta_1+\eta_2}$

10. In the case of reflection by a perfect dielectric with normal incidence, the ratio of the reflected electric field to the incident electric field is given by

    (a) $\dfrac{2\eta_1}{\eta_1+\eta_2}$ (b) $\dfrac{(\eta_1-\eta_2)}{(\eta_1+\eta_2)}$ (c) $\dfrac{(\eta_2-\eta_1)}{(\eta_1+\eta_2)}$ (d) $\dfrac{2\eta_2}{\eta_1+\eta_2}$

11. In the case of reflection by a perfect dielectric with normal incidence, the ratio of the reflected ray to the incident ray given by H_r/H_i is

    (a) $\dfrac{2\sqrt{\varepsilon_2}}{\sqrt{\varepsilon_1}+\sqrt{\varepsilon_2}}$ (b) $\dfrac{\sqrt{\varepsilon_2}}{\sqrt{\varepsilon_1}+\sqrt{\varepsilon_2}}$ (c) $\dfrac{\sqrt{\varepsilon_1}-\sqrt{\varepsilon_2}}{\sqrt{\varepsilon_1}+\sqrt{\varepsilon_2}}$ (d) $\dfrac{\sqrt{\varepsilon_2}-\sqrt{\varepsilon_1}}{\sqrt{\varepsilon_1}+\sqrt{\varepsilon_2}}$

12. In the case of reflection by a perfect conductor with normal incidence, the ratio of the reflected electric field to the incident electric field is given by

    (a) 1 (b) −1 (c) 0 (d) none

13. The angle of incidence is equal to the angle of reflection. This is known as

    (a) the law of reflection (b) Snell' law (c) both (a) and (b) (d) none

14. Surface impedance is defined as

    (a) $\dfrac{E_{\tan}}{J_s}$ (b) $E_{\tan}J_s$ (c) $\dfrac{J_s}{E_{\tan}}$ (d) none

15. The ratio of the sines of the angles of incidence and refraction is equivalent to the reciprocal of the ratio of the refractive indices. This is known as

    (a) the law of reflection (b) Snell' law (c) both (a) and (b) (d) none

16. The angle of incidence at which no reflection takes place is called

    (a) Brewster angle (b) critical angle (c) refracted angle (D) reflected angle

17. A parallel polarized wave propagates from a dielectric medium 1, $\varepsilon_{r_1}=9$ to the interface with dielectric medium 2, $\varepsilon_{r_2}=4$. The Brewster angle is

    (a) 50° (b) 66.5° (c) 33.7° (d) 23.96°

18. For parallel polarization, the sum of the Brewster angle and the angle of refraction is

    (a) 50° (b) 45° (c) 90° (d) zero

19. The Poynting vector (watts/m^2) of an electromagnetic wave is defined as

    (a) $\vec{E}\cdot\vec{H}$ (b) $\vec{E}\times\vec{H}$ (c) $\vec{E}/\vec{H}$ (d) $\varepsilon\mu\vec{E}\cdot\vec{H}$

20. The surface resistance of a copper conductor at 100 MHz with $\sigma = 5.8 \times 10^7$ S/m, and $\delta = 6.6 \times 10^{-8}$ m is
    (a) 0.26 ohms (b) 3.83 ohms (c) 38.28 ohms (d) zero

21. The complex Poynting vector (watts/m^2) is defined as
    (a) $\vec{E} \times \vec{H}$ (b) $\vec{E}^* \times \vec{H}$ (c) $\vec{E} \times \vec{H}^*$ (d) $\frac{1}{2}\left(\vec{E} \times \vec{H}^*\right)$

22. The direction of energy flow in an electromagnetic wave is
    (a) parallel to the E and H planes (b) perpendicular to the E and H planes
    (c) perpendicular to the E plane only (d) parallel to the H plane only

23. A wave propagates from a dielectric, $\varepsilon_{r_1} = 78$, to free space. The critical angle is
    (a) 83.5° (b) 6.5° (c) 0.73° (d) 45°

24. When a laser beam is launched into an optical fibre, in order to minimise the reflection losses, the incidence angle at the fibre surface is kept at
    (a) Brewster angle (b) critical angle (c) either (a) or (b) (d) none

25. A plane wave has an average Poynting vector of 10 W/m^2 in free space. The rms value of the electric field (V/m) is
    (a) 2.5 (b) 5 (c) 61.4 (d) 30.7

Answers

1. *a*	2. *b*	3. *b*	4. *c*	5. *b*	6. *a*	7. *a*	8. *b*	9. *d*	10. *c*
11. *d*	12. *b*	13. *a*	14. *a*	15. *b*	16. *a*	17. *c*	18. *c*	19. *b*	20. *a*
21. *d*	22. *b*	23. *b*	24. *a*	25. *c*					

Transmission Lines - I | 9

Learning Objectives

The main objectives this chapter are to understand thoroughly the following topics:

- Types and structures of transmission lines
- Basic concepts and applications of transmission lines
- Transmission line parameters, line equations
- Characteristic impedance, propagation constant, attenuation and phase shift constants
- Phase and group velocities
- Lossless and distortionless lines, line distortions, loading of lines etc.

9.1 Introduction

In the previous chapters, the propagation of waves through free space or materials was studied. But the generation and transmission of electromagnetic radiation at low frequencies require directive antennas of huge dimensions and expensive systems. Since the wave is not guided, a lot of power is wasted in long distance communication. For efficient energy transmission, waves should be directed or guided.

The conductors or waveguides used for transmitting electric or electromagnetic waves over long distances between the transmitter and the receiver are called transmission lines. For example, open wire lines, coaxial cables, optical fiber cables and waveguides are used to transmit signals between the transmitter and the receiver.

Ordinary electrical cables are used for carrying low frequency electrical power. However, as frequency increases, energy tends to radiate off the cable, causing power losses. In practice, waves are propagated through transmission lines that are specially designed to carry electromagnetic signals with minimal reflections and power losses at all frequencies. Hollow or dielectric slab guided waves and waveguides support waves that have frequencies above the cutoff frequency. Therefore, for transmission of electromagnetic waves at all frequencies, two-conductor transmission lines such as open wire lines or coaxial cables are used.

Generally, a transmission line is a distributed parameter network, where the length of the line is much longer than the transmitted signal wavelength. The transmission line parameters such as resistance, capacitance, inductance and conductance are uniformly distributed along the transmission line. The distributed elements are measured per unit length. Ohm's law and Kirchhoff's laws are not applicable for distributed elements. When the wave propagates along the line, the amplitudes of voltages and currents decrease exponentially depending on the attenuation constant. Also, as wave propagates, the phase of the wave changes depending on the phase shift constant. At microwave frequencies, transmission lines can be used as circuit elements such as inductors, capacitors, resonators etc. They can also be used as impedance matching devices.

In this chapter, we will study the basic concepts of two conductor transmission lines—structures, types, line parameters, distortionless lines, line distortions, loading of lines etc. We will also discuss the wave equations and characteristics of TEM waves guided by the transmission lines.

9.2 Types of Transmission Lines

There are various types transmission lines available depending on the propagation constant, distance, power transmission etc.

The important types are explained below (Fig. 9.1).

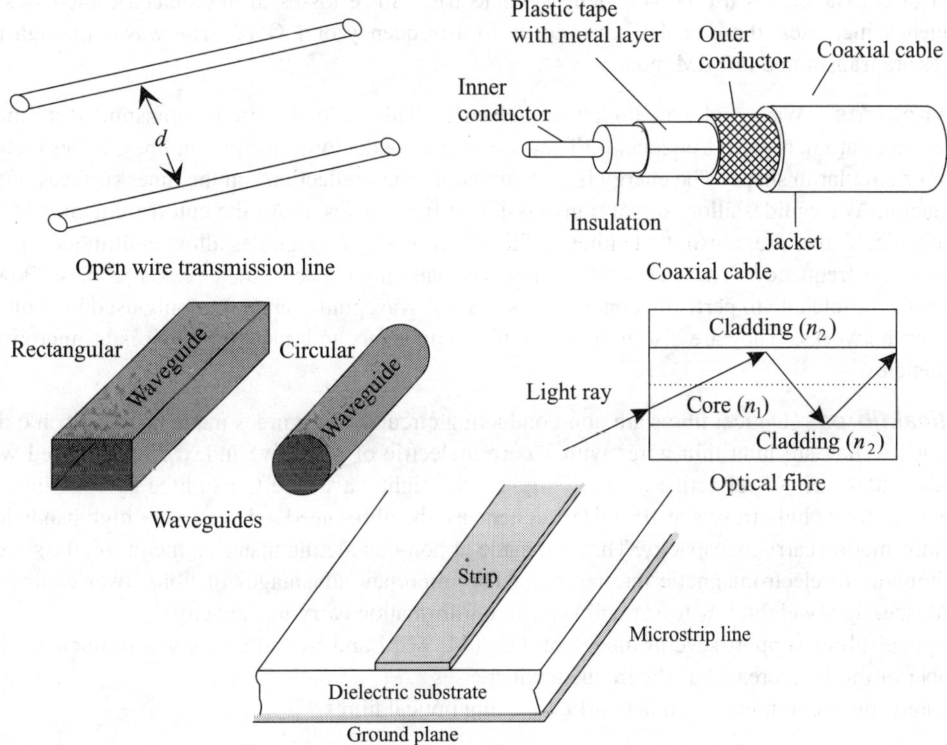

Fig. 9.1 *Types of transmission lines*

Open wire lines (parallel wire lines) An open wire line is a pair parallel conducting wires separated by a distance in free space and mounted on posts or on towers. Waves are transmitted in TEM mode. Open wire lines can be used for telephone lines, telegraphy lines and power lines.

Open wire lines are simple, low cost and easy to maintain for short distances. They have less capacitance effect compared to other types and are best suited for low frequency short and medium distance transmission. As distance increases, the cost becomes high as more number of posts and towers have to be erected. Open wire lines are badly affected by atmospheric conditions.

Short length lines at high frequency can be used as antenna feeders or matching lines. Open wires are used up to 100 MHz. Energy loss due to radiation is severe beyond 100 MHz.

Cables A cable consists of a number of twisted insulated wires in pairs protected by a lead or plastic sheath. Cables are generally laid underground or undersea, for example, telephone cables, electrical power transmission cables etc.

Coaxial cables When two conductors are placed coaxially and filled with dielectric material, the cable is called a coaxial cable. Coaxial cables can be used as TV cables and telephone cables. The dielectric inside the cables may be air, gas or solid. Coaxial cables have less attenuation, low interface and negligible radiation compared to open wire lines. It provides EM shielding against external fields as the EM fields are confined within the cable. Two strand pairs and coaxial cables are used in computer networks.

Most coaxial cables use polyethylene as dielectric. Since losses in the dielectric increases as frequency increases, these cables are used up to a frequency of 1 GHz. The waves through the cables are transmitted in TEM mode.

Waveguides Waveguides are hollow or dielectric filled conductors used to transmit electromagnetic waves at microwave frequencies. They are single conductor transmission lines, either rectangular or circular in shape. The energy is transmitted through reflections on the inner surfaces of the conductor. Waveguides allow signal transmission at frequencies above the cutoff frequency of the waveguide. Waves are transmitted either in TE or TM mode. Waveguides allow multimode signals at the same frequency. They have a higher power handling capacity than coaxial cables. Power loss is negligible due to perfectly conducting surfaces. Waveguides are commonly used in communication networks. They are also used as circuit components and matching devices at microwave frequencies.

Optical fibres Optical fibres are non-conducting circular waveguides made up of silicon oxide or silica. They are long thin wires with a core dielectric of refractive index n_1 and coated with cladding dielectric of refractive index n_2 ($n_2 < n_1$). Light waves are transmitted by total internal reflection. Since light travels at 10^5 GHz frequencies, the fibres need to have a very high bandwidth and information carrying capacity. They are made of non-conducting material, therefore, the signals are immune to electromagnetic interference. The important advantages of fibres over cables are small size, light weight, low losses and very high information carrying capacity.

Optical fibres support several modes of TE, TM, TEM and hybrid at a given frequency. The number of modes increases as the frequency increases.

Lately, most communication networks are using optical fibres.

Microstrip lines Microstrip line is a parallel plate transmission line consisting of a conducting strip and a ground plane separated by a dielectric substrate. A very thin copper sheet is used as a strip and the ground plane is fabricated using printed circuit board technology. The fields are confined in the dielectric substrate between the strip and the ground plane. Microstrip lines are lighter, more compact and flexible than other types of transmission lines, but they have low power handling capacity. They are used to transmit micowave signals and can be used as microwave components such as antennas, direction couplers, filters, circulators, etc. For example, high speed digital circuits and antennas in smart phones are designed by using microstrip line technology.

9.2.1 Transmission Line Parameters

The equivalent electric circuit of a transmission line consists of series resistance, series inductance, shunt capacitance and shunt conductance along the length. These parameters are calculated per unit length of the transmission line.

Resistance (R) A series resistance is due to the intemal resistance of the conductors of a transmission line. It is uniformly distributed along the line and depends on the conductivity and cross-sectional area of the conductors. But at high frequencies, it depends on skin depth. It is measured as loop resistance per unit length of the line. Its units are ohms/m or Ω/m.

Inductance (L) A series or loop inductance is due to the magnetic flux density produced around the conductors of a transmission line. It is uniformly distributed along the line. The flux linkages per unit current gives the inductance of the transmission line. It is measured as loop inductance per unit length of the transmission line and its units are henries/m or H/m.

Capacitance (C) Two parallel conductors or coaxial conductors of a transmission line separated by a distance d acts as a capacitor. Thus, a shunt capacitance is formed due to the electric field between the conductors. Capacitance is distributed uniformly along the line. It is measured as shunt capacitance per unit length of the transmission line. The units are farads/m or F/m.

Conductance (G) A shunt conductance is due to the leakage current between the conductors of a transmission line. Since the dielectric or insulation material around the conductors is not perfect, a small amount of current flows through the dielectric material. The conductance is also distributed uniformly along the transmission line. It is measured as shunt conductance per unit length of the transmitted line. Its units are mhos/m or S/m.

The electric equivalent circuit of the transmission line per unit length in shown in Fig.9.2.

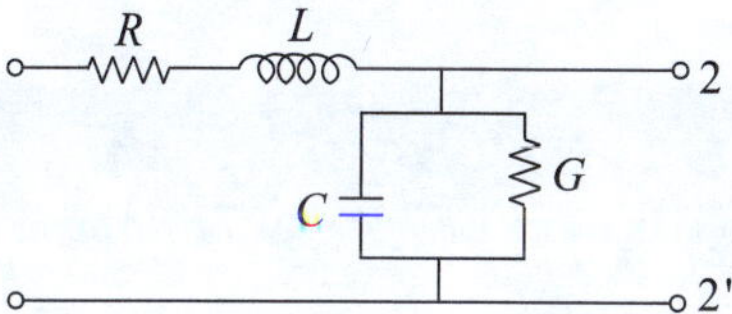

Fig. 9.2 *Equivalent circuit of a transmission line per unit length*

The parameters R, L, G and C are called primary constants of the transmission line. They are independent of the operating frequency. They are measured in either per metre or per kilometre length of the transmission line.

The series impedance Z and shunt admittance Y of the line per unit length can be expressed as

$$Z = R + j\omega L.$$

$$Y = G + j\omega C.$$

The primary constants for different types of transmission lines are discussed below.

Parallel open wire Consider a parallel open line as shown in Fig. 9.3(a). It has σ_c – conductivity of the conductors, f = frequency, a = radius of the conductor, d = separation between conductors ($d >> a$), μ_c = permeability of the conductors. The medium between the lines has σ = conductivity, μ =permeability and ε = permittivity.

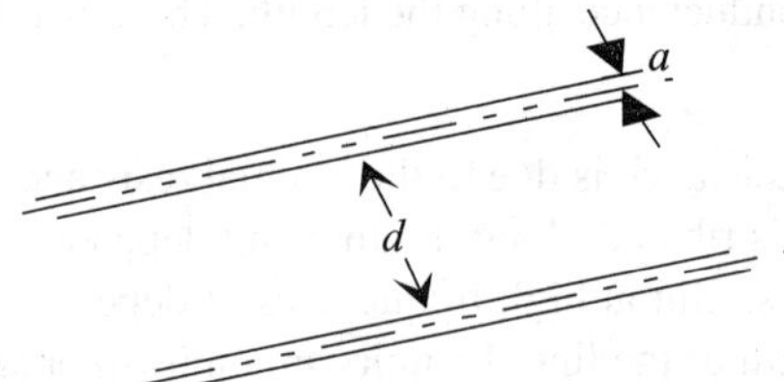

Fig. 9.3a *Open wire line*

Resistance :

$$\text{DC resistance } R_{dc} = \frac{2}{\sigma_c \pi a^2} \ \Omega/\text{m}. \tag{9.1}$$

$$\text{AC resistance } R_{ac} = \sqrt{\frac{f\mu_c}{\pi\sigma_c a^2}} = \sqrt{\frac{R_{dc} f \mu_c}{2}} \ \Omega/\text{m}. \tag{9.2}$$

Inductance :

$$L = \frac{\mu_c}{4\pi} + \frac{\mu}{\pi}\ln\left(\frac{d}{a}\right) \text{ H/m} \tag{9.3}$$

or $L = [\mu_r + 9.21\log(d/a)] \times 10^{-7}$ H/m. (since $\ln x = \frac{\log x}{\log e}$)

At high frequencies,

$$L = \frac{\mu_0}{\pi}\ln\left(\frac{d}{a}\right) \text{ H/m.}$$

Capacitance :

$$C = \frac{\pi\varepsilon}{\ln(d/a)} \text{ F/m} \tag{9.4}$$

or $C = \frac{12.07\ \varepsilon_r}{\ln(d/a)}$.

Conductance :

$$G = \frac{C\sigma}{\varepsilon} = \frac{\pi\sigma}{\ln(d/a)} \text{ S/m.}$$

Coaxial cables Consider a coaxial cable as shown in Fig. 9.3(b). It has a = radius of the inner solid conductor, b = inner surface radius of the outer conductor, c = outer surface radius of the outer conductor, σ_c = conductivity of the conductor, f = frequency, μ_c = permeability of the conductor. The medium between the conductors has σ = conductivity, μ = permeability and ε = permittivity.

Resistance :

$$R_{dc} = \frac{1}{\pi\sigma_c}\left\{\frac{1}{a^2} + \frac{1}{c^2 - b^2}\right\} \ \Omega/\text{m}. \tag{9.6}$$

$$R_{ac} = \sqrt{\frac{f\mu_c}{4\pi\sigma_c}}\left(\frac{1}{a} + \frac{1}{b}\right) \ \Omega/\text{m}. \tag{9.7}$$

Inductance :

$$L = \frac{\mu}{2\pi}\ln(b/a) \ \text{H/m}. \tag{9.8}$$

Capacitance :

$$C = \frac{2\pi\varepsilon}{\ln(b/a)} \ \text{F/m}. \tag{9.9}$$

Conductance :

$$G = \frac{C\sigma}{\varepsilon} = \frac{2\pi\sigma}{\ln(b/a)} \ \text{S/m}. \tag{9.10}$$

Note: At very high frequencies, the line parameters can be related to the medium characteristics as

$$LC = \mu\varepsilon \quad \text{and} \quad \sigma C = G\varepsilon. \tag{9.11}$$

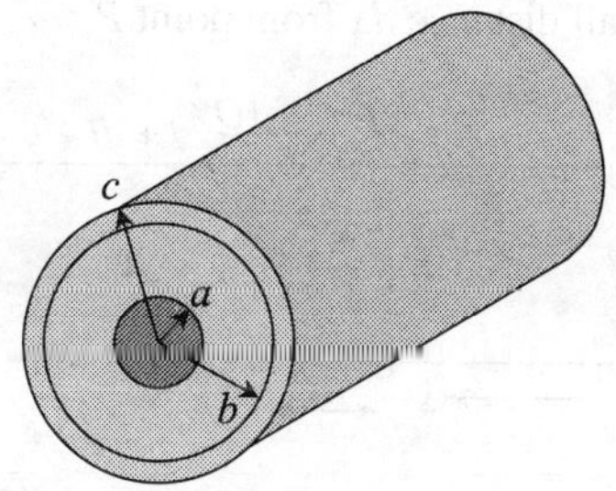

Fig. 9.3(b) *Coaxial cable*

Example 9.1 A parallel open line made of aluminium has conductivity 3.6×10^7 S/m and is kept in air. The conductor radius is 1 mm, separation between the conductors is 0.15 m, operating frequency is 1 kHz. Find the line parameters R, L, C, G.

Solution Given:
For aluminium: Conductivity $\sigma_c = 3.6\times10^7$ S/m, $\mu_c = \mu_0$, frequency $f = 1$ kHz, radius $a = 1$mm, separation between conductors $d = 0.15$ m.
For air: Conductivity $\sigma = 0$, $\varepsilon = \varepsilon_0$, $\mu = \mu_0$.
For an open wire, the line parameters are given below.

Resistance :

$$R = \frac{1}{a}\sqrt{\frac{f\mu_c}{\pi\sigma_c}} \quad \Omega/\text{m} = \frac{1}{0.001}\sqrt{\frac{1000\,\mu_0}{\pi \times 3.6 \times 10^7}} = 3.33 \quad \text{m}\Omega/\text{m}.$$

Inductance :

$$L = \frac{\mu_0}{\pi}\ln\left(\frac{d}{a}\right) \quad \text{H/m} = \frac{\mu_0}{\pi}\ln\left(\frac{0.15}{0.001}\right) = 2 \quad \mu\text{H/m}.$$

Capacitance :

$$C = \frac{\pi\varepsilon}{\ln(d/a)} \quad \text{F/m} = \frac{\pi\varepsilon_0}{\ln(150)} = 5.55 \quad \text{pF/m}.$$

Conductance :

$$G = \frac{C\sigma}{\varepsilon} = \frac{\pi\sigma}{\ln(d/a)} = 0 \quad \text{S/m}.$$

9.3 Transmission Line Equations

Consider a transmission line with two parallel conductors. Let R, L, C and G be the primary constants. Assume that these values do not vary with frequency. Consider a point P on the line at a distance x from the source as shown in Fig.9.4.

Let another point Q be at a small distance dx from point P.

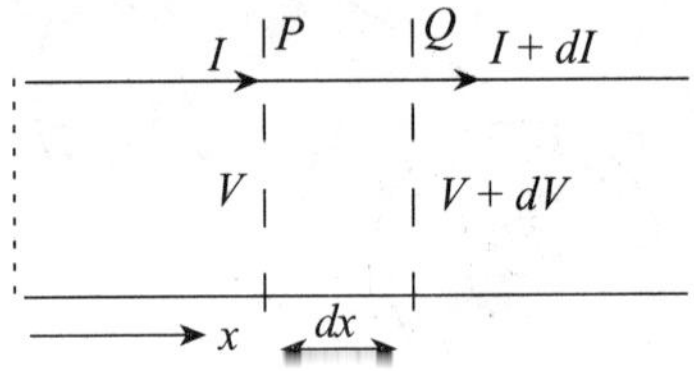

Fig. 9.4 *Voltages and currents on a transmission line*

Let V and I be the voltage and current respectively at point P. Since the voltage and currents are uniformly distributed along the line, at point Q, let the voltage be $V+dV$ and the current be $I+dI$.

For a small length dx of the line, the series impedance is $(R+j\omega L)\,dx$ and the shunt admittance is $(G+j\omega C)\,dx$.

The potential difference between P and Q is

$$V-(V+dV) = I(R+j\omega L)\,dx \tag{9.13}$$

and the current difference between P and Q is

$$I-(I+dI) = V(G+j\omega C)\,dx. \tag{9.14}$$

$$\therefore \quad \frac{-dV}{dx} = (R+j\omega L)I \tag{9.15}$$

and $$\frac{-dI}{dx} = (G + j\omega C)V. \tag{9.16}$$

Taking derivatives of Eq. (9.15),

$$\frac{-d^2V}{dx^2} = (R + j\omega L)\frac{dI}{dx}.$$

Substituting Eq. (9.16),

$$\frac{d^2V}{dx^2} = (R + j\omega L)(G + j\omega C)V \tag{9.17}$$

and similarly, $$\frac{d^2I}{dx^2} = (R + j\omega L)(G + j\omega C)I. \tag{9.18}$$

These equations are second order differential equations either in voltage or current.

Let the constant term at a given frequency be

$$(R + j\omega L)(G + j\omega C) = \gamma^2$$

or $$\gamma = \alpha + j\beta = \sqrt{(R + j\omega L)(G + j\omega C)} = \sqrt{ZY}, \tag{9.19}$$

where α = attenuation constant, β = phase shift constant and γ = propagation constant of the transmission line.

$$\therefore \quad \frac{d^2V}{dx^2} = \gamma^2 V \tag{9.20}$$

and $$\frac{d^2I}{dx^2} = \gamma^2 I. \tag{9.21}$$

Eq. (9.20) and Eq. (9.21) are called differential equations of the transmission lines. The standard solutions for these equations are

$$V = ae^{\gamma x} + be^{-\gamma x} \tag{9.22}$$

and $$I = ce^{\gamma x} + de^{-\gamma x}, \tag{9.23}$$

where a, b, c and d are the constants.

Eq. (9.22) and Eq. (9.23) represent a standing wave pattern. To find the relationship between V and I, substitute Eq. (9.22) in Eq. (9.15).

We get,

$$\frac{-d}{dx}(ae^{\gamma x} + be^{-\gamma x}) = (R + j\omega L)I.$$

$$-\gamma(ae^{\gamma x} - be^{-\gamma x}) = (R + j\omega L)I$$

or $$I = \frac{(be^{-\gamma x} - ae^{\gamma x})\sqrt{(R + j\omega L)(G + j\omega C)}}{(R + j\omega L)}.$$

$$\therefore \quad I = \sqrt{\frac{G + j\omega C}{R + j\omega L}}(be^{-\gamma x} - ae^{\gamma x}). \tag{9.24}$$

Let another constant at a given frequency, called characteristic impedance, be defined as

$$Z_0 = \sqrt{\frac{R + j\omega L}{G + j\omega C}} = \sqrt{\frac{Z}{Y}}. \tag{9.25}$$

$$\therefore \quad I = \frac{1}{Z_0}(be^{-\gamma x} - ae^{\gamma x}). \tag{9.26}$$

Eq. (9.22) and Eq. (9.23) can be written as

$$V = ae^{\gamma x} + be^{-\gamma x} \tag{9.27}$$

and
$$I = \frac{1}{Z_0}(be^{-\gamma x} - ae^{\gamma x}). \tag{9.28}$$

In terms of hyperbolic functions:

Substituting $e^{\gamma x} = \cosh \gamma x + \sinh \gamma x$ and $e^{-\gamma x} = \cosh \gamma x - \sinh \gamma x$ in Eq. (9.27) and Eq. (9.28), we get

$$V = A\cosh \gamma x + B\sinh \gamma x \tag{9.29}$$

and
$$I = \frac{-1}{Z_0}(A\sinh \gamma x + B\cosh \gamma x), \tag{9.30}$$

where $A = a + b$ and $B = a - b$.

The constants A and B can be obtained by applying initial conditions. Let V_s and I_s be the source voltage and source current respectively. At the sending end, $x = 0$, the voltage is $V = V_s$ and the current is $I = I_s$.

Then $V_s = A\cosh \gamma(0) + B\sinh \gamma(0)$.

$$\therefore \quad A = V_s \text{ and } I_s = \frac{-1}{Z_0}(A\sinh \gamma(0) + B\cosh \gamma(0)).$$

$$B = -Z_0 I_S.$$

Substituting the constants A and B in Eq. (9.79) and Eq. (9.30),

$$V = V_s \cosh \gamma x - I_s Z_0 \sinh \gamma x \tag{9.31}$$

and
$$I = I_s \cosh \gamma x - \frac{V_S}{Z_0}\sinh \gamma x. \tag{9.32}$$

These equations are called transmission line equations. They give voltage and current at a point of distance x from the sending end in terms of source voltage and current.

Note: γ and Z_0 are called secondary constants of a transmission line.

9.4 Infinite Length Line

Consider a transmission line having infinite length (very long), as shown in Fig. 9.5.

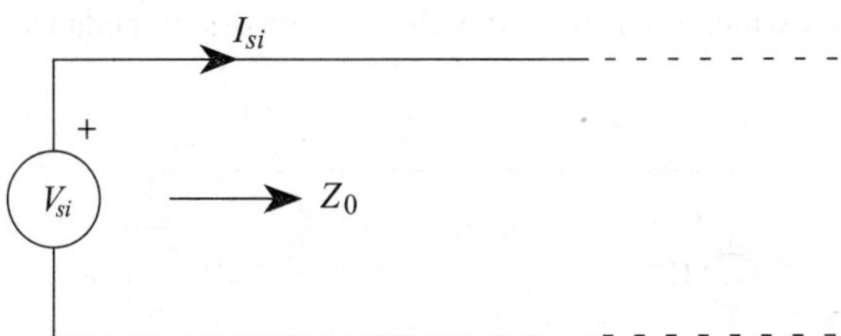

Fig. 9.5 *Infinite length line*

When a voltage source is applied to the line, a current will flow depending on the primary constants. This current will not reach the other end due to infinite attenuation.

The ratio of the voltage applied to the current flowing through the infinite line is called characteristic impedance of the line and it is denoted by Z_0.

$$\therefore \quad Z_0 = \frac{V_{si}}{I_{si}}, \tag{9.33}$$

where V_{si} and I_{si} are source voltage and source current respectively of an infinite line.

Since the wave travels only in one direction and no standing waves exist, if γ is the propagation constant, the voltage and current equations of the line are

$$V = V_{si}e^{-\gamma x}, \tag{9.34}$$

$$I = I_{si}e^{-\gamma x}, \tag{9.35}$$

where V and I are the voltage and current at a distance x from the source.

So, $e^{\gamma x} = \frac{V_{si}}{V} = \frac{I_{si}}{I}$.

The propagation constant is

$$\gamma = \frac{1}{x}\ln\left(\frac{V_{si}}{V}\right) = \frac{1}{x}\ln\left(\frac{I_{si}}{I}\right).$$

If l is the length of the line, the propagation constant at the receiving end is

$$\gamma = \frac{1}{l}\ln\left(\frac{V_{si}}{V_R}\right) \times l$$

$$\text{or} \quad \gamma = \ln\left(\frac{V_{si}}{V_R}\right) \text{ nepers (or Np).} \tag{9.36}$$

In decibels,

$$\gamma = 20\log_{10}\left(\frac{V_{si}}{V_R}\right) = 20\log_{10}\left(\frac{I_{si}}{I_R}\right) \; dB. \tag{9.37}$$

9.5 Transmission Line Terminated with Characteristic Impedance Z_0

Consider a line of length l which is terminated with characteristic impedance Z_0 as shown in Fig. 9.6.

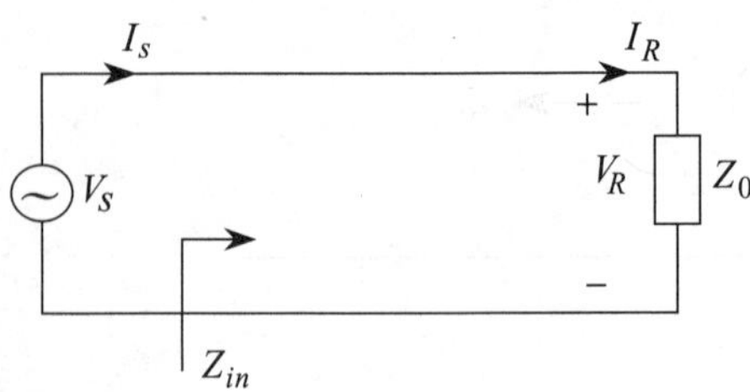

Fig. 9.6 *Transmission line with Z_0*

Let V_R = Voltage at termination,
I_R = Current at termination,
V_S = Source voltage and
I_S = Source current.

We know from Eq. (9.31) and Eq. (9.32) that the line equations are

$$V = V_S \cosh \gamma x - I_S Z_0 \sinh \gamma x \quad \text{and}$$

$$I = I_S \cosh \gamma x - \frac{V_S}{Z_0} \sinh \gamma x.$$

At distance $x = l$,

$$V = V_R \ \text{and} \ I = I_R.$$

$$V_R = V_S \cosh \gamma l - I_S Z_0 \sinh \gamma l.$$

$$I_R = I_S \cosh \gamma l - \frac{V_S}{Z_0} \sinh \gamma l.$$

From Fig. 9.6, we know that $Z_0 = \dfrac{V_R}{I_R}$

$$Z_0 = \frac{V_S \cosh \gamma l - I_S Z_0 \sinh \gamma l}{I_S \cosh \gamma l - \dfrac{V_S}{Z_0} \sinh \gamma l}$$

$$Z_0 I_S \cosh \gamma l - V_S \sinh \gamma l = V_S \cosh \gamma l - I_S Z_0 \sinh \gamma l$$

$$Z_0 I_S (\cosh \gamma l + \sinh \gamma l) = V_S (\cosh \gamma l + \sinh \gamma l)$$

$$Z_0 I_S = V_S$$

$$Z_0 = \frac{V_S}{I_S}.$$

But $\frac{V_S}{I_S} = Z_{in}$ is the input impedance of the transmission line.

$$\therefore \quad Z_{in} = Z_0. \tag{9.38}$$

Hence, the input impedance Z_{in} of a finite length line terminated with its characteristic impedance Z_0 is equal to Z_0. Thus a line is matched when the termination impedance is equal to its characteristic impedance Z_0 (not the complex conjugate of the characteristic impedance as in circuit theory).

That is, $\quad Z_R = Z_0 = Z_{in} \ \Omega.$

For an open wire line
Characteristic impedance for an open wire system is

$$Z_0 = 276 \log_{10}\left(\frac{d}{r}\right) \ \Omega, \tag{9.39}$$

where $\quad r =$ radius of the wire

$\quad d =$ separation between two wires

For a coaxial cable

$$Z_0 = \frac{138}{\sqrt{\varepsilon_r}} \log_{10}(b/a) \ \Omega, \tag{9.40}$$

where $\quad b =$ radius of the outer conductor,

$\quad a =$ radius of the inner conductor and

$\quad \varepsilon_r =$ dielectric constant.

Transmission line with source generator: Consider a line length l terminated with characteristic impedance Z_0 and fed with a generator having source voltage V_g and internal impedance Z_g. If Z_s is the input impedance line, then $Z_s = Z_0$.

The input current $I_s = \frac{V_g}{Z_g + Z_s}$, input voltage $V_s = I_s Z_s$ and average input power $P_s = |I_s||V_s|\cos\theta$.

Load voltage $V_L = V_s e^{-\gamma l}$, load current $I_L = I_s e^{-\gamma l}$ and

average load power $P_L = P_s e^{-2\alpha l} = |I_L||V_L|\cos\theta$, where θ is the angle between voltage and current.

Example 9.2 A coaxial line has an outer diameter of 8 mm and an inner diameter of 2 mm. The dielectric constant of the insulation is 2.65. Similarly, an open wire parallel line has conductor radius 2 mm and separation 8 mm. Calculate the characteristic impedance for both cases.

Solution Given: A coaxial line, outer diameter $b = 8$ mm, inner diameter $a = 2$ mm and $\varepsilon_r = 2.65$.

We know that the characteristic impedance for a coaxial cable is

$$Z_0 = \frac{138}{\sqrt{\varepsilon_r}} \log_{10}(b/a) = \frac{138}{\sqrt{2.65}} \log(8/2) = 51 \text{ ohms}.$$

Given: An open wire parallel line, separation $d = 8$ mm, radius $r = 2$ mm.

We know that the characteristic impedance for an open wire parallel line is

$$Z_0 = 276 \log_{10}(d/r) = 276 \log_{10}(8/2) = 166 \text{ ohms}.$$

9.6 Derivation of Attenuation and Phase Shift Constants

We know that the propagation constant

$$\gamma = \alpha + j\beta = \sqrt{(R + j\omega L)(G + j\omega C)}$$

$$(\alpha + j\beta)^2 = RG - \omega^2 LC + j(\omega LC + R\omega C)$$

$$\alpha^2 - \beta^2 + j2\alpha\beta = RG - \omega^2 LC + j(\omega LC + R\omega C).$$

Equating the real terms,

$$\alpha^2 - \beta^2 = RG - \omega^2 LC. \tag{9.41}$$

We know that the square magnitude of γ is

$$|\gamma|^2 = |\alpha + j\beta|^2 = \alpha^2 + \beta^2 = \sqrt{(R^2 + \omega^2 L^2)(G^2 + \omega^2 C^2)}. \tag{9.42}$$

Solving Eq. (9.41) and Eq. (9.42),

$$2\alpha^2 = (RG - \omega^2 LC) + \sqrt{(R^2 + \omega^2 L^2)(G^2 + \omega^2 C^2)}.$$

Then, $$\alpha = \sqrt{\frac{1}{2}\left[(RG - \omega^2 LC) + \sqrt{(R^2 + \omega^2 L^2)(G^2 + \omega^2 C^2)}\right]} \tag{9.43}$$

and $$\beta = \sqrt{\frac{1}{2}\left[(RG - \omega^2 LC) - \sqrt{(R^2 + \omega^2 L^2)(G^2 + \omega^2 C^2)}\right]}. \tag{9.44}$$

Attenuation constant

The real part of the propagation constant γ is called attenuation constant α.

If the line is terminated with its characteristic impedance and l is the length in kilometres of the line, the attenuation constant at the receiving end is

$$\alpha = \ln\left(\frac{V_s}{V_R}\right) \text{ nepers}$$

$$\text{or} \quad \alpha = \frac{1}{l} \times \ln\left(\frac{V_s}{V_R}\right), \quad \alpha = \frac{1}{l} \times \ln\left(\frac{I_s}{I_R}\right), \quad \alpha = \frac{1}{2l} \times \ln\left(\frac{P_s}{P_R}\right) \tag{9.45}$$

measured in nepers/km.

In decibels,

$$\alpha = \frac{1}{l} \times 20 \log_{10}\left(\frac{V_s}{V_R}\right), \quad \alpha = \frac{1}{l} \times 20 \log_{10}\left(\frac{I_s}{I_R}\right), \quad \alpha = \frac{1}{l} \times 10 \log_{10}\left(\frac{P_s}{P_R}\right) \tag{9.46}$$

measured in dB/km, where V_s, I_s and P_s are the voltage, current and power at source respectively and V_R, I_R and P_R are the voltage, current and power at load respectively.

The unit conversion is

$$1 \text{ neper} = 20 \log e = 8.686 \text{ dB} \quad \text{or} \quad 1\text{dB} = 0.15 \text{ nepers}.$$

Phase shift constant

The imaginary part of the propagation constant γ is called phase shift constant β. It is measured is radians/km. β can also be expressed in degrees/km.

$$\pi \text{ radians} = 180 \text{ degrees, or } 1 \text{ radian} = 57.3 \text{ degrees.}$$

Example 9.3 The characteristic impedance of a certain line is $850\angle{-32^\circ}$ when the frequency is 2 kHz. At this frequency, the attenuation is 0.05 nepers per km and phase function is 0.076 radians per km. Calculate the resistance, the leakage, the inductance, and the capacitance per km and velocity of propagation.

Solution Given: $Z_0 = 850\angle{-32^\circ}$ ohms, $\alpha = 0.05$ Np/km, $\beta = 0.076$ radians/km, $f = 2000$ Hz.

We know that $\gamma = \alpha + j\beta$.

$$\gamma = 0.05 + j0.076 = 0.09\angle 56.65^\circ.$$

$$\omega = 2\pi f = 4000\pi \text{ radians/s.}$$

Also, $R + j\omega L = Z_0 \times \gamma$

$$= 850\angle{-32^\circ} \times 0.09\angle 56.65^\circ$$

$$= 76.5\angle(-32 + 56.65) = 76.5\angle 24.65.$$

$$R + j\omega L = 69.5 + j31.9.$$

Equating, the real and imaginary parts,

$$R = 69.5 \ \Omega/\text{km}.$$

$$\omega L = 31.9.$$

$$L = \frac{31.9}{4000\pi} = 2.53 \times 10^{-3} \quad \text{or} \quad L = 2.53 \ \text{mH/km}.$$

Similarly, $G + j\omega L = \dfrac{\gamma}{Z_0} = \dfrac{0.09\angle 56.65^\circ}{850\angle -32}$

$$= 1.058 \times 10^{-4}\angle(56.65 + 32)$$

$$= 1.058 \times 10^{-4}\angle 88.65$$

$$= (0.0249 + j1.058) \times 10^{-4}.$$

$$G = 2.49 \times 10^{-6} \quad \text{S/km}.$$

$$\omega C = 1.058 \times 10^{-4}.$$

$$C = \frac{1.058 \times 10^{-4}}{4000\pi} = 8.42 \times 10^{-9} = 8.42 \quad \text{nF/km}.$$

Example 9.4 An open wire telephone line has $R = 10\Omega$ per km. $L = 0.0037$ H per km. $C = 0.0083 \times 10^{-6}$ farads per km and $G = 0.4 \times 10^{-6}$ mhos per km. Determine Z_0, α and β at 1000 Hz.

Solution Given: $R = 10\ \Omega$/km, $L = 0.0037$ H per km, $C = 0.0083 \times 10^{-6}$ F/km, $G = 0.4 \times 10^{-6}$ S/km, $f = 1000$ Hz, $\omega = 2\pi f = 2\pi \times 1000 = 2000\pi$ radians/s.

Now, series impedance, $Z = R + j\omega L$

$$= 10 + j2000\pi \times 0.0037 = 10 + j23.25 = 25.3\angle 66.73^\circ$$

and shunt admittance $Y = G + j\omega C = 0.4 \times 10^{-6} + j2000\pi \times 0.0083 \times 10^{-6}$

$$= (0.4 + j52.15)10^{-6} = 52.15 \times 10^{-6}\angle 89.56^\circ.$$

We know that the propagation constant

$$\gamma = \sqrt{Z \times Y}$$

$$= \sqrt{25.3\angle 66.73 \times 52.15 \times 10^{-6}\angle 89.56} = \sqrt{25.3 \times 52.1 \times 10^{-6}}\angle(66.73 + 89.56)/2.$$

$$\gamma = 0.0363\angle 78.14^\circ.$$

$$\gamma = \alpha + j\beta = (0.00746 + j0.0355) \text{ per km}.$$

Attenuation constant $\alpha = 0.00746$ Np/km.
Phase constant $\beta = 0.0355$ radians/km.

9.7 Velocity of Propagation and Group Velocity

The velocity at which a wave of single frequency propagates along a line is called phase velocity or wave velocity. As the wave propagates along the line, phase changes continuously.
The phase velocity can be expressed as

$$v_p = \lambda \times f, \tag{9.47}$$

where λ = wavelength in metres, f = frequency in Hz.

We know that $\lambda = \dfrac{2\pi}{\beta}$. (9.48)

$$\therefore \quad v_p = \frac{2\pi}{\beta} \times f$$

or $v_p = \dfrac{\omega}{\beta}$ m/s. (9.49)

In free space, phase velocity is approximately equal to the velocity of light.

$$\therefore \quad v_p = \frac{1}{\sqrt{\mu_0\,\varepsilon_0}} = 3 \times 10^8 \text{ m/s.} \tag{9.50}$$

In a medium with dielectric constant $\varepsilon = \varepsilon_0\,\varepsilon_r$, the phase velocity reduces to

$$v_p = \frac{1}{\sqrt{\mu\varepsilon}} = \frac{1}{\sqrt{\mu_0\,\varepsilon_0\,\varepsilon_r}} = \frac{3 \times 10^8}{\sqrt{\varepsilon_r}} \text{ m/s.} \tag{9.51}$$

Group velocity The velocity at which the envelope of the complex wave propagates along a line is called group velocity. It is defined as *the ratio of the change in angular frequency to the change in phase constant of the wave*.

At a given frequency, the group velocity is a derivative of angular frequency with respect to the phase constant.

$$\therefore \quad v_g = \frac{\omega_2 - \omega_1}{\beta_2 - \beta_1} = \frac{d\omega}{d\beta} \text{ m/s.} \tag{9.52}$$

9.8 Relationship between Group Velocity and Phase Velocity

We know that when a wave travels along a line, the phase velocity is

$$v_p = \frac{\omega}{\beta}.$$

Differentiating with respect to frequency,

$$\frac{dv_p}{d\omega} = \frac{\beta - \omega\frac{d\beta}{d\omega}}{\beta^2} = \frac{1}{\beta} - \frac{\omega}{\beta^2}\frac{d\beta}{d\omega}.$$

$$\frac{dv_p}{d\omega} = \frac{1}{\beta} - \frac{v_p}{\beta}\frac{d\beta}{d\omega}. \tag{9.53}$$

But we know that the group velocity is $v_g = \dfrac{d\omega}{d\beta}$.

$$\therefore \quad \beta\frac{dv_p}{d\omega} = 1 - \frac{v_p}{v_g} \tag{9.54}$$

or $$v_p = v_g\left(1 - \beta\frac{dv_p}{d\omega}\right). \tag{9.55}$$

If v_p is constant with respect to ω, then

$$\frac{dv_p}{d\omega} = 0.$$

From Eq. (9.55), $v_p = v_g$.

Thus if the phase velocity is independent of frequency, it becomes group velocity.

Example 9.5 For a 20 km long transmission line terminated with its characteristic impedence, the generator at the sending end has 10 V and a frequency of 1000 Hz. At the receiving end, the voltage is dropped to 4 V and the phase is changed to 30°. Find the values of attenuation and phase velocity.

Solution Given: Length $l = 20$ km, source voltage, $V_S = 10$ V, frequency $f = 1000$ Hz. Voltage at termination, $V_R = 4$ V and change in phase $\beta l = 30°$.

We know that,

Attenuation $$\alpha = \frac{1}{l}20\log_{10}\left(\frac{V_S}{V_R}\right) \text{ dB/km} = \frac{1}{20}\times 20\log\left(\frac{10}{4}\right)$$

$$= 0.397 \text{ dB/km} = 0.0457 \text{ nepers/km}.$$

Phase shift $$\beta = \frac{30°}{20} = 1.5° = \frac{1.5\pi}{180} = 0.02618 \text{ radians/km}.$$

Velocity of propagation $$v_p = \frac{\omega}{\beta}, \quad v_p = \frac{2\pi\times 1000}{0.02618} = 240 \text{ km/s}.$$

9.9 Lossless Transmission Line

The line equations and characteristics of a transmission line can be easily solved by considering various practical applications. Lossless transmission lines, very long transmission lines, distortionless transmission lines, and high frequency transmission lines are commonly used in many applications in addition to low frequency lossy lines. By selecting suitable primary constants, *R*, *L*, *C*, *G* and frequency *f* of a transmission line, wave characteristics can be easily analysed.

For a lossless transmission line, $R = 0$, $G = 0$. The line is made of perfect conductors and perfect dielectrics.

Condition for lossless transmission

We know that

$$\gamma = \sqrt{(R + j\omega L)(G + j\omega C)}.$$

$$Z_0 = \sqrt{\frac{R + j\omega L}{G + j\omega C}}.$$

By substituting $R = G = 0$,

$$\alpha + j\beta = \sqrt{j\omega L \times j\omega C} = j\sqrt{LC}.$$

$$Z_0 = \sqrt{\frac{j\omega L}{j\omega C}} = \sqrt{\frac{L}{C}}.$$

Thus, $\alpha = 0$, $\beta = \omega\sqrt{LC}$ and $Z_0 = \sqrt{\frac{L}{C}}$. (9.56)

Phase velocity $v_p = \frac{\omega}{\beta} = \frac{1}{\sqrt{LC}}$. (9.57)

Since $v_p = \frac{1}{\sqrt{\mu\varepsilon}}$,

$$\therefore \quad LC = \mu\varepsilon.$$

Here phase velocity and characteristic impedance depend only on L and C values.

9.10 Distortionless Transmission Line

A line is said to be a distortionless transmission line if the attenuation constant α is independent of frequency ω and β varies linearly with frequency.

Condition for distortionless transmission

We know that

$$\gamma = \sqrt{(R + j\omega L)(G + j\omega C)} = \sqrt{L\left(\frac{R}{L} + j\omega\right)C\left(\frac{G}{C} + j\omega\right)}.$$

$$\alpha + j\beta = \sqrt{LC}\sqrt{\left(\frac{R}{L} + j\omega\right)\left(\frac{G}{C} + j\omega\right)}.$$

To make the α in the above equation independent of frequency, let $\frac{R}{L} = \frac{G}{C}$. (9.58)

Then $\alpha + j\beta = \sqrt{LC}\sqrt{\left(\frac{R}{L} + j\omega\right)\left(\frac{R}{L} + j\omega\right)} = \sqrt{LC}\left(\frac{R}{L} + j\omega\right)$.

Equating real and imaginary terms,

$$\alpha = \sqrt{LC}\frac{R}{L} = R\sqrt{\frac{C}{L}} \quad \text{and} \quad \beta = \omega\sqrt{LC}. \tag{9.59}$$

Also, $Z_0 = \sqrt{\frac{R + j\omega L}{G + j\omega L}} = \sqrt{\frac{L\left(\frac{R}{L} + j\omega\right)}{C\left(\frac{G}{C} + j\omega\right)}} = \sqrt{\frac{L}{C}}$. (9.60)

Therefore for distortionless transmission,

$$\alpha = R\sqrt{\frac{C}{L}} = G\sqrt{\frac{L}{C}} \tag{9.61}$$

or $\quad \alpha = \dfrac{R}{Z_0}$ or GZ_0

and $\quad \beta = \omega\sqrt{LC}, \quad v_p = \dfrac{1}{\sqrt{LC}}. \qquad (9.62)$

Therefore, the condition for distortionless transmission is

$$\frac{R}{L} = \frac{G}{C} \quad \text{or} \quad RC = LG. \tag{9.63}$$

In a distortionless line, the attenuation, characteristic impedance and phase velocity are independent of frequency. Phase shift constant varies linearly with frequency.

Note: For a lossless line, $\alpha = 0, \ \beta = \omega\sqrt{LC}, \ Z_0 = \sqrt{\dfrac{L}{C}}$, and $v_p = \dfrac{1}{\sqrt{LC}}$.

That is, the resistive components $R = G = 0$.

Hence, the lossless line is distortionless.

9.11 High Frequency Transmission Line

We know that the propagation constant of a transmission line is

$$\gamma = \sqrt{(R + j\omega L)(G + j\omega C)} = \sqrt{j\omega L\left(1 + \frac{R}{j\omega L}\right) j\omega C\left(\frac{G}{j\omega L} + 1\right)}$$

or $\quad \gamma = j\omega\sqrt{LC}\left(1 + \dfrac{R}{j\omega L}\right)^{1/2}\left(1 + \dfrac{G}{j\omega L}\right)^{1/2}. \qquad (9.64)$

If the frequency is very high, we can assume that $\omega L >> R$ and $\omega C >> G$. Equation (9.64) can be expanded using binomial theorem as

$$\gamma = j\omega\sqrt{LC}\left(1 + \frac{R}{j2\omega L} + \frac{1}{2!}\left(\frac{R}{j\omega L}\right)^2 + \ldots\right)\left(1 + \frac{G}{2j\omega C} + \frac{1}{2!}\left(\frac{G}{j\omega C}\right)^2 + \ldots\right)$$

$$= j\omega\sqrt{LC}\left(1 + \frac{R}{j2\omega L} + \frac{G}{j2\omega C} + \frac{RG}{j^2 4\omega^2 L^2} + \frac{R^2}{2j^2\omega^2 L^2} + \frac{G^2}{2j^2\omega^2} + \cdots\right).$$

Neglecting the higher order terms of ω,

$$\gamma = j\omega\sqrt{LC}\left(1 + \frac{R}{j2\omega L} + \frac{G}{j2\omega C}\right). \tag{9.65}$$

We know that $\gamma = \alpha + j\beta = j\omega\sqrt{LC} + \frac{R}{2}\sqrt{\frac{C}{L}} + \frac{G}{2}\sqrt{\frac{L}{C}}$.

Equating imaginary and real terms:

The attenuation constant is $$\alpha = \frac{R}{2}\sqrt{\frac{C}{L}} + \frac{G}{2}\sqrt{\frac{L}{C}}. \tag{9.66}$$

The phase shift constant is $\beta = \omega\sqrt{LC}$.

The phase velocity is $v_p = \frac{\omega}{\beta} = \frac{1}{\sqrt{LC}}$.

Similarly, the characteristic impedance is

$$Z_0 = \sqrt{\frac{R + j\omega L}{G + j\omega C}}.$$

Since $\omega L >> R$ and $\omega C >> G$,

$$Z_0 = \sqrt{\frac{j\omega L}{j\omega C}} = \sqrt{\frac{L}{C}}.$$

Therefore the characteristic impedance is $$Z_0 = \sqrt{\frac{L}{C}} \tag{9.67}$$

and the attenuation constant is

$$\alpha = \frac{R}{2}\sqrt{\frac{C}{L}} + \frac{G}{2}\sqrt{\frac{L}{C}} = \frac{R}{2Z_0} + \frac{GZ_0}{2}. \tag{9.68}$$

Here also, the attenuation, characteristic impedance and phase velocity are independent of frequency. Phase shift constant varies linearly with frequency.

Hence, the high frequency transmission line is distortionless.

Example 9.6 Find the characteristic impedance, propagation constant and velocity of propagation at $f = 100$ KHz for a lossless transmission line having $L = 33\ \mu$H/m and $C = 10$ nF/m.

Solution Given: $L = 33 \times 10^{-6}$ H/m, $C = 10 \times 10^{-9}$ F/m, $f = 100 \times 10^3$ Hz,

$$\omega = 2\pi f = 2\pi \times 10^5 \text{ radians/s}.$$

For a lossless line: $R = G = 0$, $Z = j\omega L$, $Y = j\omega C$.

Characteristic impedance $Z_0 = \sqrt{\frac{Z}{Y}} = \sqrt{\frac{L}{C}} = \sqrt{\frac{33 \times 10^{-6}}{10 \times 10^{-9}}} = 57.44$ ohms.

Propagation constant $\gamma = \sqrt{ZY} = j\omega\sqrt{LC} = j2\pi \times 10^5\sqrt{33 \times 10^{-6} \times 10 \times 10^{-9}}$.

$$\gamma = j0.361.$$

$$\gamma = \alpha + j\beta = j0.361.$$

$\therefore \quad \alpha = 0$ and $\beta = j0.361$ radians/m.

Velocity of propagation $v_p = \dfrac{\omega}{\beta} = \dfrac{2\pi \times 10^5}{0.361} = 1740.5$ km/s.

9.12 Telephone Cable

A telephone cable has insulated wires twisted in pairs. It has negligible values of inductance L and conductance G. Over the range of frequencies, it is assumed that

$$\omega L << R \text{ and } G << \omega C.$$

The propagation constant is

$$\gamma = \sqrt{(R + j\omega L)(G + j\omega C)}.$$

Neglecting L and G values,

$$\gamma = \alpha + j\beta = \sqrt{jR\omega C} = \sqrt{R\omega C}\angle 45^\circ \tag{9.69}$$

or $\alpha + j\beta = \sqrt{\dfrac{R\omega C}{2}} + j\sqrt{\dfrac{R\omega C}{2}}.$

Equating real and imaginary terms,

$$\alpha = \sqrt{\frac{\omega RC}{2}} \text{ and } \beta = \sqrt{\frac{\omega RC}{2}}. \tag{9.70}$$

Velocity of propagation $v_p = \dfrac{\omega}{\beta} = \dfrac{\omega}{\sqrt{\dfrac{\omega RC}{2}}}.$

$$v_p = \sqrt{\frac{2\omega}{RC}} \tag{9.71}$$

and characteristic impedance is

$$Z_0 = \sqrt{\frac{R + j\omega L}{G + j\omega C}} = \sqrt{\frac{R}{j\omega C}} = \sqrt{\frac{-jR}{\omega C}}.$$

$$Z_0 = \sqrt{\frac{R}{\omega C}}\angle -45^\circ \text{ ohms.} \tag{9.72}$$

It is observed that α, v_p and Z_0 are dependant on frequency. This results in considerable frequency and delay distortions in cables.

9.13 Condition for Minimum Attenuation

We know that the attenuation constant α is expressed as

$$\alpha = \sqrt{\frac{1}{2}\left[(RG - \omega^2 LC) + \sqrt{(R^2 + \omega^2 C^2)(G^2 + \omega^2 C^2)}\right]}.$$

The attenuation constant depends on primary constants R, G, C and L of the line and the frequency of the wave.

To determine the primary constants at which the wave propagates with minimum attenuation on the line, the variation of attenuation should be considered separately with respect to L, C, G and R.

Variation of attenuation with respect to the inductor: Let the parameter inductor L be a variable. Then keeping all other parameters constant, the minimum attenuation occurs at $\frac{d\alpha}{dL} = 0$.

$$\therefore \quad \frac{d\alpha}{dL} = \frac{1}{2\sqrt{\left[(RG - \omega^2 LC) + \sqrt{(R^2 + \omega^2 L^2)(G^2 + \omega^2 C^2)}\right]}}$$

$$\times \frac{1}{2}\left[-\omega^2 C + \frac{2\omega^2 L(G^2 + \omega^2 C^2)}{2\sqrt{(R^2 + \omega^2 L^2)(G^2 + \omega^2 C^2)}}\right] = 0 \qquad (9.73)$$

$$= -\omega^2 C + \frac{2\omega^2 L(G^2 + \omega^2 C^2)}{2\sqrt{(R^2 + \omega^2 L^2)(G^2 + \omega^2 C^2)}} = 0$$

or $$\frac{\omega^2 L(G^2 + \omega^2 C^2)}{\sqrt{(R^2 + \omega^2 L^2)}\sqrt{(G^2 + \omega^2 C^2)}} = \omega^2 C.$$

$$\omega^2 L\sqrt{\frac{G^2 + \omega^2 C^2}{R^2 + \omega^2 L^2}} = \omega^2 C.$$

$$L\sqrt{G^2 + \omega^2 C^2} = C\sqrt{R^2 + \omega^2 L^2}. \qquad (9.74)$$

Squaring both the sides,

$$L^2(G^2 + \omega^2 C^2) = C^2(R^2 + \omega^2 L^2).$$

$$L^2 G^2 + \omega^2 C^2 L^2 = C^2 R^2 + \omega^2 C^2 L^2.$$

$$L^2 G^2 = C^2 R^2.$$

Taking only the positive values,

$$LG = CR \text{ or } \quad \frac{G}{C} = \frac{R}{L}. \qquad (9.75)$$

This condition is the same as for a distortionless line given in Eq. (9.63). Therefore, the minimum attenuation occurs at

$$L = \frac{CR}{G} \text{ H/m.} \tag{9.76}$$

Variation of attenuation with respect to capacitor

Let the parameter capacitor be a variable.

Then while keeping all other parameters constant, the minimum attenuation occurs at $\frac{d\alpha}{dC} = 0$. Similarly, if the capacitor is varying, the minimum attenuation occurs at

$$C = \frac{LG}{R} \text{ farads/m.} \tag{9.77}$$

Variation of attenuation with respect to resistor or conductor

While keeping all other parameters constant, the minimum attenuation occurs at $\frac{d\alpha}{dR} = 0$, or $\frac{d\alpha}{dG} = 0$. From the expression of attenuation, it is observed that the attenuation $\alpha = 0$ occurs only when $R = 0$ and $G = 0$.

In practice, to get minimum attenuation with R and G, these values should be kept as small as possible.

Thus, a lossless line, $\alpha = 0$, is a distortionless line when $R = 0$ and $G = 0$.

9.14 Line Distortion

The waves on the transmission line are generally in complex form. When a transmission line is used for communicating information bearing signals to long distances, the signals consist many frequency components. Since the line characteristics depend on the frequency of the signal, the presence of many components cause variation in attenuation and phase velocity of the wave. This phenomenon is called line distortion. When distortion is more, the received signal is not a true replica of the transmitted signal. For ideal transmission, the attenuation and phase velocity should be kept constant at all frequencies. The phase constant should also vary linearly with frequency and the attenuation constant should be kept as small as possible.

In practice, there are usually two types of line distortions occuring on transmission lines.

Frequency distortion Frequency distortion occurs due to variation of attenuation with frequency. Each frequency component of the wave undergoes different attenuation. At the receiver end, the wave contains different amplitudes at different frequencies resulting in frequency distortion.

If the attenuation α is made independent of frequency, frequency distortion will not exist on transmission lines.

Delay distortion Delay distortion occurs due to variation of time delay with frequency. If the time required to transmit various frequency components of the wave is not the same, then delay distortion occurs. Delay depends on phase velocity. If the phase velocity of the wave is constant with frequency, then all the frequency components of the wave will reach at the same time and delay distortion will not occur. Since $v_p = \frac{\omega}{\beta}$, if β varies linearly with ω, the phase velocity becomes constant with frequency and we can achieve a transmission without delay distortion. Delay distortion is also called phase distortion.

In general, attenuation and phase constants of the line from Eq. (9.43) and Eq. (9.44) are given by

$$\alpha = \sqrt{\frac{1}{2}\left[(RG-\omega^2LC)+\sqrt{(R^2+\omega^2L^2)(G^2+\omega^2C^2)}\right]}, \tag{9.78}$$

$$\beta = \sqrt{\frac{1}{2}\left[(RG+\omega^2LC)-\sqrt{(R^2+\omega^2L^2)(G^2+\omega^2C^2)}\right]}. \tag{9.79}$$

Thus if α is independent of frequency and β varies linearly with ω (β is a constant multiplied by ω), no distortions occur in transmission.

9.15 Conditions to Achieve Minimum Attenuation

In general, for a typical cable with both attenuation and delay distortion, the primary constants are related as

$$\frac{R}{L} >> \frac{G}{C}. \tag{9.80}$$

We can achieve the distortionless and minimum attenuation conditions in four ways.

(i) Reduce R: When the resistance is decreased to achieve less attenuation, large size conductors are required. This increases the cable size and cost. Reduction of R will also lower $|Z_0|$.

(ii) Increase G: The shunt conductance can be increased by lowering the conductor insulation or by adding shunt conductance along the line. However, this process increases losses and lowers $|Z_0|$.

(iii) Decrease C: The capacitance can be decreased by increasing the spacing between conductors. This results in increase of cable size and cost. Decreasing C also increases $|Z_0|$.

(iv) Increase L: The inductance can be increased by adding series inductance along the line. Increasing L decreases attenuation and distortion. This process is the best method to achieve minimum attenuation and distortionless conduction in the line.

9.16 Loaded Line

Increasing the inductance of a transmission line by inserting inductance in series with the line is called loading. The line is called a loaded line.

There are three methods of loading a line.

1. Continuous loading
2. Patch loading and
3. Lumped loading

In general, lumped inductors called coils are placed at suitable intervals along the line to increase the inductance.

For a loaded line, attenuation is constant and very low up to a cut off frequency f_c. It can be calculated as

$$f_c = \frac{1}{\pi\sqrt{LCd}} \text{ Hz},$$

where $L =$ inductance of the loaded line (H/m)

$C =$ capacitance of the loaded line (F/m)

$d =$ length of loaded line in metres.

Attenuation rises sharply above the cutoff frequency. The cutoff frequency can be increased by decreasing the load or keeping the coils closer. But it increases attenuation.

1. Continuous loading

In this process, a sheet of iron or magnetic material (such as permalloy or μ-metal) is wound round the copper conductor as shown in Fig. 9.7. This method increases the permeability of the surrounding space, thereby increasing the inductance up to 100 mH/km. The process of continuous loading is laborious and expensive. Also, eddy current and hysteresis losses in the magnetic material increases the primary constant R. Furthermore, cable repairs result in poor design. The advantage of continuous loading is that the cutoff frequency is very high compared to lumped loading. Continuous loading is mostly used on ocean cables.

Fig. 9.7 *Continuous loading*

2. Patch loading

In this process, the cable is loaded with sections of continuous loading separated by sections of unloaded cable. In this way, the cost of loading can be reduced. The sections are separated normally by 200 to 250 metres. Patch loading is suitable for submarine cables.

3. Lumped Loading

In this process, the loading coils are placed at suitable intervals along the cable to increase the effective inductance of the line. This method has very low attenuation up to cutoff frequency, low eddy current losses and negligible external interference. Above the cutoff frequency, the attenuation increases rapidly. The line acts as a low-pass filter. The frequency characteristics of loaded lines are shown in Fig. 9.8.

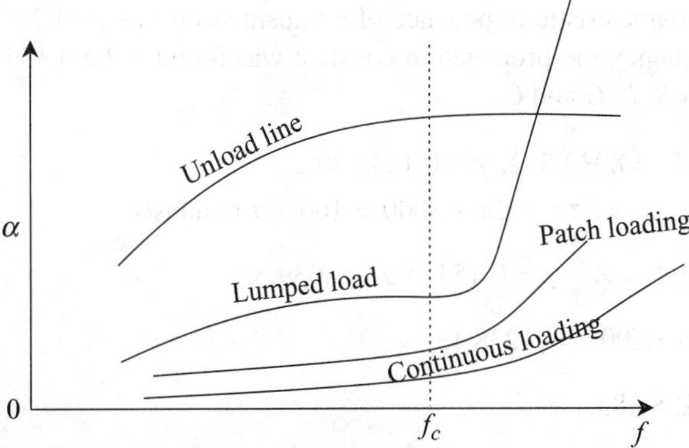

Fig. 9.8 *Frequency characteristics of the loaded lines*

In general, loaded lines provide very low distortion and perform much better than unloaded lines.

Additional Solved Problems

Example 9.7 A coaxial line has an outer diameter of 40 mm and an inner diameter of 2 mm. If the dielectric constant of the insulation is 9 and conductivity is 10^{-9} S/m. The conductors have conductivity 3.6×10^7 S/m and the operating frequency is 1 kHz. Find the line parameters R, L, C, G.

Solution Given:

Outer diameter = 40 mm and inner diameter = 2 mm; frequency $f = 1$ kHz,

radius of the inner solid conductor, $a = 1$ mm; radius of the outer conductor, $b = 20$ mm.

For an insulator: $\sigma_i = 10^{-9}$ S/m, $\varepsilon = 9\varepsilon_0$, $\mu = \mu_0$.

For aluminium: $\sigma = 3.6 \times 10^7$ S/m.

$$\text{Resistance } R = \sqrt{\frac{f\mu}{4\pi\sigma}}\left(\frac{1}{a} + \frac{1}{b}\right) \ \Omega/\text{m}$$

$$= \sqrt{\frac{1000\mu_0}{4\pi \times 3.6 \times 10^7}}\left(\frac{1}{0.001} - \frac{1}{0.02}\right) = 1.748 \text{ m}\Omega/\text{m}.$$

$$\text{Inductance } L = \frac{\mu_0}{2\pi}\ln\left(\frac{d}{a}\right) \text{ H/m} = \frac{\mu_0}{2\pi}\ln\left(\frac{0.02}{0.001}\right) = 60 \ \mu\text{H/m}.$$

$$\text{Capacitance } C = \frac{2\pi\varepsilon}{\ln(d/a)} \text{ F/m} = \frac{2\pi \times 9\varepsilon_0}{\ln(20)} = 0.167 \ \text{nf/m}.$$

$$\text{Conductance } G = \frac{2\pi\sigma_i}{\ln(d/a)} \text{ S/m} = \frac{2\pi \times 10^{-9}}{\ln(20)} = 2.1 \ \text{nS/m}.$$

Example 9.8 The characteristic impedance of a transmission line is 8039.5 Ω at a frequency of 8000 Hz. At this frequency the propagation constant was found to be $0.154\angle 90°$. Determine the values of line constant *R*, *L*, *G* and *C*.

Solution Given: $Z_0 = 8039.5\ \Omega$, $\gamma = 0.154\angle 90°$,
$f = 8$ kHz, $\omega = 2\pi f = 2\pi \times 8000 = 16000\pi$ radians/s.

We know that $R + j\omega L = Z \times \gamma = 0.154\angle 90° \times 8039.5.$

$$R + j\omega L = 1238.1\angle 90° = j1238.1.$$

$$R = 0,\ \omega L = 1238.083.$$

$$L = \frac{1238.083}{16000\pi} = 24.64 \quad \text{mH/km}.$$

Similarly, $G + j\omega C = \dfrac{\gamma}{Z_0} = \dfrac{0.154\angle 90°}{8039.5} = 1.915 \times 10^{-5}\angle 90°.$

$$G + j\omega C = j1.915 \times 10^{-5}.$$

$$G = 0,\ \omega C = 1.915 \times 10^{-5}.$$

$$C = \frac{1.915 \times 10^{-5}}{16000\pi} == 0.381 \quad \text{nF/km}.$$

Example 9.9 An open wire transmission line terminated in characteristic impedance has the following primary constants at 2 kHz: $R = 12\ \Omega$/km, $L = 4$ mH/km, $G = 1\ \mu\Omega$/km, $C = 0.005\ \mu$/km. Calculate the phase velocity and the attenuation in dB suffered by a signal in a length of 200 km.

Solution Given, $R = 12\ \Omega$/km, $L = 4$ mH/km, $G = 1\ \mu\Omega$/km,
$C = 0.005\ \mu$/km, $f = 2$ kHz. $\omega = 2\pi f = 2 \times \pi \times 2000 = 4000\pi$ radian/s.

Series impedance, $Z = R + j\omega L$

$$= 12 + j4000\pi \times 4 \times 10^{-3} = 12 + j50.26 = 51.67\angle 76.57°\ \Omega.$$

Similarly, shunt admittance $Y = G + j\omega C$

$$= 10^{-6} + j4000\pi \times 0.005 \times 10^{-6}.$$

$$Y = (1 + j62.83) \times 10^{-6} = 62.83 \times 10^{-6}\angle 89°\ \mho.$$

We know that $\gamma = \sqrt{(R + j\omega L)(G + j\omega C)} = \sqrt{Z \times Y}$

$$= \sqrt{51.67\angle 76.57 \times 62.83 \times 10^{-6}\angle 89.1}$$

$$= \sqrt{322 \times 10^{-3}\angle 165.57} = 0.0567\angle 82.78$$

$$= (7.13 + j56.3) \times 10^{-3}.$$

$$\gamma = \alpha + j\beta = 0.0713 + j0.0563.$$

Attenuation constant, $\alpha = 0.00713$ nepers/km.

Phase constant, $\beta = 0.0563$ radians/km.

Velocity of propagation $v_p = \dfrac{\omega}{\beta} = \dfrac{4000\pi}{0.0563} = 2.232 \times 10^5$ km/s.

Attenuation for 200 km:

$$\alpha = 0.00713 \times 200 = 1.426 \text{ nepers} = 1.426 \times 8.686 \text{ dB} = 12.38 \text{ dB}.$$

Example 9.10 An open wire transmission line terminated in its characteristic impedance has the following primary constants at 1 kHz. $R = 6\ \Omega$/km, $L = 2$ mH/km, $G = 0.5\ \mu$ mhos/km, $C = 0.05\ \mu$ f/km. Calculate α, β and phase velocity.

Solution Given:
$R = 6\,\Omega$/km, $L = 2$ mH/km, $G = 0.5\,\mu$mhos/km, $C = 0.05\,\mu$f/km, $f = 1$ kHz,
$\omega = 2\pi f = 2000\pi$ radians/s.

$$Z = R + j\omega L = 6 + j12.57 = 13.93\angle 64.48^\circ.$$

$$Y = G + j\omega C$$

$$= (0.5 + j2000\pi \times 0.05) \times 10^{-6}$$

$$= 315.16 \times 10^{-6}\angle 89.9^\circ.$$

$$Z_0 = \sqrt{\frac{Z}{Y}} = \sqrt{\frac{13.93\angle 64.48^\circ}{315.16 \times 10^{-6}\angle 89.9^\circ}} = 29.24\angle -12.71^\circ.$$

$$\gamma = \sqrt{ZY}$$

$$= \sqrt{13.93\angle 64.48^\circ \times 315.16 \times 10^{-6}\angle 154.36^\circ} = 0.06623\angle 77.2^\circ$$

$$= 0.0147 + j0.0645.$$

$$\gamma = \alpha + j\beta.$$

$$\alpha = 0.0147 \text{ nepers/km}; \quad \beta = 0.0646 \text{ radians/km}.$$

Phase velocity $\quad v_p = \dfrac{\omega}{\beta} = \dfrac{2000\pi}{0.0646} = 97263$ km/s.

Example 9.11 At 8 MHz, the characteristic impedance of a transmission line is $(40 - j2)\ \Omega$ and the propagation constant is $(0.01 + j0.18)$ per metre. Find the primary constants.

Solution Given:

Frequency $f = 8$ MHz; characteristic impedance $Z_0 = (40 - j2)\ \Omega$;
propagation constant $\gamma = (0.01 + j0.18)$ per metre.

$$\omega = 2\pi f = 16\pi \times 10^6 \text{ radians/s}.$$

We know that $Z_0 = \sqrt{\dfrac{R+j\omega L}{G+j\omega C}}$ and $\gamma = \sqrt{(R+j\omega L)(G+j\omega C)}$.

$\therefore \quad Z_0\gamma = R + j\omega L.$

$$R+j\omega L = (40-j2)(0.01+j0.18) = 0.76 + j7.18.$$

Equating real and imaginary parts,

$$R = 0.76\ \Omega\text{/m} \quad \text{and} \quad \omega L = 7.18.$$

$$L = \frac{7.18}{\omega} = \frac{7.18}{2\pi f} = \frac{7.18}{16\pi \times 10^6}\ L = 0.143\ \mu\text{H/m}.$$

Also, $G + j\omega C = \dfrac{\gamma}{Z_0} = \dfrac{(0.01+j0.18)}{(40-j2)} = 4.49 \times 10^{-3}\angle 89.68^\circ$

$$= 2.49 \times 10^{-5} + j4.5 \times 10^{-3}.$$

Equating real and imaginary parts,

$$G = 2.49 \times 10^{-5}\ \mho\text{/m} \quad \text{and} \quad \omega C = 4.5 \times 10^{-3}.$$

$$C = \frac{4.5 \times 10^{-3}}{16\pi \times 10^6} = 8.95 \times 10^{-11}\ C = 89.5\ \text{pF/m}.$$

Example 9.12 A telephone wire has the following constants per loop km. Resistance is 90 Ω, capacitance, 0.062 μF, inductance, 0.001 H and leakage $= 1.5 \times 10^{-6}$ mhos. The line is terminated in its characteristic impedance. A frequency of 1000 Hz is applied at the sending end. Calculate: (i) the characteristic impedance, (ii) wavelength, (ii) velocity propagation.

Solution Given: $R = 90\ \Omega$/km, $C = 0.062\,\mu$F/km, $L = 0.001$ H/km,
leakage conductance, $G = 1.5 \times 10^{-6}$ mhos/km, frequency, $f = 1000$ Hz.
Termination impedance = Characteristic impedance

Now $\quad Z = R + j\omega L = 90 + j2\pi \times 1000 \times 0.001 = 90 + j6.283$

or $\quad Z = 90.219\angle 3.99^\circ\ \Omega.$

$$Y = G + j\omega C = 1.5 \times 10^{-6} + j2\pi \times 1000 \times 0.062 \times 10^{-6}.$$

$$Y = (1.5 + j389.557) \times 10^{-6}\ \mho = 389.56 \times 10^{-6}\angle 89.78^\circ\ \mho.$$

(i) Characteristic impedance $Z_0 = \sqrt{\dfrac{Z}{Y}}$

$$= \sqrt{\frac{90.219\angle 3.99^\circ}{389.56 \times 10^{-6} \times \angle 89.78^\circ}} = \sqrt{0.2316 \times 10^6\angle -85.79^\circ}.$$

$$Z_0 = 481.24\angle -42.89^\circ\ \Omega.$$

(ii) Wavelength

We know that the propagation constant $\gamma = \alpha + j\beta = \sqrt{ZY}$.

$$\alpha + j\beta = \sqrt{90.219\angle 3.99^\circ \times 389.56 \times 10^{-6}\angle 89.78^\circ} = 0.1874\angle 46.89$$
$$= 0.128 + j0.1368.$$

$\therefore \quad \beta = 0.1368$ radians per km.

Then wavelength $\lambda = \dfrac{2\pi}{\beta} = \dfrac{2\pi}{0.1368} = 45.93$ km.

(iii) Velocity of propagation:

$$v_p = \frac{\omega}{\beta} = \frac{2\pi \times 1000}{0.1368} = 45930 \text{ km/s}.$$

Example 9.13 A coaxial line with an outer diameter of 8 mm has 50 ohm characteristic impedance. If the dielectric constant of the insulation is 1.60. Calculate the inner diameter.

Solution Given a coaxial line of characteristic impedance $Z_0 = 50\ \Omega$, outer diameter $b = 8$ mm, $\varepsilon_r = 1.6$.

We know that for a coaxial cable, the characteristic impedance is

$$Z_0 = \frac{138}{\sqrt{\varepsilon_r}}\log_{10}(b/a),$$

where $a =$ inner diameter.

$$Z_0 = \frac{138}{\sqrt{1.6}}\log_{10}(b/a) = 109.1\log_{10}(b/a) = 50.$$

$$\log_{10}(b/a) - 0.458 \quad \text{or} \quad b/a - 2.872.$$

$$a = \frac{8}{1.872} = 2.784 \text{ mm}.$$

The diameter of the inner conductor is 2.784 mm.

Example 9.14 The characteristic impedance of a 1 km long line is 100 ohms and is terminated in 200 ohms. It is fed with 10 V, having a source resistance of 50 ohms at $\omega = 0.3$ radian/s. Find the input voltage and current.

Solution Given: Length $l = 20$ km; characteristic impedance $Z_0 = 100$ ohms; termination impedance $Z_R = 200$ ohms; source resistance $Z_g = 50$ ohms; source voltage $V_g = 10$ V; $Z_0 = 100$ ohms.

We know that the input current is $\quad I_s = \dfrac{V_g}{Z_0 + Z_g} = \dfrac{10}{100 + 50} = 66.67$ mA.

Input voltage is $\quad V_s = Z_0 I_s = 66.67 \times 10^{-3} \times 100 = 6.67$ V.

Example 9.15 A coaxial cable has the following parameters: $Z_0 = 50$ ohms, length $= 20$ km. If the power input is 1 watt and the attenuation constant is 1.5 dB/km, find the output power of the cable provided it is terminated by Z_0. Also determine the output current.

Solution Given: Length $l = 20$ km; input power $P_s = 1$ watt; characteristic impedance $Z_0 = 50$ ohms; attenuation $\alpha = 1.5$ dB/km.

We know that if output power is P_0, then

$$\text{attenuation,} \quad \alpha = \frac{1}{l} 10\log_{10}\left(\frac{P_s}{P_0}\right) \text{ dB/km.}$$

$$1.5 = \frac{1}{20} \times 10\log_{10}\left(\frac{1}{P_0}\right).$$

$$P_0 = \frac{1}{10^3} = 1 \text{ mW.}$$

If output current is I_0, then

$$P_0 = |I_0|^2 Z_0, \; |I_0|^2 = \frac{P_0}{Z_0}.$$

$$I_0 = \sqrt{\frac{P_0}{Z_0}} = \sqrt{\frac{1 \times 10^{-3}}{50}} = 4.47 \text{ mA.}$$

Example 9.16 A uniform transmission line operating at 1 kHz has $Z_o = 75$ ohms and propagation constant of $(0.1 + j0.2)\ \text{m}^{-1}$. Find G and C.

Solution Given: Frequency $f = 1$ kHz; characteristic impedance $Z_0 = 75\ \Omega$; propagation constant $\gamma = 0.1 + j0.2$ per metre; $\omega = 2\pi f = 2000\pi$ radians/s.

We know that,

$$G + j\omega C = \frac{\gamma}{Z_0} = \frac{0.1 + j0.2}{75} = (1.33 + j2.667) \times 10^{-3}.$$

Equating real and imaginary parts,

$$G = 1.33 \times 10^{-5}\ \text{℧/m and} \quad \omega C = 2.667 \times 10^{-3}.$$

$$C = \frac{2.667 \times 10^{-3}}{2000\pi} = 0.424 \times 10^{-6} = 0.424\ \mu\text{F/m.}$$

Example 9.17 A transmission line has $R = 0.01\ \Omega/\text{m}$, $G = 1 \times 10^{-6}\ \Omega/\text{m}$, $L = 10 \times 10^{-6}$ H/m, $C = 1 \times 10^{-6}$ F/m. Find the attenuation constants, phase constants and phase velocity.

Solution Given: $R = 0.01\ \Omega/\text{m}$, $L = 10 \times 10^{-6}$ H/m; $C = 1 \times 10^{-6}$ F/m, $G = 1 \times 10^{-6}$ ℧/km.

Assume signal frequency $f = 1$ kHz. Then,

$$\omega = 2\pi f = 2000\pi \text{ radians/s}$$

$$Z = R + j\omega L = 0.01 + j2000\pi \times 10^{-6} = 0.01 + j0.0628 = 0.0636\angle 80.95°.$$

$$Y = G + j\omega C = (1 + j2000\pi) \times 10^{-6} = 6283 \times 10^{-6}\angle 89.99°.$$

We know that $\gamma = \sqrt{ZY}$

$$= \sqrt{0.0636\angle 80.95° \times 6283 \times 10^{-6}\angle 89.99°} = 0.02\angle 85.47 = 0.00158 + j0.0199.$$

$$\gamma = \alpha + j\beta.$$

$$\alpha = 0.00158 \text{ nepers/m}; \ \beta = 0.0199 \text{ radians/m}.$$

Wavelength $\lambda = \dfrac{2\pi}{\beta} = \dfrac{2\pi}{0.0199} = 315.73$ m.

Phase velocity $v_p = \lambda \times f = 315.73 \times 10^3 = 315.73$ km/s.

Example 9.19 Find the characteristic impedance, propagation constant and velocity of propagation at 10 kHz for a lossless transmission line having $L = 62\ \mu$H/m and $C = 37\ \mu$F/m.

Solution Given: $L = 62 \times 10^{-6}$ H/m, $C = 37 \times 10^{-6}$ F/m, $f = 1 \times 10^4$ Hz.

For a lossless line, $R = G = 0; Z = j\omega L; Y = j\omega C.$

Characteristic impedance for a lossless transmission line is

$$Z_0 = \sqrt{\frac{Z}{Y}} = \sqrt{\frac{L}{C}} = \sqrt{\frac{62 \times 10^{-6}}{37 \times 10^{-6}}} = 1.3\ \Omega \text{ ohms}.$$

Propagation constant

$$\gamma = \sqrt{ZY} = j\omega\sqrt{LC} = j2\pi \times 10^4\sqrt{62 \times 10^{-6} \times 37 \times 10^{-6}}.$$

$$\gamma = \alpha + j\beta = j3.01.$$

$$\alpha = 0, \ \beta = j3.01 \text{ radians/m}.$$

Velocity of propagation $v_p = \dfrac{1}{\sqrt{LC}} = \dfrac{\omega}{\beta} = \dfrac{2\pi \times 10^4}{3.01} = 20874$ km/s.

Example 9.20 A lossless transmission line has a capacitance of 50 μF/m and an inductance of 200 mH/m. Find the characteristic impedance for a section of a line 10 m long and 500 m long.

Solution Given: $L = 200 \times 10^{-3}$ H/m; $C = 50 \times 10^{-9}$ F/m.

For a lossless line: $R = G = 0; Z = j\omega L; Y = j\omega C.$

Characteristic impedance

$$Z_0 = \sqrt{\frac{Z}{Y}} = \sqrt{\frac{L}{C}} = \sqrt{\frac{200 \times 10^{-3}}{50 \times 10^{-6}}} = 63.24\Omega.$$

Since the characteristic impedance of a lossless line is independent of line length, the characteristic impedance for a section of a line 10 m long and 500 m long is the same as

$$Z_0 = 63.24\,\Omega.$$

Example 9.21 A high frequency line has the following primary constants: $L = 1.2$ mH/km, $C = 0.05\ \mu$F/km, $R = G =$ negligible. Determine the characteristic impedance and propagation constant of the line.

Solution Given: $L = 1.2 \times 10^{-3}$ H/m, $R = G = 0$, $z = j\omega L$, $\gamma = j\omega C$, $C = 0.05 \times 10^{-6}$ F/m.

Assume that $f = 1 \times 10^6$ Hz.

Characteristic impedance

$$Z_0 = \sqrt{\frac{Z}{Y}} = \sqrt{\frac{L}{C}} = \sqrt{\frac{12 \times 10^3}{0.05 \times 10^{-6}}} = 155\ \Omega.$$

Propagation constant

$$\gamma = \sqrt{ZY} = j\omega\sqrt{LC}$$

$$= j2\pi \times 10^6 \sqrt{1.2 \times 10^{-3} \times 0.05 \times 10^{-6}}.$$

$$\gamma = \alpha + j\beta = j48.67.$$

$$\alpha = 0; \quad \beta = j48.67 \text{ radians/km}.$$

Example 9.22 A transmission line operating at 500 MHz has $Z_0 = 80$ ohms, $\alpha = 0.04$ Np/m, $\beta = 1.5$ rad/m. Find the line parameters R, L, G and C.

Solution Given $Z_0 = 80$ ohms, $\alpha = 0.04$ nepers/m, $\beta = 1.5$ radians/m, $f = 500$ MHz.

We know that $\gamma = a + j\beta = 0.04 + j1.5$.

$$\omega = 2\pi f = 2\pi \times 500 \times 10^6 = \pi \times 10^9 \text{ radians/s}.$$

Also, $R + j\omega L = Z_0 \times \gamma = 80 \times (0.04 + j1.5) = 3.2 + j120.$

Equating real and imaginary parts,

$$R = 3.2\ \Omega/\text{m}; \quad \omega L = 120.$$

$$L = \frac{120}{\pi \times 10^9} = 38.2 \times 10^{-9} \quad \text{or} \quad L = 38.2 \text{ nH/m}.$$

Similarly, $G + j\omega L = \dfrac{\gamma}{Z_0} = \dfrac{0.04 + j1.5}{80} = 5 \times 10^{-4} + j0.01875.$

Equating real and imaginary parts,

$$G = 5 \times 10^{-4} \text{ S/m}.$$

$$\omega C = 0.01875.$$

$$C = \frac{0.01875}{\pi \times 10^9} = 5.968 \times 10^{-12} = 5.968 \text{ pF/m}.$$

Example 9.23 A lossy cable which has $R = 2.25\ \Omega$/m, $L = 1\ \mu$H/m, $C = 1$ pF/m and $G = 0$ operates at $f = 0.5$ GHz. Find the attenuation constant of the line.

Solution Given: $R = 2.25\ \Omega$/m, $L = 1 \times 10^{-6}$ H/m, $C = 1 \times 10^{-12}$ F/m, $G = 0$.
Signal frequency $f = 0.5 \times 10^9$ Hz; $\omega = 2\pi f = \pi \times 10^9$ radians/s.

$$Z = R + j\omega L = 2.25 + j\pi \times 10^9 \times 10^{-6} = 2.25 + j3141.6.$$

$$Y = G + j\omega C = j\pi \times 10^9 \times 10^{-12} = j3.1426 \times 10^{-3}.$$

Propagation constant $\gamma = \sqrt{ZY}$

$$= \sqrt{(2.25 + j3141.6) \times j3.1426 \times 10^{-3}} = \sqrt{9.869\angle 179.96}$$

$$= 3.1414\angle 89.98 = 0.0011 + j3.1414.$$

$$\gamma = \alpha + j\beta.$$

Attenuation constant $\alpha = 0.0011$ nepers/m.
Phase shift constant $\beta = 3.1414$ radians/m.

Example 9.24 A 75 ohm TV cable of length 20 km has attenuation 3 dB/km and is terminated with matched load. The dielectric constant of the cable is $\varepsilon = 5\varepsilon_0$ and $\mu = \mu_0$. If a 5 V source at 80 MHz is applied, find the voltage and current at the load.

Solution Given:

$$Z_0 = 75 \text{ ohms}, \ f = 80 \text{ MHz}, \ V_s = 5 \text{ V}, \ \varepsilon = 5\varepsilon_0 \text{ and } \mu = \mu_0$$

$$\alpha = 3 \text{ dB/km} = 0.3454 \text{ nepers/m}$$

$$\omega = 2\pi f = 2\pi \times 80 \times 10^6 = 16\pi \times 10^7 \text{ radians/s}$$

$$\beta = \omega\sqrt{\mu\varepsilon} = 2\pi \times 80 \times 10^6 \sqrt{5\mu_0 \varepsilon_0} = 3.746 \text{ radians/m}.$$

$$\gamma = a + j\beta = 0.3454 + j3.746.$$

We know that $V_R = V_s e^{-\gamma l}$

$$V_R = 5 \times e^{-(0.3454 + j3.746) \times 20} = 5 \times 10^{-3} e^{-j74.92} = 5 \times 10^{-3} \angle 332.6^\circ \text{ V}.$$

$$I_R = \frac{V_R}{Z_R} = \frac{5 \times 10^{-3}\angle 332.6}{75} = 66.67\angle 332.6^\circ\ \mu\text{A}.$$

Example 9.25 A transmission line has $R = 15\,\Omega$/m, $L = 2$ H/m, $C = 0.15 \times 10^{-6}$ F/m, $G = 1 \times 10^{-6}$ ℧/km. Find the additional inductance to give a distortionless line. Also find the propagation constant at 1 kHz.

Solution Given: $R = 15\ \Omega/\text{m}$, $L = 2$ H/m, $C = 0.15 \times 10^{-6}$ F/m, $G = 1 \times 10^{-6}$ ℧/km.

Assume that signal frequency $f = 1$ kHz, Then, $\omega = 2\pi f = 200\pi$ radians/s.

Condition for distortionless line is $L_1 G = RC$.

$$L_1 = \frac{RC}{G} = \frac{15 \times 0.15 \times 10^{-6}}{1 \times 10^{-6}} = 2.25 \text{ H/m}.$$

So, additional inductance required $L' = L_1 - L = 2.25 - 2 = 0.25$ H/m.

For distortionless line, we know that

$$\alpha = R\sqrt{\frac{C}{L_1}} = 15\sqrt{\frac{0.15 \times 10^{-6}}{2.25}} = 3.873 \times 10^{-3} \text{ nepers/m}.$$

$$\beta = \omega\sqrt{L_1 C} = 2000\pi\sqrt{2.25 \times 0.15 \times 10^{-6}} = 3.65 \text{ radians/m}.$$

Propagation constant $\gamma = \alpha + j\beta = 0.00387 + j3.65$ per m.

Example 9.26 A 50 ohm distortionless transmission line has attenuation 2.5 dB/m. The line has capacitance of 0.1 nF/m. Find *R*, *L*, *G*, v_p and voltage at 5 m if source voltage is 10 V.

Solution Given: $Z_0 = 50$ ohms, $f = 80$ MHz, $V_s = 10$ V, $C = 0.1$ nF/m, $\alpha = 2.5$ dB/m $= 0.2878$ nepers/m.

The condition for distortionless transmission is $RC = LG$.

$$\alpha = R\sqrt{\frac{C}{L}}, \quad Z_0 = \sqrt{\frac{L}{C}}.$$

$$\alpha = \frac{R}{Z_0}, \quad v_p = \frac{1}{\sqrt{LC}} \quad \text{and} \quad \beta = \omega\sqrt{LC}.$$

$$L = CZ_0^2 = 0.1 \times 10^{-9} \times 50^2 = 0.25\ \mu\text{H/m}.$$

$$R = \alpha Z_0 = 0.2828 \times 50 = 14.4\ \Omega/\text{m}.$$

$$G = \frac{RC}{L} = \frac{R}{Z_0^2} = \frac{14.4}{50^2} = 5.76 \text{ m℧/m}.$$

$$v_p = \frac{1}{\sqrt{LC}} = \frac{1}{\sqrt{0.25 \times 10^{-6} \times 0.1 \times 10^{-9}}} = 2 \times 10^8 \text{ m/s}.$$

We know that the voltage on the line is $V_1 = V_s e^{-\alpha l}$.

Voltage at 5 m is $V_1 = 10 \times e^{-0.2878 \times 5} = 10 \times e^{-1.439} = 2.37$ V.

Summary

1. There are various types of transmission lines available, depending on the propagation constant, propagation distance, power of transmission, etc. The important types are: open-wire lines transmission cables, coaxial cables, waveguides, optical fibres and strip lines.
2. The equivalent electrical circuit of a transmission line consists of a series resistance R (Ω/m), series inductance L (H/m), shunt capacitance C (F/m) and shunt conductance G (℧/m) along the length.
3. At very high frequencies, the line parameters can be related with medium characteristics as:

$$LC = \mu\varepsilon \quad \text{and} \quad \sigma C = G\varepsilon.$$

4. The voltage and current at a point at distance x from the sending end in terms of source voltage and current for a transmission line are

$$V = V_s \cos h\gamma x - I_s Z_0 \sin h\gamma x \quad \text{and} \quad I = I_s \cos h\gamma x - \frac{V_S}{Z_0} \sin h\gamma x.$$

These equations are called transmission line equations.

5. The propagation constant $\gamma = \alpha + j\beta = \sqrt{(R + j\omega L)(G + j\omega C)} = \sqrt{ZY}$ and the characteristic impedance $Z_0 = \sqrt{\dfrac{R + j\omega C}{G + j\omega C}} = \sqrt{\dfrac{Z}{Y}}$ are called the secondary constants of the transmission line.
6. The characteristic impedance for open wire system is $Z_0 = 276 \log_{10}\left(\dfrac{d}{r}\right)$ Ω, and for coaxial cable it is $Z_0 = \dfrac{138}{\sqrt{\varepsilon_r}} \log_{10}(b/a)$.
7. The expression for attenuation constant at the receiving end is $\alpha = \ln\left(\dfrac{V_s}{V_R}\right)$ nepers.

1 neper = 20 $\log e$ = 8.68 dB.

8. At a given frequency, the group velocity is a derivative of angular frequency with repect to the phase constant.

$$V_g = \frac{\omega_2 - \omega_1}{\beta_2 - \beta_1} = \frac{d\omega}{d\beta} \text{ m/s, and phase velocity is } v_p = v_g\left(1 - \beta\frac{dv_p}{d\omega}\right).$$

9. For lossless transmission line,

$$R = 0,\ G = 0,\ \alpha = 0,\ \beta = \omega\sqrt{LC},\ Z_0 = \sqrt{\frac{L}{C}} \quad \text{and} \quad v_p = \frac{\omega}{\beta} = \frac{1}{\sqrt{LC}}.$$

10. For distortion less transmission line,

$$RC = LG,\quad Z_0 = \sqrt{\frac{L}{C}},\quad \alpha = \frac{R}{Z_0} \text{ or } GZ_0,\quad \beta = \omega\sqrt{LC} \quad \text{and} \quad v_p = \frac{1}{\sqrt{LC}}.$$

11. For high frequency transmission line, $\omega L >> R$ and $\omega C >> G$.

$$\alpha = \frac{R}{2}\sqrt{\frac{C}{L}} + \frac{G}{2}\sqrt{\frac{L}{C}},\quad \beta = \omega\sqrt{LC},\ Z_0 = \sqrt{\tfrac{L}{C}} \text{ and } v_p = \frac{1}{\sqrt{LC}}.$$

12. For telephone cable, $\omega L << R$ and $G << \omega C$.

$$\alpha = \sqrt{\frac{\omega RC}{2}}, \quad \beta = \sqrt{\frac{\omega RC}{2}}, \quad v_p = \sqrt{\frac{2\omega}{RC}} \quad \text{and} \quad Z_0 = \sqrt{\frac{R}{\omega C}} \angle -45^\circ.$$

13. The condition for distortionless line is the same as the condition for minimum attenuation, and it is

$$RC = LG.$$

14. To achieve minimum attenuation in cables, the inductance of the transmission line can be increased by inserting inductance in series with the line. This is called loading, and the line is called a loaded line. Coils are placed at some points along the line to increase the effective inductance per length.

Review Questions

1 Prove that a line of finite length terminated by its characteristic impedance Z_0 is equivalent to a line of infinite length. *(May 2011)*

2 Derive the relationship between secondary constants and primary constants of a transmission line. *(Dec 2012)*

3 Draw the equivalent circuit of a two-wire transmission line and list the applications of a transmission lines. *(Dec 2012)*

4 Using expressions of wave amplitudes in a transmission line, deduce the relation for characteristic impedance of the line. *(May 2013)*

5 Derive the secondary constants for a low-loss transmission line. *(Dec 2011)*

6 Derive the following general equations for a transmission line:
$E = E_R \cosh\gamma l + I\, Z_0 \sinh\gamma l; \quad I = I_R \cosh\gamma l + \frac{E}{Z_0} \sinh\gamma l.$ *(Dec 2012)*

7 What is distortionless line? Determine the condition for distortionlessness and minimum attenuation. *(Dec 2013)*

8 Show that a line will be distortionless if $RC = LG$. *(Dec 2011)*

9 Obtain expressions for (i) attenuation constant, (ii) phase constant and (iii) characteristic impedance of a transmission line. *(Dec 2012)*

10 Show the effect terminating a transmission line with its characteristic impedance.

11 Derive the expression for propagation constant of an infinitely long transmission line. *(May 2013)*

12 Starting from the equivalent circuit, derive the transmission line equations for V and I in terms of the source parameters. *(Jan 2014)*

13 Derive the characteristic impedance of a transmission line in terms of its line constants. *(Jan 2014)*

14 What is meant by distortion? Derive the conditions for a distortionless transmission line. *(Jan 2014)*

15 Describe the losses in transmission lines. *(April 2012)*

16 What is meant by inductive loading? With the help of suitable expressions, explain the advantages of loading and also discuss the disadvantages. *(April 2012)*

17 Write a short notes on (i) lossless Transmission lines; (ii) distortionless line. *(Dec 2012)*

Problems

1. A uniform transmission line operating at 10 kHz has $Z_0 = 50$ ohms and propagation constant of $(0.1 + j0.1)$ per metre. Find R, L, G and C.
2. A transmission line has $R = 0.01$ ohm/m, $G = 1.0$ μmho/m, $L = 10\mu$ H/m, $f = 796$ Hz, $C = 1.0$ μ F/m. Find the attenuation constant, phase constant and phase velocity.
3. A high frequency line has the following primary constants: $L = 1.5$ mH/km, $C = 0.047$ μF/km, $R = G = 0$. Determine the characteristic impedance and propagation constant of the line.
4. The characteristic impedance of 200 km long line is $Z_0 = 683 - j138$ ohms and it is terminated in Z_0. Propagation constant is $\gamma = 0.0074 + j0.0356$. It is fed with 10 V, having a source resistance of 500 ohms at 1 kHz. Find the input and output voltages and currents.
5. A coaxial cable has following parameters: $Z_0 = 10$ ohms, length $= 100$ km. If the power input is 15 kW and the attenuation constant is 1.5 dB/km, find the output power of the cable provided it is terminated by Z_0. Also, determine the output current.
6. The constants of a transmission line per km are $R = 169$ Ω, $L = 12.2$ mH, $C = 10$ μF and $G = 10.25$ μmhos. At the frequency of 9000 Hz, calculate (i) the terminating impedance for which no reflection will be set up in the line, (ii) the attenuation in dB differed by the signal travelling a distance of 200 km when the line is properly terminated, and the phase velocity with which the signal would travel.
7. A parallel line made of copper of conductivity 5.8×10^7 S/m is kept in air. Conductor radius is 2 mm, separation between the conductors is 0.5 m, operating frequency is 10 kHz. Find the line parameters R, L, C, and G.
8. The primary constants of a line per loop km are $R = 160$ Ω, $C = 0.209$ μF, $L = 0.10$ mH and leakage conductance is negligible. Calculate the characteristic impedance and the propagation constant at a frequency of c0 = 15000 radians/s.
9. The parameters of the line are $R = 130$ Ω/km, $L = 3.2$ mH/km, $C = 0.2$ μF/km, $G = 4.50$ μ℧/km. Calculate the characteristic impedance at frequency $f = 1000$ Hz.
10. A transmission line has the following constants: $R = 20.8$ Ω/km, $L = 6.622$ mH/km, $C = 0.00835$ μF/km, $G = 0.08$ μS/km. Calculate Z_0, α, β and v_p at $\omega = 1000$ radians per second.
11. A telephone line has resistance of 30 ohm, inductance of $L = 50$ mH, capacitance of 0.67 μF and insulation resistance of 0.05 MΩ/km. Find the input impedance at regular frequency of 10000 radians/s if the line is very long.
12. A two-wire line has a characteristic impedance of 328.63 Ω and the conductor is 1.2 cm in diameter. Find centre-to-centre spacing in air.
13. A coaxial line with an outer diameter of 8 mm has 50 ohm characteristic impedance. If the dielectric constant of the insulation is 1.6, calculate the inner diameter and derive the formula to be used.
14. A lossless transmission line has a capacitance of 50 pF/m and an inductance of 200 nH/m. Find the characteristic impedance for sections of lines 10 m long and 500 m long.

Answers

1. $R = 5\Omega/\text{m}$, $L = 79.57$ μH/m, $C = 0.0318$ μF/m. $G = 2$ mS/m
2. $\alpha - 1.56 \times 10^{-3}$ nepers/m, $\beta = 0.01572$ radian/m, and $v_p - 318$ km/s
3. $Z_0 = 178.6$ Ω, $\gamma = j8.39 \times 10^{-6}$ ω per km

4. $V_s = 5.83$ volts, $I_s = 8.4$ mA, $V_R = 1.33\angle -48°$, volts, $I_R = 1.91$ mA
5. $P_0 = 15$ pW, $I_0 = 1.22\ \mu$A
6. i) $1220\angle -1.33°$, ii) 4dB, 2863 km/s
7. $R = 4.15$ mΩ/m, $L = 2.308\ \mu$H/m, $C = 5.037$ pF/m
8. $Z_0 = 225.9\angle -44.73°\ \Omega$, $\gamma = 0.7082\angle 45.268°$
9. $Z_0 = 322.95\angle -40.54°\ \Omega$
10. $Z_0 = 1616.8\angle -35.89°$, $\alpha = 8.02 \times 10^{-3}$ nepers/km
11. $Z_{in} = 273.42\angle -1.63°\ \Omega$
12. $d = 9.29$ cm
13. $a = 2.78$ mm
14. $Z_0 = 63.25\ \Omega$ irrespctive of line length.

Multiple-choice Questions

1. The input impedance of an infinite length transmission line is called
   (a) characteristic impedance (b) line impedance
   (c) image impedance (d) matching impedance

2. The primary constants of transmission line the order of kilometres are $R = 10.15$ ohm, $L = 393$ mH, $G = 0.29$ mhos. The value of C for distortionless transmission line is
   (a) 35. 15 mF (b) 11.22 mF (c) 393 mF (d) none of these

3. In a telephone cable, the condition for primary constants is
   (a) $\omega L \ll R,\ G \ll \omega C$ (b) $\omega L \gg R,\ G \ll \omega C$
   (c) $\omega L \ll R,\ G \gg \omega C$ (d) no condition

4. For distortionless transmission, the characteristics impedance is
   (a) $Z_0 = \sqrt{\dfrac{\omega L}{R}}$ (b) $Z_0 = \sqrt{\dfrac{L}{G}}$ (c) zero (d) $Z_0 = \sqrt{\dfrac{L}{C}}$

5. In a telephone cable, α, v_p and Z_0 are dependent on
   (a) frequency (b) velocity (c) constants (d) phase

6. Very high frequency transmission line is
   (a) noisey (b) lossy (c) distortionless (d) phaseless

7. Characteristic impedance for open wire system is
   (a) $276\log\left(\dfrac{d}{r}\right)$ (b) $276\log\left(\dfrac{d}{2r}\right)$ (c) $276\log\left(\dfrac{2r}{d}\right)$ (d) zero

8. For lossless transmission line, the primary constants are
   (a) $R = L = 0$ (b) $R = G = 0$ (c) $C = G = 0$ (d) $R = C = 0$

9. For lossless transmission line, the secondary constants are
   (a) $\alpha = \sqrt{\dfrac{\omega RC}{2}},\ \beta = 0$ (b) $\alpha = \sqrt{\dfrac{\omega RC}{2}},\ \beta = \omega\sqrt{LC}$
   (c) $\alpha = 0,\ \beta = \omega\sqrt{LC}$ (d) $\alpha = 0,\ \beta = \omega\sqrt{\dfrac{L}{C}}$

10. At very high frequency, the transmission line parameters can be related with medium characteristics as
    (a) $RC = \mu\varepsilon,\ \sigma C = G\varepsilon$ (b) $LC = \mu\varepsilon,\ \sigma C = R\varepsilon$
    (c) $LC = \mu\varepsilon,\ \sigma C = G\varepsilon$ (d) $RC = \mu\varepsilon,\ \sigma C = R\varepsilon$
11. The velocity at which the envelope of the complex wave propagates along the line is called
    (a) group velocity (b) phase velocity (c) complex velocity (d) guide velocity
12. For a lossless transmission line, the phase velocity can expressed as
    (a) $\sqrt{LC}$ (b) $\sqrt{\frac{L}{C}}$ (c) $\frac{1}{\sqrt{LC}}$ (d) $\sqrt{\frac{C}{L}}$
13. Various frequency components of a signal undergoing different attenuations is called
    (a) frequency distortion (b) delay distortion
    (c) both (a) and (b) (d) none of these
14. Transmission line will neither delay distortion, nor will there be frequency distortion if only
    (a) α is independent of ω (b) β is constant multiplied by ω
    (c) both (a) and (b) (d) (a) only
15. If $R/L = G/C$, then it is known as
    (a) distortion condition (b) distortionless condition
    (c) both (a) and (b) (d) (a) only
16. To reduce the capacitance C value, the following is resorted to:
    (a) increase in the spacing between the conductors of the transmission line
    (b) decrease in the spacing between the conductors of the transmission line
    (c) cable size reduction
    (d) all of the above
17. In a transmission line, inserting inductance in series with the line (to increase the inductance) is termed as
    (a) unloading (b) loading (c) distortionless transmission line (d) none
18. The inductance of a line can also be increased by the introduction of a loading coil at uniform intervals. This is known as
    (a) patch loading (b) lumped loading (c) continuous loading (d) none of these
19. The type of loading employing sections of continuously loaded cable separated by sections of unloaded cable is called
    (a) patch loading (b) lumped loading (c) continuous loading (d) none of these
20. In a telephone cable, the primary constants of transmission line are R = 100 ohms/ km and $C = 2\mu$F/km. at frequency 10 kHz. The characteristics impedance is
    (a) 28.21 ohms (b) 282.1 ohms (c) 795.2 ohms (d) 79.5 ohms

Answers

1. *a* 2. *b* 3. *a* 4. *d* 5. *a* 6. *c* 7. *a* 8. *b* 9. *c* 10. *c*
11. *a* 12. *c* 13. *a* 14. *a* 15. *b* 16. *a* 17. *b* 18. *b* 19. *a* 20. *a*

Transmission Lines - II | 10

Learning Objectives

The main objectives of this chapter are to understand thoroughly the following topics:

- Voltage and current distributions on the transmission line
- Open and short circuited transmission lines, input impedance, load impedance
- Reflection and refraction coefficients, VSWR
- Location of voltage maximum and voltage minimum for standing wave patterns
- Short length transmission lines used as circuit elements
- Impedance matching devices and stubs, single and double stub matching
- Smith chart, properties and applications
- Solving problems using Smith chart

10.1 Introduction

In the previous chapter, we discussed the basic concepts and characteristics of transmission lines. This chapter explores standing waves, input impedance, reflection coefficient, VSWR, impedance matching devices and stubs. It also explains how to solve problems using the Smith chart.

10.2 Standing Waves on a Transmission Line

Consider a transmission line kept along the x-axis and terminated with impedance Z_R as shown in Fig. 10.1.

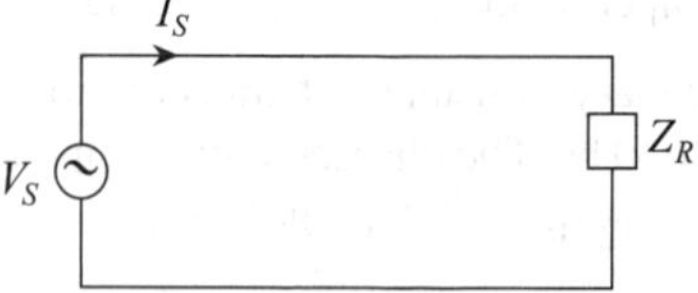

Fig. 10.1 *Transmission line terminated with Z_R*

When the source is applied on the line, the voltage and current components of the travelling wave decrease exponentially along the line with a decaying factor $e^{-\gamma x}$. This wave is called the incident wave. At the far end termination, due to impedance mismatch, the wave reflects back and travels in the opposite direction. This wave is called the reflected wave. The voltage and current components of the reflected wave again decrease exponentially along the line in the negative x direction with a decaying factor $e^{\gamma x}$. The incident and reflected waves having the same frequency combined on the line result in standing waves.

The voltage and current distributions of a standing wave at a given frequency, from Eq. (9.27) and Eq. (9.28) are:

$$V = be^{-\gamma x} + ae^{\gamma x} \tag{10.1}$$

and $$I = \frac{1}{Z_0}\left(be^{-\gamma x} - ae^{\gamma x}\right), \tag{10.2}$$

where Z_0 = characteristic impedance, a and b are constants.

The first term in Eq. (10.1) represents the incident wave and the second term represents the reflected wave.

Thus, in standing waves, two waves will travel in opposite directions between the source and the load. At some points on the line, these two waves are in-phase, while at other points, they are out of phase. The in-phase components added give maximum voltage at points called anti-nodes, while the out-of-phase components subtracted give minimum voltage at other points called nodes as shown in Fig.10.2(a).

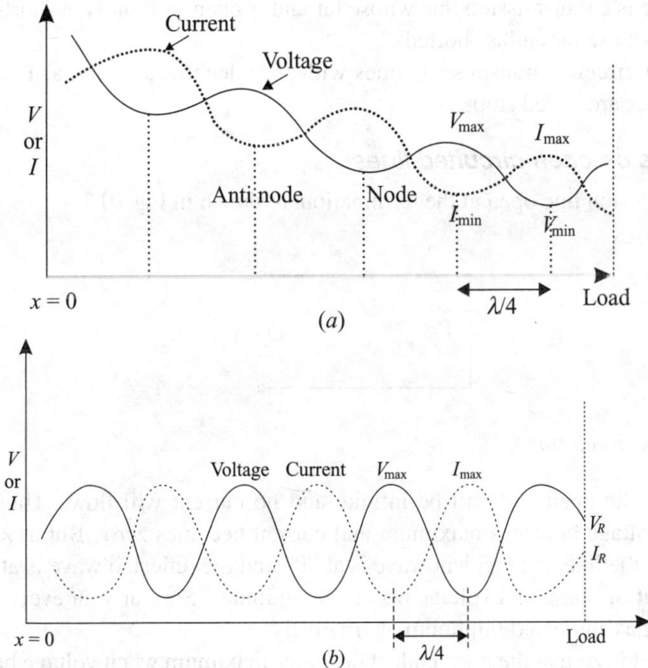

Fig. 10.2 *Voltage and current distribution of a standing wave along (a) lossy transmission line; (b) lossless transmission line*

Lossless transmission line:

For lossless line, $\alpha = 0,\ \gamma = j\beta$.

The voltage and current are

$$V = be^{-j\beta x} + ae^{j\beta x} \tag{10.3}$$

and $$I = \frac{1}{Z_0}(be^{-j\beta x} - ae^{j\beta x}). \tag{10.4}$$

The voltage and current distributions of a standing wave on a lossless transmission line are shown Fig. 10.2(b), where V_R and I_R are the voltage and current at the load.

10.3 Open and Short-Circuited Lines

The voltage and current distributions on a transmission line at any point depend on the distance of the point from the load and the termination impedance (load). The impedance of the line at any point also depends on these parameters.

Thus the voltage, current and impedances on the transmission line can be calculated for different loads, such as open-circuit end, short-circuit end and load equal to the characteristic impedance. An open-circuited line is a transmission line whose far end is open. Similarly, a short-circuited line is a transmission line whose far end is shorted.

Open and short-circuited transmission lines with short lengths are used as impedance matching devices. These lines are called stubs.

Standing waves on open-circuited lines

Consider a transmission line open at the termination as shown in Fig.10.3.

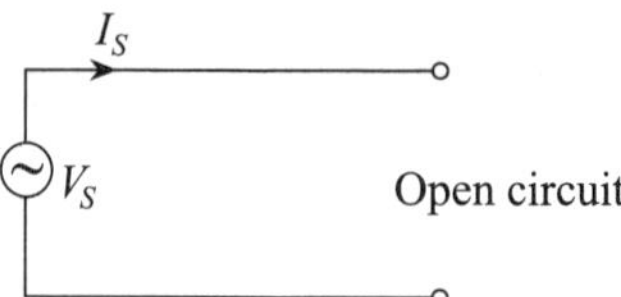

Fig. 10.3 *Open circuit line*

The impedance at the open end will be infinite and no current will flow. Thus at the open end termination, the voltage becomes maximum and current becomes zero. But at $\lambda/4$ distance from the termination on the line, the incident wave is at 90° and the reflected wave is at −90°. Thus both waves are 180° out of phase and voltage becomes minimum. Similarly, at every $\lambda/4$ distance, the voltage becomes maximum and minimum alternatively.

Also, the current is zero at the open end. It becomes maximum when voltage becomes minimum and vice versa as shown in Fig. 10.4.

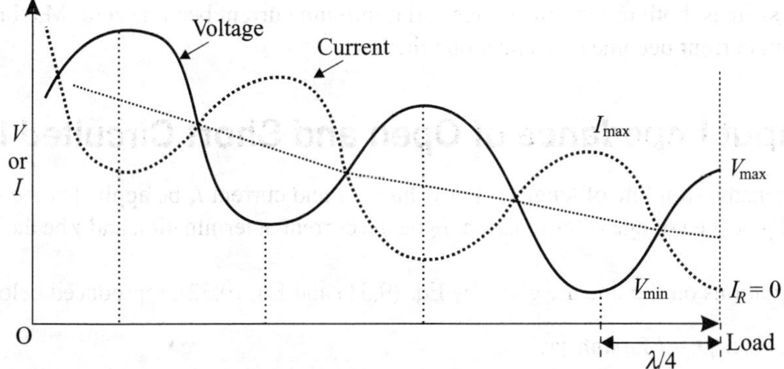

Fig. 10.4 *Voltage and current distribution of a standing wave*

Standing waves on short-circuited lines

Consider a transmission line short-circuited at termination as shown in Fig. 10.5.

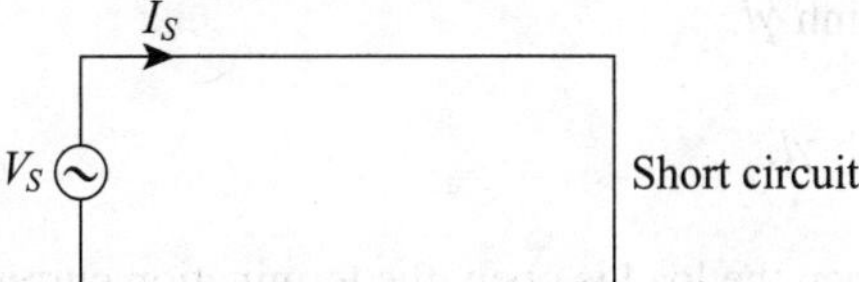

Fig. 10.5 *Short circuit line*

At short-circuit termination, since the impedance is zero, the current becomes maximum and voltage becomes zero. At $\lambda/4$ distance from the termination on the line, both the incident and reflected waves are in-phase, so the voltage becomes maximum and current becomes minimum. Similarly, at every $\lambda/4$ distance, the voltage becomes minimum and maximum alternatively. The current becomes minimum when voltage becomes maximum and vice versa, as shown in Fig.10.6.

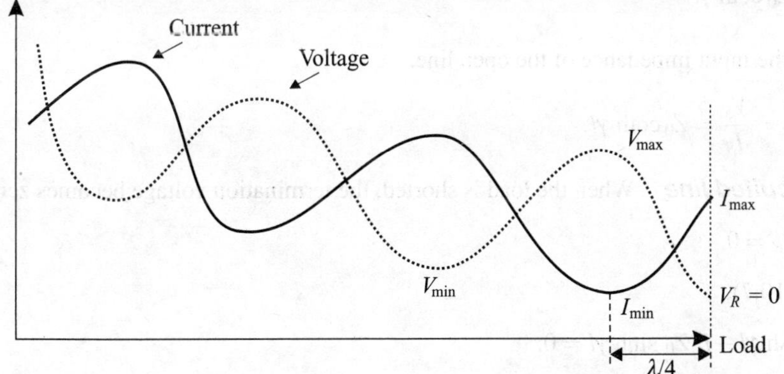

Fig. 10.6 *Voltage and current distribution of a standing wave in a short circuited line*

Note: It is observed that in an open circuit line, the voltage becomes maximum and current becomes minimum at even multiples of $\lambda/4$ distance from the load towards the source.

Similarly, in a short circuit line, the voltage becomes maximum and current becomes minimum at odd multiples of $\lambda/4$ distance from the load towards the source.

For lossless lines, both minimum voltage and minimum current become zero. Maximum voltage and maximum current become constant along the line.

10.4 Input Impedance of Open and Short Circuited Lines

Consider a transmission line of length l. Let voltage V_s and current I_s be applied at the source.

Also, let V_R be the voltage at termination, I_R be the current at termination and γ be the propagation constant.

The wave equations on the line are given by Eq. (9.31) and Eq. (9.32), reproduced below.

$$V = V_s \cosh \gamma x - I_s Z_0 \sinh \gamma x, \tag{10.5}$$

$$I = I_s \cosh \gamma x - \frac{V_s}{Z_0} \sinh \gamma x. \tag{10.6}$$

At $x = l$,

$$V_R = V_s \cosh \gamma l - I_s Z_0 \sinh \gamma l, \tag{10.7}$$

$$I_R = I_s \cosh \gamma l - \frac{V_s}{Z_0} \sinh \gamma l. \tag{10.8}$$

Open-circuited line When the load is open, the termination current becomes zero.

That is, $I_R = 0$.

From Eq. (10.8),

$$I_S \cosh \gamma l - \frac{V_S}{Z_0} \sinh \gamma l = 0.$$

or $\dfrac{V_S}{I_S} = Z_0 \coth \gamma l.$

Let Z_{oc} be the input impedance of the open line.

Then, $$Z_{oc} = \frac{V_S}{I_S} = Z_0 \coth \gamma l. \tag{10.9}$$

Short-circuited line When the load is shorted, the termination voltage becomes zero.

That is, $V_R = 0$.

From Eq. (10.7),

$$V_s \cosh \gamma l - I_s Z_0 \sinh \gamma l = 0.$$

or $\dfrac{V_s}{I_s} = Z_0 \tanh \gamma l.$

Let Z_{sc} be the input impedance of the short circuited line.

Then, $$Z_{sc} = \frac{V_s}{I_s} = Z_0 \tanh \gamma l. \tag{10.10}$$

Z_0 and γ in terms of Z_{sc} and Z_{oc}:

From Eq. (10.9) and Eq. (10.10),

$$Z_{oc}\, Z_{sc} = Z_0^2\, (\coth \gamma l)(\tanh \gamma l\,).$$

$$Z_0 = \sqrt{Z_{oc}\, Z_{sc}} \tag{10.11}$$

and $$\frac{Z_{sc}}{Z_{oc}} = \frac{Z_0 \tanh \gamma l}{Z_0 \coth \gamma l} = \tanh^2 \gamma l$$

$$\tanh \gamma l = \sqrt{\frac{Z_{sc}}{Z_{oc}}}$$

or $$\gamma = \frac{1}{l} \tanh^{-1}\left(\sqrt{\frac{Z_{sc}}{Z_{oc}}}\right). \tag{10.12}$$

Thus by measuring the short-circuit and open-circuit impedances at the termination of a transmission line, we can calculate the characteristic impedance and propagation constant of the line.

Again, from Z_0 and γ, we can compute the primary constants R, G, C and L.

Example 10.1 The open and short-circuit impedance of a transmission line at 36 kHz are $1800\angle{-60^\circ}$ ohm and $800\angle{-20^\circ}$ ohms. Calculate its Z_0.

Solution Given: Frequency, $f = 36$ kHz, $Z_{oc} = 1800\angle{-60^\circ}$, $Z_{sc} = 800\angle{-20^\circ}$.

Characteristic impedance Z_0 is given by

$$Z_0 = \sqrt{Z_{oc} \times Z_{sc}} = \sqrt{1800\angle{-60^\circ} \times 800\angle{-20^\circ}}.$$

$$Z_0 = 1200\angle{-40^\circ}\,.$$

Example 10.2 A transmission line 100 km long gave the following results for an impedance measurement at 1796 Hz. $Z_{oc} = 328\angle{-29.2^\circ}$ and $Z_{sc} = 1548\angle{6.8^\circ}$. Determine the line constants.

Solution Given:

$$Z_{oc} = 328\angle{-29.2^\circ},$$

$$Z_{sc} = 1548\angle{6.8^\circ},$$

$$f = 1796 \text{ Hz}.$$

$$\omega = 2\pi f$$

$$= 2 \times 3.14 \times 1796 = 11278.8 \text{ radians/s}.$$

$$Z_0 = \sqrt{Z_{sc} \times Z_{oc}}$$

$$= \sqrt{1548\angle{6.8^\circ} \times 328\angle{-29.2^\circ}} = 712.56\angle\frac{6.8 - 29.2}{2}$$

$$= 712.56\angle{-10.2}$$

and $\tanh \gamma l = \sqrt{\dfrac{Z_{sc}}{Z_{oc}}} = \sqrt{\dfrac{1548\angle 6.8^\circ}{328\angle -29.2^\circ}} = 2.172\angle \dfrac{6.8+29.2}{2}.$

$$\tanh \gamma l = 2.172\angle 18^\circ = 2.07 + j0.6712.$$

$$\frac{e^{\gamma l} - e^{-\gamma l}}{e^{\gamma l} + e^{-\gamma l}} = 2.07 + j0.6712.$$

By componendo and dividendo, we get

$$\frac{e^{\gamma l} + e^{-\gamma l} + e^{\gamma l} - e^{-\gamma l}}{e^{\gamma l} + e^{-\gamma l} - e^{\gamma l} + e^{-\gamma l}} = \frac{1 + 2.07 + j0.6712}{1 - (2.07 + j0.6712)}.$$

$$\frac{2e^{\gamma l}}{2e^{-\gamma l}} = \frac{3.07 + j0.6712}{-1.07 - j0.6712}.$$

$$e^{2\gamma l} = \frac{3.1425\angle 12.33}{1.263\angle -147.9}.$$

Taking ln on both sides,

$$2\gamma l = \ln(2.488\angle 160.23).$$

$$\gamma = \frac{1}{2l}\{\ln 2.488 + j160.23^\circ\}. \qquad \text{(Since } \ln(\gamma\angle\theta) = \ln\gamma + j(\theta + 2n\pi)\text{; assume } n = 0)$$

For $l = 100$ km,

$$\gamma = \frac{1}{200}(0.911 + j2.8) = 0.0147\angle 72^\circ. \qquad \text{(Since } 160.23^\circ = 2.8 \text{ radians).}$$

We know that, $R + j\omega L = \gamma \times Z_0 = 0.0147\angle 72^\circ \times 712.56\angle -10.2^\circ.$

$$R + j\omega L = 5.11 + j9.143$$

$$\omega L = 9.143.$$

$$L = \frac{9.143}{11278.8} = 0.81 \text{ mH/km}; \quad R = 5.11\ \Omega\text{/km}$$

$$G + j\omega C = \frac{\gamma}{Z_0} = \frac{0.0147\angle 72^\circ}{712.56\angle -10.2} = 2.44 \times 10^{-6} + j2.048 \times 10^{-6}.$$

$$G = 2.44 \times 10^{-6}\ \mho\text{/km}.$$

$$\omega C = 2.048 \times 10^{-6}; C = \frac{2.048 \times 10^{-6}}{11278.8} = 0.181 \text{ nF/km}.$$

10.5 Transmission Line with Load Impedance

Consider a wave is applied on a transmission line. If the line is of infinite length or if the line is terminated with characteristic impedance (Z_0), then there is no reflected wave.

If the line is terminated with a load impedance Z_R other than Z_0, some part of the energy will transfer to the load and the other part of it will be reflected back. Thus, standing waves appear on the line, resulting in loss of power.

For maximum power transfer to the load, the load impedance Z_R should be a complex conjugate of the source impedance.

Voltage and current distributions of a transmission line

Consider a transmission line of length l terminating with an impedance Z_R. Let V_R be the voltage at Z_R and I_R be the current through Z_R.

The voltage and current at any point on the transmission line are given by Eq. (9.29) and Eq. (9.30), which are reproduced below:

$$V = A \cosh \gamma x + B \sinh \gamma x, \tag{10.13}$$

$$I = \frac{-1}{Z_0}(B \cosh \gamma x + A \sinh \gamma x), \tag{10.14}$$

where A and B are constants.

At $x = l$, $V = V_R$ and $I = I_R$.

Substituting these values in Eq. (10.13) and Eq. (10.14),

$$V_R = A \cosh \gamma l + B \sinh \gamma l, \tag{10.15}$$

$$I_R = \frac{-A}{Z_0} \sinh \gamma l - \frac{B}{Z_0} \cosh \gamma l. \tag{10.16}$$

Multiplying Eq. (10.15) with cosh γl and Eq. (10.16) with sinh γl,

$$V_R \cosh \gamma l = A\cosh^2 \gamma l + B \sinh \gamma l \cosh \gamma l, \tag{10.17}$$

$$Z_0\, I_R \sinh \gamma l = -A \sinh^2 \gamma l - B \cosh \gamma l \sinh \gamma l. \tag{10.18}$$

Adding Eq. (10.17) and Eq. (10.18),

$$V_R \cosh \gamma l + Z_0\, I_R \sinh \gamma l = A\,(\cosh^2 \gamma l - \sinh^2 \gamma l).$$

$$\therefore \quad A = V_R \cosh \gamma l + Z_0 I_R \sinh \gamma l. \tag{10.19}$$

Multiplying Eq. (10.15) with sinh γl and Eq. (10.16) with cosh γl,

$$V_R \sinh \gamma l = A \sinh \gamma l \cosh \gamma l + B\sinh^2 \gamma l, \tag{10.20}$$

$$Z_0 I_R \cosh \gamma l = -A \sinh \gamma l \cosh \gamma l - B\cosh^2 \gamma l. \tag{10.21}$$

Adding Eq. (10.20) and Eq. (10.21),

$$V_R \sinh \gamma l + Z_0 I_R \cosh \gamma l = B\sinh^2 \gamma l - B\cosh^2 \gamma l$$

or $$B = -I_R Z_0 \cosh \gamma l - V_R \sinh \gamma l. \tag{10.22}$$

Substituting the values of A and B in Eq. (10.13), we get

$$V = (V_R \cosh \gamma l + Z_0 I_R \sinh \gamma l) \cosh \gamma x - (V_R \sinh \gamma l + I_R Z_0 \cosh \gamma l) \sinh \gamma x$$

$$= V_R\ (\cosh \gamma l \cosh \gamma x - \sinh \gamma l \sinh \gamma x) + Z_0 I_R\ (\sinh \gamma l \cosh \gamma x - \cosh \gamma l \sinh \gamma x).$$

$$V = V_R \cosh \gamma(l - x) + Z_0 I_R \sinh \gamma(l - x).$$

Similarly, substituting the values of A and B in Eq. (10.14), we get

$$I = I_R \cosh \gamma(l - x) + \frac{V_R}{Z_0} \sinh \gamma(l - x).$$

If $y = l - x$ is the distance from the load point, the voltage and current distribution at point y is

$$V = V_R \cosh \gamma y + Z_0 I_R \sinh \gamma y \tag{10.23}$$

$$I = I_R \cosh \gamma y + \frac{V_R}{Z_0} \sinh \gamma y. \tag{10.24}$$

Equations (10.23) and (10.24) are general line equations of voltage and current at a point of distance y from the receiving end in terms of terminal voltage and current.

Input impedance :

At $x = 0$ or $y = l$, $V = V_s$ and $I = I_s$.

The input impedance is given by

$$Z_s = Z_{\text{in}} = \frac{V_s}{I_s} = \frac{V_R \cosh \gamma l + Z_0 I_R \sinh \gamma l}{I_R \cosh \gamma l + \frac{V_R}{Z_0} \sinh \gamma l} \tag{10.25}$$

or $$Z_{\text{in}} = \left(\frac{V_R \cosh \gamma l + I_R Z_0 \sinh \gamma l}{V_R \sinh \gamma l + I_R Z_0 \cosh \gamma l} \right). \tag{10.25}$$

Z_{in} in terms of Z_R:

$$Z_{\text{in}} = Z_0 \left[\frac{\frac{V_R}{I_R} \cosh \gamma l + Z_0 \sinh \gamma l}{\frac{V_R}{I_R} \sinh \gamma l + Z_0 \cosh \gamma l} \right].$$

If $Z_R = \dfrac{V_R}{I_R}$, then $Z_{\text{in}} = Z_0 \left[\dfrac{Z_R \cosh \gamma l + Z_0 \sinh \gamma l}{Z_R \sinh \gamma l + Z_0 \cosh \gamma l} \right]$

or $$Z_{\text{in}} = Z_0 \left[\frac{Z_R + Z_0 \tanh \gamma l}{Z_0 + Z_R \tanh \gamma l} \right]. \tag{10.26}$$

If $$Z_R = Z_0, \quad \text{then } Z_{\text{in}} = Z_0. \tag{10.27}$$

Line impedance:

Line impedance is the impedance at any point on the line. From Eq. (10.23) and Eq. (10.24), the line impedance at a distance y from the load is

$$Z = \frac{V}{I} = \frac{V_R \cosh \gamma y + Z_0 I_R \sinh \gamma y}{V_R \cosh \gamma y + \frac{V_R}{Z_0} \sinh \gamma y}$$

$$= \frac{Z_R \cosh \gamma y + Z_0 \sinh \gamma y}{Z_0 \cosh \gamma y + Z_R \sinh \gamma y} = \frac{Z_R + Z_0 \tanh \gamma y}{Z_0 + Z_R \tanh \gamma y}. \tag{10.27}$$

Thus, when the load impedance is equal to the characteristic impedance, the transmission line is said to be perfectly matched or said to have proper termination.

The input impedance is also equal to the characteristic impedance.

$$\therefore \quad Z_{\text{in}} = Z_R = Z_0. \tag{10.28}$$

This is the condition for a perfectly matched transmission line.

When the transmission line is perfectly matched, no standing waves exist.

Note: For an oc line, $Z_R = \infty$, $Z_{\text{in}} = Z_{oc}$.

From Eq. (10.26), $$Z_{oc} = Z_0 \left[\frac{1 + \frac{Z_0}{Z_R} \tanh \gamma l}{\frac{Z_0}{Z_R} + \tanh \gamma l} \right].$$

$$Z_{oc} = Z_0 \left[\frac{1 + 0}{0 + \tanh \gamma l} \right] = Z_0 \coth \gamma l.$$

Similarly, for an SC line, $Z_R = 0$.

$$Z_{sc} = Z_0 \left[\frac{0 + Z_0 \tanh \gamma l}{Z_0 + 0} \right] = Z_0 \tanh \gamma l.$$

These equations for Z_{oc} and Z_{sc} are the same as Eq. (10.9) and Eq. (10.10), respectively.

10.6 Voltage and Current Distributions on a Lossless Line

We know that the voltage and current distributions at any point y on a transmission line are

$$V = V_R \cosh \gamma y + Z_0 I_R \sinh \gamma y,$$

$$I = I_R \cosh \gamma y + \frac{V_R}{Z_0} \sinh \gamma y.$$

For lossless line, $\alpha = 0,\ \gamma = j\beta$

$$V = V_R \cosh j\beta y + Z_0 I_R \sinh j\beta y,$$

$$I = I_R \cosh j\beta y + \frac{V_R}{Z_0} \sinh j\beta y,$$

or $$V = V_R \cos \beta y + jZ_0 I_R \sin \beta y, \tag{10.29}$$

$$I = I_R \cos \beta y + j\frac{V_R}{Z_0} \sin \beta y. \tag{10.30}$$

These equations represent voltage and currents at a distance y from the load on a lossless or distortionless line.

Input impedance:

At the source, $y = l,\ V = V_s$ and $I = I_s$.

$$V_s = V_R \cos \beta l + jZ_0 I_R \sin \beta l,$$

$$I_s = I_R \cos \beta l + j\frac{V_R}{Z_0} \sin \beta l.$$

$$Z_s = Z_{\text{in}} = \frac{V_s}{I_s} = \frac{V_R \cos \beta l + jZ_0 I_R \sin \beta l}{I_R \cos \beta l + j\frac{V_R}{Z_0} \sin \beta l}.$$

Or, $$Z_{\text{in}} = Z_0 \cdot \frac{Z_R + jZ_0 \tan \beta l}{Z_0 + jZ_R \tan \beta l} \tag{10.30}$$

where $Z_R = \dfrac{V_R}{I_R}$.

We know that $$\beta l = \frac{2\pi}{\lambda} l = \omega\sqrt{LC}\, l. \tag{10.31}$$

$$\therefore \quad Z_{\text{in}} = Z_0 \frac{Z_R + jZ_0 \tan(\omega l\sqrt{LC})}{Z_0 + jZ_R \tan(\omega l\sqrt{LC})}, \tag{10.32}$$

where L and C are primary constants of a lossless line.

Example 10.3 Determine the input impedance of a short-circuited lossless transmission line of length $\lambda/8$ and having $Z_0 = 600\ \Omega$.

Solution: Given: $Z_0 = 600\ \Omega,\ Z_R = 0,\ l = \pi/8$.

We know that, a for lossless line the input impedance is

$$Z_{\text{in}} = Z_0 \frac{Z_R + jZ_0 \tan \beta l}{Z_0 + jZ_R \tan \beta l}.$$

Also, $\beta l = \frac{2\pi}{\lambda}\frac{\lambda}{8} = \frac{\pi}{4}$.

$$\tan \beta l = \tan \frac{\pi}{4} = 1.$$

$$\therefore \quad Z_{\text{in}} = 600\frac{0 + j600}{600 + j.0} = j600\ \Omega.$$

$$Z_{\text{in}} = j600\ \Omega.$$

Example 10.4 Find the input impedance of a lossless transmission line for $L = 2\ \mu$H/m, $C = 80$ pF/m at a distance 0.1 m from the load impedance of a 100 ohm resistor in series with a 0.1 mH inductor at a frequency of 500 MHz.

Solution: Given: $L = 2\ \mu$H/m, $C = 80$ pF/m, distance $l = 0.1$ m, load resistance $R_L = 100\ \Omega$, load inductance $L_L = 0.1$ mH, and $f = 500$ MHz $= 500 \times 10^6$ Hz.

$$Z_R = R_L + j\omega L_L$$

$$= 100 + j2\pi \times 500 \times 10^6 \times 0.1 \times 10^{-6} = 100 + j314.15 = 329.6\angle 72.3\Omega.$$

We know that the characteristic impedance for a lossless line is

$$Z_0 = \sqrt{\frac{L}{C}} = \sqrt{\frac{2 \times 10^6}{80 \times 10^{-9}}} = 5\ \Omega.$$

Also, input impedance of a lossless transmission line is

$$Z_{\text{in}} = Z_0\frac{Z_R + jZ_0 \tan \beta l}{Z_0 + jZ_R \tan \beta l},$$

where $\beta = \omega\sqrt{LC}$

$$= 2\pi \times 500 \times 10^6 \sqrt{2 \times 10^{-6} \times 80 \times 10^{-9}} = 1256.6 \text{ radians/m}.$$

$$\therefore \quad Z_{\text{in}} = Z_0\frac{Z_R + jZ_0 \tan \beta l}{Z_0 + jZ_R \tan \beta l}$$

$$= 5\frac{100 + j314.15 + j5 \tan(1256.6 \times 0.1)}{5 + j(100 + j314.15) \tan(1256.6 \times 0.1)}$$

$$Z_{\text{in}} = 5 \times \frac{100 + j314.15 + j5(-1.39)}{5 + j329.6\angle 72.3 \times (-1.39)}$$

$$= 5 \times \frac{100 + j307.18}{5 + 458.14\angle -17.7}$$

$$= \frac{1615.23\angle 71.96}{464.24\angle -17.7} = 3.48\angle 89.66.$$

$$Z_{\text{in}} = 0.02 + j3.48\ \Omega.$$

10.7 Reflection and Reflection Coefficients

When the parameters of a transmission line are not uniform along the line or the termination impedance is different from the characteristic impedance, $(Z_R \neq Z_0)$, the incident wave is reflected back at the load.

The phenomenon of a wave being reflected at the load due to improper termination is called reflection.

Power loss occurs due to reflection. If the attenuation is less, the reflected wave again reflects back at the source. The wave travels back and forth on the line until it disappears.

Reflection coefficient

Reflection coefficient is defined as *the ratio of reflected voltage to the incident voltage or reflected current to the incident current.*

It is a vector quantity and designated by letter K.

$$\therefore \quad K = \frac{V_r}{V_i} \tag{10.33}$$

$$\text{or} \quad K = \frac{-I_r}{I_i}, \tag{10.34}$$

where V_r and V_i are the incident and reflected voltages respectively and I_r and an I_i are the incident and reflected currents respectively. The negative sign indicates that I_r is in an opposite direction to I_i.

We know that the voltage and current distributions at any point on the line, from Eq. (9.27) and Eq. (9.28), are:

$$V = ae^{\gamma x} + be^{-\gamma x}, \tag{10.35}$$

$$I = \frac{1}{Z_0}\left(be^{-\gamma x} - ae^{\gamma x}\right). \tag{10.36}$$

If y is the distance measured from the termination Z_R, then by substituting $-y$ in place of x, we get

$$V = be^{\gamma y} + ae^{-\gamma y}, \tag{10.37}$$

$$I = \frac{b}{Z_0}e^{\gamma y} - \frac{a}{Z_0}e^{-\gamma y}. \tag{10.38}$$

At load point, the first term is the incident voltage/current:

$$V_i = be^{\gamma y} \text{ and } I_i = \frac{b}{Z_0}e^{\gamma y} \tag{10.39}$$

and the second term is the reflected voltage/current:

$$V_r = ae^{-\gamma y} \text{ and } I_r = \frac{-a}{Z_0}e^{-\gamma y}. \tag{10.40}$$

To find the constants:

The conditions at the load are at $y = 0$, $V = V_R$ and $I = I_R$. Substituting these in Eq. (10.37) and Eq. (10.38), we get

$$V_R = b + a,$$

$$I_R = \frac{1}{Z_0}(b - a).$$

Solving these equations,

$$a = \frac{1}{2}(V_R - I_R Z_0), \quad b = \frac{1}{2}(V_R + I_R Z_0).$$

The voltage reflection coefficient is

$$K = \frac{V_r}{V_i} = \frac{ae^{-\gamma y}}{be^{\gamma y}} = \frac{a}{b}e^{-2\gamma y}. \tag{10.41}$$

At the termination, $y = 0$, $K = a/b$.

Substituting the values of a and b,

$$K = \frac{(1/2)(V_R - I_R Z_0)}{(1/2)(V_R + I_R Z_0)} = \frac{\frac{V_R}{I_R} - Z_0}{\frac{V_R}{I_R} + Z_0}.$$

We know that $\frac{V_R}{I_R} = Z_R$, termination impedance.

Therefore the reflection coefficient $K = \frac{Z_R - Z_0}{Z_R + Z_0}$. (10.42)

The voltage reflection coefficient K is completely dependent only on load impedance and characteristic impedance.

K is a complex quantity, always less than or equal to ± 1.

Reflection coefficient for different conditions

(a) For matched termination, i.e., $Z_R = Z_0$,

$$K = 0. \tag{10.43}$$

It shows that the reflected wave is zero.

Also $V_R = V_S$ or $V_S = Z_o I_S$ and $V_R = Z_o V_R$.

(b) For short circuited termination, i.e., $Z_R = 0$,

$$K = \frac{-Z_0}{Z_0} = -1. \tag{10.44}$$

It shows that the entire incident wave reflects back with 180° phase shift.

Also $V_R = 0$ and $I_R = \frac{2V_S}{Z_0}$. (10.45)

(c) For open circuited termination, i.e., $Z_R = \infty$,

$$K = \frac{Z_R - Z_0}{Z_R + Z_0}$$

or $$K = \frac{1 - Z_0/Z_R}{1 + Z_0/Z_R} = \frac{1-0}{1+0}.$$

$$\therefore \quad K = 1. \tag{10.46}$$

It shows that the entire incident wave reflects back with the same phase.

So, $V_R = 2V_S$ and $I_R = 0$. (10.47)

Reflection loss

Reflection loss is the power loss due to reflection. It is defined as the ratio of load power to the incident power. Reflection loss in decibels can be expressed as

$$K_l = 10\log\left[\frac{\text{Load power}}{\text{Incident power}}\right] = 20\log\left|\frac{1}{K_f}\right|,$$

where K_f is the reflection factor given by

$$K_f = \sqrt{1-K^2} = \frac{2\sqrt{Z_0 Z_R}}{Z_R + Z_0}.$$

10.8 Standing Wave Ratio

When the lossless line is not terminated with characteristic impedance, the combination of incident and reflected waves gives rise to standing waves as shown in Fig. 10.7.

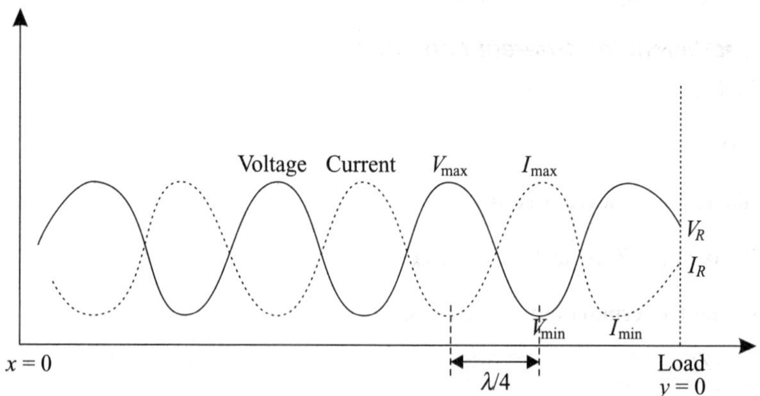

Fig. 10.7 *Standing wave pattern along a lossless high frequency transmission line* $(Z_R \neq Z_o)$

Let V_{max} be the maximum voltage, V_{min} be the minimum voltage, I_{max} be the maximum current and I_{min} be the minimum current.

The distance between two maximum or minimum points is $\lambda/2$. The maximum values occur when the incident and reflected waves are added.

i.e., $$|V_{\max}| = |V_r| + |V_i| \tag{10.48}$$

and $$|I_{\max}| = |I_r| + |I_i|. \tag{10.49}$$

The minimum values occur when the incident and reflected waves are subtracted.

i.e., $$|V_{\min}| = |V_i| - |V_r| \tag{10.50}$$

and $$|I_{\min}| = |I_i| - |I_r|. \tag{10.51}$$

Voltage standing wave ratio

The ratio of the maximum magnitude of the voltage to the minimum magnitude of the voltage is called voltage standing wave ratio. It is abbreviated as VSWR and denoted by *S*.

Thus $$\text{VSWR} = S = \frac{|V_{\max}|}{|V_{\min}|}. \tag{10.52}$$

Current standing wave ratio

The ratio of the maximum magnitude of the current to the minimum magnitude of the current is called current standing wave ratio.

It is abbreviated as ISWR.

$$\therefore \quad \text{ISWR} = \frac{|I_{\max}|}{|I_{\min}|}. \tag{10.53}$$

VSWR in terms of reflection coefficient (K):

We know that

$$S = \frac{|V_{\max}|}{|V_{\min}|}.$$

$$S = \frac{|V_r| + |V_i|}{|V_r| - |V_i|} = \frac{1 + \left|\frac{V_r}{V_i}\right|}{1 - \left|\frac{V_r}{V_i}\right|}.$$

But we know that the reflection coefficient

$$K = \frac{V_r}{V_i} \quad \text{or} \quad |K| = \left|\frac{V_r}{V_i}\right|.$$

$$\therefore \quad S = \frac{1 + |K|}{1 - |K|} \tag{10.54}$$

or $$|K| = \frac{S - 1}{S + 1}. \tag{10.55}$$

VSWR is a real quantity. It is always greater than 1.

The different conditions are:

(a) If the line is perfectly matched, $Z_R = Z_0$ and $K = 0$.

$\therefore \quad S = 1$ and there is no reflection.

(b) For an open or short circuit line, $K = \pm 1$ and $S = \infty$.

That is, when the incident wave is completely reflected back, VSWR becomes infinite.

Note 1: The range of reflection coefficient is $-1 \le K \le 1$. The range of VSWR is $0 \le S \le \infty$.

Note 2: VSWR is more preferable compared to the current standing wave ratio because VSWR at different points on the line can be easily measured by using a voltmeter. It is very difficult to measure currents on the line.

Example 10.5 A lossless transmission line of 100 Ω characteristic impedance is connected to a load of 300 Ω. Calculate the reflection coefficient and standing wave ratio.

Solution Given: $Z_0 = 100\ \Omega,\ Z_R = 300\ \Omega$.

$\therefore$ The reflection coefficient is, $K = \dfrac{Z_R - Z_0}{Z_R + Z_0} = \dfrac{300 - 100}{300 + 100} = \dfrac{200}{400} = \dfrac{1}{2} = 0.5.$

The standing wave ratio is $S = \text{VSWR} = \dfrac{1 + |K|}{1 - |K|}\ S = \dfrac{1 + 0.5}{1 - 0.5} = 3.$

10.9 Input Impedance in Terms of Reflection Coefficient

We know that the input impedance of a transmission line terminated with load impedance Z_R is given by

$$Z_{\text{in}} = Z_0 \frac{Z_R \cosh \gamma l + Z_0 \sinh \gamma l}{Z_0 \cosh \gamma l + Z_R \sinh \gamma l}.$$

In exponential form,

$$\cosh \gamma l = \frac{e^{\gamma l} + e^{-\gamma l}}{2}, \quad \sinh \gamma l = \frac{e^{\gamma l} - e^{-\gamma l}}{2}. \tag{10.56}$$

Then,

$$Z_{\text{in}} = Z_0 \frac{Z_R(e^{\gamma l} + e^{-\gamma l}) + Z_0(e^{\gamma l} - e^{-\gamma l})}{Z_0(e^{\gamma l} + e^{-\gamma l}) + Z_R(e^{\gamma l} - e^{-\gamma l})}$$

$$= Z_0 \frac{e^{\gamma l}(Z_R + Z_0) + e^{-\gamma l}(Z_R - Z_0)}{e^{\gamma l}(Z_0 + Z_R) - e^{-\gamma l}(Z_R - Z_0)}.$$

$$Z_{in} = Z_0 \cdot \frac{1 + e^{-2\gamma l}\dfrac{(Z_R - Z_0)}{(Z_R + Z_0)}}{1 - e^{-2\gamma l}\dfrac{(Z_R - Z_0)}{(Z_R + Z_0)}}. \tag{10.57}$$

But we know that

$$K = \frac{Z_R - Z_0}{Z_R + Z_0}.$$

$$\therefore \quad Z_{in} = Z_0 \frac{(1 + Ke^{-2\gamma l})}{(1 - Ke^{-2\gamma l})}. \tag{10.58}$$

Thus the input impedance of the line depends on the reflection coefficient, characteristic impedance, propagation constant and length of the line.

Note: The line impedance at a point y is

$$Z = Z_0 \frac{1 + Ke^{-2\gamma y}}{1 - Ke^{-2\gamma y}}.$$

10.10 Location of Voltage Maxima and Minima

Consider a lossless transmission line terminated with load impedance $Z_R \neq Z_O$. When the wave is sent, standing waves exist on the line with definite maxima and minima of voltage along the line as shown in Fig. 10.8. When the line is lossless at very high frequency, attenuation of the wave is zero.

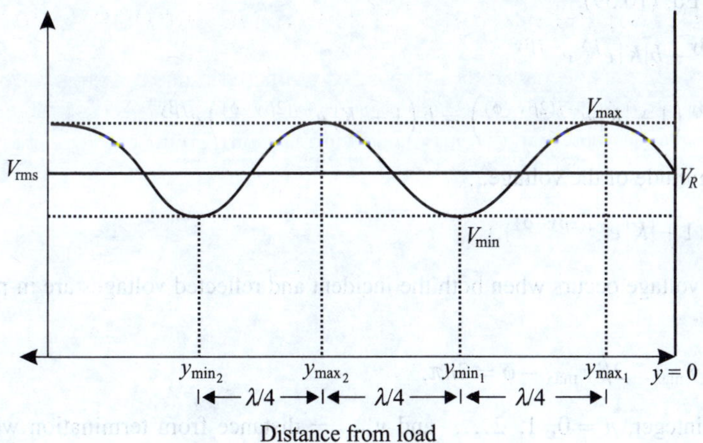

Fig. 10.8 *Location of voltage maximum and voltage minimum on standing wave pattern*

Let V_i be the incident voltage, V_r be the reflected voltage,

V_{max} = maximum voltage of the standing wave,

V_{min} = minimum voltage of the standing wave.

Let y be the distance of the line from termination.

We know that the voltage at any point on the line from termination is, from Eq. (10.36),

$$V = be^{\gamma y} + ae^{-\gamma y}.$$

Since the line is lossless, $\gamma = j\beta$,

$$V = be^{j\beta y} + ae^{-j\beta y}. \tag{10.59}$$

The reflected and incident voltages are

$$V_r = ae^{-j\beta y} \text{ and } V_i = be^{j\beta y}. \tag{10.60}$$

If K is the reflection coefficient, then

$$K = \frac{V_r}{V_i}$$

$$\text{or} \quad K = |K|\, e^{j\phi} = \frac{ae^{-j\beta y}}{be^{j\beta y}}, \tag{10.61}$$

where ϕ is the phase angle of the reflection coefficient.

At the termination, $y = 0$.

$$\therefore \quad |K|\, e^{j\phi} = \frac{a}{b}$$

$$\text{or} \quad a = b|K|\, e^{j\phi}. \tag{10.62}$$

Substituting in Eq. (10.59),

$$V = be^{j\beta y} + b|K|\, e^{j\phi}\; e^{-j\beta y}$$

$$= be^{j\beta y}\left(1 + |K|\; e^{-j(2\beta y-\phi)}\right) = b\left(1 + |K|\; e^{-j(2\beta y-\phi)}\right) e^{j\beta y}. \tag{10.63}$$

Taking the magnitude of the voltage,

$$|V| = |b|(1 + |K|\, e^{-j(2\beta y-\phi)}). \tag{10.64}$$

The maximum voltage occurs when both the incident and reflected voltages are in-phase or integer multiples of π.

$$\text{At} \quad y = y_{\max}, \;\; 2\beta y_{\max} - \phi = 2n\pi, \tag{10.65}$$

where n is an integer, $n = 0,\ 1,\ 2\ldots,$ and $y_{\max}$ = distance from termination where maximum voltage occurs.

$$\therefore \quad y_{\max} = \frac{1}{2\beta}(2n\pi + \phi). \tag{10.66}$$

Substituting in Eq. (10.64), the maximum amplitude of the voltage is

$$|V_{\max}| = |b|(1 + |K| e^{-j2n\pi})$$

$$|V_{\max}| = |b|[1 + |K|]. \tag{10.67}$$

Similarly, the minimum voltage occurs when both the incident and reflected voltages are out of phase or odd integer multiples of π.
From Eq. (10.64), at $y = y_{\min}$,

$$2\beta y_{\min} - \phi = (2n+1)\pi, \tag{10.68}$$

where n is an integer, $n = 0, 1, 2 \ldots\ldots$, and $y_{\min}$ = distance from termination where minimum voltage occurs.

$$y_{\min} = \frac{1}{2\beta}((2n+1)\pi + \phi) \tag{10.69}$$

Substituting in Eq. (10.64), the minimum amplitude of the voltage is

$$|V_{\min}| = |b|\,[1 - |K|]. \tag{10.70}$$

$$\therefore \quad \text{VSWR is} \quad S = \frac{|V_{\max}|}{|V_{\min}|} = \frac{|b|[1+|K|]}{|b|[1-|K|]} \tag{10.71}$$

$$\text{or} \quad S = \frac{1+|K|}{1-|K|}. \tag{10.72}$$

The position of first maxima and minima from the termination can be found by putting $n = 0$.
First maxima is

$$y_{\max 1} = \frac{\phi}{2\beta} = \frac{\phi\,\lambda}{2 \times 2\pi} = \frac{\phi\,\lambda}{4\pi}. \tag{10.73}$$

First minima is

$$y_{\min 1} = \frac{1}{2\beta}(\phi + \pi)$$

$$y_{\min 1} = \frac{\lambda}{2 \times 2\pi}(\phi + \pi) = \frac{\phi\,\lambda}{\pi\,4} + \frac{\lambda}{4} = \frac{\lambda\,\phi}{4\pi} + \frac{\lambda}{4}. \tag{10.74}$$

$$\therefore \quad y_{\min 1} = y_{\max 1} + \frac{\lambda}{4}. \tag{10.75}$$

Thus the first voltage maxima occurs at $y_{\max 1} = \dfrac{\phi\,\lambda}{4\pi}$ and the first voltage minima occurs at $y_{\min 1} = y_{\max 1} + \dfrac{\lambda}{4}$ from the termination.

Similarly, the position of second maxima and minima can be computed by putting $n = 1,2$ and so on.

The maximum and minimum positions of the standing wave are separated by a distance of $\lambda/4$.

Note: For resistive load, if $Z_L > Z_0$, the first voltage maximum occurs near the load, and if $Z_L < Z_0$, the first voltage minimum occurs near the load.

Significance of Z_{max} and Z_{min}

We know that the line impedance at any point y on the lossless line is

$$Z = Z_0 \frac{1 + Ke^{-j2\beta y}}{1 - Ke^{-j2\beta y}}.$$

The line impedance is complex, but for a lossless line the impedances at V_{max} and V_{min} positions are always real.

At voltage maximum (anti node position), the line impedance is called impedance maxima, Z_{max}. It can be expressed as

$$Z_{max} = \frac{V_{max}}{I_{min}} = \frac{V_{max}}{V_{min}/Z_0} = Z_0 \left| \frac{V_{max}}{V_{min}} \right|$$

$$Z_{max} = Z_0 S. \tag{10.76}$$

At voltage minimum (node position), the line impedance is called impedance minima, Z_{min}. It can be expressed as

$$Z_{min} = \frac{V_{min}}{I_{max}} = Z_0 \left| \frac{V_{min}}{V_{max}} \right| = \frac{Z_0}{S}$$

$$Z_{min} = \frac{Z_0}{S}. \tag{10.78}$$

Thus the impedance on the line varies from maximum value $Z_{max} = Z_0 S$ at the anti-node to minimum value $Z_{min} = \frac{Z_0}{S}$ at the node point.

For example, if a transmission line has VSWR, $S = 2.5$, and characteristic impedance is 50 Ω, then the maximum and minimum impedance on the line will be $Z_{max} = 50 \times 2.5 = 125$ Ω and $Z_{min} = 20$ Ω irrespective of the load impedance.

Example 10.6 A lossless line in air has a characteristic impedance of 350 Ω and is terminated by an unknown impedance. When the frequency is 1200 MHz, the standing wave ratio is 9.48 and the first voltage minima is situated at 6 cm from the load. Determine the complex reflection coefficient and the terminating impedance of the line.

Solution Given: $Z_0 = 350\ \Omega$, $f = 1200$ MHz $= 1200 \times 10^6$ Hz, VSWR $= 9.48$.

First voltage minima at 6 cm from load:

$$y_{min} = 6 \times 10^{-2} \text{ m}.$$

Wavelength λ of the wave in air is given by

$$\lambda = \frac{C}{f} = \frac{3 \times 10^8}{1200 \times 10^6} = 0.25 \text{ m}.$$

We know that $\beta = \frac{2\pi}{\lambda} = \frac{2 \times 3.14}{0.25} = 25.12$ radians.

For voltage minima, the condition is,

$$2\beta y_{\text{min}} - \phi = \pi.$$

$$2 \times 25.12 \times \frac{6}{100} - \phi = 3.14$$

$$\phi = -0.96 \text{ radians}$$

The magnitude of the reflection coefficient is given by

$$|K| = \frac{S-1}{S+1} = \frac{9.48-1}{9.48+1} = 0.80916.$$

Here, $K = |K|\, e^{j\phi} = 0.8096 e^{-j0.96}$.

$$|K| = 0.8096\angle 55^\circ.$$

Now, $K = \dfrac{Z_R - Z_0}{Z_R + Z_0}$

$$0.81\angle 55^\circ = \frac{Z_R - 350}{Z_R + 350}$$

$$(0.464 + j0.663)(Z_R + 350) = Z_R - 350$$

$$(0.464)\, Z_R + (0.464 \times 350) + j(Z_R)(0.663) + j0.633 \times 350 = Z_R - 350$$

$$Z_R[0.464 - 1 + j0.663] = -350 - 162.4 - j232.05$$

$$Z_R = \left[\frac{-512.4 - j232.05}{-0.536 + j0.663}\right] = \frac{562.5\angle -155.63}{0.8526\angle 128.95} = 659.74\angle -284.58$$

$$Z_R = 659.74\angle -284.58.$$

10.11 Impedance Transformation

Input impedance of a transmission line, depends on its length. Thus, a short length transmission line is used as an impedance matching device or impedance transformation. If the load impedance is not equal to the complex conjugate of the source impedance, a short length transmission line is added to the line to achieve maximum power transfer.

The important short length transmission lines are:

(1) The eighth wave ($\lambda/8$ length) transmission line

(2) The quarter wave ($\lambda/4$ length) transmission line

(3) The half wave ($\lambda/2$ length) transmission line.

Eighth wave transmission line

The length of the eighth wave transmission line is $\lambda/8$, where λ is the wavelength. Consider a $\lambda/8$ length transmission line terminated with impedance Z_R and characteristic impedance Z_0 as shown in Fig. 10.9.

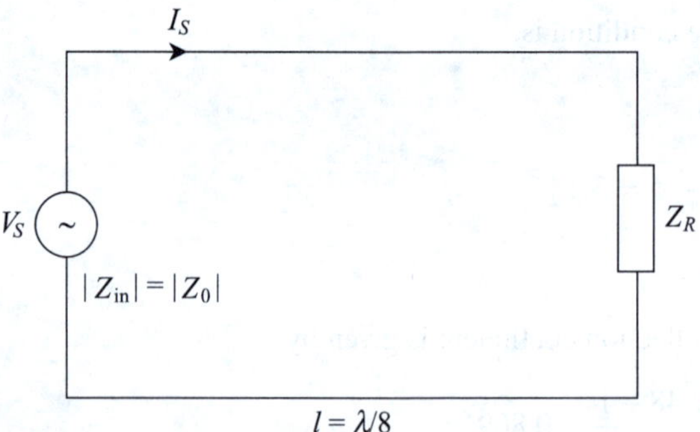

Fig. 10.9 *$\lambda/8$ length transmission line*

We know that the input impedance of a transmission line is

$$Z_{in} = Z_0 \left[\frac{Z_R + Z_0 \tanh \gamma l}{Z_0 + Z_R \tanh \gamma l}\right].$$

For lossless line, $\alpha = 0$ and $\gamma = j\beta$.

$$Z_{in} = Z_0 \left[\frac{Z_R + Z_0 \tanh j\beta l}{Z_0 + Z_R \tanh j\beta l}\right].$$

$$Z_{in} = Z_0 \frac{Z_R + jZ_0 \tan \beta l}{Z_0 + jZ_R \tan \beta l}. \tag{10.79}$$

For length $l = \lambda/8$,

$$\beta l = \frac{2\pi}{\lambda} \times \frac{\lambda}{8} = \frac{\pi}{4}.$$

$$\therefore \quad Z_{in} = Z_0 \frac{Z_R + jZ_0 \tan (\pi/4)}{Z_0 + jZ_R \tan (\pi/4)}$$

$$\text{or} \quad Z_{in} = Z_0 \left[\frac{Z_R + jZ_0}{Z_0 + jZ_R}\right]$$

$$\text{or} \quad |Z_{in}| = |Z_0| \left[\frac{Z_R + jZ_0}{Z_0 + jZ_R}\right].$$

$$\text{But} \quad |Z_R + jZ_0| = \sqrt{Z_R^2 + Z_0^2} = |(Z_0 + jZ_0)|. \tag{10.80}$$

$$\therefore \quad |Z_{in}| = |Z_0|.$$

The magnitude of the input impedance is equal to the magnitude of the characteristic impedance and is independent of load impedance.

Thus a $\lambda/8$ length transmission line is used to transform any impedance (Z_R) to a magnitude of characteristic impedance.

Quarter wave ($\lambda/4$) transmission line

Consider a $\lambda/4$ length transmission line terminated with impedance Z_R and characteristic impedance Z_0 as shown in Fig. 10.10. This $\lambda/4$ length transmission line is also called a quarter wave transformer.

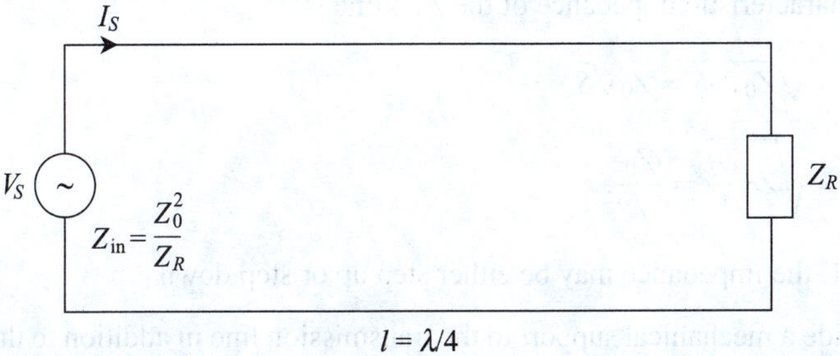

Fig. 10.10 *Quarter wavelength transmission line*

We know that the input impedance of a distortionless transmission line is $Z_{in} = Z_0 \dfrac{Z_R + jZ_0 \tan \beta l}{Z_0 + jZ_R \tan \beta l}$.

For length $l = \lambda/4$,

$$\beta l = \frac{2\pi}{\lambda} \times \lambda/4 = \pi/2.$$

$$\therefore \quad Z_{in} = Z_0 \frac{\dfrac{Z_R}{\tan \beta l} + jZ_0}{\dfrac{Z_0}{\tan \beta l} + jZ_R}.$$

Substituting, $\beta l = \pi/2$,

$$Z_{in} = Z_0 \frac{jZ_0}{jZ_R} \quad \text{or} \quad Z_0^2 = Z_{in} Z_R.$$

$$Z_0 = \sqrt{Z_{in} Z_R}. \tag{10.81}$$

Thus, a $\lambda/4$ length line is considered as a transformer that matches a load of Z_R to a source impedance of Z_{in}. The transformer characteristic impedance is equal to the geometric mean of the source and the load impedance.

Applications of a $\lambda/4$ line transformer

(1) To match the impedance between a transmission line and an antenna.
For example, if a line with characteristic impedance Z_0 is connected to a mismatched load Z_R, then for proper matching, a quarter wave transformer is inserted in between the line and load.
The characteristic impedance of the quarter wave transformer should be

$$Z_0 = \sqrt{Z_0 Z_R}. \tag{10.82}$$

(2) To step up or step down the characteristic impedance Z_0 of a transmission line.
If the load of a line is not resistive, the impedance of the line at node points is either SZ_0 or $\frac{Z_0}{S}$, irrespective of the load impedance, where S is the VSWR.

$\therefore$ The characteristic impedance of the $\lambda/4$ line is

$$Z_0' = \sqrt{Z_0 Z_0 S} = Z_0\sqrt{S}$$

or
$$Z_0' = \sqrt{Z_0 \frac{Z_0}{S}} = \frac{Z_0}{\sqrt{S}}. \tag{10.83}$$

Since $S > 1$, the impedance may be either step up or step down.

(3) It can provide a mechanical support to the transmission line in addition to the impedance. For example, the line connected between the transmission line and the ground acts as an insulator at the point of contact as shown in Fig. 10.11.

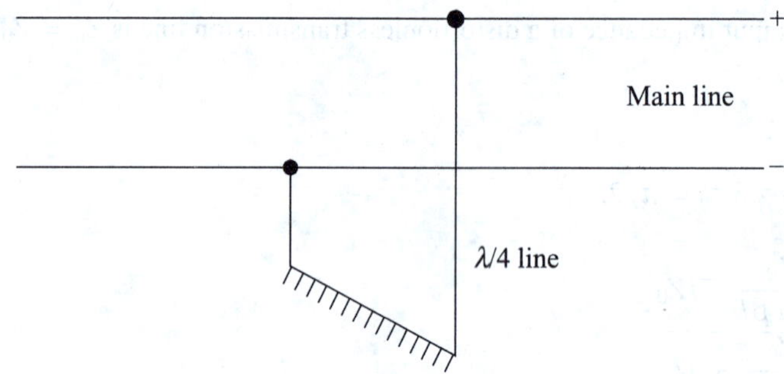

Fig. 10.11 *Quarter wave transformer as insulator*

Half wave transmission line

The length of the half wave transmission line is $\lambda/2$. We know that the input impedance of a distortionless transmission line is

$$Z_{\text{in}} = Z_0 \frac{Z_R + jZ_0 \tan \beta l}{Z_0 + jZ_R \sin \beta l}.$$

For length $l = \lambda/2$,

$$\beta l = \frac{2\pi}{\lambda} \times \lambda/2 = \pi.$$

$$\therefore \quad Z_{\text{in}} = Z_0 \frac{Z_R}{Z_0}$$

or
$$Z_{\text{in}} = Z_R. \tag{10.84}$$

Thus the input impedance of a half wave line is equal to its termination impedance as shown in Fig. 10.12.

Note: The termination impedance of the line repeats at every $\lambda/2$ distance.

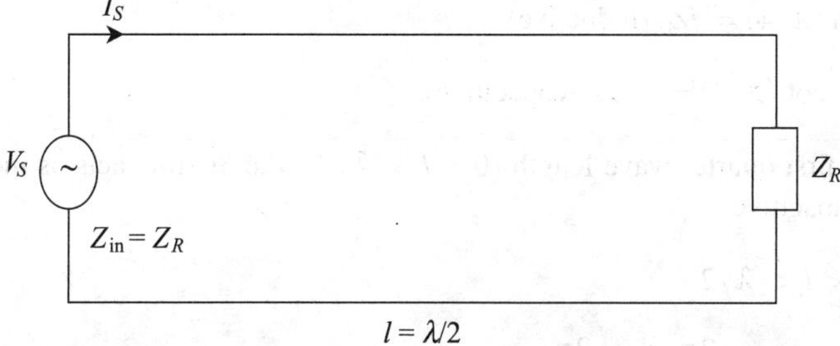

Fig. 10.12 *$\lambda/2$ length transmission line*

Application: If the load and source cannot be made adjacent, a half wavelength line may be connected at the load point for accurate measurements.

Example 10.7 Design a quarter wave transformer to match a line having impedance 75 Ω, to a load of 25 Ω.

Solution Given: $Z_R = 25\ \Omega$. line impedance $Z_{in} = 75\ \Omega$.

To match the load with line, the characteristic impedance of the quarter wave transformer should be

$$Z_0 = \sqrt{Z_{in} Z_R} = \sqrt{75 \times 25} = 43.3\ \Omega.$$

Therefore the characteristic impedance of the quarter wave transformer is 43.3 Ω.

10.12 Input Impedance of Lossless SC and OC Transmission Lines

From Eq. (10.9) and Eq. (10.10), we know that the input impedance of short circuited and open circuited lines are

$$Z_{sc} = Z_0 \tanh \gamma l \text{ and } \quad Z_{oc} = Z_0 \coth \gamma l.$$

For a lossless transmission line,

$$\gamma = j\beta,$$

$$Z_{sc} = Z_0 \tanh j\beta l = jZ_0 \tan \beta l \tag{10.85}$$

$$\text{and} \quad Z_{oc} = Z_0 \coth j\beta l = -jZ_0 \cot \beta l. \tag{10.86}$$

Since Z_0 is resistive, the input impedance for both open and short circuited lossless lines is pure reactive. Depending on the length, the transmission line can provide either a capacitive or an inductive effect.

Case (a): $l < \lambda/4$

Let $l = \lambda/8$. Then $\beta l = \frac{2\pi}{\lambda}\frac{\lambda}{8} = \frac{\pi}{4}$.

$$\therefore \quad Z_{sc} = jZ_0 \tan(\pi/4) = jZ_0 \text{ (inductive)}$$

$$\text{and} \quad Z_{oc} = -jZ_0 \cot(\pi/4) = -jZ_0 \text{ (capacitive).} \tag{10.87}$$

Therefore, in the first quarter wave length $(0 < l < \lambda/4)$, the SC line acts as inductive while the OC line acts as capacitive.

Case (b): $\lambda/4 < l < \lambda/2$

Let $l = \lambda/3$. Then $\beta l = \frac{2\pi}{\lambda}.\frac{\lambda}{3} = \frac{2\pi}{3}$.

$$\therefore \quad Z_{sc} = jZ_0 \tan\left(\frac{2\pi}{3}\right) = -j\sqrt{3}Z_0 \text{ (capacitive)}$$

$$\text{and} \quad Z_{oc} = -jZ_0 \tan\left(\frac{2\pi}{3}\right) = j\sqrt{3}Z_0 \text{ (inductive).} \tag{10.88}$$

Therefore for $(\lambda/4 < l < \lambda/2)$, the SC line acts as capacitive, while the OC line acts as inductive.

Case (c): $l = \lambda/4$

$$\beta l = \frac{2\pi}{\lambda} \times \frac{\lambda}{4} = \frac{\lambda}{2}.$$

$$\therefore \quad Z_{sc} = jZ_0 \tan(\lambda/2) = \pm\infty$$

$$\text{and} \quad Z_{oc} = -jZ_0 \tan(\lambda/2) = 0. \tag{10.89}$$

Therefore, for $= \lambda/4$, the SC line acts as open circuit while the OC line acts as short circuit.

Case (d): $l = \lambda/2$

$$\beta l = \frac{2\pi}{\lambda}.\frac{\lambda}{2} = \pi.$$

$$\therefore \quad Z_{sc} = jZ_0 \tan(\pi) = 0.$$

$$Z_{oc} = -jZ_0 \cot(\pi) = \pm\infty. \tag{10.90}$$

Therefore, for $l = \lambda/2$, the SC line acts as short circuit while the OC line acts as open circuit.

It is observed that after each quarter wavelength of the line, the nature of reactance reverses. The same reactance values repeat every half wavelength distance. Figure 10.13 shows the variation of input impedance for lossless SC and OC lines with length.

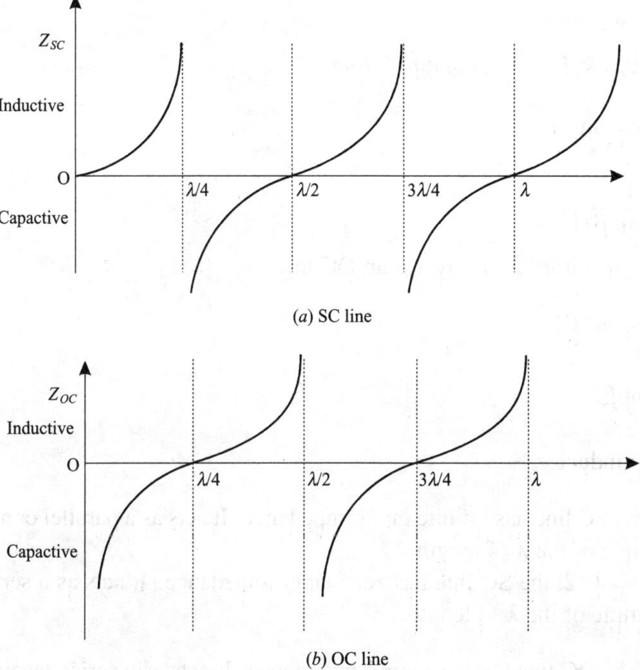

Fig. 10.13 *Input impedance variation for (a) SC line; (b) OC line.*

10.13 UHF Lines as Circuit Elements

At ultra high frequency, the transmission line becomes lossless. The short length SC and OC transmission lines can be used as circuit elements.

(*a*) *For lengths* $(0 < l < \lambda/4)$ *of an SC line*

We know that $Z_{sc} = jZ_0 \tan \beta l$.

If L_{eq} = equivalent inductance, then $j\omega L_{eq} = jZ_0 \tan \beta l$.

$$\therefore \quad L_{eq} = \frac{Z_0}{\omega} \tan \beta l. \tag{10.91}$$

Therefore the line acts as an inductor.

Similarly, for an OC line,

$$Z_{oc} = -jZ_0 \cot \beta l.$$

If C_{eq} = equivalent capacitance, then $\dfrac{1}{j\omega C_{eq}} = -jZ_0 \cot \beta l$.

$$C_{eq} = \frac{1}{\omega Z_0 \cot \beta l}. \tag{10.92}$$

The line acts as a capacitor.

(***b***) *For lengths* $(\lambda/4 < l < \lambda/2)$ *of an SC line*

$$\frac{1}{j\omega C_{eq}} = jZ_0 \tan \beta l$$

$$\text{or} \quad C_{eq} = \frac{-1}{\omega Z_0 \tan \beta l}. \tag{10.93}$$

The line acts as a capacitor. Similarly, for an OC line,

$$j\omega L_{eq} = jZ_0 \cot \beta l.$$

$$L_{eq} = \frac{Z_0}{\omega} \cot \beta l. \tag{10.94}$$

The line acts as an inductor.

(***c***) For $l = \lambda/4$, the SC line has infinite input impedance. It acts as a parallel or anti-resonant circuit at every odd multiple of the $\lambda/4$ length.

Similarly for $l = \lambda/2$, the SC line has zero input impedance. It acts as a series resonant circuit at every even multiple of the $\lambda/4$ length.

(***d***) For $l = \lambda/4$, the OC line has zero input impedance. It acts as a series resonant circuit at every odd multiple of the $\lambda/4$ length.

Similarly for $l = \lambda/2$, the OC line has infinite input impedance. It acts as a parallel or anti-resonant circuit at every even multiple of the $\lambda/4$ length.

Figure 10.14 shows the SC lines at different lengths with equivalent circuit elements and Fig. 10.15 shows the OC lines at different lengths with equivalent circuit elements.

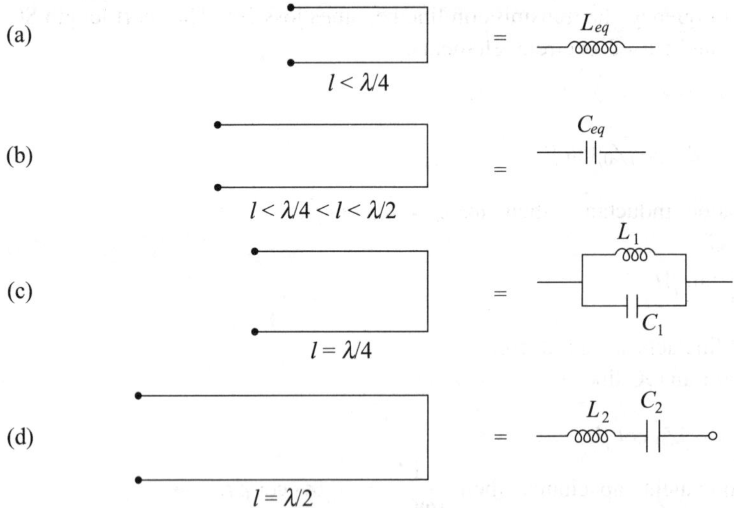

Fig. 10.14 *SC lines at different lengths with equivalent circuit elements*

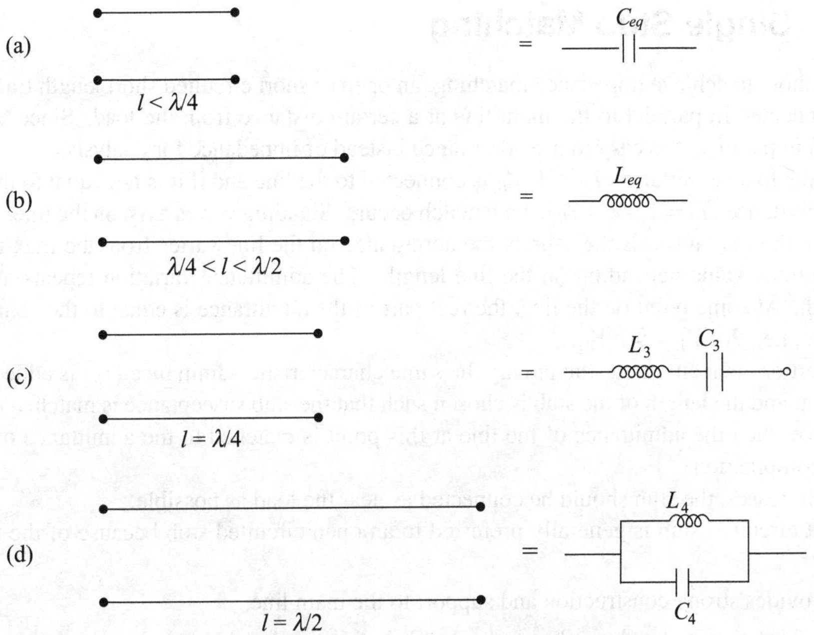

Fig. 10.15 *OC lines at different lengths with equivalent circuit elements*

10.14 Stub Matching

When a UHF line is terminated with a load impedance which is not equal to the characteristic impedance of the line, mismatch occurs. Mismatch reduces efficiency and increases power loss. To avoid mismatching, it is necessary to add impedance matching devices between the load and the line.

A quarter wave transformer can be used to achieve impedance matching but we have to cut the line to insert a transformer in between the line and the load.

The other method is to use open or short circuited short length transmission lines as a matching device, which can be connected in parallel to the line at a certain distance or distances from the load. This matching device is called stub matching.

Advantages

1) The length and characteristic impedance of the line remains the same.
2) Since the stub is added in shunt, there is no need to cut the line.
3) The susceptance of the stub can be adjusted for perfect matching.

Methods of stub matching

1) Single stub matching
2) Double stub matching

10.15 Single Stub Matching

In this method, to achieve impedance matching, an open or short circuited short length transmission line is connected in parallel to the main line at a certain distance from the load. Since the stub is connected in parallel, it is easy to use admittance instead of impedance for analysis.

When the load admittance, $Y_R = 1/Z_R$ is connected to the line and if it is not equal to the characteristic admittance $(Y_0 = 1/Z_0 \neq Y_R)$, a mismatch occurs. Standing waves exist on the line. When we move from the load towards the source, the admittance on the line varies from the max/min value to the min/max value depending on the line length. The admittance variation repeats after every $\lambda/2$ length. At some point on the line, the real part of the admittance is equal to the characteristic admittance, i.e., $\Re e[Y] = \Re e[Y_0]$.

If a short or open circuited stub having the same characteristic admittance (Y_0) is added in shunt at this point and the length of the stub is chosen such that the stub susceptance is matched to the line susceptance, then the admittance of the line at this point is matched to the admittance of the load and stub combination.

To avoid losses, the stub should be connected as near the load as possible.

A short circuited stub is generally preferred to an open circuited stub because of the following reasons.

1) It provides strong construction and support to the main line.
2) The short circuited stub can be easily established with a large metal plate.
3) Radiation loss is very less compared to an open circuited stub.

Length and location of the single stub

Consider a transmission line terminated with load admittance Y_R. Let a short circuited stub of length l_t at a distance l_s from the load be connected to the line as shown in Fig. 10.16.

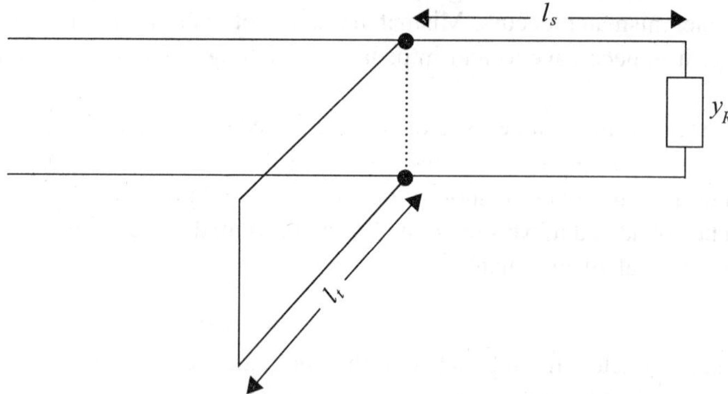

Fig. 10.16 *Single stub matching*

We know that the input impedance at any point on the line is

$$Z_{\text{in}} = Z_0 \frac{Z_R \cosh \gamma y + Z_0 \sinh \gamma y}{Z_0 \cosh \gamma y + Z_R \sinh \gamma y}$$

$$= Z_0 \frac{Z_R + Z_0 \tanh \gamma y}{Z_0 + Z_R \tanh \gamma y},$$

where y is the distance from the load.

At UHF for lossless line, $\alpha = 0,\ \gamma = j\beta$.

$$Z_{\text{in}} = Z_0 \frac{Z_R + jZ_0 \tan \beta y}{Z_0 + jZ_R \tan \beta y}.$$

Converting the impedance into admittance,

$$Y_{\text{in}} = \frac{1}{Z_{\text{in}}} = Y_0 \frac{Y_R + jY_0 \tan \beta y}{Y_0 + jY_R \tan \beta y}. \tag{10.95}$$

Since the load is variable, it has to be normalised.

Normalised load is $y_R = \dfrac{Y_R}{Y_0}$.

$\therefore$ Normalised input admittance is

$$y_{\text{in}} = \frac{Y_{\text{in}}}{Y_0} = \frac{\dfrac{Y_R}{Y_0} + j \tan \beta y}{1 + j\dfrac{Y_R}{Y_0} \tan \beta y}$$

or $$y_{\text{in}} = \frac{y_R + j \tan \beta y}{1 + jy_R \tan \beta y}. \tag{10.96}$$

Let Y_R and Y_0 be real. Separating real and imaginary terms,

$$y_{\text{in}} = \left(\frac{y_R + j\tan\beta y}{1 + jy_R\tan\beta y}\right)\left(\frac{1 - jy_R\tan\beta y}{1 - jy_R\tan\beta y}\right)$$

$$= \frac{y_R + y_R\tan^2\beta y - jy_R^2 \tan \beta y + \tan \beta y}{1 + y_R^2\tan^2\beta y}$$

$$y_{\text{in}} = \frac{y_R(1+\tan^2\beta y)}{1+y_R^2\tan^2\beta y} + j\frac{(1-y_R^2) \tan \beta y}{1+y_R^2\tan^2\beta y}. \tag{10.97}$$

For no reflection, at a distance $y = l_s$, the real part of the normalised admittance should be equal to unity.

That is, $\text{Re}[Y_{\text{in}}] = Y_0$ or $\text{Re}[y_{\text{in}}] = 1$.

$\therefore$ at $y = l_s$, $$\frac{y_R(1 + \tan^2 \beta l_s)}{1 + y_R^2 \tan^2 \beta l_s} = 1$$

or $$(y_R + y_R \tan^2 \beta l_s) = 1 + y_R^2 \tan^2 \beta l_s.$$

$$(y_R - y_R^2) \tan^2 \beta l_s = 1 - y_R.$$

$$\tan^2 \beta l_s = \frac{1-y_R}{y_R(1-y_R)} = \frac{1}{y_R}.$$

$$\therefore \quad \tan \beta l_s = \frac{1}{\sqrt{y_R}} = \sqrt{\frac{Y_0}{Y_R}}$$

$$\text{or} \quad l_s = \frac{1}{\beta} \tan^{-1} \sqrt{\frac{Y_0}{Y_R}}. \tag{10.98}$$

Therefore the location of the short circuited stub is

$$l_s = \frac{\lambda}{2\pi} \tan^{-1} \sqrt{\frac{Z_0}{Z_R}}. \tag{10.99}$$

At this location, the imaginary part of y_{in} (susceptance) from Eq. (10.97) is

$$\text{Im}\,[y_{\text{in}}] = b_s.$$

$$b_s = \frac{(1-y_R^2)\tan \beta l_s}{1+y_R^2 \tan^2 \beta l_s}. \tag{10.100}$$

Substituting the value of tan βl_s from Eq. (10.98),

$$b_s = \frac{(1-y_R^2)\dfrac{1}{\sqrt{y_R}}}{1+y_R^2 \dfrac{1}{y_R}} = \frac{1-y_R^2}{\sqrt{y_R}(1+y_R)} = \frac{1-y_R}{\sqrt{y_R}}.$$

$$\therefore \quad b_s = \frac{1-\dfrac{Y_R}{Y_0}}{\sqrt{\dfrac{Y_R}{Y_0}}} = \frac{(Y_0 - Y_R)}{\sqrt{Y_R Y_0}}. \tag{10.101}$$

Therefore at length l_s,

$$y_{\text{in}} = 1 + jb_s. \tag{10.102}$$

If a short-circuited stub is added in parallel at this point with susceptance equal to $-jb_s$, then admittance is

$$y_{\text{in}} = 1 + jb_s - jb_s.$$

$$y_{\text{in}} = 1 \text{ or } \quad Y_{\text{in}} = Y_0.$$

Thus, matching is achieved at a distance l_s from the load.
To find the length of the short circuited stub:

We know that the impedance of the short circuited stub line with length l_t is

$$Z_{sc} = jZ_0 \tan \beta l_t.$$

The susceptance is

$$-jB_s = \frac{1}{Z_{sc}} = \frac{1}{jZ_0 \tan \beta l_t} \quad \text{or} \quad B_s = Y.\cot \beta l_t$$

$$\text{or} \quad b_s = \frac{B_s}{Y_0} = \cot \beta l_t. \tag{10.103}$$

Equating with Eq. (10.101),

$$\cot \beta l_t = \frac{Y_0 - Y_R}{\sqrt{Y_R Y_0}}$$

$$\text{or} \quad \tan \beta l_t = \frac{\sqrt{Y_R Y_0}}{Y_0 - Y_R}. \tag{10.104}$$

Converting into impedances,

$$\tan \beta l_t = \frac{\dfrac{1}{\sqrt{Z_R Z_0}}}{\dfrac{1}{Z_0} - \dfrac{1}{Z_R}} = \frac{1}{\sqrt{Z_R Z_0}} \times \frac{Z_0 Z_R}{Z_R - Z_0} = \frac{\sqrt{Z_0 Z_R}}{Z_R - Z_0}. \tag{10.105}$$

$$l_t = \frac{1}{\beta} \tan^{-1}\left(\frac{\sqrt{Z_R Z_0}}{Z_R - Z_0}\right).$$

Therefore the length of the short circuited stub is

$$l_t = \frac{\lambda}{2\pi} \tan^{-1}\left(\frac{\sqrt{Z_R Z_0}}{Z_R - Z_0}\right) \text{ for } (Z_R > Z_0). \tag{10.106}$$

If $Z_R > Z_0$, to obtain a positive length:

$$l_t = \frac{\lambda}{2\pi}\left[\pi - \tan^{-1}\left(\frac{\sqrt{Z_R Z_0}}{Z_0 - Z_R}\right)\right] \text{ for } Z_R > Z_0. \tag{10.107}$$

$$(\text{since } \tan^{-1}(-x) = -\tan^{-1} x)$$

Note: The stub susceptance cancels the susceptance of the line at stub point so that the line is terminated with the characteristic impedance Z_0 at the place of stub connection.

Stub distance l_s and a stub length l_t in terms of reflection coefficient K:

We know that the input impedance at any point on the line is

$$Z_{\text{in}} = Z_0 \frac{(1 + Ke^{-2\gamma l})}{(1 - Ke^{-2\gamma l})}.$$

Converting into admittance

$$Y_{\text{in}} = Y_0 \frac{(1 - Ke^{-2\gamma l})}{(1 + Ke^{-2\gamma l})}$$

Assuming that the characteristic impedance of the line is resistive, the normalised admittance is

$$y_{\text{in}} = \frac{Y_{\text{in}}}{Y_0} = \frac{1-|K|e^{j(\phi-2\beta l)}}{1+|K|e^{j(\phi-2\beta l)}}$$

Separating real and imaginary terms, we get

$$y_{\text{in}} = g_{\text{in}} + jb_{\text{in}} = \frac{1-|K|^2}{1+|K|^2+2|K|\cos(\phi-2\beta l)} - \frac{j2|K|\sin(\phi-2\beta l)}{1+|K|^2+2|K|\cos(\phi-2\beta l)}. \tag{10.108}$$

For no reflection, at a distance $l = l_s$, the real part of the normalised admittance should be equal to unity.

$$\text{At} \quad l = l_s, \quad \frac{1-|K|^2}{1+|K|^2+2|K|\cos(\phi-2\beta l_s)} = 1.$$

$$\phi - 2\beta l_s = \cos^{-1}(-|K|). \tag{10.109}$$

$$\text{Since,} \quad \cos^{-1}(-|K|) = \pi - \cos^{-1}(|K|),$$

$$l_s = \frac{\phi + \pi - \cos^{-1}(|K|)}{2\beta}$$

$$\text{or} \quad l_s = \frac{\lambda(\phi + \pi - \cos^{-1}(|K|))}{2\pi}. \tag{10.110}$$

At this location, the imaginary part of y_{in} (susceptance) from Eq. (10.108) is

$$b_{\text{in}} = -\frac{2|K|\sin(\phi-2\beta l_s)}{1+|K|^2+2|K|\cos(\phi-2\beta l_s)}$$

Substituting the value of $\phi - 2\beta l_s$ from Eq. (10.109),

$$b_{\text{in}} = -\frac{2|K|\sin(-\pi+\cos^{-1}(|K|))}{1+|K|^2+2|K|\times|K|}$$

$$\text{since,} \quad \sin(-\pi+\cos^{-1}(-|K|)) = \sqrt{1-|K|^2}$$

$$b_{\text{in}} = -\frac{2|K|\sqrt{1-|K|^2}}{1+|K|^2+2|K|^2} = \frac{2|K|}{\sqrt{1+|K|^2}}$$

To find the length of the short circuited stub, we know that the impedance of the short circuited stub line with length l_t is

$$Z_{sc} = jZ_0 \tan\beta l_t$$

$$b_{\text{in}} = \frac{Z_0}{Z_{sc}} = \tan\beta l_t = \frac{\sqrt{1+|K|^2}}{2|K|}$$

$$l_t = \frac{1}{\beta}\tan^{-1}\frac{\sqrt{1+|K|^2}}{2|K|} \tag{10.111}$$

Disadvantages of single matching

1 Since the location and length of the single stub matching depend on frequency, if the frequency of the wave changes, the location and length of the stub should be changed. However, it is very difficult to change the stub once it is fixed.

2 In practical cases, the location of the stub has to be moved along the line for final adjustment. This tuning is possible only on open wire lines. However, it is very difficult to place a stub on coaxial cables.

Example 10.8 A lossless line of 300 Ω impedance is terminated with a load impedance of $100 + j650\ \Omega$. The frequency of operation is 60 MHz. Find the length and location of a single stub needed for impedance match.

Solution Given: $f = 60$ MHz, $Z_R = 100 + j650\ \Omega$, $Z_0 = 300\ \Omega$.

$$\text{Wavelength,}\quad \lambda = \frac{v_0}{f} = \frac{3\times 10^8}{60\times 10^6} = 5 \text{ m.}$$

$$\text{Reflection coefficient,}\quad K = \frac{Z_R - Z_0}{Z_R + Z_0} = \frac{100 + j650 - 300}{100 + j650 + 300} = \frac{-200 + j650}{400 + j650}.$$

$$K = \frac{680\angle 107.1^\circ}{763.2\angle 58^\circ} = 0.89\angle 48.7^\circ.$$

We know that the length of stub

$$l_t = \frac{\lambda}{2\pi}\tan^{-1}\frac{\sqrt{1-|K|^2}}{2|K|} = \frac{5}{2\pi}\tan^{-1}\frac{\sqrt{1-(0.89)^2}}{2\times 0.89} = 0.2 \text{ m.}$$

Location of the stub from the load

$$l_s = \frac{\phi + \pi - \cos^{-1}|K|}{2\beta} = \frac{48.7^\circ + 180^\circ - \cos^{-1}(0.89)}{4\pi}\times\lambda.$$

$$l_s = \frac{201.57^\circ}{4\times 180^\circ}\times 5,\quad l_s = 1.4 \text{ m.}$$

10.16 Double Stub Matching

To overcome the disadvantages of single stub matching, two stubs can be used at different locations. This is called double stub matching.

Consider a double stub matching system consisting of two short circuited stubs connected in parallel to the line near the load as shown in Fig. 10.17.

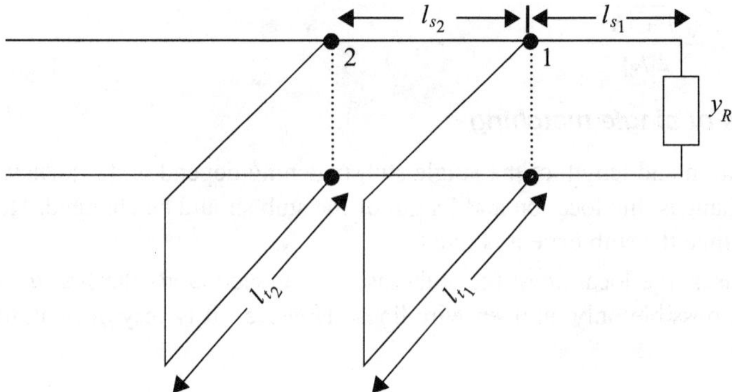

Fig. 10.17 *Double stub matching*

Let l_{s_1} = location of stub A from load, l_{s_2} = location of stub B from load, l_{t_1} = length of stub A, l_{t_2} = length of stub B, l_s = separation between stubs and Y_R = load admittance.

The characteristic admittance of the stubs should be equal to the line characteristic admittance Y_0.

In this system, the locations of the stubs have to be chosen arbitrarily. But in practice, it is very difficult to design stub matching with arbitrary locations. Due to some problems encountered in the design, the locations of stubs should be restricted.

Since the admittance repeats at every $\lambda/2$, the total distance $l_{s_1}+l_{s_2}$ can never be more than or equal to $\lambda/2$. In practice, l_{s_1} is taken in between 0.1 λ and 0.15 λ. Sometimes, $l_{s_1}=0$ (at load) can be chosen. The space between two stubs is generally taken as $\lambda/8$, $\lambda/4$ or $3\lambda/8$ distance. The total distance of the double stab matching should be kept as small as possible.

Therefore, in the design of a double stub matching, it is better to keep the locations of the stubs fixed. Impedance matching is done by finding the lengths of the stubs. When the signal frequency changes, the stub lengths can be adjusted to achieve impedance matching.

Design of double stub matching

1. Fix the locations of the stubs.
2. The normalised admittance at the location of stub 1 from the load is (from Eq. (10.96)),

$$y_A = \frac{y_R + j\tan\beta l_{s_1}}{1 + jy_R\tan\beta l_{s_1}}$$

or $$y_A = \frac{y_R(1+\tan^2\beta l_{s_1})}{1+y_R^2\tan^2\beta l_{s_1}} + j\frac{(1-y_R^2)\tan\beta l_{s_1}}{1+y_R^2\tan^2\beta l_{s_1}} = g_A + jb_A.$$

3. Connect a short-circuited stub 1 having susceptance $\pm jb_1$ to the line at 1. The admittance at 1 after stub connection is

$$y'_A = g_A + j(b_A \pm b_1).$$

4. For impedance matching, the admittance of the line at the location of stub 2 should be equal to the characteristic admittance. That is,

$$Y'_B = Y_0.$$

The normalised admittance at stub 2 should be equal to

$$y_B = 1 \pm jb_2.$$

5. Connect a short-circuited stub 2 having susceptance $\mp jb_2$ to the line at 2.
6. The stub lengths can be obtained from Eq. (10.103),

$$b_1 = \cot \beta l_{t_1}, \qquad b_2 = \cot \beta l_{t_2}. \tag{10.112}$$

10.17 Smith Chart

Phillip H. Smith in the year 1939 developed a polar chart for calculating transmission line characteristics. This chart is called Smith chart. It consists of two sets of orthogonal circles which represent the values of normalised impedance. One set of circles represents the resistive component R, called R circles, and the other set of circles represent the reactive component X, called X circles.

Derivation of R circles and X circles

Consider a transmission line having characteristic impedance Z_0 terminated with Z_R. The reflection coefficient is

$$K = \frac{Z_R + Z_0}{Z_R + Z_0} = \frac{\dfrac{Z_R}{Z_0} - 1}{\dfrac{Z_R}{Z_0} + 1} = \frac{z_R - 1}{z_R + 1},$$

where $z_R = \dfrac{Z_R}{Z_0}$ is the normalised load impedance.

$$z_R = \frac{1+K}{1-K}.$$

Since z_R and K are complex quantities,

$$R + jX = \frac{1 + K_R + jK_X}{1 - K_R - jK_X},$$

where R and X are the real and imaginary parts of z_R, K_R and K_X are the real and imaginary parts of K respectively. Separating the real and imaginary terms,

$$R + jX = \frac{(1 + K_R + jK_X)(1 - K_R + jK_X)}{(1 - K_R - jK_X)(1 - K_R + jK_X)}$$

$$= \frac{1 - K_R + jK_X + K_R - K_R^2 + jK_RK_X + jK_X - jK_RK_X - K_X^2}{(1 - K_R)^2 + K^2}.$$

$$R + jX = \frac{1 - K_R^2 - K_X^2 + j2K_X}{(1 - K_R)^2 + K_X^2}. \tag{10.113}$$

$$\therefore \quad R = \frac{1 - K_R^2 - K_X^2}{(1 - K_R)^2 + K_X^2} \tag{10.114}$$

and $$X = \frac{2K_X}{(1-K_R)^2+K_X^2}. \tag{10.115}$$

Arranging the above equations in standard form, the real part is

$$R = \frac{1-K_R^2-K_X^2}{(1-K_R)^2+K_X^2}.$$

$$R(1-K_R)^2+RK_X^2 = 1-K_R^2-K_X^2$$

$$R\left(1+K_R^2-2K_R\right)+RK_X^2 = 1-K_R^2-K_X^2$$

$$(1+R)K_X^2+(1+R)K_R^2+R-2RK_R = 1.$$

$$K_R^2+K_X^2-\frac{2R}{(1+R)^2}K_R = \frac{1-R}{1+R}.$$

$$\left(K_R-\frac{R}{1+R}\right)^2-\frac{R^2}{(1+R)^2}+K_X^2 = \frac{1-R}{1+R}$$

$$\left(K_R-\frac{R}{1+R}\right)^2+K_X^2 = \frac{1-R}{1+R}+\frac{R^2}{(1+R)^2}$$

$$\left(K_R-\frac{R}{1+R}\right)^2+K_X^2 = \frac{1}{(1+R)^2}. \tag{10.116}$$

This equation represents a family of R circles on the K plane (reflection coefficient plane) as shown in Fig. 10.18.

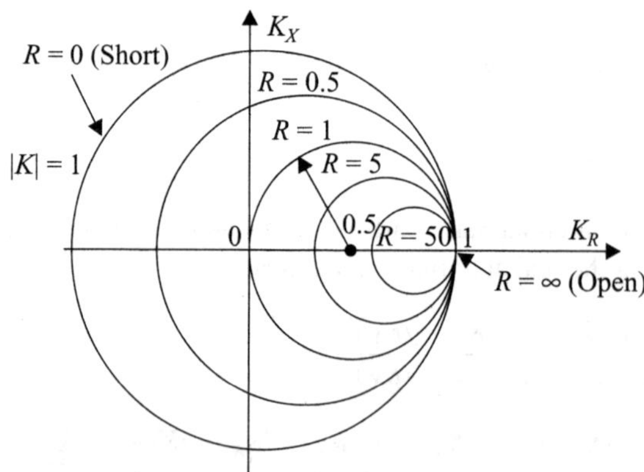

Fig. 10.18 *Family of constant R circles*

These circles are called constant R circles, having centres at $(R/(1+R),0)$ and radii of $(1/(1+R))$. A set of circles can be generated at different values of R.

At $R = 0$, the centre of the circle is at (0, 0) and the radius is 1. This is an outer circle. As R increases, the circle radius decreases. $R = \infty$, represents a point at (1, 0). All circles touch the point (1, 0). Now the imaginary part is

$$X = \frac{2K_X}{(1-K_R)^2 + K_X^2}.$$

$$(1-K_R)^2 + K_X^2 = \frac{2K_X}{X}$$

$$(1-K_R)^2 + K_X^2 - \frac{2K_X}{X} = 0$$

$$(K_R - 1) + \left(K_X - \frac{1}{X}\right)^2 - \frac{1}{X^2} = 0$$

$$(K_R - 1)^2 + \left(K_X - \frac{1}{X}\right)^2 - \frac{1}{X^2} = 0. \tag{10.117}$$

This equation represents another family of circles on the K-plane as shown in Fig. 10.19.

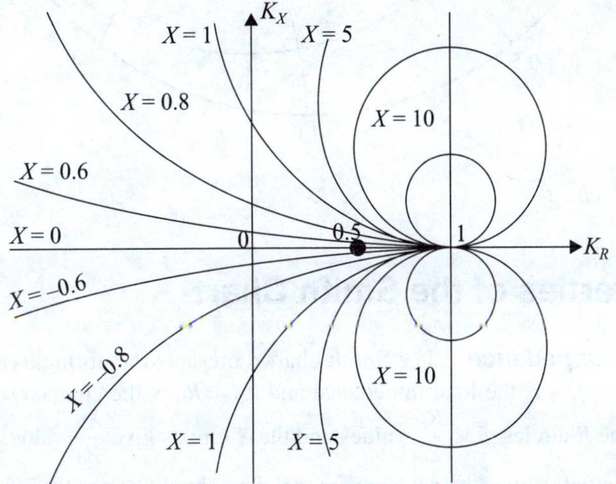

Fig. 10.19 *Family of constant X circles*

These circles are called constant X circles, having centres at $(1,\ 1/X)$ and radii of $1/X$.

$X = 0$ represents a straight line along the K_R-axis. If X is positive, then K_X is positive and the circles are generated above the $K_R = 0$ axis. If X is negative, then K_X is negative and the circles are generated below the $K_R = 0$ axis. $X = \infty$ represents a point at (1, 0). All circles touch the point (1, 0).

The complete Smith chart can be obtained by the superposition of the family of the constant-R circles and constant- X circles on the K plane as shown in Fig. 10.20.

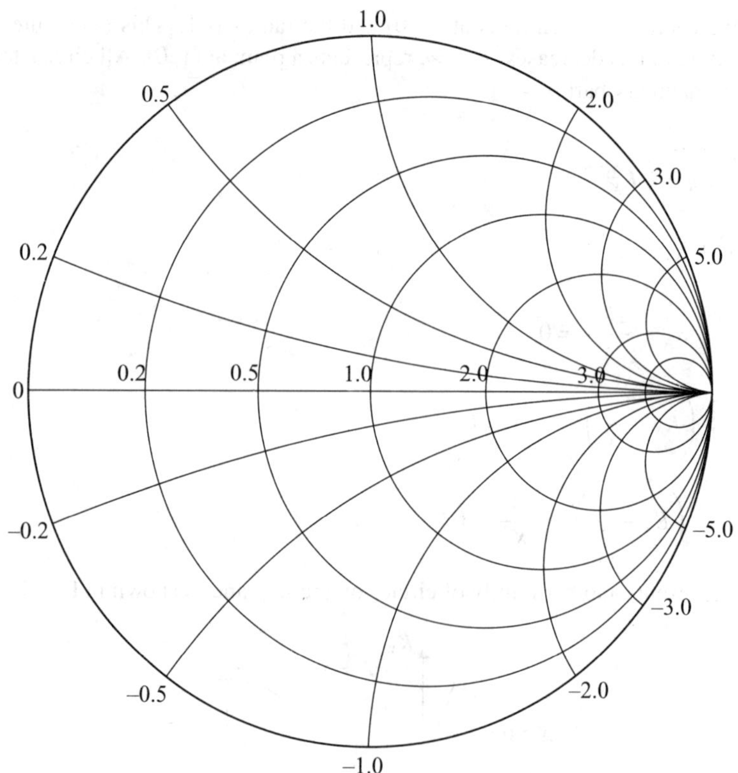

Fig. 10.20 *Smith chart*

10.18 Properties of the Smith Chart

Normalisation of impedance The Smith chart represents the normalised values of R and X circles. If $Z_R = R_R + jX_R$ is the load impedance and $Z_0 = R_0$ is the characteristic impedance of a lossless line, then the R circles give $\frac{R_R}{Z_0}$ values and the X circles give $\frac{X_R}{Z_0}$ values.

Thus, to obtain actual values, the values from the chart should be multiplied by Z_0.

Load impedance plot The intersection points of R and X circles give normalised load impedance (z_R) values. If X is positive, the point is above the $K_R = 0$ axis and if X is negative, the point is below $K_R = 0$ axis.

VSWR plot The VSWR values can be obtained by drawing S-circles on the chart, as shown in Fig. 10.21. The circles with a centre at the origin (0, 0) with radius $|z_R|$ are called S circles.

The radius of the S circle $|z_R|$ gives VSWR value $S = \left|\frac{Z_R}{Z_0}\right|$.

If M is the intersection point of an S circle with the horizontal axis AB, the normalised resistance at M is equal to the value of VSWR. Therefore, $S =$ OM.

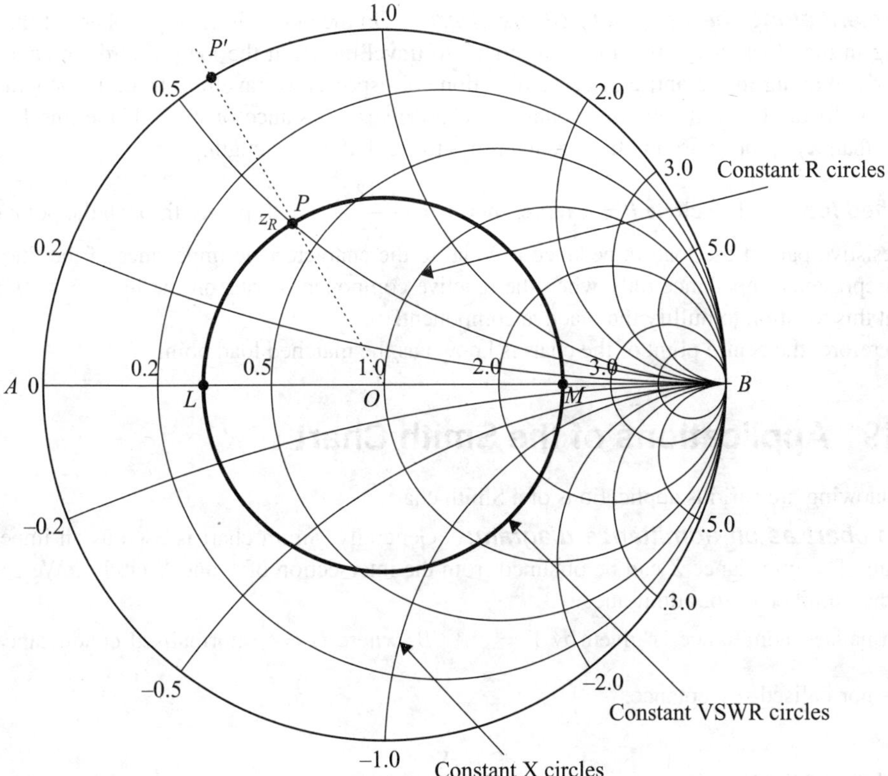

Fig. 10.21 *S circle on Smith chart*

Reflection coefficient values Draw the line OP and extend it to the outer circle, It cuts the outer circle at P$'$. The angles are indicated on the outer circle. The angle (θ) of the line OP gives the angle of the reflection coefficient. The value (K_R, K_X) at the point P gives the magnitude of K. $|K|$ can be obtained from the K scale provided in the chart. The length of OP on the K scale gives the magnitude of K. Also $|K|$ can be calculated directly as $|K| = \dfrac{\text{OP}}{\text{OP}'}$.

Location of voltage maximum and minimum: There are two intersection points of the S circle with the horizontal axis AB. The point at the left side of the centre represents voltage minima (V_{min}) and the point at the right side of the centre represents voltage maxima (V_{max}). Thus the points M and L respectively gives the positions voltage maximum and voltage minimum.

The location of the first V_{max} can be obtained from the wavelength scale on the outer circle. The arc AP$'$ gives the distance of V_{min} from the load. Similarly, the arc P$'$AB gives the distance of the first V_{max} from the load.

Open and short-circuited line: At point B on the right side end of the horizontal axis, both R and X are infinite which represents an open circuit termination of the line. Similarly, at point A on the left side end of the horizontal axis, both R and X are zero which represent short circuit termination of the line.

Movement along the periphery of the chart: On the outer circle or periphery of the chart, moving in the clockwise direction corresponds to travelling from the load towards the generator. Similarly, moving in the anti-clockwise direction corresponds to travelling from the generator towards the load. The full rotation around the chart gives a distance of $\lambda/2$. If the line length is greater than $\lambda/2$, rotate around the circle n times to reach the line length.

Matched load: The circle $R = 1$ represents $z_R = \dfrac{\Re e[Z_R]}{Z_0} = 1$. It passes through the point (1, 0). The resistive part of the load impedance is equal to the characteristic impedance of the line. This circle represents impedance only when the reactive component varies on the line. A stub can be used at this location to nullify the reactive component.

Therefore, the centre point of the chart is known as the matched load point.

10.19 Applications of the Smith Chart

The following are various applications of a Smith chart.

Smith chart as an admittance diagram: Generally, Smith chart is used as an impedance diagram. The impedance Z can be obtained from the intersection of R and X circles. We can also draw the Smith chart for admittance.

Normalised admittance is given by $Y = G + jB$, where $G = \dfrac{1}{R}$ normalised conductance, and $B = \dfrac{1}{X}$ normalised susceptance.

$$\therefore \quad \text{At termination, } y_R = \frac{1}{z_R} = \frac{1}{R + jX} = \frac{R - jX}{R^2 + X^2}.$$

$$\text{So,} \quad G - jB = \frac{R}{R^2 + X^2} - j\frac{X}{R^2 + X^2},$$

$$\text{also,} \quad y_R = \frac{1}{z_R} = \frac{1 - K}{1 + K}.$$

$$G - jB = \frac{1 - K}{1 + K}.$$

The G and B circles on the K plane can be drawn similar to R and X circles.

A Smith chart with G and B circles is called an admittance diagram. The admittance diagram is the mirror image of the impedance diagram; all measurements will be taken in the reverse direction.

Converting impedance into admittance: We know that for a lossless quarter wave transformer, if Z_R is the characteristic impedance and Z_0 is the termination impedance, then the input impedance is

$$Z_{\text{in}} = \frac{Z_0^2}{Z_R}.$$

$$\frac{Z_{in}}{Z_0} \times \frac{Z_R}{Z_0} = 1 \text{ or } \quad z_{in} \times z_R = 1.$$

$$y_R = z_{in}.$$

Thus, in order to find the admittance on the chart, we first locate the impedance point and rotate it to a distance $\lambda/4$ (quarter wave) towards the generator. The $\lambda/4$ distance is an opposite point on the chart. Hence, the point opposite to the impedance point on the circle gives the admittance point.

Determination of an input impedance: Consider a transmission line of length l terminated with a load impedance Z_R. Denote the normalised load impedance z_R as point P on the Smith chart as shown in Fig. 10.22. With centre O and radius OP, draw the S circle. Extend the OP line to the outer circle which cuts at point P$'$. Rotate towards the generator (clockwise) up to a length l/λ. Denote this point as N'. Draw the line ON$'$ which cuts the S circle at point N. This point N represents the normalised input impedance z_{in}.

The angle NOP gives the electric length βl of the line.

Determination of the load impedance: Consider a transmission line of length l. Given the VSWR and the location of the first V_{min} point from the load. To locate the load impedance, first draw the S circle with centre O and radius VSWR. Locate point A on the outer circle at the left side end of the horizontal axis as the position of V_{min}.

Move towards the load (clockwise direction) to a given length l/λ on the wavelength scale and locate the point P. Draw the line OP which cuts the S circle at point P. The location of the point P gives the normalised load impedance.

Input impedance and admittance of an SC (short-circuited) line We know that the input impedance of an SC line is purely reactive and $R = 0$. The short circuit termination represents the position of V_{min}, i.e., a point A on the outer circle at left side end of the horizontal axis (i.e., $R = 0$ on the R circle). From point A, move towards the generator (anti-clockwise direction) to a given length l/λ on the wavelength scale and locate the point P. The location of point P gives normalised input impedance of the SC line. The opposite point Q gives the normalised input admittance.

Input impedance and admittance of an OC (open-circuited) line: We know that the input impedance of an OC line is purely reactive and $R = \infty$. The open circuit termination represents the position of V_{max}, i.e., a point B on the outer circle at a right side end of the horizontal axis (i.e., $R = \infty$ on R-circle). From point B, move towards the generator to a given length l/λ on the wavelength scale and locate the point P. The location of the point P gives normalised input impedance of the OC line. The opposite point Q gives the normalised input admittance.

Determination of locations and lengths of stubs by Smith chart: Design of impedance matching can be easily done by using Smith charts. The locations and lengths of single and double stub matching can be obtained by locating the admittances on the chart. Since the stubs are connected in parallel, it is much easier to combine admittance in parallel than impedances. Also, short-circuited stubs are preferred over open-circuited stubs to avoid radiation losses.

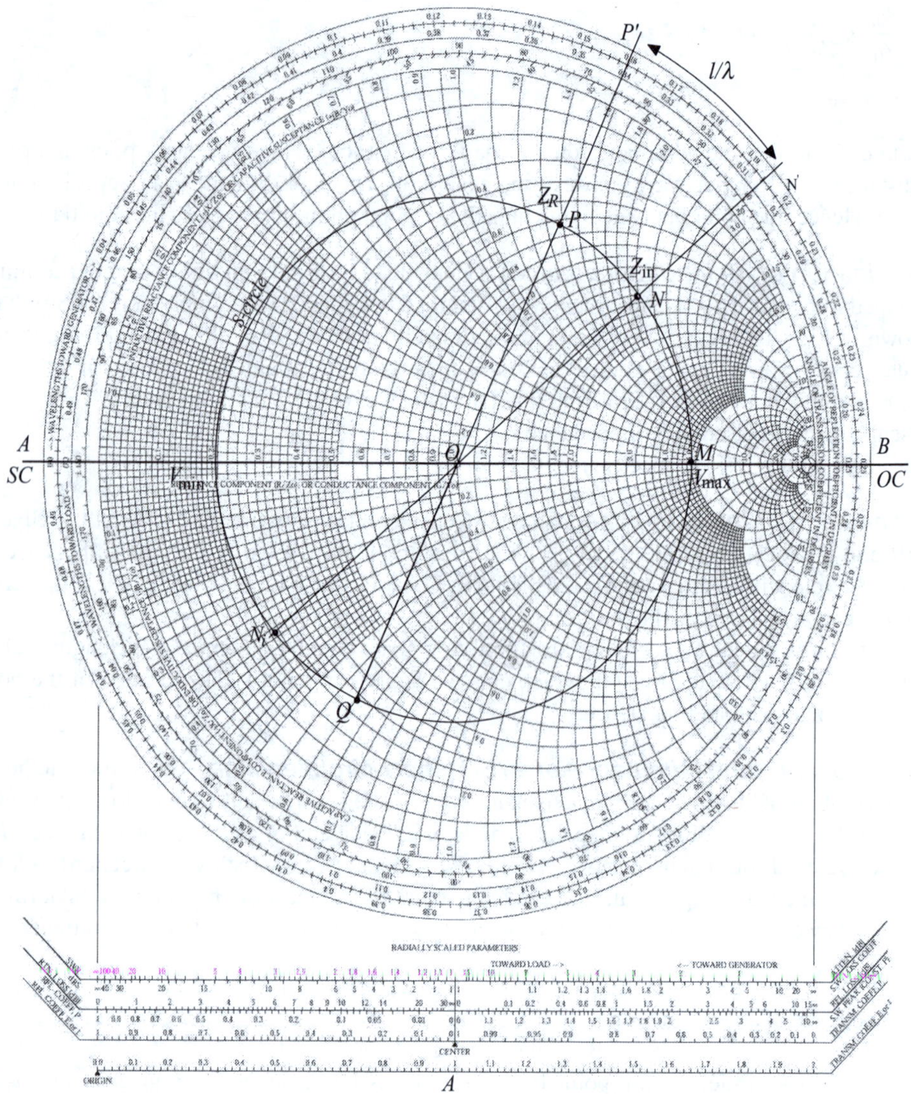

Fig. 10.22 *Smith chart for determination of impedances*

10.20 Single Stub Matching Using Smith Chart

Consider a transmission line of length l terminated with a load impedance Z_R. Assume that a single stub having a characteristic impedance same as that of the line characteristic impedance and length l_t is connected in parallel to the main line at a distance l_s from the load.

Steps for using Smith chart:

1) Locate the normalised load impedance (z_R) point as P on the chart shown in Fig. 10.23.
2) Locate the normalised load admittance point Q which is opposite to P.

3) Draw the S circle with centre O and radius OP. The length OP on the horizontal axis gives VSWR value before the stub is connected.
4) Locate the point T at the intersection of the S circle with the $R = 1$ circle. There are two intersection points. The point nearer to the load will be taken as the point T.
5) Extend the line OQ on to the outer circle to a point Q′ and extend the line OT on to the outer circle to a point T′. The distance Q′T′ measured on a wavelength scale towards the generator gives the location of the stub on the line from the load.
6) The value of the intersection point T gives the susceptance of the line at the stub location. To nullify the line susceptance, a stub of the same negative susceptance value should be connected. Locating point C with the negative line susceptance on the outer circle gives the susceptance of the stub.
7) The input admittance of the SC stub represents the point B (i.e., $R = \infty$ on the R circle). Measure the distance on the wavelength scale from point B to C. This distance gives the length of the SC stub.
8) Similarly, the input admittance of the OC stub represents the point A (i.e., $R = 0$ on R circle). Measure the distance on the wavelength scale from point A to C. This distance gives the length of the OC stub.

Figure 10.23 shows the Smith chart for determination of length and location of a single stub matching.

Example 10.9 A line characteristic impedance 300 Ω is terminated with a load of $175 + j207$ Ω. An electrical signal of 200 MHz is transmitted along the line in free space.

Determine the following using the Smith chart:

a) Voltage standing wave ratio (VSWR)

b) Load admittance

c) Distance between the load and the first voltage minimum along the transmission line.

Solution Given: Characteristic impedance, $Z_0 = 300$ Ω,
Load impedance $Z_R = Z_L = 175 + J207$ Ω, frequency $f = 200$ MHz.

$$\therefore \quad \text{Signal wavelength } \lambda = \frac{c}{f} = \frac{3 \times 10^8}{200 \times 10^6} = 1.5 \text{ m.}$$

Steps for using the Smith chart:

1. Normalised load impedance is

$$z_L = \frac{Z_L}{Z_0} = \frac{175 + j207}{300} = 0.58 + j0.69.$$

The point P is located on the Smith chart shown in Fig. 10.24 at the intersection of the circles $R = 0.58$ and $X = 0.69$.

2. Taking the centre as point O and the radius as OP, the impedance circle is drawn. The distance between the centre and point M where the impedance circle crosses the horizontal axis on the right side of the chart gives the voltage standing wave ratio.

VSWR = OM= 2.8.

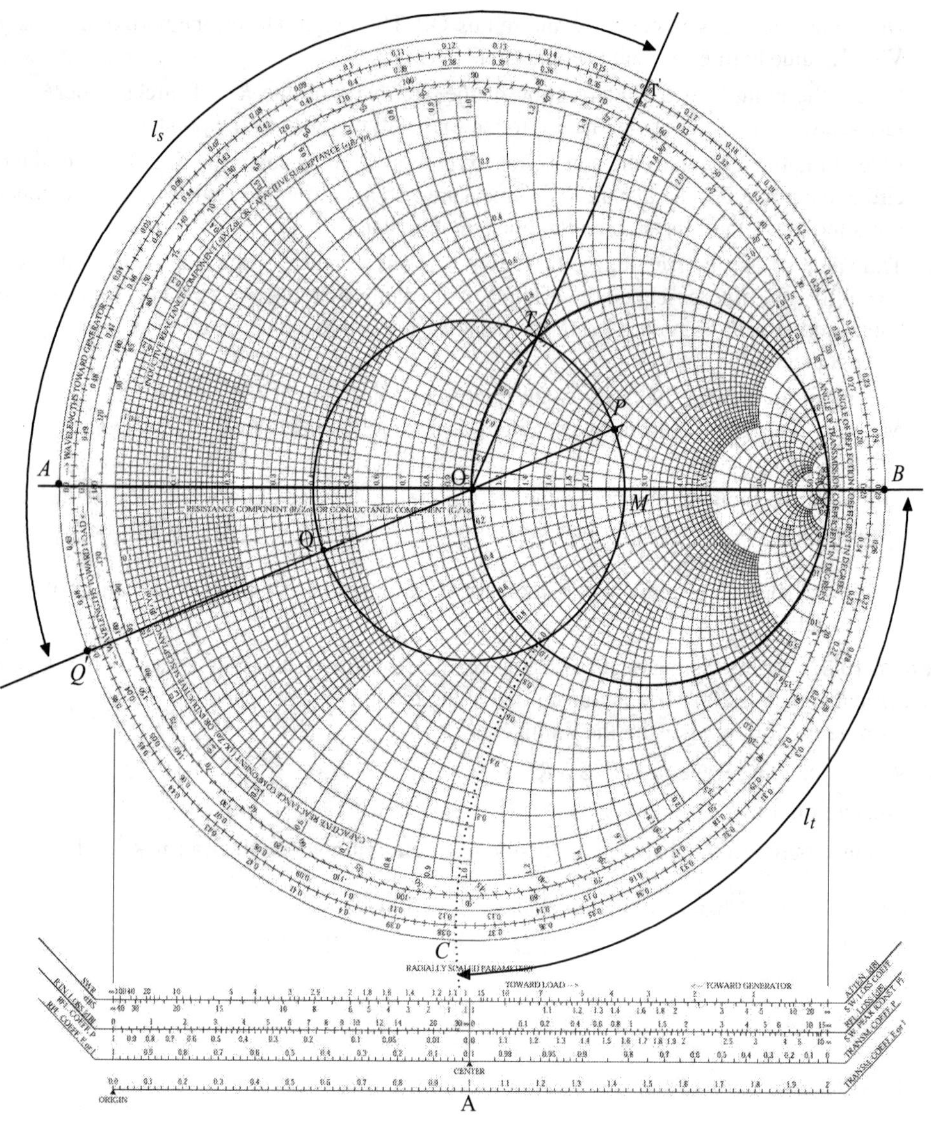

Fig. 10.23 *Smith chart for determination of VSWR, admittance and* $y_{\min}$.

3. The point Q opposite to the normalised impedance point on the impedance circle is the admittance of the load.
   Normalised admittance at point Q is $y_l = 0.72 - j0.85$.
   Load admittance, $Y_L = Y_0 y_l$.

$$Y_L = \frac{1}{300}(0.72 - j0.85) = 2.38 - j2.8 \text{ m℧}.$$

4. The first voltage minimum from the load lies along the horizontal axis to the left side end of the chart, i.e., at point A.

The line OP is extended to the outer circle at point P′. This represents load point.

The position of P′ in terms of the wavelength is 0.114 λ.

The distance between load point P′ and the first voltage minimum point A towards the generator is measured on the wavelength scale. That is,

$$P'A = 0.5\lambda - 0.114\lambda = 0.386\ \lambda.$$

Hence, the distance between the load and the first voltage minimum is

$$y_{min} = 0.386 \times 1.5 \text{ m.} = 0.58 \text{ m.}$$

Figure 10.24 shows the Smith chart for determining VSWR, admittance and y_{min}.

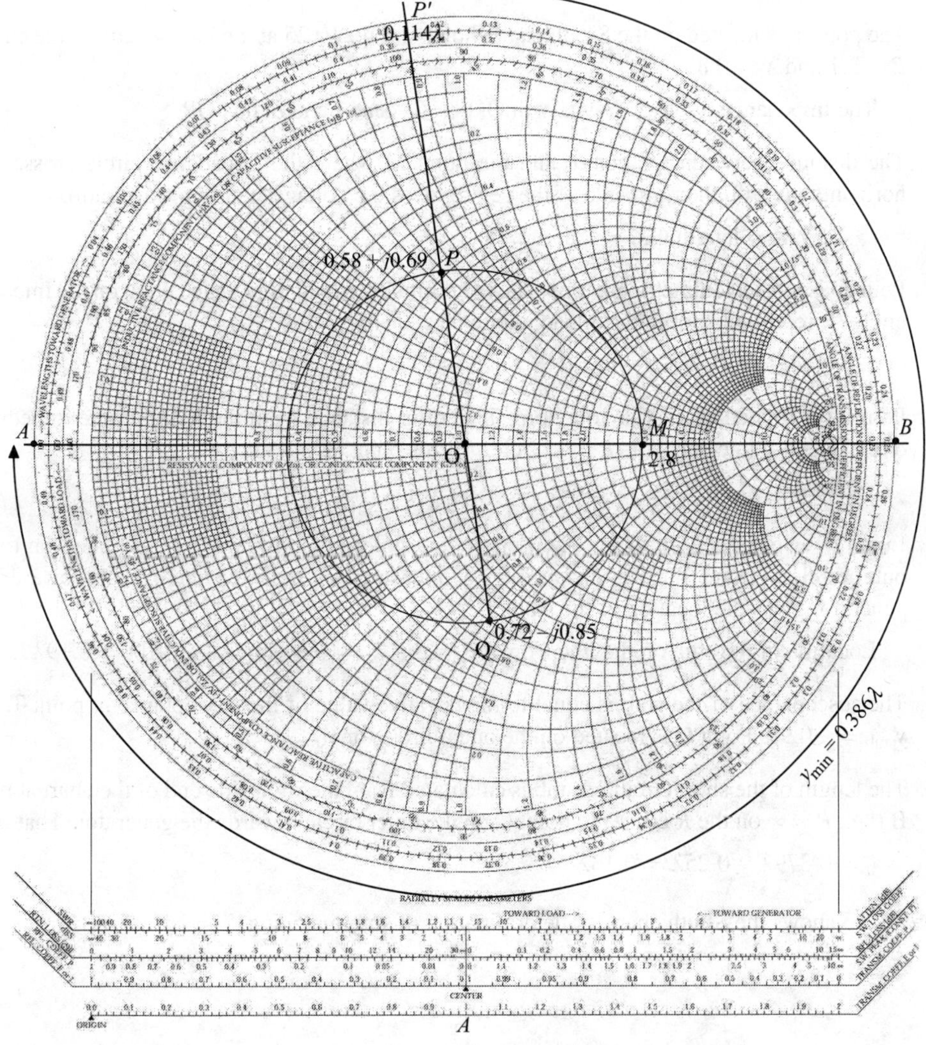

Fig. 10.24 *Smith chart for determination of length and location of a single stub.*

Example 10.10 Find the location and length of a stub matching for a transmission line having characteristic impedance 50 Ω and terminated with $105 + j40\ \Omega$ using the Smith chart.

Solution Let a short circuited stub of length l_t at a distance l_s from the load be connected parallel to the main line.

Given: Characteristic impedance, $Z_0 = 50\ \Omega$, load impedance, $Z_R = Z_L = 105 + j40\ \Omega$.

Normalised load impedance is

$$z_R = \frac{Z_R}{Z_0} = \frac{105 + j40}{50} = 2.1 + j0.8.$$

Steps for using the Smith chart:

1. The point P is located on the Smith chart shown in Fig. 10.25 at the intersection of the circles $R = 2.1$ and $X = 0.8$.

   The impedance circle is drawn with $O(1 + j0)$ centre and radius OP.

2. The distance between the centre and the point M where the impedance circle crosses the horizontal axis on the right side of the centre gives the voltage standing wave ratio.

   VSWR = length OM = 2.5.

3. Point Q is located diametrically opposite the normalised load impedance point on the impedance circle. It represents normalised load admittance. That is,

   $y_L = 0.415 - j0.158.$

4. Point Q is rotated clockwise to a point T (shorter distance from the load) on the impedance circle where it intersects the $R = 1$ circle. At this point, admittance

   $y = 1 + j0.95.$

5. Line OQ is extended on to the outer circle to a point Q′ and line OT is extended on to the outer circle to point T′. The distance Q′ T′ measured on the wavelength scale towards the generator gives the location of the stub.

   Location of the stub on the line from the load is $l_s = (0.5 - 0.47 + 0.16)\lambda = 0.19\lambda$.

6. The susceptance of the stub is equal to the negative value of the susceptance at point T, i.e., $y_{\text{stub}} = -0.95$. Point C is located on the outer circle for y_{stub}.

7. The length of the short circuited stub is calculated from the right side end of the chart at point B (i.e., $R = \infty$ on the R circle) to point C on the outer circle towards the generator. That is,

   $l_t = 0.379\lambda - 0.25\lambda = 0.13\lambda.$

Figure 10.25 shows the Smith chart for determination of stub matching.

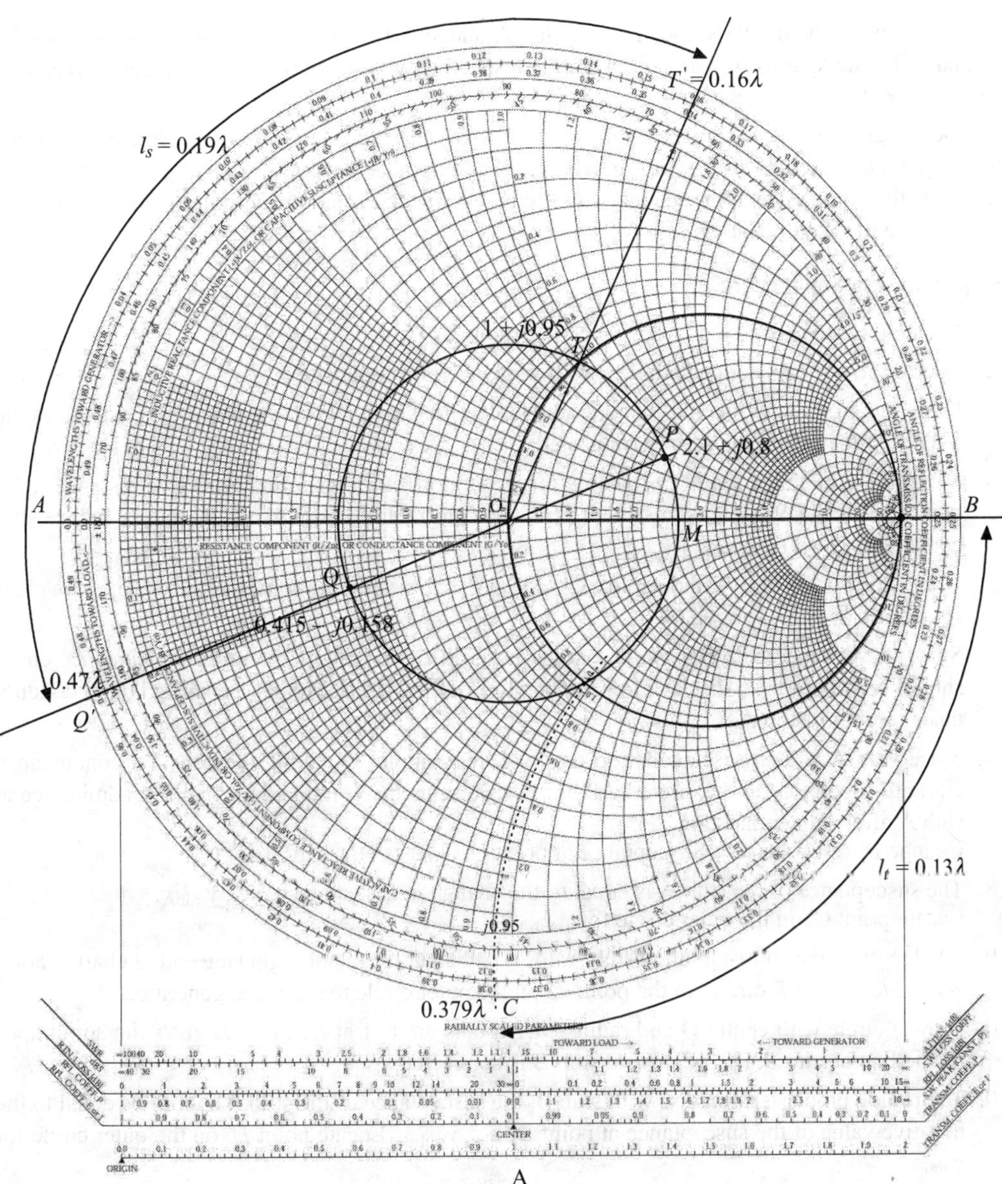

Fig. 10.25 *Smith chart for determination of stub matching*

10.21 Double Stub Matching Using Smith Chart

Consider a transmission line length l terminated with a load impedance Z_R. Assume that two stubs A and B having a characteristic impedance same as that of the line characteristic impedance are connected in parallel nearer to the main line.

Let l_{s_1} and l_{s_2} be the fixed locations of stub A and stub B respectively from the load. Also, let l_{t_1} and l_{t_2} be the lengths of stub A and stub B respectively. Assume that the distance between two stubs is $l_s < \lambda/2$.

The design of a double stub matching is different from a single stub matching. Here the locations of the stubs are fixed as shown in Fig. 10.27. Matching is achieved by adjusting the stub lengths such that the admittance at the second stub is equal to the characteristic admittance. To accomplish this, we use the Smith chart as follows.

Steps for using Smith chart:

1. Locate the normalised load impedance (z_R) point as P.
2. Locate the normalised load admittance point Q which is opposite P.
3. Draw the S circle with centre O and radius OP. The length OP on the horizontal axis gives the VSWR value before the stub is connected.
4. Move the admittance point Q towards the generator (clockwise direction) to a distance l_{s_1} at point Q_1 on the S circle. This point represents normalised admittance at stub A before stub connection.
5. Draw circle 1 for unity conductance ($y = 1$ or $R = 1$ circle) on the right-hand side of the chart for stub B.
6. Since the distance between the two stubs is l_s, all the points of the unity conductance circle should be shifted by l_s distance towards the load. Draw circle 2 to get the shifted unity conductance circle for stub A.
7. Since stub A adds a susceptance to the load, move point Q_1 along the constant conductance circle to a point C on the circle 2. Point C represents the combined normalised admittance at stub A after stub connection.
   Change in susceptance = Susceptance at point C − susceptance at point Q_1.
8. The susceptance of the stub A is equal to the change in susceptance, i.e., y_{stubA}.
   Locate point C′ on the outer circle for y_{stubA}.
9. The length of the short-circuited stub A, l_{t_1}, is calculated from the right side end of chart at point B (i.e., $R = \infty$ on R circle) to the point C′ on the outer circle towards the generator.
10. Draw a circle with centre O and radius OC. It cuts circle 1 at D. Point D gives the admittance on the line at stub B, i.e., y_{s_2} before stub B is connected.
10. To provide proper termination with stub B, the susceptance of the stub B should be equal to the negative value of the susceptance at point D, i.e., y_{stubB}. Locate point D' on the outer circle for y_{stubB}.
12. The length of the short circuited stub B, l_{t_2}, is calculated from the right-side end of the chart at point B (i.e., $R = \infty$ on R circle) to the point D′ on the outer circle towards the generator.

Example 10.11 A lossless transmission line $Z_0 = 300$ ohms has a load impedance of $75 + j225$ ohms. Using a Smith chart, determine what lengths of shorted stubs must be attached to the line to achieve proper matching to maximise the received power.

Solution For proper matching, consider two short circuited stubs connected in parallel to the main line. Let stub A be connected at the load and stub B be connected at a distance $\lambda/4$ from stub B.

Let l_{t_1} and l_{t_2} be the lengths of stub A and stub B respectively.

Characteristic impedance of the stubs should be equal to line impedance $Z_0 = 300$ ohms.

Steps for using Smith chart:

Given: Characteristic impedance, $Z_0 = 300\ \Omega$, load impedance, $Z_L = 75 + j225\ \Omega$,

$$\text{Normalised load admittance, } y_L = \frac{Y_L}{Y_0} = \frac{Z_0}{Z_L} = \frac{300}{75 + j225} = 0.4 - j1.2\ \mho.$$

1. The point Q located on the chart as shown in Fig. 10.26, represent $y_L = 0.4 - j1.2$, which is the normalised admittance at stub A before stub connection.
2. Circle 1 for unity conductance ($y = 1$ or $R = 1$ circle) is drawn on the right-hand side of the chart.
3. Since the distance between the two stubs is $\lambda/4$, all the points of the unity conductance circle should be shifted by 0.25 λ towards the load. Circle 2 is drawn for shifted unity conductance circle for stub A on the left hand side of the chart.
4. Since stub A adds a susceptance to the load, point Q is moved along the constant conductance circle to a point C on the circle 2. The point C represents the combined normalised admittance at stub A after stub connection, i.e., $y_{L_1} = 0.4 - j0.5$.

   Change in susceptance $= y_{L_1} - y_L = 0.4 - j0.5 - (0.4 - j1.2) = j0.7$.

5. The susceptance of stub A is equal to the change in susceptance, i.e., $y_{\text{stubA}} = j0.7$. Locate point C′ on the outer circle for y_{stubA}.
6. The length of the short-circuited stub A, l_{s_1}, is calculated from the right side end of chart at point B (i.e., $R = \infty$ on R circle) to point C′ on the outer circle towards the generator.
   Length of the stub A, $l_{t_1} = C'B = 0.098\lambda + 0.25\lambda = 0.348\lambda$.
7. A circle with centre O and radius OC is drawn. It cuts circle 1 at D. Point D gives the admittance on the line at stub B, i.e., $y_{s_2} = 1 + j1.2$ before stub B is connected.
8. To provide proper termination with stub B, the susceptance of the stub B should be equal to the negative value of the susceptance at point D, i.e., $y_{\text{stubB}} = -j1.2$. Point D′ is located on the outer circle for y_{stubB}.
9. The length of the short circuited stub B, l_{t_2}, is calculated from the right side end of the chart at point B (i.e., $R = \infty$ on the R circle) to point D′ on the outer circle towards generator.
   Length of the stub B, $l_{t_2} = D'B = 0.36\lambda - 0.25\lambda = 0.11\lambda$.

Figure 10.26 shows the Smith chart for determination of double stub matching.

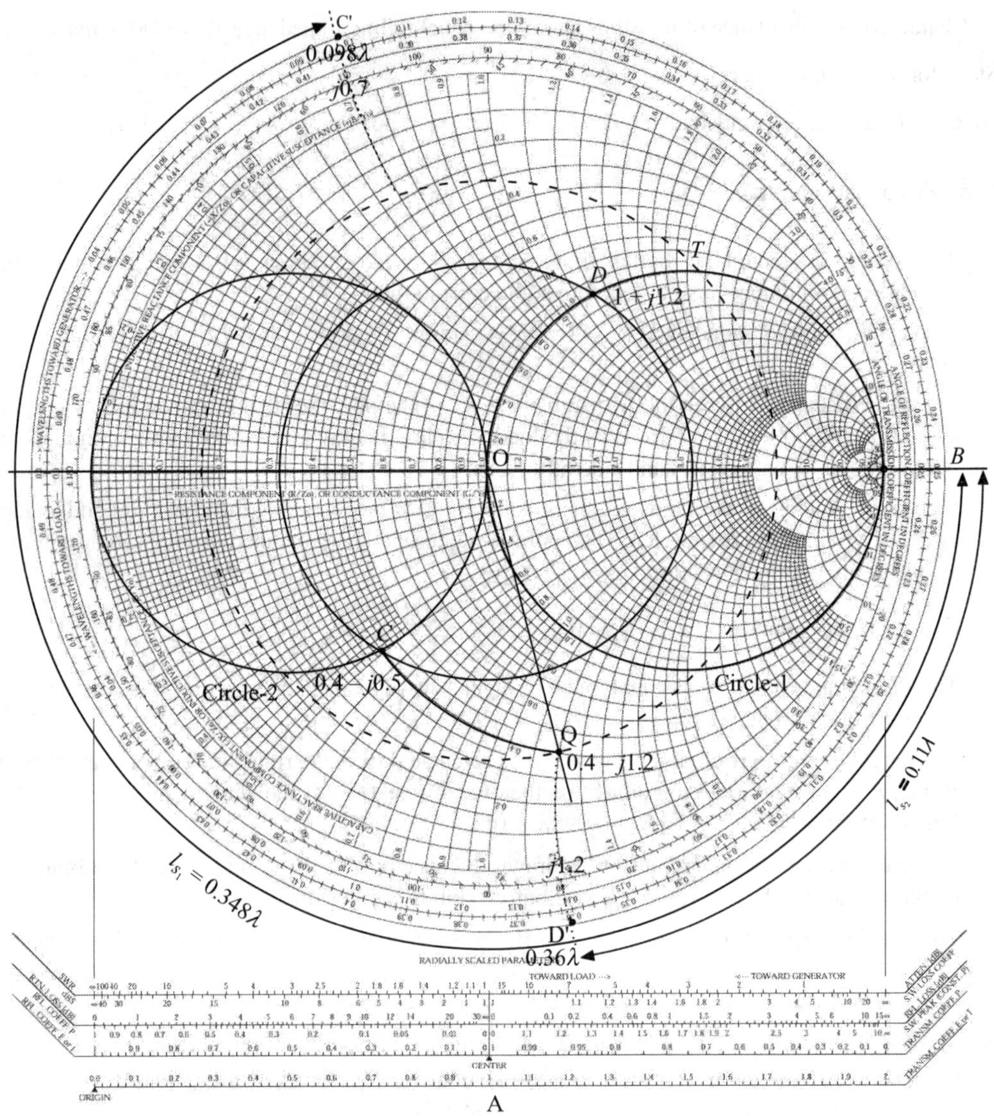

Fig. 10.26 *Smith chart for determination of double stub matching.*

Additional Solved Problems

Example 10.12 Find the characteristic impedance of a line at 1600 Hz, if the following measurements have been made on the line at 1600 Hz, $Z_{oc} = 750\ \Omega$ and $Z_{sc} = 500\ \Omega$.

Solution Given:

Operating frequency, $f = 1600$ Hz.

Open-circuited impedance, $Z_{oc} = 750\ \Omega$.
Short-circuited impedance, $Z_{sc} = 500\ \Omega$.
The characteristic impedance of a line is given by

$$Z_0 = \sqrt{Z_{oc} \times Z_{sc}} = \sqrt{750 \times 500} = \sqrt{375000}.$$

The characteristic impedance is $Z_0 = 612.37\ \Omega$.

Example 10.13 $Z_{oc} = 900\angle{-30°}\Omega$, $Z_{sc} = 400\angle{-10°}\Omega$. Calculate the Z_0 and γ of a 12 km long line.

Solution Given: $Z_{oc} = 900\angle{-30°}\Omega$, $Z_{sc} = 400\angle{-10°}\Omega$.

a) The characteristic impedance is given by

$$Z_0 = \sqrt{Z_{oc} \times Z_{sc}} = \sqrt{900\angle{-30} \times 400\angle{-10}} = 600\angle{-20°}\ \Omega.$$

Hence, the characteristic impedance, $Z_0 = 600\angle{-20°}\ \Omega$.

We know that $\tanh \gamma l = \sqrt{\dfrac{Z_{sc}}{Z_{oc}}}$.

$$\tanh \gamma l = \sqrt{\frac{400\angle{-10}}{900\angle{-30}}} = 0.67\angle 10.$$

$$\frac{e^{\gamma l} - e^{-\gamma l}}{e^{\gamma l} + e^{-\gamma l}} = 0.66 - j0.116.$$

By componendo and dividendo principle,

$$\frac{2e^{\gamma l}}{2e^{-\gamma l}} = \frac{1 + 0.66 - j0.116}{1 - (0.66 - j0.116)}.$$

$$e^{2\gamma l} = 4.269 - j1.798 = 4.632\angle{-22.8}.$$

Taking ln of both sides,

$$2\gamma l = \ln(4.632\angle{-22.8°}).$$

$$\gamma = \frac{1}{2l}(\ln(4.632) + j(-22.8°))$$

or $$\gamma = \frac{1}{2 \times 12}(1.533 - j0.398).$$

$$\gamma = \alpha + j\beta = 0.0638 - j0.01658 \text{ napers/km}.$$

Example 10.14 A two wire line has a characteristic impedance of 300 Ω and is fed to a 90 Ω resistor at 200 MHz. A quarter wave line is used as a tube 0.25 cm in diameter. Find the centre-to-centre spacing in air.

Solution Given: $Z_1 = 300\ \Omega$, $Z_2 = 90\ \Omega$, tube diameter = 0.25 cm, radius $r = 0.125$ cm.

For a quarter wave line, we know that

$$Z_0 = \sqrt{Z_1 \times Z_2} = \sqrt{300 \times 90} = 164.32\ \Omega.$$

Also, we know that the characteristic impedance of a parallel wire line is,

$$Z_0 = 276 \log (d/r) \text{ ohms},$$

where d = centre-to-centre spacing.

$$\therefore \quad 164.32 = 276 \log \left[\frac{d}{0.125}\right].$$

$$\log \left[\frac{d}{0.125}\right] = 0.595.$$

$$\frac{d}{0.125} = 3.936 \quad \text{or} \quad d = 0.49 \text{ cm}.$$

Example 10.15 A 60 ohm lossless line is connected to a source with 10 V, $Z_g = 50 - j40$ and terminated with a load of $j40$ ohms. If the line is 100 m long and $\beta = 0.25$ rad/m, calculate Z_{in} and voltage at (i) the sending end, (ii) the receiving end, (iii) 4 m from the load end and (iv) 3 m from the source.

Solution Given:
Length, l = 100 m, characteristic impedance, $Z_0 = 60$ ohms, termination impedance, $Z_R = j40$ ohms, source resistance, $Z_g = 50 - j40$ ohms, source voltage, $V_g = 10$ V, $\alpha = 0$, $\beta = 0.25$ rad/m,

$$\text{input current is } I_s = \frac{V_g}{Z_0 + Z_g} = \frac{10}{60 + 50 - j40} = 85.4\angle 20 \text{ mA}.$$

We know that the source impedance for a lossless transmission line is

$$Z_s = Z_0 \left(\frac{Z_R + jZ_0 \tan \beta l}{Z_0 + jZ_R \tan \beta l}\right).$$

(i) At the sending end, $l = 100$ m, $\beta l = 0.25 \times 100 = 25$ rad $= 1432.4°$.

$$Z_s = 60 \left(\frac{j40 + j60 \tan 1432.4}{60 + j^2 40 \tan 1432.4}\right) = j29.38\ \Omega.$$

Input voltage is $V_s = Z_s I_s = j29.38 \times (85.4\angle 20) \times 10^{-3} = 2.5\angle 110$ V.

(ii) At the receiving end, $l = 0$.

$$Z_s = Z_R = j40\ \Omega.$$

$$V_R = Z_R I_s = j40 \times (85.4\angle 20) \times 10^{-3} = 3.416\angle 110 \text{ V}.$$

(iii) At 4 m from the load end: $l = 4$, $\beta l = 0.25 \times 4 = 1$ rad $= 57.3°$.

$$Z_1 = 60\left(\frac{j40 + j60\tan 57.3}{60 + j^2 40\tan 57.3}\right) = -j435.53\ \Omega.$$

$$V_1 = Z_1 I_s = -j435.53 \times 85.4\angle 20 \times 10^{-3} = 37.2\angle -70° \text{ V}.$$

(iv) At 3 m from the source end: $l = 97$, $\beta l = 025 \times 97 = 2425$ rad $= 1389.4°$.

$$Z_2 = 60\left(\frac{j40 + j60\tan 1389.4}{60 + j^2 40\tan 1389.4}\right) = -j0.303\ \Omega.$$

$$V_2 = Z_2 I_s = -j0.303 \times (85.4\angle 20) \times 10^{-3} = 0.0258\angle -70° \text{ V}.$$

Example 10.16 An open wire unloaded line, 75 km long, is operated at a frequency of 1000 Hz. The open circuit impedance is found to be $330\angle -30°\ \Omega$ and the short circuit impedance is $540\angle 7°\ \Omega$. Calculate the parameters of line.

Solution Given: Length of the unloaded line, $l = 75$ km, $f = 1000$ Hz, $Z_{oc} = 330\angle -30°\ \Omega$, $Z_{sc} = 540\angle 7°\ \Omega$.

We know that $Z_0 = Z_{sc} \times Z_{oc}$

$$= \sqrt{540\angle 7° \times 330\angle -30°} = 422.14\angle -11.5.$$

Also, $Z_{sc} = Z_0 \tanh \gamma l$.

$$\tanh \gamma l = \frac{Z_{sc}}{Z_0} = \frac{540\angle 7°}{422.14\angle -23°}.$$

$$= 1.28\angle 30° = 1.108 + j0.64.$$

$$\frac{e^{\gamma l} - e^{-\gamma l}}{e^{\gamma l} + e^{-\gamma l}} = 1.108 + j0.64.$$

By componendo and dividendo principle,

$$\frac{2e^{\gamma l}}{2e^{-\gamma l}} = \frac{1 + 1.108 - j0.64}{1 - 1.108 - j0.64}.$$

$$e^{2\gamma l} = \frac{2.108 + j0.64}{-0.108 - j0.64}.$$

$$e^{2\gamma l} = \frac{2.203\angle 16.89°}{0.649\angle -99.58°} = 3.394\angle 116.47°.$$

Taking ln of both sides,

$$2\gamma l = \ln(3.394\angle 116.47^\circ).$$

Since, $\ln(r\angle\theta) = \ln r + j\theta^\circ$,

$$\gamma = \frac{1}{2l}(\ln(3.394) + j(116.47^\circ))$$

$$= \frac{1}{2l}(1.222 + j2.033) = \frac{1.222 + j2.033}{150} = 0.00815 + j0.0135.$$

$$\gamma = 0.0157\angle 58.88^\circ.$$

$$\beta = 0.00815 \text{ rad/km} \quad \text{and} \quad \alpha = 0.0135 \text{ nepers/km}.$$

Also we know that $= R + j\omega L$.

$$R + j\omega L = 0.0157\angle 58.88^\circ \times 422.14\angle -10.5^\circ = 6.627\angle 47.38$$

$$R + j\omega L = 4.487 + j4.877.$$

$$R = 4.487\ \Omega\text{/km} \quad \text{and} \quad \omega L = 4.877.$$

$$2\pi f \times L = 4.877.$$

$$L = \frac{4.877}{2\pi f} = \frac{4.877}{2 \times \pi \times 1000} = 0.776 \text{ mH/km}.$$

$$\frac{\gamma}{Z_0} = G + j\omega C.$$

$$G + j\omega C = \frac{0.0157\angle 58.88^\circ}{422.14\angle -10.5^\circ} = 3.72 \times 10^{-5}\angle 70.38^\circ$$

$$G + j\omega C = 12.5 \times 10^{-6} + j35 \times 10^{-6}.$$

$$G = 12.5\ \mu\mho\text{/km and} \quad \omega C = 3.5 \times 10^{-5}.$$

$$2\pi f C = 3.5 \times 10^{-5}.$$

$$C = \frac{3.5 \times 10^{-5}}{2\pi f} = \frac{3.5 \times 10^{-5}}{2 \times \pi \times 1000}$$

$$\therefore \quad C = 5.576 \text{ nF/km}.$$

Example 10.17 The input impedance of a short circuited lossy transmission line of length 2 m and characteristic impedance 75 Ω is $45 + j225$ Ω.

(a) Find α and β of a line.

(b) Determine the input impedance if the short circuit is replaced by a load impedance of $67.5 - j45$ Ω.

Solution Given: Input impedance of a short circuited line, $Z_{sc} = 45 + j225\ \Omega$, characteristic impedance, $Z_0 = 75\ \Omega$, length $l = 2$ m.

(a) We know that $Z_{sc} = Z_0 \tanh \gamma l$.

$$\tanh \gamma l = \frac{Z_{sc}}{Z_0} = \frac{45 + j225}{75} = 0.6 + j3.$$

$$\frac{e^{\gamma l} - e^{-\gamma l}}{e^{\gamma l} + e^{-\gamma l}} = 0.6 + j3.$$

By componendo and dividendo principle,

$$\frac{2e^{\gamma l}}{2e^{-\gamma l}} = \frac{1 + 0.6 - j3}{1 - 0.6 - j3}.$$

$$e^{2\gamma l} = \frac{1.6 + j3}{0.4 - j3} = 1.12\angle 114.3.$$

Taking ln of both sides,

$$2\gamma l = \ln(1.12\angle 144.3°).$$

Since, $\ln(r\angle\theta) = \ln r + j\theta°$,

$$\gamma = \frac{1}{2l}(\ln(1.12) + j(144.3°)) = \frac{0.1133 + j2.518}{2 \times 2} = 0.028 + j0.63.$$

$$\gamma = 0.0157\angle 58.88°;\ \beta = 0.63 \text{ rad/km and } \quad \alpha = 0.028 \text{ nepers/km}.$$

(b) Given: Load impedance $Z_R = 67.5 - j45\ \Omega$.
We know that the input impedance is

$$Z_{\text{in}} = Z_0 \left(\frac{1 + Ke^{-2\gamma l}}{1 - Ke^{-2\gamma l}} \right),$$

where $K = \dfrac{Z_R - Z_0}{Z_R + Z_0} = \dfrac{67.5 - j45 - 75}{67.5 - j45 + 75} = 0.3\angle -82°.$

$$e^{2\gamma l} = 1.12\angle 114.3.$$

$$Z_{\text{in}} = 75 \left(\frac{1 + \dfrac{0.3\angle -82}{1.12\angle 144.3}}{1 - \dfrac{0.3\angle -82}{1.12\angle 144.3}} \right) = 75 \left(\frac{1 + 0.2678\angle 226.3}{1 - 0.2678\angle 226.3} \right) = 52.35\angle 22.66.$$

Input impedance is $Z_{\text{in}} = 48.3 + j20\ \Omega$.

Example 10.18 A dipole antenna is fed by a lossless transmission line having $Z_o = 60\ \Omega$. The source impedance is 600 Ω. If the length of the line is 0.1λ, determine antenna impedance.

Solution Given: Length of the line $l = 0.1\lambda$, characteristic impedance, $Z_0 = 60\ \Omega$, source impedance $Z_S = 600\ \Omega$.

Let Z_R be the antenna impedance. We know that the source impedance for a lossless transmission line is

$$Z_S = Z_0 \left(\frac{Z_R + jZ_0 \tan \beta l}{Z_0 + jZ_R \tan \beta l} \right),$$

or $$Z_R = Z_0 \left(\frac{Z_S - jZ_0 \tan \beta l}{Z_0 - jZ_S \tan \beta l} \right).$$

Now $$\tan \beta l = \tan \left[\frac{2\pi}{\lambda} (0.1\lambda) \right] = \tan (0.2\pi) = 0.726.$$

$$\therefore \quad Z_R = 60 \left[\frac{600 - j60(0.726)}{60 - j600(0.726)} \right] = 60 \left[\frac{600 - j43.56}{60 - j435.6)} \right]$$

$$= 17.06 + j80.3 = 82.09\angle 78^\circ.$$

$\therefore$ Antenna impedance is $Z_R = 82.09\angle 78^\circ$.

Example 10.19 A transmission line of 100 Ω characteristic impedance is connected to a load of 400 Ω. Calculate the reflection coefficient and standing wave ratio.

Solution Given: Characteristic impedance, Z_0= 100 Ω, load, $Z_R = 400\ \Omega$.

We know that reflection coefficient K is given by

$$K = \frac{Z_R - Z_0}{Z_R + Z_0} = \frac{400 - 100}{400 + 100}.$$

$\therefore$ Reflection coefficient, $K = 3/5 = 0.6$.

The standing wave ratio is $\text{VSWR} = \dfrac{1 + |K|}{1 - |K|} = \dfrac{1 + 0.6}{1 - 0.6} = \dfrac{1.6}{0.4} = 4.$

Example 10.20 An UHF transmission line of $Z_0 = 150$ ohms is terminated with an unknown load. The VSWR measured in the line is 5 and the position of minimum current nearest the load is one-fifth wavelength away. Calculate the value of the load impedance.

Solution Given: Characteristic impedance, $Z_0 = 150$, VSWR $= S = 5$, position of minimum current $y_{max} = \lambda/5$.

Position of the first minimum current or voltage maximum can be obtained by

$$2\beta y_{max} - \phi = 0.$$

$$2.\frac{2\pi}{\lambda}.\frac{\lambda}{5} - \phi = 0$$

or $\phi = \frac{4\pi}{5} = 2.51$ radians or 144°.

We know that the magnitude of the reflection coefficient is

$$|K| = \frac{S-1}{S+1} = \frac{5-1}{5+1} = 0.666.$$

The reflection coefficient K is

$$K = |K|e^{j\phi} = 0.667\angle 144°.$$

Also we know that $K = \frac{Z_R - Z_0}{Z_R + Z_0}$.

$$0.667\angle 144° = \frac{Z_R - 150}{Z_R + 150}$$

$$-0.5388 + j0.3916 = \frac{Z_R - 150}{Z_R + 150}$$

$$Z_R(-0.5388 + j0.3916 - 1) = (-150 + 150 \times 0.5388 - j150 \times 0.3916)$$

$$Z_R(-1.5388 + j0.3916) = -69.18 - j58.74$$

$$Z_R = \frac{90.753\angle -139.66}{1.588\angle 165.72}.$$

The load impedance, $Z_R = 57.15\angle -305.38$ or $Z_R = 57.15\angle 54.62\ \Omega$.

Example 10.21 An open wire transmission line having $Z_0 = 650\angle -12°\ \Omega$ is terminated in Z_0 at the receiving end. If this line is supplied from a source of internal resistance 300 Ω, calculate the reflection factor and reflection loss at the sending end terminals.

Solution Given: $Z_i = 300\ \Omega$, $Z_0 = 650\angle -12°\ \Omega$.
We know that the reflection factor is

$$K_f = \frac{2\sqrt{Z_i Z_0}}{(Z_i + Z_0)}.$$

$$\therefore \quad K_f = \frac{2\sqrt{300 \times 650\angle -12°}}{(300 + 6358 - j135)} = \frac{883\angle -6}{945.5\angle -8.22} = 0.934\angle 2.22°.$$

The reflection loss,

$$K_e = 20\log_{10}\left|\frac{1}{K_f}\right| = 20\log\left(\frac{1}{0.934}\right).$$

$$K_e = 0.59 \text{ dB}.$$

$\therefore$ The reflection loss is 0.59 dB.

Example 10.22 A 80 Ω distortionless line connects a signal of 50 kHz to a load of 140 Ω. The load power is 75 mW. Calculate:

(i) Voltage reflection coefficient,

(ii) VSWR,

(iii) Position of V_{max}, I_{max}, V_{min} and I_{min}.

Solution Given: $Z_R = 140\ \Omega$, $Z_0 = 80\ \Omega$.

(i) Voltage reflection coefficient

$$K = \frac{Z_R - Z_0}{Z_R + Z_0} = \frac{140-80}{140+80} = 0.273.$$

$$\therefore \quad K = 0.273, \quad \phi = 0.$$

(ii) VSWR

$$\text{VSWR} = \frac{1+|K|}{1-|K|} = \frac{1+0.273}{1-0.273} = 1.75.$$

(iii) Position of V_{max}, I_{max}, V_{min} and I_{min}

The condition for maximum voltage occurs at y_{max} from the load, i.e.,

$$2\beta y_{max} - \phi = 2n\pi.$$

For first maximum, $n = 0$,

$$2\beta y_{max} - \phi = 0$$

$$2\beta y_{max} - 0 = 0, \quad y_{max} = 0.$$

Therefore, the first voltage maximum, V_{max}, and current minimum, I_{min}, occur at the load position.

The first voltage minimum, V_{min}, occurs at a distance of $\lambda/4$ from V_{max}.

$$\therefore \quad y_{min} = y_{max} + \lambda/4 = 0 + \lambda/4 = \lambda/4.$$

We know that the wavelength, $\lambda = \dfrac{v_0}{f} = \dfrac{3\times 10^8}{50\times 10^3} = 6$ km.

$$\therefore \quad y_{min} = \frac{v_0}{4\times f} = \frac{3\times 10^8}{4\times 50\times 10^3} = 1.5 \text{ km}.$$

Therefore, the first voltage minimum and current maximum, I_{max} occur at a distance of 1.5 km from the load. These values repeat every $\dfrac{\lambda}{2} = 3$ km distance from the load.

Given: The power at the load is $P = 75$ mW.

We know that $P = \frac{V_{\max}^2}{ZR}$ and $I_{\min} = \frac{V_{\max}}{Z_R}$.

$V_{\max}$ at the load is $\sqrt{P \times Z_R} = \sqrt{75 \times 10^{-3} \times 140} = 3.24$ volts and $I_{\min} = \frac{3.24}{140} = 0.23$ mA.

Example 10.23 Design a quarter wave transformer to match a line having impedance of 300 Ω to a load of 600 Ω.

Solution Given: $Z_0 = 300\ \Omega$, $Z_L = 600\ \Omega$.

The quarter wave transformer should have a sending end impedance of

$$Z_S = \frac{Z_0^2}{Z_L} = \frac{300 \times 300}{600} = 150\ \Omega.$$

The impedance will be matched if the Z_S of a $\lambda/4$ transformer is 150 Ω.

Example 10.24 Calculate the reflection coefficient and VSWR of a 50 Ω line terminated with (i) matched load, (ii) short circuit, (iii) $+j50\ \Omega$ load, (iv) $-j50\ \Omega$ load.

Solution

(i) Matched load

$$Z_0 = 50\ \Omega, \quad Z_R = 50\ \Omega.$$

Reflection coefficient $K = \frac{Z_R - Z_0}{Z_R + Z_0} = \frac{50 - 50}{50 + 50} = 0$

and VSWR $S = \frac{1 + |K|}{1 - |K|} = \frac{1 + 0}{1 - 0}$

$\therefore \quad S = 1$

(ii) Short circuit

$$Z_R = 0,\ Z_0 = 50\ \Omega.$$

$$K = \frac{Z_R - Z_0}{Z_R + Z_0} = \frac{0 - 50}{0 + 50} = -1.$$

$$S = \frac{1 + |K|}{1 - |K|} = \frac{1}{0} = \infty.$$

(iii) $+j50\ \Omega$ load

$\therefore \quad Z_R = +j50\ \Omega,\ Z_0 = 50\ \Omega.$

$$K = \frac{+j50 - 50}{j50 + 50} = \frac{-1 + j}{1 + j} = \frac{1.414\angle 135^\circ}{1.414\angle 45^\circ}$$

$\therefore \quad K = 1\angle 90^\circ.$

$$S = \frac{1+|K|}{1-|K|} = \frac{1+1}{1-1} = \frac{2}{0} = \infty.$$

(iv) $-j50\ \Omega$ load

$\therefore \quad Z_R = -j50\ \Omega,\ Z_0 = 50\ \Omega.$

$$K = \frac{-j50-50}{j50+50} = \frac{1.414\angle -135^\circ}{1.414\angle -45^\circ} = 1\angle -90^\circ.$$

$$\therefore \quad S = \frac{1+1}{1-1} = \frac{2}{0} = \infty.$$

Example 10.25 An aerial of $(200 - j300)\ \Omega$ is to be matched with 500 Ω lines. The matching is to be done by means of a low loss 600 Ω stub line. Find the position and length of the stub line used if the operating wave length is 20 metres.

Solution Given: $Z_R = 200 - j300\ \Omega = 360.55\angle -56.31$, $Z_0 = 500\ \Omega$, $\lambda = 20$ m.

We know that the reflection coefficient is

$$K = \frac{Z_R - Z_0}{Z_R + Z_0} = \frac{200 - j300 - 500}{200 - j300 + 500} = \frac{-300 - j300}{700 - j300} = 0.2068 - j0.5172.$$

$\therefore \quad K = 0.557\angle -110.8.$

So, $|K| = 0.557$ and $\phi = -110.8^\circ$.

We know that the position of the stub line l_s in terms of the reflection coefficient is

$$l_s = \frac{\lambda}{2\pi}(\phi + \pi - \cos^{-1}(|K|).$$

Substituting the values of λ, ϕ and $|K|$, we get,

$$l_s = \frac{20}{2\pi}(-110.8 + \pi - \cos^{-1}(0.557))$$

$$= \frac{10}{180}(-110.8 + 180 - 56.15) = \frac{10 \times 12.05}{180} = 0.669.$$

The stub position from the aerial is

$$l_s = 0.669 \text{ metres}.$$

Also the length of the stub line l_s in terms of the reflection coefficient $|K|$ is

$$l_t = \frac{\lambda}{2\pi}\tan^{-1}\frac{\sqrt{1-|K|^2}}{2|K|}.$$

Substituting the values of λ and $|K|$, we get,

$$l_t = \frac{20}{2\pi} \tan^{-1} \frac{\sqrt{1-(0.557)^2}}{2(0.557)} = \frac{10}{180} \times 36.7 = 2.039 \text{ metres.}$$

Example 10.26 A 50 Ω transmission line is terminated by an unknown impedance. The VSWR is 4 and the first voltage minimum is formed 2 cm from the load end. The frequency of operation is 1 GHz. Design a single stub line matching the above conditions.

Solution Given: $Z_0 = 50\ \Omega$, VSWR, $S = 4$,

$$f = 1 \text{ GHz} = 1 \times 10^9 \text{ Hz}, \lambda = \frac{C}{f} = \frac{3 \times 10^8}{10^9} = 0.3 \text{ m.}$$

Then, reflection coefficient, $|K| = \frac{S-1}{S+1} = \frac{4-1}{4+1} = 0.6.$

The first voltage minima is obtained at $y_{\min} = 2 \text{ cm} = 2 \times 10^{-2}$ m.

The condition for first minima is

$$2\beta y_{\min} - \phi = \pi.$$

$$\beta = \frac{2\pi}{\lambda} = \frac{2\pi}{0.3} = \frac{20\pi}{3}$$

$$\therefore \quad 2 \times \frac{20\pi}{3} \times 2 \times 10^{-2} - \phi = \pi.$$

$$\phi = 0.267\pi - \pi = -0.733\pi.$$

$$\phi = -2.304 \text{ radians} = -131.94^\circ.$$

We know that the length of the stub is

$$l_t = \frac{\lambda}{2\pi} \tan^{-1} \frac{\sqrt{1-|K|^2}}{2|K|}$$

$$= \frac{0.3}{2\pi} \tan^{-1} \frac{\sqrt{1-0.6^2}}{2 \times 0.6} = \frac{0.3 \times 33.69}{2 \times 180} = 2.8 \text{ cm.}$$

The location of the stub is

$$l_s = \frac{\lambda}{4\pi}(\phi + \pi - \cos^{-1}(0.6))$$

$$= \frac{0.3}{4 \times 180}(-131.94 + 180 - \cos^{-1}(0.6)) = -0.12 \text{ cm.}$$

Since l_s is negative, the next location is

$$l_s = -0.12 + \frac{\lambda}{2} = -0.12 + \frac{0.3 \times 10^2}{2} = 14.88 \text{ cm.}$$

Problems Using Smith Chart

Example 10.27 A 50 Ω line is terminated with a receiving end impedance of $100 + j121$. The wavelength of the electrical signal along the line is 2.5 m. Determine the following from Smith chart:

a) Standing wave ratio (VSWR)
b) Load admittance
c) Impedance of the transmission line at voltage maximum and minimum
d) Distance between load and first voltage maximum for a transmission line.

Solution Given: Characteristic impedance $Z_0 = 50\ \Omega$, load impedance $Z_L = 100 + j121\ \Omega$, signal wavelength $\lambda = 2.5$ m.

Steps for using Smith chart:

1. Normalised load impedance is $z_L = \dfrac{Z_L}{Z_0} = \dfrac{100 + j121}{50} = 2 + j2.42.$

   The point P is located on the Smith chart shown in Fig 10.27 at the intersection of the circles $R = 2$ and $X = 2.42$. The impedance circle is drawn with $O(1 + j0)$ as centre and radius as OP. The distance between the centre and the point M where the impedance circle crosses the horizontal axis on the right side of the chart gives the voltage standing wave ratio VSWR = length OM = 5.

   $\therefore \quad S = 5.$

2. The point Q diametrically opposite to the normalised load impedance point on the impedance circle is the normalised admittance of the load.

   Normal load admittance is $y_L = 0.22 - j0.25$.

   Load admittance $Y_L = y_L \times Y_0 = \dfrac{0.02 - j0.25}{50}$.

   $\therefore \quad Y_L = 4.4 - j5\ \text{m}\mho.$

3. Impedance at the first voltage maximum is $Z_{\max} = Z_0 S = 50 \times 5 = 250\ \Omega$.

   Impedance at the first voltage minimum is $Z_{\min} = \dfrac{Z_0}{S} = \dfrac{50}{5} = 10\ \Omega$.

4. The first voltage maximum from the load lies along the horizontal axis to the right side end of the chart, i.e., at point B.

   The line OP is extended to the outer circle at point P′. This represents load point.
   The position of P′ in terms of the wavelength is $0.208\ \lambda$.
   The distance between load point P′ and the first voltage maximum point B towards the generator measured on wavelength scale is

   $$\text{P}'\text{B} = 0.25\lambda - 0.208\lambda = 0.042\lambda.$$

   Hence the distance between load and the first voltage maximum is $y_{\max} = 0.042 \times 2.5 = 0.105$ m.

Figure 10.27 shows the Smith chart for determination of VSWR, admittance, impedance and y_{max}.

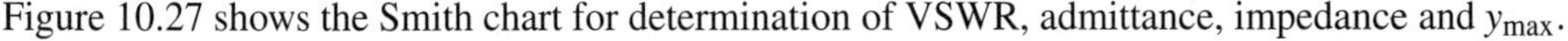

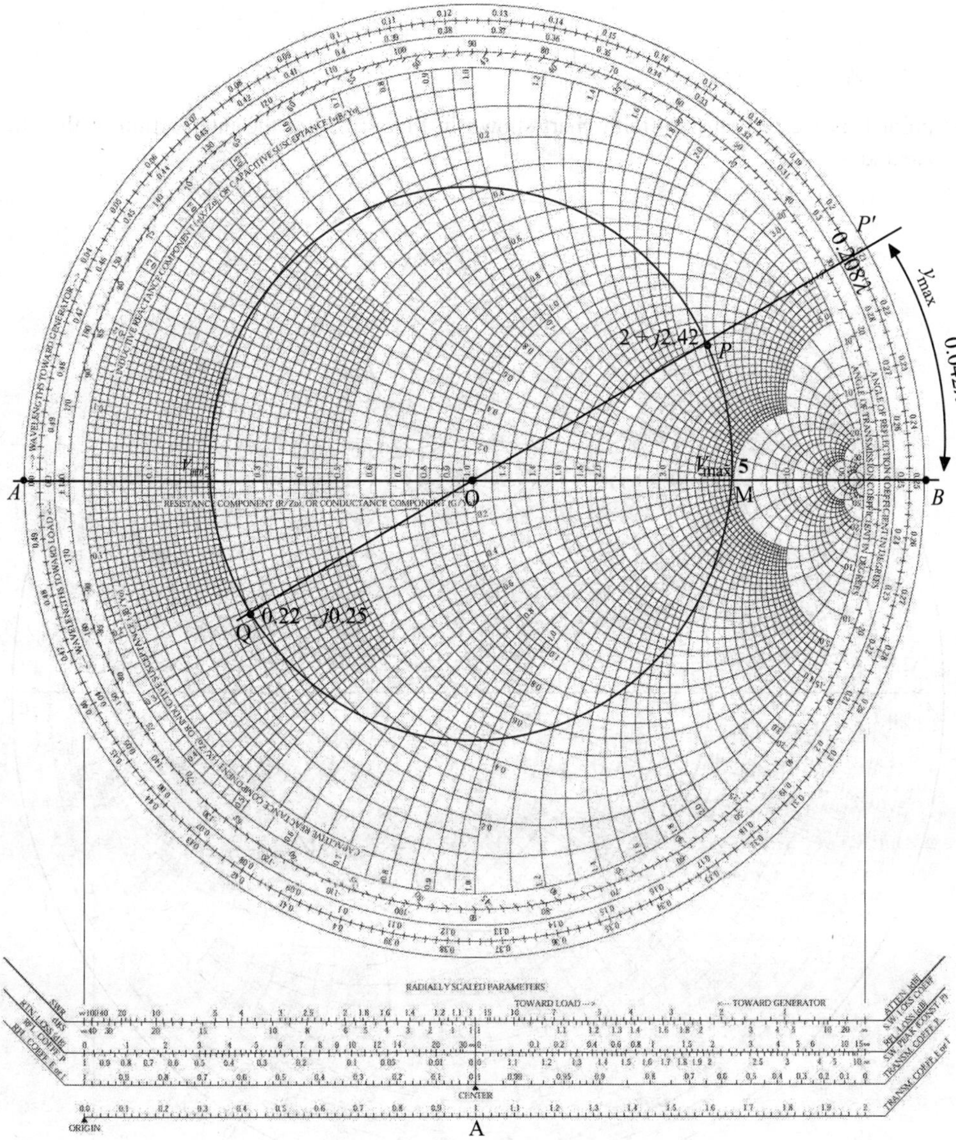

Fig. 10.27 *Smith chart for determination of VSWR, admittance, impedance and y* $_{max}$

Example 10.28 A lossless transmission line of length 0.434 λ and $Z_0 = 100\ \Omega$ is terminated at an impedance of $260 + j180$. Find (a) VSWR, (b) reflection coefficient, (c) input impedance, (d) location of voltage maximum on the line using Smith Chart. Assume the line is placed in free space.

Solution Given: Characteristic impedance, $Z_0 = 100\ \Omega$,

Load impedance, $Z_L = 260 + j180$, line length $l = 0.434\lambda$.

Steps for using Smith chart:

Normalised load impedance is

$$z_L = \frac{Z_L}{Z_0} = \frac{260 + j180}{100} = 2.6 + j1.8.$$

The point P is located on the Smith chart shown in Fig. 10.28 at the intersection of the circles $R = 2.6$ and $X = 1.8$.

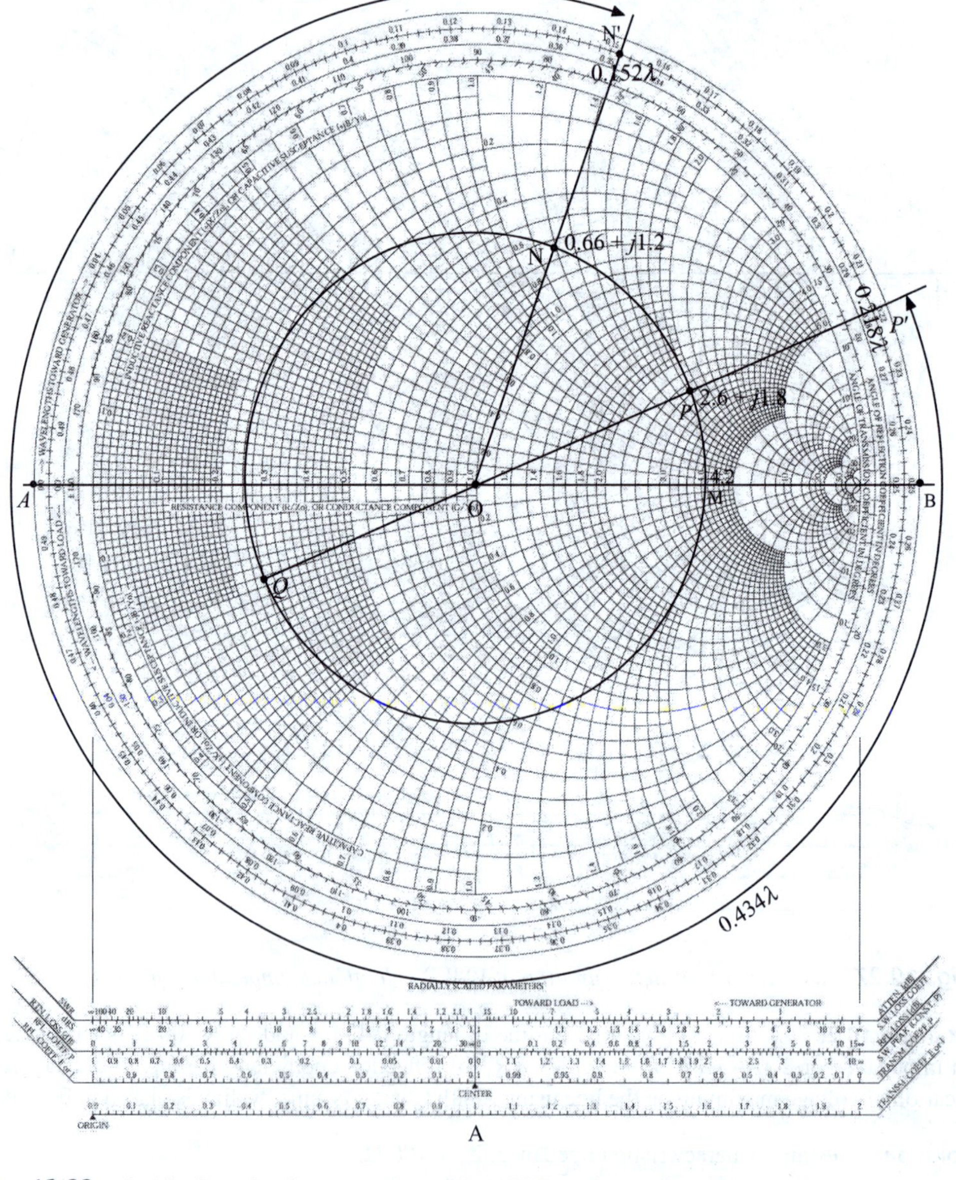

Fig. 10.28 *Smith chart for determination of K, VSWR and impedance*

1. VSWR: The voltage standing wave ratio S can be obtained by drawing a circle with $O(1+j0)$ as centre and radius OP which is called the impedance circle (S circle). The distance between the centre and the point M where the impedance circle crosses the horizontal axis on the right side of the centre gives the voltage standing wave ratio.

$$\text{VSWR} = \text{OM} = 4.2.$$

2. Voltage reflection coefficient: The reflection coefficient K is calculated by extending the line OP to meet the outer circle at P′. The lengths OP and OP′ are measured. The magnitude of voltage reflection coefficient is obtained by

$$|K| = \frac{\text{OP}}{\text{OP}'} = \frac{5.1}{8.5} = 0.6.$$

The phase angle θ of the reflection coefficient at the load can be obtained directly by measuring the angle at point P′ in degrees on the outer circle, i.e., $\theta = 22°$.

$\therefore \quad K = 0.6\angle 23°$.

3. Input impedance of the line: The input impedance of a line of length $l = 0.434\lambda$ can be calculated by extending the line OP to the outer circle at point P′ to represent the load point. This position is $0.218\ \lambda$.

   Move the load point P′ towards the generator (clockwise direction) to a distance $0.434\ \lambda$ to point N′. This position is $0.434\ \lambda + 0.218\lambda = 0.652\lambda$.

   The position of point N′ is $0.652\ \lambda - 0.5\lambda = 0.152\lambda$.

   The straight line ON′ is drawn and it cuts the impedance circle at point N. The impedance at point N represents normalised input impedance

$$Z_{\text{in}} = 0.66 + j1.2\ \Omega$$

   and the input impedance is

$$Z_{\text{in}} = z_{\text{in}} \times Z_0 = (0.66 + j1.2) \times 100 = 66 + j120\ \Omega.$$

4. Location of voltage maximum: The point B (i.e., $R = \infty$ on the R circle) at the right side end of the chart represents voltage maxima (V_{max}). From load point P, moving to a point towards the generator, location of the voltage maxima from load is $= 0.25\lambda - 0.218\lambda = 0.032\lambda$.

Figure 10.28 shows the Smith chart for determination of VSWR and input impedance.

Example 10.29 Determine the input impedance and VSWR for a $1.25\ \lambda$ long transmission line at the sending end with a characteristic impedance, $Z_0 = 50\ \Omega$ and a load impedance $Z_L = 30 + j40\ \Omega$ using a Smith chart. Assume that the line is placed in free space.

Solution Given: Characteristic impedance, $Z_0 = 50\ \Omega$,

Load impedance, $Z_L = 30 + j40\ \Omega$, length of the transmission line $= 1.25\lambda$.

Steps for using Smith chart:

Normalised load impedance is $z_L = \dfrac{Z_L}{Z_0} = \dfrac{30 + j40}{50} = 0.6 + j0.8$.

1. The point P is located on the Smith chart shown in Fig. 10.29 at the intersection of the circles $R = 0.6$ and $X = 0.8$.

2. The impedance circle is drawn with $O(1 + j0)$ as centre and radius OP.
   The distance between the centre and the point M where the impedance circle crosses the horizontal axis on the right side of the chart gives the voltage standing wave ratio.

   VSWR = length OM = 3.

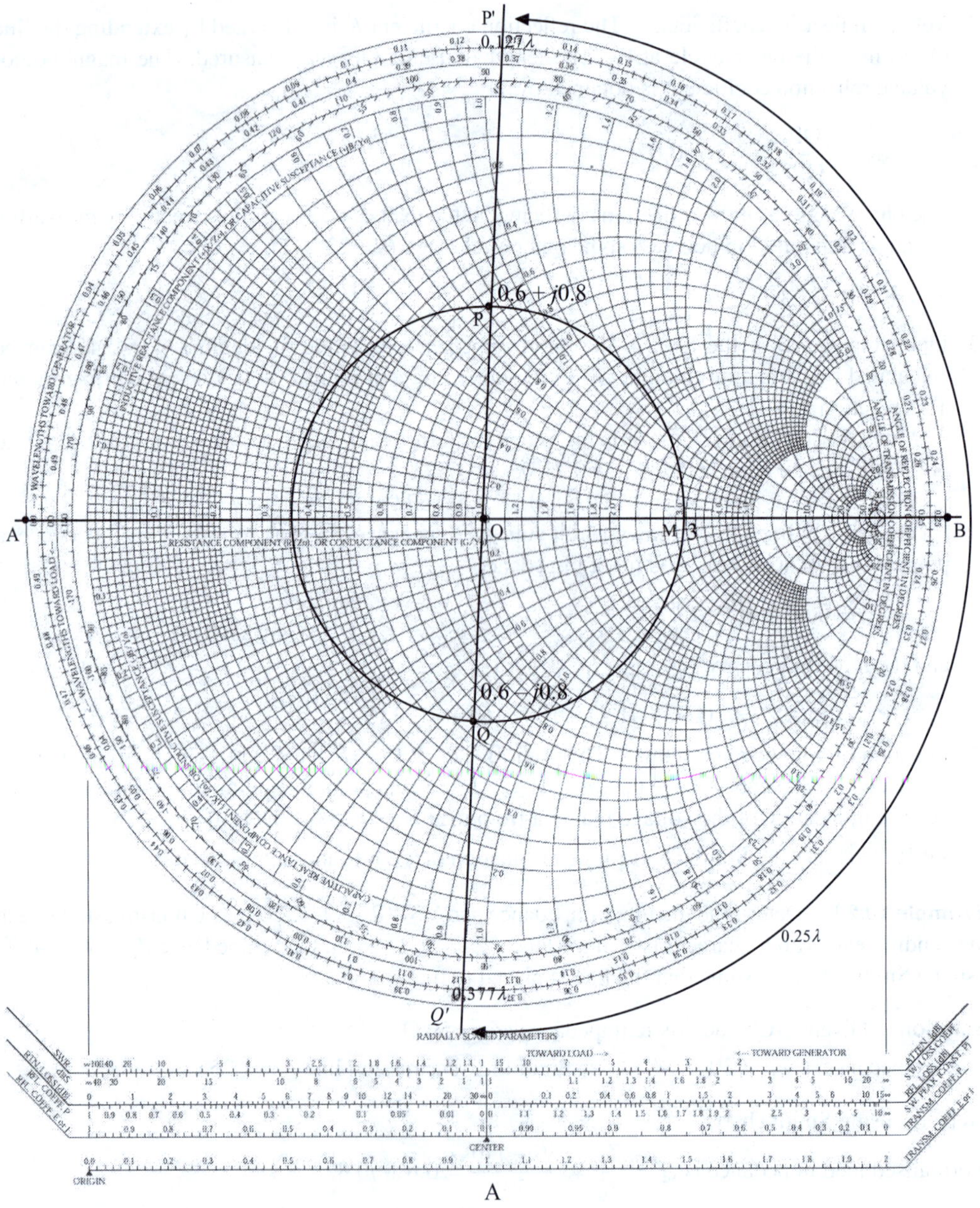

Fig. 10.29 *Smith chart for determination of VSWR and input impedance.*

3. Extending the line OP to the outer circle at point P′ represents the load point. This position is $0.127\,\lambda$.

   Move the load point P′ towards the generator (clockwise direction) to a distance $1.25\,\lambda$ at point N′. This position is $0.127\,\lambda + 0.25\lambda = 0.377\lambda$.

   The straight line ON′ is drawn and it cuts the impedance circle at point N. The impedance at point *N* is the normalised input impedance $z_{\text{in}} = 0.6 - j0.8$.

   Therefore the input impedance is $Z_{\text{in}} = z_{\text{in}} \times Z_0 = (0.6 - j0.8) \times 50 = 30 - j40\ \Omega$.

Note: Since the length of the transmission line is integer multiples of $\frac{\lambda}{4}\,\frac{5\lambda}{4}$, the input impedance is

$$Z_{\text{in}} = \frac{Z_0^2}{Z_L} = \frac{50^2}{30 + j40} = 30 - j40\Omega.$$

Figure 10.29 shows the Smith chart for determination of VSWR and input impedance.

Example 10.30 Consider a line with $Z_0 = 100\ \Omega$ terminated by an unknown impedance. When a signal with frequency 100 MHz is transmitted, the VSWR is 2.5 and the first voltage minimum occurs at 16 cm from termination. Determine the termination impedance by using Smith chart. Assume the line is placed in free space.

Solution Given: Voltage standing wave ratio = 2.5, characteristic impedance $Z_0 = 100\ \Omega$, frequency = 100 MHz

$$\text{Wavelength } \lambda = \frac{c}{f} = \frac{3 \times 10^8}{100 \times 10^6} = 3 \text{ m}.$$

Steps for using Smith chart:

1. The VSWR point, M = 2.5 on the horizontal axis of the chart is located as shown in Fig. 10.30. The impedance circle is drawn with $O(1 + j0)$ as centre and the radius OM = 2.5.
2. The point A is located on the horizontal axis to the left side of the chart. This represents the first voltage minimum point.
3. Given the distance between the load and the first voltage minimum is 16 cm,

   $$y_{\text{min}} = 16 \text{ cm} = \frac{0.16}{3}\lambda = 0.053\lambda.$$

   Move from point A towards the load (anti-clockwise direction) to a length $0.053\,\lambda$ at point P′. This position is $0.5\,\lambda - 0.053\lambda = 0.447\lambda$.
4. The straight line OP′ is drawn and it cuts the impedance circle at point P. The impedance at point P is the normalised load impedance

   $$z_L = \frac{Z_L}{Z_0} = 0.44 - j0.29.$$

   Hence, the termination load impedance is

   $$Z_L = z_L \times Z_0 = (0.44 - j0.29) \times 100 = 44 - j29\ \Omega.$$

Figure 10.30 shows the Smith chart for determination of VSWR and load impedance.

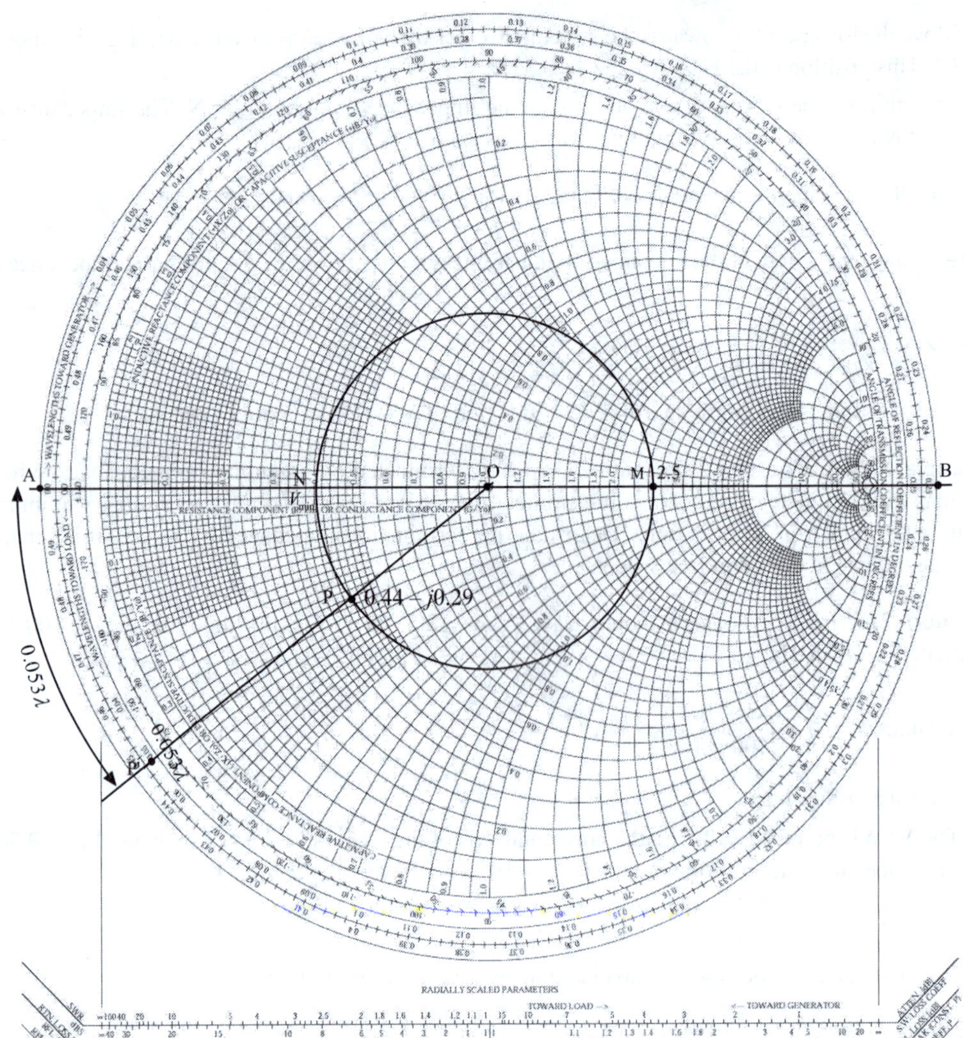

Fig. 10.30 *Smith chart for determination of VSWR and input impedance.*

Example 10.31 Determine the VSWR, characteristic impedance of a quarter wave transformer and the distance at which the transformer must be placed from the load to match a 75 Ω transmission line to a load $Z_L = 25 - j50\ \Omega$ using the of Smith chart.

Solution Given: Characteristic impedance $Z_0 = 75\ \Omega$, load impedance $Z_L = 25 - j50\ \Omega$.

Steps for using Smith chart:

1. Normalised load impedance is

$$z_L = \frac{Z_L}{Z_0} = \frac{25 - j50}{75} = 0.33 - j0.66.$$

The point P is located on the Smith chart shown in Fig. 10.31 at the intersection of the circles $R = 0.33$ and $X = 0.66$.

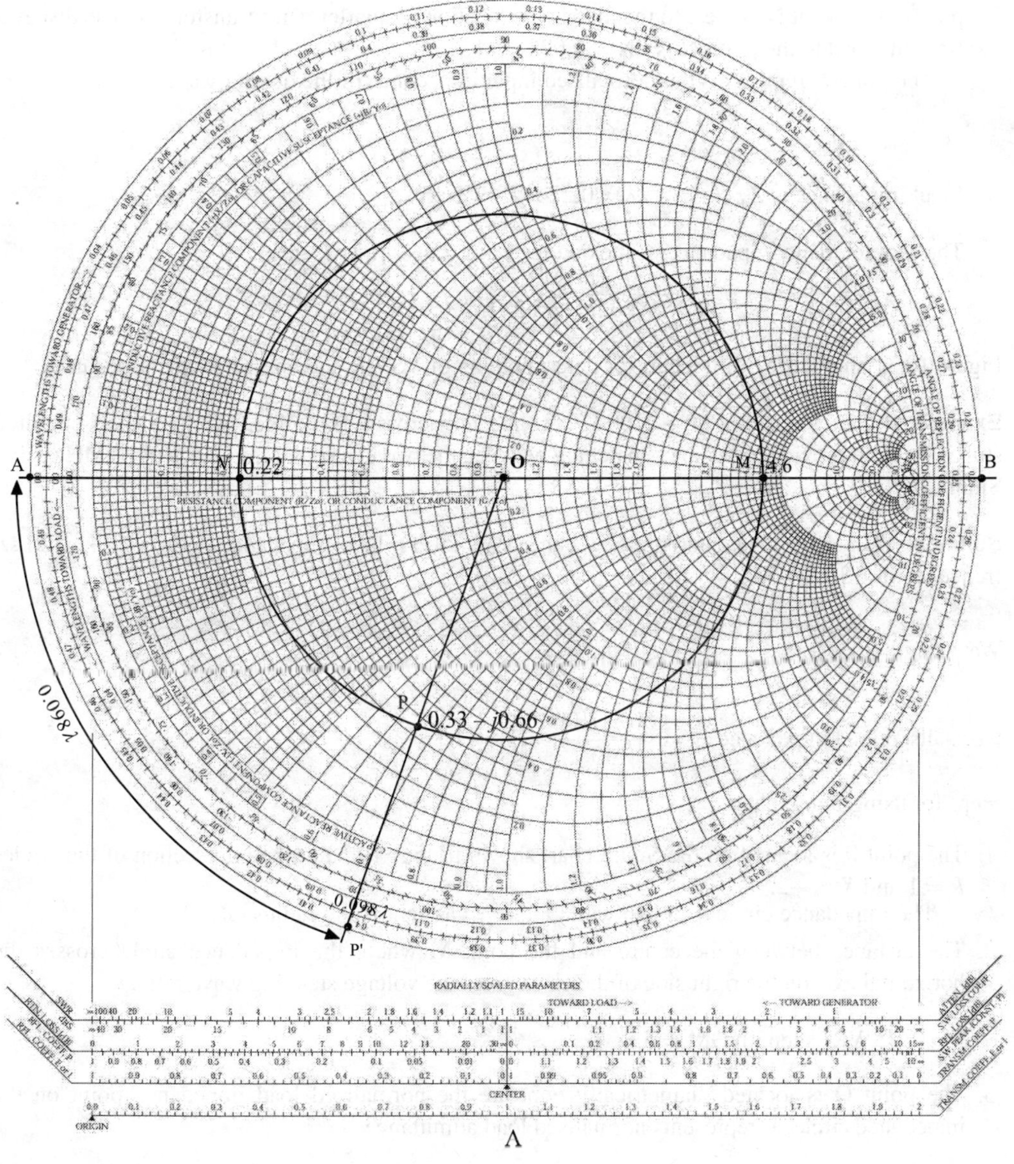

Fig. 10.31 *Smith chart for determination of load impedance*

2. The impedance circle is drawn with $O(1+j0)$ as centre and radius OP.

   The distance between the centre and the point M where the impedance circle crosses the horizontal axis on the right side of the chart gives the voltage standing wave ratio.

   $$\text{VSWR} = \text{length OM} = 4.6.$$

3. For impedance matching with a quarter wave transformer, the impedance of the line at the location of the transformer should be purely resistive.

   Point P is rotated on to the $X = 0$ line along impedance circle. There are two intersection points. The point N nearest to the load is the location of quarter wave transformer. The distance from the load to the point N is $\text{PN} = 0.098\lambda$.

   The point N represents the normalised input impedance of the quarter wave transformer,

   i.e., $z_{\text{in}} = 0.22$.

   Input impedance is $Z_{\text{in}} = z_{\text{in}} \times Z_0 = 0.22 \times 75 = 16.5\ \Omega$.

   The characteristic impedance of the quarter wave transformer is given by

   $$Z_o' = \sqrt{Z_0 \times Z_{\text{in}}} = \sqrt{75 \times 16.5} = 35.2\ \Omega.$$

Figure 10.31 shows the Smith chart for determination of VSWR and characteristic impedance.

Example 10.32 A load of $50 - j100\ \Omega$ is connected across a $50\ \Omega$ line. Design a short-circuited stub in order to provide impedance matching between the two at a signal frequency of 30 MHz using Smith chart.

Solution Given: Characteristic impedance $Z_0 = 50\ \Omega$, load impedance $Z_L = 50 - j100\ \Omega$, frequency $= 30$ MHz.

$$\text{Wavelength } \lambda = \frac{c}{f} = \frac{3 \times 10^8}{30 \times 10^6} = 10 \text{ m}.$$

$$\text{Normalised load impedance } z_L = \frac{50 - j100}{50} = 1 - j2.$$

Steps for using Smith chart:

1. The point P is located on the Smith chart shown in Fig. 10.32 at the intersection of the circles $R = 1$ and $X = -2$.

   The impedance circle is drawn with $O(1+j0)$ as centre and radius OP.
2. The distance between the centre and the point M where the impedance circle crosses the horizontal axis on the right side of the chart gives the voltage standing wave ratio.

   $$\text{VSWR} = \text{length OM} = 5.8.$$

3. The point Q is located diametrically opposite the normalised load impedance point on the impedance circle. It represents normalised load admittance.

   $$y_L = 0.2 + j0.4.$$

4. The point Q is rotated clockwise to a point T (shorter distance from load) on the impedance circle where it intersects the $R = 1$ circle. At this point, admittance $y = 1 + j0.2$.

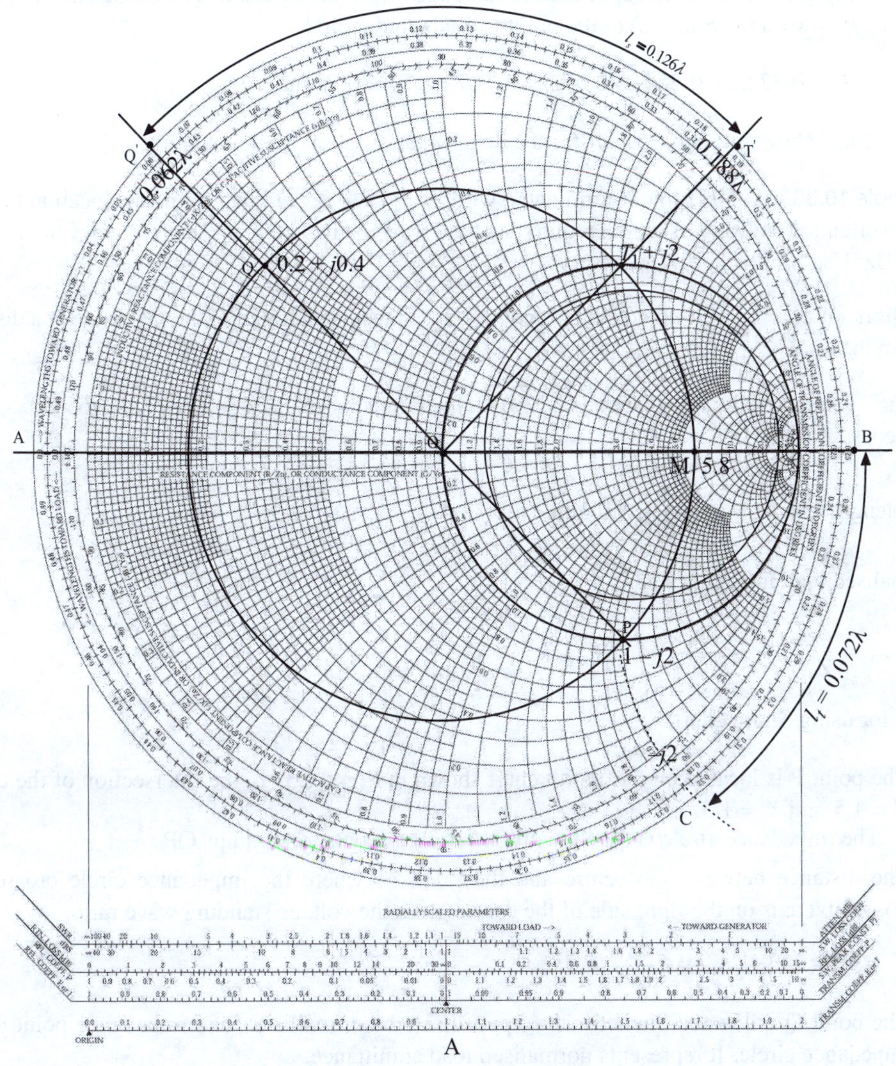

Fig. 10.32 *Smith chart for determination of VSWR and characteristic impedance.*

5. The line OQ is extended on to the outer circle to a point Q′ and the line OT is extended on to the outer circle to a point T′. The distance Q′T′ measured on the wavelength scale towards the generator gives the location of the stub.

   Location of the stub on the line from the load is $l_s = (0.188 - 0.062)\,\lambda = 0.126\lambda$.

   $l_s = 0.126\lambda = 0.126 \times 10 = 1.26$ m.

6. The susceptance of the stub is equal to the negative value of the susceptance at point T, i.e., $y_{stub} = -j0.2$. Point C is located on the outer circle for y_{stub}.
7. The length of the short circuited stub is calculated from the right side end of the chart at point B (i.e., $R = \infty$ on the R circle) to the point C on the outer circle.

$$l_t = 0.322\lambda - 0.25\lambda = 0.072\lambda = 0.072 \times 10 = 0.72 \text{ m}.$$

Figure 10.32 shows the Smith chart for the determination of SC stub matching

Example 10.33 A 50 Ω line is terminated with $Z_L = 75 + j65$ Ω. Determine the location to insert a short circuited stub and its length required to match the load to the line. The frequency of operation is 1 GHz. Use Smith chart if necessary.

Solution Let a short circuited stub of length l_t be connected parallel to the main line at a distance l_s from the load.

Given: Characteristic impedance, $Z_0 = 50$ Ω, load impedance, $Z_R = Z_L = 75 + j65$ Ω, frequency $= 1$ GHz.

$$\text{Wavelength } \lambda = \frac{c}{f} = \frac{3 \times 10^8}{1 \times 10^9} = 0.3 \text{ m}.$$

Normalised load impedance is

$$z_R = \frac{Z_R}{Z_0} = \frac{75 + j65}{50} = 1.5 + j1.3.$$

Steps for using Smith chart:

1. The point P is located on the Smith chart shown in Fig. 10.33 at the intersection of the circles $R = 1.5$ and $X = 1.3$.
   The impedance circle is drawn with $O(1 + j0)$ as centre and radius OP.
2. The distance between the centre and the point M where the impedance circle crosses the horizontal axis on the right side of the centre gives the voltage standing wave ratio.

   VSWR = length OM = 3.

3. The point Q is located diametrically opposite to the normalised load impedance point on the impedance circle. It represents normalised load admittance.

   $$y_L = 0.38 - j0.33.$$

4. The point Q is rotated clockwise to a point T (shorter distance from load) on the impedance circle where it intersects the $R = 1$ circle. At this point, admittance

   $$y = 1 + j1.16.$$

5. The line OQ is extended on to the outer circle to a point Q′ and the line OT is extended on to the outer circle to a point T′. The distance Q′T′ measured on wavelength scale towards the generator gives the location of the stub.

Location of the stub on the line from the load is $l_s = (0.056 + 0.166)\,\lambda = 0.222\lambda$.

$l_s = 0.222\lambda = 0.222 \times .3 = 6.6$ cm.

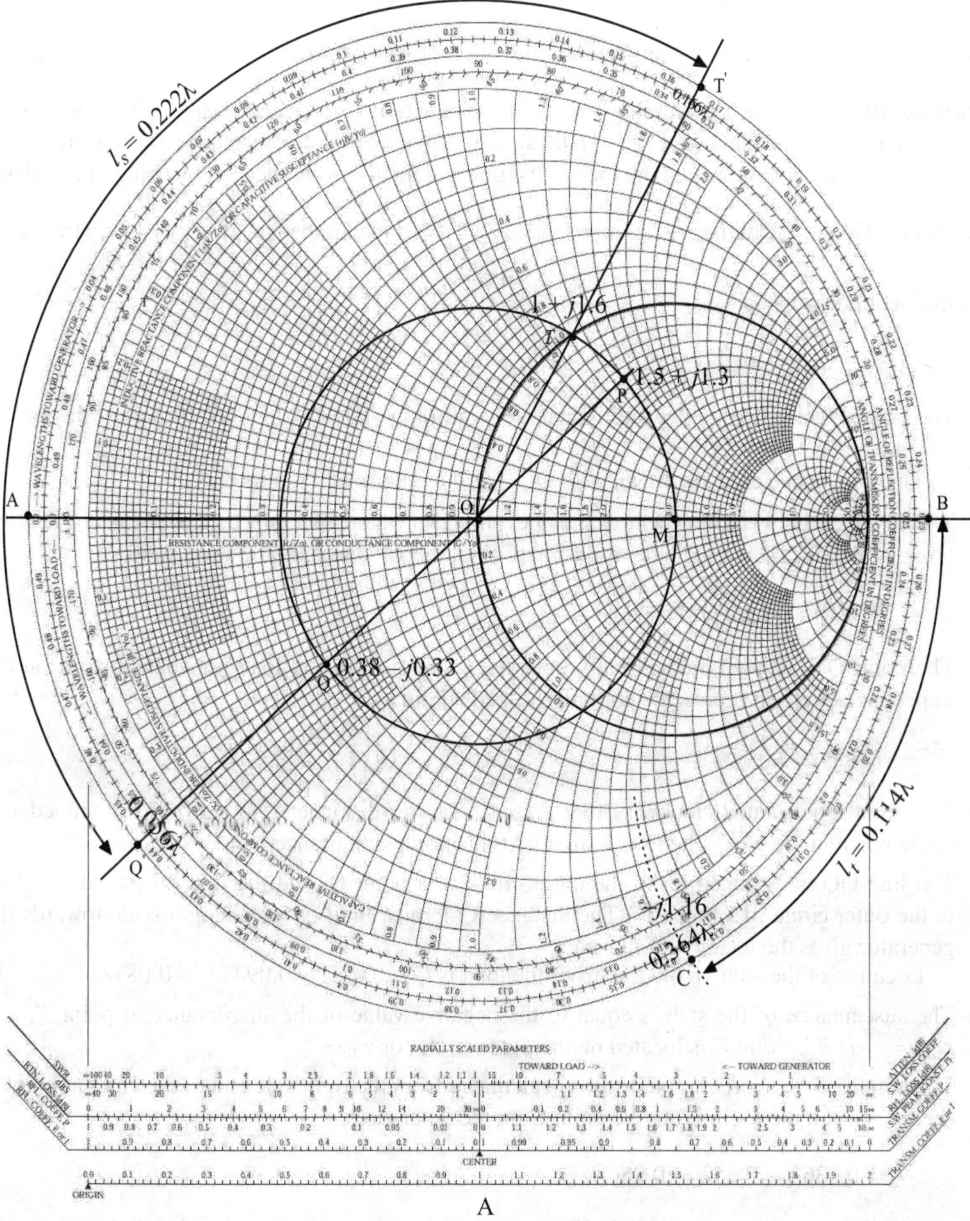

Fig. 10.33 *Smith chart for determination of SC stub matching*

6. The susceptance of the stub is equal to the negative value of the susceptance at point T, i.e., $y_{stub} = -j1.16$. Point C is located on the outer circle for y_{stub}.

7. The length of the short circuited stub is calculated from the right side end of the chart at point B (i.e., $R=\infty$ on the R circle) to the point C on the outer circle towards the generator.

$$l_t = 0.346\lambda - 0.25\lambda = 0.114\lambda = 0.114 \times 0.3 = 3.4 \text{ cm}.$$

Figure 10.33 shows the Smith chart for determination of stub matching.

Example 10.34 For a transmission line with a characteristic impedance $Z_0 = 75\ \Omega$ and a load with a capacitive impedance $Z_L = 45 - j100\ \Omega$, determine VSWR, the distance a shorted stub must be placed from the load to match the load to the line and the length of the stub by using Smith chart.

Solution Given: Characteristic impedance $Z_0 = 75\ \Omega$, load impedance $Z_L = 45 - j100\ \Omega$.

Normalised load impedance $z_L = \dfrac{Z_L}{Z_0} = \dfrac{45 - j100}{75} = 0.6 - j1.33.$

Steps for using Smith chart:

1. The point P is located on the Smith chart shown in Fig. 10.34 at the intersection of the circles $R = 0.6$ and $X = -1.33$.
   The impedance circle is drawn with $O(1 + j0)$ as centre and radius OP.
2. The distance between the centre and the point M where the impedance circle crosses the horizontal axis on the right side of the chart gives the voltage standing wave ratio.

$$\text{VSWR} = \text{length OM} = 5.$$

3. The point Q is located diametrically opposite to the normalised load impedance point on the impedance circle. It represents normalised load admittance

$$y_L = 0.28 + j0.62.$$

4. The point Q is rotated clockwise to a point T (shorter distance from load) on the impedance circle where it intersects the $R = 1$ circle. At this point, admittance $y = 1 + j1.75$.
5. The line OQ is extended on to the outer circle to a point Q′ and the line OT is extended on to the outer circle to a point T′. The distance Q′T′ measured on wavelength scale towards the generator gives the location of the stub.
   Location of the stub on the line from the load is $l_s = (0.181 - 0.094)\,\lambda = 0.087\lambda$.
6. The susceptance of the stub is equal to the negative value of the susceptance at point T, i.e., $y_{\text{stub}} = -j1.75$. Point C is located on the outer circle for y_{stub}.
7. The length of the short circuited stub is calculated from the right side end of the chart at point B (i.e., $R = \infty$ on the R circle) to the point C on the outer circle.

$$l_t = 0.336\lambda - 0.25\lambda = 0.086\lambda.$$

Figure 10.34 shows the Smith chart for determination of SC stub matching.

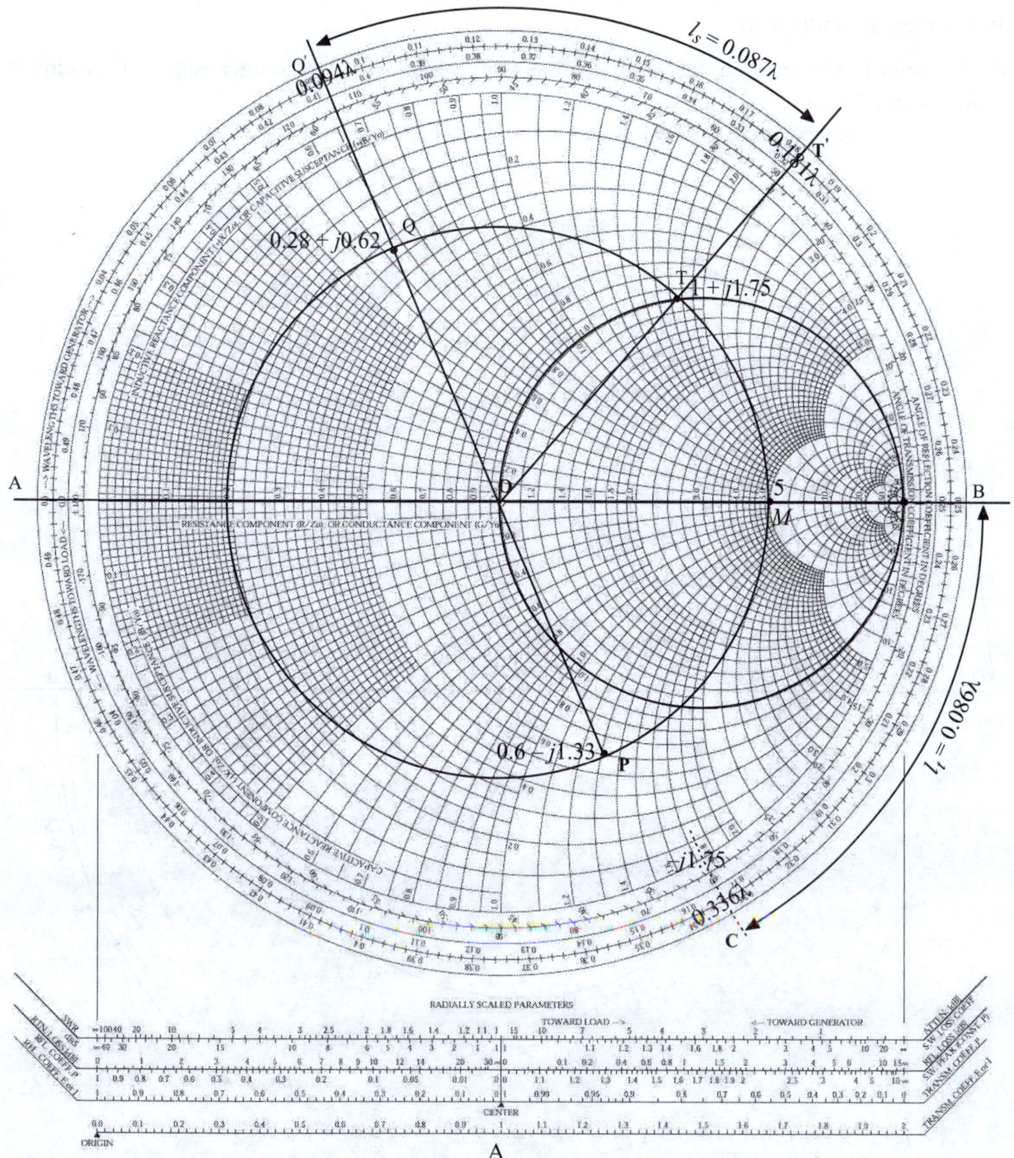

Fig. 10.34 *Smith chart for determination of stub matching*

Example 10.35 A lossless transmission line is terminated by a 300 Ω load. Find a suitable stub location as close to the load as possible and determine the length of a 50 Ω stub required using the Smith chart. Is the stub inductive or capacitive?

Solution Given: $Z_0 = 50\ \Omega \quad Z_R = 300\ \Omega.$

Normalised load impedance $z_L = \dfrac{Z_L}{Z_0} = \dfrac{300}{50} = 6.$

Steps for using Smith chart:

1. The point P is located on the Smith chart shown in Fig. 10.35 at the intersection of the circles $R = 6$ and $X = 0$.

   The impedance circle is drawn with $O(1 + j0)$ as centre and radius OP.

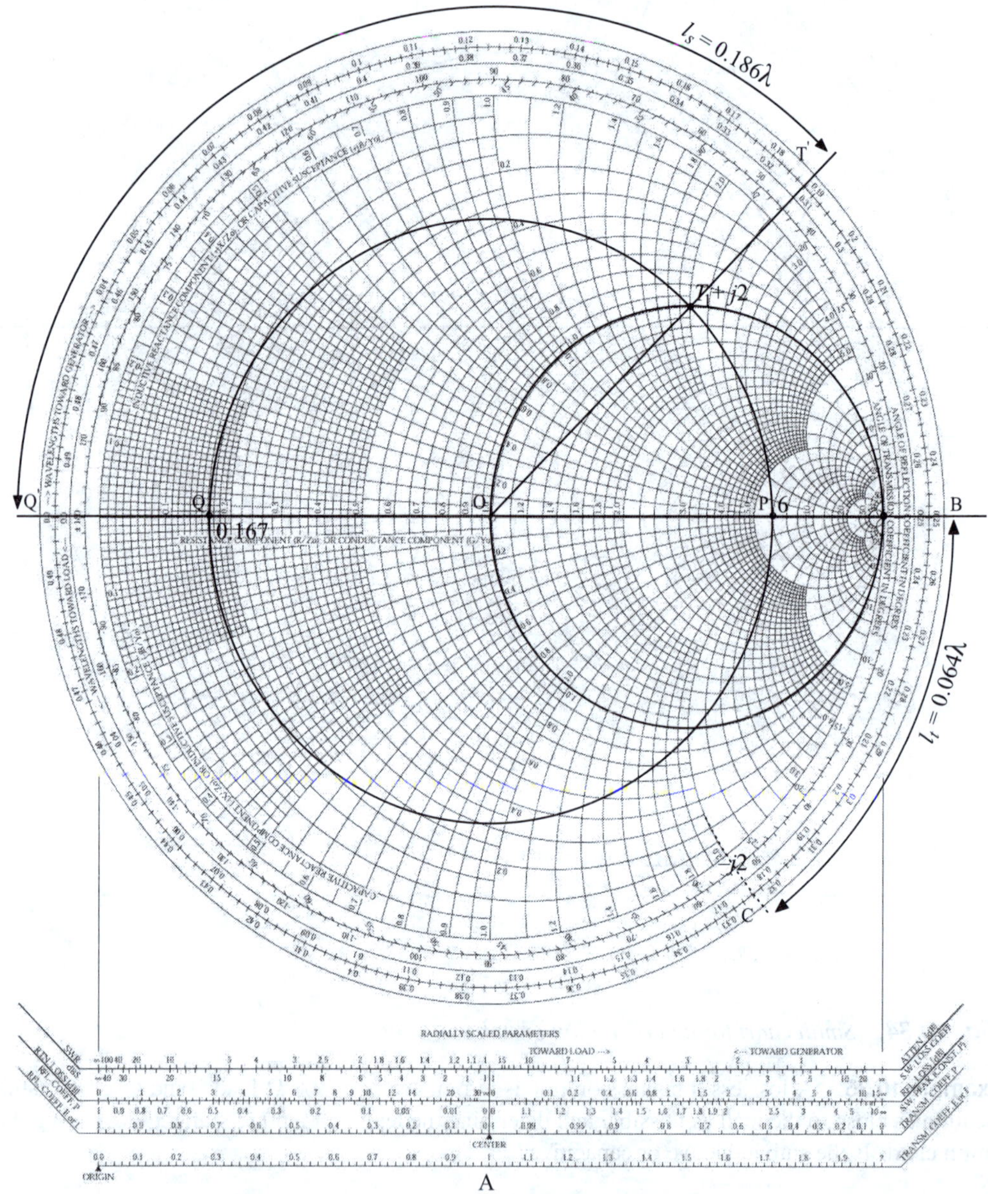

Fig. 10.35 *Smith chart for determination of SC stub matching*

2. The distance between the centre and the point M where the impedance circle crosses the horizontal axis on the right side of the chart, i.e., OM, is the voltage standing wave ratio.

   VSWR = length OM = 6.

3. The point Q is located diametrically opposite to the normalised load impedance point on the impedance circle. It represents normalised load admittance.

   $y_L = 0.166.$

4. The point Q is rotated clockwise to a point T (shorter distance from load) on the impedance circle where it intersects the $R = 1$ circle. At this point, admittance

   $y = 1 + j2.$

5. The line OQ is extended on to the outer circle to a point Q′ and the line OT is extended on to the outer circle to a point T′. The distance Q′ T′ measured on wavelength scale towards the generator gives the location of the stub.
   Location of the stub on the line from the load is $l_s = 0.186\lambda$.

6. The susceptance of the stub is equal to the negative value of the susceptance at point T, i.e., $y_{\text{stub}} = -j2$. Point C is located on the outer circle for y_{stub}.

7. The length of the short circuited stub is calculated from the right side end of the chart at point B (i.e., $R = \infty$ on R circle) to the point C on the outer circle.

   $l_t = 0.25\lambda - 0.186\lambda = 0.064\lambda.$

   Stub reactance is $b_s = -j2 \times Y_0 = -\dfrac{j2}{50} = -j0.04$ mhos.
   Therefore, the stub is capacitive.

Figure 10.35 shows the Smith chart for determination of single stub matching.

Example 10.36 A lossless transmission line $Z_0 = 300\ \Omega$ has a load impedance of $150 + j225\ \Omega$. Using a Smith chart, design a double stub matching such that the two stubs are 0.1λ apart.

Solution For proper matching, consider two short circuited stubs connected in parallel to the main line. Let stub A be connected at the load and stub B connected at a distance of 0.1λ from stub B.
Let l_{t_1} and l_{t_2} be the lengths of stub A and stub B respectively.
Characteristic impedance of the stubs should be equal to line impedance $Z_0 = 300\ \Omega$.

Steps for using Smith chart:

Given: Characteristic impedance $Z_0 = 300\ \Omega$, load impedance $Z_L = 150 + j225\ \Omega$,

Normalised load admittance $y_L = \dfrac{Y_L}{Y_0} = \dfrac{Z_0}{Z_L} = \dfrac{300}{150 + j225} = 0.62 - j0.92\ \mho$.

1. The point Q located on the chart and represented by $y_L = 0.62 - j0.92\ \mho$ is the normalised admittance at stub A before stub connection.
2. Circle 1 for unity conductance ($y = 1$ or $R = 1$ circle) is drawn on the right-hand side of the chart.

3. Since the distance between the two stubs is 0.1λ, all the points of the unity conductance circle should be shifted by $0.1\ \lambda$ towards the load. Circle 2 is drawn with shifted unity conductance circle for stub A.

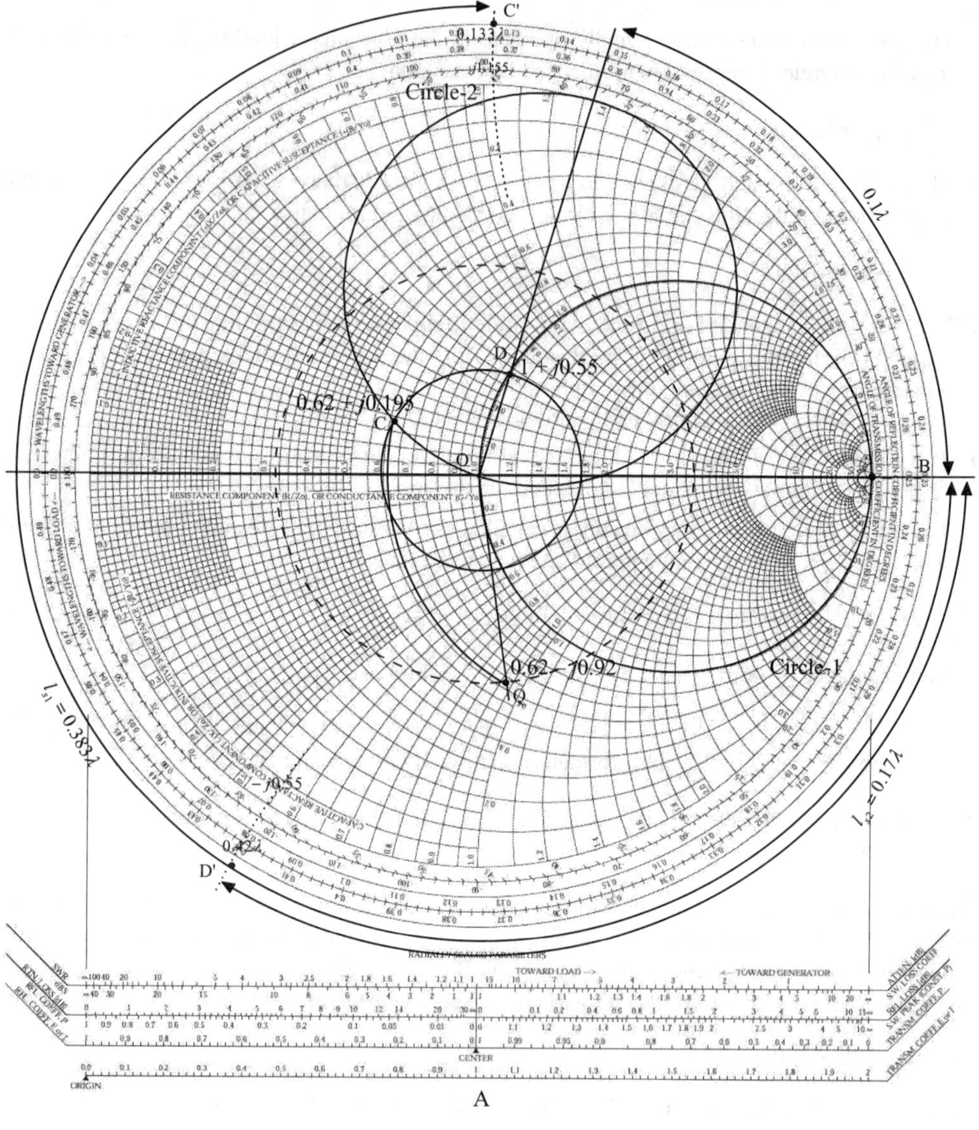

Fig. 10.36 *Smith chart for determination of single stub matching*

4. Since stub A adds a susceptance to the load, point Q is moved along the constant conductance circle to a point C on circle 2. Point C represents the combined normalised admittance at stub A after stub connection, i.e., $y_{L_1} = 0.62 + j0.195\ \mho$.

Change in susceptance $= y_{L_1} - y_L = 0.62 + j0.195 - (0.62 - j0.92) = j1.115.$

5. The susceptance of the stub A is equal to the change in susceptance, i.e., $y_{\text{stubA}} = j1.115$. Locate point C' on the outer circle for y_{stubA}.
6. The length of the short circuited stub A, l_{s_1} is calculated from the right side end of chart at point B (i.e., $R = \infty$ on the R circle) to the point C′ on the outer circle towards the generator.

   Length of the stub A, $l_{t_1} = \text{C}'\text{B} = 0.133\lambda + 0.25\lambda = 0.383\lambda$.

7. A circle with centre O and radius OC is drawn. It cuts circle 1 at D. Point D gives the admittance on the line at stub B, i.e., $y_{s_2} = 1 + j0.55$ before stub B is connected.
8. To provide proper termination with stub B, the susceptance of the stub B should be equal to the negative value of the susceptance at point D, i.e., $y_{\text{stubB}} = -j0.55$.
   Locate point D' on the outer circle for y_{stubB}.
9. The length of the short circuited stub B l_{s_1} is calculated from the right side end of the chart at point B (i.e., $R = \infty$ on the R circle) to the point D′ on the outer circle towards the generator.

   Length of the stub B, $l_{t_2} = \text{D}'\text{B} = 0.42\lambda - 0.25\lambda = 0.17\lambda$.

Figure 10.36 shows the Smith chart for determination of double stub matching.

Example 10.37 A 50 Ω line is connected to a load $60 + j75$ Ω. Stub A and stub B are connected in parallel with the line at distances $\lambda/8$ and $\lambda/4$ respectively from the load. Find the shortest lengths of two stubs for a match. The stubs are of 50 Ω characteristic impedance.

Solution Let l_{t_1} and l_{t_2} be the lengths of stub A and stub B respectively from the load.

Given: Characteristic impedance $Z_0 = 50$ ohms.

Load impedance $Z_R = 60 + j75$ Ω for a transmission line with double stub matching as shown in Fig. 10.37.

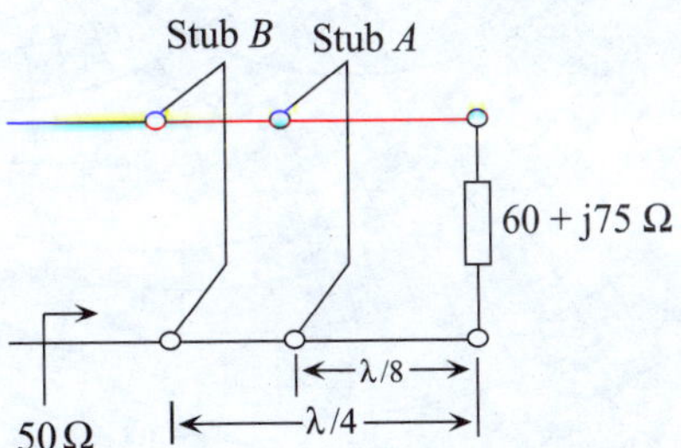

Fig. 10.37 *Transmission line with double stub matching*

$$\text{Normalised load admittance } y_L = \frac{Y_L}{Y_0} = \frac{Z_0}{Z_L} = \frac{50}{60 + j75} = 0.32 - j0.4.$$

Steps for using Smith chart:

1. The point Q located on the chart as shown in Fig. 10.38 represents, $y_L = 0.32 - j0.4$ mhos. The impedance circle is drawn with $O(1 + j0)$ as centre and radius OQ.

2. At a distance $\lambda/8$ from the load, stub A is connected. The admittance point Q is moved towards the generator (clockwise direction) to a distance $l_{s_1} = 0.125\lambda$ at point Q_1 on the S circle. This point represents normalised admittance at stub A before stub connection.

   The admittance is $y_L = 0.32 + j0.4$.

3. Circle 1 for unity conductance ($y = 1$ or $R = 1$ circle) is drawn on the right-hand side of the chart for stub B.

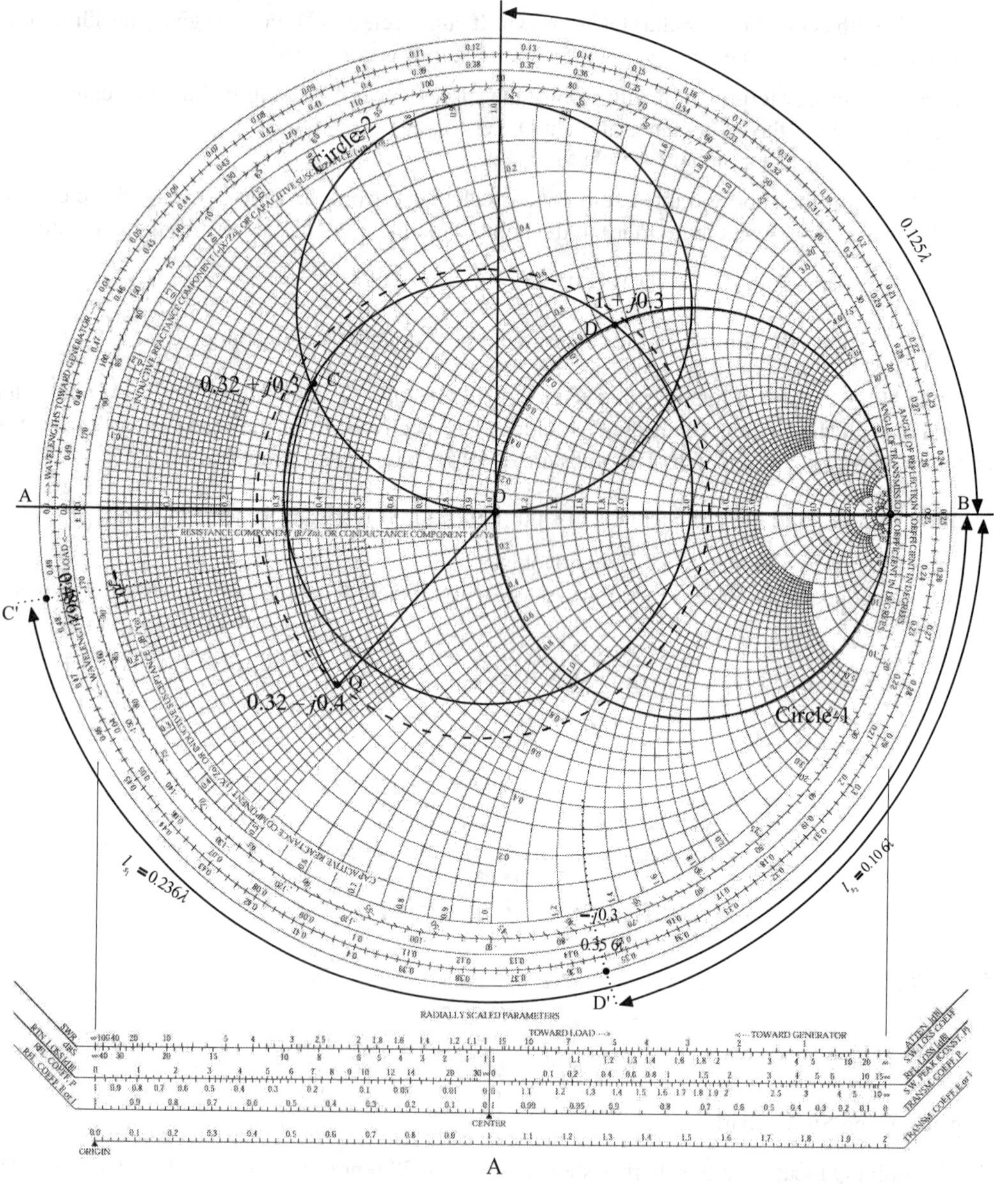

Fig. 10.38 *Smith chart for determination of double stub matching*

4. Since the distance between the two stubs is $l_s = \lambda/8$, all the points of the unity conductance circle should be shifted by 0.125λ distance towards the load. Draw circle 2 as the shifted unity conductance circle for stub A.

5. Since stub A adds a susceptance to the load, move point Q_1 along the constant conductance circle to a point C on circle 2. The point C represents the combined normalised admittance at stub A after stub connection, i.e., $y_{L_1} = 0.32 + j0.3\ \mho$.

   Change in susceptance $= y_{L_1} - y_L = 0.32 + j0.3 - (0.32 + j0.4) = -j0.1$.

6. The susceptance of the stub A is equal to the change in susceptance, i.e., $y_{\text{stubA}} = -j0.1$ Locate point C′ on the outer circle for y_{stubA}.

7. The length of the short circuited stub A, l_{t_1} is calculated from the right side end of the chart at point B (i.e., $R = \infty$ on the R circle) to the point C′ on the outer circle towards the generator.

   Length of the stub A, $l_{t_1} = C'B = 0.486\lambda - 0.25\lambda = 0.236\lambda$.

8. Draw a circle with centre O and radius OC It cuts circle 1 at D. Point D gives the admittance on the line at stub B, i.e., $y_{s_2} = 1 + j1.3$ before stub B is connected.

9. To provide proper termination with stub B, the susceptance of the stub B should be equal to the negative value of the susceptance at point D, i.e., $y_{\text{stubB}} = -j1.3$.

   Locate point D′ on the outer circle for y_{stubB}.

10. The length of the short circuited stub B, l_{t_2} is calculated from the right side end of the chart at point B (i.e., $R = \infty$ on the R circle) to the point D′ on the outer circle towards the generator.

    Length of the stub B, l_{t_2} =D′B= $0.356\lambda - 0.25\lambda = 0.106\lambda$.

Figure 10.38 shows the Smith chart for determination of double stub matching.

Summary

1. The voltage and current distributions of standing wave at a given frequency can be expressed as

$$V = V_s \cosh \gamma x - I_s Z_0 \sinh \gamma x, \quad I = I_s \cosh \gamma x - \frac{V_s}{Z_0} \sinh \gamma x.$$

2. The input impedance of the open-circuited line is $Z_{oc} = \frac{V_s}{I_s} = Z_0 \coth \gamma l$ and for the short-circuited line, it is $Z_{sc} = \frac{V_s}{I_s} = Z_0 \tanh \gamma l$.

3. Z_0 and γ in terms of Z_{sc} and Z_{oc} are given by $Z_0 = \sqrt{Z_{oc} Z_{sc}}$, $\gamma = \frac{1}{l} \tanh^{-1}\left(\sqrt{\frac{Z_{sc}}{Z_{oc}}}\right)$.

4. The input impedance is given by $Z_{\text{in}} = Z_0 \left[\frac{Z_R + Z_0 \tanh \gamma l}{Z_0 + Z_R \tanh \gamma l}\right]$.

   The condition for perfectly matched transmission line is $Z_{\text{in}} = Z_R = Z_0$.

5. For lossless line, $\alpha = 0$. The input impedance $Z_{\text{in}} = Z_0 \dfrac{Z_R + jZ_0 \tan(\omega l\sqrt{LC})}{Z_0 + jZ_R \tan(\omega l\sqrt{LC})}$.
6. The reflection coefficient is defined as the ratio of reflected voltage or current to the incident voltage or current. It can be expressed as $K = \dfrac{Z_R - Z_0}{Z_R + Z_0}$.
7. The ratio of the maximum magnitude of voltage to the minimum magnitude of voltage is called voltage standing wave ratio. It is given by

$$\text{VSWR} = \text{S} == \frac{|V_{\max}|}{|V_{\min}|} \qquad \text{S} = \frac{1+|K|}{1-|K|}.$$

8. The input impedance in terms of the reflection coefficient is $Z_{\text{in}} = Z_0 \dfrac{(1+Ke^{2\gamma l})}{(1-Ke^{2\gamma l})}$.
9. The position of the first maxima and minima from the termination can be obtained from:

$$y_{\max 1} = \frac{\phi}{2\beta} = \frac{\phi\lambda}{4\pi} \quad \text{and} \quad y_{\min 1} = y_{\max} + \frac{\lambda}{4}.$$

10. The impedance on the line varies from a maximum value $Z_{\max} = Z_0 S$ at antinode to a minimum value $Z_{\min} = \dfrac{Z_0}{S}$ at node point.
11. A short length transmission line is added to the main line to achieve maximum power transfer. The important short length transmission lines are:

    (i) The eight-wave ($\lambda/8$ length) transmission line; $|Z_{\text{in}}| = |Z_0|$
    (ii) The quarter-wave ($\lambda/4$ length) transmission line; $Z_0 = \sqrt{Z_{\text{in}} Z_R}$
    (iii) The half-wave ($\lambda/2$ length) transmission line; $Z_{\text{in}} = Z_R$

12. At ultra high frequencies, the transmission line becomes lossless. The short length SC and OC transmission lines can be used as circuit elements.

    For ($0 < l < \lambda/4$), the SC line acts as inductive while the OC line acts as capacitive
    For ($\lambda/4 < l < \lambda/2$), the SC line acts as capacitive while the OC line acts as inductive.
    For $l = \lambda/4$, the SC line acts as open-circuit while the OC line acts as short-circuit.
    For $l = \lambda/2$, the SC line acts as short-circuit while the OC line acts as open circuit.

14. Single stub matching and double stub matching methods are used to achieve impedance matching.
15. The location and length of a short-circuited stab are

$$l_s = \frac{\lambda}{2\pi} \tan^{-1} \sqrt{\frac{Z_0}{Z_R}} = \frac{\phi + \pi - \cos^{-1}|K|}{2\beta}, \quad \text{and}$$

$$l_l = \frac{1}{\beta} \tan^{-1} \left(\frac{\sqrt{Z_R Z_0}}{Z_R - Z_0} \right) = \frac{\lambda}{2\pi} \tan^{-1} \frac{\sqrt{1-|K|^2}}{2|K|}.$$

16. In double stub matching, the space between two stubs is generally fixed at $\lambda/8$, $\lambda/4$ or $3\lambda/8$ distance.

17. The Smith chart graphically represents the normalized values of R and X circles on the reflection coefficient plane. It is used to obtain line characteristics such as admittance, input impedance, VSWR, K, location of voltage maximum and voltage minimum etc. It is also used in the design of impedance matching devices using SC and OC lines.

Review Questions

1. Determine the reflection coefficient of a transmission line when (i) Z_L = Short circuit (ii) $Z_L = Z_0$. *(Aug 2010)*
2. Using the general line equations, obtain an expression for the input impedance of a line. *(Feb 2008)*
3. Sketch input impedance versus line length for short- and open-circuited lines with $0 < l < \lambda$. *(April 2013)*
4. Discuss single and double stub matching. *(Dec 2012, April 2013)*
5. What is stub? Why are short circuited stubs preferred to open circuited ones? Explain. *(Dec 2010, Aug 2014)*
6. Derive the expression for the input impedance of a lossless line. Hence evaluate Z_{sc} and Z_{oc}. Sketch their variation with line length. *(Dec 2012, Aug 2014)*
7. What are the applications of Smith chart? *(April 2013)*
8. Explain the significance and utility of $\lambda/8$, $\lambda/4$ and $\lambda/2$ lines. *(Dec 2012)*
9. Explain the principle of impedance matching with a quarter wave transformer. *(Dec 2012, Aug 2014)*
10. Discuss why impedance matching is necessary and show with a suitable diagram. Calculate how a single stub is used for impedance matching. *(May 2012, Jan 2014)*
11. A lossless transmission line of phase constant β radians/metre and characteristic impedance Z_0 is 1 metre long and terminated by an impedance Z_L. Derive from first principles an expression for input impedance. *(May 2011)*
12. What are the various methods impedance matching? Explain the principle of quarter wave transformer. *(Dec 2013, Jan 2014)*
13. Derive the expression for the current, voltage and impedance relations along a high frequency transmission line. Explain characteristic impedance. Deduce an expression for the suitable length of a short circuited stub section to match a loss free feeder line with resistance load. *(April 2013)*
14. Explain reflection coefficient and voltage standing wave ratio of a transmission line. *(May 2012)*
15. Describe the characteristics and applications of $\lambda/2$ and $\lambda/4$ radio frequency (lossless) transmission line elements. *(Dec 2012)*
16. Explain the principle of the design of a Smith chart used for transmission line calculations. *(Aug 2014)*

17. Define the following terms used in connection with transmission line. (*May 2011*)
    (i) Propagation constant
    (ii) Characteristic impedance
    (iii) Reflection coefficient
    (iv) Voltage standing wave ratio
18. Derive a relation between reflection coefficient and characteristic impedance. (*Dec 2012, Aug 2014*)
19. What is meant by stub? Explain the single stub matching concept. (*Aug 2014*)

Problems

1. A transmission line of length 0.4 km of characteristic impedance of 100 Ω is terminated in a load impedance of $(200 + j80)$ ohms. Find (i) voltage reflection coefficient, and (ii) voltage standing wave ratio.
2. A transmission line is a lossless and is 25 m long. It is terminated in a load of $Z = (40 + j30)$ ohms at a frequency of 10 MHz. The inductance and capacitance of line are $L = 300\text{nH/m}$ and $C = 40\text{pF/m}$. Find the input impedance at the source and at the midpoint of the line.
3. A load of $(50 - j100)$ ohms is connected across a 50 Ω line. Design a short-circuited stub in order to provide impedance matching between the two at a signal frequency of 30 MHz.
4. A low loss line with characteristic impedance of 50 ohms is connected to a load of $(100 + j150)$ ohms. Calculate the reflection coefficient and the maximum and minimum line impedances.
5. Calculate the characteristic impedance of a transmission line if the following measurements have been made at 1600 MHz on the line, $Z_{oc} = 550\ \Omega$ and $Z_{sc} = 660\ \Omega$.
6. A lossless transmission line with $Z_0 = 50$ ohms is to be matched to a load impedance of 300 ohms with a quarter wave transformer. Find characteristic impedance of the quarter wave transformer.
7. A lossless transmission line with $Z_0 = 300$ ohms connected to an unknown load produces VSWR $= 2.8$. Voltage minimum occurred at 0.12λ from the load. Find the load impedance.
8. A lossless transmission line with $Z_0 = 300$ ohms connected to an unknown load produces VSWR $= 4.48$ at 200 MHz. Voltage minimum is located at 6 cm from the load. Find the load impedance.
9. A lossless transmission line with $Z_0 = 50$ ohms is 30 m long and operates at 2 MHz. The line is terminated with a load of $(60 + j40)$ ohms. If the velocity is $v_p = 0.6C$ (where C is the free space velocity) on the line, find the following: (i) the reflection coefficient, (ii) the standing wave ratio, and (iii) input impedance.
10. A short-circuited lossy transmission line of length 5 m and characteristic impedance of 150 Ω has input impedance of $(90 + j450)$ ohms. Find out propagation constant.
11. Find Z_{oc} and Z_{sc} if the characteristic impedance of a 20 m length lossless line at 1.2 MHz is 1000 Ω.
12. A two-wire line has a characteristic impedance of 600 Ω and is to feed a 1800 Ω resistor at 200 MHz. Determine the shortest length and location of a single SC stub for matching
13. A two-wire RF line has a characteristic impedance of 50 ohms and is used to feed 300 MHz to a load of 500 ohms. Find the position on the line at which the matching stub is to be connected.

14. A uniform lossless line with characteristic impedance of 100 ohms is connected to a load of $(200 - j150)$ ohms. A short-circuited stub of the same characteristic impedance is connected in parallel with the line at a distance d_1 from the load. Find the shortest values of d_1 and the stub length d_2 for a match.
15. A transmission line of 100 Ω characteristic impedance is connected to a load of 400 Ω. Calculate the reflection coefficient and standing wave ratio using Smith chart.
16. A lossless 100 ohm line is connected to a load 100 $(-j100)$ ohms. Stub A is connected in parallel with the line $3/16\lambda$ from the load and stub B, $\lambda/4$ from the load. Find the shortest lengths of the two stubs for a match using Smith chart. The stubs are of 100 ohm characteristics resistance.
17. A lossless transmission line of $Z_0 = 50$ ohms has a load impedance of $60 + j25$ ohms. Using Smith chart, find two possible stub locations within 0.5λ from the load that will permit a match to be achieved. Determine the length of the stub required in each case if Z_0 of the stub is also 50 ohms.
18. With the aid of Smith chart, calculate the position and length of a short-circuited stub matching a $(150 + j120)$ ohms load to a 300 ohm transmission line. Assuming that the load impedance remains constant, find the VSWR on the main line when the frequency is (i) increased by 10% and (ii) doubled.

Answers

1. $K = 0.412\angle 23.730$, $S = 2.4$
2. $Z_{in} = 57.17\angle 41.11$, $53.43\angle 39.18°$
3. $l_s = 3.75$ m, $l_t = 0.738$ m
4. $K = 0.33$, $Z_{max} = 100$ Ω, $Z_{min} = 25$ Ω
5. $Z_0 = 602.5$ Ω
6. $Z_0 = 122.5$ Ω
7. $Z_R = 658.5\angle -32.8$ ohms
8. $Z_R = 101.9\angle -47.5$ ohms
9. $K = 0.352\angle 56°$, $S = 2.09$, $Z_{in} = 24.01\angle 3.22$ ohms
10. $0.0116 + j0.252$ per metre
11. $Z_{oc} = -j\ 1819$ Ω, $Z_{oc} = -j550\Omega$
12. $l_s = 2.5$ m, $l_t = 0.17$ m
13. 0.201 m from load
14. $d_1 = 0.129\lambda$, $d_2 = 0.106\lambda$
15. $K = 0.6$, $S = 4$

Multiple-choice Questions

1. The phenomenon of setting up a reflected wave at the load due to improper termination or due to impedance irregularity in a line is called
   (a) refraction (b) diffraction (c) reflection (d) none of the above
2. The ratio of reflected current to incident current is known as
   (a) VSWR (b) reflection coefficient
   (c) reflection loss (d) none of the above
3. The magnitude of the voltage when the incidence and reflected waves are in phase is
   (a) minimum (b) maximum (c) zero (d) none of the above

4. We can find the first voltage maximum from the load by the following equation

(a) $2\beta y_{\max} - \phi = 2n\pi$ (b) $2\beta y_{\min} - \phi = (2n+1)\pi$

(c) both (a) and (b) (d) none of the above

5. On the Smith chart, if resistance increases from the top to the bottom part, then the top part of the transmission line is

(a) short circuit (b) open circuit

(c) transmission line with impedance (d) none of the above

6. It is possible to connect a section of an open or short circuited line in shunt with the main line at some point or points to effect impedance matching. This is called

(a) loading (b) stub matching

(c) frequency distortion (d) none of the above

7. Stub matching used for fixed frequencies is

(a) single stub (b) double stub

(c) both (a) and (b) (d) none of the above

8. Using Smith chart, we can find the following parameters

(a) VSWR (b) load admittance

(c) length of single and double stub (d) all of the above

9. When the transmission line is terminated with impedance Z_R, then there will be no reflections. This type of transmission line behaves like a

(a) open circuit (b) short circuit

(c) line terminated with characteristic impedance (d) none of the above

10. $V = V_s \cosh \gamma x - I_s Z_0 \sinh \gamma x$ is the voltage equation on the line, when x is measured from

(a) source to load (b) load to source

(c) from centre to load (d) constant parameter

11. For a lossless line the characteristic impedance $Z_0 = 50$ ohms and load imped- ance $Z_R = 60 + j75\Omega$, the normalised load admittance is

(a) $1.\ 2 - j1.5$ (b) $0.\ 32 - j0.4$

(c) $12 - j15$ (d) $0.016 - j0.013$

12. For a lossless line, the characteristic impedance is Z_0 and VSWR is S. Then the maximum impedance on the line is

(a) $Z_0 S$ (b) $Z_0/2S$ (c) $Z_0 S^2$ (d) Z_0/S

13. The current equation at a distance y from the load on a lossless or distortionless line is

(a) $I_R \cos\beta y - j\frac{V_R}{Z_0}\sin\beta y$ (c) $I_R \cos\beta y + j\frac{V_R}{Z_0}\sin\beta y$

(b) $I_s \sin\beta y + j\dfrac{V_s}{Z_0}\cos\beta y$ (d) $I_s \sin\beta y - j\dfrac{V_s}{Z_0}\sin\beta y$

14. On the Smith chart, full rotation around the chart gives a distance of

(a) $\lambda/2$ (b) $\lambda/4$ (c) λ (d) 2λ

15. For a transmission line, VSWR is 4. Then the reflection coefficient is
    (a) 0.4 (b) 0.6 (c) 1 (d) 0.5

16. For a lossless line, the position of single stub line l_s in terms of reflection coefficient $K\angle\varphi$ at a given frequency is,
    (a) $\frac{\lambda}{2\pi}(\varphi+\pi-\cos^{-1}|K|)$ (b) $\frac{\lambda}{2\pi}(\varphi+\pi+\cos^{-1}|K|)$
    (c) $\frac{1}{2\pi}(\varphi+\pi+\cos^{-1}|K|)$ (d) $\frac{1}{2\pi}(\varphi+\pi-\cos^{-1}|K|)$

17. For a given transmission line of length l , the relationship between propagation constant, Z_{oc} and Z_{sc} is
    (a) $\coth\gamma l=\sqrt{\frac{Z_{sc}}{Z_{oc}}}$ (b) $\tanh\gamma l=\sqrt{\frac{Z_{oc}}{Z_{sc}}}$
    (c) $\tanh\gamma l=\sqrt{\frac{Z_{sc}}{Z_{oc}}}$ (d) $\tanh\gamma l=\sqrt{Z_{sc}Z_{oc}}$

18. For a given transmission line of length l, the relationship between characteristic impedance Z_0, Z_{oc} and Z_{sc} is
    (a) $Z_0=\sqrt{Z_{oc}Z_{sc}}$ (c) $Z_0=\sqrt{\frac{Z_{sc}}{Z_{oc}}}$
    (b) $Z_0=\sqrt{\frac{Z_{oc}}{Z_{sc}}}$ (d) $Z_0=Z_{sc}Z_{oc}$

19. A $\lambda/4$ length transmission line having characteristic impedance Z_0 is terminated with impedance Z_R. The input impedance is
    (a) $Z_{\text{in}}=\sqrt{Z_0Z_R}$ (b) $Z_{\text{in}}=\frac{Z_0^2}{Z_R}$
    (c) $Z_{\text{in}}=\frac{Z_R^2}{Z_0}$ (d) $Z_{\text{in}}=\sqrt{\frac{Z_0}{Z_R}}$

20. For a lossless line, the characteristic impedance is 100 ohms and VSWR is 20. Then the minimum impedance on the line is
    (a) 20 ohms (b) 5 ohms (c) 0.25 ohms (d) 2.5 ohms

Answers

1. *c*	2. *a*	3. *b*	4. *a*	5. *a*	6. *b*	7. *a*	8. *d*	9. *c*	10. *a*
11. *b*	12. *a*	13. *c*	14. *a*	15. *b*	16. *a*	17. *c*	18. *a*	19. *b*	20. *b*

Appendix-A

Symbols of Quantities

Quantity	Symbol	Units
Force	$\vec{F}$	newton, N
Charge	Q, e	coulomb, C
Permittivity of free space	ε_0	farad/m, F/m
Permittivity of the medium	ε	farad/m, F/m
Relative permittivity of the medium, dielectric constant	ε_r	constant
Electric field intensity	$\vec{E}$	force per coulomb or volt per metre, F/C, V/m
Volume charge density	ρ_v	charge per m^3, C/m^3
Bounded charge density	ρ_b	charge per m^3, C/m^3
Work done or energy	W	joule, J
Electric potential	V	volts, V
Electric flux	ψ	coulomb, C
Electric flux density	$\vec{D}$	coulomb per metre^2, C/m^2
Line charge density	ρ_l	coulomb per metre, C/m
Surface charge density	ρ_s	coulomb per metre^2, C/m^2
Polarization	$\vec{P}$	coulomb per metre^2, C/m^2
Electric dipole moment	$\vec{p}$	coulomb-metre, C-m
Torque	$\vec{T}$	newton-metre, N-m
Capacitance	C	farad, F
Velocity vector	$\vec{u}$	metre per second, m/s
Distance vector	$\vec{R}$	metre, m
Energy density	w	joules per metre^3, J/m^3
Current	I	ampere, A

contd...

contd...

Current density	$\vec{J}$	ampere per metre^2, A/m^2
Conductivity	σ	siemen per metre, S/m
Electron mobility	μ_e	meter^2/volt, m^2/V
Electric susceptibility	χ_e	constant
Magnetic field intensity	$\vec{H}$	ampere per metre, A/m
Magnetic flux	ϕ	weber, Wb
Magnetic flux density	$\vec{B}$	weber per metre^2, Wb/m^2
Vector magnetic potential	$\vec{A}$	weber per metre, Wb/m
Surface current density, sheet current	K	ampere per metre, A/m
Scalar magnetic potential	V_m	weber per metre, Wb/m
Magnetic dipole moment	$\vec{m}$	ampere metre^2, A-m^2
Current element	Idl	ampere metre, A-m^2
Magnetization	$\vec{M}$	amperes per metre, A/m
Magnetic susceptibility	χ_m	constant
Self inductance	L	henry, H
Mutual inductance	M	henry, H
Permeability	μ	henry per metre, H/m
Reluctance	$\mathbb{R}$	ampere turn for weber, AT/Wb
Angular velocity	ω	radian per second, r/s
Linear frequency	f	hertz, Hz
Magneto motive force	NI	ampere-turn, AT
Poynting vector	$\vec{P}$	watt/m^2
Average power flow	p_{av}	watt/m^2
Total average power	P_{av}	watt
Energy stored in electric field	W_e	joule
Energy stored in magnetic field	W_m	joule

Appendix-B Material Properties

i) Relative permittivities amd dielectric strengths

Material	Relative permittivity (ε_r)	Dielectric strength, E (MV/m)
Air	1.0006	3
Petroleum oil	2.1	12
Paraffin	2.1	30
Mineral water	2.3	15
Paper	2–4	15
Polystyrene	2.7	20
Rubber	2.3–4	21–25
Soil (dry)	2.8	25
Bakelite	5	25
Glass	4–10	30
Mica	6	200
Quartz (fused)	5	30
Porcelain	6	–
Silicon	12	–
Germanium	16	–
Water	81	–
Barium titanate	1200	7.5

ii) Material relative permeabilities

Material type	Material	Relative permeability (μ_r)
Diamagnetic	Bismuth, silver, lead, mercury, copper, water	~0.9999
Non-magnetic	Vacuum	1
Paramagnetic	Air, aluminum, palladium, tungsten, platinum	~1
Ferromagnetic	Permalloy powder (Iron with 2% Mo, 81%Ni)	130
Ferromagnetic	Cobalt	250
Ferromagnetic	Nickel	600

contd...

...contd

Ferromagnetic	Ferroxcube -3 (Mn-An-ferrite powder)	1,500
Ferromagnetic	Mild steel (0.2% C)	2,000
Ferromagnetic	Iron (0.2% impurity)	5,000
Ferromagnetic	Silicon iron (2% Si)	7,000
Ferromagnetic	Permalloy (78.5%Ni)	1,00,000
Ferromagnetic	Mumetal (75% Ni, 5% Cu, 2% Cr)	1,00,000
Ferromagnetic	Purified iron (99.95%)	2,00,000
Ferromagnetic	Superalloy (5% Mo, 79% Ni)	10,00,000

iii) Material conductivities

Material type	Material	Conductivity, σ (Mhos/m)
Insulators	Quartz, polystyrene, sulphur, mica, paraffin, rubber, wax	$> 10^{-15}$
Insulators	Porcelain, glass	$\sim 10^{-12}$
Insulators	Bakelite	$\sim 10^{-9}$
Insulators	Distilled water	$\sim 10^{-4}$
Bad insulators	Dry, sandy soil, silicon	$\sim 10^{-3}$
Bad insulators	Marshy soil	$\sim 10^{-2}$
Bad insulators	Fresh water	$\sim 10^{-2}$
Bad insulators	Animal fat	4×10^{-2}
Bad conductors	Animal body	0.2
Bad conductors	Animal blood	0.7
Bad conductors	Germanium	2
Conductors	Sea water	4
Conductors	Ferrite	100
Conductors	Silicon	1000
Conductors	Carbon	30000
Good conductors	Graphite, cast iron, mercury, nichrome, stainless steel	$\sim 10^{6}$
Good conductors	Tin, bronze, brass, zinc, iron	$\sim 10^{7}$
Good conductors	Tungsten	1.8×10^{7}
Good conductors	Aluminum	3.5×10^{7}
Good conductors	Gold	4.1×10^{7}
Good conductors	Copper	5.8×10^{7}
Good conductors	Silver	6.1×10^{7}
Super conductors	Hg (< 4K), Nb (< 9.2K)	Infinite

Index